Pro/ENGINEER Wildfire for Designers

Sham Tickoo
Professor
Department of Mechanical Engineering Technology
Purdue University Calumet
Hammond, Indiana
U.S.A.

CADCIM Technologies
(www.cadcim.com)
U.S.A.

CADCIM Technologies
Excellence de Technology

CADCIM Technologies

Pro/ENGINEER Wildfire for Designers
Sham Tickoo

Published by CADCIM Technologies, 525 St Andrews Drive, Schererville, IN 46375 USA.

ISBN 0-9663537-6-5

Cover designer: *CADCIM Technologies*
Copy editor: *Vivek Kumar*
Cover illustration: *Created by Deepak Maini and Abhinav Rai using Pro/ENGINEER and SolidWorks*
Technical editors: *Abhinav Rai, Deepak Maini, Gurpreet Singh*
Typeface: *10/12 New Baskerville Bt*

www.cadcim.com

DEDICATION

To teachers, who make it possible to disseminate knowledge to enlighten the young and curious minds of our future generations

To students, who are dedicated to learning new technologies and making the world a better place to live

THANKS

To the faculty and students of the MET department of Purdue University Calumet for their cooperation

To Abhinav Rai, Deepak Maini, Gurpreet Singh, Puja Manchanda, and Saket Singhal of CADCIM Technologies for their valuable help

Teaching Aids for Faculty

The following teaching aids are available to faculty:

1. All Part files (.prt), Assembly files (.asm), and Drawing files (.drw) used for tutorials and exercises in this book.
2. Free chapters for download.
3. PowerPoint presentations for every chapter of the book.
4. Instructor's Guide with answers to review questions and solution to exercises.
5. Course outlines.
6. Students projects.
7. Free online technical support by contacting cadsoft@vsnl.com

To access the web site that contains these teaching aids, please contact the author, Prof. Sham Tickoo, at the following address:

stickoo@calumet.purdue.edu
or
tickoo@cadcim.com

Engineering Services Offered by CADCIM Technologies

The following are the services offered by CADCIM Technologies:

Solid Modeling, MAX/VIZ

CADCIM Technologies specializes and undertakes the following solid modeling related projects:

- Conversion of 2D drawings to 3D solid/surface models using Mechanical Desktop (MDT), Autodesk Inventor, Pro/ENGINEER, I-DEAS, CATIA, SolidWorks, and AutoCAD.
- Conversion of 3D models to 2D drawings.
- Conversion of raster drawings into vector drawings.
- Generation of assembly models.
- Animation and rendering using 3D Studio VIZ/MAX.
- Animation of assemblies using Autodesk Inventor and SolidWorks.

Training and Technical Documentation

CADCIM Technologies offer training on the following software packages:
• AutoCAD, • Mechanical Desktop (MDT), • Autodesk Inventor, • Pro/ENGINEER • SolidWorks, and • 3D Studio VIZ/MAX.

CADCIM Technologies also offers services in developing technical documents, study/training materials, and brochures.

For more information, please visit www.cadcim.com

Table of Contents

Chapter 4: Datums

Chapter 5: Options Aiding Construction of Parts-I

Chapter 6: Options Aiding Construction of Parts-II

Chapter 7: Advanced Modeling Tools-I

Chapter 8: Advanced Modeling Tools-II

Chapter 9: Assembly Modeling

Chapter 10: Generating, Editing, and Modifying the Drawing Views

Chapter 11: Dimensioning the Drawing Views

Chapter 12: Other Drawing Options

Chapter 13: Projects

Appendices

Index

Chapter for free download

The following chapter is available on the author's Website for free download. To download the free chapter, logon to

www.cadcim.com or

http://technology.calumet.purdue.edu/met/tickoo/students/students.htm

Chapter 14: Surface Modeling *Chapter For Free Download*

Author's Web Sites

For Faculty: Please contact the author at **stickoo@calumet.purdue.edu** or **tickoo@cadcim.com** to access the web site that contain the PowerPoint presentations, solid models used in the text book, Instructor's Guide, and other related material.

For Students: You can download solid modeling exercises, tutorials, and special topics by accessing the author's web site at **http://technology.calumet.purdue.edu/met/tickoo/students/students.htm** or **www.cadcim.com**

Preface

Pro/ENGINEER WILDFIRE

Pro/ENGINEER, developed by Parametric Technology Corpration, is one of the world's fastest growing solid modeling softwares. It is a parametric feature-based solid modeling tool and it not only unites the 3D parametric features with 2D tools but also addresses every design-through-manufacturing process. Based mainly on the solid modeling users feedback, this solid modeling tool is remarkably user-friendly and it allows you to be productive from day one.

This solid modeling tool allows you to easily import the standard format files with an amazing compatibility.

The 2D drawing views of the components are automatically generated in the Drawing mode. The drawing views that can be generated include detailed, orthographic, isometric, auxiliary, section, and so on. You can use any predefined drawing standard files for generating the drawing views. You can display the model dimensions in the drawing views or add reference dimensions whenever you want. The bidirectional associative nature of this software ensures that any modification made in the model is automatically reflected in the drawing views and any modification made in the dimensions in drawing views automatically updates the model.

Pro/ENGINEER Wildfire for Designers is a book that is written with the intent of helping the people who are into 3D design. This book is written with the tutorial point of view, with learn-by-doing as the theme. The mechanical engineering industry examples and tutorials are used in this book to ensure that the user can relate his knowledge of this book with the actual mechanical industry designs. The main features of the book are as follows:

- **Tutorial approach.**
  The author has adopted the tutorial point-of-view, with learn-by-doing as the theme, throughout the book. This approach will guide the users through the process of creating the model in the tutorial.

- **Real-World Projects as Tutorials.**
  The author has used the real-world mechanical engineering projects as tutorials in this book so that the reader can correlate the tutorials in this book with the real-time models in the mechanical engineering industry.

- **Tips and Notes.**

  The additional information related to the topics is provided to the users in the form of tips and notes.

- **Learning Objectives.**

  The first page of every chapter provides in brief the topics that will be covered in that chapter. This will help the users to easily refer to a topic.

- **Tools section.**

  Every chapter begins with the tools section that provides the detailed explanation of the Pro/ENGINEER tools.

- **Self-Evaluation Test, Review Questions, and Exercises.**

  Every chapter ends with a Self-Evaluation Test so that the users can assess their knowledge of the chapter. The author has given the answers of the Self-Evaluation Tests so that the users can compare their answers with the correct answers. The Review Questions and Exercises are also given at the end of each chapter and can be used by the Instructors as test questions and exercises in the classroom.

- **Heavily illustrated text.**

  The text in this book is heavily illustrated with the help of around 500 line diagrams and 700 photos that support the tools section and tutorials.

Organization of the Text

This textbook uses real-world design engineering projects that are explained in a step-by-step method so that the user can learn by doing them. The following flowchart suggests the possible sequence of chapters in this book to learn Autodesk Inventor.

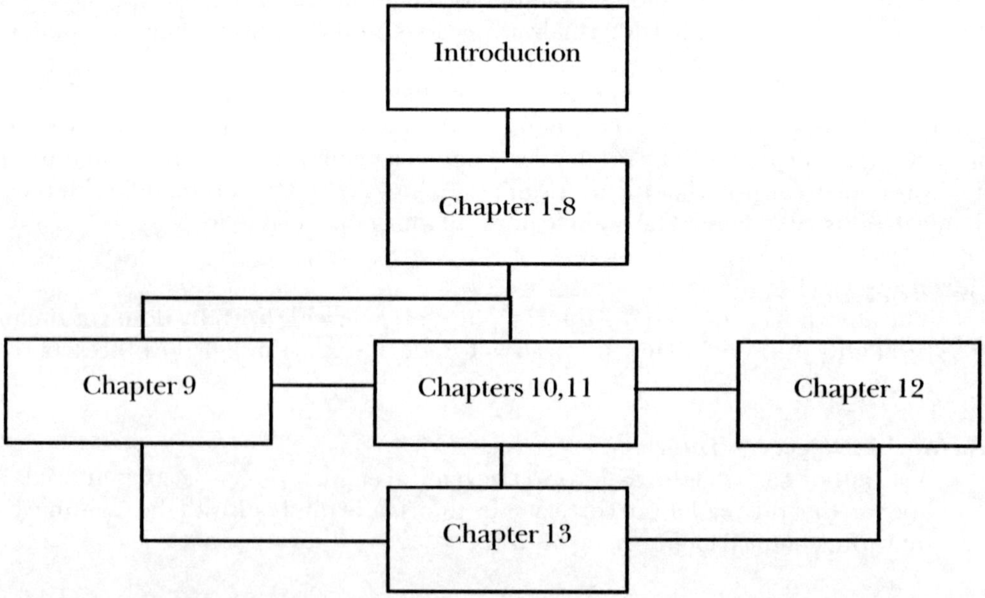

The Introduction provides the basic introduction to this solid model tool. In addition, the various important terms used in this textbook are also explained in detail. Chapters 1 through 8 explain the basic and advanced techniques of creating solid models. Chapters 9 through 12 discuss the techniques of assembling the solid models and generating their drawing views. Chapter 13 has the projects that require the knowledge of part modeling, assembly modeling, and drawing view generation. As a result, to teach this chapter, the instructor needs to teach all the chapters.

Features of the Text

Learning Objectives

Each chapter begins with Learning Objectives. These are the points that in brief explain that tools and options that the reader will be able to work with after completing the chapter.

Tutorials

These are real-world mechanical automobile engineering components whose modeling is explained in a step-by-step method.

Self-Evaluation Test

These are the questions given at the end of each chapter. Answers to these questions are also given at the end of the chapter so that the user can check his answers with the correct answers.

Review Questions

These are the questions that are also given at the end of the chapter. However, the answer to these questions are not given in the book and so the instructor can use these questions as test questions. The teaching faculty will get the answer to these questions in the Intructor's Guide that is provided by the Authors.

Note

Authors have provided a lot of additional information to the users about the topic being discussed in the form of notes.

Tip

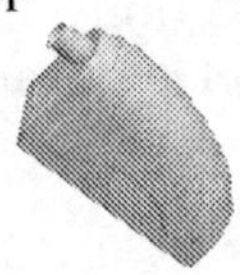

Special information and techniques are provided in the form of tips that allow the users to increase their efficiency.

About the Author

Sham Tickoo is a professor of Manufacturing Engineering Technology at Purdue University Calumet. He has been in the education and training since 1981 and joined Purdue University in the year 1987. Since then he has been teaching drafting and design, AutoCAD, AutoLISP, 3ds max, 3ds viz, Mechanical Desktop, Pro/Engineer, Autodesk Inventor, etc.

Prof. Tickoo is also one of the best known authors of CAD/CAM books in the world market. He has authored a number of books on the latest CAD/CAM technology. He is one of the very few authors in the world whose books are translated and sold in the countries like Italy, China, Philippines, and Russia. The list of some of his books is given below:

AutoCAD: A Problem Solving Approach, Delmar Publishers, USA
AutoCAD LT: A Problem Solving Approach, Delmar Publishers, USA
Customizing AutoCAD, Delmar Publishers, USA
3D Studio VIZ, Goodheart-Willcox Publishers, USA
3D Studio MAX, Goodheart-Willcox Publishers, USA
Mechanical Desktop Instructor, McGraw-Hill, USA
Autodesk Inventor for Designers, CADCIM Technologies, USA
Pro/ENGINEER for Designers, CADCIM Technologies, USA
Designing with Pro/ENGINEER, dreamtech PRESS, India
AutoCAD with Applications, Tata McGraw-Hill, India
Understanding AutoCAD, Tata McGraw-Hill, India
Advanced Techniques in AutoCAD, Tata McGraw-Hill, India
AutoCAD 2000 Tecniche Avanzate, APOGeO education, Italy
AutoCAD 2000 Fondamenti, APOGeO education, Italy
AutoCAD 2000: A Problem Solving Approach, China Machine Press, China
Customizing AutoCAD 2000, China Machine Press, China
AutoCAD 2002, Russian Edition

In addition to his academic career, he has also been employed as a design engineering, quality control engineer, software developer, and has also worked as a consultant for a number of companies in US and Canada.

Prof. Tickoo has also developed a software package called "SMLayout" that has been registered by Autodesk Inc. as a third party software product. This software package is currently being used by some companies in Canada, United States, and Columbia. SMLayout is written in the AutoLISP programming language. It generates flat layout drawing of some complicated geometric shapes like: transitions, cones, cone-cylinder intersections, sphere-cylinder intersections, and cylinder-cylinder intersections. This software package addresses the needs of steel fabricators and manufacturers of sheet metal products. The programs also provide a dimensioning option for automatic dimensioning of the drawing.

In March 2000, Prof. Tickoo was issued a US patent for a product called "Self Adjusting Cargo Organizer" that he designed for vehicles.

Introduction

PRO/ENGINEER WILDFIRE

Welcome to Pro/ENGINEER Wildfire. If you are a new user of Pro/ENGINEER software package, you are going to join hands with thousands of users of this high-end CAD/CAM/CAE tool worldwide. If you are a user of the previous releases of this software, you are going to upgrade your designing skills with the tremendous improvement in this latest release.

Pro/ENGINEER Wildfire is a powerful program used to create complex designs with a great precision. The design intent of any three-dimensional (3D) model or an assembly is defined by its specification and its use. You can use the powerful tools of Pro/ENGINEER Wildfire to capture the design intent of any complex model by incorporating intelligence into the design.

To make the designing process simple and quick, this software package has divided the steps of designing into different modules. This means that each step of designing is completed in a different module. For example, generally a design process consists of the following steps:

- Sketching using the basic sketch entities.
- Converting the sketch into features and parts.
- Assembling different parts and analyzing them.
- Documentation of the parts and the assembly in terms of drawings views.
- Manufacturing the final part and assembly.

All these steps are divided into different modes of Pro/ENGINEER Wildfire, namely, the Sketch Mode, Part Mode, Assembly Mode, Drawing Mode, and Manufacturing Mode.

In spite of making various modifications in a design, the parametric nature of this software helps to preserve the design intent of a model with tremendous ease. Once you understand the feature-based, associative, and parametric nature of Pro/ENGINEER Wildfire, you can appreciate its power as a solid modeler. It allows you to work in a 3D environment and calculates the mass properties directly from the created geometry. You can switch to various display modes like wireframe, shaded, hidden, and no hidden at any time with ease as it only changes the appearance of the model.

FEATURE-BASED NATURE OF PRO/ENGINEER WILDFIRE

Pro/ENGINEER Wildfire is a feature-based solid modeling tool. A feature is defined as the smallest building block and any solid model created in Pro/ENGINEER Wildfire is an integration of a number of these building blocks. Each feature can be edited individually to bring in any change in the solid model. The use of the feature-based property provides greater flexibility to the parts created. For example, consider the part shown in Figure I-1. It consists of a counterbore hole at the center and six counterbore holes around it at some Bolt Circle Diameter (BCD).

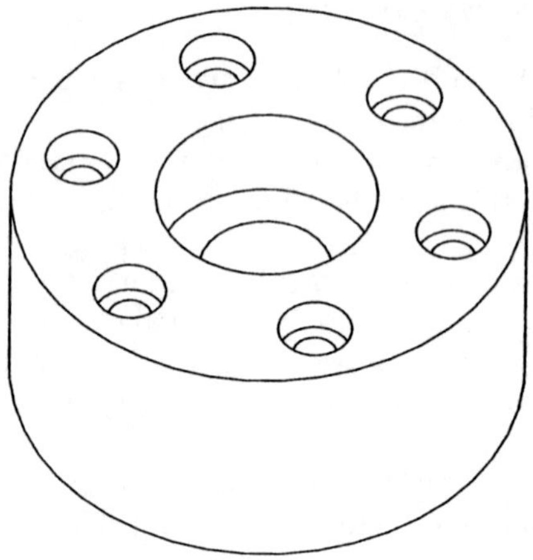

Figure I-1 *Model displaying the counterbore holes*

Now, consider a case where you have to change all the outer counterbore holes to drill holes keeping the central counterbore hole and the BCD for the outer holes same. Also, you have to change the number of holes from six to eight. In a nonfeature-based software package, you will have to delete the entire part and then create a new part as per the new specifications; whereas, Pro/ENGINEER Wildfire allows you to make this modification by just modifying some values in the same part, see Figure I-2. This shows that the solid parts created in Pro/ENGINEER Wildfire are a combination of various features that can be modified individually at any time.

BIDIRECTIONAL ASSOCIATIVE NATURE OF PRO/ENGINEER WILDFIRE

There is a bidirectional associativity between all modes of Pro/ENGINEER Wildfire. The bidirectional associative nature of a software package is defined as its ability to ensure that if any modification is made in a particular model in one mode, the modification is reflected in the same model in other modes also. For example, if you make any change in a model in the Part mode and regenerate it, the changes will be highlighted in the Assembly mode also. Similarly, if you make any change in a part in the Assembly mode, after regeneration, the change will be highlighted in the Part mode also. This bidirectional associativity also correlates

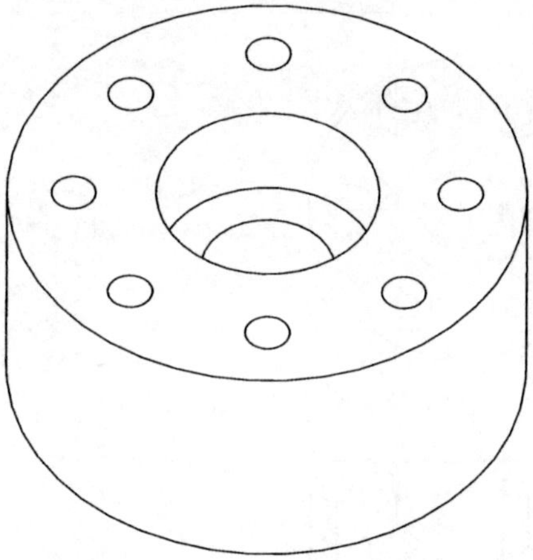

Figure I-2 *Model after making the modifications*

the two-dimensional (2D) drawing views generated in the Drawing mode and the solid model created in the Part mode of Pro/ENGINEER Wildfire. This means that if you modify the dimensions of the 2D drawing views in the Drawing mode, the change will be automatically reflected in the solid model and also in the assembly after regeneration. Likewise, if you modify the solid model in the Part mode, the changes will also be seen in the 2D drawing views of that model in the Drawing mode. Thus, bidirectional associativity means that if modification is made to any one application, it changes the output of all the other modes related to the model. This nature relates the various modes available in Pro/ENGINEER.

Figure I-3 shows the drawing views of the part shown in Figure I-1 generated in the Drawing mode. The views show that the part consists of a counterbore hole at the center and six counterbore holes around it.

Now, when the part is modified in the Part mode, the modifications are automatically reflected in the Drawing mode, as shown in Figure I-4. The views in this figure show that all the outer counterbore holes are converted into the drilled holes and the number of holes is increased from six to eight.

Figure I-5 shows the Crosshead assembly. It is clear from the assembly that the diameter of the hole is more than what is required (shown using dotted lines). In an ideal case, the diameter of the hole should be equal to the diameter of the bolt.

The diameter of the hole can be easily changed by opening the file in the Part mode and making the necessary modifications in the part. This modification is reflected in the assembly shown in Figure I-6. This is due to the bidirectional associative nature of Pro/ENGINEER.

Similarly, if you change the dimensions of the model in the Drawing mode, the modifications will be automatically reflected in the Part mode. This shows that all the modes are related to each other. Thus, it becomes very easy to modify your model at any time. This makes the application software more user-friendly.

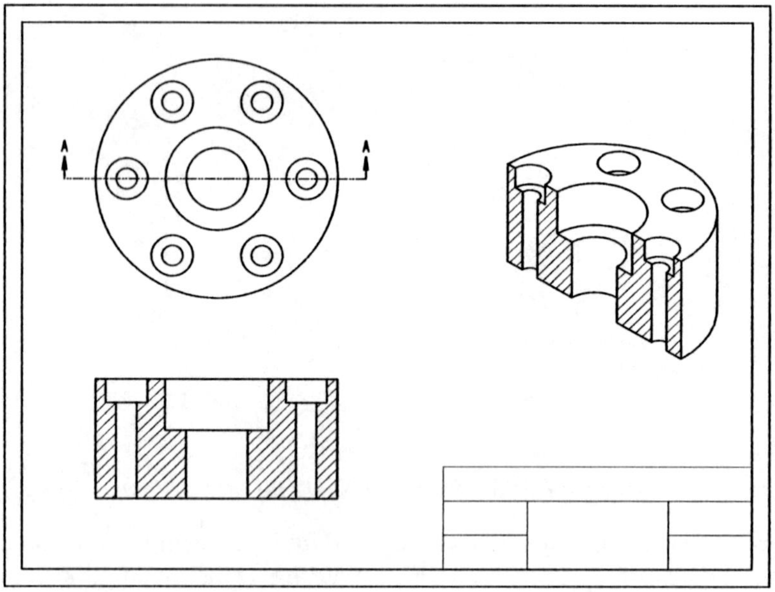

Figure I-3 Drawing views of the model before modifications

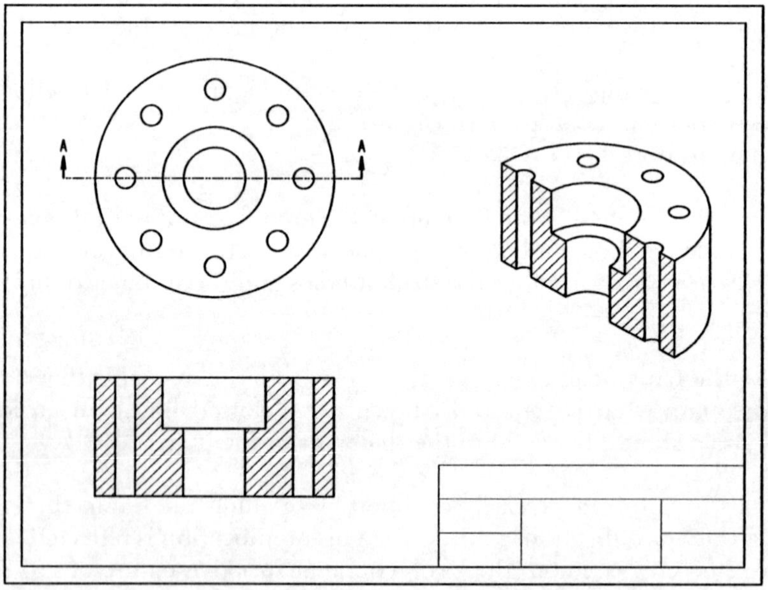

Figure I-4 Drawing views of the model after modifications

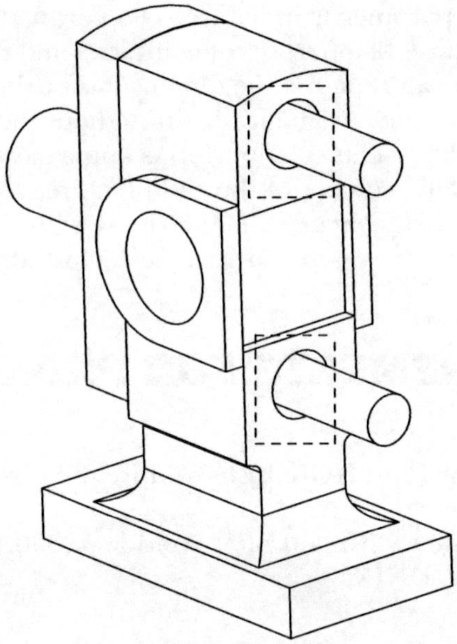

Figure I-5 *Diameter of the hole and the bolt*

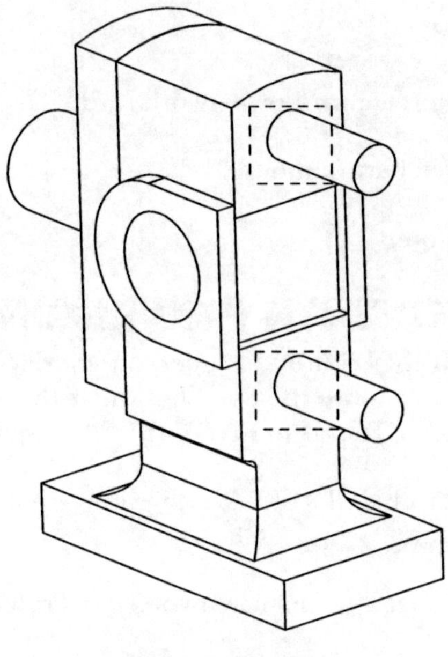

Figure I-6 *Figure after modifying the diameter of the hole*

PARAMETRIC NATURE OF PRO/ENGINEER WILDFIRE

Pro/ENGINEER Wildfire is parametric in nature. The parametric nature of Pro/ENGINEER Wildfire means that the features of a part become interrelated if they are drawn by taking the reference of each other. You can redefine the dimensions or the attributes of a feature at any time. The changes will propagate automatically throughout the model. Thus, they develop a relationship among themselves. This relationship is known as the parent-child relationship. So if you want to change the placement of the child feature, you can make alterations in the dimensions of the references and hence change the design as per your requirement. The parent-child relationship will be discussed in detail while discussing the datums in later chapters.

SYSTEM REQUIREMENTS FOR PRO/ENGINEER WILDFIRE

The system requirements for Pro/ENGINEER Wildfire are given below.

1. Operating System: Windows NT 4.0 SP5, Windows 2000, Windows XP Home Edition, XP Professional Edition, UNIX.
2. Processor 233MHz or higher.
3. Memory: 128MB RAM minimum requirement (512MB or higher recommended).
4. Swap Memory: 256MB minimum (1024MB or higher recommended).
5. Hard disk space: 1.12GB.
6. An ethernet adapter interface card or network card.
7. Microsoft approved three button mouse.
8. Microsoft Internet Explorer 6.0.

GETTING STARTED WITH PRO/ENGINEER WILDFIRE

Once you have Pro/ENGINEER Wildfire installed on your system, there are two options to start it. The first option is to choose the **Start** button at the lower left corner of the screen and select **Start** > **Programs** > **PTC** > **Pro ENGINEER** > **proewildfire** as shown in Figure I-7.

The second option to start Pro/ENGINEER Wildfire is by double-clicking on its shortcut icon on the desktop of the computer.

Figure I-8 shows the screen that appears when you start Pro/ENGINEER Wildfire.

IMPORTANT TERMS AND DEFINITIONS

Some important terms that will be used in this book while working with Pro/ENGINEER

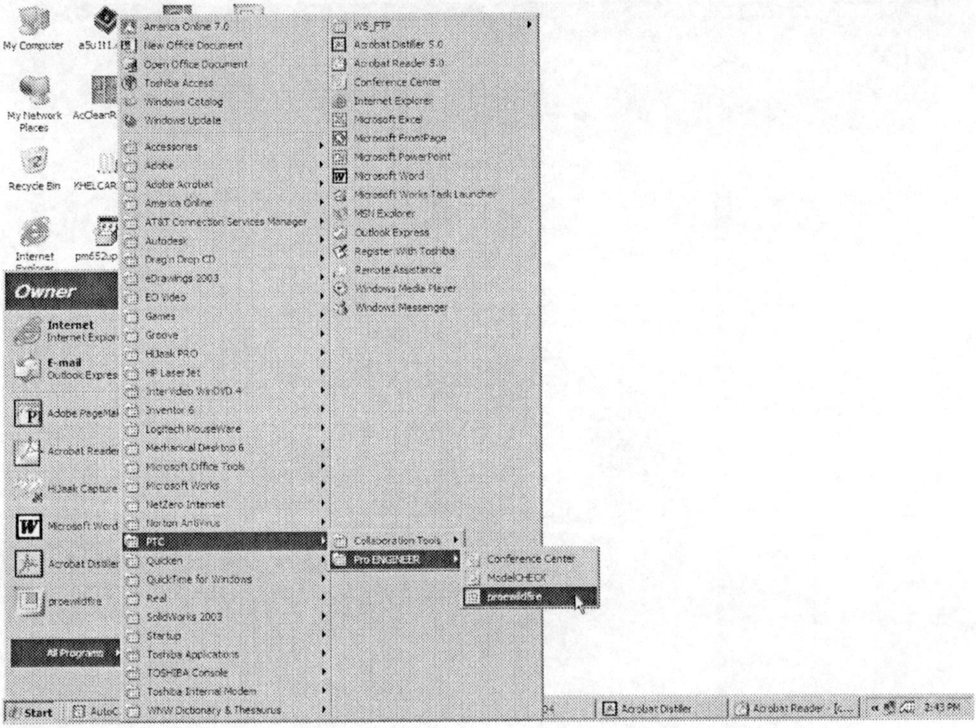

Figure I-7 Windows screen with task bar and application icons

Wildfire are discussed next.

Entity

An element of a section geometry is called an entity. The entity can be an arc, line, circle, point, conic, coordinate system, and so on. When one entity is divided at a point then the total number of entities is said to be two.

Dimension

It is the measurement of one or more entities.

Constraint

Constraints are logical operations that are performed on the selected geometry to make it more accurate in defining its position and size with respect to the other geometry.

Parameter

It is defined as a numeric value or any definition that defines a feature. For example, all the dimensions in a sketch are parameters. The parameters can be modified at any time.

Relation

A relation is an equation that relates two entities.

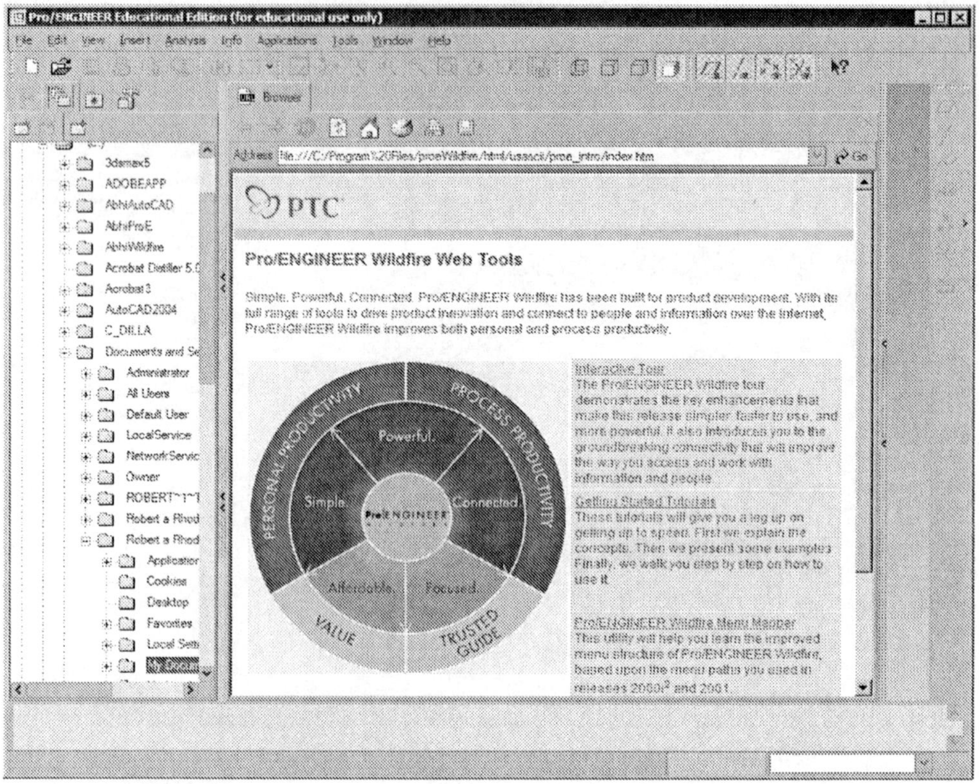

Figure I-8 Initial screen appearance after starting Pro/ENGINEER Wildfire

Weak Dimensions and Weak Constraints

Weak dimensions and weak constraints are temporary dimensions or constraints that appear in gray color. These are automatically applied to the sketch when it is drawn using the **Intent Manager**. They are removed from the sketch without any confirmation from the user. The weak dimensions or the weak constraints should be changed to strong dimensions or constraints if they seem to be useful for the sketch. This only saves an extra step of dimensioning the sketch or applying constraints to the sketch.

Strong Dimensions and Strong Constraints

Strong dimensions and strong constraints appear in yellow color. These dimensions and constraints are neither removed automatically nor applied automatically. All the dimensions added manually to a sketch are strong dimensions.

FILE MENU OPTIONS

When you choose **File** from the menu bar, various options are displayed in the **File** menu. These options are discussed next.

Working Directory

A working directory is a directory on your system where you can save the work done in the current session of Pro/ENGINEER Wildfire. You can set any directory existing on your system as the working directory. Before starting work in Pro/ENGINEER Wildfire, it is important to specify the working directory. If the working directory is not selected before saving an object file then the object file will be saved in a default directory set by Pro/ENGINEER Wildfire. This default directory is set at the time of installing Pro/ENGINEER Wildfire. If the working directory is selected before saving the object files that you create, it becomes easy to organize them. In Pro/ENGINEER Wildfire, the working directory can be set in two ways:

> **Tip**: *When several strong dimensions or constraints conflict, Pro/ENGINEER make the constraints and dimensions appear in red color, and prompts you to delete one or more.*

Using the Navigator

When you start a Pro/ENGINEER Wildfire session, the Navigator is displayed on the left of the graphics window. This Navigator can be used to select a folder and set it as the working directory.

Right-click on the folder that you need to set as the working directory. The shortcut menu appears as shown in Figure I-9. From this shortcut menu, choose the **Make Working Directory** option to set the selected folder as the working directory. To make a new folder, choose the **New Folder** option from the shortcut menu.

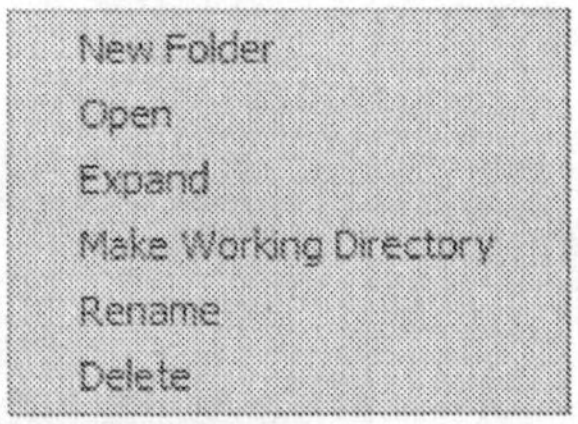

Figure I-9 *Shortcut menu*

Using the Select Working Directory dialog box

In order to specify a working directory, choose **File > Set Working Directory** from the menu bar. The **Select Working Directory** dialog box is displayed as shown in Figure I-10. Using this dialog box you can set any directory as the working directory. The various options in this dialog box are discussed next.

> **Tip**: *The* ***Select Working Directory*** *dialog box has some of the properties of the MS Windows operating system. You can set the working directory using this dialog box by browsing through directories and folders. However, you cannot rename a file, directory, or a folder name in this dialog box. You can create a new directory using this dialog box.*

Look In drop-down list

The **Look In** drop-down list displays all drives present on your computer along with a **Favorites** folder as shown in Figure I-10. When the **Select Working Directory** dialog box is invoked, by default, it displays the contents of a default directory. However, the default directory that appears every time you open this dialog box can be changed. This is discussed later. The **Favorites** folder contains all the directories saved as favorites. The saving of the favorite directories will be discussed later.

Name

The **Name** edit box displays the name of the directory selected in the **Select Working Directory** dialog box. You can either select a directory from the **Look In** drop-down list

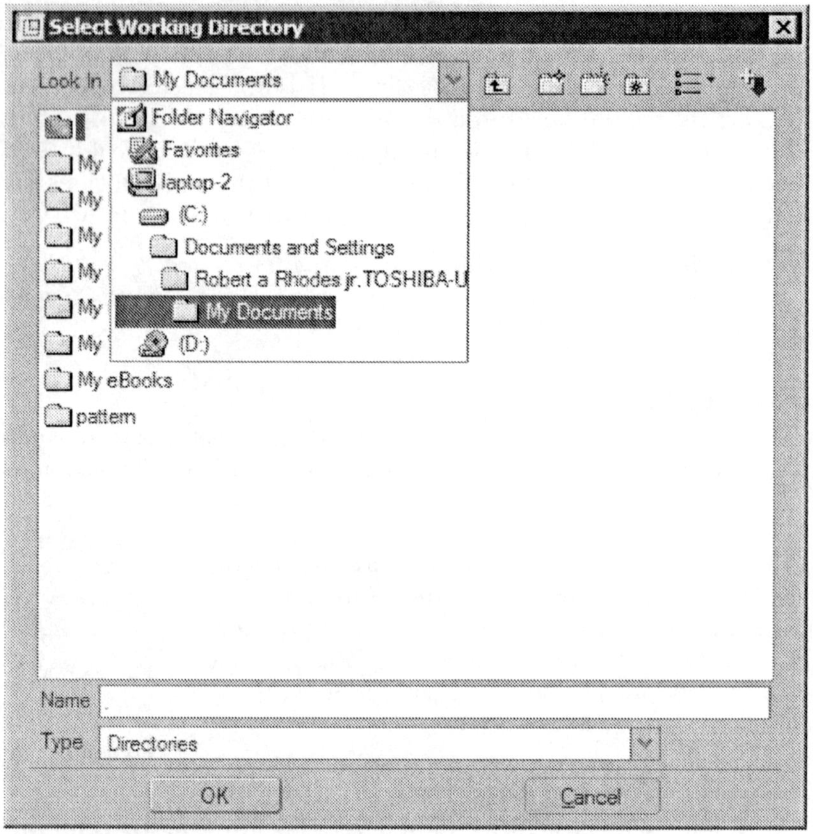

Figure I-10 Select Working Directory dialog box

or enter the path of any existing directory in this edit box.

Type drop-down list

The **Type** drop-down list has two options: **Directories** and **All Files**. If you select the **Directories** option, all the directories present are listed and if you select **All Files** then all the files along with the directories are listed in the dialog box.

Up One Level

The **Up One Level** button allows you to move one level up in the directory. When you choose this button, a directory is displayed that is one level above the current directory. This button is generally available in most of the dialog boxes of Windows operating system and has the same function.

Working Directory

This is used when you have already set the working directory. You may browse through the directories in the **Select Working Directory** dialog box, but when you choose this button, the directory selected previously as working directory is displayed in the **Look In** drop-down list.

New Directory

The **New Directory** button is used to create a new directory that can be selected as a working directory. When you choose the **New Directory** button, the **New Directory** dialog box is displayed. You are prompted to enter the name of the new directory you want to create.

Favorites

The **Favorites** button is used to save the location of the directories that are to be used frequently. You just have to specify the working directories to be used frequently and save the location of those directories by selecting the **Favorites** button.

When you want to select one of the favorite working directories, you can select the **Favorites** folder from the **Look In** drop-down list available in the **Select Working Directory** dialog box. In this folder there is a subfolder named **Personal Favorites**. When you double-click on this folder, all the directories that were selected as favorites are displayed.

When you choose the **Favorites** button, a menu is displayed. The options in this menu are discussed next.

Save location

The **Save location** option is used to save the current directory in the **Favorites** folder. This option is available only when the directory selected is not already saved in the **Favorites** folder.

Remove location

The **Remove location** option removes the directory from the **Favorites** folder. This option is available only when the directory selected is already saved in **Favorites** folder.

Browse favorites

The **Browse favorites** option allows you to browse through your favorite directories that you saved using the **Save location** option.

Display Configuration

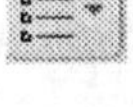

When you choose the **Display Configuration** button, a menu is displayed. The options in this menu are discussed next.

List

The **List** option is used to view the contents of the current directory or drive, which includes files and folders in the form of a list.

Details

The **Details** option is used to view the contents of the current folder or drive in the form of a table, which indicates the name, size, and date on which it is modified.

Commands and Settings

The **Command and Settings** button can be used to customize the **Select Working Directory** dialog box. This button when chosen displays a menu. The options in this menu are discussed next.

'Look In' Default

The **'Look In' Default** option allows you to set a directory as a default directory. When you select this option, the **'Look In' Default** dialog box is displayed. Figure I-11 shows this dialog box with the options in the drop-down list. In the drop-down list, there are four options. If you select the **Default** option, whenever the **File Open** dialog box is invoked it displays the directory that is set as default. If you select the **Working Directory** option in this drop-down list, then whenever the **File Open** dialog box is invoked it displays the working directory that is set. If you select the **In Session** option, then whenever you select **Open** from the **File** menu in the **File Open** dialog box the **In Session** folder is selected by default. Similarly, the **Pro/Library** can be set as the working directory.

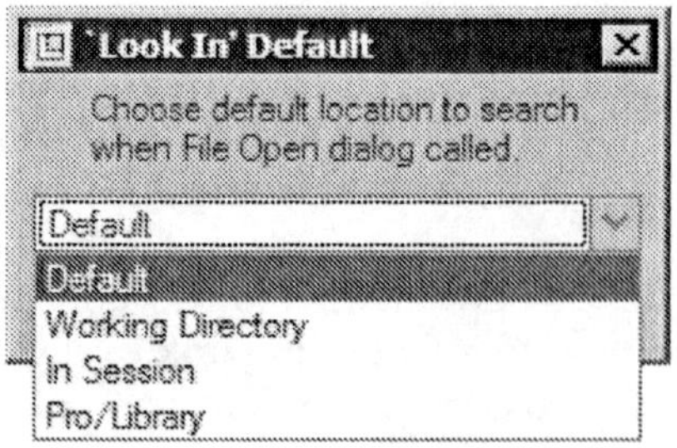

Figure I-11 'Look In' Default dialog box

All Versions

This option when selected displays all the versions of an object file. In Pro/ENGINEER Wildfire, the file once saved will generate a new version of it with an extension 1. An object file is not copied on another object file but a new version of it is created. Therefore, every time you save an object using the **Save** option, a new version of it is created on the disk in the current working directory.

By default, in the **Select Working Directory** dialog box, the **Directories** option is displayed in the **Type** drop-down list. From this list if you select the **All Files** option and then select the **All Versions** option, all the versions of the object files are displayed in the dialog box.

Note

An object in Pro/ENGINEER Wildfire is defined as a file that is created using any of its modes like Part, Drawing, Sketch, and so on.

New

In order to create a new object, select **New** from the **File** menu or choose the **Create a new object** button from the **Top Toolchest**. The **New** dialog box is displayed as shown in Figure I-12. The dialog box displays the various modes available in Pro/ENGINEER Wildfire. By default, the Part mode radio button is selected. A default name of the object file is displayed in the **Name** edit box. You can also enter the name of the object file as desired.

When you select the Part, Assembly, and Manufacturing mode radio buttons in this dialog

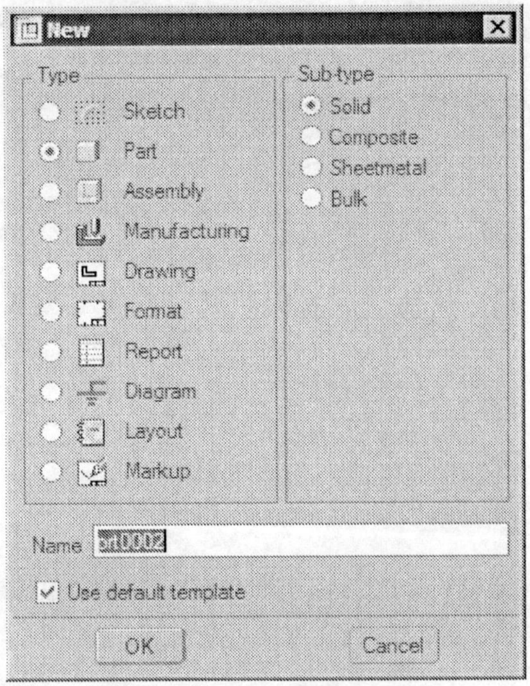

***Figure I-12 New** dialog box*

box, their subtypes are displayed under the **Sub-type** area of this dialog box.

Accept the default name of the Part mode file, and choose the **OK** button in the **New** dialog box. The three default datum planes are displayed on the graphics window and some of the toolbars become active. Also, the **Model Tree** appears on the left of the screen as shown in Figure I-13. The options available in the **New** dialog box are discussed next.

Use default template

The **Use default template** check box is selected to start a new file using an existing default template file that includes three default datum planes and a coordinate system. This default template file creates the model in inches. The check box is selected by default. If you clear this check box and choose the **OK** button from the **New** dialog box, the **New File Options**

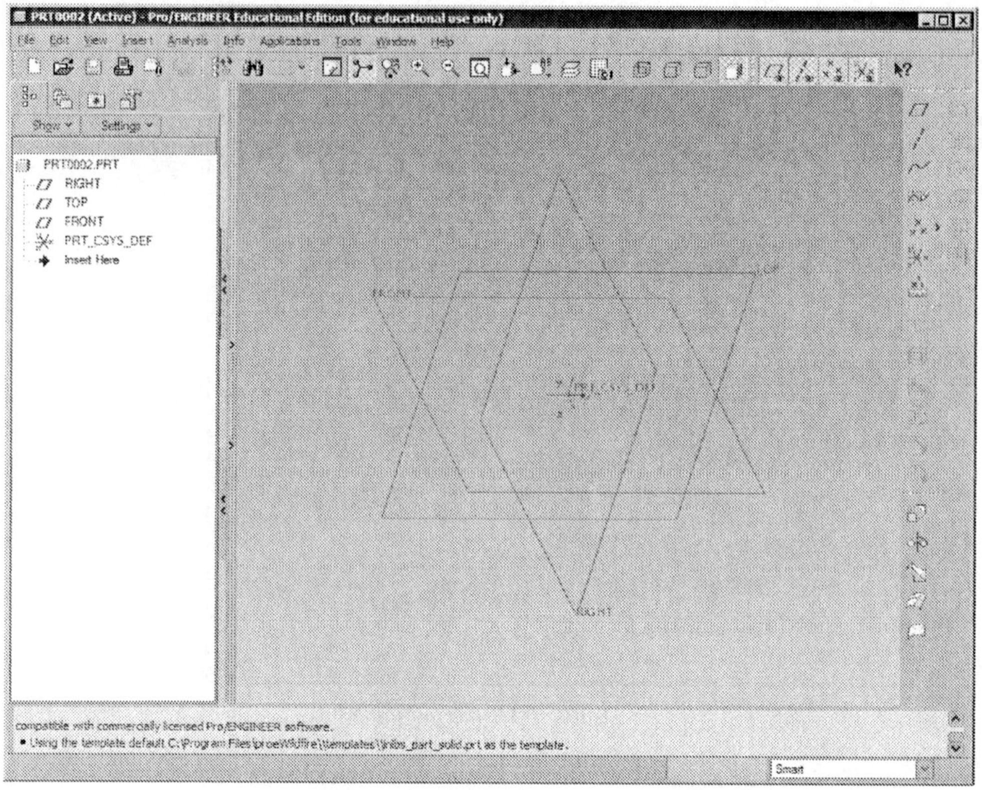

***Figure I-13** The initial screen appearance after entering the Part mode*

dialog box is displayed as shown in Figure I-14. Using the **New File Options** dialog box you can select the predefined templates provided in Pro/ENGINEER Wildfire or a user-defined template created and saved earlier. You can also open an empty template provided in the **New File Options** dialog box in which you have to create the datum planes and the coordinate system manually.

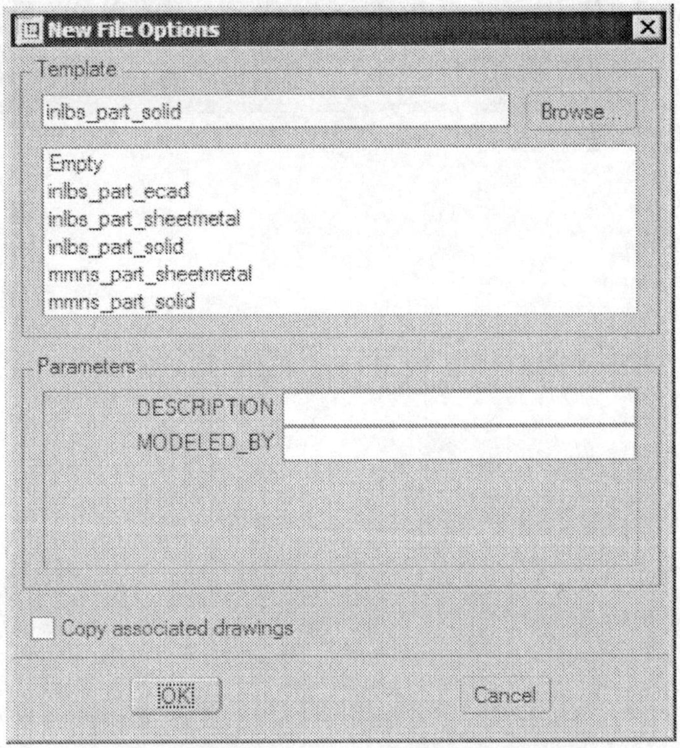

***Figure I-14 New File Options** dialog box*

Open

The **Open** option is used to open an existing object file. When you choose the **Open** option from the **File** menu or choose the **Open an existing object** button from the **File** toolbar, the **File Open** dialog box is displayed as shown in Figure I-15. The working directory you had selected is displayed in the **Look In** drop-down list. The various options in this dialog box are discussed next.

Look In

In the **Look In** drop-down list, various browsing options for selecting the directories are available. You can browse through these folders to search for the object file you want to open.

In Session

The **In Session** option displays all the object files that are in the current session. The object files that you open in Pro/ENGINEER Wildfire in the current session are stored in its temporary memory. This temporary memory is a folder named **In Session**. Once you exit Pro/ENGINEER Wildfire, the contents of this folder are deleted automatically. However, the original files are not removed from their actual location.

Commands and Settings

The **Commands and Settings** option can be used to customize the **File Open** dialog box.

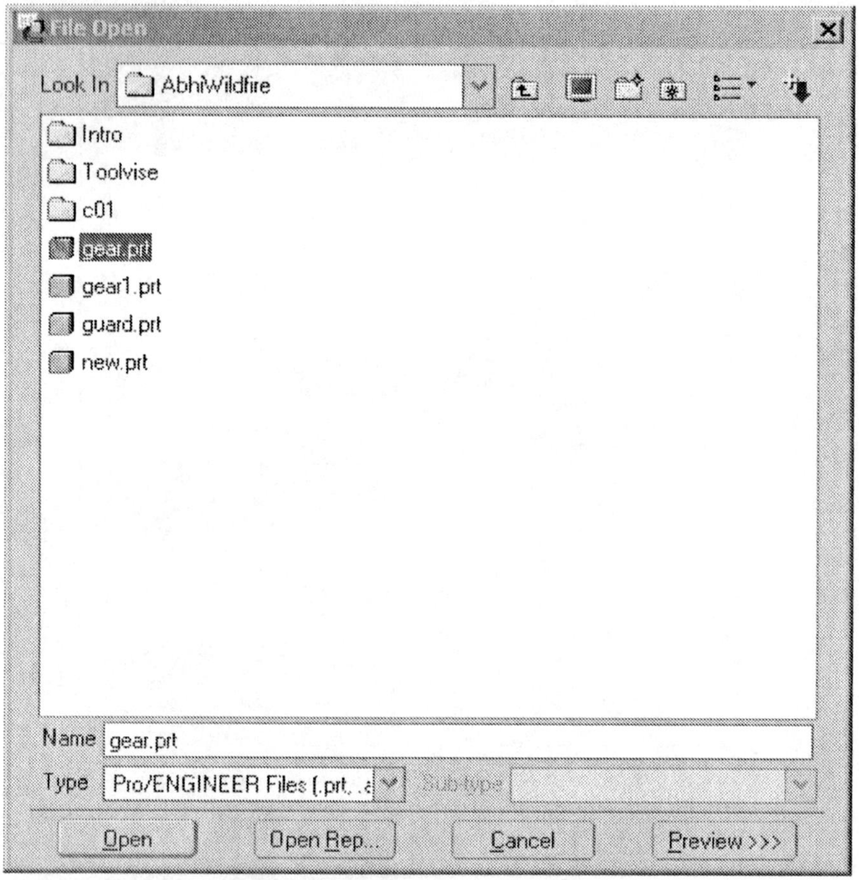

***Figure I-15 File Open** dialog box*

When you choose this button a shortcut menu is displayed. The options under this menu are similar to those available in the **Select Working Directory** dialog box.

Sort By

The **Sort By** option displays all files present in the directory in order so that it becomes easy to search a file. When you choose the **Sort By** option, a cascading menu is displayed. In the cascading menu, there are two options: **Model Name** and **Markup/Instance Name**. If you select the **Model Name** option, the file list will be sorted alphabetically by model name in the **File Open** dialog box. The **Markup/Instance Name** option sorts the file list by specific markups or instance names in the **File Open** dialog box.

Retrieve Drawing as View Only

The **Retrieve Drawing as View Only** option is used to open the drawing as view only. Generally when you open a drawing file, its part model is listed in the current session. You can verify this by opening the current session. But when you open a drawing file using this option, its part model is not listed in the current session.

The modification of an object in view only mode is not possible unless you choose **File > Retrieve Models** from the menu bar. The **Retrieve Drawing as View only** option is used to open drawings in the drawing mode, that is, the object files having the extension *.drw*.

Name

In the **Name** edit box you can enter the name of the existing object file you want to open.

Type

The **Type** drop-down list contains the file formats of the various modes available in Pro/ENGINEER Wildfire. This drop-down list also has other file formats that can be imported in Pro/ENGINEER Wildfire. By default, in this drop-down list the **Pro/ENGINEER Files** option is selected, and hence the files created in any mode of Pro/ENGINEER can be opened. However, if you select a specific mode from this drop-down list, only the files created in that mode are displayed. For example, if you select **Part** in the drop-down list, then only the *.prt* files are displayed. This makes the selection and opening of the files easy.

Preview

The **Preview** button is used to see the preview of the model. When you choose this button, the **File Open** dialog box expands and a preview is displayed on the right of the dialog box. In this window, you can see the preview of the model selected. This feature of the **File Open** dialog box helps in previewing the model before actually opening it.

Note

*Assembly files with the file extension .asm can also be previewed using the **Preview** button. If you are not able to see the preview of the assembly files on the preview displayed on the right side of the **File Open** dialog box then choose the **Refit** button present on the top of the preview area to resize the assembly according to the preview area.*

*There is no command prompt in Pro/ENGINEER Wildfire. However, you are provided with prompts in the **Message Area**. Whenever you have to enter any numerical value or text, a **Message Input Window** is displayed in the **Message Area**.*

Erase

The **Erase** option is used to delete the files from the temporary memory known as the **In Session** folder. To invoke this option, choose **File > Erase** from the menu bar. The cascading menu is displayed with two options; **Current** and **Not Displayed**. As discussed earlier, all the files opened in a session of Pro/ENGINEER Wildfire are saved in the temporary memory. You can use the **Erase** option to erase files from the temporary memory. The options that are displayed in the cascading menu are discussed next.

Current

The **Current** option is used to erase the file that is opened and displayed on the graphics screen. When the **Current** option of the cascading menu is chosen, the system prompts you to confirm erase file. The **Erase Confirm** dialog box is displayed as shown in Figure I-16.

Not Displayed

This option is used to erase the files present in the **In Session** folder but not available presently on the screen. The **Erase Not Displayed** dialog box is displayed as shown in Figure I-17. You can select the files from this dialog box to erase them.

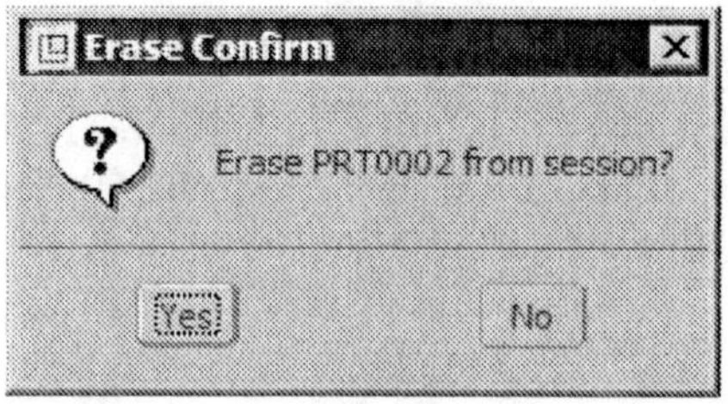

***Figure I-16 Erase Confirm** dialog box*

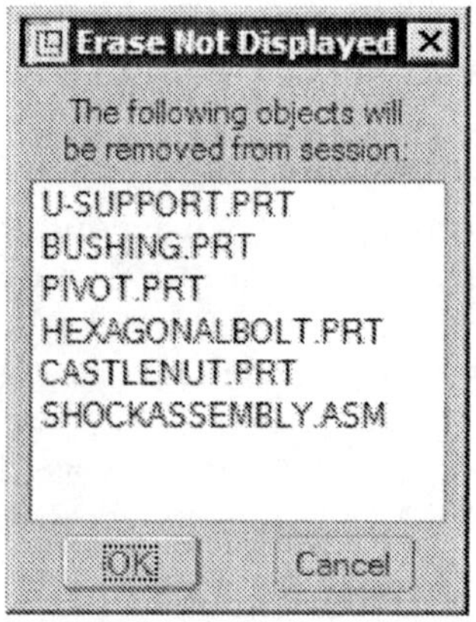

***Figure I-17 Erase Not Displayed** dialog box*

Delete

This option removes the selected file permanently from the hard disk. To invoke this option, choose **File > Delete** from the menu bar. The cascading menu is displayed with two options; **Old Versions** and **All Versions**.

Old Versions

This option is used to delete all the old versions of the current file. When you choose the **Old Versions** option, you are prompted to enter the name of the object file of which the old versions have to be deleted. When the **Message Input Window** is displayed, enter the object file name in this window. All the versions of that file will be deleted from the hard disk except the latest version.

All Versions

This option is used to delete all the versions including the current file from the hard disk. When you choose the **All Versions** option, a warning is displayed stating that performing this function can result in loss of data. This option is chosen when the file is opened and is displayed on the graphics window.

Save

The **Save** option is used to save the objects present in the **In Session** folder or an object on the graphics screen. When you choose the **Save** option from the **File** menu or the **Save the active object** button on the **File** toolbar, you are prompted to specify the name of the object file. The name is displayed in the **Message Input Window** that is displayed when you choose this button.

Note

Remember that all the object files saved using the ***Save*** *option in the* ***File*** *menu should be present in the* ***In Session*** *folder. If the file to be saved is not present in the* ***In Session*** *folder then Pro/ENGINEER Wildfire will display an error informing you that the file does not exist in the current session.*

It should be noted here that the file opened currently on the screen may not be in the ***In Session*** *folder but is in the current session and hence will be saved without any error.*

Save a Copy

The **Save a Copy** option is used to save a copy of the current object in the same working directory or in some other directory. When you invoke this option from the **File** menu or select the **Save a copy of the active object** button from the **File** toolbar, the **Save a Copy** dialog box is displayed and you need to specify the new name of the object file to be saved as a copy and the name of the target directory in the **Save a Copy** dialog box. You can browse through the directories and select the target directory. The file will be saved in the selected directory.

Using this option you can also export the file in other file formats like IGES, STEP, STL, and so on. After specifying the name of the new file and the target directory choose the file format in which you want to export the file from the **Type** drop-down list in the **Save a Copy** dialog box.

Backup

The **Backup** option of the **File** menu creates a backup copy of an object file that is in memory. When you choose this option, the **Backup** dialog box is displayed as shown in Figure I-18. In the **Model Name** edit box the name of the file that you want to take a backup of is displayed. In the **Backup To** edit box the name of the directory is specified where the object will be saved as a backup. If you backup an assembly or a drawing object, Pro/ENGINEER Wildfire saves all its dependent files in the specified directory. For example, if you backup a drawing file in a different directory then the part file of the drawing will also be created automatically in the same directory where the drawing backup is created.

Rename

The **Rename** option in the **File** menu renames the active object that is on the screen. When you choose this option, the **Rename** dialog box is displayed as shown in Figure I-19.

MANAGING FILES IN Pro/ENGINEER

As discussed earlier, a new file is generated whenever you save an object. The number of files generated is directly proportional to the number of times you save that object. Hence, these files occupy a lot of disk space. The latest version of the object is of use and should be stored. Latest version implies to the highest number that is suffixed with the file name of that object. The rest of the files called old versions should be deleted from the hard disk if they are not required.

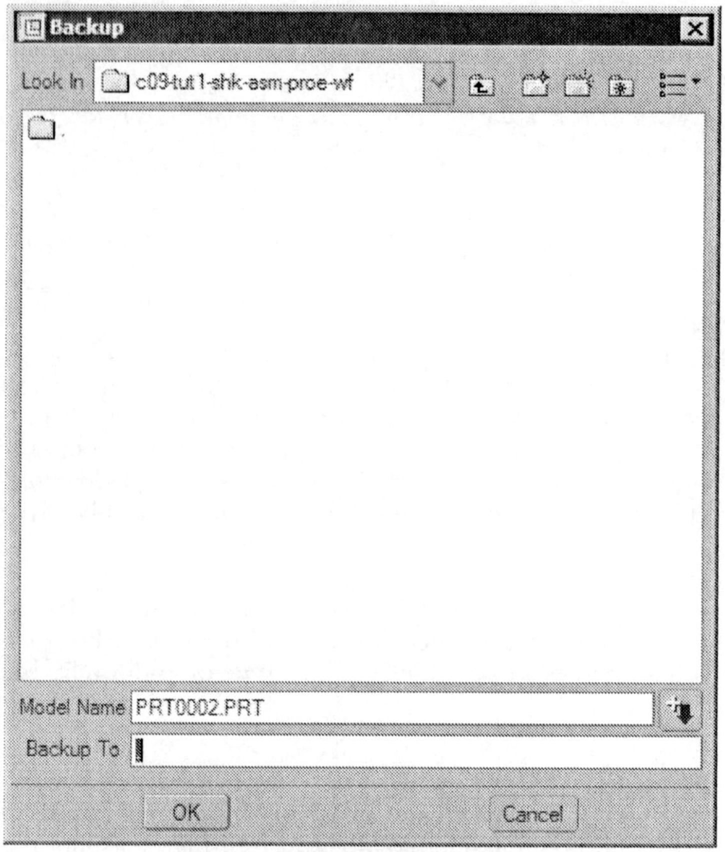

***Figure I-18 Backup** dialog box*

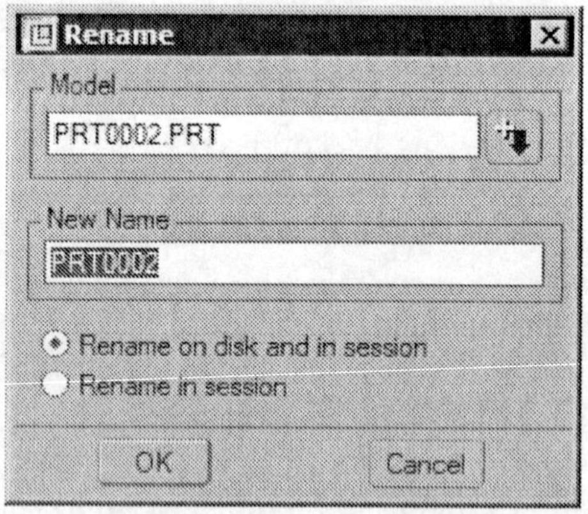

***Figure I-19 Rename** dialog box*

Note

*To save disk space you should keep deleting the old versions of a file. This is done using the **Delete** option in the **File** menu.*

Tip: *Suppose you open an assembly that has a component named Nut. Close the assembly and now open another assembly that has a component named Nut. Now, there are chances that the second assembly that you choose to open may open with the Nut that was present in the previous assembly. This is because the component with the file name Nut was already present in the memory of Pro/ENGINEER (In session) before opening the second assembly .*

To avoid this error of assemblies, you should erase the In session memory of Pro/ENGINEER before opening the second assembly.

MENU MANAGER

From this release of Pro/ENGINEER the **Menu Manager** is not available when you enter the Part mode, Assembly mode, or the Drawing mode. The **Menu Manager** is displayed with some selected options of feature creation. There are menus and submenus cascaded in the **Menu Manager**. In the **Menu Manager**, all the options are available to complete the desired task using Pro/ENGINEER Wildfire.

While using the **Menu Manager**, always complete the option selected by choosing **Done** or **Done Sel** after the current task is over. This is important when you are in the Drawing mode of Pro/ENGINEER. If you are directly selecting one option after another, then it is easy to loose track of commands or options in the **Menu Manager**.

Note
From this release of Pro/ENGINEER most of the features are created using the ***Dashboard****. The* ***Dashboard*** *is displayed above the* ***Message Area*** *and it contains all the options to complete the selected operation on the model.*

MODEL TREE

The **Model Tree** stores and displays all the features in a chronicle. You can select any desired feature of a model or an assembly from the model tree and apply different operations on the selected feature. You can select the feature by right-clicking on it. The shortcut menu will be displayed as shown in Figure I-20. Move the cursor on the shortcut menu and choose an option from the menu pressing the right mouse button. You can also select the feature using the left mouse button and then right-click to invoke the shortcut menu.

When you create a new object file, the **Model Tree** appears and is attached to the graphics window by default. But in this case with the **Model Tree** attached to the graphics window, the graphics window becomes small in size. You can hide the **Model Tree** by clicking the sash on the right edge of the Navigator as shown in Figure I-20. The Model Tree can slide out or slide in, thus increasing the area on the graphics window. It can also be stretched horizontally to cover the graphics window.

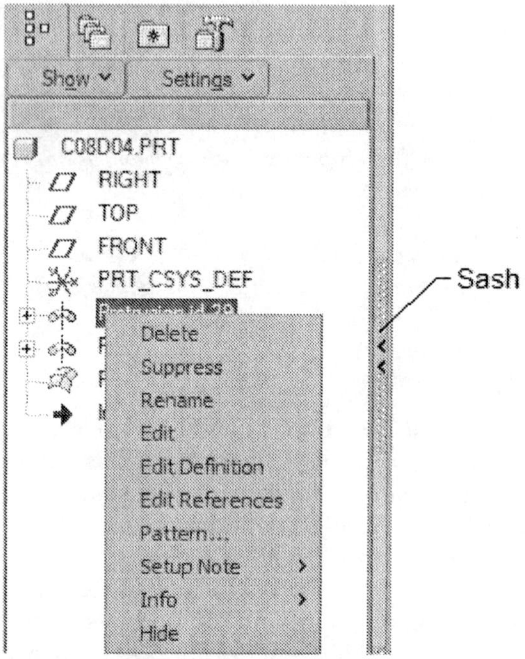

***Figure I-20** **Model Tree** with the shortcut menu*

Tip: *By looking at the **Model Tree** of a model, you can understand the method and approach used by the designer to create the model. If you have the **Model Tree** of a model you can modify the model. Generally when you import a model of a different file format in Pro/ENGINEER, you do not get the features of the model in the **Model Tree,** and therefore, you are not able to modify that model.*

UNDERSTANDING THE FUNCTION OF THE MOUSE BUTTONS

While working with Pro/ENGINEER Wildfire, it is important to understand the function of the three buttons of the mouse to make efficient use of this device. The various combinations of the keys and three buttons of the mouse are listed below.

1. Figure I-21 shows the function of the left mouse button. The left mouse button is used to make a selection. Using CTRL+left mouse button you can add or remove items from the selection set. Using SHIFT+left mouse button you can select a chain of entities or a set of surfaces.

2. Figure I-22 shows the function of the right mouse button. The right mouse button is used to invoke shortcut menus and to query select the items. When you bring the cursor on an item, it is highlighted in cyan color. Now if you hold down the right mouse button, a shortcut menu is displayed. Choose the **Pick From List** option from the shortcut menu.

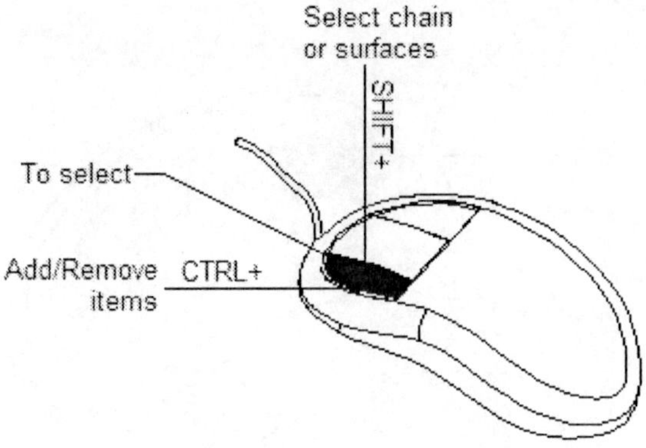

Figure I-21 Functions of the left mouse button

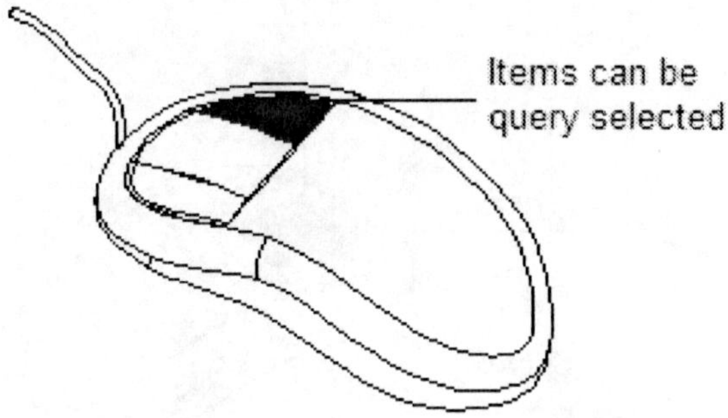

Figure I-22 Functions of the right mouse button

The **Pick From List** dialog box is displayed. You can make selections from this dialog box.

3. Figure I-23 shows the function of the middle mouse button in the 3D mode. The middle mouse button is used to spin the model in the graphics window and view it from different directions.

   CTRL+middle mouse button is used to dynamically zoom in and zoom out by moving the mouse. When you press and hold the CTRL key and move the mouse up, the view is reduced, and thus you zoom out. When the mouse is moved down, the view is enlarged, and thus you zoom in.

   When you use CTRL+middle mouse button and move the mouse horizontally, the model is turned.

   Shift+middle mouse button is used to pan the object on the screen.

4. Figure I-24 shows the function of the middle mouse button in the 2D mode (sketcher environment and Drawing mode). The middle mouse button is used to place dimensions on the graphics screen.

   It is also used to confirm an option or to abort the creation of an entity.

   The middle mouse button is used to pan in the Sketch mode and the Drawing mode.

Note
When you spin the model with the ***Spin Center*** *on, the model is spinned about the spin center. If the spin center is turned off, then the model is spinned about the cursor.*

TOOLBARS

Before starting work on Pro/ENGINEER Wildfire, it is very important for you to understand

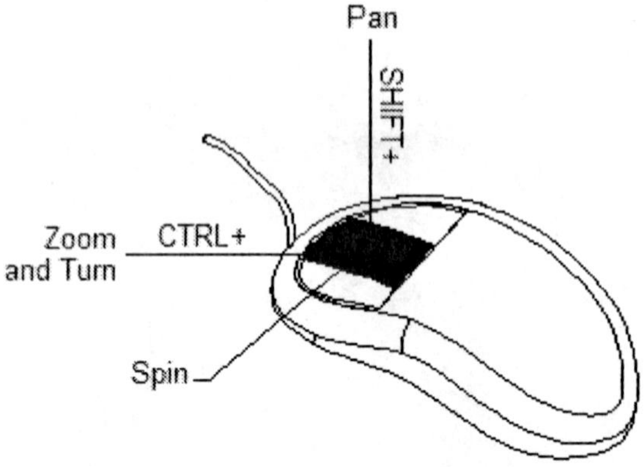

Figure I-23 *Functions of the middle mouse button in the 3D mode*

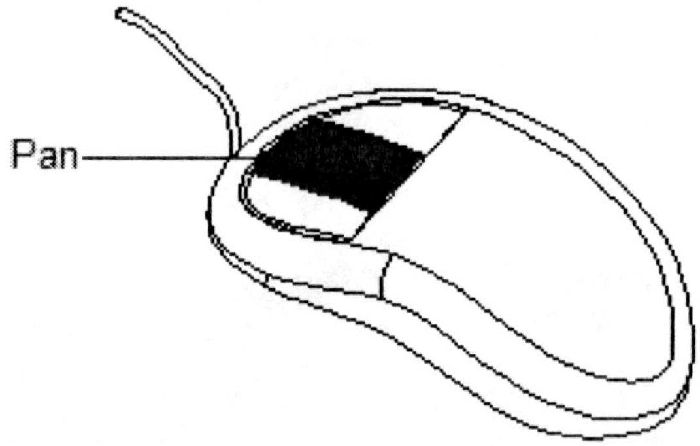

Figure I-24 *Functions of the middle mouse button in the 2D mode*

the default toolbars and buttons on the main window. Figure I-25 shows various default screen components in Pro/ENGINEER Wildfire.

In Pro/ENGINEER Wildfire, the toolbars are at two locations: the **Top Toolchest** and the **Right Toolchest**. The area on the top of the graphics window where the toolbars exist is called the **Top Toolchest** and the area on the right of the graphics window where the toolbars exist is called the **Right Toolchest.** The toolbars that initially appear on the screen are highlighted in the figure shown. Here you will notice that all the toolbar buttons are not available. These buttons are available after you open a new or existing file. Only those buttons are active that are required for the current session. As you proceed with the procedure to enter one of the modes provided by Pro/ENGINEER Wildfire, you will notice that the toolbar buttons required by that mode will be active. To make the designing easy and user-friendly, this software package has provided you with a number of toolbars. Different modes of Pro/ENGINEER Wildfire display different toolbars. Some of the frequently used toolbars are shown in Figure I-25.

Tip: *To add buttons to the toolbar and add toolbars to the Toolchest, choose* ***Tools > Customize Screen*** *from the menu bar. The* ***Customize*** *dialog box is displayed. This dialog box is used to customize the main window of Pro/ENGINEER Wildfire.*

Right-click on the ***Right Toolchest*** *to invoke a shortcut menu that has all the toolbar names that are available for the current session. Select any toolbar name to display that toolbar in the selected toolchest.*

Datum Display Toolbar

Figure I-26 shows the **Datum Display** toolbar. It has four buttons that are used to control the visibility of datum features like the datum planes, datum axes, and datum points. You can also use this toolbar to control the visibility of the coordinate systems.

File Toolbar

The **File** toolbar has six buttons. The buttons available in this toolbar are used to save a file,

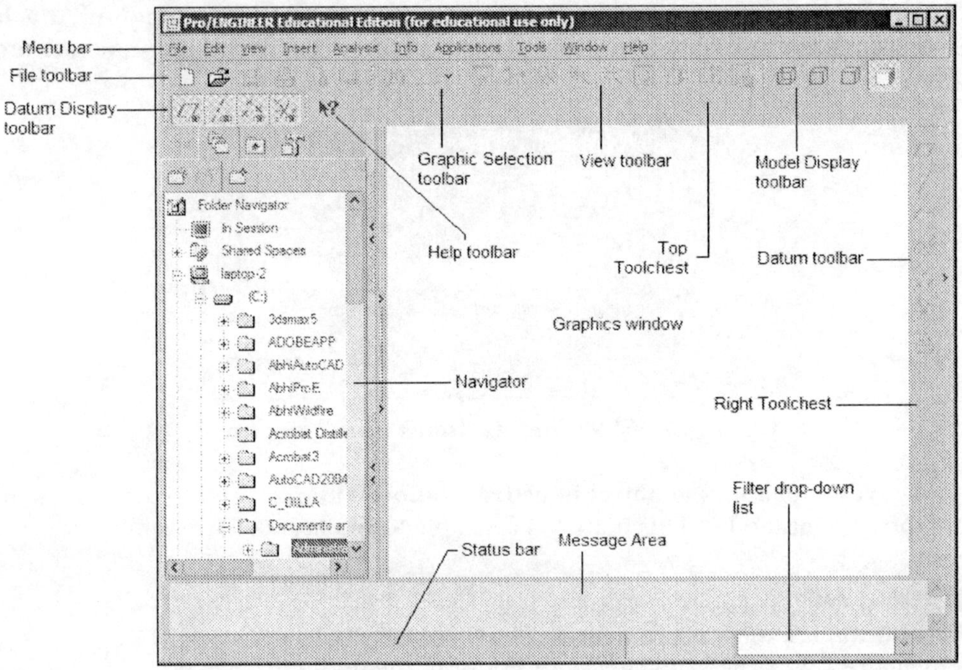

Figure I-25 Items on the main window

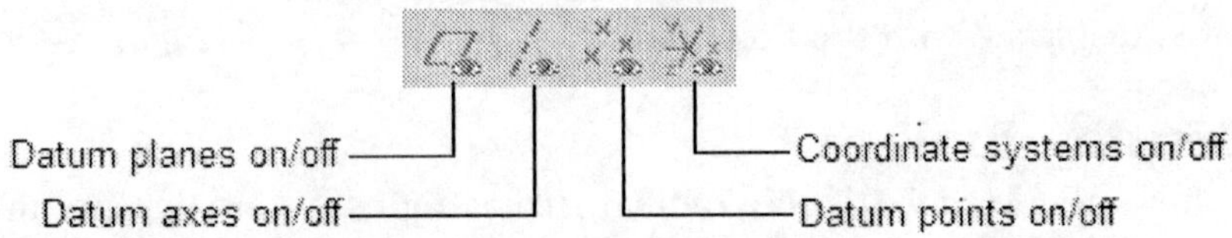

***Figure I-26 Datum Display** toolbar*

print the current file, open a new file, or open an existing file. The description of the buttons provided under this toolbar is shown in Figure I-27.

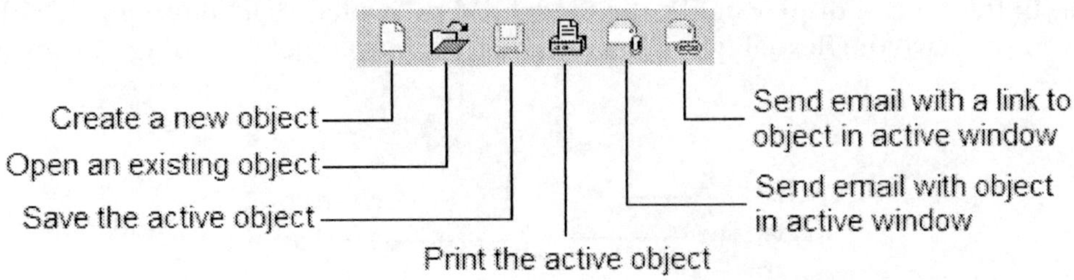

***Figure I-27 File** toolbar*

With this release of Pro/ENGINEER, there are two buttons added to the **File** toolbar. The **Send email with object in active window** button is used to send an email with the current model attached with mail. The attachment may be zipped by using the **Send As Attachment** dialog box which is displayed when you choose this button. The **Send As Attachment** dialog box is shown in Figure I-28.

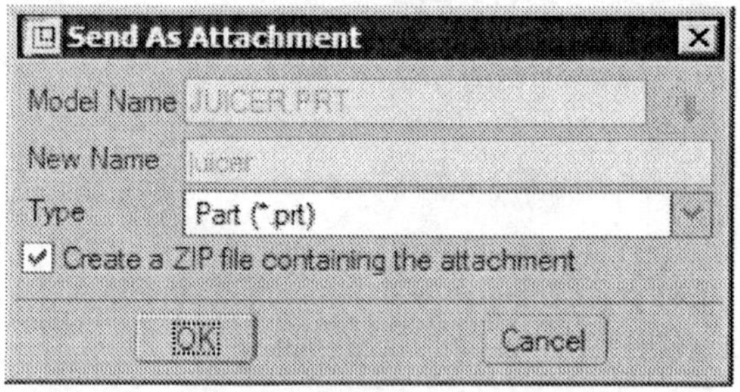

***Figure I-28 Send As Attachment** dialog box*

The **Send email with a link to object in active window** button is used to send an email with a link to the current model. This button is available to the users of Windchill.

Help Button

The **Help** button is known as the **Context Sensitive help** button, shown in Figure I-29. When you choose this button, the question mark symbol (?) is attached with the cursor. Depending on the location in the main window where you click, the **Pro/ENGINEER Help** window is displayed with the help topics.

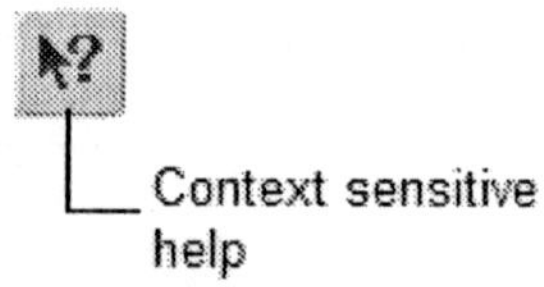

***Figure I-29 Help** button*

Model Display Toolbar

Figure I-30 shows the **Model Display** toolbar. It has four buttons that are used to set the display mode for viewing a model. The **Wireframe** button is used to display the model as a wireframe model. The visible and the invisible edges of the model will be displayed in this display mode. The **Hidden Line** button is used to display the visible lines of the model in bright color and the hidden lines in dull color. However, if you plot the model in this display mode, you will notice that the hidden lines are shown by dashed lines in the plot. The **No Hidden** button displays only the visible edges of the model. The last one is the **Shading** button. In this type of display mode, the model is displayed as shaded using a default color. However, you have the flexibility to alter the color of your model according to your needs.

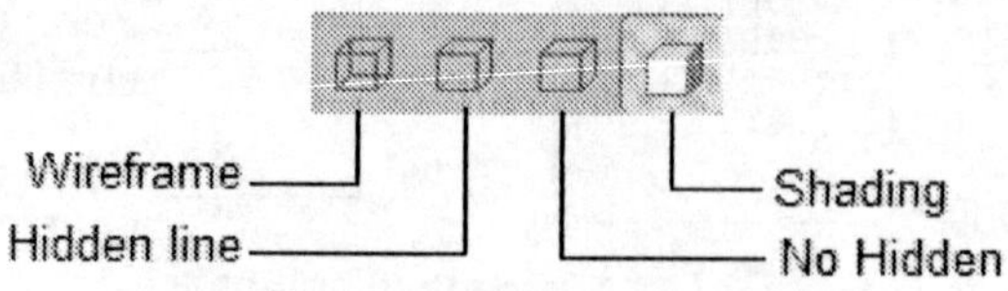

***Figure I-30 Model Display** toolbar*

View Toolbar

The **View** toolbar shown in Figure I-31 has ten buttons. The first button **Redraw the current**

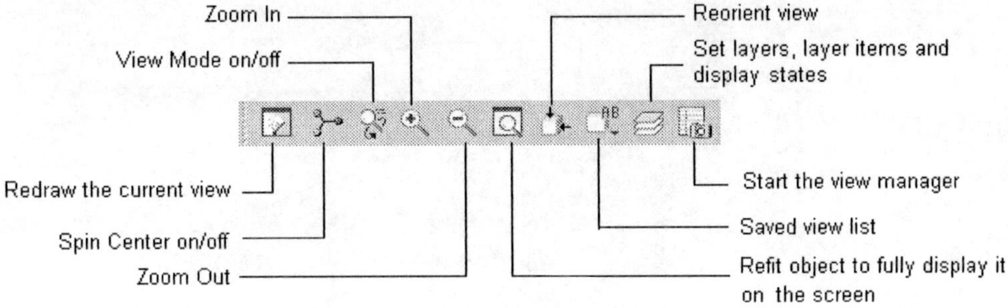

***Figure I-31 View** toolbar*

view is used for repainting the screen, meaning to remove any temporary information from the graphics screen. The second button **View Mode on/off** when chosen turns off the visibility of datums and allows you to spin the model using the middle mouse button. The third button toggles the visibility of the spin center. The fourth and fifth buttons are used to zoom in and zoom out the object. These options are discussed in Chapter 1. The **Refit object to fully display it on the screen** button is used to fit the model on the screen. The **Reorient view** button is used to orient the model. The next button is **Saved view list**. This button is used to display the saved views. The **Set layers, layer items and display states** button displays the layers in the navigation area.

Note

*The default system of units in Pro/ENGINEER Wildfire is in Inch Pound Seconds. To change the units choose **Edit > Setup** to display the **PART SETUP** menu in the **Menu Manager**. From this menu choose the **Units** option to invoke the **Units Manager** dialog box. Using this dialog box you can set the system of units available in Pro/ENGINEER Wildfire.*

NAVIGATOR

The Navigator is present on the left of the graphics window and can slide in or out on the graphics window. To make the Navigator slide in or out, you need to select the sash present on its right edge. The Navigator is shown in Figure I-32. It has the following functions:

1. When you browse files uisng the Navigator, the browser expands and the files in the selected folder are displayed in the browser.

2. When you open a model, the **Model Tree** is displayed in the navigation area.

3. The buttons on the top of the Navigator are used to display the different items in the navigation area. The **Model Tree** button is used to display the model tree in the navigation area. This button is available only when a model is opened on the screen. The **Folder Browser** button is used to display the folders that are in the local system of the user. The **Favorites** button is used to display the contents of the **Personal Favorites** folder. The **Connections** button is used to connect you to the network. This button is used only when you are connected to the internal or the external network.

4. Any other locations if available to your system can also be accessed by using the Navigator.

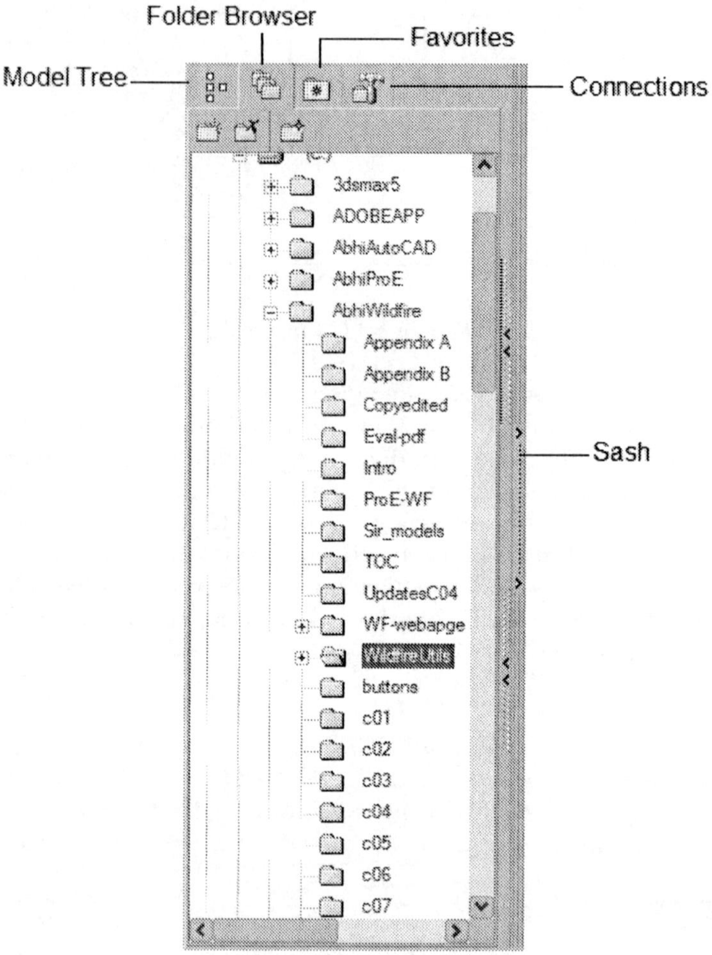

Figure I-32 Navigator

BROWSER

The Browser is present on the right of the navigator. It can slide in and out of the navigator and can also be stretched horizontally. When the Pro/ENGINEER Wildfire session is started, the browser is displayed on the screen. Figure I-33 shows the browser that displays some part files. The functions of the browser are listed next.

1. It is used to preview the Pro/ENGINEER Wildfire files and browse the file system.

2. A Pro/ENGINEER file can be opened by using the browser by double-clicking on the file or by dragging it to the graphics window. When you open a file using the browser, the model is displayed on the screen and the browser is closed automatically.

3. By default, the homepage of this browser is set to ptc.com. But you can view other Web sites and access ftp sites too.

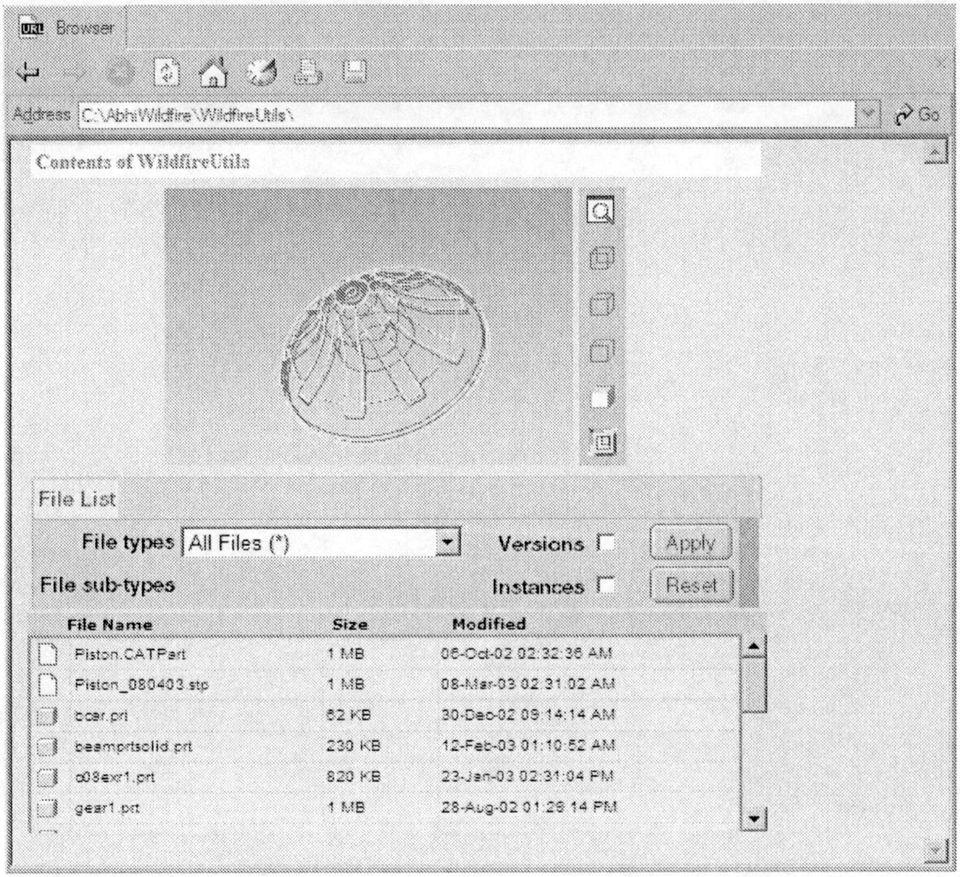

Figure I-33 *Pro/ENGINEER Wildfire Browser*

4. You can connect to your client's computer and jointly work with them using the Browser.

Chapter 1

Creating Sketches in the Sketch Mode-I

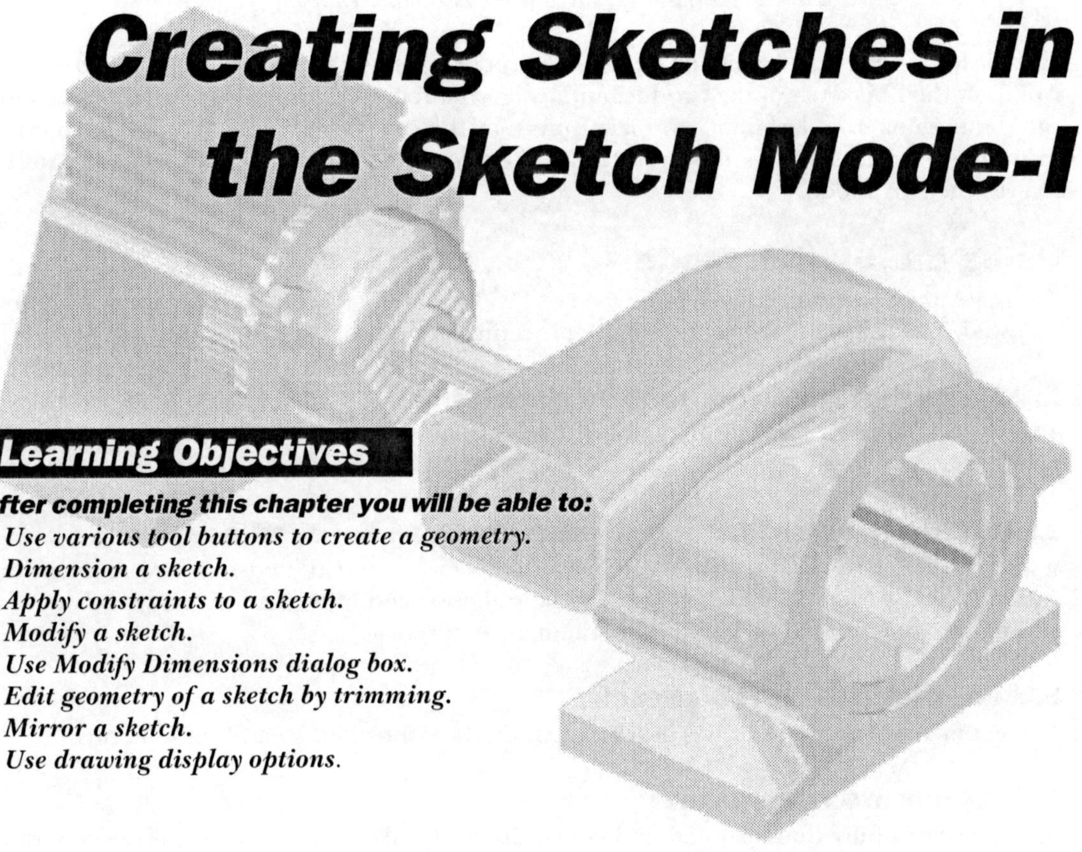

Learning Objectives

After completing this chapter you will be able to:

- *Use various tool buttons to create a geometry.*
- *Dimension a sketch.*
- *Apply constraints to a sketch.*
- *Modify a sketch.*
- *Use Modify Dimensions dialog box.*
- *Edit geometry of a sketch by trimming.*
- *Mirror a sketch.*
- *Use drawing display options.*

THE SKETCH MODE

Almost all the models designed in Pro/ENGINEER consists of datums, sketched features, and placed features. Generally, for creating datums and placed features, you do not require sketches. However, to create a three-dimensional (3D) feature, it is necessary to draw its two-dimensional (2D) sketch. When you enter the Part mode and select the options to create any sketched feature, the system automatically takes you to the sketcher environment. In the sketcher environment, the sketch of the feature is created, dimensioned, and constrained. Then you return to the Part mode to create the required feature. The sketches created in the Sketch mode are stored in the *.sec* format.

Note
You will learn about datums and placed features in later chapters.

The Sketch mode is used when the design of a product is at its development stage. The designer can draw the 2D sketch of the product and assign the required dimensions to it. By assigning the dimensions, the designer can make sure that the 2D sketch of the product or model is satisfying the necessary conditions. He can then continue for the 3D model of the design, that is, the Part mode.

Using the Sketch Mode

To create any section in the Sketch mode of Pro/ENGINEER, certain basic steps have to be followed. The following steps outline the procedure to use the Sketch mode:

1. Sketch the required section geometry

The various sketcher tools available in this mode can be used to sketch the required section geometry.

2. Add the constraints and dimension the sketched section

While sketching the section geometry, weak constraints and dimensions are automatically added to the section. The sketch can also be dimensioned and constrained manually. After adding the dimensions you can modify them as required.

3. Add relations to the sketch

The geometry of various entities of the sketch can be controlled by adding relations.

4. Regenerate the section

If the sketch is fully dimensioned and constrained, the sketch is automatically regenerated. Throughout this book, it is assumed that you are sketching in the Sketch mode with the **Intent Manager** on. Pro/ENGINEER has the capability to analyze the section, and if the section is not complete for any reason, the section will not be regenerated. You will know about these reasons as you go through this chapter.

Tip: *In Pro/ENGINEER, the phrase "Sketch is regenerated" means that the sketch is fully defined in terms of dimensions and constraints and is accepted by Pro/ENGINEER.*

Entering the Sketch Mode

To enter the Sketch mode, select **New** from the **File** menu or choose the **Create a new object** button from the **File** toolbar. The **New** dialog box is displayed, as shown in Figure 1-1, with various Pro/ENGINEER modes available. Select the **Sketch** radio button to open a new file in the Sketch mode. When you select the **Sketch** radio button, a default name of a sketch file appears in the **Name** edit box. You can change the sketch name as required. Choose the **OK** button to enter the Sketch mode.

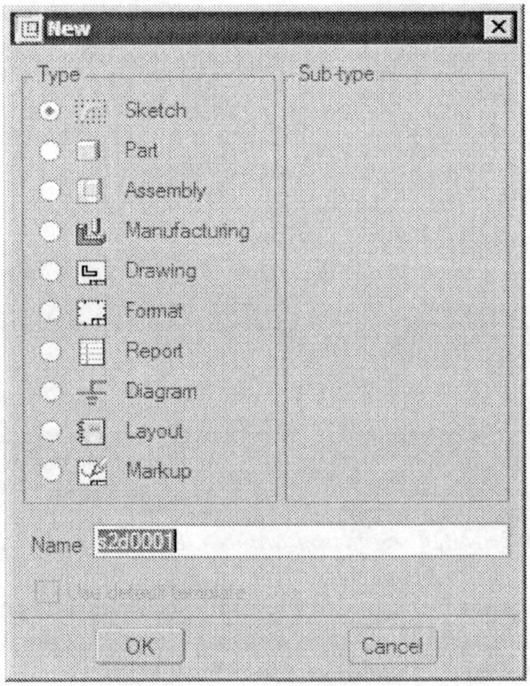

***Figure 1-1** **New** dialog box*

THE SKETCHER ENVIRONMENT

When you enter the Sketch mode, the initial screen appearance is similar to the one shown in Figure 1-2. This figure also shows the **Sketcher Tools** toolbar that is on the right of the graphics screen. The buttons available in this toolbar are used to draw a sketch. The various drawing options are also available in the **Sketch** menu in the menu bar. When you enter the sketcher environment, the **Intent Manager** is on by default. When you are in the selection mode, the shortcut menus can be invoked by holding down the right mouse button on the graphics screen. These shortcut menus depend on the item selected and the options in the shortcut menus vary accordingly. These shortcut menus also contain the options to draw the sketches.

Note

The selection mode in the sketcher environment is discussed later in this chapter.

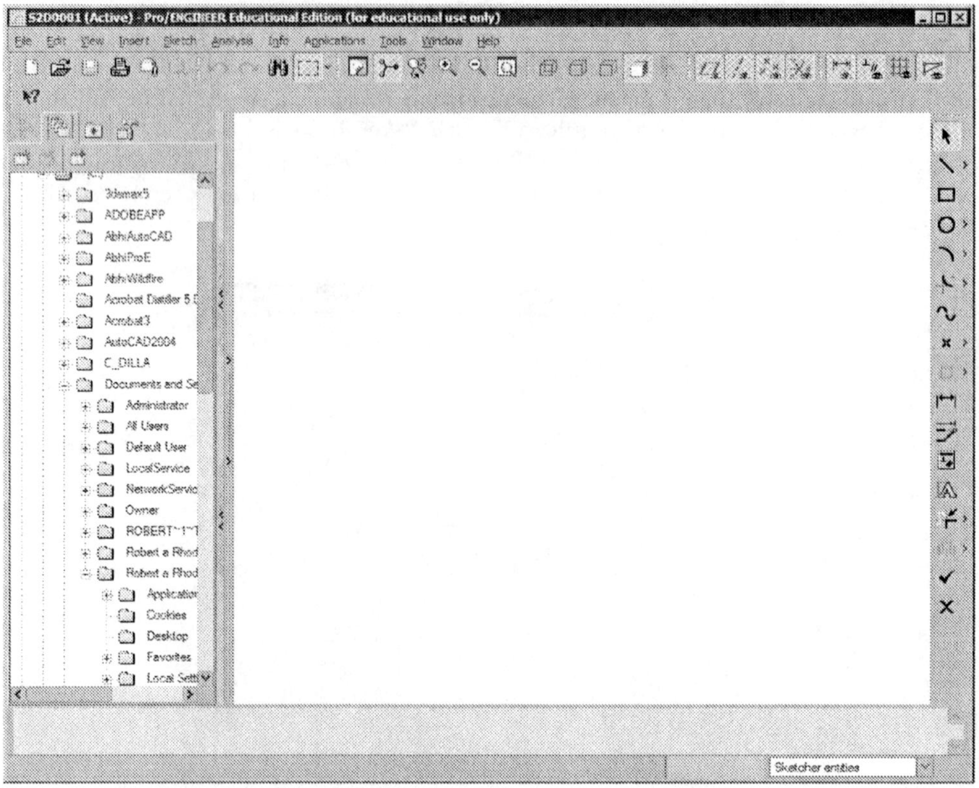

***Figure 1-2** Initial screen appearance after entering the Sketch mode*

The Navigator is displayed on the left of the graphics screen. It covers the graphics screen, and therefore the drawing area is decreased. Make the Navigator to slide in by clicking the Navigator sash. Now, the area available for sketching is increased.

Note

The function of the navigator is discussed in Introduction.

WORKING WITH THE SKETCH IN THE SKETCH MODE

When you enter the sketcher environment, the **Select items** button is selected by default. If this button is selected, the sketcher environment is said to be in the selection mode. In the selection mode, you can select entities from the sketch to edit them and to invoke the shortcut menu. The options in the shortcut menu can be used to apply various operations on the selected item.

Note

The sketch is saved as a .sec file extension. While drawing in the Part mode, if you save a drawing in the sketcher environment, the sketch is saved in the ***.sec*** *file extension.*

DRAWING A SKETCH USING THE SHORTCUT MENU

You can create a simple sketch by using the options available in the shortcut menu. You can draw lines, arcs that start at the endpoint of any existing geometry, and circles by using the left mouse button.

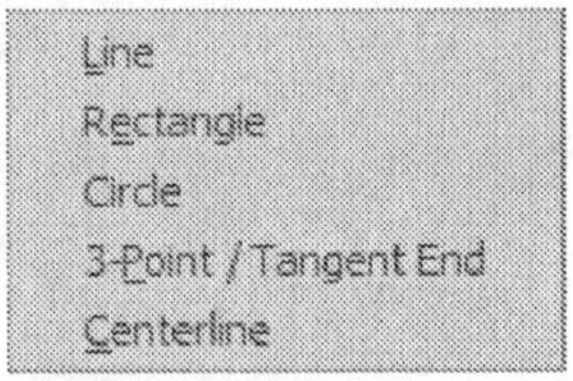

Figure 1-3 *Shortcut menu*

To invoke the shortcut menu, hold down the right mouse button on the graphics screen. Note that, once the shortcut menu is displayed, the right mouse button can be released. The shortcut menu that is displayed when you are in the selection mode is shown in Figure 1-3. These options in the shortcut menu are displayed only when there is no entity drawn on the graphics screen. This menu allows you to draw a sketch without selecting the tools from the **Sketcher Tools** toolbar. For example, you can draw a line, a circle, and an arc without selecting these tools from the **Sketcher Tools** toolbar. The procedure to draw various entities by using this shortcut menu is discussed next.

Drawing a Line Using the Shortcut Menu

The following steps explain the procedure to sketch a line using the shortcut menu:

1. Invoke the shortcut menu by holding down the right mouse button on the graphics screen.

2. Choose the **Line** option from the shortcut menu.

3. Using the left mouse button, specify a point on the graphics screen from where you want the line to start. A white rubber-band line appears, one end of which is fixed at the point you specified and the other is attached to the cursor. Now, move the cursor on the screen to a desired point where you want the line to end.

4. Specify the endpoint of the line by pressing the left mouse button. The line ends at this point. Note that line creation does not end at this point. The next rubber-band line yellow in color is attached to the cursor. The endpoint of the last line is the start point of this new line. This process will continue until you terminate line creation.

5. Press the middle mouse button to end the continuous line creation. The yellow rubber-band line disappears. Press it again to exit the tool. The lines drawn appear in yellow color.

Note

The above explanation for drawing lines is not for a single line creation. When you draw a single line, then the color of the line after you have drawn it is red.

After drawing a line when you press the middle mouse button to end the line creation, the line drawn is highlighted in red color. In the sketcher environment, the red color of an entity indicates that it is selected. If you press the DELETE key then the line will be erased from the graphics screen. After drawing a line, weak dimensions that appear in gray color are applied to the sketch. The weak dimensions are applied automatically to the sketcher entities as you draw them.

Drawing a Rectangle Using the Shortcut Menu

The following steps explain the procedure to sketch a rectangle using the shortcut menu:

1. Invoke the shortcut menu by holding down the right mouse button on the graphics screen.

2. Choose the **Rectangle** option from the shortcut menu.

3. Using the left mouse button, specify a point on the graphics screen to locate the first corner of the rectangle. A yellow rubber-band rectangle appears, one corner of which is fixed at the point you specified and the diagonally opposite corner is attached to the cursor.

4. Move the cursor and specify the location of the opposite corner on the graphics screen by pressing the left mouse button. The rectangle is created and appears in yellow color.

Note
If you want to abort sketching an entity, you can do so by pressing the middle mouse button.

Drawing a Circle Using the Shortcut Menu

The following steps explain the procedure to sketch a circle using the shortcut menu:

1. Invoke the shortcut menu by holding down the right mouse button on the graphics screen.

2. Choose the **Circle** option from the shortcut menu.

3. Specify the center of the circle you want to draw by pressing the left mouse button on the graphics screen. A white rubber-band circle appears that has its center at the specified point. The rubber-band circle is attached to the cursor.

4. You can now move the cursor away from the center point to give the circle the required size.

5. Once you get the appropriate size of the circle, press the left mouse button. The circle appears in yellow color.

Drawing an Arc Using the Shortcut Menu

You can draw two types of arcs using the shortcut menu. The first one is drawn by selecting three points on the graphics screen and the second one by drawing it tangent to an existing entity. The following steps explain the procedure to sketch an arc using the shortcut menu:

1. Invoke the shortcut menu by holding down the right mouse button on the graphics screen.

2. Choose the **3-Point / Tangent End** option from the shortcut menu.

3. Using the left mouse button, select the endpoint of an entity from where you want to start the arc. A circle with four quadrants formed by two lines appears at the selected point. This is called the **Target** symbol.

As you move the cursor out of one of the quadrant tangentially, a yellow rubber-band arc appears with one end attached to the cursor and the other end tangent to the entity.

If you want to draw the arc by specifying three points, then move the cursor out of the quadrant perpendicular to endpoint of the entity.

4. Now, move the cursor on the graphics screen to size the arc. If you are drawing an arc by specifying three points then the rubber-band arc will not appear.

5. When you get the required size of the arc, press the left mouse button to complete the arc. A yellow colored arc is sketched. You can abort the arc by pressing the middle mouse button.

Note

The color of the entities displayed depends on the system settings of the colors you set. The colors referred to above are the default system colors.

DRAWING A SKETCH USING THE SKETCHER TOOLS TOOLBAR

The **Sketcher Tools** toolbar that is available on the **Right Toolchest** when you are in the sketcher environment contains the tools to draw a sketch, dimension it, and modify the dimensions. In this section, you will learn to draw various drawing entities using the tools available in the **Sketcher Tools** toolbar.

Drawing a Point

Points are used mainly for dimensioning the vertices that are missing due to fillets. For example, if you create fillets at the corners of a sketch but want to show the dimensions of the sketch. In this case, you can draw points on the corners and then create the fillet at the corners. Now, since a point is present at the corner, you can easily dimension the sketch. The following steps explain the procedure to sketch a point:

1. Choose the **Create points** button from the **Sketcher Tools** toolbar. When you choose this button, the system prompts you to select a location for the point on the graphics screen.

2. As soon as you select a point by pressing the left mouse button, the point is placed at the desired location on the graphics screen.

Note

*To increase the number of visible command prompt lines in the **Message Area**, select the top sash of the **Message Area** using the left mouse button and drag it toward the screen.*

When you draw a single point no dimensions appear but when you draw two points then they are dimensioned with each other.

Drawing a Line

To create lines, there are three tool buttons available in the **Sketcher Tools** toolbar. To view the tools available to draw lines, choose the black arrow on the right of the **Create 2 point lines** button. The flyout appears with the three tool buttons. The first button is **Create 2 point lines**. This button is used to create a line by selecting two points on the graphics screen. The second button on the flyout is **Create lines tangent to 2 entities**. This button is used to create a tangent between two entities. The third button is **Create 2 point centerlines**. This button is used to create a centerline by selecting two points on the graphics screen. The centerline is used for creating revolved features, mirroring, and so on.

The procedures to create lines using the various tools are discussed next.

Drawing a line using the Create 2 Point lines button

The following steps explain the procedure to create a line using the **Create 2 point lines** button:

1. Choose the **Create 2 point lines** button. Select a point on graphics screen to start the line by pressing the left mouse button. A rubber-band line appears from the selected point with the other end attached to the cursor.

2. The system prompts you to specify the endpoint. Move the cursor on the graphics screen to give the desired length to the line. Press the left mouse button to specify the endpoint of the line. The line appears in yellow color. However, the rubber-band line continues to draw the second line.

3. Repeat step 2 until all the lines are drawn. End line creation by pressing the middle mouse button. To abort line creation, use the middle mouse button.

Drawing a line using the Create lines tangent to 2 entities button

The **Create lines tangent to 2 entities** button is used to draw a tangent between two entities such as arcs, circles, splines, or a combination of these. The following steps explain the procedure to draw a tangent using this button:

1. Choose the arrow on the right of the **Create 2 points line** button and then choose the **Create lines tangent to 2 entities** button from the **Sketcher Tools** toolbar. You are prompted to select two different arcs, circles, or splines.

2. Select the first entity from where the tangent line will be drawn. The color of the entity changes to red. You are prompted to select the second entity. As soon as you select the second entity, a line that is tangent to both the selected entities is drawn.

Note

Whenever you are prompted to select an entity in the sketcher environment, the ***SELECT*** *dialog box is displayed. You can ignore this dialog box because it appears automatically and disappears without any confirmation.*

Drawing a centerline using the Create 2 point centerlines button

You can draw horizontal, vertical, or inclined centerline using the **Create 2 point centerlines** button. This button is available in the flyout that is displayed when you choose the black arrow on the right of the **Create 2 point lines** button. The centerline in a sketch is used as axis of rotation, for mirroring entities, for alignment, and for dimensioning.

The following steps explain the procedure to draw a centerline:

1. Choose the black arrow on the right of the **Create 2 point lines** button and then choose the **Create 2 point centerlines** button from the **Sketcher Tools** toolbar. You are prompted to select the start point.

2. Using the left mouse button, select a point on the graphics screen to specify the start point. You are prompted to select the end point.

3. Select an endpoint to create the centerline. A centerline of infinite length is drawn.

Drawing a Rectangle

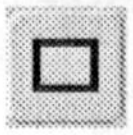

The following steps explain the procedure to sketch a rectangle using the **Create rectangle** button:

1. Choose the **Create rectangle** button from the **Sketcher Tools** toolbar. You are prompted to select two points to indicate the diagonal of box. Select the first point.

2. As you select the first point by pressing the left mouse button, a yellow rubber-band box appears with the cursor attached to the opposite corner of the box. Move the cursor to the desired location on the graphics screen to size the diagonal of the rectangle. Press the left mouse button to select the second point for the diagonal of the rectangle.

Drawing a Circle

To draw a circle there are four tool buttons and one button to draw an ellipse in the **Sketcher Tools** toolbar. To view the tool buttons available to draw circles and ellipses, choose the black arrow on the right of the **Create circle by picking the center and a point on the circle** button. The flyout appears with five tool buttons.

The procedures to create circle and ellipse using the various tool buttons are discussed next.

Drawing a circle using the Create circle by picking the center and a point on the circle button

As the name suggests, the **Create circle by picking the center and a point on the circle** button is used to draw a circle by specifying the center of the circle and a point on the circle. The following steps explain the procedure to draw a circle using the **Create circle by picking the center and a point on the circle** button from the **Sketcher Tools** button:

1. Choose the **Create circle by picking the center and a point on the circle** button. You are prompted to select the center of the circle.

2. Specify the center point for the circle on the graphics screen by pressing the left mouse button.

3. You are prompted to select a point on the circle. A yellow rubber-band circle appears with the center at the specified point and the cursor attached to its circumference. Move the cursor to size the circle. Press the left mouse button to complete circle creation. You are again prompted to select the center of the circle.

4. Repeat steps 2 and 3 until you have drawn all the required circles. If you want to abort circle creation before completing it, press the middle mouse button.

Drawing a Construction Circle

A construction circle is a circle that is used to align entities, is used for diametrical or radial dimensioning, and is used to reference entities. Figure 1-4 shows an application of construction circle. In the sketch of a flange, centers of the circles lie on a particular bolt circle diameter (BCD) that is defined using a construction circle.

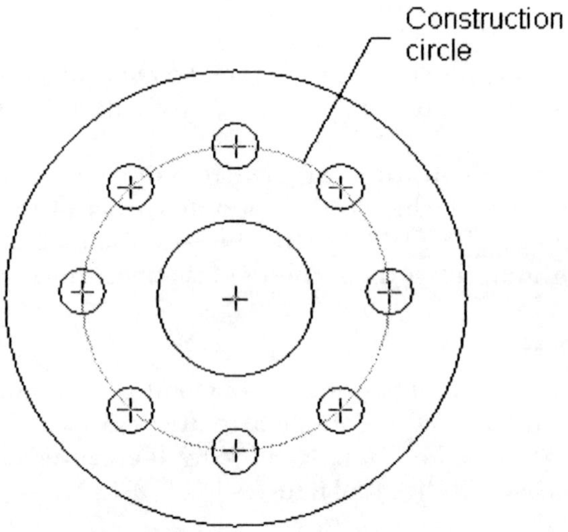

Figure 1-4 *Sketch of a flange*

To create a construction circle, draw a circle and then hold down the right mouse button on it to invoke the shortcut menu as shown in Figure 1-5.

Choose the **Construction** option from the shortcut menu. Pick a point on the screen to remove the circle from the selection set. The circle appears yellow in color and with a dotted line indicating that it is a construction circle.

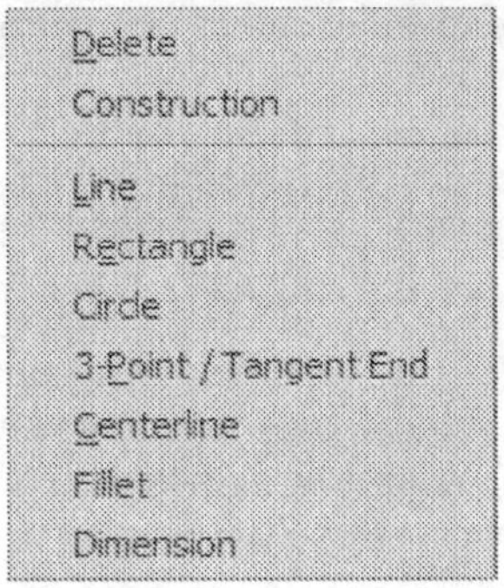

Figure 1-5 Construction
option in the shortcut menu

Tip: *To convert a construction circle back to a solid entity, select the construction circle and hold down the right mouse button to invoke the shortcut menu. Choose the **Solid** option from this shortcut menu.*

Drawing a circle using the Create Concentric circle button

The following steps explain the procedure to draw a concentric circle using the **Create Concentric circle** button:

1. Choose the **Create Concentric circle** button from the flyout in the **Sketcher Tools** toolbar. You are prompted to select an arc to determine the center. You can select an arc or a circle to specify the center point.

2. Press the left mouse button to select an arc or a circle to determine the concentricity of the circle to be drawn. Move the mouse to size the circle.

3. After sizing the circle, finish circle creation by pressing the left mouse button. Press the middle mouse button to end circle creation or to abort it.

Drawing a circle using the Create circle by picking its 3 points button

The following steps explain the procedure to draw a circle using the **Create circle by picking its 3 points** button:

1. Choose the **Create circle by picking its 3 point** button from the flyout in the **Sketcher Tools** toolbar. You are prompted to specify the first point on the circle.

2. Select the first point at the desired location on the graphics screen by pressing the left mouse button. You are prompted to select the second point on the circle. Move the cursor to select the second point on the graphics screen.

3. As soon as you select the second point, a yellow rubber-band circle appears with the cursor attached to it. You are prompted to select the third point. Move the mouse to size the circle. A circle is drawn when you select the third point by pressing the left mouse

button. You will again be prompted to select first point on the circle to draw the next circle.

4. Repeat step 2 until you have drawn all the circles. Press the middle mouse button to end circle creation or to abort circle creation before the circle is completed.

Drawing a circle using the Create a circle tangent to 3 entities button

The **Create a circle tangent to 3 entities** button is used to draw a circle that is tangent to three existing entities. This tool references other entities to draw a circle. The circle drawn using this tool is drawn irrespective of the points selected on the entity. The following steps explain the procedure to draw a circle using the **Create a circle tangent to 3 entities** button:

1. Choose the **Create a circle tangent to 3 entities** button from the flyout in the **Sketcher Tools** toolbar. You are prompted to select the start location on an arc, circle, or line.

2. Select the first entity using the left mouse button. The color of the entity changes to red. You are prompted to select an end location on an arc, circle, or line. Select the second tangent entity. Next, you are prompted to select the third location on an arc, circle, or line. Select the third tangent entity.

   As you select all the three entities, a circle that is tangent to all the three entities is drawn. You are again prompted to select the first entity for the second circle if required to be drawn.

3. Repeat step 2 until you have drawn all the circles. To end the creation of circle using this option or to abort it, press the middle mouse button.

Drawing an ellipse using the Create a full ellipse button

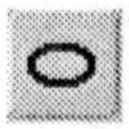

The following steps explain the procedure to draw an ellipse using the **Create a full ellipse** button:

1. Choose the **Create a full ellipse** button from the flyout in the **Sketcher Tools** toolbar. You are prompted to specify the center of the ellipse.

2. Select the center point at the desired location on the graphics screen by pressing the left mouse button. A yellow rubber-band ellipse appears with the cursor attached to the ellipse. Move the cursor on the graphics screen to size the ellipse.

3. An ellipse is drawn when you select the second point by pressing the left mouse button.

Drawing an Arc

To draw an arc there are five tool buttons in the **Sketcher Tools** toolbar. To view these tool buttons, choose the black arrow on the right of the **Create an arc by 3 points or tangent to an entity at its endpoint** button. The flyout appears with five tool buttons.

The procedures to draw arcs using various tool buttons in the flyout are discussed next.

Drawing an arc using the Create an arc by 3 points or tangent to an entity at its endpoint button

The **Create an arc by 3 points or tangent to an entity at its endpoint** button is used to draw arcs tangent from the endpoint of an existing entity or by defining three points on the graphics screen.

When you choose this tool button to draw an arc from an endpoint, the **Target** symbol is displayed as soon as you select the endpoint. This **Target** symbol is in the form of a circle that is divided into four quadrants. The following steps explain the procedure to draw an arc from the endpoint of an existing entity by using this tool button:

1. Choose the **Create an arc by 3 points or tangent to an entity at its endpoint** button from the **Sketcher Tools** toolbar. You are prompted to select the start point of the arc.

2. Specify three points on the graphics screen to draw an arc. If you want to draw an arc from the endpoint of an existing entity, select the endpoint of that entity. As soon as you select the endpoint, the **Target** symbol appears at the endpoint of the entity. Move the cursor along the tangent direction through a small distance. A yellow rubber-band arc appears with one end attached to the endpoint and the other end attached to the cursor. Note that when you move the cursor out of the **Target** symbol perpendicular to the endpoint, the arc is drawn by specifying three points. In this case the rubber-band arc does not appear as shown in Figure 1-6.

   On the other hand, if you move the cursor out horizontally from one of the quadrants of the **Target** symbol, the arc is drawn tangent to the endpoint as shown in Figure 1-7.

3. Move the cursor to the desired position on the graphics screen to size the arc. Use the left mouse button to complete the arc.

Tip: *If you do not want to draw a tangent arc, move the cursor out of the* ***Target*** *symbol perpendicular to the endpoint.*

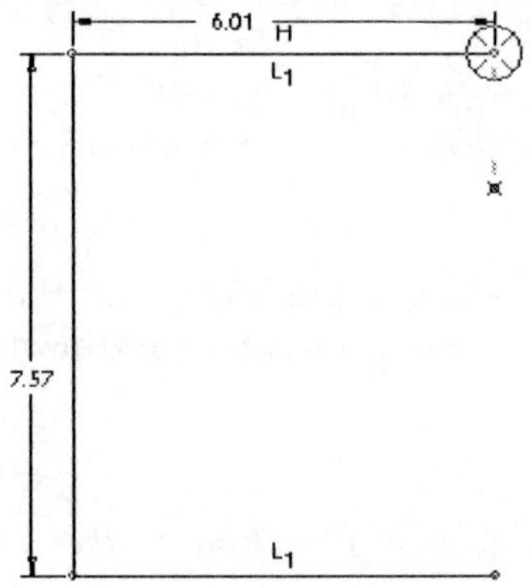

Figure 1-6 *The* ***Target*** *symbol to draw three point arc*

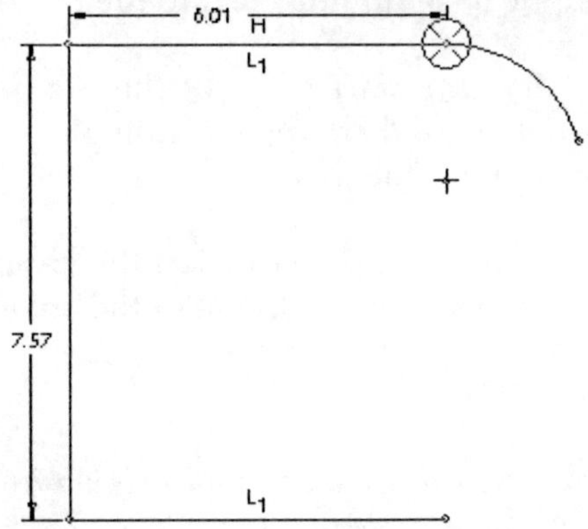

Figure 1-7 *The* ***Target*** *symbol to draw tangent arc*

Drawing an arc using the Create concentric arc button

The **Create concentric arc** button is used to draw an arc concentric to an existing arc. You will have to select an entity to which the arc will be concentric. The entity selected must be an arc or a circle. The following steps explain the procedure to draw an arc using this tool button:

1. Choose the **Create concentric arc** button from the flyout in the **Sketcher Tools** toolbar. You are prompted to select an arc to determine the center of the arc to be created. The entity to be selected should be an arc or a circle.

2. As soon as you select an entity, you are prompted to select the start point of the arc. Select the start point of the arc. A yellow rubber-band arc will appear with one end attached to the start point and center point at the center of the selected arc. The size of the arc will change as you move the cursor. Move the mouse to size the arc. You are prompted to select the endpoint of the arc.

3. Specify the endpoint by pressing the left mouse button.

4. Repeat steps 2 and 3 until you have drawn the required number of arcs. You can end arc creation by pressing the middle mouse button.

Drawing an arc using the Create an arc by picking its center and endpoints button

The following steps explain the procedure to draw an arc using the **Create an arc by picking its center and endpoints** button:

1. Choose the **Create an arc by picking its center and endpoints** button from the flyout in the **Sketcher Tools** toolbar. You are prompted to select the center of the arc.

2. Select a center point for the arc on the graphics screen by pressing the left mouse button. A yellow colored center mark appears at that point on the graphics screen. Now, you are prompted to select the start point of the arc. As you move the cursor, a dotted circle appears and is attached to the cursor.

3. Select the start point of the arc on the circumference of the dotted circle. A yellow rubber-band arc appears from the start point. The size of this arc changes dynamically as you move the mouse.

4. You are prompted to select the endpoint of the arc. Move the mouse to size the arc, and then select the endpoint of the arc using the left mouse button. An arc is drawn between the two points selected.

Note

Note that you can draw only one arc with one center. If you want to draw another arc you will have to select the center again.

Drawing an arc using the Create an arc tangent to 3 entities button

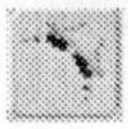

The **Create an arc tangent to 3 entities** button is used to draw an arc that is tangent to three selected entities. The following steps explain the procedure to draw an arc using this tool button:

1. Choose the **Create an arc tangent to 3 entities** button from the flyout in the **Sketcher Tools** toolbar. You are prompted to select the start location on an arc, circle, or line.

2. As soon as you select the first entity, the color of the entity changes to red. Now, you are prompted to select the end location on an arc, circle, or line.

3. Select the second entity. You are prompted to select a third location on an arc, circle, or line. Select a third entity. An arc is created instantly when all the three entities are selected. The arc drawn is tangent to all the three entities selected.

4. Repeat steps 2 and 3 until you have drawn the required number of arcs. If you want to abort arc creation, you can press the middle mouse button.

Drawing an arc using the Create a conic arc button

The **Create a conic arc** button is used to draw a conic arc. The following steps explain the procedure to draw a conic arc using this tool button:

1. Choose the **Create a conic arc** button from the flyout in the **Sketcher Tools** toolbar. You are prompted to specify the first endpoint of the conic entity.

2. Specify a point on the screen. You are prompted to specify the second endpoint of the conic arc.

3. Specify the second endpoint. A centerline is drawn between the two points. Now, you are prompted to specify the shoulder point of the conic arc. Specify a point on the screen. The conic arc is drawn.

Note

If you delete the centerline of the conic arc, the arc will not be deleted.

Remember that if the conic arc is the only entity on the graphics screen, then you cannot delete its centerline.

DIMENSIONING THE SKETCH

After you draw a sketch, the next step involves the dimensioning of the sketch. The basic purpose of dimensioning in Pro/ENGINEER is to control the size of the sketch and to locate it with some reference. In Pro/ENGINEER, a sketch cannot be regenerated unless it is fully dimensioned and constrained. The phrase "the sketch cannot be regenerated" means that the sketch cannot be accepted by Pro/ENGINEER.

By default, the sketched entities are dimensioned and constrained automatically while sketching or as soon as you sketch them. However, sometimes you need to add additional dimensions to the sketch. The **Create defining dimension** button is used to manually dimension the entities. You can also choose **Sketch > Dimension** to display a cascaded menu as shown in Figure 1-8. The menu contains the various dimensioning options.

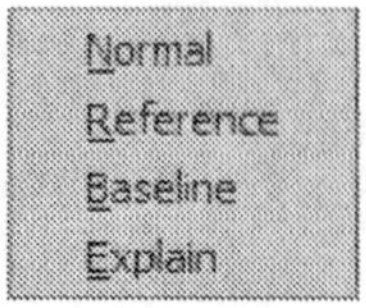

Figure 1-8 *Dimensioning options*

Note

*When the **Intent Manager** is off, the dimensions and constraints are not automatically applied. You need to manually add the dimensions by choosing the **Dimension** option from the **SKETCHER** menu. The **DIMENSION** submenu appears with various options to dimension the sketch.*

Dimensioning a Sketch Using the Create defining dimension button or the Normal Option

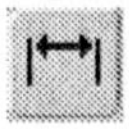

The **Create defining dimension** button or the **Normal** option are used for normal dimensioning of the sketch. The following steps explain the procedure to dimension a sketch:

1. Choose the **Create defining dimension** button from the **Sketcher Tools** toolbar. Select the entity you want to dimension by pressing the left mouse button. The color of the entity changes from yellow to red.

2. Place the dimension at the desired place by pressing the middle mouse button. The dimension appears in yellow color. You can modify the dimension values using the modifying options discussed later in this chapter.

The remaining options in the cascaded menu are not used much while sketching and therefore they are discussed in Chapter 2.

DIMENSIONING THE BASIC SKETCHER ENTITIES

The procedure to dimension the sketcher entities such as lines, arcs, circles, revolved sections, and so on is discussed next.

Linear Dimensioning a Line

You can dimension a line by selecting its endpoints or by selecting the line. After selecting the two endpoints or the line, press the middle mouse button to place the dimension. If the line is inclined and you select the two endpoints to dimension, then the location where you press the middle mouse button is important. The location on the graphics screen where you press the middle mouse button to place the dimension defines the orientation of the dimension that is displayed on the screen.

Figure 1-9 explains the three possible orientations of dimensions that can be displayed when you dimension a line.

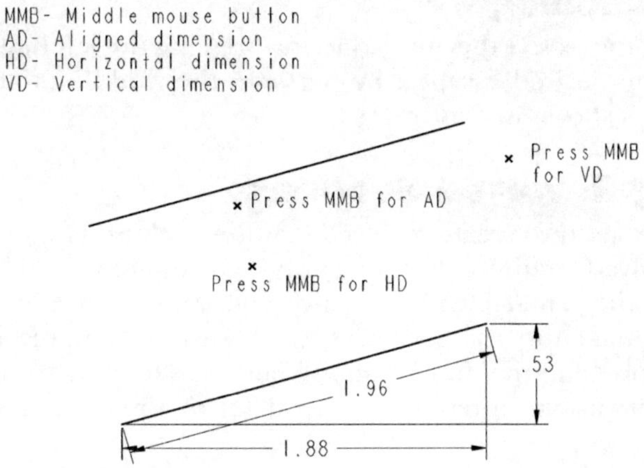

Figure 1-9 *Approximate locations to press the middle mouse button to achieve the different dimensions*

Note

It is not possible to dimension a line in all the three orientations at the same time in the sketcher environment. The dimensions in Figure 1-9 are only for explanation.

Angular Dimensioning an Arc

To add angular dimension to an arc, select both the ends of the arc by pressing the left mouse button and then select a point on the arc. Next, place the dimension at the desired point by pressing the middle mouse button. The dimension appears as shown in Figure 1-10. You can modify the dimension using a tool button that is discussed later.

Diameter Dimensioning

For diameter dimensioning, select the entity twice by pressing the left mouse button. Then place the dimension at the desired place by pressing the middle mouse button. The diameter dimension appears as shown in Figure 1-11. The same diameter dimensioning technique is also used for arcs.

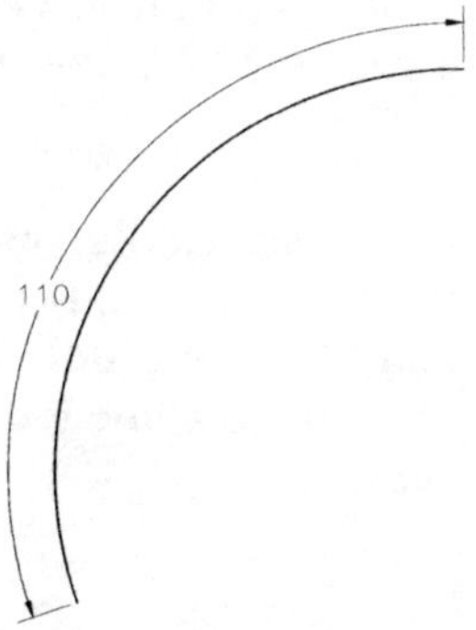

Figure 1-10 *Angular dimensioning technique*

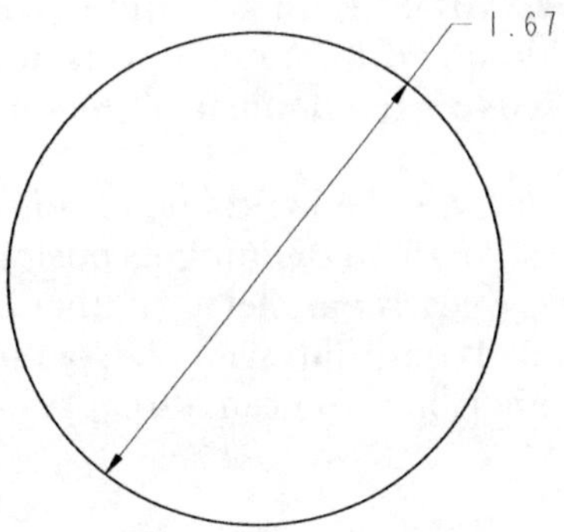

Figure 1-11 *Diameter dimensioning technique*

Radial Dimensioning

For radial dimensioning, select the entity once by pressing the left mouse button. Then place the dimension at the desired location by pressing the middle mouse button. The radial dimension appears as shown in Figure 1-12.

Dimensioning Revolved Sections

Revolved sections are used to create revolved features such as flanges, couplings, and so on. To dimension a revolved section, select the entity to be dimensioned by using the left mouse button. Next, select the centerline about which you want the section to be revolved. Once again select the original entity that you want to dimension. Now, place the dimension at the desired location by pressing the middle mouse button. The dimension appears as shown in Figure 1-13. This dimension represents the diameter of a revolved section.

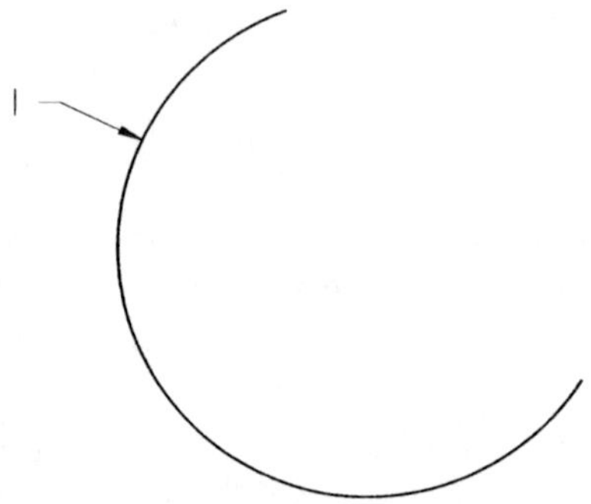

Figure 1-12 *Radial dimensioning technique*

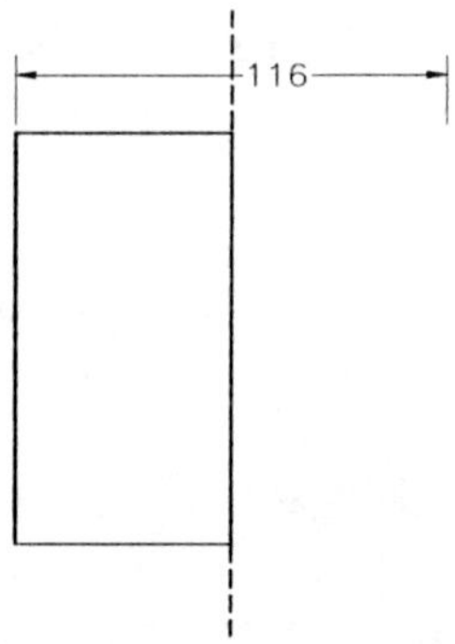

Figure 1-13 *Dimensioning technique for revolved sections*

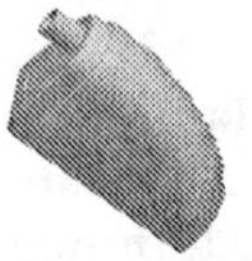

Tip: *You can also first select the centerline, then the entity to dimension, and then again the centerline to add dimension to revolved sections.*

WORKING WITH CONSTRAINTS

In Pro/ENGINEER, the entities in a sketch have to be fully specified in terms of size, shape, orientation, and location. This is achieved by setting constraints. Using constraints in the sketch reduces the number of dimensions in that sketch.

Constraints are the logical operations that are performed on the selected geometry to make it more accurate in defining its position with respect to the other geometry. For example, if a line is very nearly parallel to another line, Pro/ENGINEER snaps the line parallel and displays the parallel constraint symbol. Now, if you confirm the line creation, the line is drawn parallel to the other line. You can also apply constraints manually.

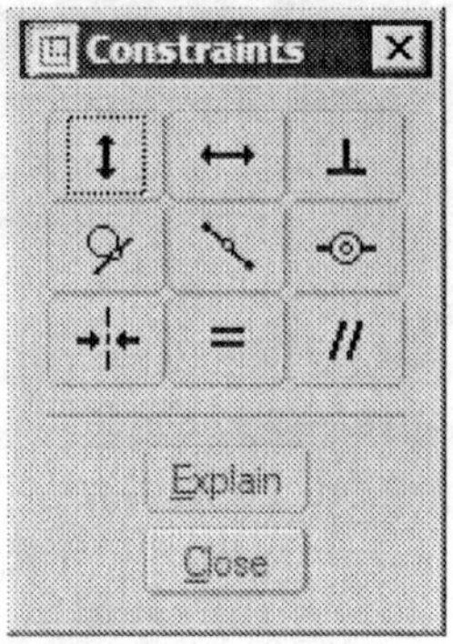

Figure 1-14
***Constraints** dialog box*

There are two types of constraints in Pro/ENGINEER, **Geometry** constraints and **Assembly** constraints. Here, you will learn about the **Geometry** constraints and the **Assembly** constraints will be discussed in later chapters.

To apply constraints manually, choose the **Impose sketcher constraints on the section** button from the **Sketcher Tools** toolbar to display the **Constraints** dialog box. This dialog box is shown in Figure 1-14.

This dialog box is used to apply constraints manually. Although the constraints are applied automatically as you draw the sketch, you can use this dialog box if you want to manually apply additional constraints. The constraints that are applied automatically are weak constraints. Weak constraints appear in gray color. Weak constraints can be made strong. This is discussed later in this chapter. The various options in the **Constraints** dialog box are discussed next.

Make line or two vertices vertical

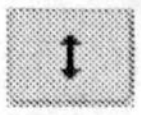

This constraint forces the selected line segment to become a vertical line. This constraint also forces the two vertices to be placed along a vertical line.

Make line or two vertices horizontal

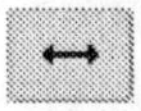

This constraint forces the selected line segment or two vertices that are apart by some distance to become horizontal or to lie in a horizontal line.

Make two entities perpendicular

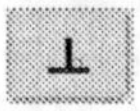

This constraint forces the selected entity to become normal to another selected entity.

Make two entities tangent

This constraint forces the two selected entities to become tangent to each other. You are prompted to select two entities that you want to make tangent to each other.

Place point on the middle of the line

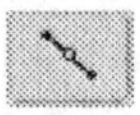

This constraint forces a selected point or vertex to lie on the middle of a line.

Create same points, points on entity or collinear constraint

This constraint performs three functions. This constraint can be used to force the two selected points to become coincident, to constrain a point on the selected entity, and to make two selected entities collinear, so that they lie on the same line. This constraint aligns two vertices or entities.

Make two points or vertices symmetric about a centerline

This constraint makes a section symmetrical about the centerline. When you select this constraint, you are prompted to select a centerline and two vertices to make them symmetrical.

Create Equal Lengths, Equal Radii, or Same Curvature constraint

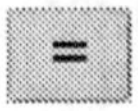

This constraint forces any two selected entities to become of equal dimension. When you select this constraint, you are prompted to select two lines to make their length the same, or two arcs, circles, or ellipses to make their radii equal. You can also force the curvature of a spline to become equal to a selected line or arc.

Make two lines parallel

This constraint is used to force two lines to become parallel. When selected, this constraint prompts you to select two entities that you want to make parallel. The two selected entities become parallel to each other.

Explain Option

The **Explain** option of the **Constraints** dialog box provides information about the constraints that are applied to a sketch. The constraints in the sketch are displayed as symbols. When you choose the **Explain** button, you are prompted to select the constraint or dimension on which you want the explanation. Select the symbol using the left mouse button. The information about the selected constraint is displayed in the **Message Area**.

Note

This option is generally helpful when you view a sketch drawn by some other person. By using the ***Explain*** *option you can obtain information about the various constraints applied in the sketch.*

Disabling the Constraints

The need to disable a constraint arises when you are drawing an entity. For example, if you draw a circle at some distance apart from a circle. While drawing it, the system tends to apply the equal radius constraint when the sizes of the two circles become equal. If at this moment you do not want to apply the equal radius constraint, right-click to disable the equal radius constraint. When you right-click to disable a constraint, an orange / line appears across the symbol. To enable the constraint, right-click once again.

Converting a Weak Constraint into a Strong Constraint

As discussed earlier, when you draw a sketch, some weak dimensions are automatically applied to the sketch. As you proceed to complete the sketch, these dimensions are automatically deleted from the sketch without any confirmation.

Select a weak dimension or a weak constraint from the graphics screen. The selected dimension or constraint is highlighted in red. Press and hold the right mouse button to invoke the shortcut menu as shown in Figure 1-15. Choose the **Strong** option from the shortcut menu.

You can also choose **Edit** > **Convert To** > **Strong** from the menu bar. The color of the selected dimension is changed from gray to yellow, indicating that the selected constraint or dimension is made permanent.

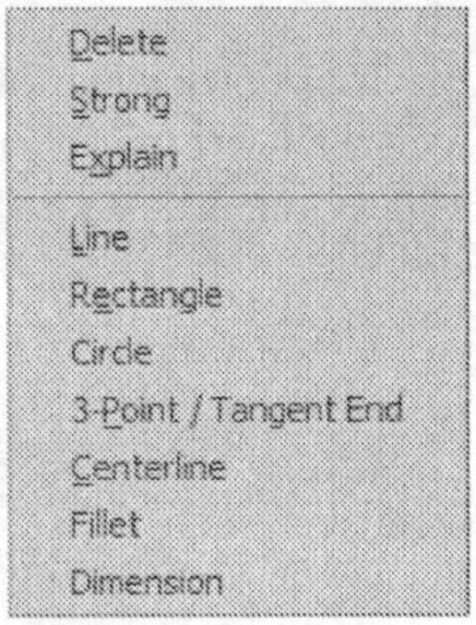

Figure 1-15 *Shortcut menu to convert the weak dimensions to strong*

MODIFYING THE DIMENSIONS OF A SKETCH

There are four ways to modify the dimensions of a sketch. These methods are discussed next.

Using the Modify Dimensions Dialog Box

You can select a dimension or more than one dimension from the sketch to modify. When you select a dimension(s) from a sketch, it is highlighted in red. If you want to select more than one dimension, hold down the CTRL key and select the dimensions by pressing the left mouse button. You can also use CTRL+ALT+A or define a window to select all the dimensions in the sketch. Choose the **Modify the values of dimensions, geometry of splines, or text entities** button from the **Sketcher Tools** toolbar to modify. The **Modify Dimensions** dialog box shown in Figure 1-16 is displayed.

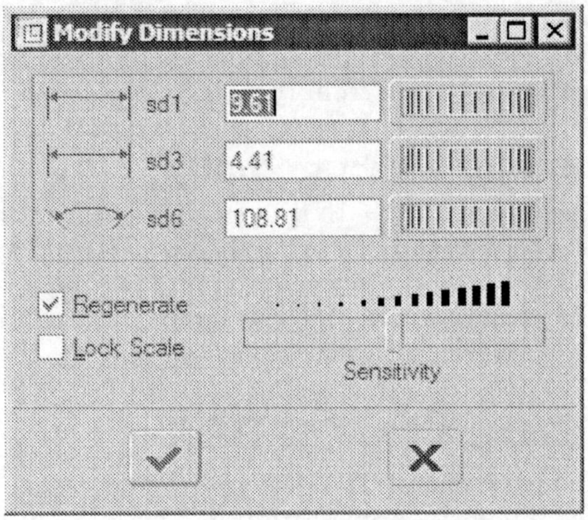

***Figure 1-16 Modify Dimensions** dialog box*

To modify dimensions using this dialog box, you can either enter a value in the edit box or use the thumbwheel that is available on the right of the edit box. The **Sensitivity** slider is used to set the sensitivity of the thumbwheel.

By default, the **Regenerate** check box is selected and any modifications in the dimensions are automatically updated in the sketch. If you want to delay the modification process of the sketch based on the new value of the selected dimension, you need to clear this check box. If this check box is cleared, the dimensions will not be modified until you exit this dialog box. This means that Pro/ENGINEER allows you to do multiple modifications before updating the sketch.

Note

*It is recommended that you clear the **Regenerate** check box and then modify the dimensions if you have to modify more than one dimension.*

The **Lock Scale** check box is used to lock the scale of all the selected dimensions. After locking the scale, if you modify a dimension, all the other dimensions will also be modified with the same scale. To lock a dimension more than one dimension should be selected.

Using the Edit Menu

The **Modify** option is available in the **Edit** menu and it can also be used to modify the dimensions. When you choose the **Modify** option from the **Edit** menu, a check mark appears to the left of the **Modify** option in the **Edit** menu. Now, you can select a dimension from the sketch to modify. When you select a dimension, the **Modify Dimensions** dialog box is displayed. By default, the **Regenerate** check box is selected. Therefore, the sketch will be regenerated dynamically as you modify the dimension. Note that you can modify only a single dimension using this option.

Modifying a dimension by double-clicking

You can also modify a dimension by double-clicking on it. When you double-click on a dimension, the pop-up text field appears. Enter a new dimension value in this field and press ENTER or use the middle mouse button. Remember that you can select a dimension only when you choose **Select items** button from the **Right Toolchest**.

Modifying dimensions dynamically

In the sketcher environment, Pro/ENGINEER is always in the selection mode, unless you have invoked some other tool. When you bring the cursor to an entity, the color of the entity changes to cyan. Now, if you hold down the left mouse button, a hand appears on the entity and you can modify the entity by dragging the mouse. You will notice that as the entity is modified, the dimensions referenced to the selected entity are also modified.

RESOLVE SKETCH DIALOG BOX

While applying constraints or dimensions, the system may sometimes prompt you to delete one or more highlighted dimensions or constraints. This is because while adding dimensions or constraints some strong dimensions or constraints conflict with the added dimensions or constraints. As soon as the conflict occurs the **Resolve Sketch** dialog box is displayed as shown in Figure 1-17 and the constraint or the dimension under conflict is displayed in orange color. On the graphics screen, when you select a dimension or constraint from the **Resolve Sketch** dialog box, the dimension or constraint selected is enclosed in a yellow box.

The buttons available in the **Resolve Sketch** dialog box are discussed next.

***Figure 1-17 Resolve Sketch** dialog box*

Undo

When you choose the **Undo** button, the section is brought back to the state that was just before the conflict occurred.

Delete

The **Delete** button is used to delete a selected dimension or constraint that is enclosed within a yellow box. Select the dimension or the constraint to delete from the **Resolve Sketch** dialog box.

Dim > Ref

When you choose the **Dim > Ref** button, the selected dimension is converted to a reference dimension.

Note

The reference dimensions are used only for reference. They do not participate in feature creation.

Explain

When you choose the **Explain** button, the system provides you with information about the selected constraint or dimension. The information is displayed in the **Message Area**.

DELETING THE SKETCHER ENTITIES

To delete a sketched entity, select it by defining a window by dragging the cursor around the entity. The color of the selected entity changes to red. Right-click on the graphics screen and hold down the right mouse button until a shortcut menu appears. Now, choose the **Delete** option from this menu. The selected item will be deleted.

You can also delete an item by selecting it and pressing the DELETE key when the selected item turns red in color.

To delete more than one item from the graphics screen, press CTRL and use the left mouse button to select the entities to be deleted. Press the DELETE key. All the selected entities are deleted. You can also define a box to select the entities.

Note

It is necessary to be in the selection mode while selecting the items. The term "items" used in this chapter refers to dimensions and entities.

To restore the last deleted item, choose the **Undo Modify Dimensions** button. This button is available in the **Edit** toolbar of the **Top Toolchest**.

TRIMMING THE SKETCHER ENTITIES

When creating a design, there are a number of places where you need to remove the unwanted and extended entities. You can do this by choosing the tool buttons available for trimming. These tool buttons are available in the **Sketcher Tools** toolbar. You can trim entities using three tools. These tools are discussed next.

Dynamically trim section entities button

This tool button deletes the selected entities. After choosing the **Dynamically trim section entities** button, when you move the cursor over an entity, the entity is highlighted in cyan color. Press the left mouse button to trim the entity. This tool button also trims entities that extend beyond the required point of intersection.

Trim entities (cut or extend) to other entities or geometry button

The **Trim entities (cut or extend) to other entities or geometry** button is used to trim two entities at their corners. Note that when you trim entities using this option, the portion from where you select the entities is retained and the other portion is trimmed. The following steps explain the procedure to trim entities using this tool button:

1. Choose the black arrow on the right of the **Dynamically trim section entities** button to display the flyout. From this flyout, choose the **Trim entities (cut or extend) to other entities or geometry** button. You are prompted to select two entities to be trimmed.

2. Using the left mouse button, select the two entities on the sides you want to keep after trimming, see Figure 1-18. These two entities must be intersecting entities. The entities are trimmed from the point of intersection.

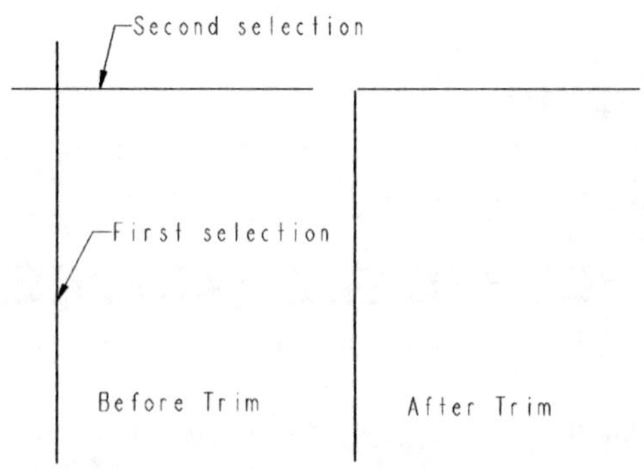

***Figure 1-18** Trimming the lines*

Divide

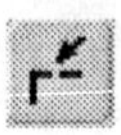

The **Divide an entity at the point of selection** button is used to divide an entity into any number of parts or entities by specifying the points on the entity.

This button is available on the flyout that is displayed when you choose the black arrow that is on the right side of the **Dynamically trim section entities** button.

The following steps explain the procedure to divide an entity:

1. Choose **Divide an entity at the point of selection** button from the flyout. You are prompted to select an entity to be divided.

2. Press the left mouse button to select the entity at the point where you want to divide it. The entity is divided into two different entities. They can now be treated as two separate entities.

3. Repeat step 2 until you have divided the entities in the required number of parts.

MIRRORING THE SKETCHER ENTITIES

The **Mirror selected entities** button is used to mirror sketched geometries about a centerline. This tool button helps to reduce the time consumed for creation of symmetrical geometries and the process of dimensioning the symmetrical entities.

The following steps explain the procedure to mirror a sketched geometry:

1. Sketch a geometry using the tool buttons. Sketch a centerline about which you need to mirror the geometry.

2. Select the entities that you need to mirror. The selected entities turn red in color.

3. Choose the **Mirror selected entities** button from the **Sketcher Tools** toolbar. You are prompted to select the centerline about which you need to mirror. Select the centerline using the left mouse button. The selected entities are mirrored about the centerline.

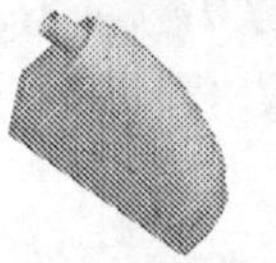

Tip: *In case of symmetrical parts, you can save the time involved in dimensioning the sketch by dimensioning half of the section and then mirroring it. Pro/ENGINEER will assume that the mirrored half has the same dimensions as the sketched half.*

DRAWING DISPLAY OPTIONS

While working with complex sketches, you need to increase the display of a particular portion of a sketch so that you can work on the minute details of the sketch. For example, if you are drawing a sketch of a piston, you have to work on the minute details of the grooves for the piston rings. To work on these minute details, you have to enlarge the display of these grooves. You can enlarge or reduce the drawing display using the various drawing display options provided in Pro/ENGINEER. Some of these drawing display options are discussed next. The remaining drawing display options will be discussed in later chapters.

Zoom In

This option enlarges the view of the drawing on the screen. When you choose the **Zoom In** button, you will be prompted to define a box. The area that you will enclose inside the box will be enlarged and displayed on the graphics screen. Note that when you enlarge the view of the drawing, the original size of the entities is not changed. To exit the zoom tool right-click in the graphics screen.

Zoom Out

This option reduces the view of the drawing on the screen, thus increasing the drawing display area. Choose this button once to zoom out. The display of the sketch on the graphics screen is reduced by some factor.

Refit object to fully display it on the screen

This option reduces or enlarges the display such that the entities that comprise the sketch are fitted inside the current display. Note that the dimensions may not necessarily be included in the current display.

Redraw the current view

While working with complex sketches, some unwanted temporary information is retained on the screen. The unwanted information include the shadows of the deleted sketched entities, dimensions, and so on. This unwanted information can be removed from the graphics screen using the **Redraw the current view** button. This option will be extensively used while designing in Pro/ENGINEER.

Note

*To remove the temporary information you can also choose **View > Repaint** from the menu bar. You can also use the shortcut keys to repaint the screen. The shortcut is CTRL+R.*

If you have a mouse that has a middle mouse button wheel, then scrolling the wheel will zoom in and out.

One more way to zoom in and out is to use the middle mouse button and the CTRL key. When you use CTRL+middle mouse button and drag the mouse upwards the sketch is zoomed out and when you drag the mouse down, the sketch is zoomed in.

To pan the sketch, use SHIFT+middle mouse button.

TUTORIALS

Tutorial 1

In this tutorial, you will draw the sketch for the model shown in Figure 1-19. The sketch is shown in Figure 1-20. **(Expected time: 30 min)**

The following steps outline the procedure to create this sketch:

a. Start Pro/ENGINEER Wildfire session.

b. Set the working directory and create a new object file.

c. Draw lines using the tool buttons.

d. Draw an arc and a circle.

e. Dimension the sketch and then modify the dimensions of the sketch.

f. Save the sketch.

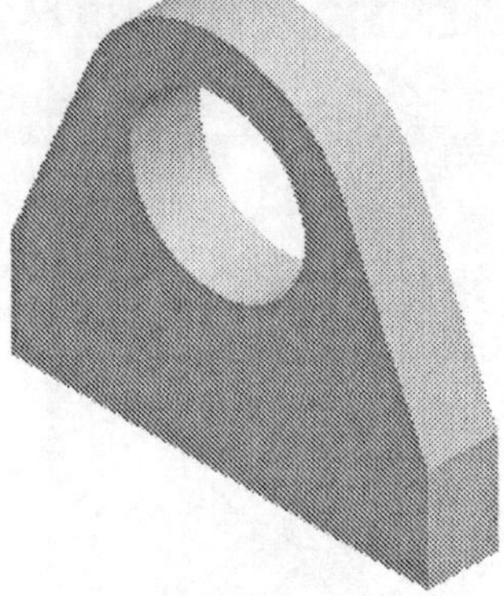

Figure 1-19 Model for Tutorial 1

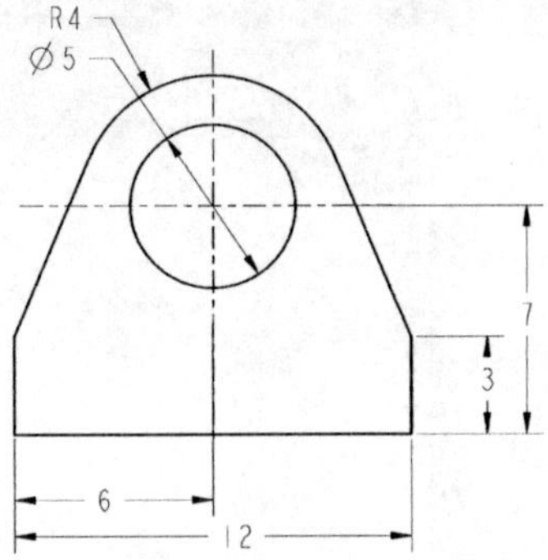

Figure 1-20 Sketch of the model

Starting Pro/ENGINEER

1. Start Pro/ENGINEER Wildfire by double-clicking on the Pro/ENGINEER Wildfire icon on the desktop of your computer or by using the **Start** menu.

Setting the Working Directory

When the Pro/ENGINEER session is started, the first task is to set the working directory. A working directory is a directory on your system where you can save the work done in the current session of Pro/ENGINEER. You can set any directory existing on your system as the working directory. Since this is the first tutorial of this chapter, you need to create a folder named *c01*, if it does not exist.

1. Choose the **Set Working Directory** option from the **File** menu. The **Select Working Directory** dialog box is displayed as shown in Figure 1-21.

2. Browse and select *C:\ProE-WF*. It is assumed that the **ProE-WF** folder exists.

3. Choose the **New Directory** button in the **Select Working Directory** dialog box. The **New Directory** dialog box is displayed.

4. Type **c01** in the **New Directory** edit box. Choose **OK** from the dialog box. You have created a folder named *c01* in *C:\ProE-WF*.

5. Choose **OK** from the **Select Working Directory** dialog box. You have set the working directory to *C:\ProE-WF\c01*. A message is displayed in the **Message Area** that the directory successfully changed to *C:\ProE-WF\c01* directory.

Creating New Object File

Any sketch drawn in the Sketch mode is saved with the *.sec* file extension. This file format is one of the file formats available in Pro/ENGINEER.

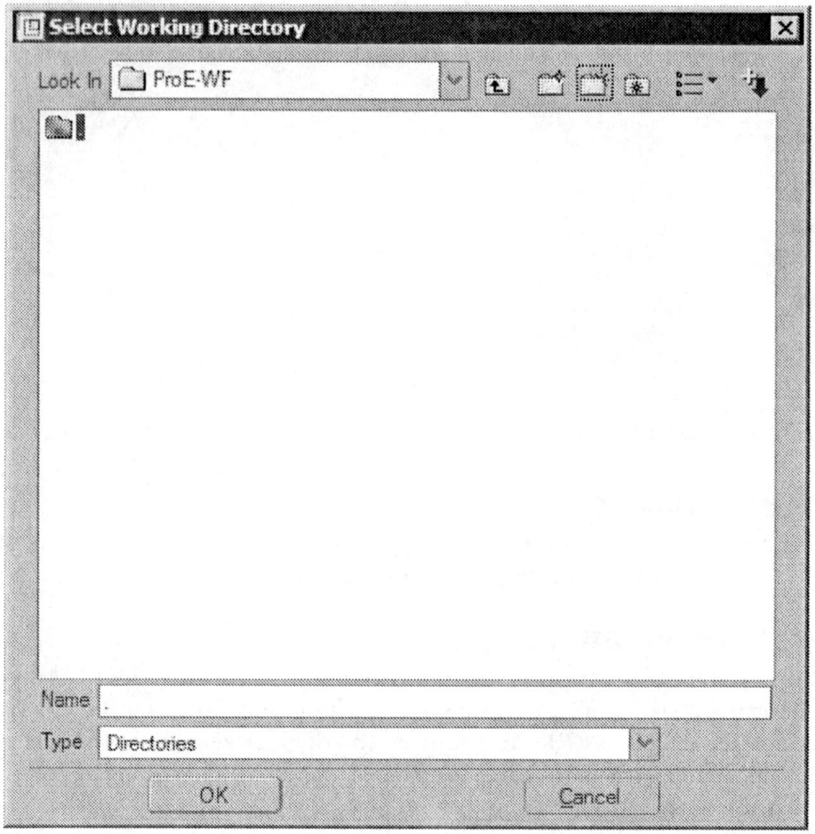

***Figure 1-21 Select Working Directory** dialog box*

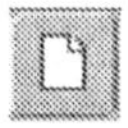

1. Choose the **Create a new object** button from the **File** toolbar. The **New** dialog box is displayed. Select the **Sketch** radio button from the **Type** area of the **New** dialog box. A default name of the sketch appears in the **Name** edit box.

2. Enter **c01tut1** in the **Name** edit box. Choose the **OK** button.

   You are in the sketcher environment of the Sketch mode. When you enter the sketcher environment, the Navigator is displayed to the left on the graphics screen.

3. Slide in the navigator by clicking on the sash present on its right edge. Now, the drawing area is increased.

Drawing the Lines of the Sketch

Start drawing the sketch with the right vertical line.

1. Choose the **Create 2 point lines** button from the **Sketcher Tools** toolbar.

2. Specify the start point to the right on graphics screen by pressing the left mouse button.

One end of the line is attached to the cursor. Move the cursor down to get an approximate size of the line.

Notice that when the cursor is moved vertically downwards, a red colored constraint named **V** appears on the graphics screen next to the line. This shows that if you draw a line now, the vertical constraint will be applied to the line.

3. Press the left mouse button to specify the endpoint of the line. The vertical constraint **V** is applied to the line and the symbol **V** appears in yellow. The color of the constraint indicates that this constraint is strong. This means that you cannot change the orientation of this line until you delete the constraint that is applied on the line.

   Another rubber-band line is attached to the cursor with the start point as the endpoint of the last line.

4. Move the cursor horizontally toward the left; a horizontal rubber-band line extends to the left as you move the mouse.

5. After you get the desired size of the line, press the left mouse button to end the line. Notice that a horizontal constraint named **H** that is yellow in color is applied to the line.

6. Move the cursor upwards on the graphics screen. A vertical rubber-band line extends as you move the mouse. As you move the cursor upwards, notice that at a particular point where the length of the left vertical line is equal to the length of the right vertical line, $\mathbf{L_1}$ symbol is displayed on both the vertical lines. This symbol suggests that the equal length constraint is applied to the two vertical lines.

7. When the $\mathbf{L_1}$ constraint appears on the vertical line, press the left mouse button to specify the endpoint of the vertical line. Notice that the $\mathbf{L_1}$ constraint is displayed in gray color as shown in Figure 1-22. This suggests that it is a weak constraint. The rubber-band line is still attached to the cursor.

   You can also apply the constraints later. But to save an extra step of adding the constraints, you will use the constraints that are applied automatically while drawing.

8. Move the cursor to size the line and specify the endpoint of the left inclined line, see Figure 1-22.

9. Press the middle mouse button to end line creation. You will notice that gray colored dimensions are applied to the sketch, see Figure 1-22. The color of these dimensions indicates that these dimensions are weak dimensions. These dimensions are automatically deleted anytime while you are completing the sketch or when you are adding dimensions and constraints manually. When system deletes weak dimensions, it does not confirm their deletion.

10. The line option is still active. Move the cursor close to the top end of the right vertical line. You will notice that as you bring the cursor close to the top end, the cursor snaps to

that point. Select the point by pressing the left mouse button.

11. Size the inclined line and specify the endpoint of the right inclined line. Press the middle mouse button to end line creation.

Figure 1-23 shows all the lines that you have drawn. Now, the arc and circle will be drawn.

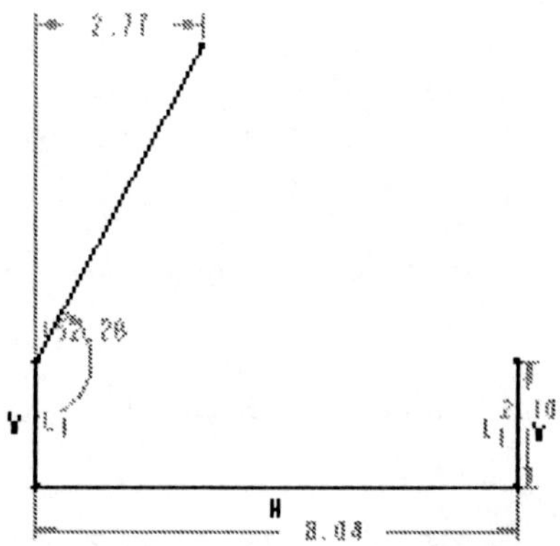

Figure 1-22 *Lines with weak dimensions*

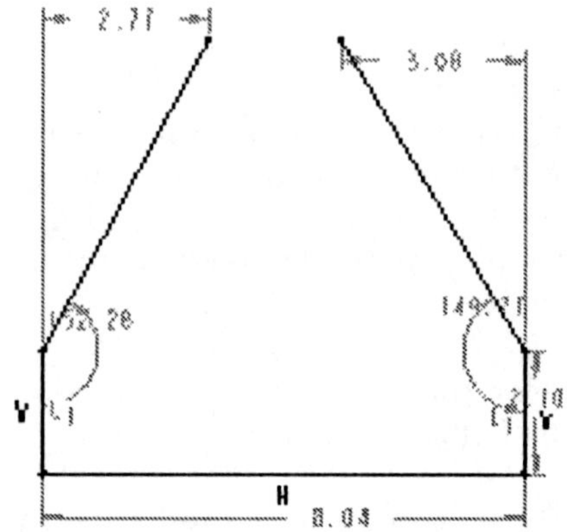

Figure 1-23 *Partial sketch with weak dimensions*

Drawing the Arc

1. Choose the **Create an arc by 3 points or tangent to an entity at its endpoint** button from the **Sketcher Tools** toolbar. You are prompted to select the start point of the arc.

2. Select the endpoint of the left inclined line by pressing the left mouse button. The **Target** symbol appears in green color.

3. Move the cursor along the tangent direction through a small distance. A rubber-band arc that is tangent to the endpoint of the line appears. As you move the cursor to the endpoint of the right inclined line, at a particular point the tangent constraint is applied at both the ends of the arc. This is evident by the symbol **T** that appears on the endpoints of the right inclined line.

4. As the tangent constraint appears, use the left mouse button to end arc creation. You will notice that the tangent constraint with a symbol **T** appears at the endpoints of the arc as evident from Figure 1-24. Press the middle mouse button to end arc creation.

The tangent constraint **T** will appear in yellow, which suggests that it is a strong constraint and the tangency of the inclined line with the arc cannot be modified until you delete the tangent constraint.

Note that in Figure 1-23 there are some weak dimensions that are not displayed in Figure 1-24. This is because the weak dimensions are deleted without confirming their deletion. Hence, after drawing the arc some weak dimensions got deleted automatically.

Note

If the tangent constraint symbol is not displayed on any of the inclined lines, apply the constraint manually using the ***Constraints*** *dialog box that is displayed when you choose the* ***Impose sketcher constraints on the section*** *button from the* ***Sketcher Tools*** *toolbar.*

Drawing the Circle

1. Choose the black arrow on the right of the **Create circle by picking the center and a point on the circle** button to display the flyout. From this flyout, choose the **Create concentric circle** button. You are prompted to select an arc.

2. Select the arc by pressing the left mouse button. Move the mouse and a circle appears.

3. By pressing the left mouse button, select a point inside the sketch to draw the circle.

4. Press the middle mouse button to end circle creation. The sketch is complete and appears similar to that shown in Figure 1-25.

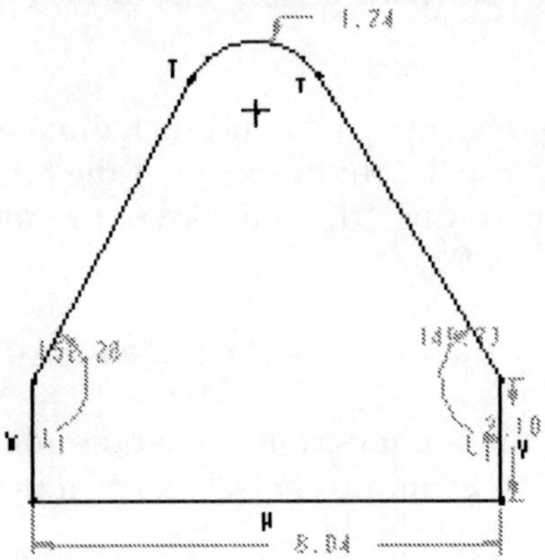

Figure 1-24 *Sketch with arc*

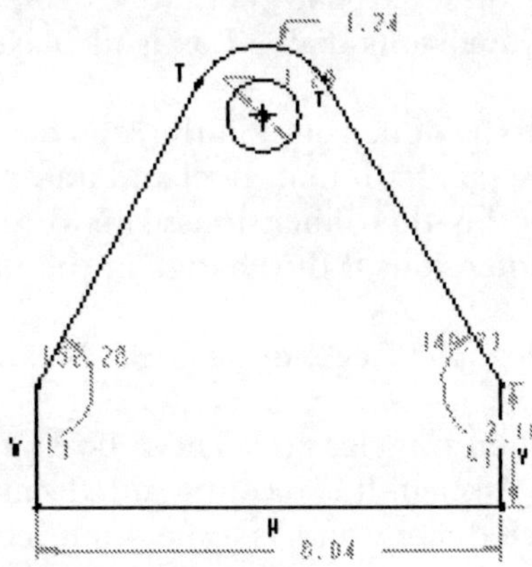

Figure 1-25 *Sketch with all the entities, weak dimensions, and weak constraints*

Dimensioning the Sketch

The right vertical line, the bottom horizontal line, the arc, and the circle are dimensioned automatically and weak dimensions are applied to them. You will use these dimensions. Hence, there is no need to dimension these entities again.

1. Choose the **Create defining dimension** button from the **Sketcher Tools** toolbar.

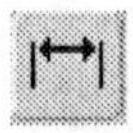

2. Select the center of the circle and then select the bottom horizontal line by pressing the left mouse button. Both, the center and the line turn red in color.

3. Place the dimension on the right of the sketch by pressing the left mouse button.

4. Select the center of the circle and then select the left vertical line by pressing the left mouse button. Both the center and the vertical line turn red in color.

5. Press the middle mouse button to place the dimension below the sketch, refer to Figure 1-26.

Modifying the Dimensions

The sketch is dimensioned with default values. You need to modify these values to the given values.

1. Choose the **Select items** button.

2. Select all the dimensions by defining a window.

Note

You can also use CTRL+ALT+A to select the whole sketch with dimensions.

3. When all the dimensions turn red in color, choose the **Modify the values of dimensions, geometry of splines, or text entities** button. The **Modify Dimensions** dialog box is displayed.

   All the dimensions in the sketch are displayed in this dialog box and each dimension has a separate thumbwheel and an edit box. You can use the thumbwheel or the edit box to modify the dimensions. It is recommended that you use the edit boxes to modify the dimensions if the change in the dimension value is large.

4. Clear the **Regenerate** check box and then modify the values of the dimensions.

   When you clear this check box, any modification in a dimension value does not update the sketch. It is recommended that you clear the **Regenerate** check box when more than one dimension has to be modified.

   Notice that the dimension you select in the **Modify Dimensions** dialog box gets enclosed in a yellow box on the graphics screen.

5. Modify all the dimensions as shown in Figure 1-26. After modifying all the dimensions, choose the **Regenerate the section and close the dialog** button from the **Modify Dimensions** dialog box. A message **Dimension modifications successfully completed** is displayed in the **Message Area**.

   The sketch is complete and is shown in Figure 1-26.

Saving the Sketch

The sketch will now be saved. You have to save the sketch because you may need the sketch later in the Part mode in order to create a 3D model.

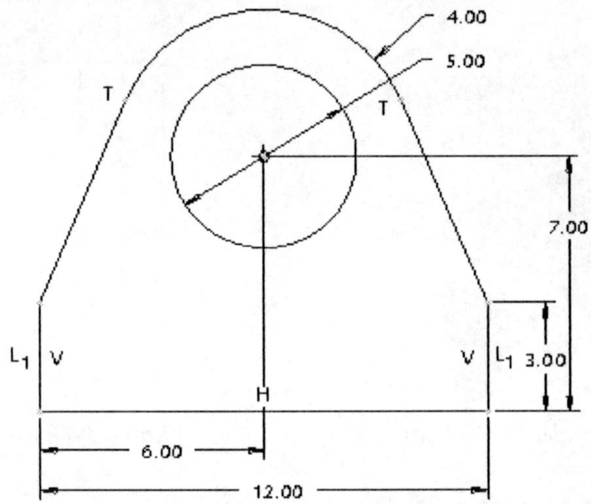

Figure 1-26 Sketch with dimensions and constraints for Tutorial 1

Tip: *You can modify the location of the dimensions as they appear on the screen by selecting and dragging them to a new location..*

1. Choose the **Save the active object** button from the **File** toolbar. The **Message Input Window** is displayed with the name of the sketch that you had entered earlier.

2. Press ENTER. The sketch is saved.

3. After saving the sketch, choose the **Continue with the current section** button to exit the sketch.

Tutorial 2

In this tutorial, you will draw the sketch for the model shown in Figure 1-27. The sketch is shown in Figure 1-28. For your reference, all the entities in the sketch are labeled alphabetically. **(Expected time: 30 min)**

The following steps outline the procedure for creating this sketch:

a. Create a new object file.

b. Draw the sketch using the tool button for creating lines.

c. Dimension the required entities and then modify the dimensions of the sketch.

d. Save the sketch.

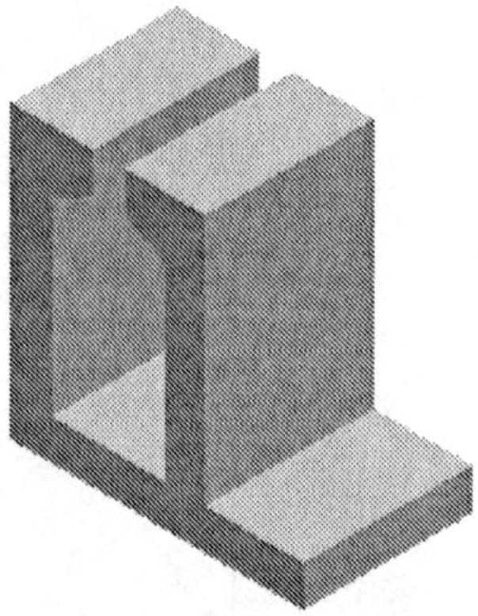

Figure 1-27 Model for Tutorial 2

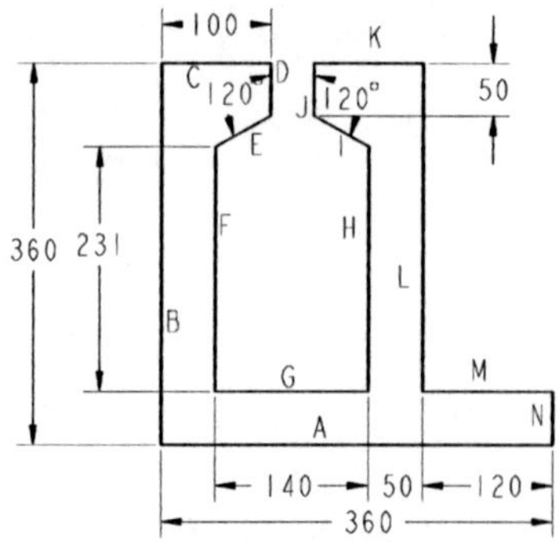

Figure 1-28 Sketch of the model

Setting the Working Directory

The working directory was selected in Tutorial 1, and therefore there is no need to select the working directory again. But if you still want to select the working directory, follow the steps given next.

1. Open the navigator by clicking the top arrows on the left edge of the Pro/ENGINEER main window. The navigator slides out.

2. Click on the plus symbol adjacent to the *ProE-WF* folder in the navigator. The contents of the *ProE-WF* folder are displayed.

3. Now right-click on the *c01* folder to display a shortcut menu. From this shortcut menu, choose the **Make Working Directory** option. The working directory is set to *c01*.

4. Close the Navigator by clicking sash on the right edge of the navigator. The Navigator slides in.

Creating New Object File

1. Choose the **Create a new object** button from the **File** toolbar. The **New** dialog box is displayed. Select the **Sketch** radio button from the **Type** area of the **New** dialog box. A default name of the sketch appears in the **Name** edit box.

2. Enter **c01tut2** in the **Name** edit box and choose **OK**. You are in the sketcher environment of the Sketch mode.

Drawing the Sketch

The sketch consists of only lines. For ease of understanding, all the lines in the sketch are labelled alphabetically.

1. Choose the **Create 2 point lines** button from the **Sketcher Tools** toolbar. Select a point close to the lower right corner of the graphics screen by pressing the left

mouse button and start drawing the horizontal line A. Here, you will notice that as you draw line A, the **H** symbol is displayed on the line. This shows that the line is horizontally constrained. Move the cursor toward the left and specify the endpoint of the line.

2. Move the cursor vertically upwards so that the **V** constraint appears on the line. When you get the appropriate size of the line, press the left mouse button to specify the endpoint of line B. Line B is completed.

3. Move the cursor to the right on the graphics screen and press the left mouse button to specify the endpoint of line C.

4. Now, to draw line D, move the cursor down and press the left mouse button to specify the endpoint of line D.

5. Line E to be drawn is inclined. Move the cursor to size the line and press the left mouse button to specify the endpoint of line E.

6. The next line you need to draw is line F. Move the cursor vertically downwards and press the left mouse button to specify the endpoint of line F.

7. Now, to draw line G, move the cursor horizontally and press the left mouse button to specify the endpoint of line G.

8. Move the cursor vertically upwards and press the left mouse button to specify the endpoint of line H.

9. Now, continue drawing the remaining lines that are shown in Figure 1-29. When the sketch is completed, end line creation by pressing the middle mouse button. Notice that the sketched entities are dimensioned automatically as you draw them. These dimensions are weak dimensions and appear in gray color.

Applying the Constraints to the Sketch

Constraints are applied to the sketch to maintain the design intent of the feature and this might sometimes result in less dimensions in the sketch.

1. Choose the **Impose sketcher constraints on the section** button from the **Sketcher Tools** toolbar. The **Constraints** dialog box is displayed.

2. Choose the **Create Equal Lengths, Equal Radii, or Same Curvature constraint** button and select lines F and H. The equal length constraint $\mathbf{L_2}$ is applied to both the lines. The constraint labels such as L_2 or L_3 vary from sketch to sketch.

3. Select lines C and K. The equal length constraint is applied to both the lines.

4. Now, select lines J and N. The equal length constraint is applied to both the lines.

5. Select lines A and B. The equal length constraint is applied to both the lines.

6. Choose the **Make line or two vertices horizontal** button from the **Constraints** dialog box. You are prompted to select a line or two vertices. 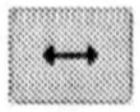

7. Select the vertex that is joining lines L and M and the vertex that is joining lines G and H. Both the vertices are aligned horizontally as shown in Figure 1-29.

Dimensioning the Sketch

Weak dimensions are already applied to the sketch while drawing. You need to dimension only the angle between lines D and E and lines J and I.

1. Choose the **Create defining dimension** button. 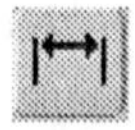

2. Select lines D and E using the left mouse button. The selected lines turn red in color. Now, press the middle mouse button to place the dimension close to the vertex where lines D and E join.

3. Similarly, dimension the angle between lines J and I.

   Figure 1-30 shows the sketch after applying dimensions. If your sketch does not have all the dimensions shown in Figure 1-30, apply them using the **Create defining dimension** button.

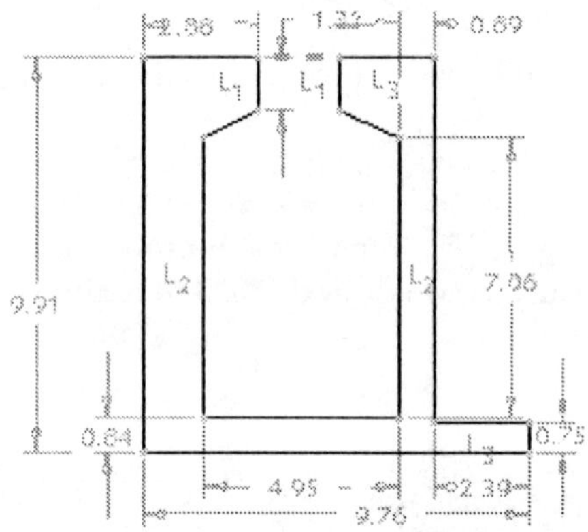

Figure 1-29 Sketch with weak dimensions and weak constraints

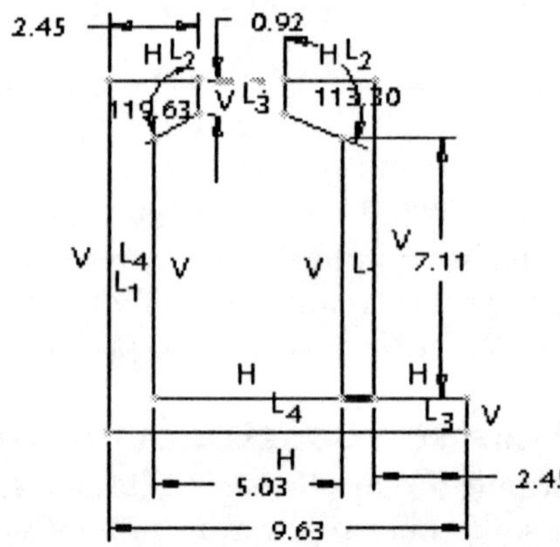

Figure 1-30 Sketch after dimensioning

Modifying the Dimensions

The dimensions that are applied to the sketch need modification in dimension values.

1. Choose the **Select items** button. 

2. Select all the dimensions by defining a window.

3. When all the dimensions turn red in color, choose the **Modify the values of dimensions, geometry of splines, or text entities** button. The **Modify** 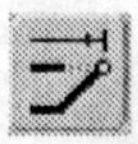

Dimensions dialog box is displayed.

4. Clear the **Regenerate** check box and then modify the values of the dimensions. When you clear this check box, the sketch is not regenerated while you modify the dimensions.

   Notice that the dimension you select in the **Modify Dimensions** dialog box is enclosed in a yellow box on the graphics screen.

5. When all the dimensions are modified, choose the **Regenerate the section and close the dialog** button from the **Modify Dimensions** dialog box. A message **Dimension modifications successfully completed** is displayed in the **Message Area**.

   The completed sketch is shown in Figure 1-31.

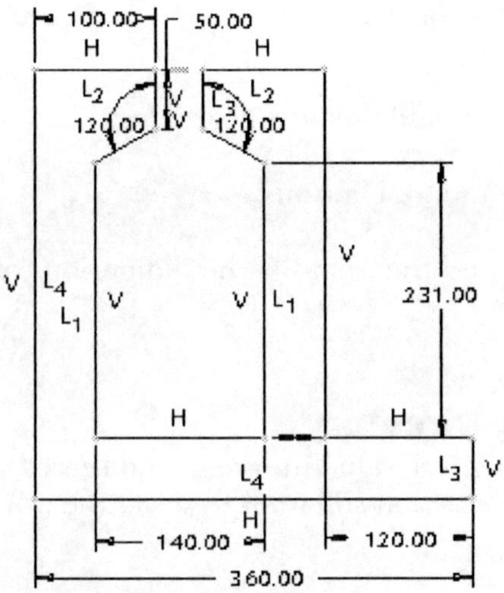

Figure 1-31 Complete sketch with dimensions and constraints

6. Save the sketch as discussed earlier. After saving the sketch, choose the **Continue with the current section** button to exit the Sketch mode. 

Note

You can also modify dimensions individually. But, individual modification of dimensions is recommended only when either there is a minor change in the dimension value or when only one dimension is required to be modified.

Tutorial 3

In this tutorial, you will draw the sketch for the model shown in Figure 1-32. The sketch is shown in Figure 1-33. For your reference, all the entities in the sketch are labeled alphabetically.

(Expected time: 30 min)

The following steps outline the procedure for creating this sketch:

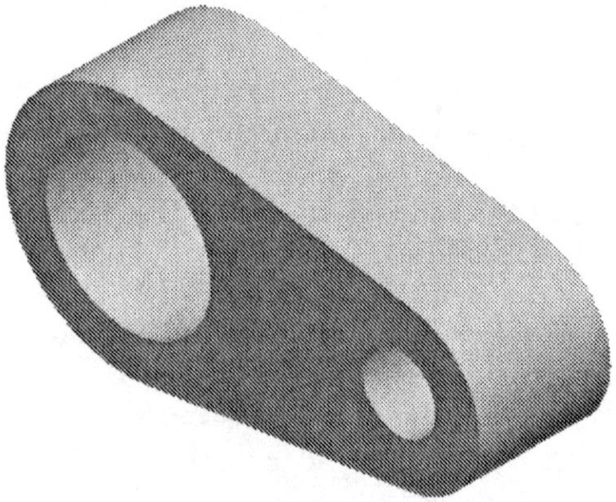

Figure 1-32 *Model for Tutorial 3*

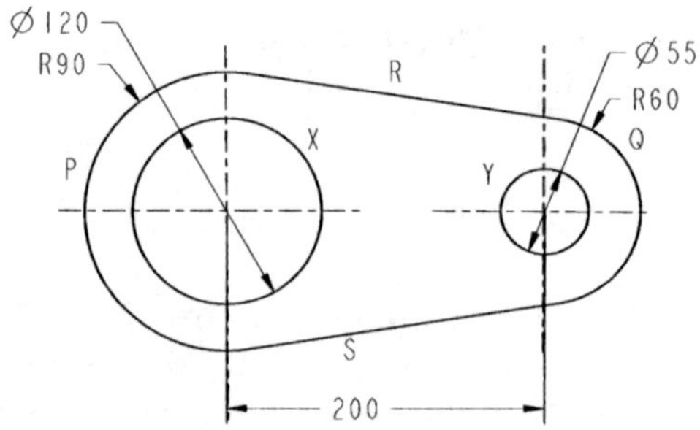

Figure 1-33 *Sketch of the model*

a. Set the working directory and create a new object file.

b. Draw the sketch using the tool buttons.

c. Dimension the sketch and then modify the dimensions of the sketch.

d. Save the sketch.

Setting the Working Directory

The working directory was selected in Tutorial 1, and therefore there is no need to select the working directory again. But if you still want to select the working directory, follow the steps given next.

1. Open the Navigator by sliding it out. Click on the plus symbol adjacent to the *ProE-WF* folder in the Navigator. The contents of the *ProE-WF* folder are displayed.

2. Now right-click on the *c01* folder to display a shortcut menu. From this shortcut menu, choose the **Make Working Directory** option. The working directory is set to *c01*. Close the Navigator.

Creating New Object File

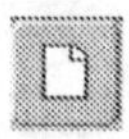

1. Choose the **Create a new object** button from the **File** toolbar. The **New** dialog box is displayed. Select the **Sketch** radio button from the **Type** area of the **New** dialog box. A default name of the sketch appears in the **Name** edit box.

2. Enter **c01tut3** in the **Name** edit box. Choose the **OK** button.You will enter the sketcher environment of the Sketch mode.

Drawing the Circles

1. Choose **Create circle by picking the center and a point on the circle** button

from the **Sketcher Tools** toolbar.

2. Specify the center of the circle by pressing the left mouse button.

3. Move the cursor to size the circle. Press the left mouse button to complete the circle.

4. Draw another circle whose center is collinear with the center of the previous circle.

   Figure 1-34 shows the two collinear circles drawn using the **Create circle by picking the center and a point on the circle** button from the **Sketcher Tools** toolbar.

Drawing the Tangent Lines

1. Choose the **Create lines tangent to 2 entities** button from the flyout in the **Right Toolchest**. You are prompted to select the start location on the arc or a circle.

2. Select the left circle at the top by pressing the left mouse button. A rubber-band line appears whose one end is attached to the circle and the other end is attached to the cursor.

3. Press the left mouse button on the top of the right circle. The tangent that connects the two circles is drawn.

4. Similarly, draw a tangent by selecting the two circles at the bottom.

   Figure 1-35 shows the sketch after drawing the tangent lines.

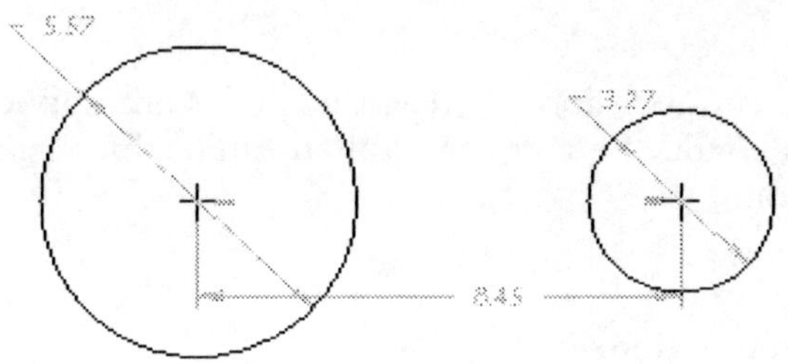

Figure 1-34 *The two circles with weak dimensions and constraints*

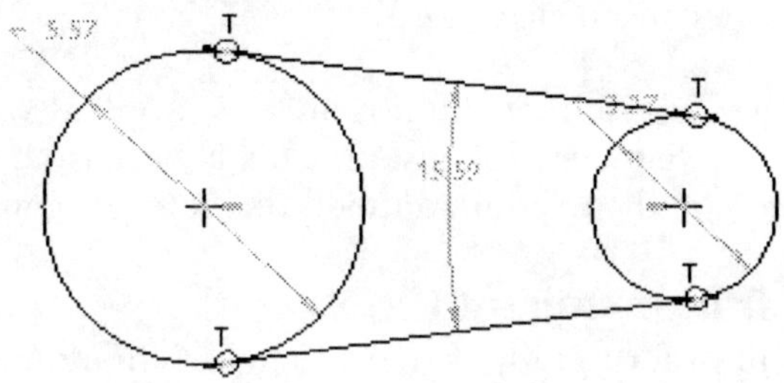

Figure 1-35 *Circles joined by lines and the tangent constraint applied to them*

Trimming the Circles

As evident from Figure 1-35, the tangents that are drawn intersect the circles at the point where they meet the circle. Therefore, the part of the circle that is not required can be dynamically trimmed.

1. Choose the **Dynamically trim section entities** button from the **Sketcher Tools** toolbar.

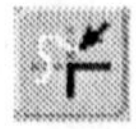

2. Select the two circles individually to trim them at the locations shown in Figure 1-36.

   Figure 1-37 shows the two circles after deleting the unwanted portion of the circle.

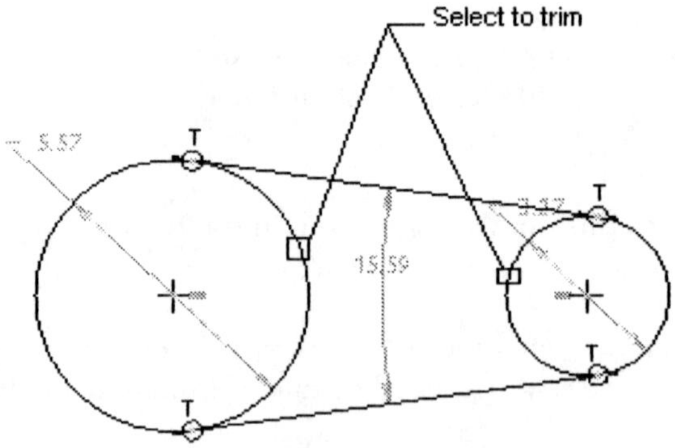

Figure 1-36 Locations to trim

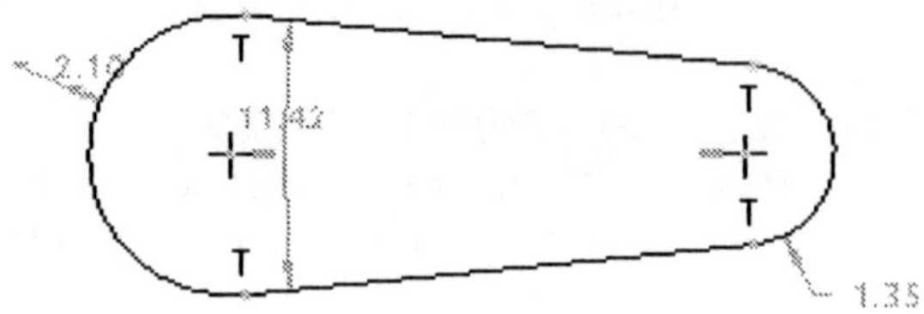

Figure 1-37 Sketch after trimming

Drawing the Circles

1. Choose the black arrow on the right of the **Create circle by picking the center and a point on the circle** button to display the flyout. From this flyout, choose the **Create concentric circle** button. You are prompted to select an arc.

2. Select arc P and create circle X concentric to the arc. Similarly, select arc Q to create a concentric circle Y.

   Notice that the two arcs are applied radius dimension whereas the circles are applied diameter dimension. This is because by default, the arcs are applied radius dimension and circles are applied diameter dimension.

Dimensioning the Sketch

In order to fully define a sketch it should be dimensioned.

1. Choose the **Create defining dimension** button.

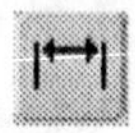

2. Select the centers of the two circles and place the dimension at the bottom of the sketch.

Modifying the Dimensions

1. Choose the **Select items** button.

2. Select all the dimensions by defining a window.

Note

You can also use CTRL+ALT+A from the keyboard to select all the entities and items in the sketch.

3. When all the dimensions turn red in color, choose the **Modify the values of dimensions, geometry of splines, or text entities** button. The **Modify Dimensions** dialog box is displayed.

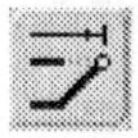

4. Clear the **Regenerate** check box and then modify the values of the dimensions.

   You will notice that the dimension you edit in the **Modify Dimensions** dialog box is enclosed by a yellow box on the graphics screen.

5. When all the dimensions are modified, choose the **Regenerate the section and close the dialog** button from the **Modify Dimensions** dialog box. A message **Dimension modifications successfully completed** is displayed in the **Message Area**.

   The sketch is completed and is shown in Figure 1-38.

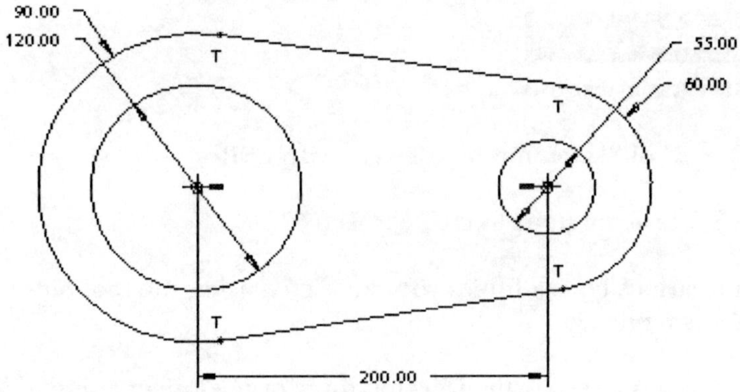

Figure 1-38 Sketch with dimensions and constraints

6. Save the sketch as discussed earlier. After saving the sketch, choose the **Continue with the current section** button to exit the Sketch mode.

Self-Evaluation Test

Answer the following questions and then compare your answers to the answers given at the end of this chapter.

1. Dimensions and constraints are automatically applied to a sketch when you draw it. (T/F)

2. While working in the Sketch mode of Pro/ENGINEER, the sketch need not be a closed loop for successful regeneration. (T/F)

3. If the **Intent Manager** is on and you draw a line, the cursor snaps to the endpoint of the previous line. (T/F)

4. You can convert a weak constraint to strong by using the shortcut menu that is displayed when you right-click on the weak constraint. (T/F)

5. While drawing a circle, first you need to specify the diameter of the circle it. (T/F)

6. The __________ menu in the menu bar has the **Modify** option in it.

7. __________ can be modified to modify the shape of the sketch.

8. **Intent Manager** is __________ by default when you enter the Sketch mode.

9. For the Sketch mode the symbolic dimension is represented by __________ symbol.

10. The Sketch mode file is saved as a __________ file extension.

Review Questions

Answer the following questions:

1. What is the need of Sketch mode in Pro/ENGINEER?

2. What are the four basic steps to create a sketch?

3. What are the various types of lines you can sketch using the tool buttons available in the **Sketcher Tools** toolbar?

4. Why is it important to select the working directory before creating a new file?

5. Write all the steps involved in creating a sketch that is accepted by Pro/ENGINEER.

6. You can dynamically modify the geometry of a sketch. (T/F)

7. You can use the **Create rectangle** button from the **Sketcher Tools** toolbar to draw a square. (T/F)

8. The __________ button is used to exit the sketcher environment. (T/F)

9. You cannot undo a previous operation in the sketcher environment. (T/F)

10. You can also use the options to draw a sketch from the **Sketch** menu in the menu bar. (T/F)

Exercises

Exercise 1

In this exercise, you will draw the sketch for the model shown in Figure 1-39. The sketch is shown in Figure 1-40. **(Expected time: 30 min)**

Figure 1-39 Solid model for Exercise 1

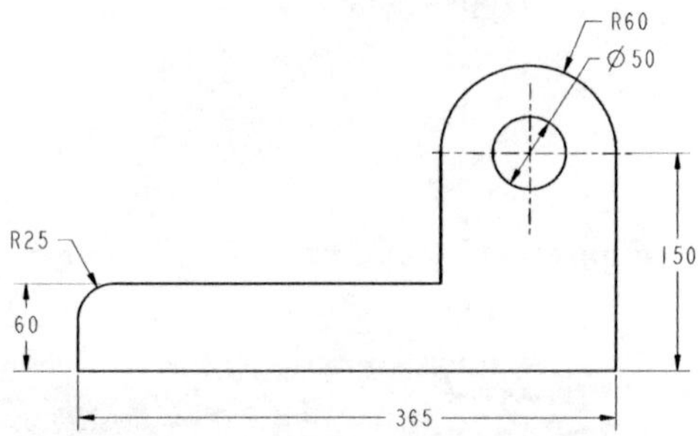

Figure 1-40 Sketch of the model

Exercise 2

In this exercise, you will draw the sketch for the model shown in Figure 1-41. The sketch is shown in Figure 1-42. **(Expected time: 30 min)**

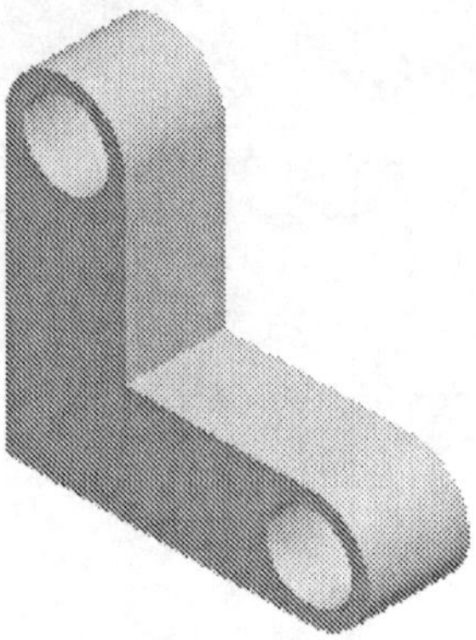

Figure 1-41 Solid model for Exercise 2

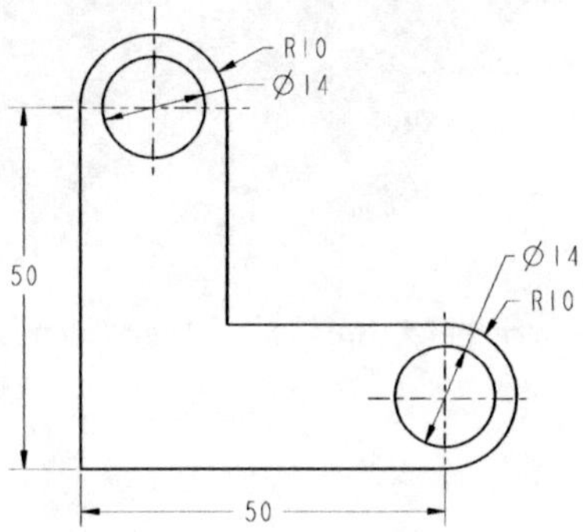

Figure 1-42 Sketch of the model

Exercise 3

In this exercise, you will draw the sketch for the model shown in Figure 1-43. The sketch is shown in Figure 1-44. **(Expected time: 30 min)**

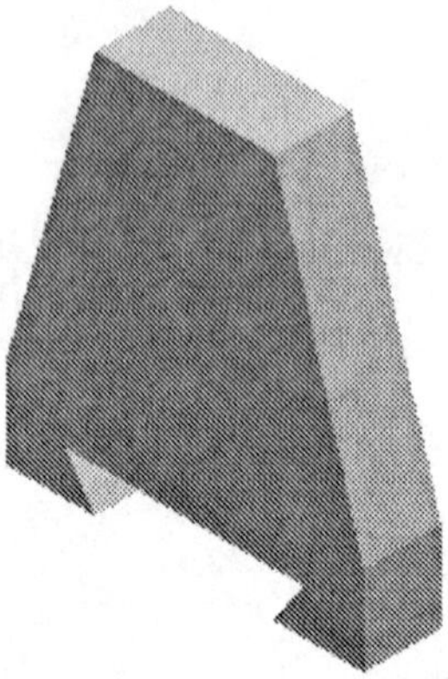

Figure 1-43 Solid model for Exercise 3

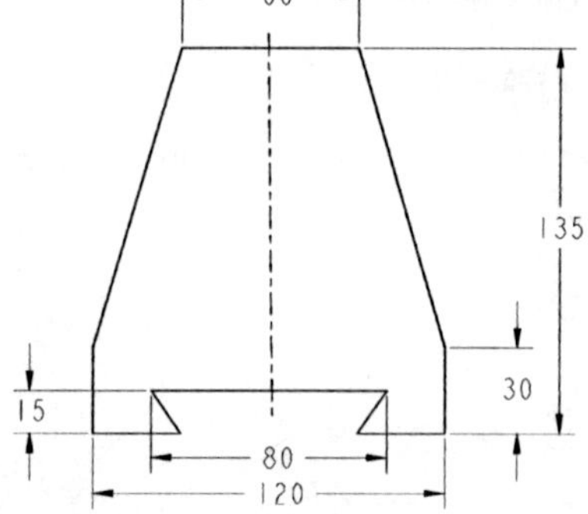

Figure 1-44 Sketch of the model

Exercise 4

In this exercise, you will draw the sketch for the model shown in Figure 1-45. The sketch is shown in Figure 1-46. **(Expected time: 30 min)**

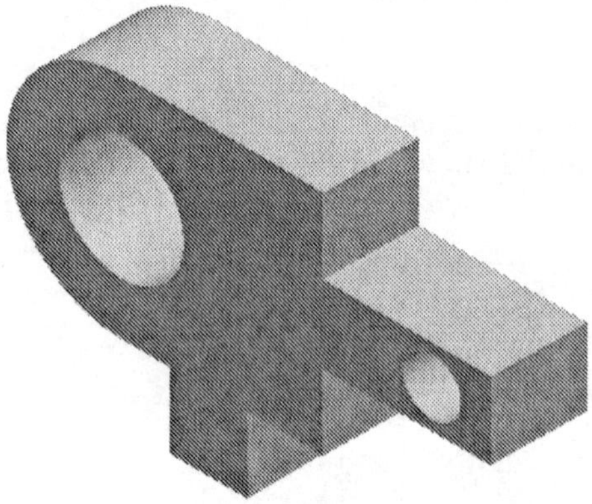

Figure 1-45 Solid model for Exercise 4

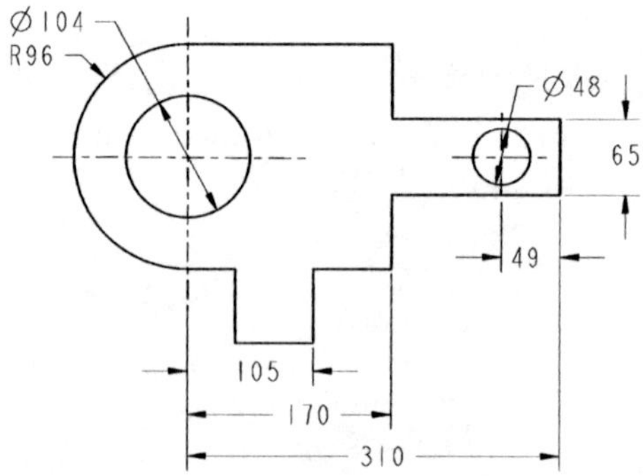

Figure 1-46 Sketch of the model

Answers to Self-Evaluation Test

1 - T, **2** - T, **3** - T, **4** - T, **5** - F, **6** - **Edit**, **7** - Dimensions, **8** - on, **9** - sd, **10** - *.sec*

Chapter 2

Creating Sketches in the Sketch Mode-II

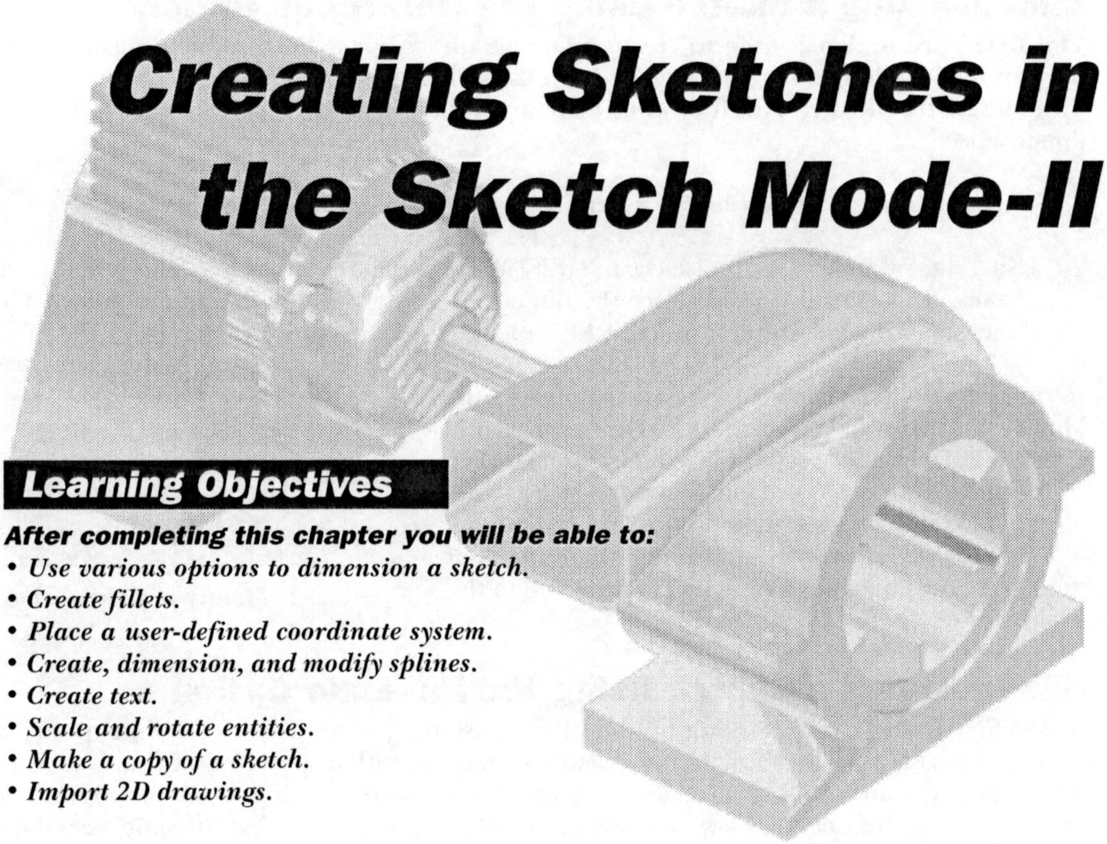

Learning Objectives

After completing this chapter you will be able to:

- *Use various options to dimension a sketch.*
- *Create fillets.*
- *Place a user-defined coordinate system.*
- *Create, dimension, and modify splines.*
- *Create text.*
- *Scale and rotate entities.*
- *Make a copy of a sketch.*
- *Import 2D drawings.*

DIMENSIONING THE SKETCH

In Chapter 1 you learned how to dimension a sketch using the **Normal** option or using its equivalent tool, that is, **Create defining dimension** button in the **Sketcher Tools** toolbar. In this chapter, you will learn the other options to dimension a sketch.

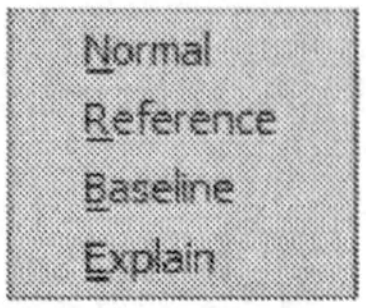

***Figure 2-1** Dimensioning options*

Choose **Sketch** > **Dimension** from the menu bar. The cascading menu is displayed as shown in Figure 2-1. The options that are available to dimension a sketch are shown in the figure.

Dimensioning a Sketch Using the Reference Option

The **Reference** option is used to create reference dimensions. These dimensions are used only for information and are not used in manufacturing the part. These dimensions appear with the suffix "REF". The following steps explain the procedure to create reference dimensions:

1. Choose **Sketch** > **Dimension** > **Reference** from the menu bar.

2. Using the left mouse button, select an entity in the sketch to dimension. The yellow color of the entity changes to red. Place the dimension using the middle mouse button. The dimension appears with the suffix "REF" after the dimension value.

If you select a reference dimension to modify, the **Modify Ref Dim** dialog box is displayed as shown in Figure 2-2. The value of a reference dimension is driven by the normal dimensions and therefore, you cannot modify the value of a reference dimension. Therefore, if you make any change in the normal dimensions of the sketch, the change is automatically reflected in the value of the reference dimension.

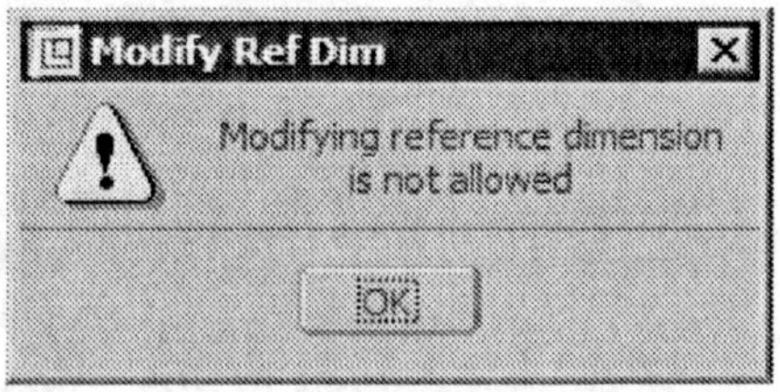

***Figure 2-2** **Modify Ref Dim** dialog box*

Dimensioning a Sketch Using the Baseline Option

In Pro/ENGINEER, the **Baseline** option of dimensioning is used to create dimensions in terms of horizontal and vertical location values of an entity with respect to a specified baseline. This type of dimensioning in a drawing can be used for writing a CNC program to manufacture a component. This type of dimensioning is also used for dimensioning assemblies from where a designer can just get an idea of the size (length and width) of the assembly. These assemblies are then developed by designers.

The **Baseline** option can be used to dimension lines, conics, arcs, and so on. The following steps explain the procedure to create dimensions using the **Baseline** option:

1. Choose **Sketch** > **Dimension** > **Baseline** from the menu bar.

2. Select the entity that will act as the baseline (origin or reference). Press the middle mouse button to place the dimension. The dimension **0.00** is displayed where you place the

dimension. Note that since the location value of the baseline is taken as the origin, the dimension value of the baseline entity will become 0.00. The dimension values of all the other entities dimensioned with reference to the baseline will be measured from this origin.

Depending upon the entity selected to act as the baseline, the horizontal or the vertical dimension value of the location of the entity will be placed. For example, if you select a vertical line, the vertical value of its location will be placed. Similarly, if you select a horizontal line, the horizontal value of its location will be placed.

For arcs, circles, and splines there are two options to dimension using the **Baseline** option. When you select a circle center or an arc for baseline dimensioning and press the middle mouse button, the **Dim Orientation** dialog box is displayed as shown in Figure 2-3. You are prompted to select the dimension orientation. The dimension is placed according to the orientation selected.

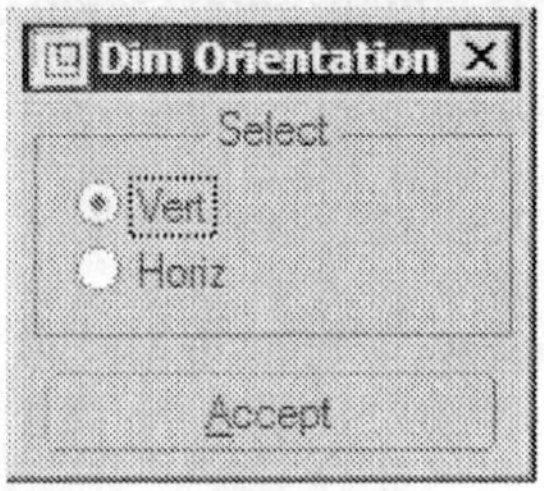

***Figure 2-3 Dim Orientation** dialog box*

3. Next, choose the **Create defining dimension** button from the **Sketcher Tools** toolbar. Select the baseline dimension that was placed earlier and then select the entity to dimension. Now, press the middle mouse button to place the dimension.

Depending upon the baseline dimension and the entity selected, the dimension will be placed. Figure 2-4 shows a sketch dimensioned using the above-mentioned method. In this figure, the two baselines are dimensioned using the **Baseline** option. Therefore, the dimensions of these lines are displayed as 0.00. The remaining lines are dimensioned by using the **Create defining dimension** button by first selecting the baseline dimension and then the entity to dimension.

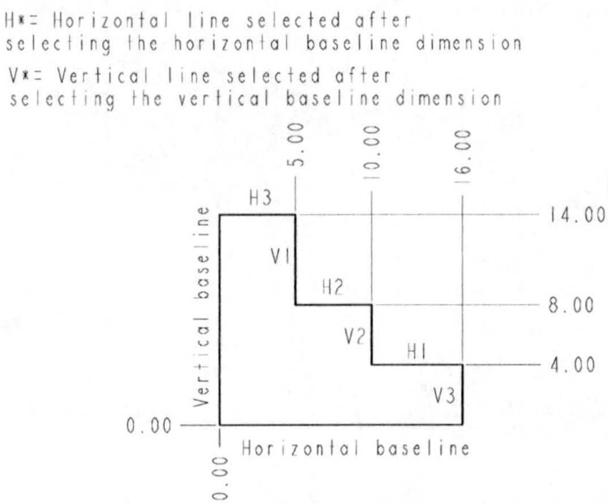

***Figure 2-4** Baseline dimensioning of a sketch*

Dimensioning a Sketch Using the Perimeter Option

The **Perimeter** option specifies the perimeter of the entities or the loop of entities selected. In this type of dimensioning, a dimension is selected from a dimensioned sketch that will vary with the change in the perimeter. If you modify the perimeter value of the sketch, the modification is reflected in the dimension of the entity selected to vary. Also, the sketch is modified corresponding to the variable dimension. The following steps explain the procedure to dimension a sketch using the **Perimeter** option:

1. Draw a sketch as shown in Figure 2-5 and dimension it using the **Create defining dimension** button.

2. Modify the dimensions to the dimension values shown in Figure 2-5.

3. Press CTRL+left mouse button to select the three lines (a), (b), and (c) shown in Figure 2-5. All the three lines turn red in color.

4. Now, choose **Edit > Convert To > Perimeter** from the **Edit** menu in the menu bar.

   You are prompted to select a dimension that will be driven by the perimeter dimension. Use the left mouse button to select the dimension of line (a) as the variable dimension, see Figure 2-5.

5. Now, if you modify the perimeter value, the dimension selected to vary will be proportionately modified, thus altering the geometry of the sketch. Remember that the dimension selected to vary cannot be modified.

   Figure 2-6 shows the perimeter modified to 100 and the consequent change in the variable dimension.

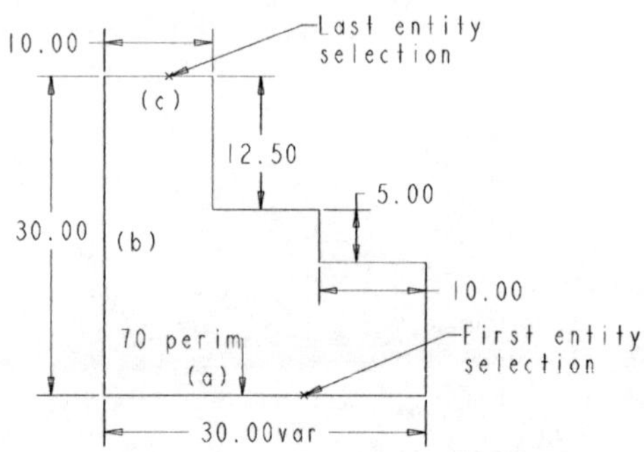

***Figure 2-5** The **Perimeter** option*

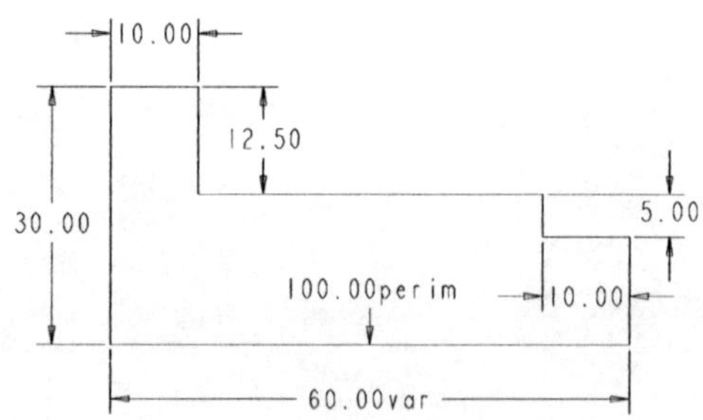

***Figure 2-6** The **Perimeter** option*

Note

If you delete the variable dimension, the perimeter dimension is also deleted.

Replacing the Dimensions of a Sketch Using the Replace Option

The **Replace** option is used to replace a dimension from a sketch. To use this option you must have a dimensioned sketch. The following steps explain the procedure to dimension a sketch using the **Replace** option:

1. Choose **Edit > Replace** from the menu bar. You will be prompted to select a dimension to be replaced.

2. Select the dimension to be replaced using the left mouse button. The selected dimension is erased. Select an entity to dimension using the left mouse button and place the dimension at the desired place. The previous dimension is replaced by a new dimension.

CREATING FILLETS

In the sketcher environment, you can create two types of fillets.

1. Circular fillets
2. Elliptical fillets

Creating Circular Fillets

A circular fillet is the arc that is used to join two lines, a line and an arc, or two arcs. The fillet is controlled by its radius or diameter dimension. The fillet that is created depends on the location where you select the entities to create the fillet.

Figure 2-7 shows two lines that do not join at the corners and Figure 2-8 shows the circular fillet created between them. The circular fillet that is created is an arc with its endpoints tangent to the two lines. This is evident from the **T** symbol that is automatically applied to the endpoints of the fillet arc.

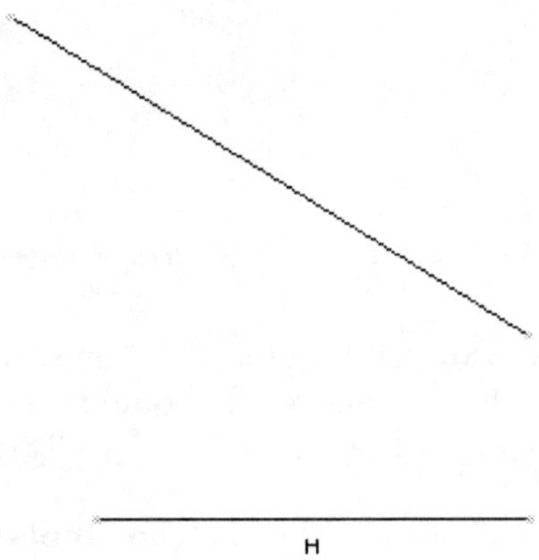

Figure 2-7 *Two lines that do not join*

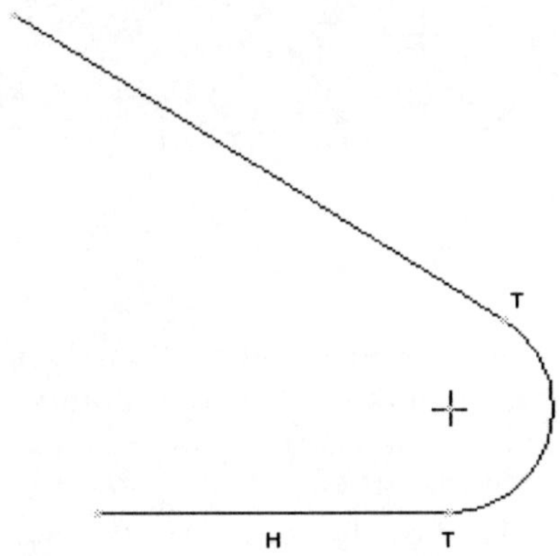

Figure 2-8 *Fillet created between the two lines*

Figure 2-9 shows two lines that join at the corner and Figure 2-10 shows the circular fillet created that joins the two lines by an arc. The corner is automatically deleted.

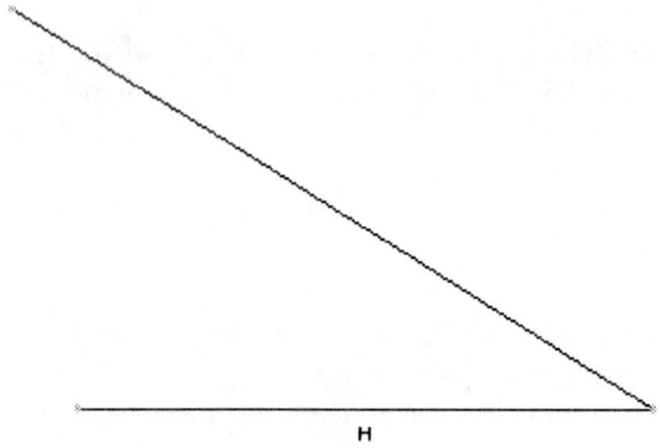

Figure 2-9 *Two lines joining at the corner*

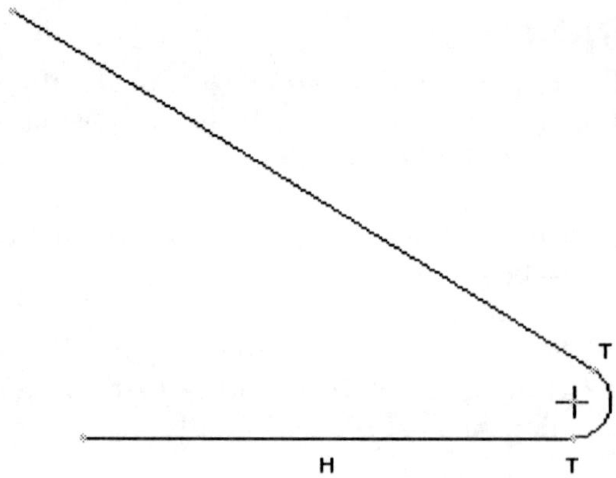

Figure 2-10 *Filleted corner*

Figure 2-11 shows two sets of arcs and Figure 2-12 shows the circular fillet created between the two arcs. The location where you select the arcs to create the fillet is important. The fillet is created tangent to the selection points on the arcs. Here, the endpoints of the arcs are selected to create the fillet.

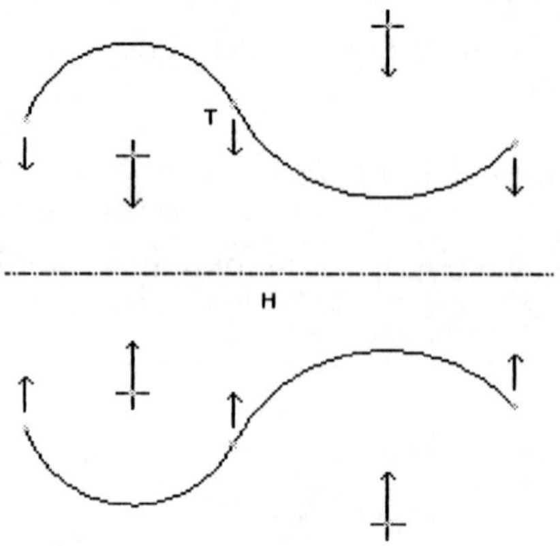

Figure 2-11 *Two sets of arcs*

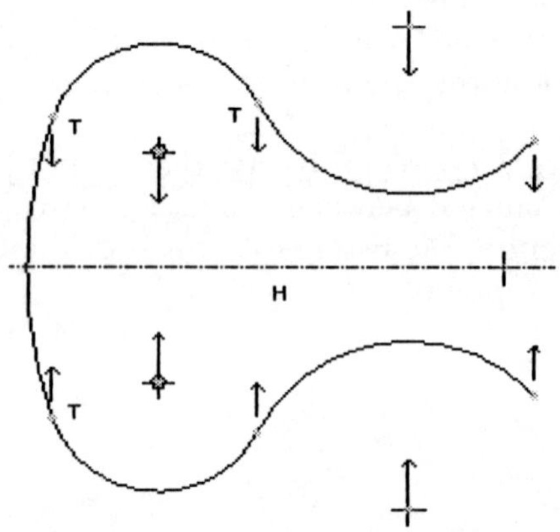

Figure 2-12 *Fillet created by selecting the endpoints*

If the points of selection on the two arcs are away from the endpoints of the arcs, then the fillet is created at the selection points. The portion of the arc that is left should be manually deleted or trimmed. This case of creating a fillet is explained in Figures 2-13 and 2-14.

The **Create a circular fillet between two entities** button in the **Sketcher Tools** toolbar is used to create circular fillets. The following steps explain the procedure to create a circular fillet:

1. Choose the **Create a circular fillet between two entities** button from the **Sketcher Tools**

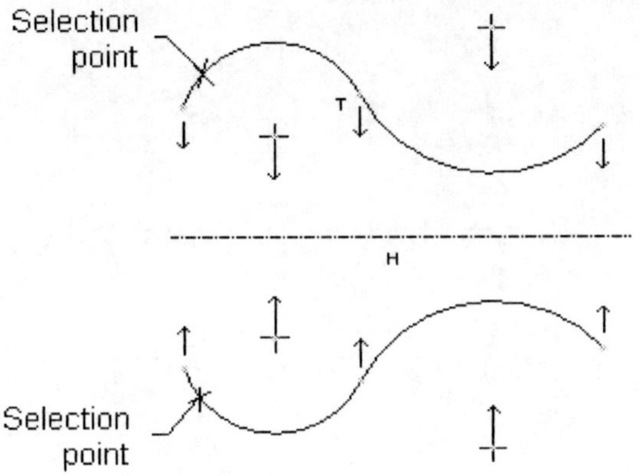

Figure 2-13 *Points to be selected on arcs*

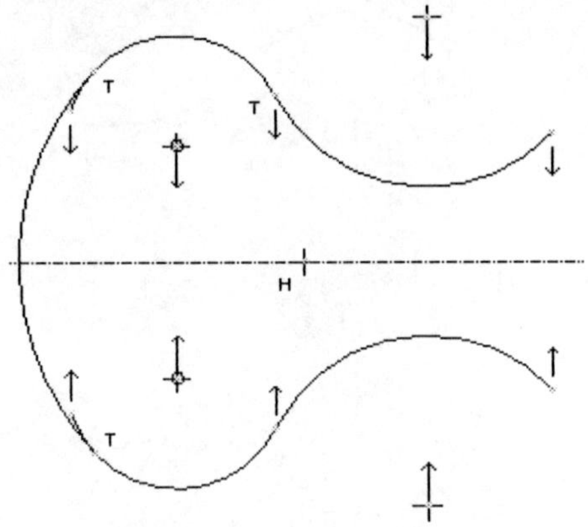

Figure 2-14 *Filleted arcs*

toolbar. You are prompted to select two entities.

2. Select the first entity for filleting by using the left mouse button. The yellow color of the first entity changes to red. Now, select the second entity. As soon as you select the second entity, a possible fillet is drawn between the two selected entities.

3. Repeat step 2 until the required number of possible fillets are drawn.

Note
You can create a fillet between any two entities except between two parallel lines and between a centerline and an entity.

Creating Elliptical Fillets

An elliptical fillet is the arc in the form of an ellipse that joins two lines, two arcs, or a line and an arc. The geometry of the elliptical fillet depends on the location where you select the entities to create a fillet.

The advantage of elliptical fillets over circular fillets is that elliptical fillets have the geometry of an ellipse. As you know, an ellipse is controlled by dimensions in two directions. Therefore, when an elliptical fillet is dynamically modified, its geometry can be controlled in either the x-direction or the y-direction. Hence, the elliptical fillet when modified, results in more curved geometric shapes than a circular fillet.

Figures 2-15 and 2-16 illustrate the elliptical fillet. Notice, that a strong tangent constraint **T** is automatically applied when you create a fillet.

An elliptical fillet is created by choosing the **Create an elliptical fillet between two entities** button in the **Sketcher Tools** toolbar. The following steps explain the procedure to create elliptical fillets:

1. Choose the black arrow on the right of the **Create a circular fillet between two entities**

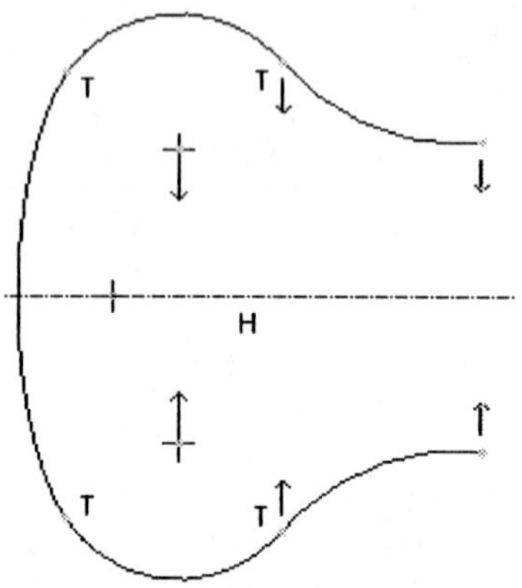

Figure 2-15 Arcs to be filleted

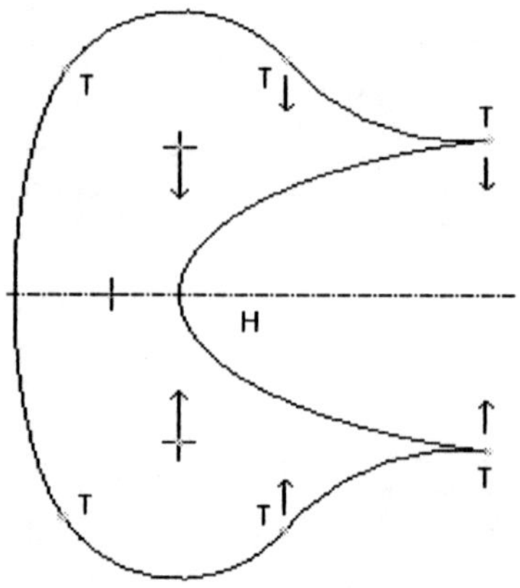

Figure 2-16 Elliptical fillet created

button to display a flyout. Choose the **Create an elliptical fillet between two entities** button from this flyout. You are prompted to select two entities.

2. Select the first entity by using the left mouse button. The color of the entity changes to red.

3. Select the second entity. As soon as you select the second entity, the elliptical fillet is created. The shape of the elliptical fillet depends upon the specified points. After the fillet is created, you are again prompted to select two entities for elliptical fillet.

4. Repeat steps 2 and 3 until you have created all the fillets.

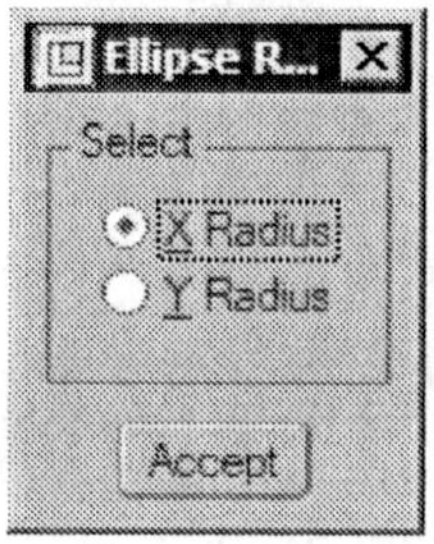

***Figure 2-17** Ellipse Rad dialog box*

When you select an elliptical fillet to dimension, the **Ellipse Rad** dialog box is displayed as shown in Figure 2-17. There are two radio buttons in this dialog box. The **X Radius** radio button when selected dimensions the elliptical fillet radially in the X-direction. The **Y Radius** radio button when selected dimensions the elliptical fillet radially in the Y-direction.

CREATING A REFERENCE COORDINATE SYSTEM

The **Create reference coordinate system** button in the **Sketcher Tools** toolbar is used to create a coordinate system and add it to the sketch you create. The coordinate system acts as reference for dimensioning. You can dimension the splines using the coordinate system. Thus, it provides you flexibility to modify the spline points by specifying the different coordinates with respect to the coordinate system.

The user-defined coordinate system is used in blend features to align different sections in a blend. It is also used in the Assembly mode and Manufacturing mode of Pro/ENGINEER.

The following steps explain the procedure to create a coordinate system:

1. Choose **Create reference coordinate system** button on the flyout that is displayed when you choose the black arrow on the right of the **Create points** button in the **Sketcher Tools** toolbar. You are prompted to select the location for the coordinate system. The coordinate system symbol is attached to the cursor.

2. Place the coordinate system at the desired points on the screen by pressing the left mouse button. The coordinate system will be placed at as many places as you select on the graphics screen. You can end coordinate system creation by using the middle mouse button.

Note

If you add a coordinate system to a sketch, it must be dimensioned unless the coordinate system is placed at the endpoints of a line, an arc, a spline, or at the center of an arc or a circle. In other words, a coordinate system must be referenced to an entity in a sketch.

WORKING WITH SPLINES

Splines are curved entities that can pass through an infinite number of intermediate points. Generally, splines are used to define the outer surface of a model. This is because the splines can provide different shapes to curves and the flexibility to modify the surfaces that result from the splines. Splines find application in automobile and aeroplane body designing.

Creating a Spline

To draw a spline, choose the **Create a spline curve** button from the **Sketcher Tools** toolbar.

The following steps explain the procedure to create a spline:

1. Choose the **Create a spline curve** button from the **Sketcher Tools** toolbar. You are prompted to select the location for spline.

2. Use the left mouse button to select the start point for the spline. Similarly, select additional points on the graphics screen passing through which the spline will be drawn.

Dimensioning of Splines

When a spline is drawn, the weak dimensions are automatically applied to the spline. A spline can be dimensioned manually in the following ways:

1. Dimensioning the endpoints.

2. Radius of curvature dimensioning.

3. Tangency dimensioning.

4. Coordinate dimensioning.

5. Dimensioning the interpolation points.

Dimensioning the endpoints

To dimension a spline by selecting the endpoints:

1. Choose the **Create defining dimension** button from the **Sketcher Tools** toolbar.

2. Select the two endpoints of the spline and place the horizontal or vertical dimension by pressing the middle mouse button.

   Figure 2-18 shows a spline that is dimensioned by selecting the endpoints.

Radius of curvature dimensioning

The radius of curvature of a spline can be dimensioned only if its tangency is defined. In other words, radius of curvature of a spline can be dimensioned only if the spline is tangent to an entity. For radius of curvature dimensioning of a spline:

1. Choose the **Create defining dimension** button from the **Sketcher Tools** toolbar.

2. Select the endpoint of the spline where tangency is defined.

3. Press the middle mouse button to place the dimension.

   Figure 2-19 shows the radius of curvature dimensioning of a spline.

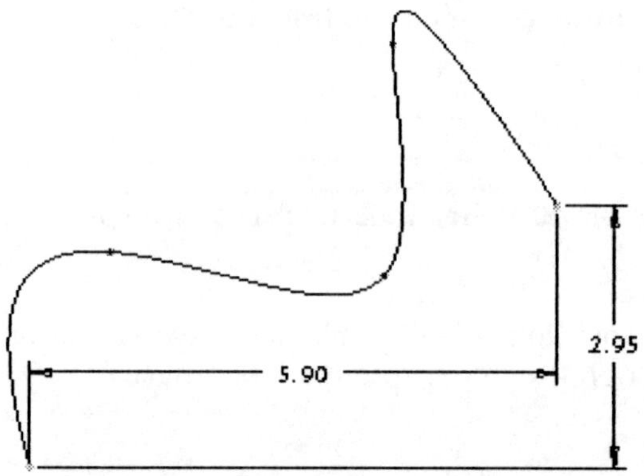

Figure 2-18 Endpoint dimensioning

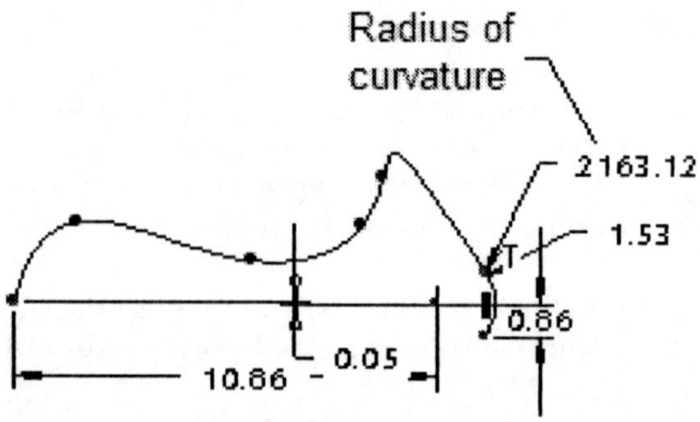

Figure 2-19 Radius of curvature

Tangency or Angular Dimensioning

A spline can be dimensioned tangentially. This type of dimensioning is also called angular dimensioning. To angular dimension a spline and a line tangent to it:

1. Choose the **Create defining dimension** button from the **Sketcher Tools** toolbar.

2. Select the spline by using the left mouse button.

3. Select the entity tangent to the spline by using the left mouse button.

4. Select the vertex of the spline that is to be dimensioned tangentially.

5. Press the middle mouse button to place the dimension.

Coordinate dimensioning

The spline can be dimensioned with respect to a user-defined coordinate system. Choose the **Create reference coordinate system** button from the flyout in the **Sketcher Tools** toolbar. The coordinate system is attached to the cursor. Place the coordinate system on the graphics screen. Now, the spline can be dimensioned with respect to the coordinate system.

Dimensioning the interpolation points

A spline can be dimensioned by dimensioning its interpolation points or vertices. This type of dimensioning is used when the designer wants the spline to be standard for all designs. This is because the exact curve can be duplicated if the interpolation points or the vertices of a spline are dimensioned.

Tip: *A dimension can be moved by pressing and holding the left mouse button on the dimension and moving it. The dimension text is replaced by an orange colored box and the cursor is replaced by a hand. You can drag the dimension to the desired location on the graphics screen and release the left mouse button to place the dimension at that point.*

Modifying a Spline

In Pro/ENGINEER Wildfire, a spline can be modified in the following ways:

1. Moving the points or vertices of the spline.

2. Adding points or vertices to a spline.

3. Deleting points or vertices of a spline.

4. Creating a control polygon and moving its control points.

5. Modifying the dimensions of the spline.

Moving the points or vertices of the spline

When you draw a spline, the points that you select on the graphics screen to create the spline are called vertices. The vertices of a spline can be dynamically modified. To modify a spline, select a vertex on the spline and when the hand appears, drag the vertex to modify the shape of the spline.

You can also use the dashboard to modify a spline. To invoke the dashboard, select the spline and hold the right mouse button down to invoke the shortcut menu. Choose the **Modify** option from the shortcut menu. The modify spline dashboard is displayed with the options and tool buttons to modify a spline at the bottom of the graphics screen.

The vertices of the spline appear in black on the graphics screen. Drag the vertices to modify the shape of the spline. Choose **OK** from the dashboard to exit it.

Adding points or vertices to a spline

To add points or vertices on a spline, invoke the dashboard. Now, right-click on the spline to invoke the shortcut menu. Choose the **Add Point** option; a point is added to the spline where the spline was selected. The new point appears in black. Note that the **Add Point** option is available only when the dashboard is displayed at the bottom of the graphics screen. You cannot increase the length of the spline by adding vertex before the start point and after the endpoint of the spline.

Deleting points or vertices of a spline

To delete a point or a vertex, invoke the dashboard. Now, select the vertex and hold down the right mouse button to invoke the shortcut menu. Choose the **Delete Point** option from the shortcut menu. The selected point is deleted. You can continue deleting vertices or points from a spline until only two vertices are left in the spline.

Creating a control polygon and moving its control points

When you draw a spline, it is associated with a control frame. The vertices of its frame are called control points. To create a control polygon, choose the **Create control polygon** button from the dashboard. The control polygon is displayed on the graphics screen. The control points of this polygon can be moved by dragging to modify the spline shape.

Modifying the dimensions of the spline

The dimensions of the spline can be modified to obtain its different shapes. Since the dimensions control the shape of the entities, therefore to modify the shape the dimensions can be modified.

Tip: *To dynamically modify the shape of the sketch, you need to select the sketch and when the hand symbol appears, move the mouse to modify the sketch. Remember that if the selected entity is constrained then you cannot modify it. You can modify it only by first disabling the constraint.*

CREATING TEXT IN SKETCHER ENVIRONMENT

There are various instances when a designer needs a text command to write text on the model. For example, for creating a label, model number, company name, and so on. In Pro/ENGINEER, you can write this text in the sketcher environment. Each alphabet or number that is created acts as an entity and can be modified as any other entity in the sketcher environment.

In the sketcher environment, the text is created using the **Create text as a part of a section** button from the **Sketcher Tools** toolbar. The following steps explain the procedure to create text in the sketcher environment:

1. Choose the **Create text as a part of a section** button from the **Sketcher Tools** toolbar. You are prompted to select the start point of line to determine the text height and orientation.

2. Specify the start point on the screen by using the left mouse button. You are prompted to select the second point of line to determine the text height and orientation.

3. Note that, to write the text upright, the second point should be above the start point and in a straight line. If the second point is below the start point, the text will be written upside down. Specify the second point on the screen by using the left mouse button. The **Text** dialog box is displayed as shown in Figure 2-20.

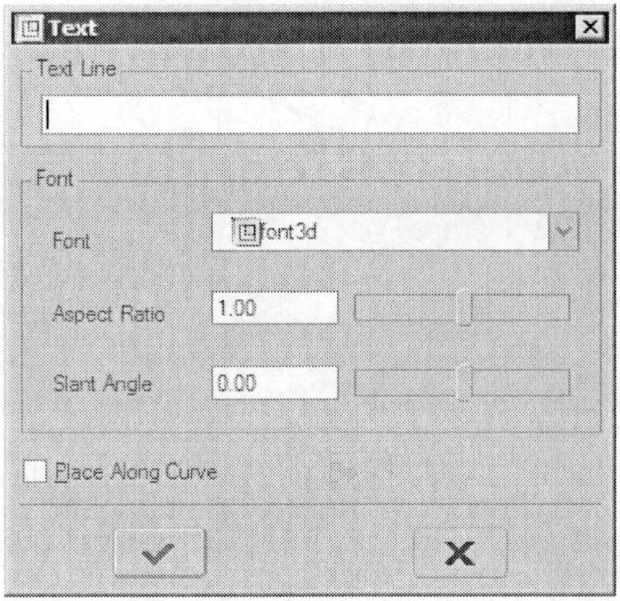

***Figure 2-20* Text** *dialog box*

After specifying the second point, a construction line is drawn having height equal to the distance between the two points. The height and orientation of the text depends on the height and angle of the construction line. If the construction line is drawn at an angle, then the text is written at that angle.

4. In the **Text Line** edit box, enter the text, which can be up to 79 characters. As you enter the text, the same text is displayed dynamically on the graphics screen. You can modify the font, aspect ratio, and slant angle using the slider bars.

5. Choose the **Close the dialog** button in the **Text** dialog box to exit it.

Note

In the later chapters of this book, you will learn that to enter the sketcher environment there are various methods other than entering through the Sketch mode.

SCALING AND ROTATING ENTITIES

The sketches can be scaled or rotated by using the **Scale and rotate selected entities** button from the **Sketcher Tools** toolbar. To invoke this button, select the sketch and then choose the arrow on the right of the **Mirror selected entities** button. This button is available on the flyout that is displayed. After choosing this tool button, the sketch that is composed of various entities now acts as a single entity. The sketch appears orange in color and is enclosed in a boundary as shown in Figure 2-21.

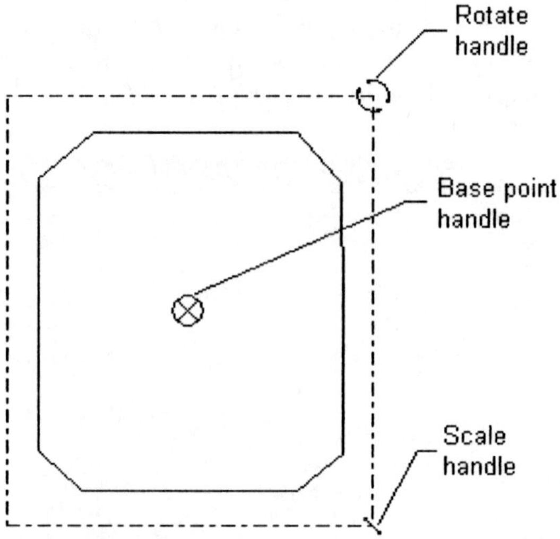

Figure 2-21 *Original entities and the copied entities with three handles*

There are three handles that facilitate in scaling and rotating the selected sketch. The rotate handle is used to dynamically rotate the selected entities. The scale handle is used to dynamically scale the selected entities. The base point handle is used to pick the sketch and place it at any other location on the graphics screen. To change the location of any of the three handles, right-click on the handle. The handle symbol is attached to the cursor. Place the symbol at the desired location. The following steps explain the procedure to scale and rotate a sketch:

1. Select the sketch to be rotated and scaled, and then choose the black arrow on the right of the **Mirror selected entities** button from the **Sketcher Tools** toolbar.

2. From the flyout, choose the **Scale and rotate selected entities** button. The **Scale Rotate** dialog box is displayed as shown in Figure 2-22. This dialog box contains the **Scale** and **Rotate** edit boxes. The sketch can be scaled and rotated dynamically or by entering a value in these edit boxes.

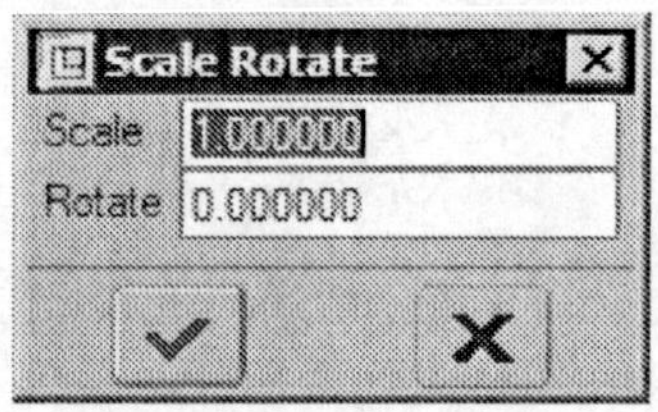

Figure 2-22 Scale Rotate *dialog box*

3. To dynamically rotate the sketch, select the rotate handle and then move the cursor. The sketch is rotated as you move the cursor. You can also enter the rotation angle in the **Rotate** edit box.

To scale the sketch, select the scale handle and then move the cursor. As you move the cursor, the sketch is scaled dynamically on the graphics screen. You can also enter the scale value in the **Scale** edit box.

4. After the sketch is rotated and scaled, choose the **Accept the changes and close the dialog** button in the **Scale Rotate** dialog box.

COPYING DRAWING ENTITIES IN SKETCH MODE

The **Make a copy of selected entities** button in the **Sketcher Tools** toolbar is used to copy the entities in the sketcher environment. To invoke this button, select the sketch and then choose the arrow on the right of the **Mirror selected entities** button. Choose the **Make a copy of selected entities** button from the flyout that is displayed. The **Scale Rotate** dialog box is displayed. Using this dialog box, you can simultaneously copy and rotate the selected entities. The following steps explain the procedure to copy the selected entities:

1. Select the entities to be copied and then choose the black arrow on the right of the **Mirror selected entities** button from the **Sketcher Tools** toolbar. From the flyout, choose the **Make a copy of selected entities** button.

   The copy of the selected entity is placed at the top left corner on the graphics screen. This copy appears orange in color. There are three handles in the copied entities that facilitate in sizing and orienting them. Figure 2-23 shows the copied sketch enclosed in a boundary with the three handles.

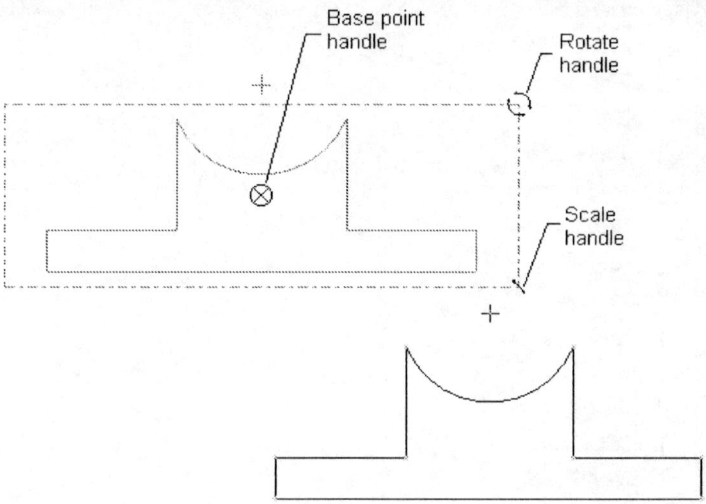

Figure 2-23 *Original entities and the copied entities with three handles*

   The **Scale Rotate** dialog box is also displayed. Use this dialog box to specify the scale factor and the rotation angle.

2. Select the base point handle; the copied entities are attached to the cursor. Place the entities where needed.

IMPORTING 2D DRAWINGS IN THE SKETCH MODE

The two-dimensional (2D) drawings when opened in the sketcher environment can be saved in the *.sec* format. The *.sec* file can then be converted to a solid model. The **Data from file** option in the **Sketch** menu in the menu bar is used to import the 2D sketches. This option saves time in drawing the same or similar section again. The file formats from which the data can be imported are shown in Figure 2-24.

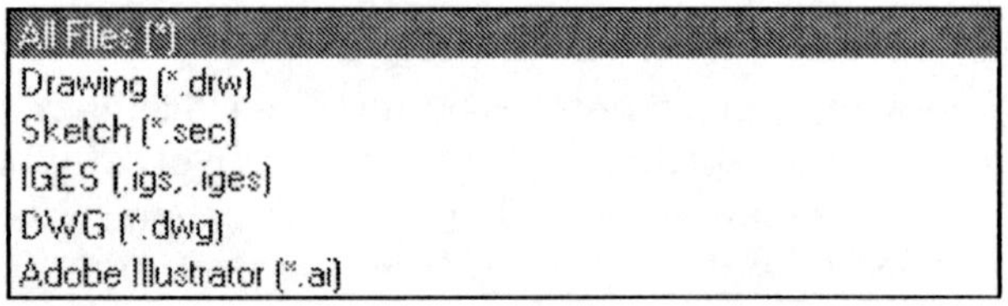

Figure 2-24 File formats

When you choose **Edit > Data from file** from the menu bar, the **Open** dialog box is displayed as shown in Figure 2-25. Use this dialog box to select and open the file.

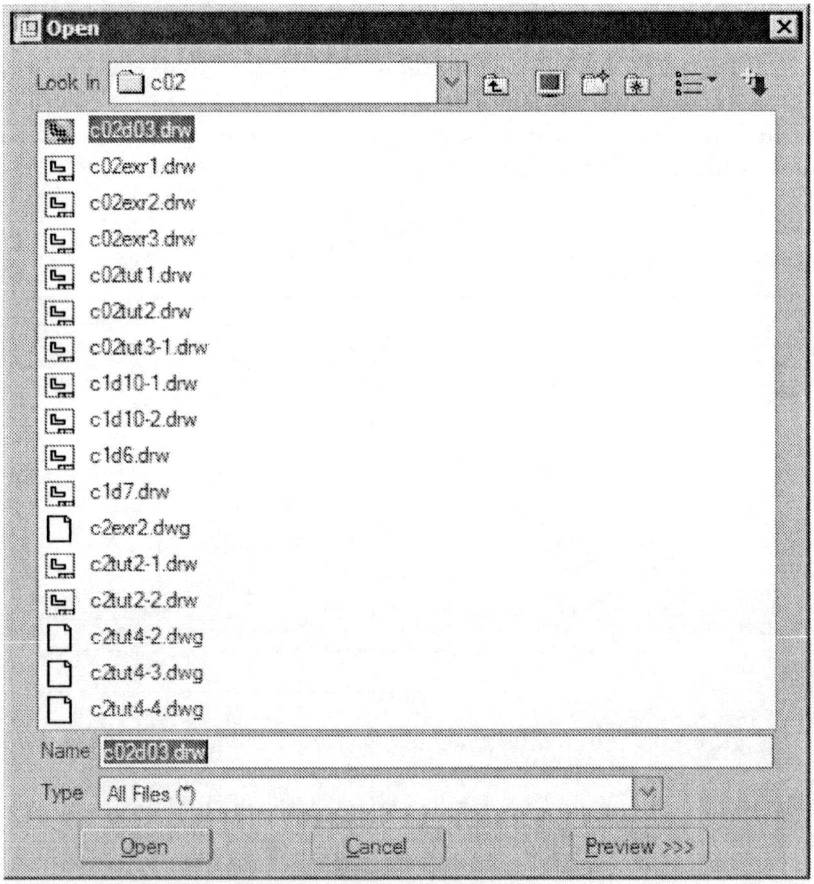

***Figure 2-25 Open** dialog box*

In the **Open** dialog box, if you select a drawing file that is created in the Drawing mode of Pro/ENGINEER Wildfire, the draft entities in that file are imported. The selected drawing is opened in a sub window. You are prompted to select the entities to copy from the sub window. Select the draft entities and then press the middle mouse button. The selected entities are enclosed in a boundary and the **Scale Rotate** dialog box is displayed. Use this dialog box to set the scale and orientation of the sketch. Note that if the *.drw* file does not consist of draft entities, no data is imported.

In the **Open** dialog box, if you select a *.sec* file that is created in the sketcher environment, the sketch is either displayed on the graphics screen enclosed in a boundary or is directly placed on the graphics screen. If there are no entities in the sketcher environment, the selected sketch will be automatically placed on the graphics screen. If there are some entities, the sketch is enclosed in a boundary and the **Scale Rotate** dialog box is displayed. Use this dialog box to set the scale and orientation of the sketch.

The section imported using the **Data from file** option in the current sketch is an independent copy. The imported section is no longer associated with the source section. The units, dimensions, grid parameters, and accuracy are acquired from the current sketch.

Tip: *The display of vertices of the section, the display of dimensions, and the display of constraints can be turned on or off from the* ***Sketcher*** *toolbar. This toolbar is in the* ***Top Toolchest****.*

TUTORIALS

Tutorial 1

In this Tutorial you will place an existing sketch that you had drawn in Tutorial 3 of Chapter 1. After placing the sketch, draw the keyway as shown in Figure 2-26.

(Expected time: 15 min)

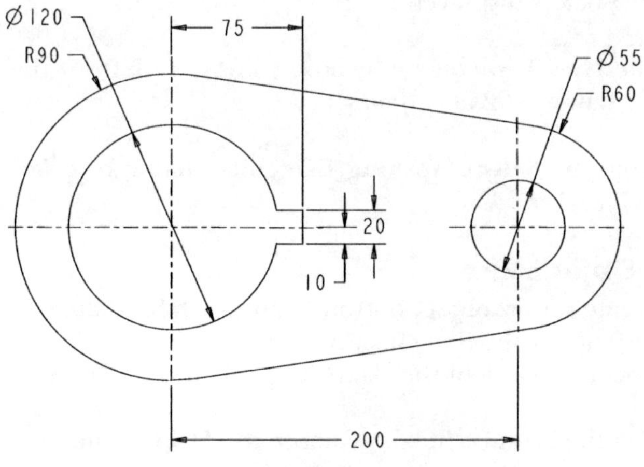

Figure 2-26 *Sketch for Tutorial 1*

The following steps outline the procedure for completing this tutorial.

a. Start Pro/ENGINEER Wildfire.

b. Set the working directory and create a new object file.

c. Import the section by using the **Data from file** option.

d. Draw the keyway and dimension it.

e. Modify the dimensions.

f. Save the sketch and exit the sketcher environment.

Starting Pro/ENGINEER

1. Start Pro/ENGINEER Wildfire by double-clicking on the Pro/ENGINEER Wildfire icon on the desktop of your computer or by using the **Start** menu.

Setting the Working Directory

When the Pro/ENGINEER session is started, the first task is to set the working directory. As mentioned earlier, working directory is a directory on your system where you can save the work done in the current session of Pro/ENGINEER. You can set any directory existing on your system as the working directory. Since this is the first tutorial of this chapter, you need to create a folder named **c02** in the *C:\ProE-WF* folder, if it does not exist.

1. Choose the **Set Working Directory** option from the **File** menu. The **Select Working Directory** dialog box is displayed.

2. Browse and select *C:\ProE-WF*.

3. Choose the **New Directory** button in the **Select Working Directory** dialog box. The **New Directory** dialog box is displayed.

4. Type **c02** in the **New Directory** edit box. Choose **OK** from the dialog box. You have created a folder named *c02* at *C:\ProE-WF*.

5. Choose **OK** from the **Select Working Directory** dialog box. You have set the working directory to *C:\ProE-WF\c02*.

Creating New Object File

1. Choose the **Create a new object** button from the **File** toolbar. The **New** dialog box is displayed. Select the **Sketch** radio button from the **Type** area of the **New** dialog box. A default name of the sketch appears in the **Name** edit box.

2. Enter **c02tut1** in the **Name** edit box. Choose the **OK** button.

You are in the sketcher environment of the Sketch mode. When you enter the sketcher environment, the navigator is displayed to the left on the graphics screen. Slide in the navigator by clicking on the top arrows present on its right edge. Now, the drawing area is increased.

Importing the Section

1. Choose **Sketch > Data from file** from the menu bar. The **Open** dialog box is displayed with the working directory as the current directory.

2. Choose the **Up One Level** button and then open the *c01* directory. Make sure that the **sketch (*.sec)** option is selected in the **Type** drop-down list. Select *c01tut3.sec* and choose the **Open** button from the **Open** dialog box.

   The sketch is placed on the graphics screen as shown in Figure 2-27.

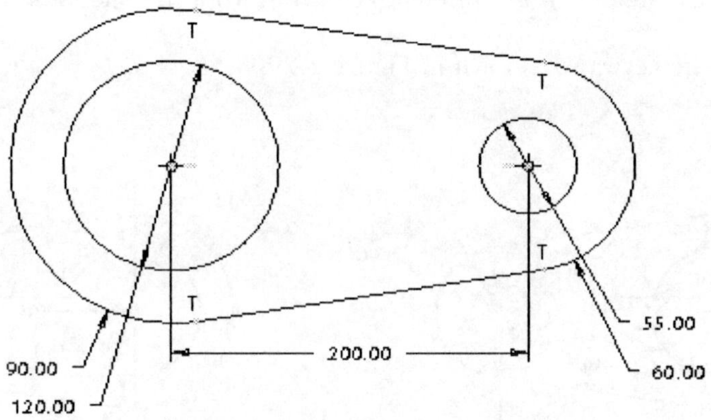

Figure 2-27 *Sketch placed in the current file*

Drawing the Keyway

To create the keyway, you will sketch the keyway and then the portion of the circle that lies between the horizontal lines will be removed.

1. Choose the **Create 2 points line** button from the **Sketcher Tools** toolbar. 

2. Draw the keyway as shown in Figure 2-28. The weak dimensions and constraints are applied automatically to the sketch of the keyway.

   The horizontal lines of the keyway and the circle intersect at the points where the lines meet the circle. The intersection points are displayed in orange color. The portion of the circle that lies between the two horizontal lines of the keyway is separated from the circle.

3. Choose the **Zoom In** button from the **View** toolbar in the **Top Toolchest**. The cursor is converted into a magnifying glass symbol. 

4. Draw a window around the keyway to zoom in. Now, the display area of the keyway is enlarged.

5. Choose the **Dynamically trim section entities** button from the **Sketcher Tools** toolbar.

6. Press the left mouse button to select the part of the circle that lies between the two intersection points. The part of the circle that was unwanted is deleted.

7. Choose the **Refit object to fully display it on the screen** button from the **View** toolbar to view the full sketch.

Dimensioning the Keyway

The dimensions that will be applied to the keyway are shown in Figure 2-26.

1. Choose the **Create defining dimensions** button from the **Sketcher Tools** toolbar.

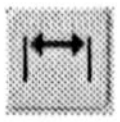

2. Dimension the keyway as shown in Figure 2-29.

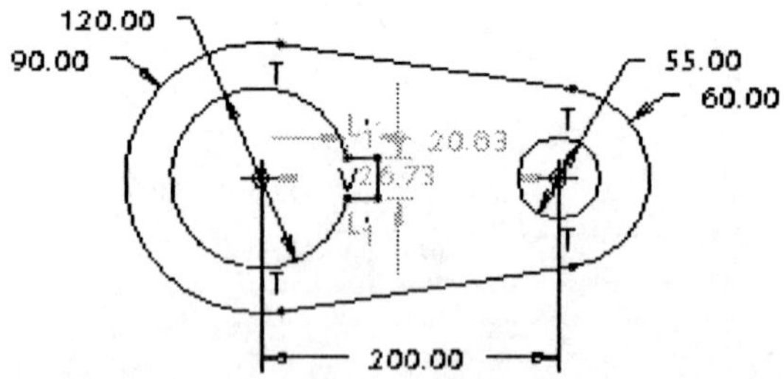

Figure 2-28 Sketch of the keyway with weak dimensions and constraints

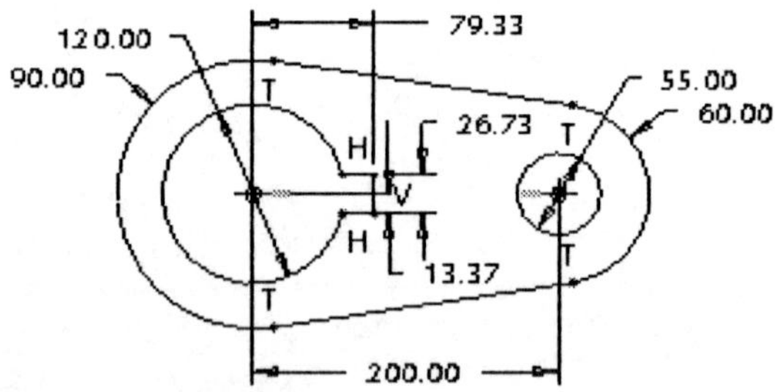

Figure 2-29 Sketch after dimensioning the keyway

Modifying the Dimensions

The dimensions of the keyway will be modified to the given dimension values.

1. Select the three dimensions of the keyway by using CTRL+left mouse button.

2. Choose the **Modify the values of dimensions, geometry of splines, or text entities** button from the **Sketcher Tools** toolbar. The **Modify Dimensions** dialog box is displayed.

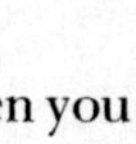

3. Clear the **Regenerate** check box and then modify the values of the dimensions. When you clear the check box, the sketch does not regenerate as you modify the dimensions.

The dimension that you edit in the **Modify Dimensions** dialog box is enclosed in a yellow box in the sketch.

4. Modify all the dimensions. Refer to Figure 2-26 for dimension values.

5. After all the dimensions are modified, choose the **Regenerate the section and close the dialog** button from the **Modify Dimensions** dialog box. The message **Dimension modifications successfully completed** is displayed in the **Message Area**.

   The sketch after modifying all the dimension values of the sketch is shown in Figure 2-30.

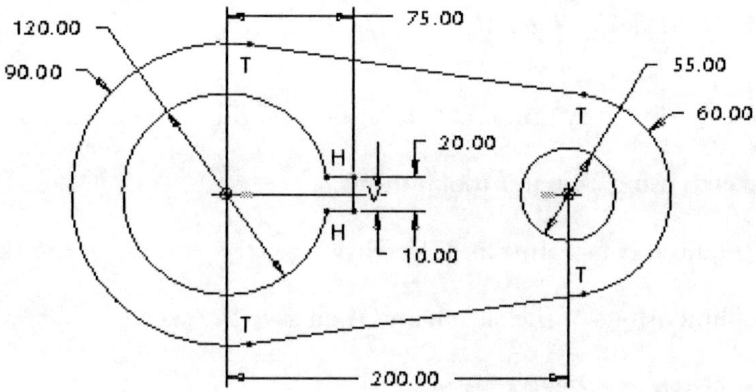

***Figure 2-30** Sketch after modifying the dimensions*

Saving the Sketch

You have to save the sketch because you may need the sketch later.

1. Choose the **Save the active object** button from the **File** toolbar. The **Message Input Window** is displayed with the name of the sketch that you had entered earlier.

2. Press ENTER or choose the check mark button on the window. The sketch is saved.

3. Choose the **Continue with the current section** button from the **Sketcher Tools** toolbar.

Tutorial 2

In this tutorial, you will draw the sketch for the model shown in Figure 2-31. The sketch is shown in Figure 2-32. **(Expected time: 30 min)**

The following steps outline the procedure for creating this sketch:

Figure 2-31 Model for Tutorial 2

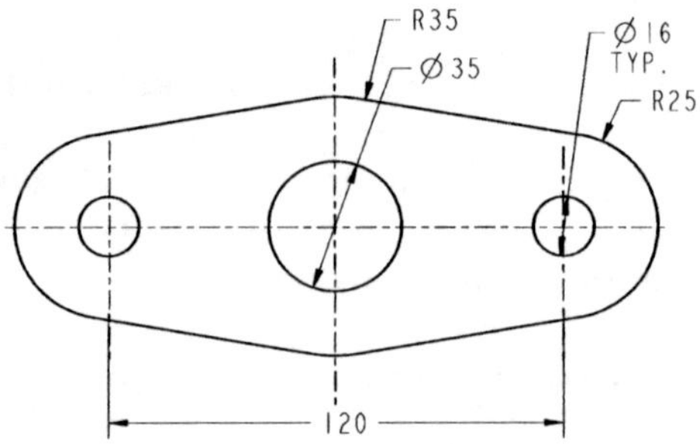

Figure 2-32 Sketch of the model

a. Set the working directory and create a new object file.

b. Draw the sketch using sketcher tool buttons.

c. Apply the required constraints and dimensions to the entities in the sketch.

d. Modify the dimensions of the sketch and then save the sketch.

Setting the Working Directory

The working directory was selected in Tutorial 1, and therefore there is no need to select the working directory again. But if you still want to select the working directory, follow the steps given next:

1. Open the Navigator by clicking the top arrows on the left edge of the Pro/ENGINEER main window. The Navigator slides out.

2. Click on the plus symbol adjacent to the *ProE-WF* folder in the Navigator. The contents of the *ProE-WF* folder are displayed.

3. Now right-click on the *c02* folder to display a shortcut menu. From this shortcut menu, choose the **Make Working Directory** option. The working directory is set to *c02*.

4. Close the Navigator by clicking the sash on the right edge of the navigator. The Navigator slides in.

Creating New Object File

1. Open a new object file in the Sketch mode. Name the file as **c02tut2**.

Drawing the Sketch

To draw the outer loop, you need to draw three circles and then draw lines tangent to the three circles.

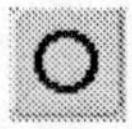

1. Choose the **Create circle by picking the center and a point on the circle** button. Draw the three circles as shown in Figure 2-33.

2. Choose the black arrow on the right of the **Create 2 points line** button to display a flyout. Choose the **Create lines tangent to 2 entities** button from this flyout.

3. Select the left circle at the top and then select the middle circle at the top by pressing the left mouse button. A tangent is drawn from the top of the left circle to the top of the middle circle.

4. Select the right circle at the top and the middle circle at the top by pressing the left mouse button. A tangent is drawn from the top of the right circle to the top of the middle circle.

5. Similarly, using the **Create lines tangent to 2 entities** button, draw the other tangents from the bottom of the circles as shown in Figure 2-34.

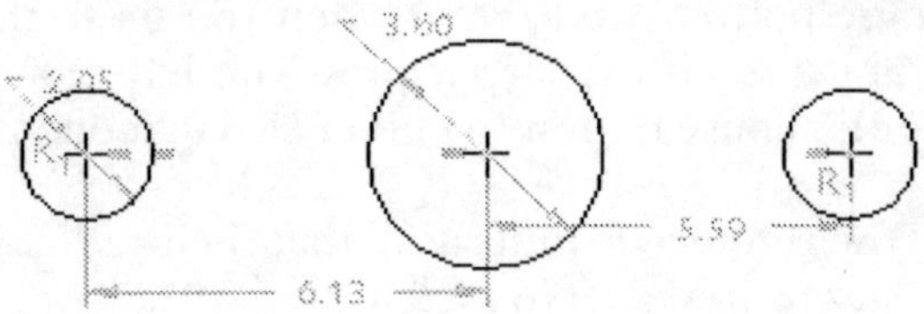

Figure 2-33 *Three circles with weak dimensions and constraints*

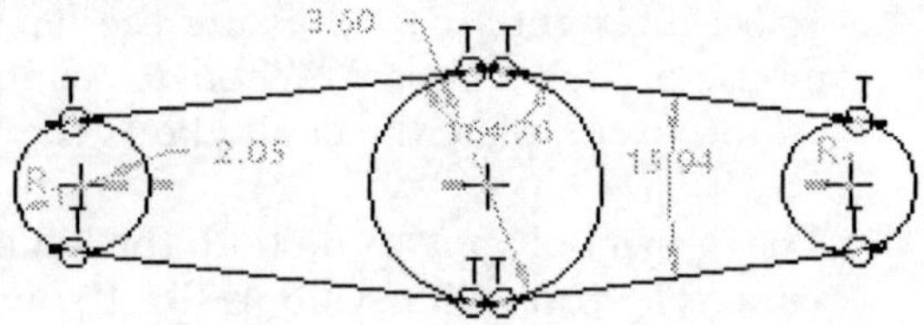

Figure 2-34 *Lines joining the three circles with weak dimensions*

Trimming the Circles

The inner portions of the circles are not required. Therefore, you need to trim them.

1. Choose the **Dynamically trim section entities** button from the **Sketcher Tools** toolbar.

2. Bring the cursor close to the right portion of the left circle. The right part of the circle turns cyan in color. Select the right portion of the circle by pressing the left mouse button to delete it.

3. Similarly, trim the part of the middle circle and the right circle that is not required. The sketch after trimming the circles is shown in Figure 2-35.

Drawing the Circles

1. Choose the black arrow on the right of the **Create circle by picking the center and a point on the circle** button to display the flyout. From this flyout, choose

the **Create concentric circle** button. You are prompted to select an arc.

> **Tip**: *While drawing a concentric circle, sometimes the circle snaps to the other circle or arc and it becomes difficult to draw a circle of the size you need. The snapping of the circle to the other circle or arc can be disabled. While drawing the circle when the circle snaps to the other circle, use TAB+right-click to disable the snapping or to disable the equal radii constraint that the system tends to apply.*

2. Select the left arc by pressing the left mouse button. Notice that when you move the mouse, a circle appears. Select a point inside the sketch to complete the circle. Press the middle mouse button.

> **Note**
> *You may need to zoom in to select the top arc in the next step.*

3. Select the top arc by pressing the left mouse button. Notice that when you move the mouse a circle appears. Select a point inside the sketch to complete the circle. Press the middle mouse button to end circle creation.

4. Select the right arc by pressing the left mouse button. Notice that when you move the mouse a circle appears. When $\mathbf{R_2}$ symbol appears on the circle, press the left mouse button to complete the circle. Press the middle mouse button to end circle creation.

   The $\mathbf{R_2}$ symbol appears on both the left and the right circle indicating that their radius is same. The constraint symbols like R_1 or R_2 vary from sketch to sketch.

   The sketch after drawing the three circles is shown in Figure 2-36.

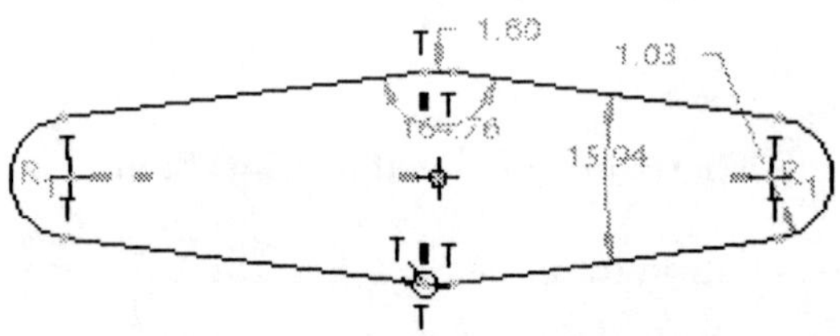

Figure 2-35 Sketch after trimming the circles

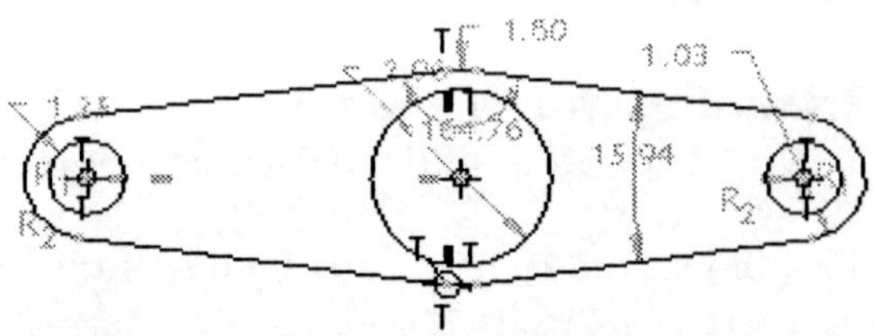

Figure 2-36 Sketch after drawing the three circles

Applying the Constraints

1. Choose the **Impose sketcher constraints on the section** button from the **Sketcher Tools** toolbar. The **Constraints** dialog box is displayed.

2. Choose the **Create Equal Lengths, Equal Radii, or Same Curvature constraint** button from the **Constraints** dialog box.

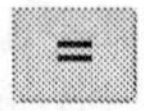

3. Select the left arc and then select the right arc to apply the equal radius constraint.

4. Select the left circle and the right circle to apply the equal radius constraint.

5. Use the CTRL+middle mouse button to zoom in and now select the top arc and the bottom arc to apply the equal radius constraint. If the constraint is already applied, the **Resolve Sketch** dialog box is displayed as shown in Figure 2-37.

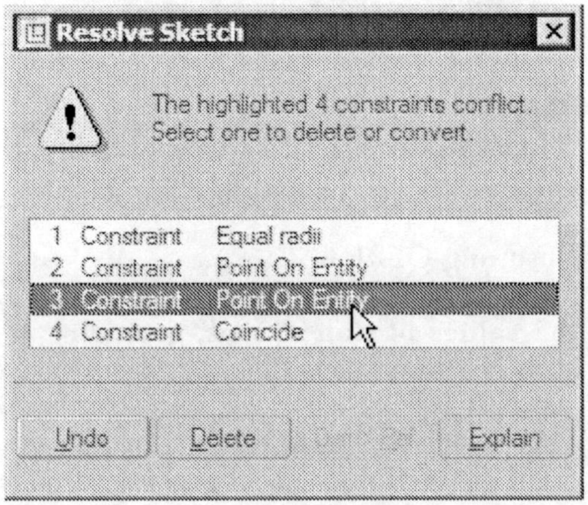

Figure 2-37 ***Resolve Sketch*** *dialog box*

6. Select the third constraint in the **Resolve Sketch** dialog box and choose the **Delete** button. The constraint that was conflicting is deleted.

7. Choose the **Make two lines parallel** constraint button from the **Constraints** dialog box.

8. Press the left mouse button to select the tangent line that connects the left circle and the middle circle at the top.

9. Press the left mouse button to select the tangent line that connects the right circle and the middle circle at the bottom.

   It is evident from the parallel constraint symbol that the parallel constraint is applied to the two tangent lines in the sketch.

10. Similarly, apply the parallel constraint to the other set of tangent lines. The sketch after applying the constraints is shown in Figure 2-38.

11. Choose the **Refit object to fully display it on the screen** button from the **View** toolbar to view the full sketch on the graphics screen.

Dimensioning the Sketch

Pro/ENGINEER applies the weak dimensions to the sketch automatically. These dimensions are not the dimensions that are needed because these dimensions will not help to machine the model. Therefore, you will dimension the sketch with the dimensions that will be used to machine the model.

1. Choose the **Create defining dimensions** button from the **Right Toolchest**.

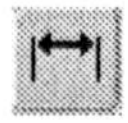

2. Select the center of right and left circles. The centers of the circles turn red in color. Now, using the middle mouse button, place the dimension below the sketch.

   The rest of the weak dimensions are needed and therefore they will be modified.

Modifying the Dimensions

All the constraints and dimensions are applied to the sketch and now dimensions will be modified.

1. Select all the dimensions using CTRL+ALT+A.

2. Choose the **Modify the values of dimensions, geometry of splines, or text entities** button. The **Modify Dimensions** dialog box is displayed.

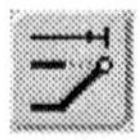

3. Clear the **Regenerate** check box and then modify the values of the dimensions. When you clear the check box, the sketch does not regenerate as you modify the dimensions.

   The dimension that you edit in the **Modify Dimensions** dialog box is enclosed in a yellow box in the sketch.

4. Modify all the dimensions. Refer to Figure 2-32 for dimension values.

5. After all the dimensions are modified, choose the **Regenerate the section and close the dialog** button from the **Modify Dimensions** dialog box. The message **Dimension modifications successfully completed** is displayed in the **Message Area**.

   The sketch after modifying the dimension values is shown in Figure 2-39.

Saving the Sketch

1. Choose the **Save the active object** button from the **File** toolbar and save the sketch.

Exiting the Sketch Mode

1. Choose the **Continue with the current section** button and exit the Sketch mode.

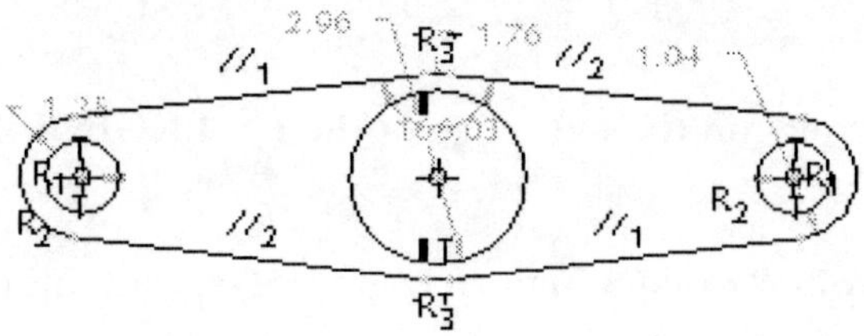

Figure 2-38 *Sketch with constraints*

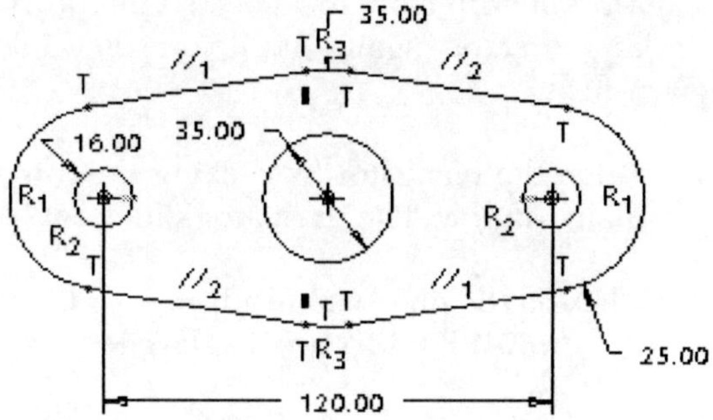

Figure 2-39 *Sketch after modifying the dimensions*

Tutorial 3

In this tutorial, you will draw the sketch for the model shown in Figure 2-40. The sketch is shown in Figure 2-41. **(Expected time: 30 min)**

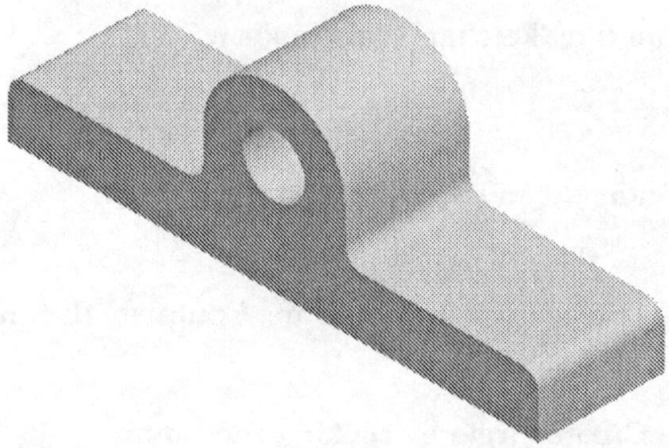

Figure 2-40 *Model for Tutorial 3*

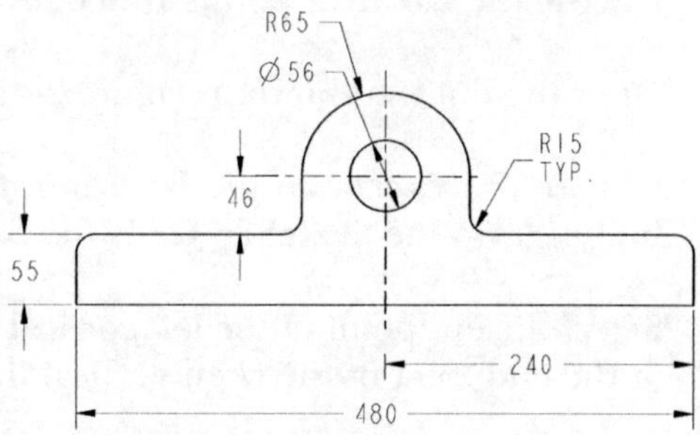

Figure 2-41 *Sketch of the model*

The following steps outline the procedure for creating this sketch:

a. Set the working directory and create a new object file.

b. Draw the sketch using sketcher tools.

c. Dimension the sketch.

d. Modify dimensions of the sketch.

e. Save the sketch and exit the Sketch mode.

Setting the Working Directory

The working directory was selected in Tutorial 1, and therefore there is no need to select the working directory again. But if you still want to select the working directory, follow the steps given next:

1. Open the navigator by clicking the top arrows on the left edge of the Pro/ENGINEER main window. The navigator slides out.

2. Click on the plus symbol adjacent to the **ProE-WF** folder in the navigator. The contents of the **ProE-WF** folder are displayed.

3. Now right-click on the **c02** folder to display a shortcut menu. From this shortcut menu, choose the **Make Working Directory** option. The working directory is set to **c02**.

4. Close the navigator by clicking the top arrows on the right edge of the navigator. The navigator slides in.

Creating New Object File

1. Open a new object file in the Sketch mode. Name the file as **c02tut3**.

Drawing the Sketch

1. Choose the **Create 2 points line** button from the **Sketcher Tools** toolbar. 

2. Draw the lines as shown in Figure 2-42.

3. Choose the **Create an arc by 3 points or tangent to an entity at its endpoint** button from the **Sketcher Tools** toolbar. 

4. Select the endpoint of the left vertical line as the start point of the arc. Complete the arc at the endpoint of the right vertical line.

5. Choose the black arrow on the right of the **Create circle by picking the center and a point on the circle** button to display the flyout. From this flyout, choose the **Create concentric circle** button. You are prompted to select an arc. 

6. Select the arc by pressing the left mouse button. A red rubber-band circle appears. Size the circle by moving the cursor and press the left mouse button to complete it.

   The sketch after drawing the circle is shown in Figure 2-43.

Note

*Choose the **Toggle display of dimensions on/off** button from the **Sketcher** toolbar in the **Top Toolchest** to turn the dimensions on or off.*

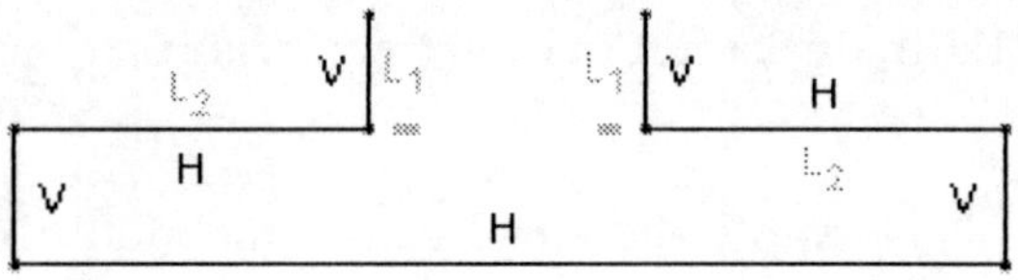

Figure 2-42 Lines in the sketch with the dimensions turned off for clarity

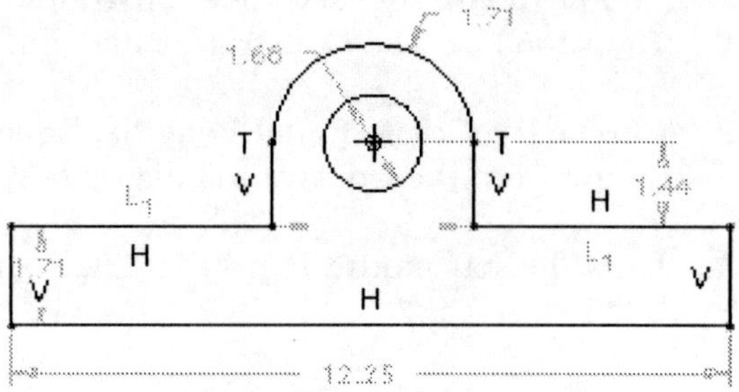

Figure 2-43 Sketch after drawing the arc and the circle

Note

Pro/ENGINEER does not have options like midpoint, endpoint, or center of. However, while drawing a sketch, these options are applied in the form of weak constraints as you sketch. For example, endpoint of any entity snaps the cursor when a new entity is to be drawn. The middle point constraint appears when you approximately bring the cursor near to the middle point of the line to draw a line.

Filleting the Corners

1. Choose the **Create a circular fillet between two entities** button from the **Sketcher Tools** toolbar. You are prompted to select the two entities to be filleted. The corners that you need to fillet are shown in Figure 2-44.

2. Select the two entities one by one using the left mouse button. The corner of the selected lines are filleted. Fillet all the required corners shown in Figure 2-44.

   The sketch after filleting is shown in Figure 2-45.

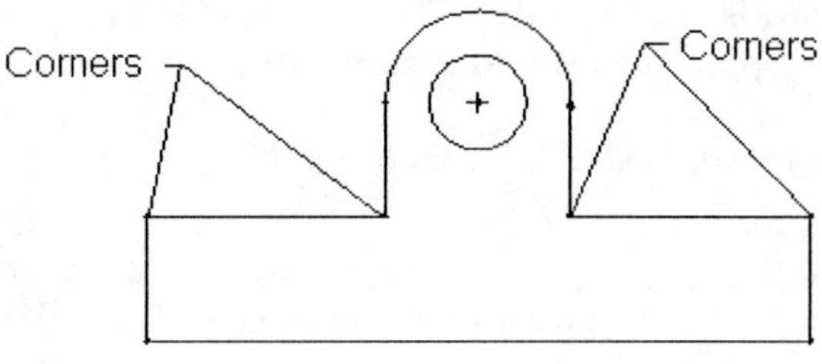

Figure 2-44 Fillet corners

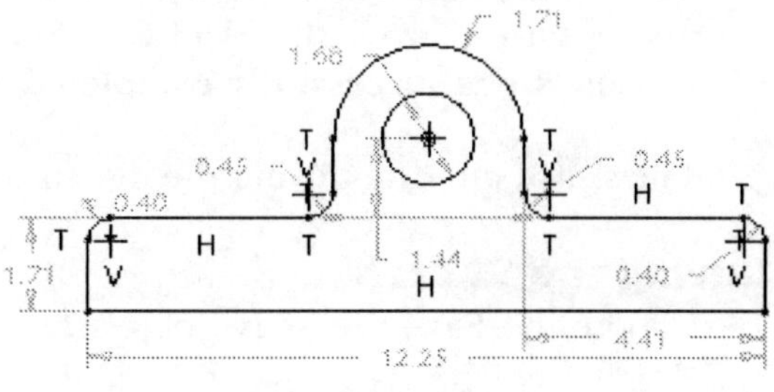

Figure 2-45 Sketch after filleting

Applying the Constraints

1. Choose the **Impose sketcher constraints on the section** button from the **Sketcher Tools** toolbar. The **Constraints** dialog box is displayed.

2. Choose the **Create Equal Lengths, Equal Radii, or Same Curvature constraint** button from the **Constraints** dialog box.

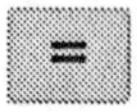

3. Press the left mouse button to select the fillets. Apply the equal constraint to all the fillets.

Dimensioning the Sketch

The weak dimensions are applied to the sketch automatically. These are not the required dimensions and therefore you need to manually dimension the sketch.

1. Choose the **Create defining dimensions** button from the **Right Toolchest**.

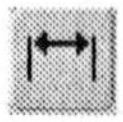

2. Dimension the sketch as shown in Figure 2-46.

Modifying the Dimensions

You need to modify the dimension values that are assigned to the sketch.

1. Select all the dimensions using CTRL+ALT+A.

2. Choose the **Modify the values of dimensions, geometry of splines, or text entities** button. The **Modify Dimensions** dialog box is displayed.

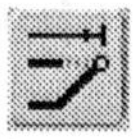

3. Clear the **Regenerate** check box and then modify the values of the dimensions. When you clear the check box, the sketch does not regenerate as you modify the dimensions.

   The dimension that you edit in the **Modify Dimensions** dialog box is enclosed in a yellow box in the sketch.

4. Modify all the dimensions. Refer to Figure 2-41 for dimension values.

5. After all the dimensions are modified, choose the **Regenerate the section and close the dialog** button from the **Modify Dimensions** dialog box. The message **Dimension modifications successfully completed** is displayed in the **Message Area**.

   The sketch after modifying the dimension values is shown in Figure 2-47.

Saving the Sketch

1. Choose the **Save the active object** button from the **File** toolbar and save the sketch.

Exiting the Sketch Mode

1. Choose the **Continue with the current section** button and exit the Sketch mode.

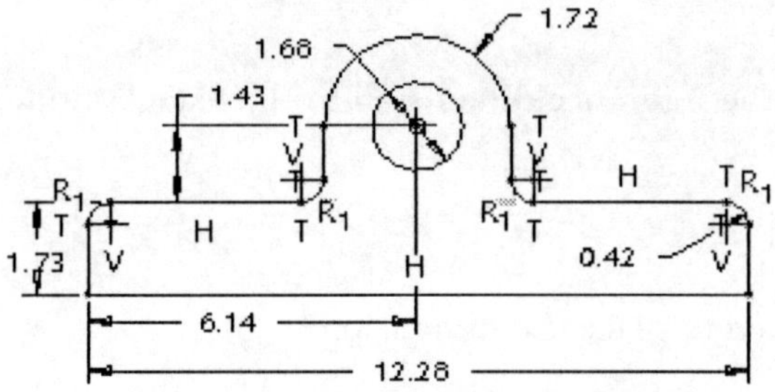

Figure 2-46 Sketch after dimensioning

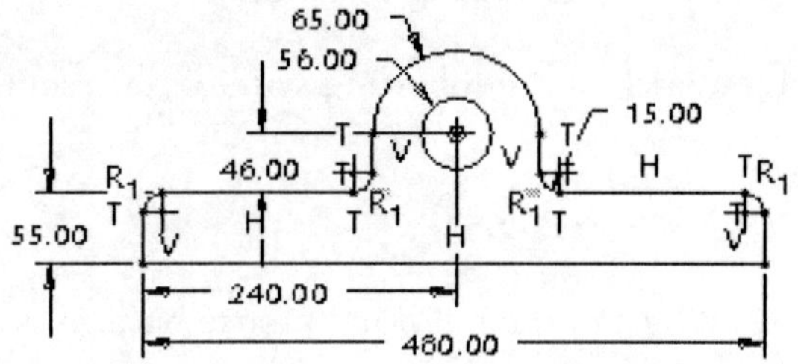

Figure 2-47 Sketch after modifying the dimensions

Self-Evaluation Test

Answer the following questions and then compare your answers to the answers given at the end of this chapter.

1. The variable dimension created using the **Perimeter** option can be modified directly by selecting it using the left mouse button. (T/F)

2. If the variable dimension is deleted then the perimeter dimension is also deleted. (T/F)

3. When you modify a weak dimension it becomes strong. (T/F)

4. You can dimension the length of a centerline. (T/F)

5. The font of the text written in the sketcher environment is fixed and cannot be modified. (T/F)

6. The __________ of a spline can be dynamically modified.

7. The __________ dialog box is used to modify dimensions.

8. The display of dimensions and constraints can be turned on/off by choosing the __________ and __________ buttons respectively from the **Sketcher** toolbar.

9. The __________ button is used to rotate selected entities.

10. You can delete the entities by selecting them and then using the __________ key.

Review Questions

Answer the following questions:

1. Which of the following symbol represents the reference dimension in the Sketch mode?

   (a) d (b) sd
   (c) ref (d) kd

2. Which of the following mouse buttons is used to place the dimension?

   (a) left (b) middle
   (c) right (d) mouse is not used for dimensioning

3. Which of the following is the default font for the text in the sketcher environment?

   (a) font (b) filled
   (c) font3D (d) isofont

4. Which of the following toolbars is used to toggle the display of dimensions and constraints in the sketcher environment?

   (a) **Sketcher Tools** (b) **Sketcher**
   (c) **File** (d) **Edit**

5. In which type of dimensioning the **Dim Orientation** dialog box is displayed while dimensioning the arcs and circles?

   (a) **Normal** (b) **Perimeter**
   (c) **Baseline** (d) None of the above

6. While placing a section in a new sketch, use the right mouse button to place the section. (T/F)

7. The grids can be turned on or off by using the **Sketcher** toolbar. (T/F)

8. The reference dimensions appear with the suffix REF. (T/F)

9. You can modify the dimensions dynamically. (T/F)

10. The modification of a spline is possible by moving its control points. (T/F)

Exercises

Exercise 1

In this exercise, you will draw the sketch for the model shown in Figure 2-48. The sketch is shown in Figure 2-49. **(Expected time: 30 min)**

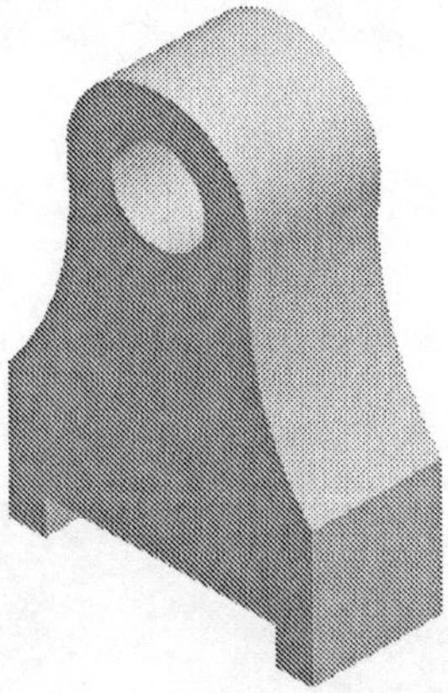

Figure 2-48 Solid model for Exercise 1

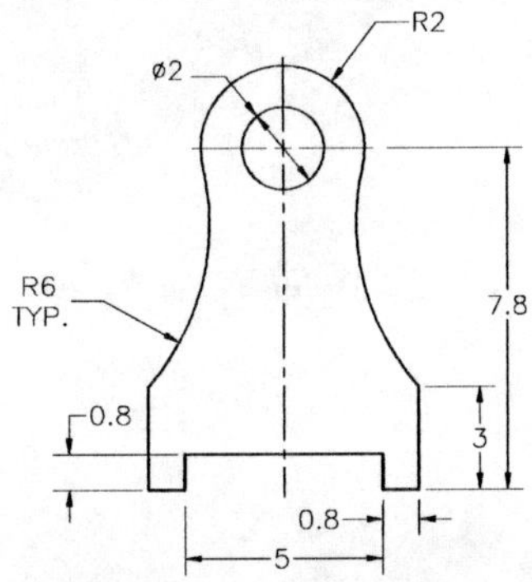

Figure 2-49 Sketch of the model

Exercise 2

In this exercise, you will draw the sketch for the model shown in Figure 2-50. The sketch is shown in Figure 2-51. **(Expected time: 15 min)**

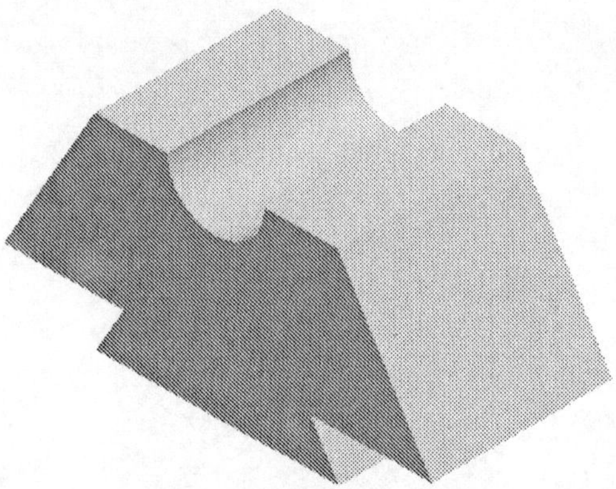

Figure 2-50 Solid model for Exercise 2

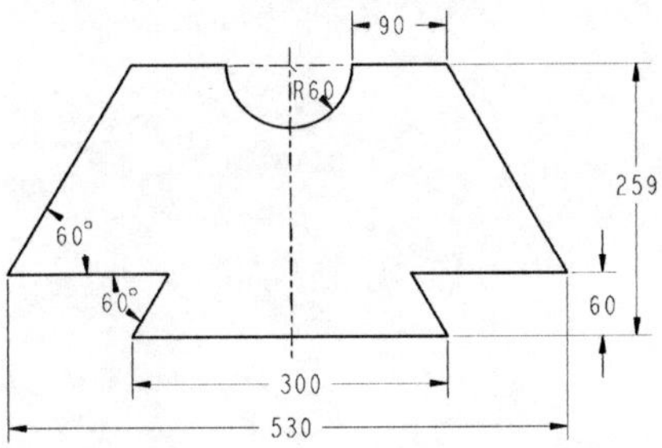

Figure 2-51 Sketch of the model

Exercise 3

In this exercise, you will draw the sketch for the model shown in Figure 2-52. The sketch is shown in Figure 2-53. **(Expected time: 30 min)**

Figure 2-52 Solid model for Exercise 3

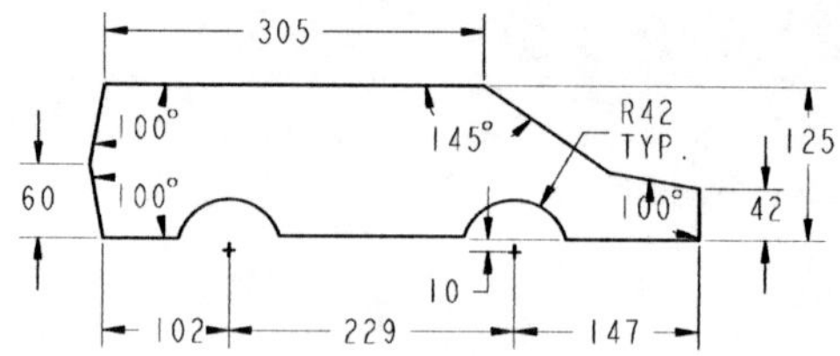

Figure 2-53 Sketch for the model

Answers to the Self-Evaluation Test

1 - F, **2** - T, **3** - T, **4** - F, **5** - F, **6** - shape, **7** - **Modify Dimensions**, **8** - **Toggle display of dimensions on/off, Toggle display of constraints on/off**, **9** - **Scale and rotate selected entities**, **10** - DELETE.

Chapter 3

Creating Base Features

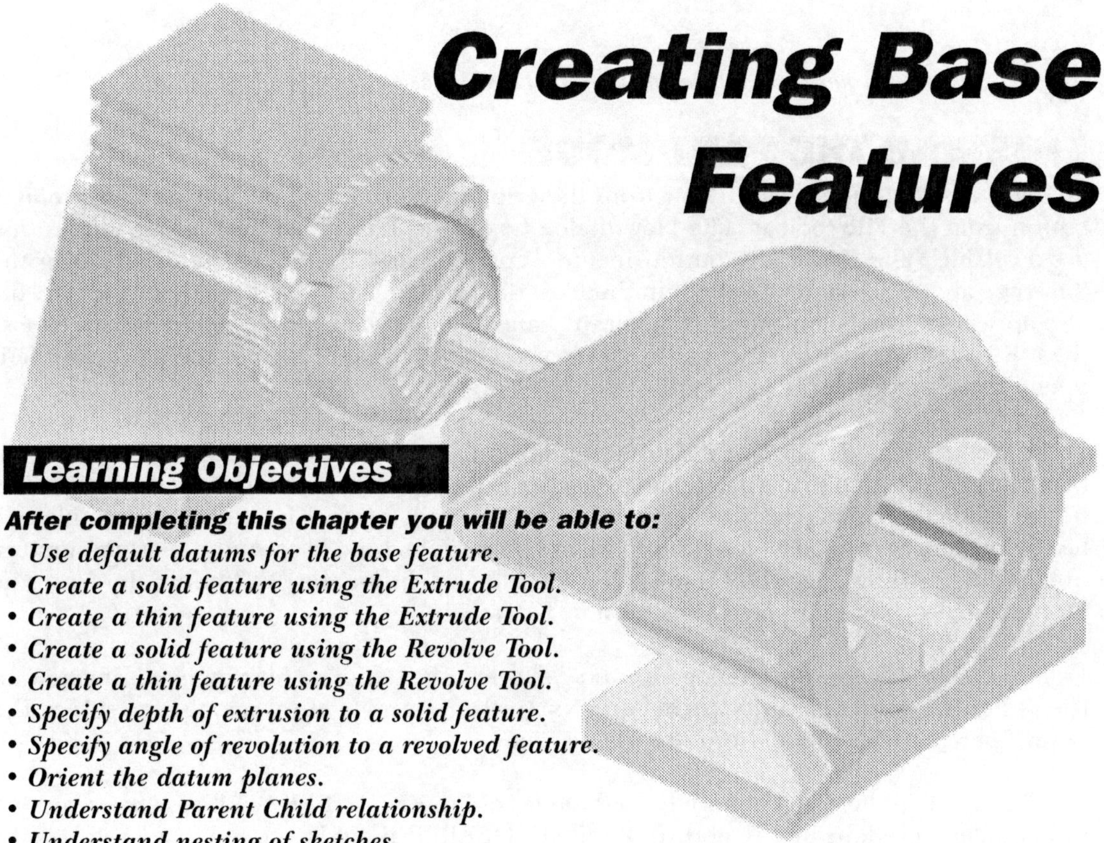

Learning Objectives

After completing this chapter you will be able to:

- *Use default datums for the base feature.*
- *Create a solid feature using the Extrude Tool.*
- *Create a thin feature using the Extrude Tool.*
- *Create a solid feature using the Revolve Tool.*
- *Create a thin feature using the Revolve Tool.*
- *Specify depth of extrusion to a solid feature.*
- *Specify angle of revolution to a revolved feature.*
- *Orient the datum planes.*
- *Understand Parent Child relationship.*
- *Understand nesting of sketches.*

CREATING BASE FEATURES

The base feature is the first solid feature created while creating a model in the Part mode. The base features are created using the datum planes. However, they can also be created without using the datum planes. But in this case, you do not have proper control over the orientation of feature and direction of feature creation.

While creating the base feature of a model, the designer should be extra careful in selecting the attributes to create it. This is because if the base feature itself is created wrong then the features created on it are also created wrong. This results in waste of time and effort. Although Pro/ENGINEER provides you with the options to redefine a feature, doing so also consumes additional time and effort. You need to enter the Part mode to create the base feature.

Note

It is recommended that you set the working directory before you open a new file.

ENTERING THE PART MODE

To enter the Part mode, select **New** from the **File** menu or choose the **Create a new object** button from the **File** toolbar. The **New** dialog box is displayed with the various modes that are available. The **Part** radio button in the **Type** area and the **Solid** radio button in the **Sub-type** area is selected by default in the **New** dialog box. The default name of the part file also appears in the **Name** edit box. You can change the part name as desired and then choose the **OK** button to enter the Part mode. When you choose the **OK** button, the new part file is opened and you enter the Part mode.

In the **New** dialog box, the **Use default template** check box is selected by default. This means that you have selected to use the default template provided by Pro/ENGINEER. This template has certain parameters related to the part file that you will create. The units of this model will be **Inches lbm Second**. The length is in inches, mass in lb, time in seconds, and temperature in fahrenheit. If this check box is not selected, and you choose the **OK** button, the **New File Options** dialog box is displayed as shown in Figure 3-1.

From the **New File Options** dialog box you can select the template file you need. If you want the default system of units to be **mmNs** (millimeter Newton sec), then select the **mmns_part_solid** template from this dialog box.

The Part mode is the most commonly used modes of Pro/ENGINEER Wildfire. This is because solid modeling is done in this mode of Pro/ENGINEER. It should be noted that a solid model is the base of a product development cycle. Product development cycle refers to the development of a product from scratch to its prototype. If you have created a solid model then it can further be used to generate its drawing views, for generating numerically controlled (NC) machining codes, analysis of the solid model, and so on.

The two-dimensional (2D) sketch drawn in the Sketch mode can be converted into a three-dimensional (3D) model in the Part mode. The Part mode contains the same sketcher environment with similar options to sketch as those available in the Sketch mode. There are some sketcher options in the Part mode that are not available in the Sketch mode because

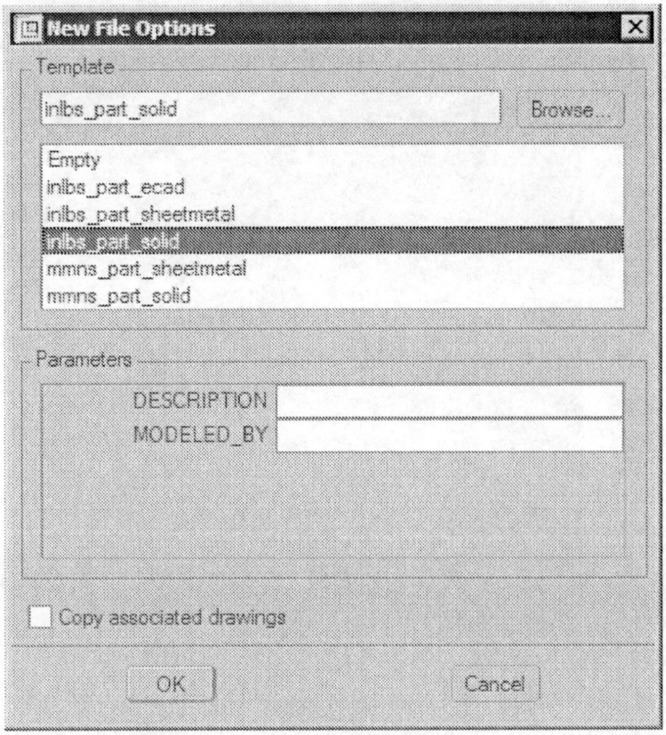

***Figure 3-1 New File Options** dialog box*

they do not have any use in the Sketch mode.

Figure 3-2 shows you the initial screen appearance on entering the Part mode, with the **Model Tree**, the three default datum planes, and the various toolbars.

THE DEFAULT DATUM PLANES

Generally, the first feature in the Part mode is the three default datum planes. These datum planes are further used to create the base feature. These datum planes act as a plane on which you can draw a 2D sketch and then convert it to a 3D model by protrusion. Generally, the base feature you create is referenced with the default datum planes.

Note

*Although it is said that the three default datum planes are the first feature in the Part mode, in the **Model Tree** the **RIGHT**, **TOP**, and **FRONT** datum planes appear as separate features. If you delete any one of them, only that datum plane is deleted.*

Tip: *It is recommended that you always use the datum planes to create a base feature. This is because the model created using the datum planes can be easily oriented. The uses of datum planes are discussed in Chapter 4.*

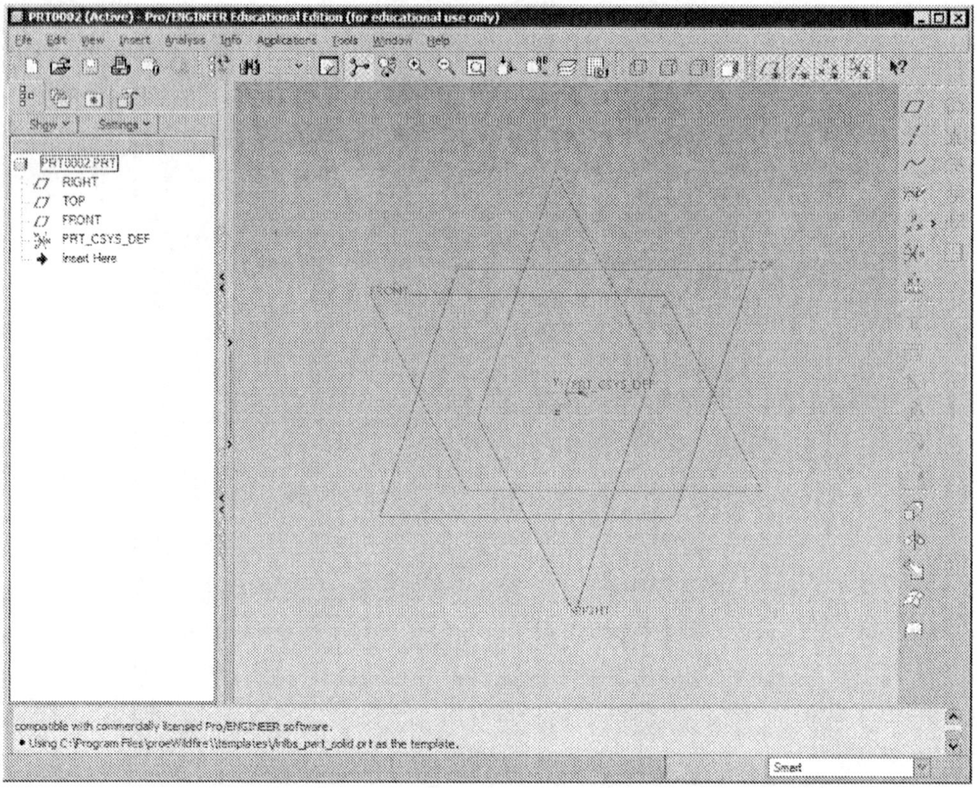

***Figure 3-2** The initial screen appearance after entering the Part mode*

These three default datum planes are mutually perpendicular to each other. They are not referenced to each other and are individual features. When a solid model is created, the datum planes adjust their size to the size of the model. You can create any number of datum planes for your requirement. The creation of additional datum planes is discussed in Chapter 4.

In the releases prior to Release $2000i^2$ of Pro/ENGINEER, you had to create the default datum planes. But in recent releases, a default template is provided, which contains the three default datum planes and is opened by default. However, if you do not need the default datum planes, then you need to clear the **Use default template** check box in the **New** dialog box at the time of creating a new file and select the **Empty** template.

Note

In this chapter, you will learn to use the three default datum planes to create the base features. It is important to remember the feature-based nature of Pro/ENGINEER while you are working on this chapter. The feature-based nature of Pro/ENGINEER has been discussed in the Introduction.

It is important to create a model in the correct orientation. Its correct orientation is the orientation in which it will be later used either in assembly or other modes. For example, the Casting component of the Plummer block is modeled in such a way such that it always stands vertical.

CREATING A PROTRUSION

Protrusion is defined as the process of adding material defined by a sketched section. In Pro/ENGINEER, there are various options of adding material such as **Extrude**, **Revolve**, **Sweep**, and so on. These options can be selected from the **Insert** menu in the menu bar. Figure 3-3 shows the options available in the **Insert** menu.

The **Base Features** toolbar shown in Figure 3-4 can also be used to invoke the protrusion options. This toolbar is available in the **Right Toolchest** when you are in the Part mode.

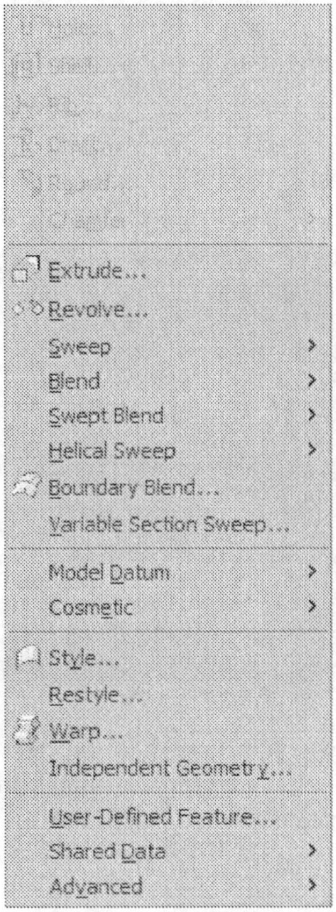

*Figure 3-3 Options in the **Insert** menu*

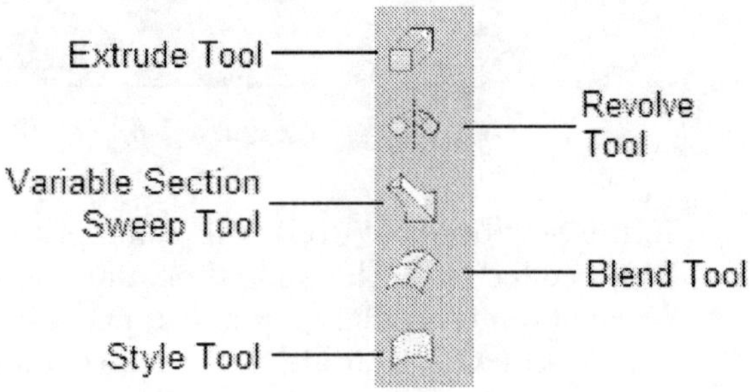

*Figure 3-4 **Base Features** toolbar*

Extruding a Sketch

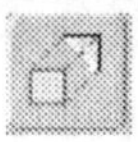

The **Extrude Tool** button in the **Base Features** toolbar adds material defined by a sketch drawn on a sketching plane. The material is added in the direction of feature creation.

The procedure to create a base feature using the **Extrude Tool** button is explained next.

1. When you choose the **Extrude Tool** button from the **Base Features** toolbar or choose **Insert > Extrude** from the menu bar, a dashboard is displayed below the graphics window as shown in Figure 3-5. This dashboard has all the options to define the extrude feature.

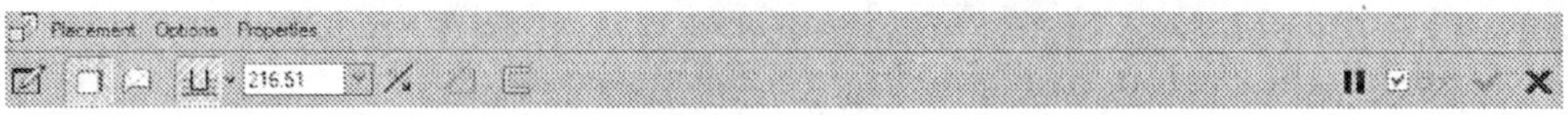

Figure 3-5 *Extrude dashboard*

2. Choose the **Create a section or red**efine the existing section button from the Extrude dashboard. The **Section** dialog box is displayed as shown in Figure 3-6. You are prompted to select a sketching plane. Select the datum plane named **FRONT** from the graphics window.

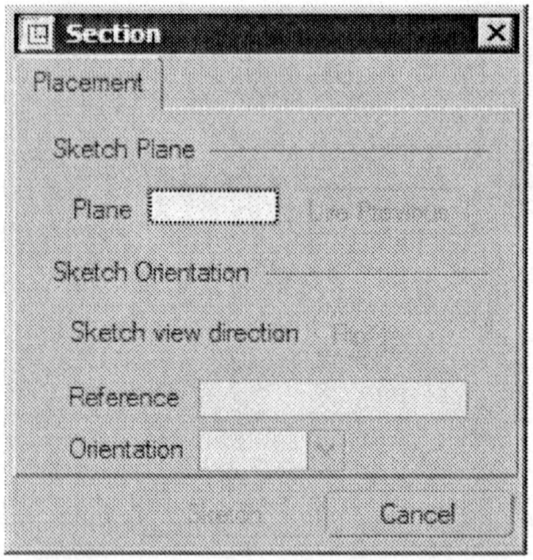

Figure 3-6 *The* ***Section*** *dialog box*

In the **Section** dialog box, the name of the plane that you have selected appears in the **Plane** collector. At the same time, the reference plane is also selected automatically. The name of reference datum plane appears in the **Reference** collector of the **Section** dialog box. Pro/ENGINEER also sets the orientation of the reference plane automatically.

3. After selecting the sketching plane and the reference plane, choose the **Sketch** button in the **Section** dialog box.

4. Now, you have entered the sketcher environment. You will notice that the **References** dialog box is displayed on the top right corner of the screen. The status displayed under the **Reference status** area is **Fully Placed**. This suggests that the references are selected by default. The **FRONT** datum plane is the sketching plane and is oriented parallel to the screen.

5. Close the **Model Tree** by clicking the sash on its right edge. The drawing space on the graphics screen increases by closing the **Model Tree** and the appearance of the graphics

screen is similar to that shown in Figure 3-7.

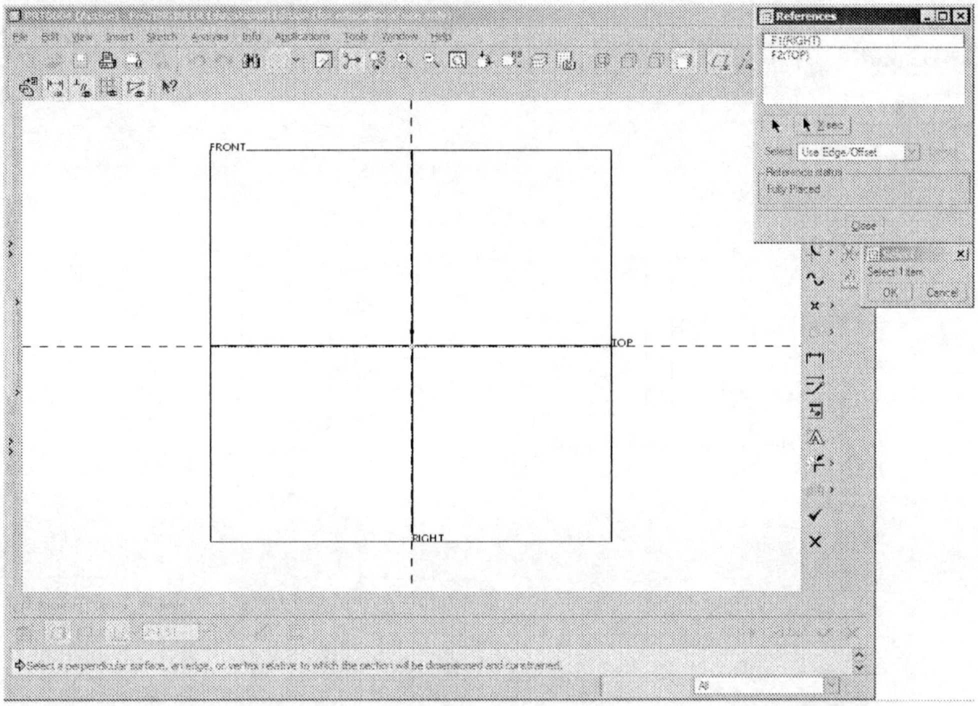

***Figure 3-7** The graphics screen after entering the sketcher environment and the **References** dialog box*

6. Use the **Create 2 points line** button to draw the right half of the I-section and then draw a vertical center line aligned with the **RIGHT** datum plane. Select all the lines in the right half of the I-section and choose the **Mirror selected entities** button. You will be prompted to select a center line. Select the center line to mirror the right half and to create the I-section as shown in Figure 3-8. Assume the dimensions. After the I-section is sketched, choose the **Continue with the current section** button from the **Sketcher Tools** toolbar.

7. When you choose the **Continue with the current section** button from the **Sketcher Tools** toolbar, the dashboard is enabled again below the graphics window. The model is created by assuming some default attributes and the model is displayed in yellow color. The attributes that a model has are all available on the dashboard and are discussed later in the chapter.

8. On the dashboard, there is a drop-down list containing some default value. This value is the depth of extrusion of the sketch that you have created. Enter an appropriate value in this drop-down list and press ENTER.

9. From the dashboard, choose the **Build feature** button. The required 3D model is displayed on the graphics window. However, it appears as a 2D entity. As a result, you need to change the display such that you can view the depth of the

model also. Change the display of the part drawn from the **View** toolbar by choosing the **Saved view list** button and selecting the **Default View** option from the drop-down list that appears. Figure 3-8 shows that the I-section after extrusion to certain depth becomes a 3D solid.

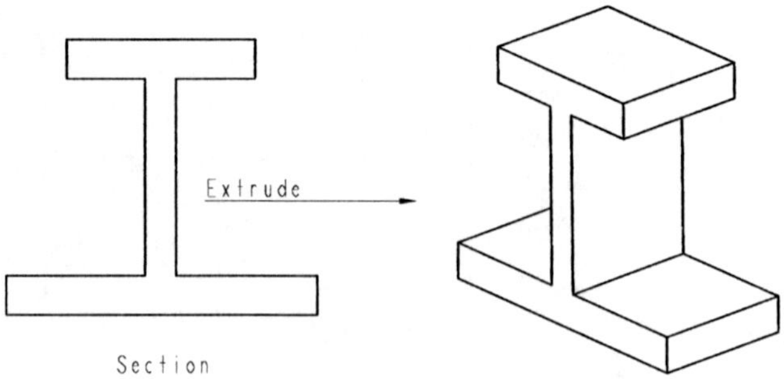

***Figure 3-8** The I-section extruded to a certain depth*

Figures 3-9 and 3-10 show some examples of **Extrude Tool** button.

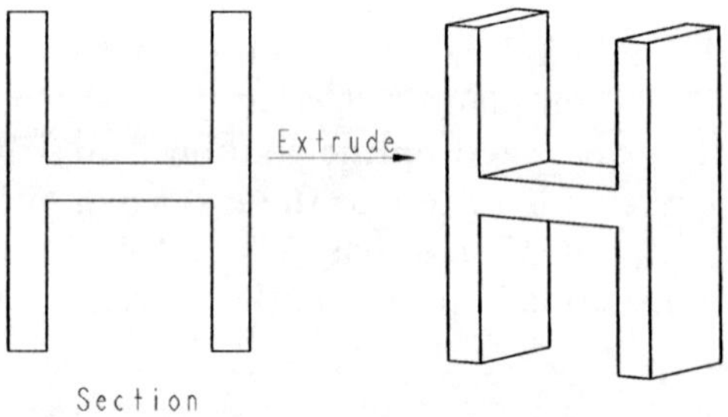

***Figure 3-9** Model created using the **Extrude Tool** button*

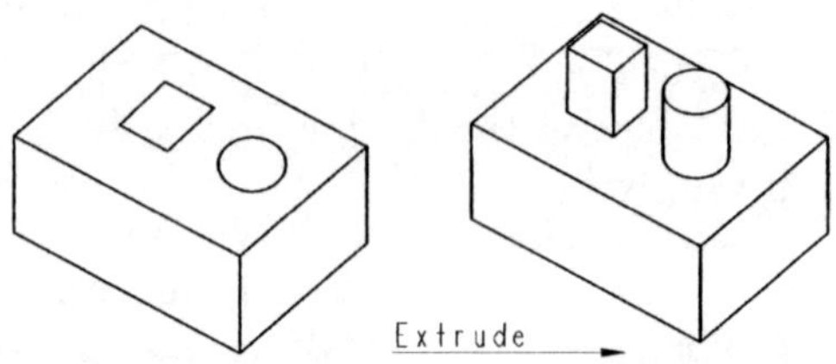

***Figure 3-10** Model created using the **Extrude Tool** button*

The above steps explain how to construct a 3D model using the **Extrude Tool** button. The Extrude dashboard, the **Section** dialog box, and the **Reference** dialog box that you came across while creating the 3D model from an I-section are discussed next.

The Extrude Dashboard

The options and the tool buttons in the Extrude dashboard shown in Figure 3-11 are used to extrude the sketch and to specify certain attributes related to the model. These attributes can be assigned to the model after the sketch of the model is drawn or before drawing the

Figure 3-11 Extrude dashboard

sketch. The tabs and the tool buttons that are available in the dashboard are discussed next.

Placement tab

When you choose the **Placement** tab, the slide-up panel is displayed as shown in Figure 3-12. The collector in this slide-up panel displays **No Items** because the section has not been drawn yet. When you choose the **Create a section or redefine the existing section** button from this slide-up panel, the **Section** dialog box is displayed.

Figure 3-12 ***Placement*** *tab slide-up panel*

Options tab

When you choose the **Options** tab, the slide-up panel is displayed as shown in Figure 3-13. This slide-up panel is used to specify whether you want the sketch to extrude to one side of the sketching plane or to both sides of the sketching plane. This slide-up panel has **Side 1** and **Side 2** drop-down lists. The options in the **Side 1** drop-down list are explained next:

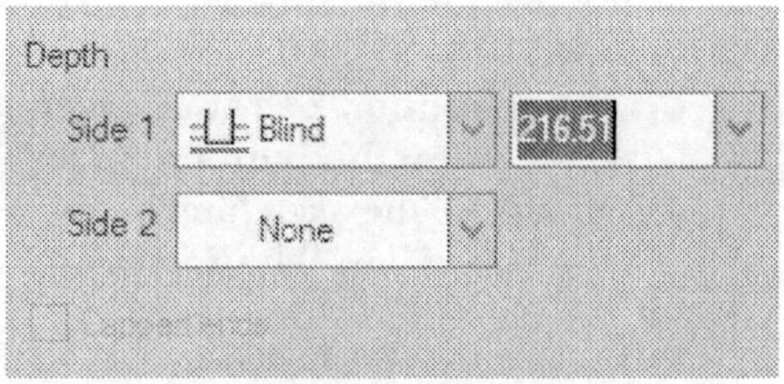

Figure 3-13 ***Options*** *tab slide-up panel*

Blind

The **Blind** option is one of the most commonly used option to define the extrusion depth of the sketch by specifying a particular depth value. When you select this option from the **Side 1** drop-down list, a default value appears in the dimension box that is adjacent to the **Side 1** drop-down list. You can enter a value in this dimension box and press ENTER. The material is added in the first direction shown by the yellow arrow. The depth of extrusion given by this value can be modified if needed.

Symmetric

If you choose the **Symmetric** option from the **Side 1** drop-down list, the material is added equally in both the directions of the sketching plane. When you select this option, the **Side 2** drop-down list becomes inactive. This means that you cannot choose any option from the **Side 2** drop-down list.

To Selected

The **To Selected** option allows you to select a point, surface, plane, or a curve up to which the section is extruded.

The options in the **Side 2** drop-down list are similar to the options available in the **Side 1** drop-down list. The **None** option in this drop-down list allows you to extrude the sketch in the first direction only. The **Blind** and **To Selected** options give the depth of extrusion to the sketch in the second direction.

Tip: *The features created using the **To Selected** option do not have a dimension associated with them, and hence, they cannot be modified by changing the dimension value. However, changing the terminating surface changes the depth of the feature. You will understand this better when you learn the modification of an existing feature.*

*The extrusion depth given using the **Symmetric** option does not appear when you generate the dimensions in the drawing views of the model. The drawing views of the model are generated in the Drawing mode of Pro/ENGINEER.*

Properties tab

When you choose the **Properties** tab, the slide-up panel is displayed. This slide-up panel displays the feature identity in the **Name** collector. The **i** button in the slide-up panel when selected opens the browser and all the information about the feature you are creating is displayed in the browser. The browser has been discussed in the Introduction.

Create a section or redefine the existing section button

Choose this button to display the **Section** dialog box. This button allows you to select the sketching plane and set its orientation with the help of the **Section** dialog box. The **Section** dialog box is discussed later in the chapter.

Extrude as solid button

This button is chosen by default. Choose this button to create a solid by adding material to the section. When you select a sketch to extrude using this button, the sketch should be a single closed loop.

Extrude as surface button

This button is used to create a surface by adding material to the section. Using this button, you can create surface models. The sketch that is drawn for the surface model need not be a closed loop.

Note

The tabs and the tool buttons that are available on the Extrude dashboard can be used before drawing the sketch or after drawing the sketch. After the sketch is completed and you exit the sketcher environment, you can dynamically specify the depth of extrusion on the model.

Change depth direction of extrude to other side of sketch

This button is used once the sketch is completed. When you choose this button, it toggles the direction of extrusion with reference to the sketch plane. The direction of extrusion is defined as the direction in which the feature is created with respect to the sketching plane. This direction is displayed by a yellow arrow on the graphics screen. You will notice that when you choose this button, the yellow arrow points in the reverse direction, suggesting that the direction of feature creation has been reversed.

The direction of the yellow arrow depends on the type of extrude. If material has to be added to a feature then the arrow by default points in the direction toward the user, that is, out of the screen. If material has to be removed from a feature then by default the arrow points into the screen.

If you spin the model using the middle mouse button then the direction of the arrow can be easily recognized.

Remove Material

The **Remove Material** button is available on the dashboard only after the base feature is created. This is because this button is used to remove material from an existing feature. Therefore, when you create a base feature this button is not available. The use of this button is explained in later chapters.

Thicken Sketch

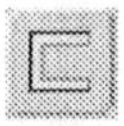

The **Thicken Sketch** button is used to create thin parts with specified thickness. When you choose this button, the **Change direction of extrude between one side, other side, or both sides of sketch** button and a dimension box appear on the right of this button.

The **Change direction of extrude between one side, other side, or both sides of sketch** button is used to set the direction with respect to the sketch where the thickness should be added. This button appears on the dashboard only when the sketch is completed and you exit the sketcher environment. This button serves three functions. It is used to set the thickness direction to either side of the sketch and even to both sides of the sketch symmetrically. When the sketch is completed, the preview of the model is displayed on the graphics window. By default, the thickness to the sketch is applied on one side. Now, when you choose the **Change direction of extrude between one side, other side, or both sides of sketch** button, the thickness is applied to the other side of the sketch. When you choose this button for the second time, the thickness is applied symmetrically to both the sides of the sketch.

The dimension box that appears when you choose the **Thicken Sketch** button is used to enter the thickness value of the feature to be created. This edit box is available only when you exit the sketcher environment.

Figure 3-14 shows the arrow pointing away from the side of the section wall where the material will be added. If you want to change the direction of the arrow, choose the **Change direction of extrude between one side, other side, or both sides of sketch** button. Figure 3-15 shows the thickness of the material added to one side of the wall pointed by the arrow.

You can choose the **Change direction of extrude between one side, other side, or both sides of sketch** button such that the material is added symmetrically to both the sides of the section wall as shown in Figure 3-16. Figure 3-17 shows the model created using the **Thicken Sketch** button.

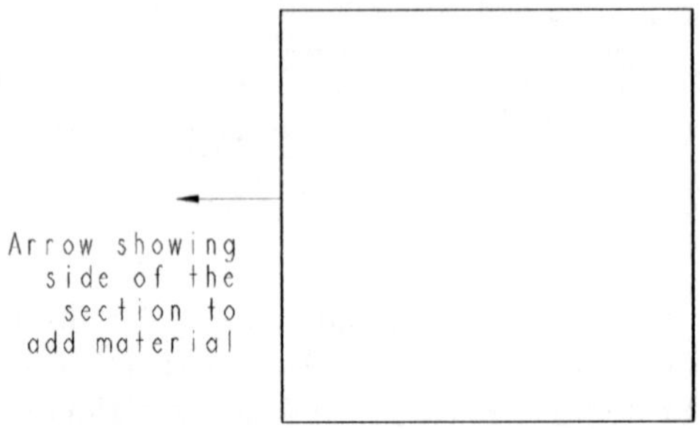

Figure 3-14 *The sketch drawn to thicken*

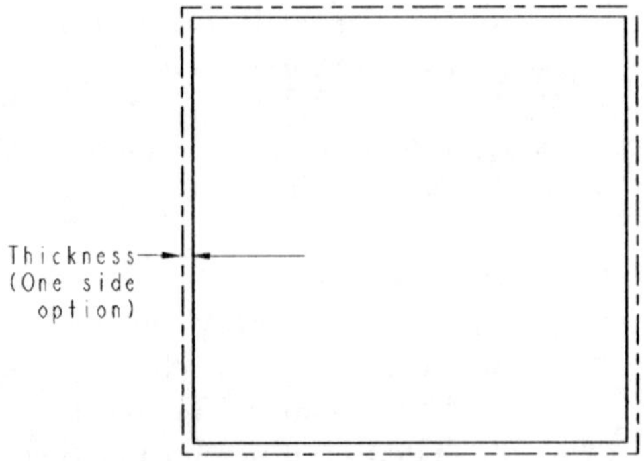

Figure 3-15 *Adding thickness to one side of the sketch*

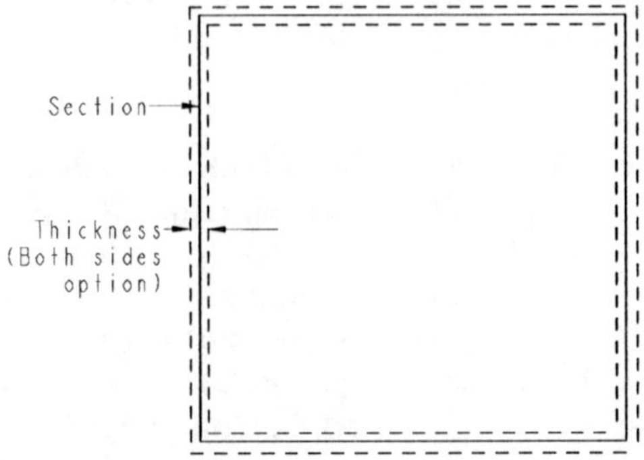

Figure 3-16 *Adding thickness to both the sides of the sketch*

Figure 3-17 *Model created using the* ***Thicken Sketch*** *button*

Pause tool button

This button is used to pause the current feature creation. After you choose this tool button, you can access other available tools such as **Datum Plane** tool. To resume the current feature creation tool, you can choose the **Resumes the previously paused tool** button that appears in place of the pause tool.

Geometry Preview/Feature Preview

This button has two functions. When you select the check box on this button, the geometry of the feature that you have created can be previewed with dimensions. When you choose the **Feature Preview** button, the system allows you to preview the feature.

Build feature

This button is used to confirm the feature creation and exit the current feature creation tool.

Close button

This button is used to abort the feature creation tool that is invoked. When you choose this button, the dashboard is closed.

The Section dialog box

The **Section** dialog box is displayed when you choose the **Create a section or redefine the existing section button** from the Extrude dashboard. The **Section** dialog box is shown in Figure 3-18. This dialog box is used to select the sketching plane and to set its orientation. As soon as you select the sketching plane, the reference plane and its orientation are selected automatically.

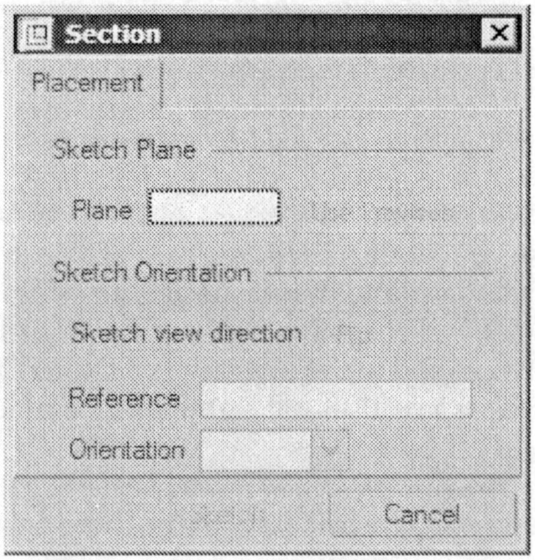

Figure 3-18 The ***Section*** *dialog box*

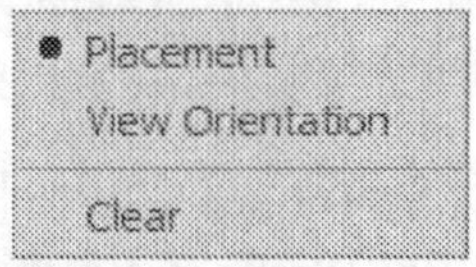

Figure 3-19 Shortcut menu invoked when the ***Section*** *dialog box is displayed*

If you want to change the sketching plane that is already selected, hold down the right mouse button on the graphics window to display a shortcut menu shown in Figure 3-19. There are three options in this shortcut menu. In the figure, the **View Orientation** option is selected (this is the reason that in the **Section** dialog box, the **Reference** collector appears yellow, indicating that it is active). If you choose the **Clear** option, the reference plane and its orientation, which were selected automatically, are cleared and now you can manually select the reference plane and its orientation. Similarly, to change the sketching plane, select the **Placement** option in the shortcut menu. Again invoke the shortcut menu and choose the **Clear** option.

However, after you select the sketching plane, no matter which reference plane is selected automatically, you can select the reference plane manually. The reference you select manually replaces the old reference. In other words, if the **Plane** collector or any other collector is yellow in color, it indicates that you can select the reference from the model. This point should be remembered because this is true with other dialog boxes available in Pro/ENGINEER and this you will learn later in this book.

Tip: *You can also remove the selected item using the **Section** dialog box. To remove a selected item in the **Section** dialog box, select the item in the dialog box and right-click to invoke a shortcut menu. From this shortcut menu, choose **Remove**.*

The **Flip** button in the **Section** dialog box can be used after you select the sketching plane and its orientation. This button is used to set the direction of viewing the sketching plane. When you choose this button, the yellow arrow that is displayed on the sketching plane changes its direction.

Tip: *Remember that if you are creating a protrusion (a material addition process) then no matter what the direction of viewing you choose, the protrusion takes place in the direction toward the user, that is, out of the screen.*

In the **Section** dialog box, the options in the **Orientation** drop-down list are used to specify the orientation of the horizontal or vertical reference for the sketching plane. Generally, your view is normal to the sketching plane you have selected. In order to orient the sketching plane normal to the viewing direction, you have to specify a plane or a planar surface that is perpendicular to the sketching plane. For example, if you select the **RIGHT** datum plane as the reference plane and then select the **Top** option from the drop-down list, then the **RIGHT** datum plane will be oriented to the top while sketching. The options in this drop-down list are common to other feature creation tools and are available whenever you need to draw a sketch.

Before you proceed further, you need to understand the three default datum planes that are displayed when you open a new part file. You can view the feature that you have created from different directions using the **Saved view list** drop-down list shown in Figure 3-20. This drop-down list is available in the **View** toolbar present in the **Top Toolchest**. In this drop-down list there are some standard preset views that are provided by Pro/ENGINEER. If you see the feature from the right then your viewing direction is normal to the **RIGHT** datum plane. If you see the feature from the top then your viewing direction is normal to the **TOP** datum plane. Similarly, if you see the feature from the front then your viewing direction is normal to the **FRONT** datum plane.

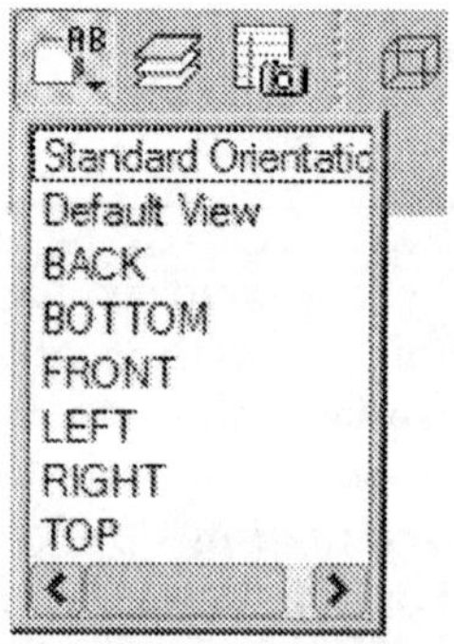

***Figure 3-20** The **Saved view list** drop-down list*

The options in this drop-down list are used to set the orientation of the model on the graphics window.

The options in the **Orientation** drop-down list are very important and you need to select the reference plane very carefully, especially for the base feature. The importance of selecting the reference plane is explained using Figure 3-21. This figure shows two cases where the sketching plane for both the solid models was the same, the same sketch was extruded, and the same option **Top** was chosen from the **Orientation** drop-down list in the **Section** dialog box but different reference planes were selected. The same sketch was used to extrude for both the models and the top curve in the sketch was on the top while sketching. Note, the difference in

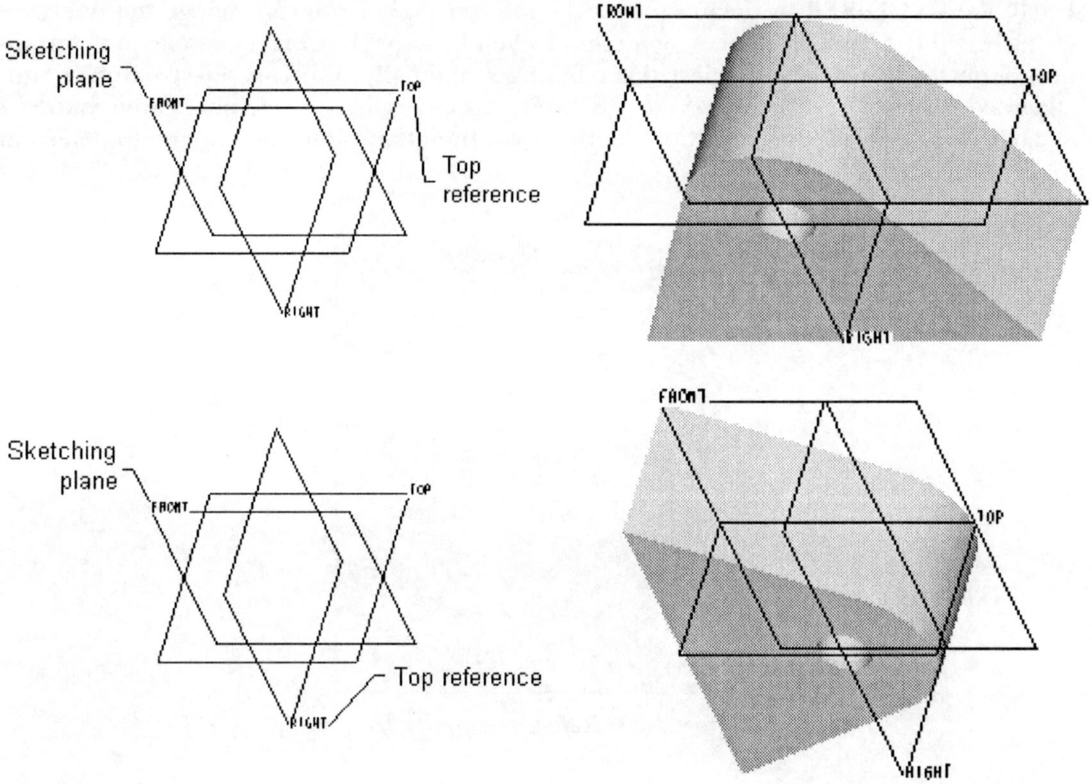

***Figure 3-21** Importance of selecting reference planes*

orientation of the resultant models in their default trimetric orientations. The model can be oriented in its default orientation using CTRL+D or by selecting the **Default View** option from the **Saved view list** drop-down list.

Once a plane has been selected for sketching and the reference plane is oriented, you can enter the sketcher environment. To enter the sketcher environment, choose the **Sketch** button in the **Section** dialog box. Now, you enter the sketcher environment. The **References** dialog box is displayed on the right of the main window.

Tip: *The **Orientation** dialog box that is displayed when you choose the **Reorient view** button from the **View** toolbar is used to orient the model on the graphics window. This dialog box provides some advanced options to orient the model and also allows you to save the orientation of the model. Using the options in this dialog box, you can dynamically orient a model, orient by selecting references, and orient by using other options.*

References Dialog Box

When you enter the sketcher environment, the **References** dialog box is displayed as shown in Figure 3-22. This dialog box is used to specify the references for dimensioning the sketch. Pro/ENGINEER by default selects the references, but you can change the references by deleting the default references. Sometimes when Pro/ENGINEER is not able to determine the references, you need to select the references manually. You can select datum planes, edges, axes, datum curves, and so on as references. Generally, two references are required in a sketch. But if you select a vertex or a datum axis, then they alone are enough to determine the references.

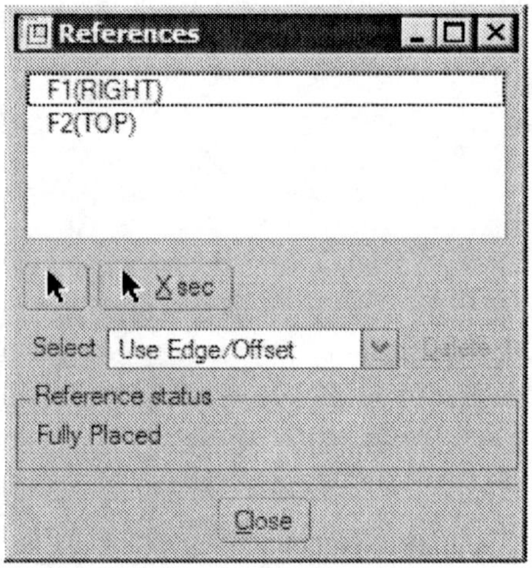

*Figure 3-22 **References** dialog box*

When you draw a sketch, the weak dimensions to the sketch are applied with references selected in the **References** dialog box. For example, if you select the two datum planes as references then the sketch is by default dimensioned with the two datum planes. If you force a dimension and dimension the sketch with an axis then the axis is also displayed in the **References** dialog box. The **References** dialog box in the sketcher environment can be invoked by choosing **Sketch > References** from the menu bar.

Revolving a Sketch

The **Revolve Tool** button allows you to revolve the sketched section through the specified angle about a center line. By revolving a sketch, you can add material (protrusion) and you can even remove material (cut). Here, you will learn to use the **Revolve Tool** to add material. The revolved feature can be revolved on one side of the sketching plane or on both the sides of the sketching plane. This and the other attributes can be specified to the sketch by using the Revolve dashboard. The Revolve dashboard is shown in Figure 3-23 and it is displayed when you choose the **Revolve Tool** button from the **Base Features** toolbar in the **Right Toolchest**.

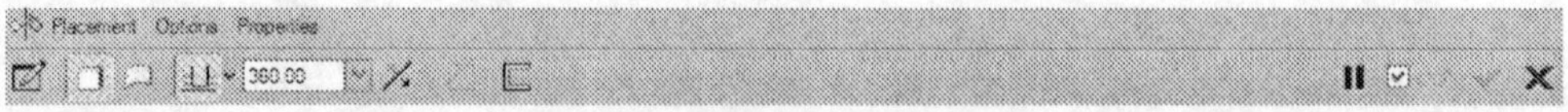

Figure 3-23 Revolve dashboard

Some of the points to be kept in mind while creating a revolve feature are given next.

1. If the revolve feature is a base feature, then the section drawn should be a closed section for revolving the sketch as a solid.

2. The section sketch of a revolved feature will not be completed until you have drawn a center line.

3. The section drawn should be on one side of the center line.

4. If there are more than one center lines in the sketch, then Pro/ENGINEER uses the center line that is drawn first and considers it the axis of rotation.

Revolving a Sketch as a Solid

The **Revolve as solid** button in the Revolve dashboard is used to revolve the sketch as a solid. To revolve a sketch as a solid, the sketch should be a closed loop. Figures 3-24 and 3-25 explain the use of **Revolve as solid** tool to revolve a sketch.

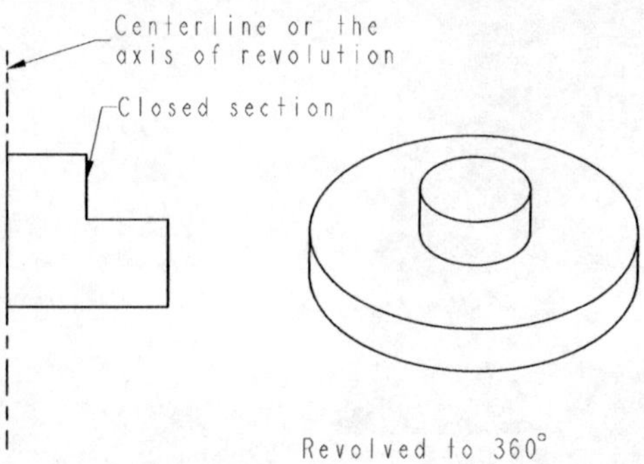

Figure 3-24 *Revolving a sketch as a solid*

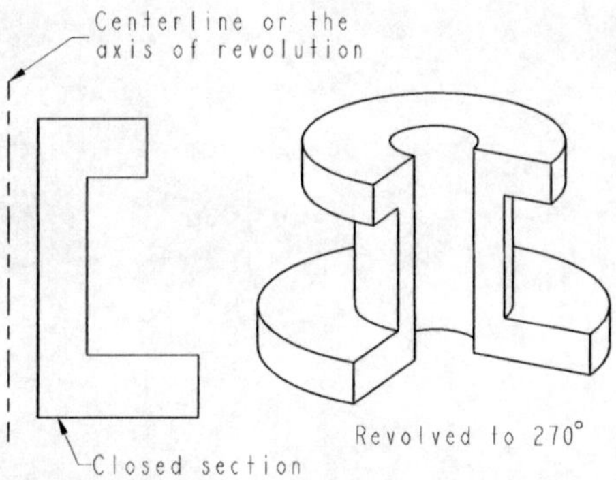

Figure 3-25 *Revolving a sketch as a solid*

Revolving a Sketch with Thickness

The **Thicken Sketch** button is used in combination with the **Revolve as solid** button to create revolved features having a certain thickness. Unlike in revolving the sketch as a solid, when you revolve a sketch with thickness, the section need not be a closed loop. When you exit the sketcher environment after the section is created, you can specify the side of the section where the material will be added. After specifying the side for the material addition, you can enter the thickness value in the edit box that appear on the dashboard.

Revolving the sketch with thickness is explained in Figures 3-26 and 3-27.

UNDERSTANDING THE ORIENTATION OF DATUM PLANES

Consider a case where you need to revolve a sketch at an angle of revolution of 270-degrees.

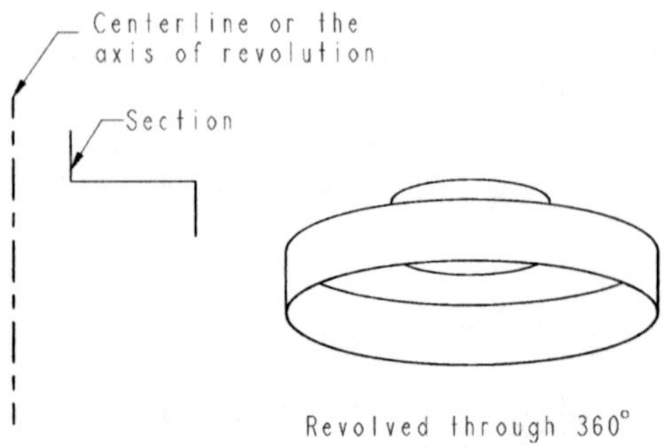

Figure 3-26 Revolving a sketch with certain thickness

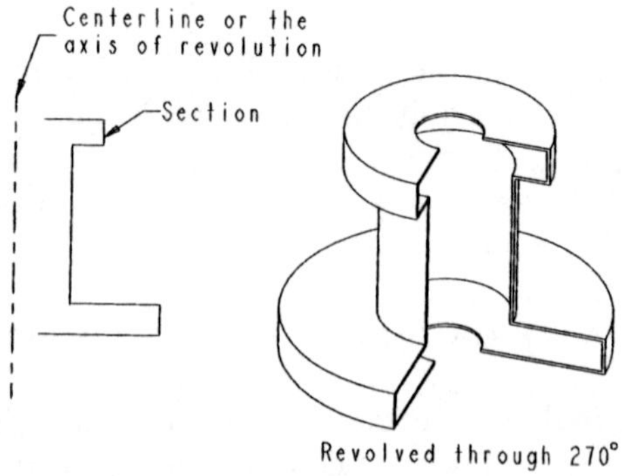

Figure 3-27 Revolving a sketch with certain thickness

The trimetric view of the solid model is shown in Figure 3-28. Now, when you approach to create the solid model, the first step is to decide the datum plane on which you need to create the sketch.

Figure 3-28 Default trimetric view of a shaded model

As evident from the model, the axis of revolution of the model is perpendicular to the **TOP** datum plane. Therefore, the **RIGHT** datum plane will be selected to be at the top while drawing the sketch. Now, only two options are left for selecting the sketching plane, the **FRONT** and the **TOP** datum planes. You can draw the section of the revolve model on any of the two planes. This is because when you view the model, you notice that the cross-section of the model is parallel to the **RIGHT** datum plane as well as to the **FRONT** datum plane.

The drawing of the cross-section sketch on the two datum planes to achieve the desired orientation of the model is discussed next.

Case-1

Drawing the sketch on the RIGHT datum plane

1. When the **Section** dialog box is displayed, select the **RIGHT** datum plane as the sketching plane.

   The selected datum plane is highlighted in red and then the reference plane is selected by default. The **TOP** reference plane is selected by default and its orientation is also set. The yellow colored arrow also appears on the sketching plane. This arrow indicates the direction of feature creation. You can change the direction of arrow but at this stage accept the default direction of arrow.

2. From the **Orientation** drop-down list, select the **Top** option.

   Now, when you draw the sketch, the **RIGHT** datum plane will be parallel to the screen and the **Top** datum plane will be at the top.

3. Choose the **Sketch** button from the **Section** dialog box. You enter the sketcher environment.

4. Draw the sketch of the cross-section, dimension it, and then modify the dimensions as shown in Figure 3-29.

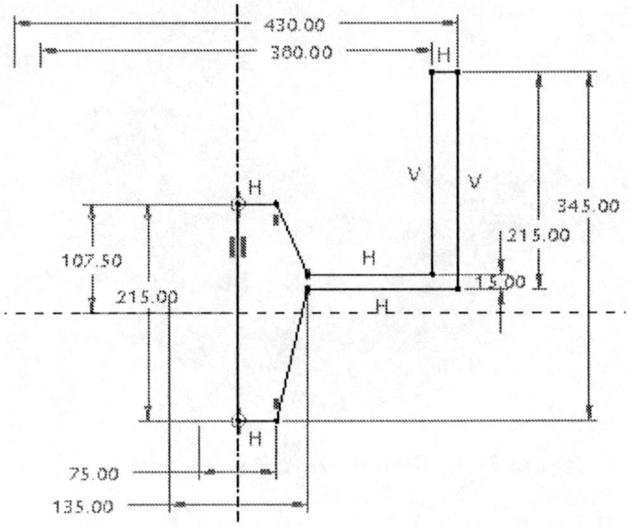

Figure 3-29 *Sketch with dimensions and constraints*

5. Exit the sketcher environment by choosing the **Continue with the current section** button.

6. The model appears as translucent and in yellow color. The dashboard under the graphics window again becomes active.

7. In the drop-down list that is present on the dashboard, select a value of **270**.

Tip: *You might have difficulty in visualizing the cross-section of a revolved model that has an angle of revolution of 360-degrees. This is because the cross-section of the model is not visible. Therefore, you cannot decide the sketch to be drawn for the cross-section of the model.*

Whenever you come across a revolve model, just imagine that if you cut the model along its axis of revolution and remove one quarter of the revolved feature then the section obtained in the one quarter of the revolved feature is the section of the revolved model. Therefore, you need to draw this section in the sketcher environment to create the desired model.

8. From the **Saved view list** button, choose the **Default View** option. The model orients in its default orientation, that is, trimetric view as shown in Figure 3-30.

   But the view that is shown in Figure 3-30 is not the view that is needed. The model is not oriented correctly. This means that the direction of feature creation that was selected was not correct in order to get the desired orientation.

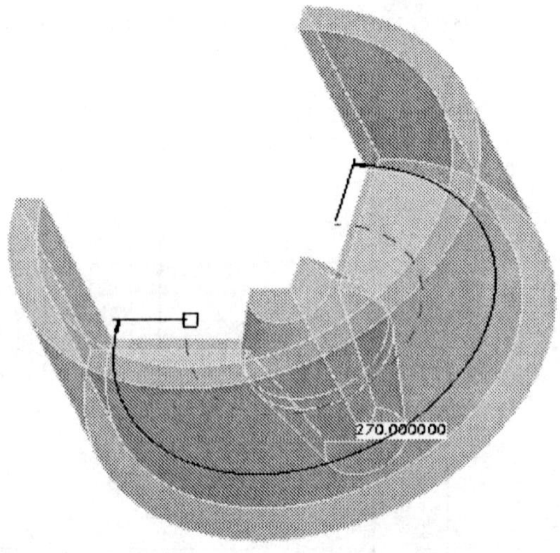

Figure 3-30 *Default trimetric view of the model*

Note
The model is created but its orientation is not the desired one. This means that at the time of selecting the sketching plane, when the yellow arrow appeared, it was pointing in the wrong direction.

Figure 3-31 shows you the sketch that you have drawn on the **RIGHT** datum plane. The sketch is always revolved in the clockwise direction. Remember that the arrow shows the direction in which you will be viewing the sketching plane while in the sketcher environment. Considering the mentioned facts, to orient the sketching plane correctly, the

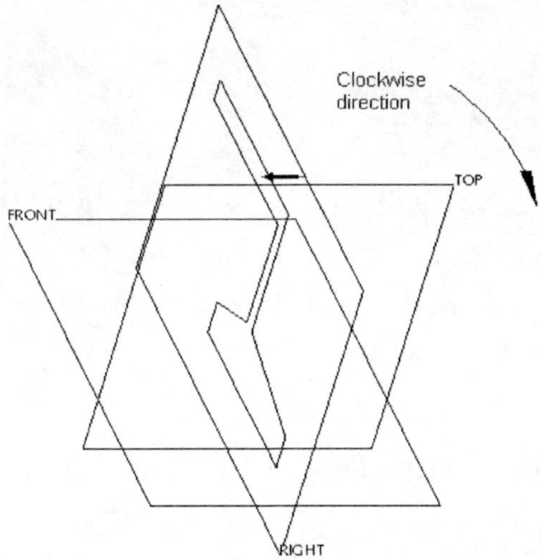

Figure 3-31 *Sketch drawn on the* ***RIGHT*** *datum plane*

arrow direction should have been flipped by using the **Flip** button in the **Section** dialog box. Now, since you have not achieved the desired orientation of the model, exit this feature creation tool. Again invoke the **Revolve Tool** button, and this time after you select the sketching plane, choose the **Flip** button to reverse the direction of viewing the sketch. Remember that the case discussed here is when you select the **RIGHT** datum plane as the sketching plane.

Tip: *By default, when Pro/ENGINEER creates a revolved model, the material is added in the clockwise direction. You can revolve the model in counterclockwise direction by using the* ***Change angle direction of revolve to other side of sketch*** *button from the Revolve dashboard. Also, to revolve the model in the anti-clockwise direction you can enter a negative value for the angle of revolution.*

Figure 3-32 shows the sketch on the sketching plane and the arrow direction reversed. In this case, the sketch when rotated in clockwise direction results in the desired orientation of the model as shown in Figure 3-33.

Case 2

Drawing the sketch on the FRONT datum plane

In this case, you will select the **FRONT** datum plane as the sketching plane and the **TOP** datum plane will be selected as reference plane. It is advisable to decide the direction of the arrow at this stage so that you create the model shown in Figure 3-28 in the required orientation.

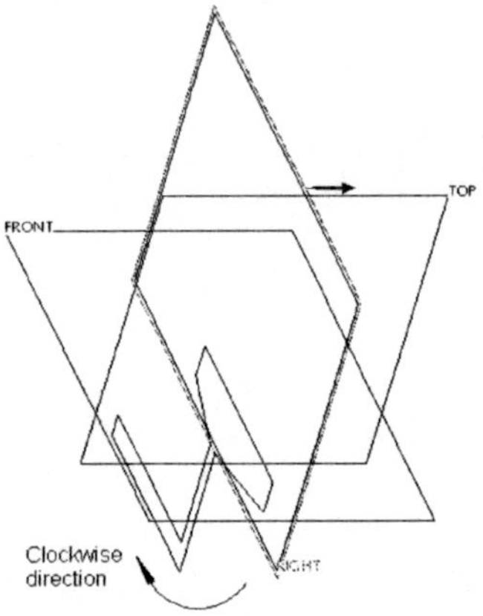

***Figure 3-32** Sketch drawn on the **RIGHT** datum plane*

***Figure 3-33** Desired trimetric view of the model*

Figures 3-34 shows the sketch drawn on the **FRONT** datum plane and the default direction of viewing the sketching plane. The arrow shows the direction of feature creation. Remember that the arrow points in the direction along which you will view the plane in the sketcher environment. Figure 3-35 shows the default trimetric view of the solid model created in this case. This is not the desired default orientation of the model. You can still get the desired orientation of the model by choosing the **Change angle direction of revolve to other side of the sketch** button from the Revolve dashboard, but, you should understand the default direction of the yellow arrow on the sketching plane. The desired orientation of the model is shown in Figure 3-36.

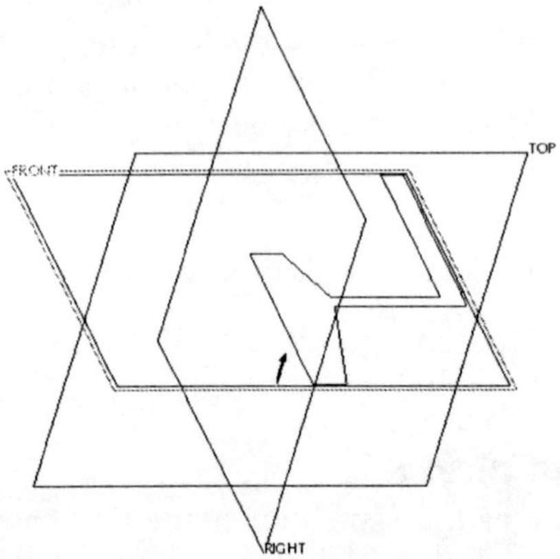

***Figure 3-34** Sketch drawn on the **FRONT** datum plane*

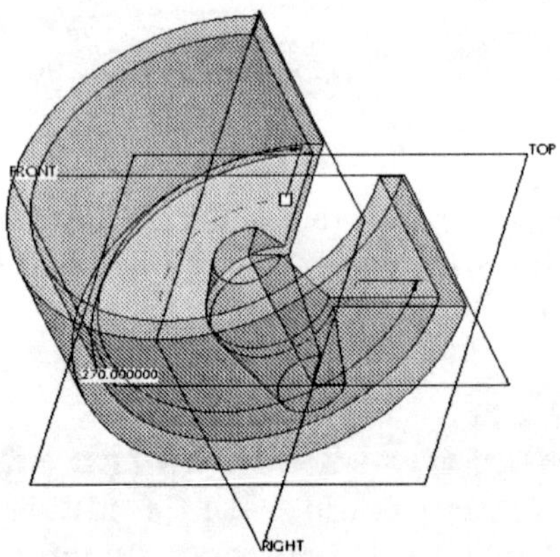

***Figure 3-35** Default trimetric view of the model*

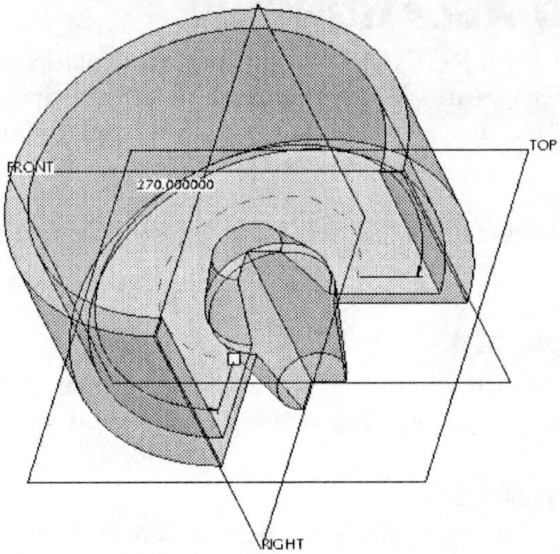

Figure 3-36 Desired trimetric view of the model

Figure 3-37 shows the sketch drawn on the **FRONT** datum plane and the arrow is showing the direction of viewing the sketching plane. The direction of the yellow arrow is reversed by choosing the **Flip** button in the **Section** dialog box. Figure 3-38 shows the default trimetric view of the model in this case.

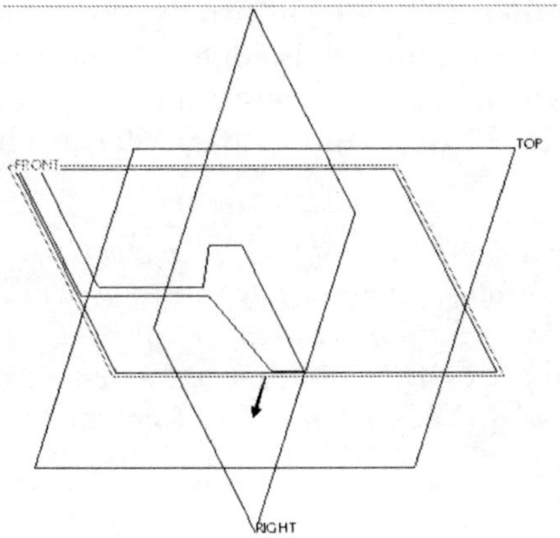

Figure 3-37 Sketch drawn on the ***FRONT*** *datum plane*

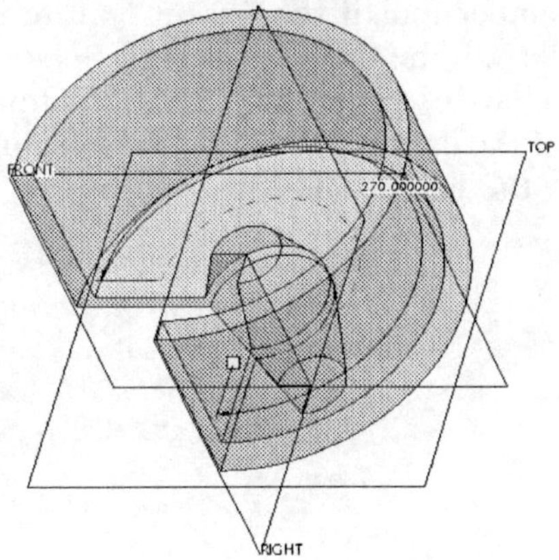

Figure 3-38 Default trimetric view of the model

In Figure 3-37, the sketch is revolved in clockwise direction to create the model shown in Figure 3-38.

From the two cases discussed, the conclusion is that the arrow that is displayed when you select the sketching plane is of great importance in orienting the model. Also, the direction of feature creation can be changed once it is created, using the **Change angle direction of revolve to other side of the sketch** button from the Revolve dashboard.

PARENT-CHILD RELATIONSHIP

Every model created in Pro/ENGINEER is composed of features that in some way or the other are related to other features in the model. The feature that occurs first in the **Model Tree** is called the parent feature, and any feature(s) that is related to this feature is called the child feature(s).

There are two types of relationships that can exist between two features, Implicit relationship and Explicit relationship.

Implicit Relationship

This type of relationship exists when the two features are related through equations. These equations are formed using relations. Relations are explained in Chapter 8.

Explicit Relationship

The explicit relationship is developed when a feature is used as a reference to create another feature. For example, one of the planar surfaces of a feature is used to create another feature, or an edge of a feature is used to dimension the other feature. In this case, the first feature is called the parent feature and the second feature is called the child feature of the first feature. Another example of this type of relationship is, when a hole is referenced to the edges of the surface it is placed on, or to the edges of some other surface. In this case, the hole is the child feature of the feature it is referenced to.

Remember that if the parent feature is modified then the child feature is also modified. Similarly, if the parent feature is deleted then the child feature is also deleted. For example, if you have used the three default datum planes to create the base feature and you delete any one of the datum planes, the base feature is also deleted. In Chapter 5, you will learn how to break the parent-child relationship.

Tip: *If you want to check the parent-child relationship of features in a model, choose **Info > Parent/Child** from the menu bar. You are prompted to select a feature. After you select the feature, the **Reference Information Window** is displayed. All the child features of the selected feature are displayed in the **Children of Current Feature** area and all the parent features are displayed in the **Parents of Current Feature** area.*

NESTING OF SKETCHES

When one or more profiles are drawn in a single sketch in order to nest more than one feature of a model in a single feature, it is called nesting of sketches. These sketches are drawn in the sketcher environment. Figure 3-39 shows a sketch in which the two circles are created on the base profile.

Advantages of Nesting the Sketches

1. One of the advantages of nesting of sketches is that the number of features used to create a model is reduced. In Figure 3-40, the model has two features. The base feature is the

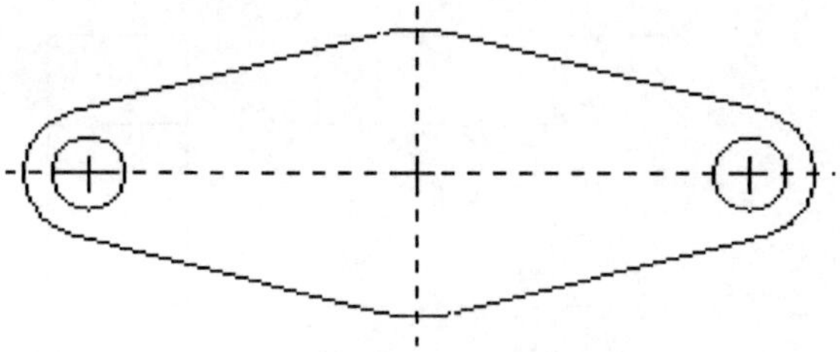

Figure 3-39 Nested sketch

Figure 3-40 Solid model of the sketch

base plate and the second feature are the holes. But, when you nest the two sketches to create the model, you are using only one feature to create the model.

2. There is no parent-child relationship that exists.

3. The depth of the hole is equal to the depth of extrusion of the base feature. This is because the two circles and the base profile are in a single sketch. This depends on the designer whether he wants the depth of the hole to be equal to the depth of extrusion.

Disadvantages of Nesting of Sketches

1. In nesting, since the two features on the model are combined into one feature, therefore there is no flexibility in editing the features of a model.

2. If at a later stage, the designer needs to convert the circular holes into elliptical holes with depth as half the depth of extrusion of the base feature, it consumes a lot of time to edit the model.

After understanding the advantages and disadvantages of nesting of sketches, it is recommended to divide a model into separate features. Draw all the features as individual features so that the model created is flexible. However, it depends on the need of the designer and the need of the model how the model is approached for creating.

TUTORIALS

Tutorial 1

In this tutorial you will create the model shown in Figure 3-41. The dimensions for the model are shown in Figure 3-42. **(Estimated time: 30 min)**

The following steps outline the procedure for creating this model:

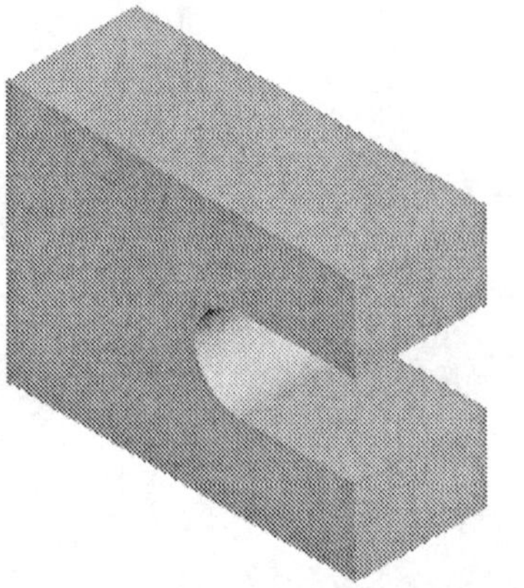

Figure 3-41 Isometric view of the model

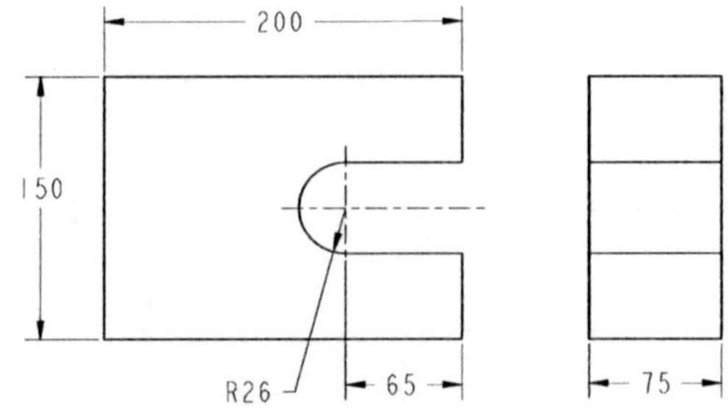

Figure 3-42 Front view and the right-side view of the model with dimensions

a. Set the working directory and create a new object file in the Part mode.

b. First examine the model and then determine the type of protrusion for the model.

c. Select the sketching plane for the model and orient it parallel to the screen.

d. Draw the sketch using the sketching tools and apply constraints and dimensions.

e. Exit the sketcher environment and define the model attributes.

Setting the Working Directory

When Pro/ENGINEER session is started, the first task is to set the working directory. A working directory is a directory on your system where you can save the work done in the current session of Pro/ENGINEER. You can set any directory existing on your system as the working directory. Since this is the first tutorial of this chapter, you need to create a folder named *c03*, if it does not exist.

1. Choose **File > Set Working Directory** from the menu bar. The **Select Working Directory** dialog box is displayed.

2. Browse and select ***C:\ProE-WF***. It is assumed that the *ProE-WF* folder exists.

3. Choose the **New Directory** button in the **Select Working Directory** dialog box. The **New Directory** dialog box is displayed.

4. Type **c03** in the **New Directory** edit box. Choose **OK** from the dialog box. You have created a folder named *c03* in *C:\ProE-WF*.

5. Choose **OK** from the **Select Working Directory** dialog box. You have set the working directory to *C:\ProE-WF\c03*.

Creating New Object File

Solid models are created in the Part mode of Pro/ENGINEER. The file extension for the files created in this mode is *.prt*.

1. Choose the **Create a new object** button from the **File** toolbar. The **New** dialog box is displayed. The **Part** radio button is selected by default in the **Type** area and the **Solid** radio button is selected by default in the **Sub-type** area of the **New** dialog box.

Note

*Before choosing the **OK** button from the **New** dialog box, make sure that the **Use default template** check box is selected. If this check box is cleared then you will need to select the template using which you will create the model.*

2. Enter the file name as **c03tut1** in the **Name** edit box and choose the **OK** button.

   The three default datum planes are displayed on the graphics screen. The **Model Tree** also appears on the left of the graphics screen in the Navigator.

3. Close the **Model Tree** by clicking on the sash present on the right edge of the Navigator. Now, the drawing area is increased.

Selecting the Protrusion Option

The given solid model is created by extruding the sketch to a distance of 75. Therefore, the sketch will be extruded as a solid to create the model. There are two methods to invoke the **Extrude** option. The first method is to use the menu bar present on the top of the screen and the second method is to use the **Extrude Tool** button present on the **Base Features** toolbar.

1. Choose **Insert > Extrude** from the menu bar or choose the **Extrude Tool** button present on the **Base Features** toolbar. The Extrude dashboard is displayed below the graphics window.

   All the attributes that are needed to create the model will be defined after the sketch is drawn.

Selecting the Sketching Plane

To create the sketch for the extruded feature, you first need to select the sketching plane for the model. The **FRONT** datum plane will be selected as the sketching plane. The sketching plane is selected such that the direction of extrusion of the solid model is perpendicular to it. From the isometric view of the model shown in Figure 3-41, it is evident that the direction of extrusion of the model is perpendicular to the **FRONT** datum plane.

1. Choose the **Create a section or redefine the existing section** button from the Extrude dashboard. The **Section** dialog box is displayed.

2. Select the **FRONT** datum plane as the sketching plane. As you select the sketching plane, the reference plane and its orientation are set automatically. The reference plane is

selected in order to orient the sketching plane.

The yellow arrow that appears on the sketching plane indicates the direction of viewing the sketch.

In the **Section** dialog box, the **Reference** box displays **RIGHT:F1(DATUM PLANE)**. This indicates that the **RIGHT** datum plane is selected as the reference plane. In the **Orientation** drop-down list, the **Right** option is selected by default. This means that while drawing the sketch the **RIGHT** datum plane will be on the right.

The **RIGHT** datum plane will be perpendicular to the sketching plane and the sketching plane will be parallel to the screen.

3. Choose the **Sketch** button in the **Section** dialog box.

Specifying References

Now, you have entered the sketcher environment and the **References** dialog box is displayed on the top right corner of the screen. You will be prompted to select a perpendicular surface, an edge, or a vertex relative to which the section will be dimensioned or constrained. Pro/ENGINEER does not use any coordinate system, and therefore it becomes necessary for the user to locate the sketch with some reference. The reference can consist of part surfaces, datums, edges, or axes.

The status displayed in the **Reference status** area is **Fully Placed**. This indicates that the references required for the sketch are automatically defined.

1. Choose the **Close** button from the **References** dialog box to exit it.

   The **FRONT** datum plane is the sketching plane and is parallel to the graphics screen. This can be verified by performing the next step.

2. Spin the datum planes by holding the middle mouse button down and dragging the cursor. Now, from this view of the datum planes, it is clear that the sketching plane was parallel to the graphics screen and the other two datum planes were perpendicular to the sketching plane.

   The above step is just to understand the orientation of the three default datum planes when you enter the sketcher environment.

3. Choose the **Orient the sketching plane parallel to the screen** button from the **Sketcher** toolbar in the **Top Toolchest**. The sketching plane and the other two perpendicular datum planes are reoriented on the screen.

Drawing the Sketch

You need to draw the sketch of the solid model that will later be extruded to create the 3D model.

1. Choose the **Create rectangle** button from the **Sketcher Tools** toolbar.

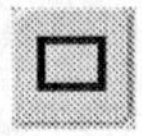

2. Draw a rectangle by defining its lower left corner and the upper right corner. The rectangle is created and weak dimensions are applied to it.

   You will notice that strong vertical and horizontal constraints are applied to the lines composing the rectangle. This is because drawing a rectangle is itself a constraint to the lines composing the rectangle.

3. Choose the **Select items** button from the **Sketcher Tools** toolbar.

4. Select the right vertical line. The line turns red in color. Press DELETE to delete the right vertical line. The sketch after deleting the vertical line is shown in Figure 3-43.

5. Now, draw the lines and the arc. Some weak dimensions and constraints are applied to the sketch as shown in Figure 3-44.

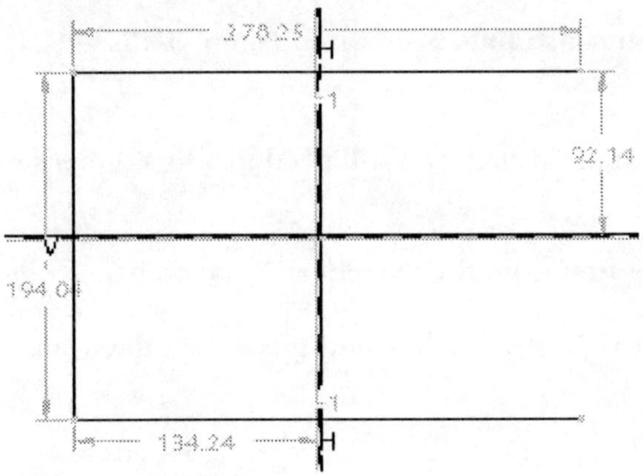

Figure 3-43 *Outer loop of the sketch with weak dimensions*

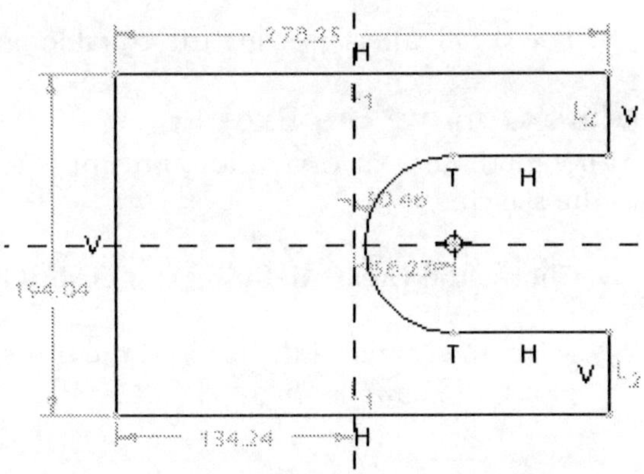

Figure 3-44 *Sketch before modifying the weak dimensions with all datums turned off*

Note

As evident from Figure 3-44, strong constraints are also applied to the sketch. These strong constraints are applied while sketching. However, if these strong constraints are not applied when you draw the sketch, you need to apply these constraints manually.

*In Figure 3-44, the center of the arc and the **TOP** datum plane are aligned by default. In case the center of the arc is not aligned to the **TOP** datum plane, you need to align them. To align the center, choose the **Create same points, points on entity or collinear constraint** button from the **Constraints** dialog box. Select the center of the arc and then select the **TOP** datum plane. Now, the center and the datum plane are aligned.*

*You can turn off the display of datums by selecting the respective tool buttons from the **Datum Display** toolbar present in the **Top Toolchest**.*

Applying Constraints to the Sketch

You need to apply equal length constraints to the sketch in order to maintain the design intent of the model.

1. Choose **Impose sketcher constraints on the section** button from the **Sketcher Tools** toolbar. The **Constraints** dialog box is displayed.

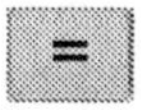

2. Choose the **Create Equal Lengths, Equal Radii, or Same Curvature constraint** button and select the two vertical lines on the right of the sketch. The weak constraint L_2 was applied when the two vertical lines were drawn. Now, the equal length constraint $\mathbf{L_2}$ is applied to both the lines. The constraint labels like L_2 or L_3 vary from sketch to sketch.

3. Select the two horizontal lines that are connected through an arc to apply on them the equal length constraint. The equal length constraint $\mathbf{L_2}$ is applied to both the lines and the constraint symbol $\mathbf{L_2}$ on the two right vertical lines is changed to $\mathbf{L_1}$.

   The sketch after applying the equal length constraints is shown in Figure 3-45.

Dimensioning the Sketch

Although some weak dimensions are applied to the sketch, you still need to add a dimension to the sketch.

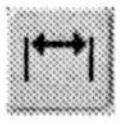

1. Choose the **Create defining dimension** button from the **Sketcher Tools** toolbar.

2. Select the center of the arc and the upper right vertical line and place the dimension as shown in Figure 3-46.

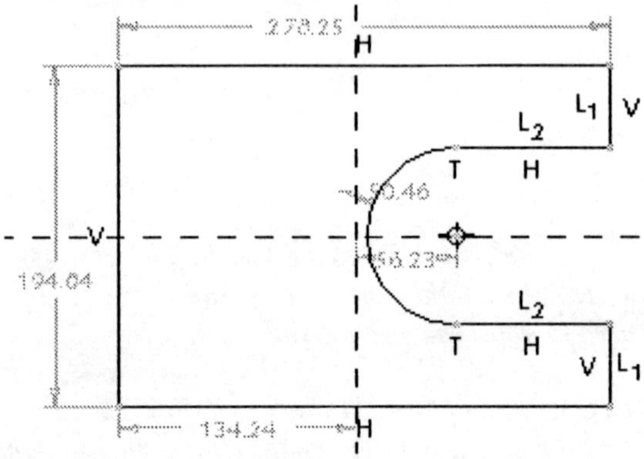

Figure 3-45 Equal length constraints applied to the sketch

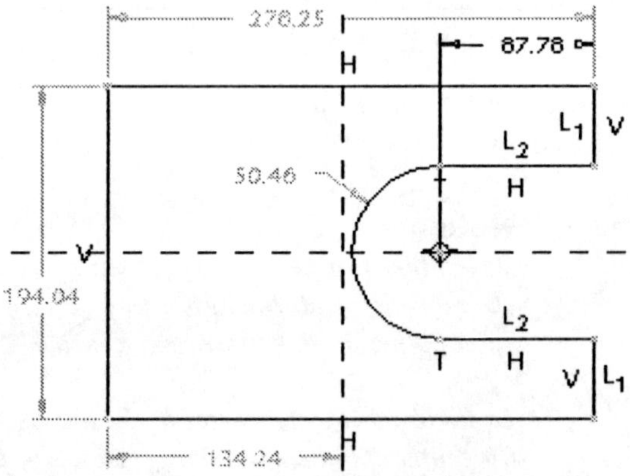

Figure 3-46 Dimension added to the sketch

You need to dimension only these entities because the rest of the weak dimensions are useful dimensions and will be modified directly.

Note

If the dimensions in your sketch are different from those shown in Figure 3-46, add the missing dimensions and delete the dimensions that are not required.

Modifying the Dimensions

You need to modify the dimension values of the sketch. You will notice that the default dimensions shown in Figure 3-46 also include the length and width of the rectangle and the distance of the sketched section from the selected references. It is recommended that you draw the base feature symmetrical with the other two datum planes. Hence, the distance from the **RIGHT** datum plane is 100 (200 divided by 2 is equal to 100).

1. Select the sketch and dimensions using CTRL+ALT+A.

2. Choose the **Modify the values of dimensions, geometry of splines, or text entities** button from the **Sketcher Tools** toolbar. The **Modify Dimensions** dialog box is displayed.

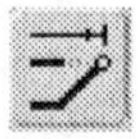

   All the dimensions in the sketch are displayed in this dialog box and each dimension has a separate thumbwheel and an edit box. You can use the thumbwheel or the edit box to modify the dimensions. It is recommended to use the edit boxes to modify the dimensions if the change in the dimension value is large.

3. Clear the **Regenerate** check box. If you clear this check box, then any modification in a dimension value does not update the sketch during the modification. The dimensions will be modified after you exit the **Modify Dimensions** dialog box. It is recommended to clear the **Regenerate** check box when more than one dimension has to be modified.

4. Modify all the dimensions one by one as shown in Figure 3-46. You will notice that the dimension you select in the **Modify Dimensions** dialog box is enclosed in a yellow box on the graphics screen.

5. After modifying all the dimensions, choose the **Regenerate the section and close the dialog** button from the **Modify Dimensions** dialog box. A message **Dimension modifications successfully completed** is displayed in the **Message Area**.

6. Choose the **Continue with the current section** button to exit the sketcher environment.

Specifying the Model Attributes

Next, you need to specify the attributes to create the model.

1. Choose the **Saved view list** button from the **View** toolbar in the **Top Toolchest**. Choose the **Default View** option from the drop-down list.

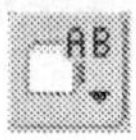

   The default trimetric view is displayed. This display gives you a better view of the sketch in the 3D space. The model is displayed in yellow color and is translucent. The yellow colored arrow is also displayed on the model, indicating the direction of extrusion. The model may not fit fully in the graphics window.

2. Press CTRL+middle mouse button and drag the mouse upwards on the graphics window. Notice a red rubber-band line attached to the cursor. The length of the rubber-band line gives an idea of the extent you need to zoom the model. After the model is in the graphics window, release the middle mouse button and the CTRL key.

   The model appears as shown in Figure 3-48. All the attributes that are selected by default in the Extrude dashboard will be used to create the model. You need to change only the depth of extrusion.

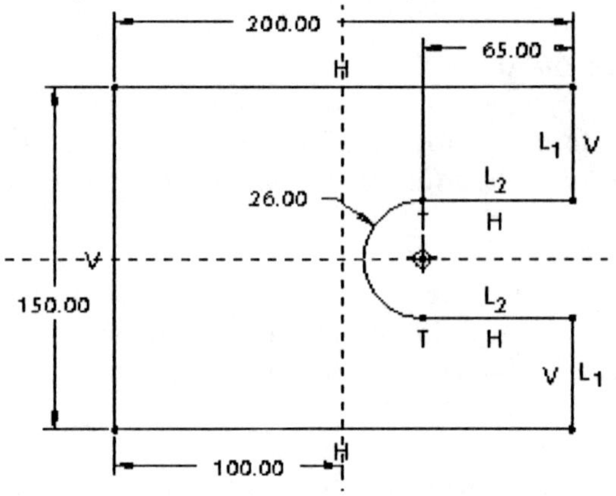

Figure 3-47 *Sketch after modifying the dimensions*

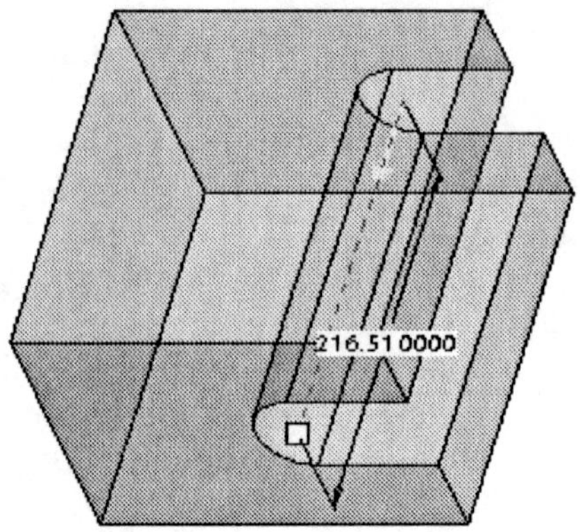

Figure 3-48 *Arrow showing the direction of feature creation*

3. In the edit box present on the Extrude dashboard, enter the depth of extrusion equal to **75** and press ENTER. The model in the graphics window is displayed with the specified depth of extrusion.

4. Choose the **Build feature** button in the Extrude dashboard. The message "**PROTRUSION has been created successfully.**" is displayed in the **Message Area**.

5. Again choose the **Default View** option from the **Saved view list** drop-down list. The model appears as shown in Figure 3-49.

Note

In Figure 3-49, the display of datum planes is turned off by choosing the ***Datum planes on/off*** *button.*

Saving the Model

1. Choose the **Save** option from the **File** menu or choose the **Save the active object** button from the **File** toolbar. The **Message Input Window** is displayed with the name of the object file that you had specified earlier.

2. Press ENTER or choose the green check mark on the **Message Input Window** to save.

Figure 3-49 Default trimetric view of the model

Closing the Current Window

The given model is created and is also saved. Now, you can close the current window.

1. Choose **File** > **Close Window** or choose the **Window** > **Close** from the menubar.

Tutorial 2

In this tutorial you will create the model shown in Figure 3-50. The dimensions are shown in Figure 3-51. **(Expected time: 30 min)**

Figure 3-50 Isometric view of the solid model

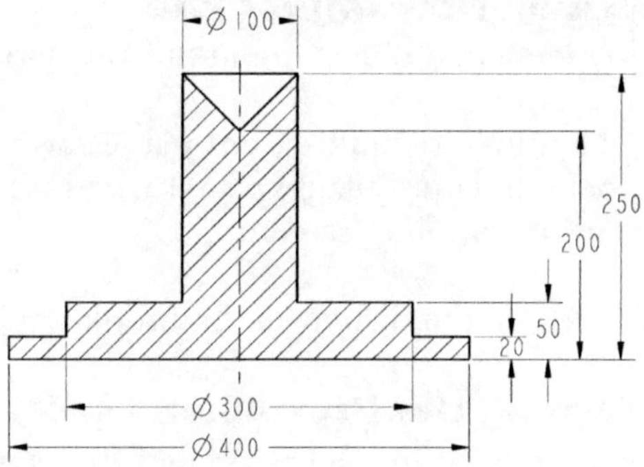

Figure 3-51 Front sectioned view of the solid model

The following steps outline the procedure for creating this model:

a. Create a new object file in the Part mode.

b. First examine the model and then determine the type of protrusion for the model.

c. Select the sketching plane for the model.

d. Draw the sketch for the revolve feature and a center line to revolve it using the sketching tools and apply dimensions.

e. Exit the sketcher environment and define the model attributes.

Setting the Working Directory

The working directory was selected in Tutorial 1, and therefore there is no need to select the working directory again. But if you still want to select the working directory, follow the steps given next:

1. Open the **Navigator** by clicking the top arrows on the left edge of the Pro/ENGINEER main window. The Navigator slides out.

2. Choose the **Folder Browser** button in the Navigator to view the folders.

3. Click on the plus symbol adjacent to the *ProE-WF* folder in the navigator. The contents of the **ProE-WF** folder are displayed.

4. Now right-click on the *c03* folder to display a shortcut menu. From this shortcut menu, choose the **Make Working Directory** option. The working directory is set to *c03*.

5. Close the Navigator by clicking the sash on the right edge of the Navigator. The Navigator slides in.

Creating New Object File

1. Open a new object file in the Part mode. Name the file as **c03tut2**.

   The three default datum planes are displayed on the graphics screen. However, if the default datum planes were turned off in the previous tutorial, then they will not appear on the graphics screen.

2. Turn on the display of datum planes by selecting the **Datum planes on/off** button.

Selecting the Protrusion Option

The given solid model is created by revolving the sketch through an angle of 360-degree about an axis. Therefore, the **Revolve Tool** button will be used to create the model. There are two methods to invoke the **Revolve** option. The first method is to use the menu bar present on the top of the screen and the second method is to use the **Base Features** toolbar present on the **Right Toolchest**.

1. Choose **Insert > Revolve** from the menu bar or choose the **Revolve Tool** button from the **Base Features** toolbar. The Revolve dashboard is displayed below the graphics window.

Some of the attributes like rotation angle and the direction of rotation of the sketch will be defined after the sketch is created.

Selecting the Sketching Plane

To create the sketch of the model, you first need to select the sketching plane for the model. Note that the axis of revolution of the revolved feature is normal to the **TOP** datum plane in the model. Therefore, any of the other two datum planes other than the **TOP** datum plane can be selected as the sketching plane. Here you will select the **FRONT** datum plane as the sketching plane.

1. Choose the **Create a section or redefine the existing section** button from the Revolve dashboard. The **Section** dialog box is displayed.

2. Select the **FRONT** datum plane as the sketching plane. As you select the sketching plane, the reference plane and its orientation are set automatically. The reference plane is selected in order to orient the sketching plane.

   In the **Section** dialog box, the **Reference** box displays **RIGHT:F1(DATUM PLANE)**. This indicates that the **RIGHT** datum plane is selected as the reference plane. In the **Orientation** drop-down list, the **Right** option is selected by default. This means that while drawing the sketch the **RIGHT** datum plane will be on the right.

   The **RIGHT** datum plane will be perpendicular to the sketching plane and the sketching plane will be parallel to the screen.

3. Choose the **Sketch** button in the **Section** dialog box.

Specifying References

Now, you have entered the sketcher environment and the **References** dialog box is displayed on the top right corner of the screen. You will be prompted to select a perpendicular surface, an edge, or a vertex relative to which the section will be dimensioned or constrained. Pro/ENGINEER does not use any coordinate system, and therefore it becomes necessary for the user to locate the sketch with some reference. The reference can consist of part surfaces, datums, edges, or axes.

The status displayed in the **Reference status** area is **Fully Placed**. This indicates that the references required for the sketch are automatically defined.

1. Choose the **Close** button from the **References** dialog box to exit it.

Drawing the Sketch

You need to draw the sketch of the revolved feature. The sketch that will be drawn is the cross-section of the revolved feature. The sketch will be revolved about the center line.

1. Choose the **Create 2 point lines** button from the **Sketcher Tools** toolbar. Draw the sketch as shown in Figure 3-52. The sketch should be a closed loop and the bottom horizontal line should be aligned to the **TOP** datum plane.

As you draw the sketch, weak dimensions and strong constraints are applied to the sketch.

2. Choose the black arrow on the right of the **Create 2 point lines** button to display the flyout. From this flyout, choose the **Create 2 point centerlines** button. Draw the center line for the axis of revolution. The center line should be drawn such that it is aligned with the **RIGHT** datum plane, see Figure 3-52.

Dimensioning the Sketch

The weak dimensions are automatically applied to the sketch. Since the model is a revolved feature, therefore, you need to manually apply linear diameter dimensions to the sketch. The linear diameter dimensions are applied using the center line that was drawn in the sketch.

Tip: *Linear diameter dimensioning is necessary for all revolved features. This is because mostly all the revolved models are machined on a lathe. Hence, while machining a revolved model it is necessary that the operator of the machine has a drawing of the model that is diametrically dimensioned.*

1. Choose the **Create defining dimension** button from the **Sketcher Tools** toolbar.

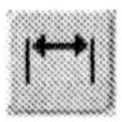

2. Select the center line, the first right vertical line, and then again the center line.

3. Now, use the middle mouse button to place the dimension on top of the sketch. The diameter dimension is placed.

   The method for diametrically dimensioning a sketch has been discussed in Chapter 1.

4. Select the center line, the second right vertical line, and then again the center line.

5. Place the dimension below the sketch.

6. Select the center line, the third right vertical line, and then again the center line. Now, place the dimension below the previous dimension.

   Dimension the remaining entities in the sketch as shown in Figure 3-53.

Modifying the Dimensions

When you dimension a sketch, default dimension values are applied to the sketch. You need to modify the dimension values of the dimensions.

1. Select the sketch and dimensions using CTRL+ALT+A.

2. Choose the **Modify the values of dimensions, geometry of splines, or text entities** button from the **Sketcher Tools** toolbar. The **Modify Dimensions** dialog box is displayed.

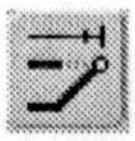

   All the dimensions in the sketch are displayed in this dialog box and each dimension has

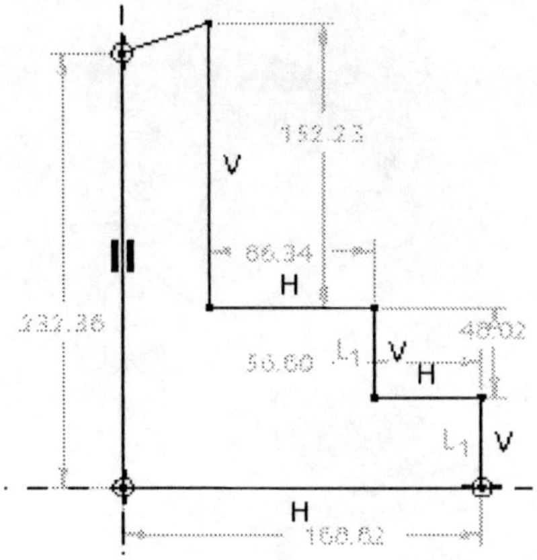

Figure 3-52 *Sketch with weak dimensions*

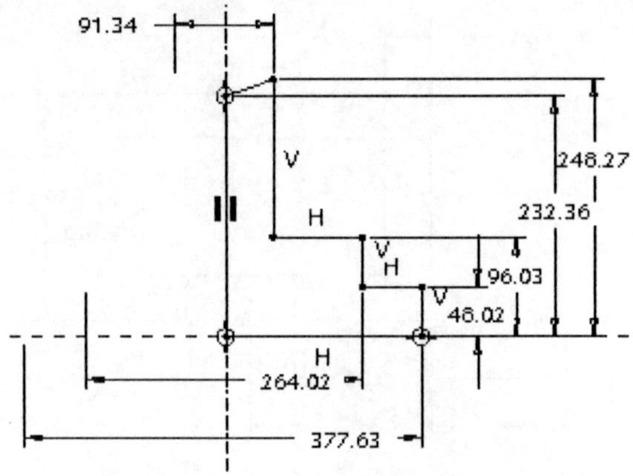

Figure 3-53 *Sketch after dimensioning*

a separate thumbwheel and an edit box. You can use the thumbwheel or the edit box to modify the dimensions. It is recommended that you use the edit boxes to modify the dimensions if the change in the dimension value is large.

3. Clear the **Regenerate** check box and then modify the values of the dimensions as shown in Figure 3-54. If you clear this check box, then any modification in a dimension value does not update the sketch. It is recommended that you clear the **Regenerate** check box when more than one dimension has to be modified.

   You will notice that the dimension you select in the **Modify Dimensions** dialog box is enclosed in a yellow box on the graphics screen.

4. After modifying all the dimensions, choose the **Regenerate the section and close the dialog** button from the **Modify Dimensions** dialog box. The message **Dimension modifications successfully completed** is displayed in the **Message Area**.

5. Choose the **Continue with the current section** button to exit the sketcher environment.

Specifying the Model Attributes

When you exit the sketcher environment, the Revolve dashboard below the graphics window is enabled again. Using this dashboard, you will specify the angle of revolution for the revolved feature.

1. Choose the **Saved view list** button from the **View** toolbar. Choose the **Default View** option from the drop-down list.

Note

If the model is not fully displayed in the graphics window, you may need to zoom out.

The feature will orient in its default orientation as shown in Figure 3-55. This display

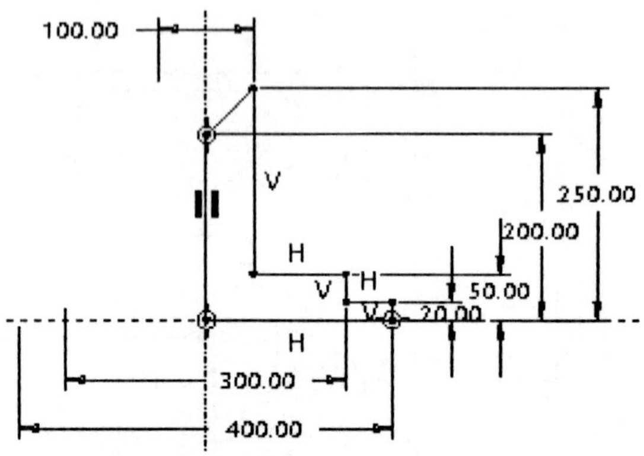

Figure 3-54 Sketch after modifying the dimensions

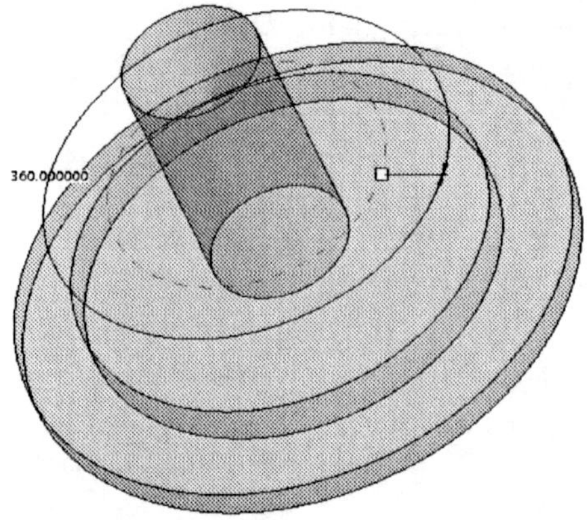

Figure 3-55 The preview of the model in default trimetric view

gives you a better view of the sketch in the 3D space. Model appears in yellow color and is translucent. The drag handle is also available on the model which can be used to modify the angle of revolution dynamically.

All the attributes that are needed to create the solid model are selected by default in the Revolve dashboard.

2. Choose the **Build feature** button in the Extrude dashboard. The model appears as shown in Figure 3-56. The message "**PROTRUSION has been created successfully.**" is displayed in the **Messsage Area**.

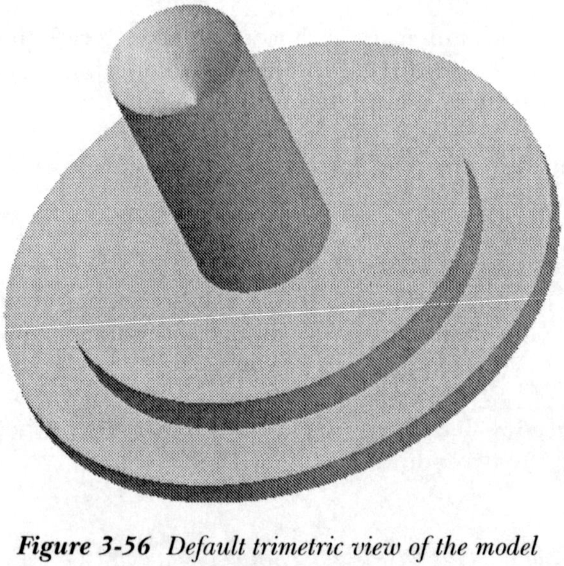

Figure 3-56 Default trimetric view of the model

Saving the Model

1. Choose the **Save** option from the **File** menu or choose the **Save the active object** button from the **File** toolbar. The **Message Input Window** is displayed with the name of the object file that you had specified earlier.

2. Press ENTER or choose the green check mark on the **Message Input Window** to save the file.

Closing the Current Window

The given model is completed and is also saved. Now you can close the current window.

1. Choose **File** > **Close Window** from the menu bar.

Note

If you need to view the **Model Tree** *of the model you have created then you need to open it. Click on the sash present on the left edge of the graphics window. The* **Model Tree** *slides out. To again close it, click on the sash again. The* **Model Tree** *slides in.*

Tutorial 3

In this tutorial you will create the model shown in Figure 3-57. Figure 3-58 shows the dimensions. **(Expected time: 30 min)**

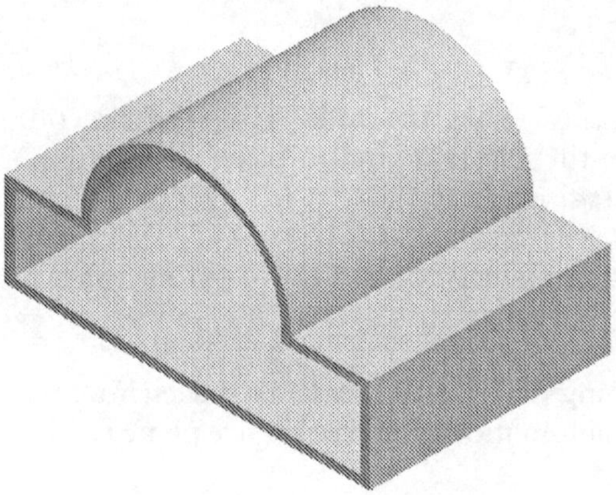

Figure 3-57 *Isometric view of the model*

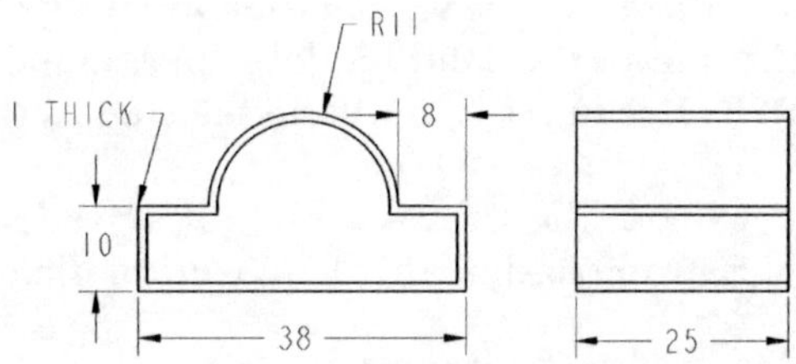

Figure 3-58 *Front view and the right-side view*

The following steps outline the procedure for creating this model:

a. Create a new object file in the Part mode.

b. First examine the model and then determine the type of protrusion for the model.

c. Select the sketching plane for the model.

d. Draw the sketch using the sketching tools and apply dimensions.

e. Specify the attributes to create the model and save the model.

The working directory was selected in Tutorial 1, and therefore there is no need to select the working directory again. But if you still want to select the working directory, use the **Navigator** to select it.

Creating New Object File

1. Open a new object file in the Part mode. Name the file as **c03tut3**.

2. If the default datum planes were not turned off in the previous tutorial, they will appear on the graphics screen. Turn on the display of datum planes.

Selecting the Protrusion Option

The given thin solid model is created after extruding the sketch to a distance of 25. Therefore, the **Thicken Sketch** tool button on the Extrude dashboard will be used to create the model. To invoke the Extrude dashboard, you need to select the **Extrude Tool** button.

1. Choose **Insert > Extrude** from the menu bar or choose the **Extrude Tool** button from the **Base Features** toolbar. The Extrude dashboard is displayed below the graphics window.

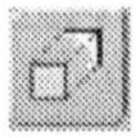

Most of the attributes that are needed to create the model will be defined after the sketch is drawn.

Selecting the Sketching Plane

From the isometric view of the model shown in Figure 3-57, it is evident that the direction of extrusion of the solid model is perpendicular to the **FRONT** datum plane. Therefore, the **FRONT** datum plane will be selected as the sketching plane.

1. Choose the **Create a section or redefine the existing section** button from the Extrude dashboard. The **Section** dialog box is displayed.

2. Select the **FRONT** datum plane as the sketching plane. As you select the sketching plane, the reference plane and its orientation are set automatically. The reference plane is selected in order to orient the sketching plane.

   In the **Section** dialog box, the **Reference** box displays **RIGHT:F1(DATUM PLANE)**. This indicates that the **RIGHT** datum plane is selected as the reference plane. In the **Orientation** drop-down list, the **Right** option is selected by default. This means that while drawing the sketch the **RIGHT** datum plane will be on the right.

   The **RIGHT** datum plane will be perpendicular to the sketching plane and the sketching plane will be parallel to the screen.

3. Choose the **Sketch** button in the **Section** dialog box.

Specifying References

Now, you have entered the sketcher environment and the **References** dialog box is displayed on the top right corner of the screen. You will be prompted to select a perpendicular surface, an edge, or a vertex relative to which the section will be dimensioned or constrained.

The status displayed in the **Reference status** area is **Fully Placed**.

1. Choose the **Close** button from the **References** dialog box to exit it.

Drawing the Sketch

Using the sketcher tools, you need to draw the sketch of the thin extruded model.

1. Draw the arc shown in Figure 3-59 using the **Create an arc by 3 points or tangent to an entity at its endpoint** button from the **Sketcher Tools** toolbar. Complete the sketch as shown in Figure 3-59.

   While drawing the sketch, weak dimensions are applied to the sketch. These dimensions appear gray in color.

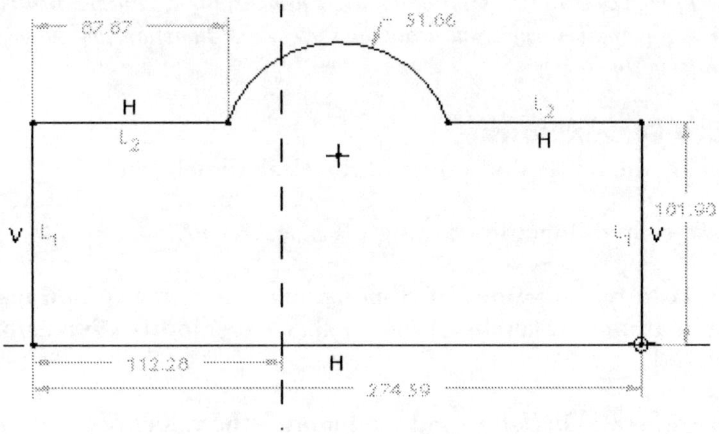

Figure 3-59 *Sketch for the thin extruded model*

Applying Constraints to the Sketch

Some weak constraints are applied to the sketch while drawing, but you need to apply the constraints using the **Constraints** dialog box.

1. Choose the **Impose sketcher constraints on the section** button from the **Sketcher Tools** toolbar. The **Constraints** dialog box is displayed.

2. Choose the **Create same points, points on entity or collinear constraint** button from the **Constraint** dialog box.

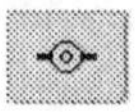

3. Select the center of the arc and then select the **RIGHT** datum plane. The center of the arc is aligned with the **RIGHT** datum plane.

4. Select the bottom horizontal line and then select the **TOP** datum plane. Now, the line is aligned with the plane. If the two horizontal lines on the top are aligned with the **TOP** datum plane, skip this point.

5. Choose the **Create Equal Lengths, Equal Radii, or Same Curvature constraint** button and select the two horizontal lines on the top to apply the equal length constraint.

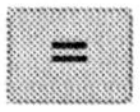

6. Choose the **Make two entities perpendicular** button and select the left horizontal line and the arc. The perpendicular constraint symbol is applied. Similarly, make the right horizontal line and the arc perpendicular. If these constraints are already applied, the **Resolve Sketch** dialog box will be displayed. Choose **Undo** from this dialog box.

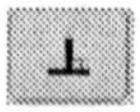

   The sketch after applying the constraints is shown in Figure 3-60.

Note

When you apply constraints, some of the weak dimensions are automatically deleted. Also, some constraints are applied when you draw the sketch, and therefore you do not need to apply those constraints again.

Modifying the Dimensions

You need to modify the dimension values of the weak dimensions.

1. Select the sketch and dimensions using CTRL+ALT+A.

2. Choose the **Modify the values of dimensions, geometry of splines, or text entities** button from the **Sketcher Tools** toolbar. The **Modify Dimensions** dialog box is displayed.

3. Clear the **Regenerate** check box and then modify the values of the dimensions as shown in Figure 3-61. When you clear the **Regenerate** check box, any modification in a dimension value does not update the sketch. As mentioned earlier, it is recommended that you clear the **Regenerate** check box when more than one dimension has to be modified.

   You will notice that the dimension you select in the **Modify Dimensions** dialog box is enclosed in a yellow box on the graphics screen.

4. After modifying all the dimensions, choose the **Regenerate the section and close the dialog** button from the **Modify Dimensions** dialog box. A message **Dimension modifications successfully completed** is displayed in the **Message Area**.

5. Choose the **Continue with the current section** button to exit the sketcher environment.

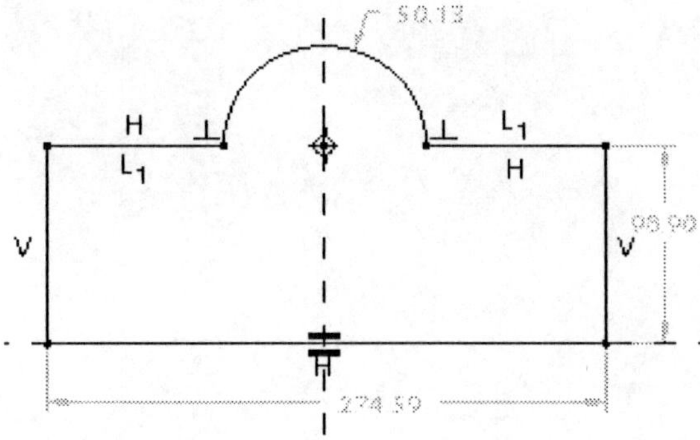

Figure 3-60 Sketch after applying constraints

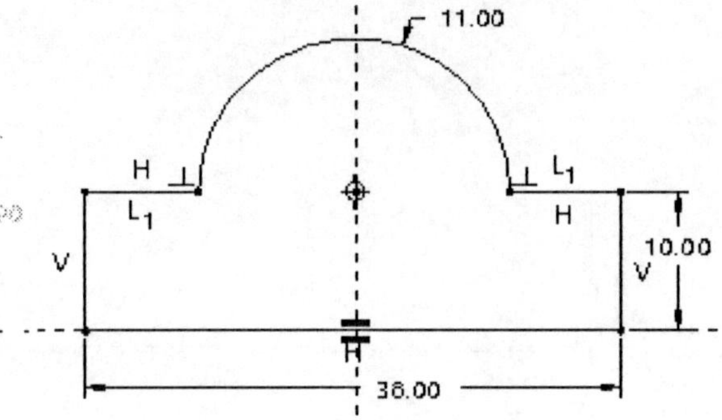

Figure 3-61 Figure after modifying the dimensions

Specifying the Model Attributes

The attributes that will create the model needs to be selected from the Extrude dashboard.

1. Choose the **Thicken Sketch** button from the Extrude dashboard. The model changes to a thin model of certain thickness as shown in Figure 3-62.

2. Choose the **Saved view list** button from the **View** toolbar. Choose the **Default View** option from the drop-down list.

   The default trimetric view of the model is displayed. The model appears yellow in color and is translucent. If the model does not fully fit on the screen then you need to zoom out. This display gives you a better view of the model in the 3D space.

   On the Extrude dashboard, there are two drop-down lists. The drop-down list that is present on the left is used to enter the depth of extrusion. The drop-down list that is present on the right appears only when the **Thicken Sketch** button is selected. This drop-down list is used to specify the thickness value of the thin model.

3. Enter a value of **25** in the left edit box and press ENTER.

4. Enter a value of **1** in the right edit box and press ENTER. The model appears as shown in Figure 3-63. The model is completed and now you need to exit the Extrude dashboard.

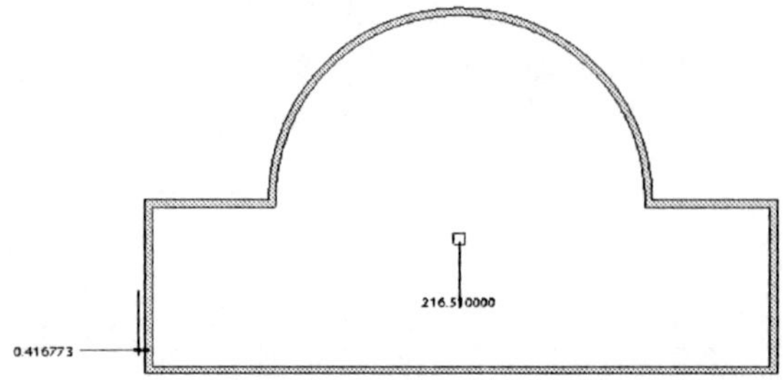

Figure 3-62 Arrow showing the direction of material addition with respect to the boundary of the section

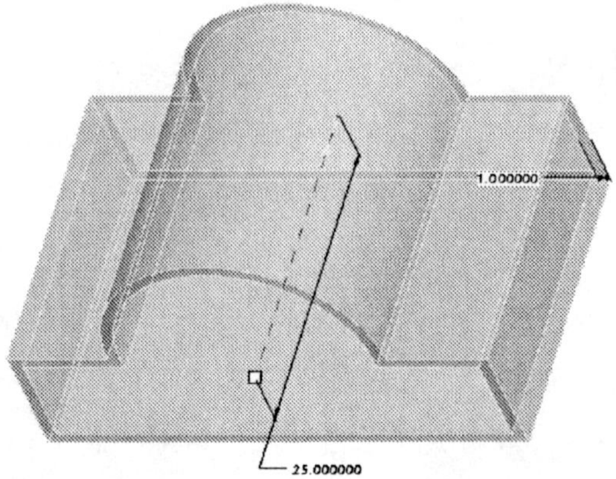

Figure 3-63 The default trimetric view of the model

5. Choose the **Build feature** button from the Extrude dashboard. The message "**PROTRUSION has been created successfully.**" will be displayed in the **Message Area** and the model will be displayed as shown in Figure 3-64.

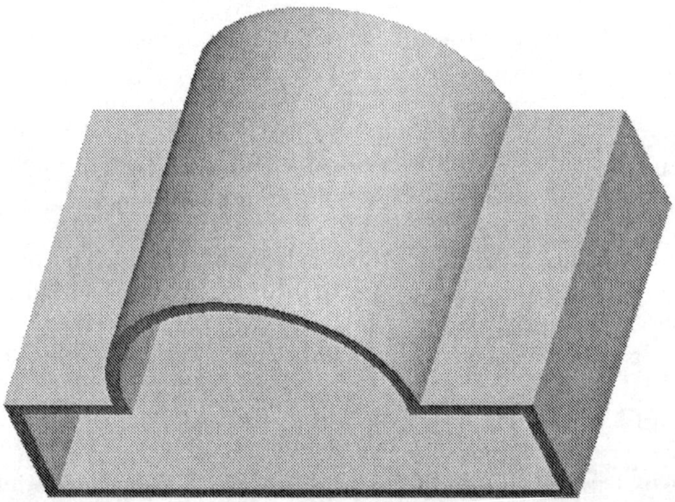

Figure 3-64 Default trimetric view of the solid model

Saving the Model

1. Choose the **Save** option from the **File** menu or choose the **Save the active object** button from the **Top Toolchest**. The **Message Input Window** is displayed with the name of the object file that you had specified earlier.

2. Press ENTER or choose the green check mark on the **Message Input Window** to save the file.

Closing the Current Window

The given model is completed and is also saved. Now you can close the current window.

1. Choose **File > Close Window** from the menu bar.

Tutorial 4

In this tutorial you will create the model shown in Figure 3-65. Figure 3-66 shows the dimensions of the model. **(Expected time: 30 min)**

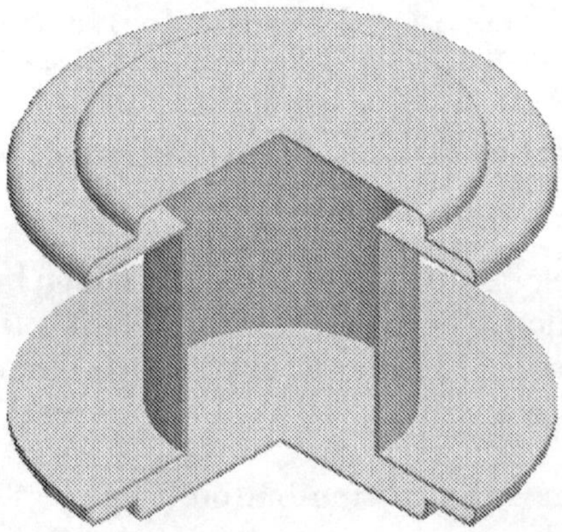

Figure 3-65 Isometric view of the solid model

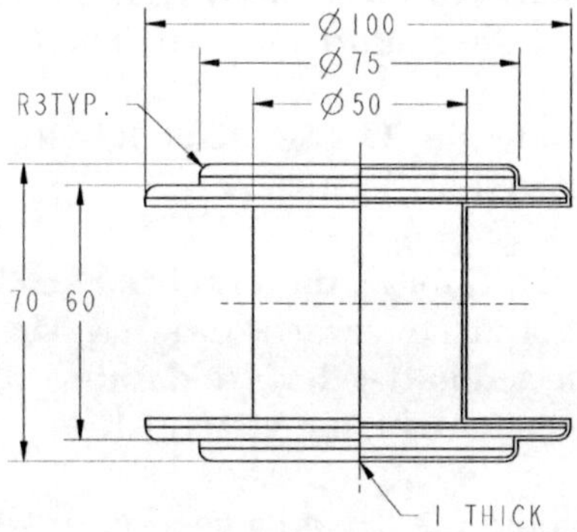

Figure 3-66 Front view of the solid model

The following steps outline the procedure for creating the given model:

a. Create a new object file in the Part mode.

b. First examine the model and then determine the type of protrusion for the model.

c. Select the sketching plane for the model.

d. Draw the sketch using sketching tools, apply dimensions, and modify dimension values.

e. Specify the attributes of the model and then save it.

The working directory was selected in Tutorial 1, and therefore there is no need to select the working directory again. But if you still want to select the working directory, use the Navigator to select it.

Creating New Object File

1. Open a new object file in the Part mode. Name the file as **c03tut4**.

2. If the default datum planes were not turned off in the previous tutorial, they will appear on the graphics window. If the datum planes are not displayed, turn them on using the **Datum planes on/off** button.

Note

Throughout the book, graphics screen is referred to the screen that appears in the sketcher environment and graphics window is referred to the screen that appears after exiting the sketcher.

Selecting the Protrusion Option

The given revolved thin model is created after revolving the sketch to a given angle. Therefore, the **Thicken Sketch** button in the Revolve dashboard will be used to create the model. There are two methods to invoke the **Revolve** option. The first method is to use the menu bar present on the top of the screen and the second method is to use the **Base Features** toolbar.

1. Choose **Insert > Revolve** from the menu bar or choose the **Revolve Tool** button from the **Base Features** toolbar. The Revolve dashboard appears below the graphics window.

2. Choose the **Thicken Sketch** button from the dashboard. The **Revolve as solid** button is selected by default.

   In this tutorial, the **Thicken Sketch** button is selected before drawing the sketch because the sketch that will be drawn will be an open section. Therefore, with the default attributes selected in the Revolve dashboard, it is not possible to draw an open sketch that can be accepted by Pro/ENGINEER.

3. Choose the **Create a section or redefine the existing section** button from the Revolve dashboard. The **Section** dialog box is displayed.

Selecting the Sketching Plane

Note that in Figure 3-65, the imaginary axis of revolution of the revolved feature is normal to the **TOP** datum plane. Therefore, the **TOP** datum plane will be selected to be at the top while drawing the sketch. Now, you need to decide the sketching plane from the datum planes **RIGHT** and **FRONT**. Any of the two datum planes can be selected as the sketching plane. The **RIGHT** datum plane will be selected as the sketching plane.

1. Select the **RIGHT** datum plane as the sketching plane. As you select the sketching plane, the reference plane and its orientation are set automatically.

   In the **Section** dialog box, the **Reference** box displays **TOP:F2(DATUM PLANE)**. This indicates that the **TOP** datum plane is selected as the reference plane. But you need to change the orientation of the plane.

2. From the **Orientation** drop-down list, choose the **Top** option.

   The **TOP** datum plane will be perpendicular to the sketching plane and the sketching plane will be parallel to the screen.

3. Change the direction of the yellow arrow that appears on the sketching plane by choosing the **Flip** button. Now, the direction of viewing the sketching plane is reversed.

4. Choose the **Sketch** button in the **Section** dialog box.

Specifying References

Now, you have entered the sketcher environment and the **References** dialog box is displayed on the top right corner of the screen. You will be prompted to select a perpendicular surface, an edge, or a vertex relative to which the section will be dimensioned or constrained. The status displayed in the **Reference status** area is **Fully Placed**.

1. Choose the **Close** button from the **References** dialog box to exit it.

Drawing the Sketch

You need to draw the sketch of the thin extruded model. The sketch can be a closed or an open loop. Here, you will draw an open sketch.

1. Draw a center line and then draw the sketch as shown in Figure 3-67. The center line is the axis of revolution. To fillet the corners, choose the **Create a circular fillet between two entities** button from the **Right Toolchest.**

   While drawing the sketch, weak constraints are applied to the sketch.

Applying Constraints to the Sketch

Weak constraints are applied to the sketch while drawing but you need to apply the constraints using the **Constraints** dialog box.

1. Choose the **Impose sketcher constraints on the section** button from the **Sketcher Tools** toolbar. The **Constraints** dialog box is displayed. 
2. Choose the **Create Equal Lengths, Equal Radii, or Same Curvature constraint** button from the **Constraints** dialog box. 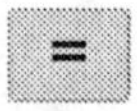
3. One by one select the two horizontal lines with the constraint symbol $\mathbf{L_1}$ shown in Figure 3-68 to apply the equal length constraint.
4. One by one select the two horizontal lines with the constraint symbol $\mathbf{L_2}$ shown in Figure 3-68 to apply the equal length constraint.
5. Select the vertical lines with the constraint symbol $\mathbf{L_3}$ shown in Figure 3-68 to apply the equal length constraint.
6. Select all the fillets to apply equal radii constraint.

Applying Dimensions to the Sketch

The weak dimensions are automatically applied to the sketch. Remember that since the model is a revolved feature, therefore, you need to apply linear diameter dimensions to the sketch manually. The linear diameter dimensions are applied using the center line that was drawn in the sketch.

1. Choose the **Create defining dimensions** button from the **Sketcher Tools** toolbar. 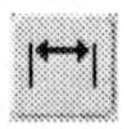

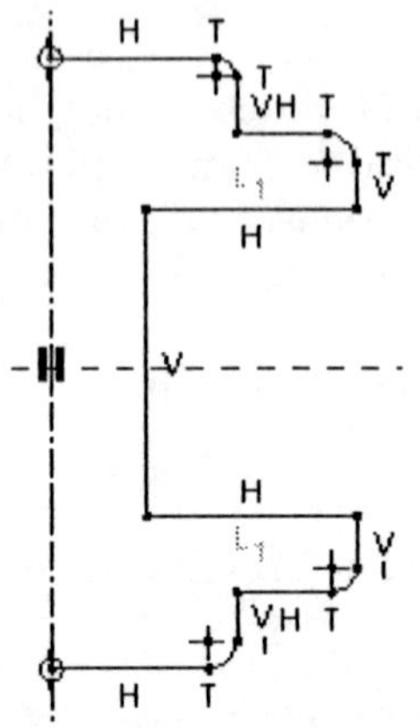

Figure 3-67 Sketch for the revolved feature with the weak dimensions turned off for clarity

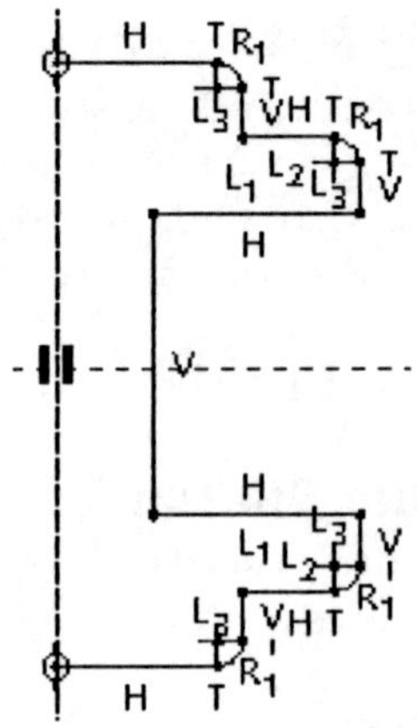

Figure 3-68 Sketch after applying constraints with the weak dimensions turned off for clarity

2. Dimension the sketch as shown in Figure 3-69.

Modifying the Dimensions

You need to modify the dimension values of the weak dimensions.

1. Select the sketch and dimensions using CTRL+ALT+A.
2. Choose the **Modify the values of dimensions, geometry of splines, or text entities** button from the **Sketcher Tools** toolbar. The **Modify Dimensions** dialog box is displayed.

3. Clear the **Regenerate** check box and then modify the values of the dimensions as shown in Figure 3-70.

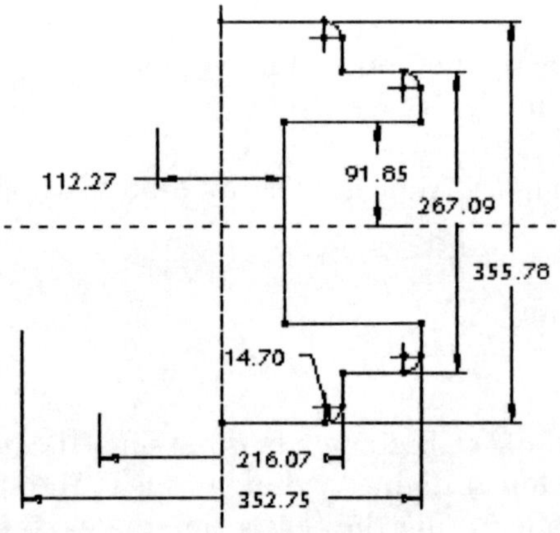

Figure 3-69 Sketch after dimensioning with constraints turned off for clarity

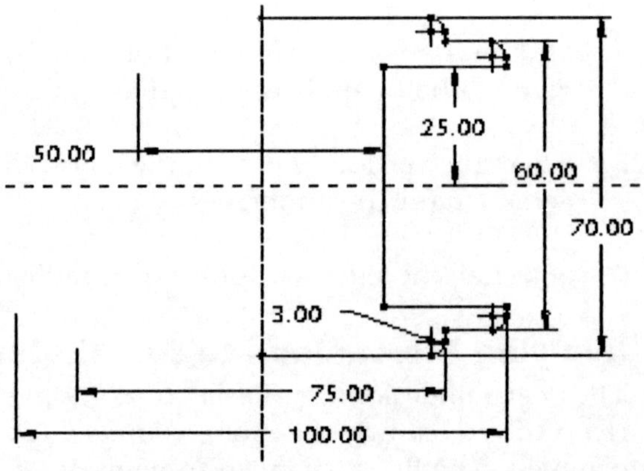

Figure 3-70 Sketch after modifying the dimensions with constraints turned off for clarity

You will notice that the dimension you select in the **Modify Dimensions** dialog box is enclosed in a yellow box on the graphics screen.

4. After modifying all the dimensions, choose the **Regenerate the section and close the dialog** button from the **Modify Dimensions** dialog box. The message **Dimension modifications successfully completed** is displayed in the **Message Area**.

5. Choose the **Continue with the current section** button to exit the sketcher environment.

Specifying the Model Attributes

After completing the sketch, you need to specify the side to which thickness of material will be applied. The thickness can be applied outside the section, inside the section, or symmetrically to both the sides of the section boundary. Here, you will apply the thickness inside the section.
When you exit the sketcher environment, the model assumes some default attributes and therefore it does not display in the graphics window.

1. In the drop-down list that is present at the right on the Revolve dashboard, enter the thickness value as **1** and press ENTER.

2. In the edit box that is present at the left on the Revolve dashboard, the default value of angle of revolution is **360**. Enter a value of **270** and press ENTER.

3. Choose the **Saved view list** button from the **View** toolbar. Choose the **Default View** option from the drop-down list.

The trimetric view of the model after specifying all the attributes is shown in Figure 3-71.

3. Choose the **Build feature** button from the Revolve dashboard. The default trimetric view of the model is shown in Figure 3-72.

Saving the Model

1. Choose the **Save** option from the **File** menu or choose the **Save the active object** button from the **Top Toolchest**. The **Message Input Window** is displayed with the name of the object file that you had specified earlier.

2. Press ENTER to save the file.

Closing the Current Window

The given model is completed and is also saved. Now you can close the current window.

1. Choose **File** > **Close Window** from the menu bar.

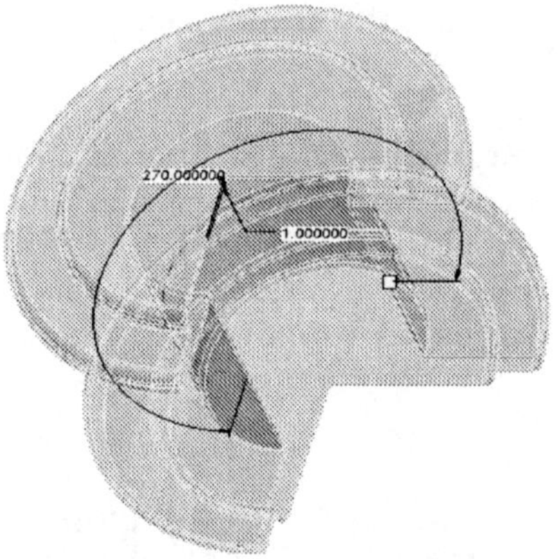

Figure 3-71 Model after specifying attributes

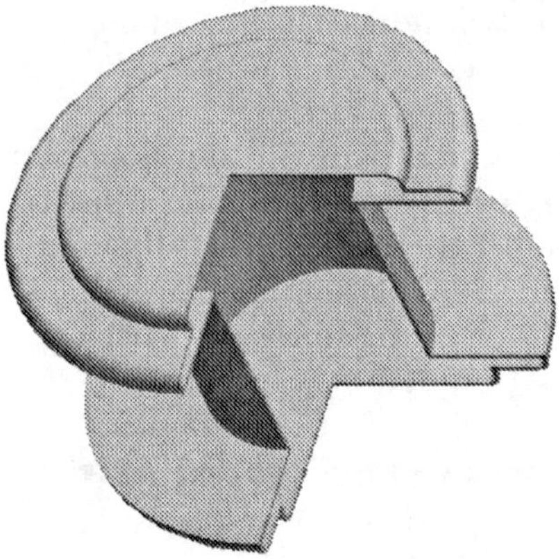

Figure 3-72 Default trimetric view of the model

Self-Evaluation Test

Answer the following questions and then compare your answers to the answers given at the end of this chapter.

1. All the features created in the Part mode are called the base features. (T/F)

2. You can extrude a sketch to both the sides of the sketching plane symmetrically. (T/F)

3. Even if you do not specify the references for sketching when you are in the sketcher environment, you can still continue to draw the sketch. (T/F)

4. The **Section** dialog box is used to select the sketching plane. (T/F)

5. The arrow displayed on the sketching plane when you select the sketching plane shows the direction of feature creation and the direction in which you will view the sketching plane. (T/F)

6. The __________ button is selected by default when the Extrude dashboard is displayed.

7. __________ button in the Revolve dashboard is used to create a thin revolve model.

8. If the material has to be added to the part then the arrow by default points in the direction __________.

9. After you exit the sketcher environment, the __________ appears on the model to dynamically modify the extrusion depth or the angle of revolution.

10. The **Revolve Tool** button revolves the sketched section about a __________ to the specified angle.

Review Questions

Answer the following questions:

1. By default a sketch is revolved about a center line in the ____________ direction.

   (a) Clockwise (b) Anti-clockwise
   (c) Right (d) None of the above

2. Which of the following menus in the menu bar contains the **Extrude** option?

   (a) **Edit** (b) **View**
   (c) **Insert** (d) **File**

3. How many datum planes are available when you enter the Part mode?

   (a) 4 (b) 3
   (c) 2 (d) None

4. Which of the following toolbar is used to turn off the grid display in the sketcher environment?

   (a) **View** (b) **Sketcher**
   (c) **Model Display** (d) **File**

5. Which of the following keyboard and mouse button combination is used to change the orientation of the model to the default orientation?

   (a) CTRL+D (b) CTRL+left mouse button
   (c) CTRL+right mouse button (d) None of the above

6. The features created using the Blind option do not have a dimension associated with them, and hence they cannot be modified by changing the dimension value. (T/F)

7. It is recommended not to use the default datum planes to create the base feature. (T/F)

8. The section drawn for revolving as a solid should be a closed loop. (T/F)

9. The revolved section should have a center line. (T/F)

10. A revolved section can be drawn on both the sides of the center line. (T/F)

Exercises

Exercise 1

Create the model shown in Figure 3-73. The dimensions of the model are shown in Figure 3-74.

(Expected time: 20 min)

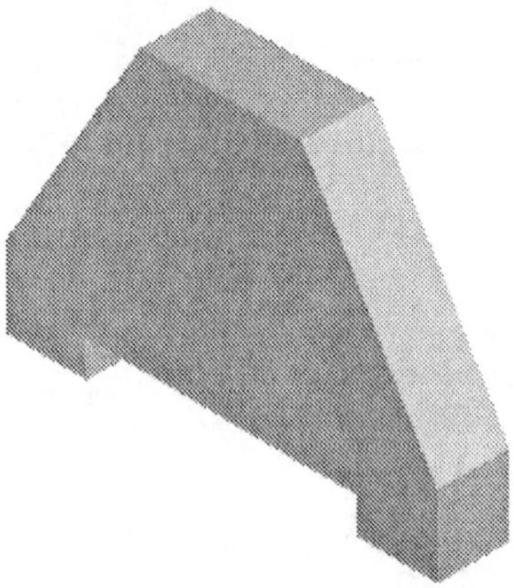

Figure 3-73 *Isometric view of the model*

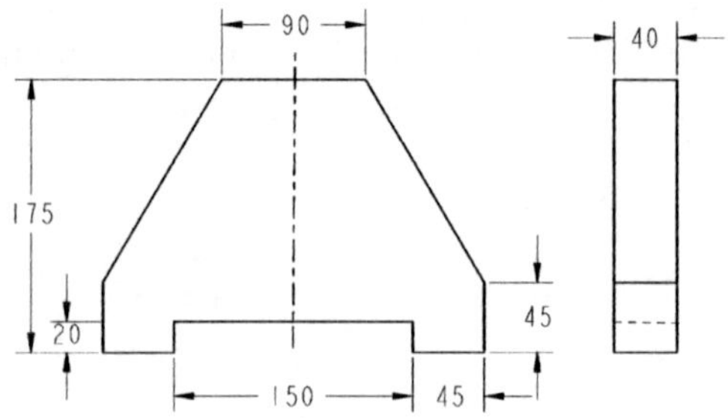

Figure 3-74 *Front and right-side views of the model*

Exercise 2

Create the model shown in Figure 3-75. The dimensions of the model are shown in Figure 3-76.

(Expected time: 30 min)

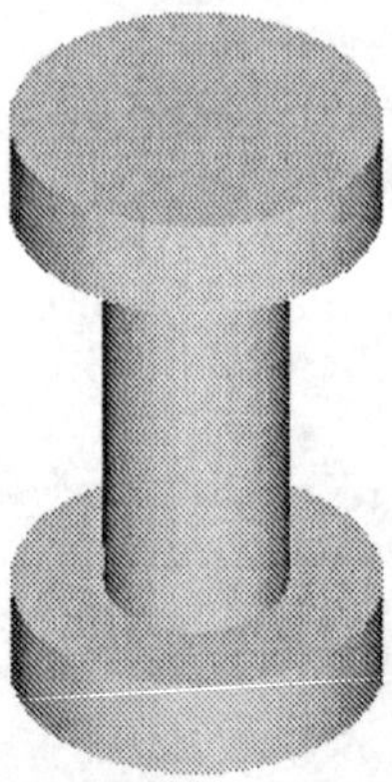

Figure 3-75 *Isometric view of the model*

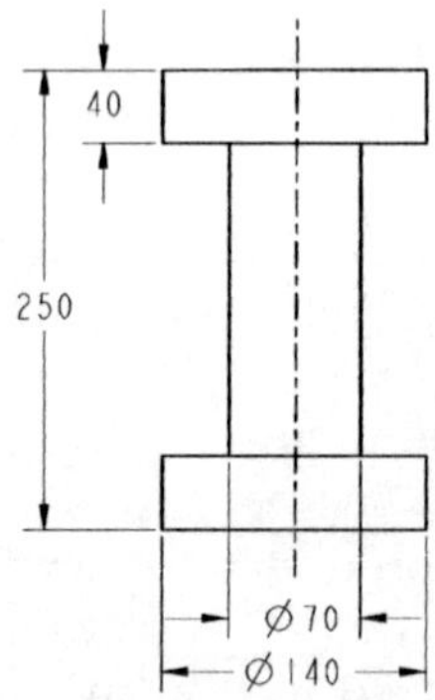

Figure 3-76 *Front view of the model*

Exercise 3

Create the model shown in Figure 3-77. The dimensions of the model are shown in Figure 3-78.

(Expected time: 30 min)

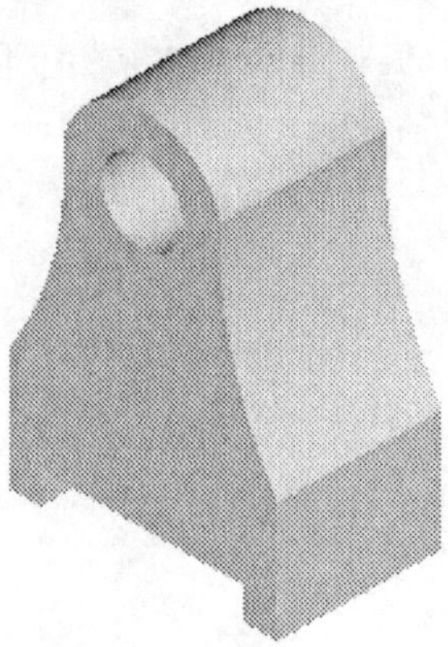

Figure 3-77 Isometric view of the model

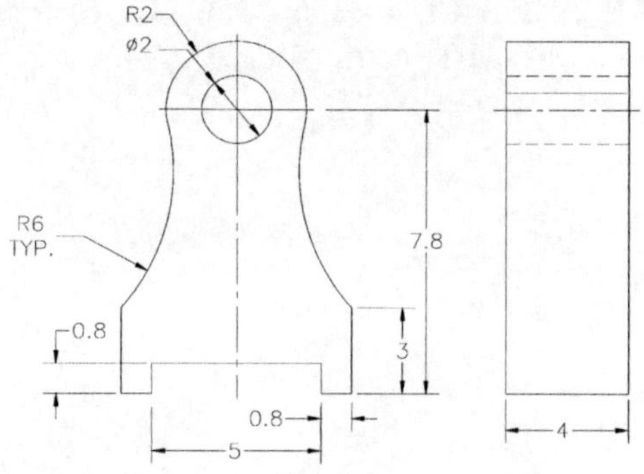

Figure 3-78 Front view of the model

Exercise 4

Create the model shown in Figure 3-79. The dimensions of the model are shown in Figure 3-80.

(Expected time: 20 min)

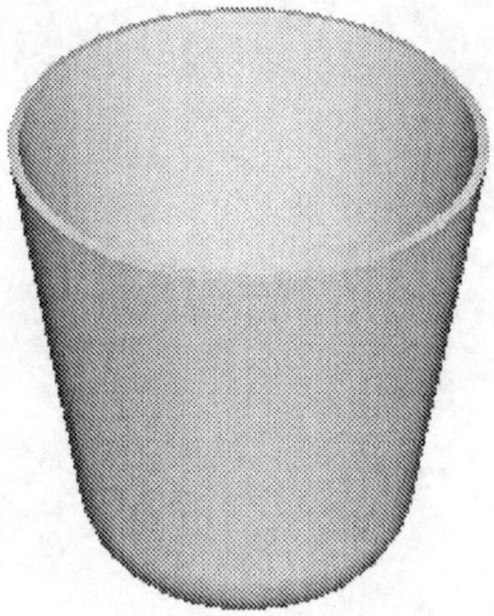

Figure 3-79 Isometric view of the model

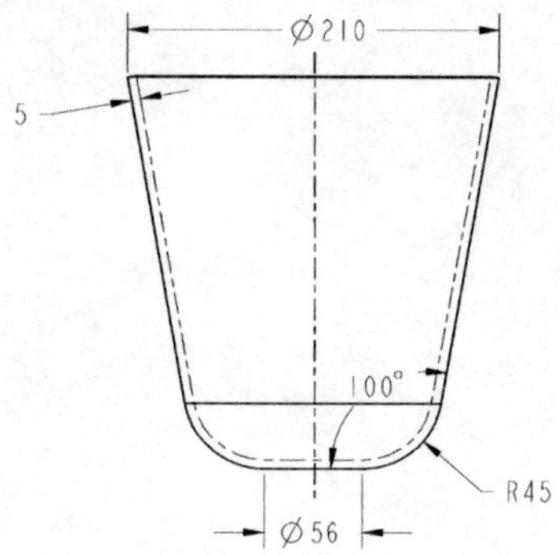

Figure 3-80 Front view of the model

Answers to the Self-Evaluation Test

1 - F, **2** - T, **3** - T, **4** - T, **5** - T, **6** - **Extrude as solid**, **7** - **Thicken Sketch**, **8** - out of the screen, **9** - handles, **10** - center line.

Chapter 4

Datums

Learning Objectives

After completing this chapter you will be able to:

- *Understand the three default datum planes.*
- *Understand selection methods in Pro/ENGINEER Wildfire.*
- *Create the datum planes using different constraints available.*
- *Create datum planes on-the-fly.*
- *Create datum axes using the different constraints available.*
- *Create the datum points.*
- *Create extrude and revolve cuts.*

DATUMS

Datums are imaginary features with no mass or volume and are available to help you in creating models. They act as reference for sketching a feature, orienting the model, assembling components, and so on. Remember that the datums play a very important role in creating complex models in Pro/ENGINEER and therefore you must have a good understanding of datums. Datums are considered to be features but not model geometry. In Pro/ENGINEER, datums exist as datum plane, datum curve, datum point, datum coordinate system, datum graph, better to mention all the datums.

Default Datum Planes

When you enter the Part mode or the Assembly mode, the three datum planes are by default displayed on the graphics window. These datum planes are mutually perpendicular to each other. These are known as the default datum planes. The only difference between the default datum planes of the Part mode and those of the Assembly mode lies in the names of the datum planes.

The default datum planes in the Part mode are named **FRONT**, **TOP**, and **RIGHT**. In case of the Assembly mode, the default datum planes are named **ASM_FRONT**, **ASM_TOP**, and **ASM_RIGHT**. However, the names of the default datum planes can be changed as required.

To change the names, choose **Edit > Setup**. The **PART SETUP** menu is displayed. Choose the **Name** option from this menu. You are prompted to select a feature to change the name. Select the datum plane you want to rename. When the **Message Input Window** appears, enter the desired name in this window.

The **SETUP PLANE** menu can also be used to set the material, units, and other parameters related to the model.

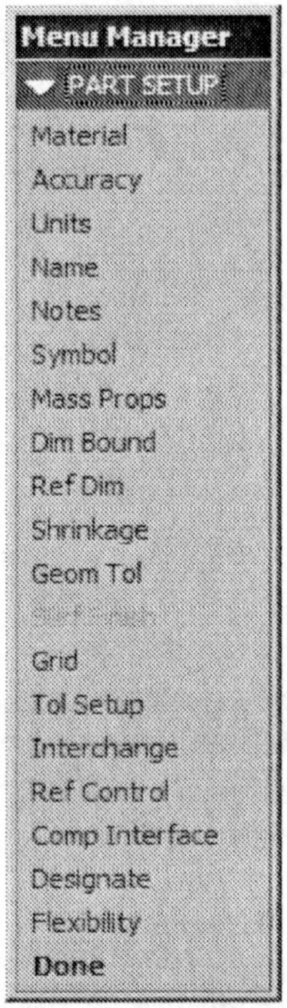

***Figure 4-1** PART SETUP menu*

Note

*The default units of a part are **Inch lbm Second**. The length is measured in inches, mass in lbm, time in seconds, and temperature in Fahrenheit. There are other system of units that are available and you can change the units as desired. However, the units are set if the part has to undergo some type of analysis or if you need to generate its drawing views and plot the drawings to a scale. The units do not affect if you need to create a solid model only.*

NEED FOR DATUMS IN MODELING

Generally, most of the engineering components or designs consist of more than one feature. First the base feature of the model is created and then the other features of the model are created. Since all the features of a model cannot be drawn on a single plane, therefore, to draw the rest of the features sometimes additional planes have to be created or selected. Also, most of the times, the three default datum planes are not enough to create a complex model having many features. For example, Figure 4-2 shows a simple model that consists of two features that require two different planes.

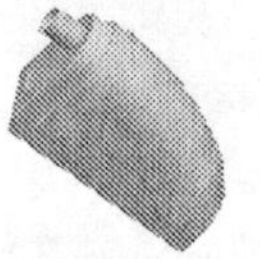

Tip: *Whenever you come across any solid model, first try to visualize the number of features in that model and then decide which feature in the model you consider as the base feature.*

In Figure 4-2 any of the two features that are defined on two different planes can be considered as the base feature. However, in this discussion, the base feature that is decided to be created is shown in Figure 4-3. After creating the base feature, the next feature has to be created. For the next feature, a sketching plane has to be defined. Therefore, an additional plane will be created on which you can draw the sketch for the second feature.

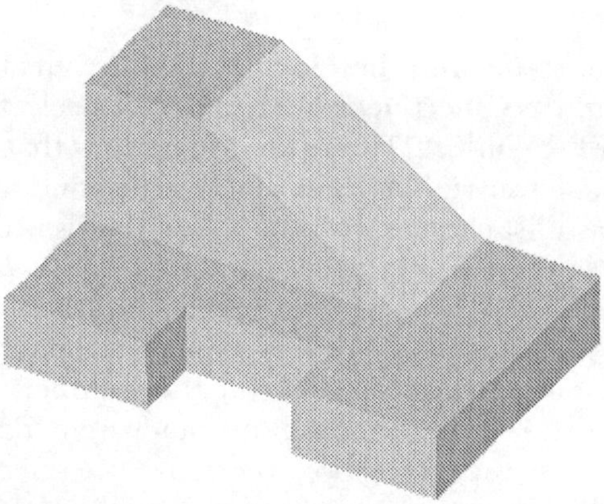

Figure 4-2 *Model having two extruded features*

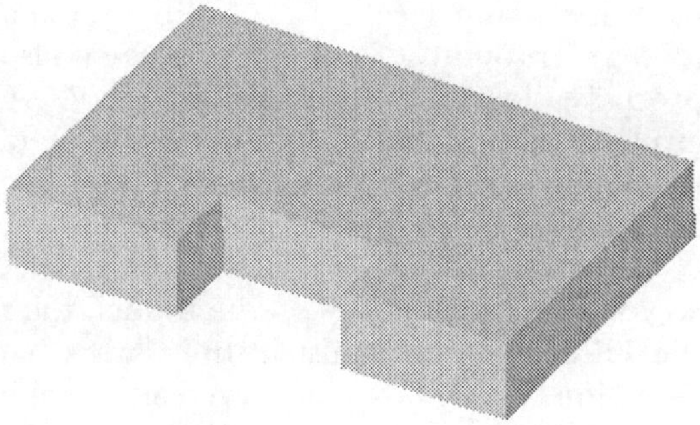

Figure 4-3 *Base feature of the model*

As shown in Figure 4-4, the plane that is used to create the base feature is highlighted by a mesh. To create the second feature, a new plane is created which is shown in Figure 4-5. The sketch of the second feature is drawn on this plane and this is the reason, its front face is coplanar with the datum plane.

To create various types of datums, you need to select the references on the model. Based on the selection you make on the model, Pro/ENGINEER applies constraints to create the datum. There are some selection methods that are introduced with Pro/ENGINEER Wildfire.

Selection Method in Pro/ENGINEER Wildfire

The selection method in Pro/ENGINEER Wildfire is divided into two types:

1. Selection
2. Collection

Selection

In Pro/ENGINEER Wildfire, selection refers to the action when you first select the entities

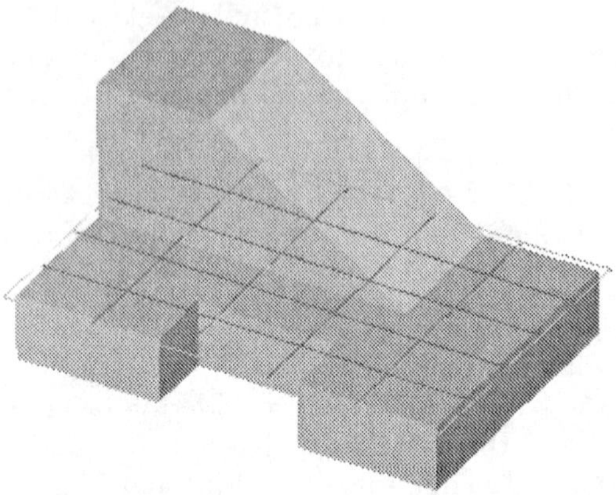

Figure 4-4 Plane selected for the base feature

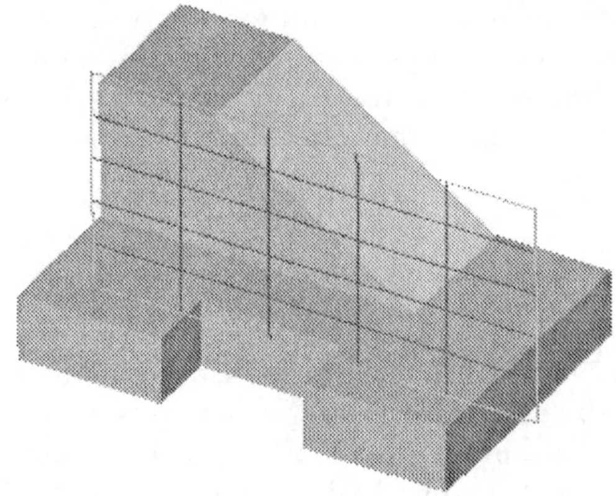

Figure 4-5 Plane selected for the second feature

like edge, plane, face, axis, coordinate system, and so on, and then invoke a feature creation tool. This property of a design package is also known as Object Action. To make your selection process easier, filters are available in Pro/ENGINEER Wildfire. Filters are the options that are available in the **Filter** drop-down list located at the lower right corner of the main window in the **Status** bar. The options in the **Filter** drop-down list change with the mode that is active. The options in this drop-down list when you are in the Part mode are shown in Figure 4-6.

As you select the entities on the model, the number of selected entities appear to the left of the **Filter** drop-down list in the **Status** bar. When you double-click on the number, the **Selections** dialog box is displayed as shown in Figure 4-7. All the selections that you made on the model are available in this dialog box. You can use the dialog box to remove the selected entities. These selections appear highlighted in red color on the model. To remove the selection on the model, press the CTRL key and then select the entity in red.

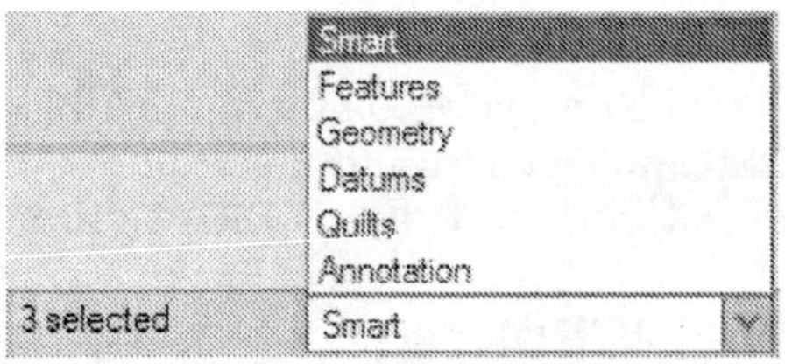

Figure 4-6 ***Filter*** *drop-down list with filters*

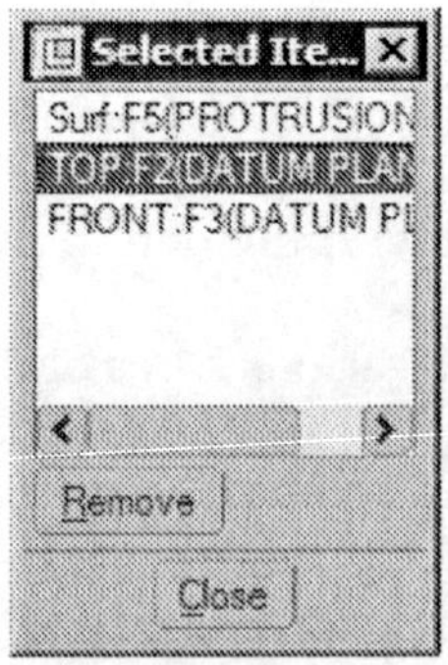

Figure 4-7 ***Selected Items*** *dialog box*

Collection

In Pro/ENGINEER Wildfire, collection refers to the action when you have invoked a feature creation tool and then select the entities to collect the references in order to create that feature. To make your selection process easy, filters can be used. The filters available in the **Filter** drop-down list depend on the feature creation tool that you have invoked.

DATUM OPTIONS

After discussing the default datum planes, which are the first feature in the Part mode, you must know the various features created using the datum options. Datums are also considered as features that have no geometry. Figure 4-8 shows the **Datum** toolbar. Figure 4-9 shows the method of invoking various types of datum features from the menu bar.

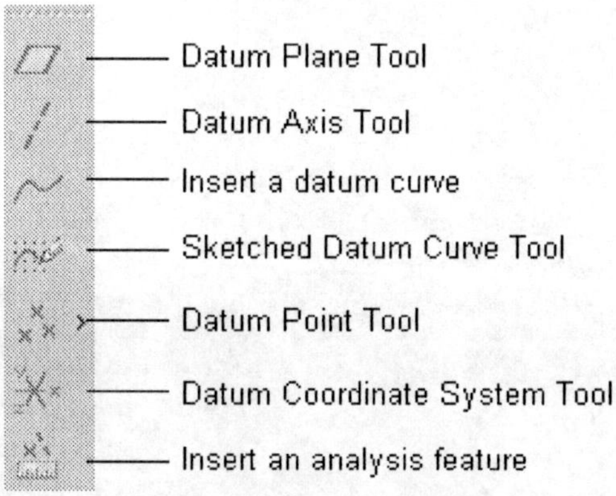

***Figure 4-8 Datum** toolbar*

Datum Planes

You can create datum planes other than the three default datum planes by using the menu bar or the **Datum** toolbar. The datum planes can be created at anytime when required. The display of the datum planes can be turned on or off by using the **Datum planes on/off** button from the **Datum Display** toolbar. Before discussing the procedure to create datum planes by using the various options, it is important for you to understand the use of datum planes. Some of the uses of datum planes in Pro/ENGINEER Wildfire are listed below.

1. Datum planes are used as sketching planes to create sketches for the various features of a model.

2. Datum planes are used as reference planes for sketching.

3. Datum planes are used as references for placing holes and for assembly.

4. Datum planes are used as a reference for mirroring features, copying features, for creating a cross-section, as well as for orientation of references.

Pro/ENGINEER provides you with various constraints to create additional datum planes. Additional datum planes are created with the help of constraints and the filters available in the **Filter** drop-down list. Datums can even be created while you are in the sketcher environment. When you choose **Insert > Model Datum > Plane** from the menu bar or

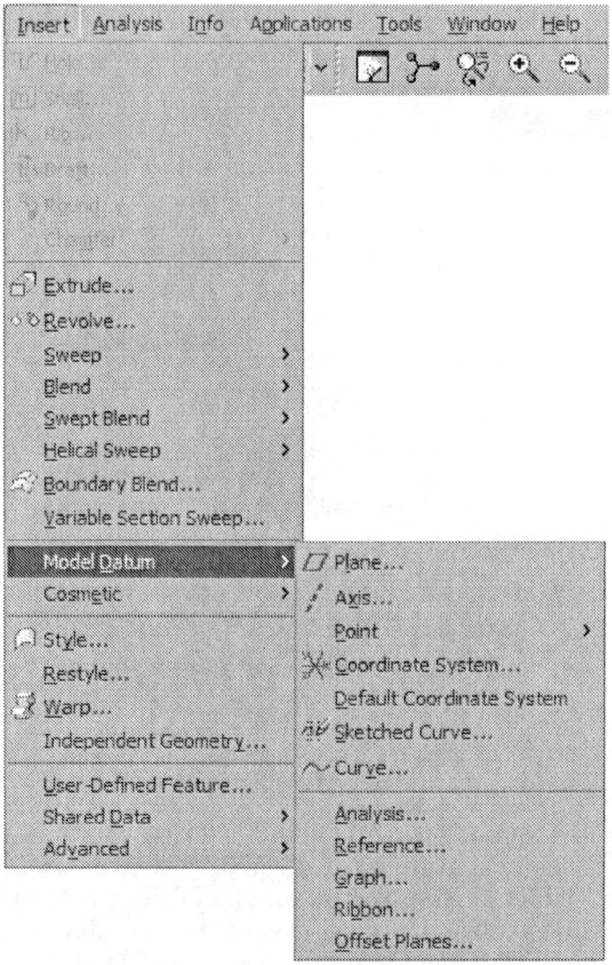

***Figure 4-9** Invoking the datum options from the **Insert** menu in the menu bar*

Datum Plane Tool button from the **Datum** toolbar, the **DATUM PLANE** dialog box is displayed as shown in Figure 4-10. This dialog box is new and is introduced in Pro/ENGINEER Wildfire. Separate buttons for creating datum axis, datum curve, datum point, datum coordinate system, and so on are available in the **Datum** toolbar.

Tip: *Generally, for the base feature creation, the three default datum planes are used and as the part becomes complex or in other words as the number of features increases, the need for additional datum planes arises.*

Figure 4-10 shows the various options available in the **DATUM PLANE** dialog box to create datum planes. The options available in the various tabs of the **DATUM PLANE** dialog box are discussed next.

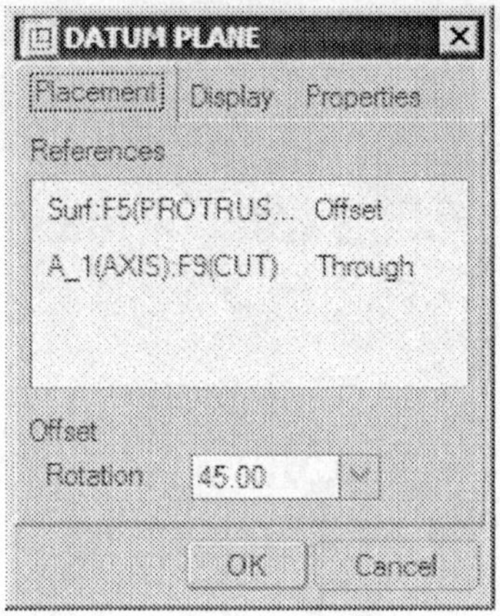

*Figure 4-10 DATUM PLANE dialog box with **Placement** tab*

Placement tab

The **Placement** tab in the **DATUM PLANE** dialog box is selected by default. Under this tab, the **References** area displays the references that are selected to create the datum plane. The constraints are displayed on the right of the references. These constraints are applied automatically based on the reference you select. To change a constraint, select the constraint in the dialog box; you will notice that a drop-down list appears to the right of the name of the constraint. From this drop-down list, select a different constraint. Generally, this drop-down list is available only for the first constraint in the dialog box. The various constraints that are used to create a datum plane are discussed later in the chapter.

The **Offset** area is available only when the **Offset** constraint is used to create a datum plane. The **Rotation** edit box is used to enter the rotation angle of the new datum plane. The angle can also be set dynamically on the model by using the drag handle displayed on the model.

Display tab

The **Display** tab of the **DATUM PLANE** dialog box is shown in Figure 4-11. The **Flip** button in this dialog box is used to change the normal direction of the datum plane being created. Every datum plane has two sides, one is brown and the other is black. By default, a sketch is protruded toward the brown side. Therefore, the **Flip** button is used to specify the side of the datum plane. The arrow points in the direction of brown side. This means that the yellow arrow is used to specify the brown side of the datum plane.

The **Adjust Outline** check box when selected, allows you to set the size for the datum plane.

Properties tab

The **Properties** tab of the **DATUM PLANE** dialog box is shown in Figure 4-12. This tab when selected, allows you to name the datum plane you are creating. By default, the first datum plane you create is named DTM1 and then every datum plane created is successively named.

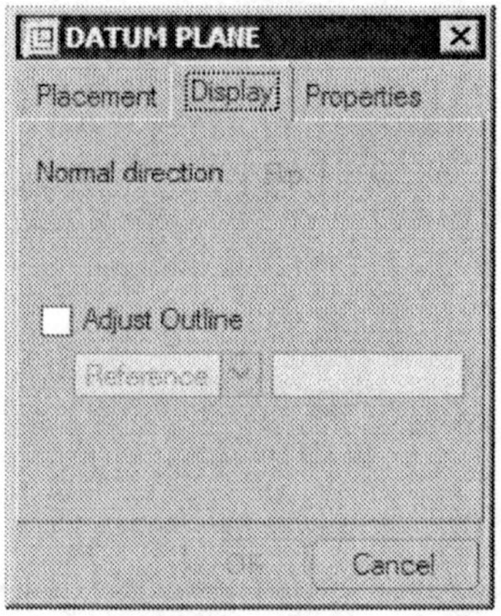

***Figure 4-11 DATUM PLANE** dialog box with **Display** tab*

***Figure 4-12 DATUM PLANE** dialog box with **Properties** tab*

Creating Datum Planes

When you create a datum plane, the constraints are applied automatically based on the selection you make on the model. These constraints appear in the **DATUM PLANE** dialog box. Sometimes, while applying constraints to define a datum plane, a single constraint is enough to define the datum plane and sometimes you need more than one constraint to define a datum plane. When only one constraint is used to create a datum plane, it is called standalone constraint. The standalone constraints are sufficient by themselves to constrain a datum plane definition. The constraints that are used to create a datum plane are discussed next.

Through constraint

The **Through** constraint is used to create a datum plane through any specified axis, edge, curve, point/vertex, plane, cylinder, or coordinate system. This constraint can be used in combination with other constraints that are discussed next.

However, the **Through** constraint can also be used as a standalone constraint when you select a plane. Figure 4-13 shows the datum plane constraint combinations using the **Through** constraint. Datum planes can be created using any of the combinations shown in the figure. The possible combinations of datum plane creation are referred to as **Yes** and the combinations that are not possible are referred to as **No** in the figure.

DATUM PLANE CONSTRAINT COMBINATIONS (USING THROUGH CONSTRAINT)		Through			Normal	Parallel	Angle	Tangent	Standalone Constraints
		Axis/Edge/ Curve	Point/ Vertex	Cylinder	Plane	Plane	Plane	Cylinder	
Through	Axis/Edge/ Curve	Yes	Yes	Yes	Yes	Yes	Yes	Yes	No
	Point/ Vertex	Yes	Yes	Yes	Yes	Yes	No	Yes	No
	Plane	No	No	No	No	No	No	No	Yes
	Cylinder	Yes	Yes	Yes	Yes	Yes	Yes	Yes	No

Figure 4-13 *Datum plane constraint combinations using the **Through** constraint*

While reading the table shown in Figure 4-13, first preference is given to the text written in the first column and then the text in the first row should be read. For example, if you want to make a datum plane that is passing through a cylinder and normal to a plane then look for **Through** in the first column and then for **Cylinder** in the second column. Now, look for **Normal** in the first row and for **Plane** in the second row. After finding both the combinations trace them in the respective column and row till they intersect. You will find **Yes**. This suggests that the creation of a datum plane that passes through a cylinder and is normal to a plane is possible. While reading the table shown in Figure 4-13, remember that the constraints that are not standalone have to be applied in pairs.

The options in the drop-down list shown in Figure 4-14 are used to make selection on the model. This drop-down list is located at the bottom right corner of the main window. The options in this drop-down list change depending on the operation being performed. For example, when you are creating a datum plane, the options shown in Figure 4-14 are available. Once you exit the datum plane creation, the options in this drop-down list change.

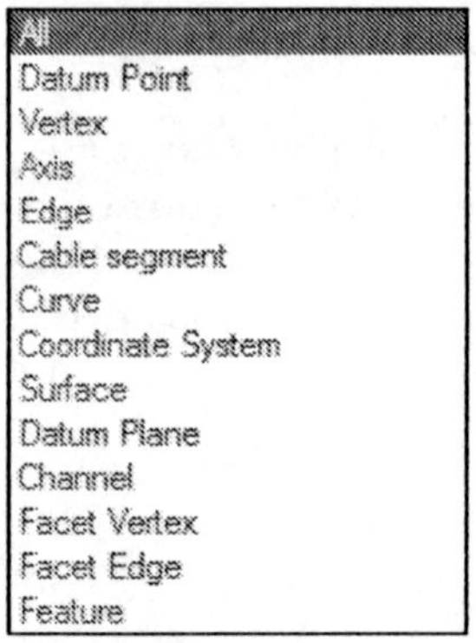

Figure 4-14 ***Filter*** *drop-down list with options to make selection*

Figure 4-15 shows that the cylindrical surface and the default datum planes are used to create a datum plane at an angle to the selected default datum plane and passing through the center of the cylindrical surface as shown in Figure 4-16. When you select the cylindrical surface, it is highlighted in red, indicating that it is selected. To make the second selection, select the **Plane** filter from the **Filter** drop-down list. Now, press the CTRL button and then select the datum plane. This method of selecting by pressing the CTRL button is known as collection. Select the **Offset** option from the drop-down list on the right of the reference in the **DATUM PLANE** dialog box and enter the angle value in the **Rotation** dimension box. The datum plane is created.

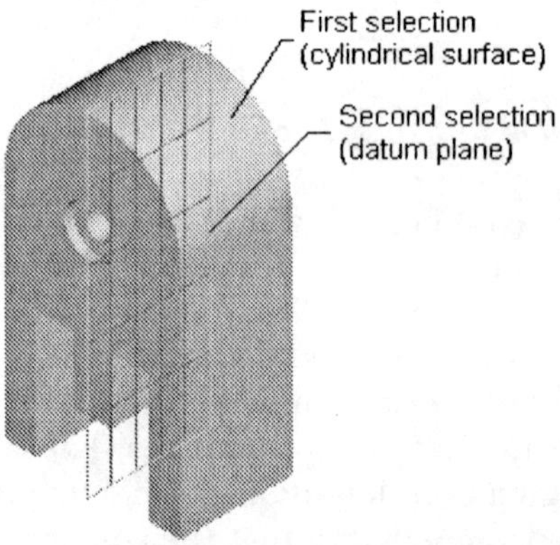

Figure 4-15 *Selecting a cylindrical surface and a default datum plane to create a datum plane*

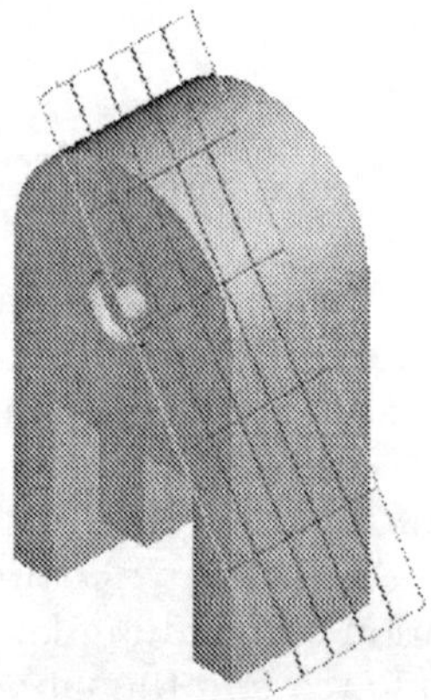

Figure 4-16 *Resultant datum plane passing through the center of a cylinder and at an angle*

Note

The first reference to constrain a datum plane is selected using the left mouse button. The second reference is selected using the CTRL+left mouse button.

Make sure you change the option in the ***Filter*** *drop-down list when you invoke any other tool.*

Normal constraint

The **Normal** constraint is used to create a datum plane normal to any specified axis, edge, curve, or plane. This constraint is used in combination with other constraints. The **Normal** constraint cannot be used as a standalone constraint. Figure 4-17 shows the datum plane constraint combinations using the **Normal** constraint.

DATUM PLANE CONSTRAINT COMBINATIONS (USING NORMAL CONSTRAINT)		Through			Normal	Parallel	Angle	Tangent	Standalone Constraints
		Axis/Edge/ Curve	Point/ Vertex	Cylinder	Plane	Plane	Plane	Cylinder	
Normal	Axis/Edge/ Curve	Yes	Yes	Yes	No	No	No	No	No
	Plane	Yes	Yes	Yes	Yes	Yes	No	Yes	No

Figure 4-17 *Datum plane constraint combinations using the* ***Normal*** *constraint*

The possible combinations of datum plane creation are referred to as **Yes** and the combinations that are not possible are referred to as **No**.

Figure 4-18 shows a planar face and a cylindrical face of a solid model. The planar face is selected as the normal surface and the cylindrical face is selected to be tangent to the datum plane. The following steps explain the procedure to create this datum plane:

1. Invoke the **DATUM PLANE** dialog box. Select the first reference shown in Figure 4-18. The **Offset** constraint is automatically applied and is displayed in the **References** collector.

2. Left-click on the **Offset** constraint in the collector; a drop-down list appears. Select the **Normal** constraint from this drop-down list.

3. Now, use CTRL+left mouse button to select the second reference shown in Figure 4-18. The **Through** constraint is displayed in the **References** collector.

4. Change this constraint to **Tangent** from the drop-down list that appears when you click on the constraint in the **References** collector. The datum plane that is created is shown in Figure 4-19.

5. Choose the **OK** button from the **DATUM PLANE** dialog box.

Parallel constraint

The **Parallel** constraint is used to create a datum plane parallel to any specified datum plane

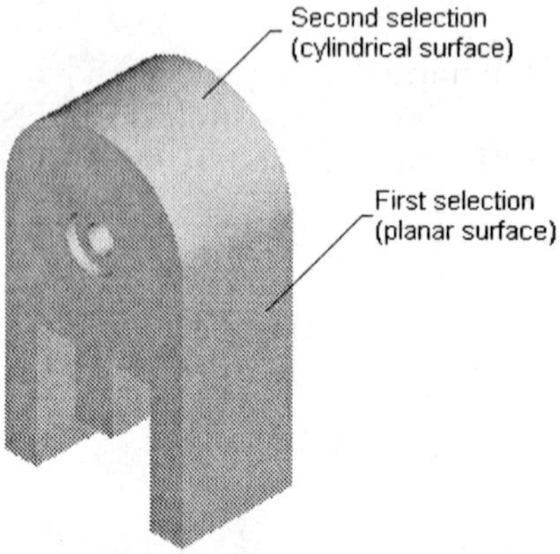

Figure 4-18 Selecting a planar surface and a cylindrical surface to create a datum plane

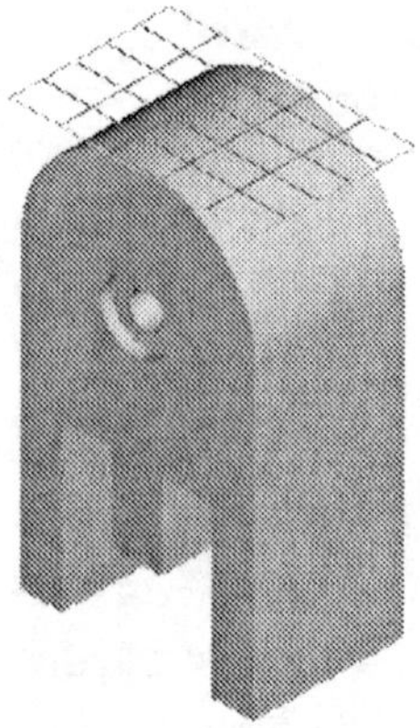

Figure 4-19 Resultant datum plane

or face. This option is used in combination with other constraints. The **Parallel** constraint cannot be used as a standalone constraint. Figure 4-20 shows various datum plane constraint combinations using the **Parallel** constraint. The possible combinations of datum plane creation are referred to as **Yes** and the combinations that are not possible are referred to as **No** in the figure.

DATUM PLANE CONSTRAINT COMBINATIONS (USING PARALLEL CONSTRAINT)		Through			Normal	Parallel	Angle	Tangent	Standalone Constraints
		Axis/Edge/ Curve	Point/ Vertex	Cylinder	Plane	Plane	Plane	Cylinder	
Parallel	**Plane**	**Yes**	**Yes**	**Yes**	**No**	**No**	**No**	**Yes**	**No**

***Figure 4-20** Datum plane constraint combinations using the **Parallel** constraint*

Figure 4-21 shows the selection of a default datum plane and an axis to create a datum plane. The resultant datum plane is parallel to the selected datum plane and passes through the axis as shown in Figure 4-22. The following steps explain the procedure to create this datum plane:

1. Invoke the **DATUM PLANE** dialog box. Select the first reference shown in Figure 4-21. The **Offset** constraint is automatically applied and is displayed in the **References** collector.

2. Left-click on the **Offset** constraint in the collector; a drop-down list appears. Select the **Parallel** constraint from this drop-down list.

3. Now, use CTRL+left mouse button to select the second reference shown in Figure 4-21. The **Through** constraint is displayed in the **References** collector.

4. Choose the **OK** button from the **DATUM PLANE** dialog box. The datum plane that is created is shown in Figure 4-22.

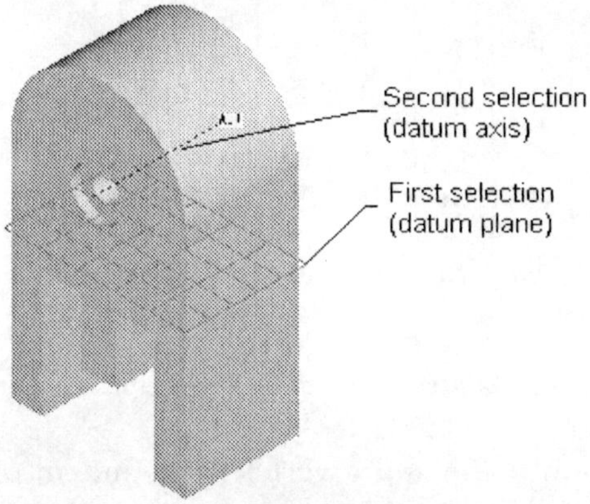

Figure 4-21 Selecting a datum plane and an axis to create a datum plane

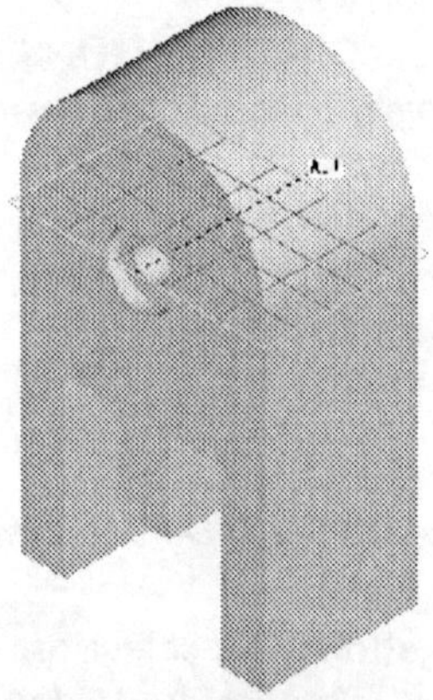

Figure 4-22 Resultant datum plane

Offset constraint

The **Offset** constraint is used to create a datum plane at an offset distance to any specified plane or coordinate system. This option is used in combination with other constraints. However, the **Offset** constraint can be used as a standalone constraint when you select a plane to offset from. This constraint is used to create a datum plane at some specified parameters.

When the **Offset** constraint is applied, the **Translation** dimension box appears. This dimension box list is used to specify an offset distance. In the case of angular planes the **Translation** dimension box changes to the **Rotation** dimension box in which you need to specify the angle. An arrow appears on the model that shows the positive direction of the offset distance or angle.

Figure 4-23 shows various datum plane constraint combinations using the **Offset** constraint. The possible combinations of datum plane creation are referred to as **Yes** and the combinations that are not possible are referred to as **No** in the figure.

DATUM PLANE CONSTRAINT COMBINATIONS (USING OFFSET CONSTRAINT)		Through			Normal	Parallel	Angle	Tangent	Standalone Constraints
		Axis/Edge/ Curve	Point/ Vertex	Cylinder	Plane	Plane	Plane	Cylinder	
Offset	Plane	No	No	No	No	No	No	No	Yes
	Coord System	No	No	No	No	No	No	No	Yes

Figure 4-23 *Datum plane constraint combinations using the* ***Offset*** *constraint*

Figure 4-24 shows the selection of a default datum plane and a vertex to define an offset datum plane. The resultant datum plane is at an offset to the selected datum plane and passes through the vertex as shown in Figure 4-25. The following steps explain the procedure to create this datum plane:

1. Invoke the **DATUM PLANE** dialog box. Select the first reference shown in Figure 4-24. The **Offset** constraint is automatically applied and is displayed in the **References** collector. The **Translation** dimension box appears.

2. Now, use CTRL+left mouse button to select the second reference, which is a vertex shown in Figure 4-24. The **Through** constraint is displayed in the **References** collector.

3. Choose the **OK** button from the **DATUM PLANE** dialog box. The datum plane that is created is shown Figure 4-25.

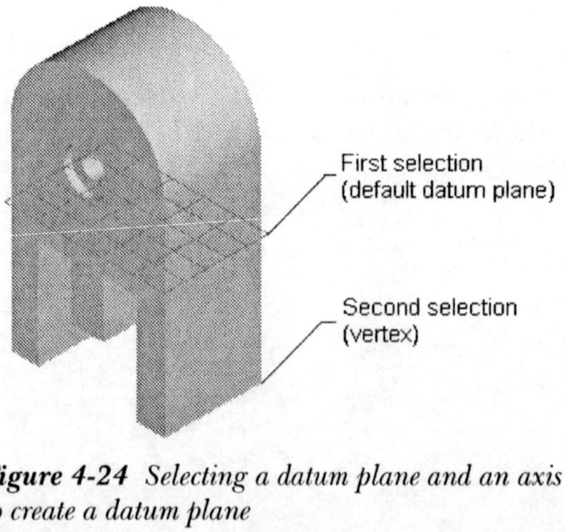

Figure 4-24 *Selecting a datum plane and an axis to create a datum plane*

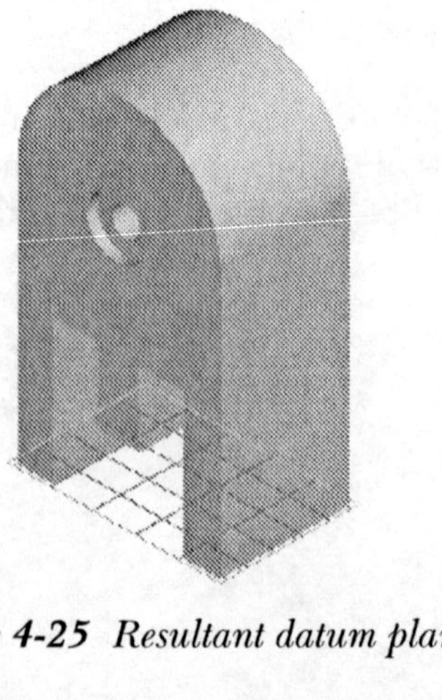

Figure 4-25 *Resultant datum plane*

Note

*The **OK** button in the **DATUM PLANE** dialog box is enabled only when the datum plane you are creating is fully constrained.*

Tangent constraint

The **Tangent** constraint creates datum planes tangent to cylindrical features. This constraint is also used with the other constraints to create various types of datum planes. Figure 4-26 shows the datum plane constraint combinations using the **Tangent** constraint. The possible combinations of datum plane creation are referred to as **Yes** and the combinations that are not possible are referred to as **No** in the figure.

DATUM PLANE CONSTRAINT COMBINATIONS (USING TANGENT CONSTRAINT)		Through			Normal	Parallel	Angle	Tangent	Standalone Constraints
		Axis/Edge/ Curve	Point/ Vertex	Cylinder	Plane	Plane	Plane	Cylinder	
Tangent	Cylinder	Yes	No	No	Yes	Yes	No	No	No

Figure 4-26 *Datum plane constraint combinations using the **Tangent** constraint*

Note

To remove a selected reference, select the reference in the dialog box and press DELETE .

Datum Planes Created On-The-Fly

The term **On-the-Fly** refers to the creation of a datum plane when the system prompts you to select or create a plane. At this step, if you choose the **Datum Plane Tool** button from the **Datum** toolbar, the datum plane created will be called datum plane on-the-fly. When you create a datum plane on-the-fly, the datum plane is neither visible on the graphics window nor is displayed under the default display of the **Model Tree** once the feature is completed. This option is provided by Pro/ENGINEER in order to avoid cluttering of datum planes in a complex model.

You can try making a datum plane on-the-fly by following these steps:

1. Choose the **Extrude Tool** button; the Extrude dashboard is displayed.

2. Choose the **Create a section or redefine the existing section** button; the **Section** dialog box is displayed. Now, you need to select a sketching plane. You can select an existing datum plane, a face as the sketch plane, or you can create a datum plane on-the-fly.

3. Choose the **Datum Plane Tool** button from the **Datum** toolbar. Select the references and constraints to create the datum plane. This datum plane when created is selected as the sketching plane automatically.

When you need to select the references to orient the sketching plane, at this step also you can create a datum plane on-the-fly. Once, the datum plane on-the-fly is created, it cannot be referenced by other features.

The datum plane created on-the-fly does not appear in the **Model Tree** under the default display. But it can be c in the **Model Tree** when you click the plus sign (+) that appears to the left of the feature you have created. The datum plane and the feature that you have created are grouped and appears as a grouped feature in the **Model Tree**. This means that the datum on-the-fly that was created during feature creation belongs to the feature that is created.

In Pro/ENGINEER, creating datum planes on-the-fly is very important and is of much use when you create rotational patterns. Patterns are discussed in Chapter 6.

Datum Axes

Datum axis is an imaginary axis that is created in Pro/ENGINEER to help you in creating a model. Datum axes can be created manually. They are also created automatically when any cylindrical feature is created. The display of a datum axis can be turned on or off by using the **Datum axes on/off** button from the **Datum Display** toolbar. The uses of datum axes are discussed next:

1. Datum axes act as reference for feature creation.

2. They are used in creating a datum plane along with different constraint combinations.

3. They are used in placing features co-axially.

4. They are also used to create radial patterns. You will learn to create patterns in Chapter 6.

Datum axes are named by default in Pro/ENGINEER. The default name of a datum axis is **A_(Number)**, where **Number** represents the number of datum axis. However, the default name of the datum axes can be changed in the same way as that of the datum planes.

When you choose **Insert > Model Datum > Axis** from the menu bar or the **Datum Axis Tool** button from the **Datum** toolbar, the **DATUM AXIS** dialog box appears as shown in Figure 4-27. The **DATUM AXIS** dialog box is introduced with this release of Pro/ENGINEER and the various options in this dialog box are discussed next.

Placement tab

When you invoke the **DATUM AXIS** dialog box, the **Placement** tab is selected by default. Under this tab, the **References** display box displays the references that are selected to create the datum axis. The constraints are displayed on the right of the references. These constraints are applied automatically based on the reference you select. The various constraints that are

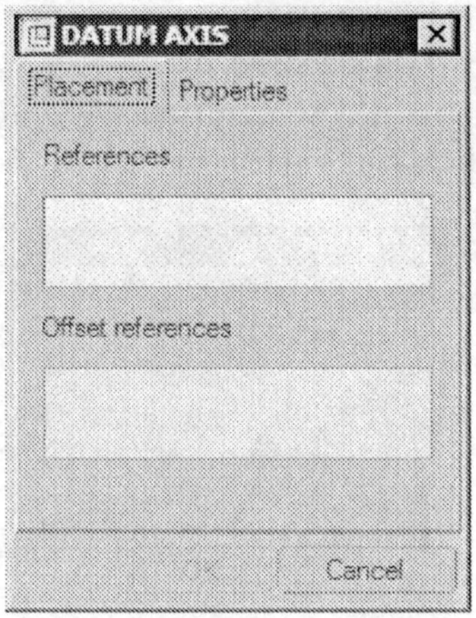

Figure 4-27 ***DATUM AXIS*** *dialog box with the* ***Placement*** *tab*

used to create a datum axis are discussed later in the chapter.

The **Offset References** collector is available only when the **Normal** constraint is used in the **References** collector to define a datum axis. To define the datum axis, you need to select two references. These references can be an edge, datum plane, face, or axis. The references can be specified dynamically on the model by dragging the handles displayed on the datum axis. The handles on the datum axis are displayed only when you select a reference. You can also select the references using the **DATUM AXIS** dialog box.

When you are defining the offset distances, there are three drag handles that are available on the model as shown in Figure 4-28. These handles appear like a white square. The middle handle is used to move the position of the axis you are creating. The other two handles are used to specify the references for dimensioning.

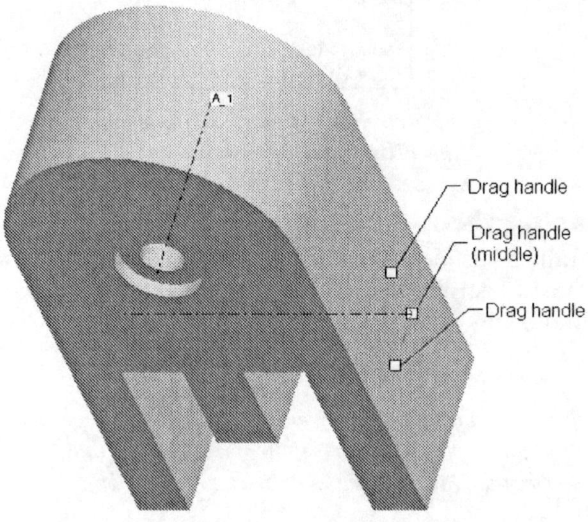

Figure 4-28 *Solid model with the three drag handles*

Properties tab

The **Properties** tab of the **DATUM AXIS** dialog box is shown in Figure 4-29. This tab, when selected, allows you to name the datum axis you are creating. By default, the first datum axis you create is named A_1 and then every datum axis created is successively named.

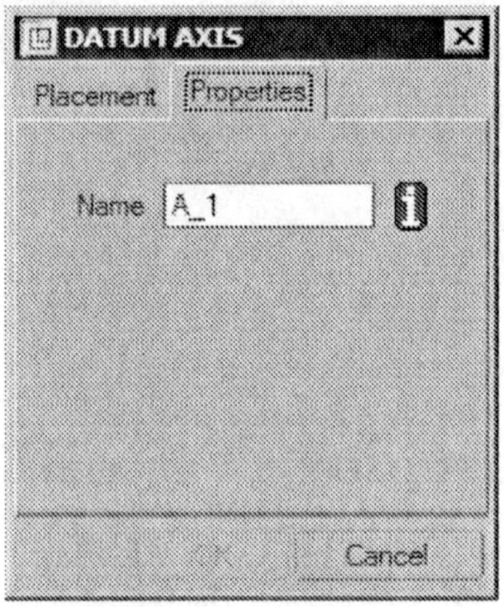

***Figure 4-29** **DATUM AXIS** dialog box with the **Properties** tab*

As mentioned earlier, a datum axis is created using constraints which are applied automatically, and filters available in the **Filter** drop-down list located in the **Status** bar. While creating a datum axis, the filters that are available are shown in Figure 4-30. The constraints that are used to create a datum axis are discussed next.

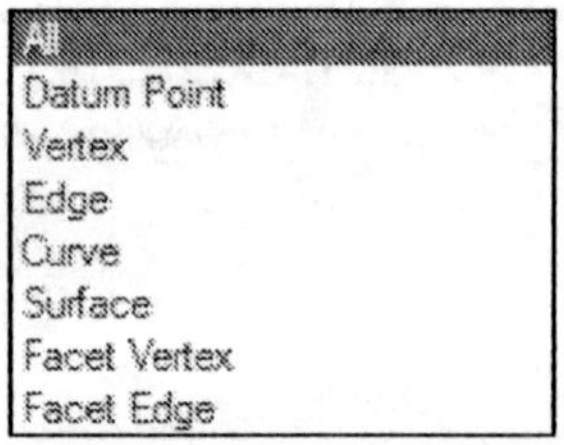

***Figure 4-30** Filters available in the **Filters** drop-down list*

Datum axis passing through an edge

The **Through** constraint is used to create a datum axis passing through any selected edge. The selected edge must be straight for the creation of a datum axis. In Figure 4-31, **A_1** is the datum axis created using this option.

Note

While creating a datum axis, you may need to create datum points in order to create the datum axis passing through them. Therefore, in the various cases that are discussed next you may need to create datum points. Datum points are discussed later in this chapter.

Datum axis normal to a plane

The **Normal** constraint is used to create a datum axis normal to any selected face or datum plane. When you select a face or a datum plane, you are prompted to select two references to place the axis. Notice that three drag handles appear on the model. Use the middle handle to change the location of the axis on the selected face. After you specify its placement location, you need to select two edges, axes, datums, or faces to specify the linear dimension for the placement of the datum axis. Click in the **Offset References** collector. The collector turns yellow in color. When you select the first edge for the placement dimensions of the axis, the offset value is displayed in the **Offset References** collector. You can accept the default dimension or change it to the required value. Similarly, select the second edge for dimensioning by pressing CTRL+left mouse button and enter the dimension value. In Figure 4-32, the preview of the datum axis is shown. In this figure note that the drag handles are displayed. It should be noted that the references for dimensioning can also be selected dynamically using the drag handles.

Note

*When you select a reference for creating the datum axis and you are prompted to select the placement references, you need to click in the **Offset References** collector to make it yellow in color. The yellow color of this area in the **DATUM AXIS** dialog box indicates that now you can select the references to place the axis. Until the **Offset References** collector is made yellow in color, you will not be able to select the placement references.*

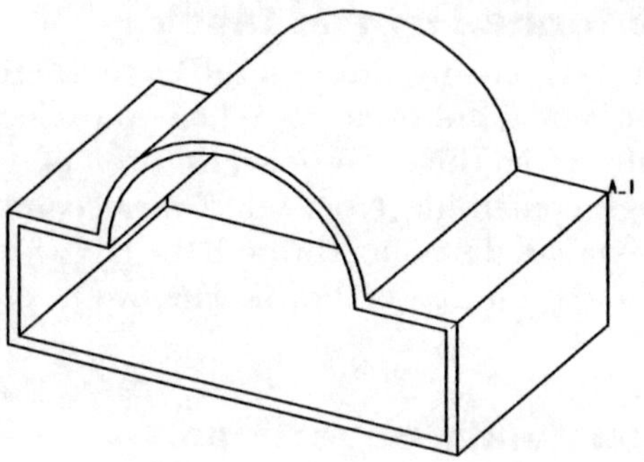

Figure 4-31 *Datum axis created along the edge*

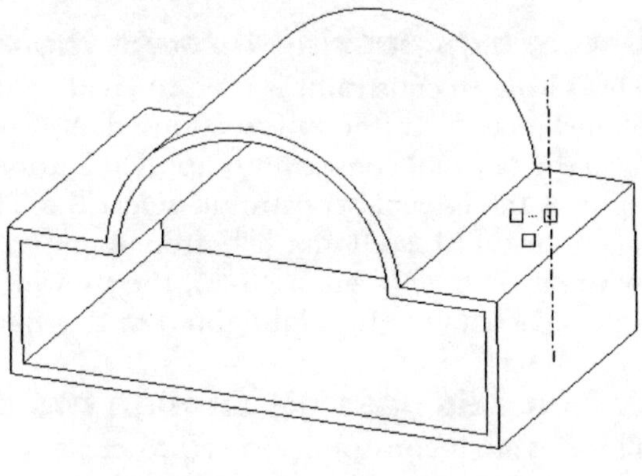

Figure 4-32 *Datum axis created normal to the plane*

Datum axis passing through a datum point and normal to a plane

The **Normal** constraint creates a datum axis passing through a datum point and normal to any face or datum plane. When you select a face or a datum plane to which the datum axis will be normal, you are prompted to select two references to place the axis. Use CTRL+left mouse button to select the datum point to create an axis passing through it. Choose the **OK** button to exit the **DATUM AXIS** dialog box. In Figure 4-33, the preview of the datum axis and the datum point is shown.

Datum axis passing through the center of a round surface

The **Through** constraint is used to create a datum axis passing through a cylindrical or a round surface. Select the round surface through which you need to pass the datum axis. The axis is automatically created around the revolved surface through an imaginary axis. In Figure 4-34, the preview of the datum axis is shown. The cylindrical surface is highlighted in this figure.

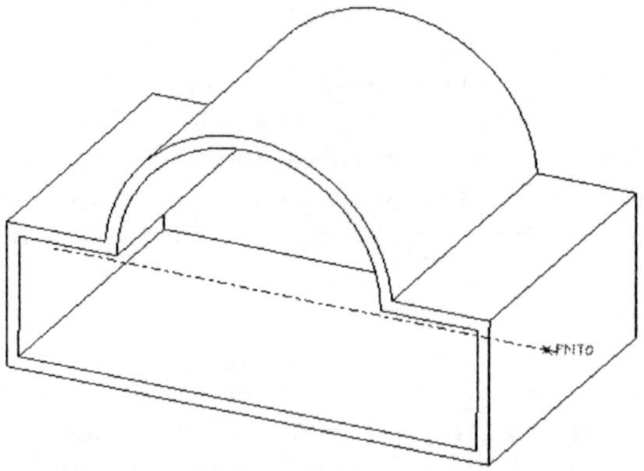

Figure 4-33 Datum axis passing through the datum point and normal to the plane

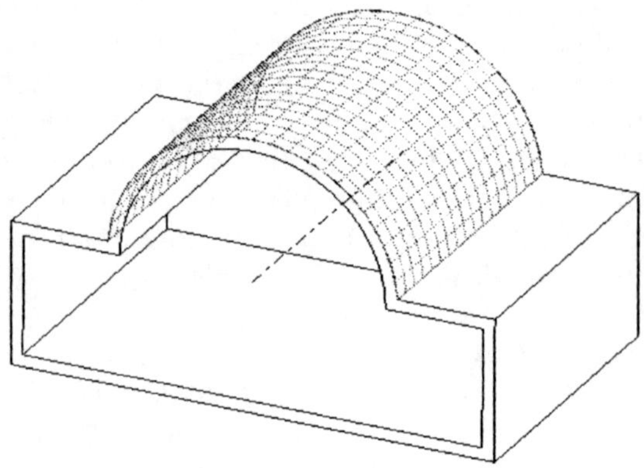

Figure 4-34 Datum axis passing through a cylinder

Datum axis passing through the edge formed by two faces

The **Through** constraint is used to create a datum axis passing through the edge where two planes meet or at the intersection edge of two faces or datum planes. When you select the face, the **Normal** constraint is applied automatically. From the drop-down list present on the right of the **Normal** constraint, select the **Through** constraint. Now, select the second face using the CTRL+left mouse button. A datum axis is created passing through the edge formed by the two faces. In Figure 4-35, the preview of the datum axis is shown. The two faces that were selected are also highlighted in this figure.

Datum axis passing through two datum points or vertices

The **Through** constraint is used to create a datum axis between two datum points or edge vertices. Select the first vertex and then using CTRL+left mouse button, select the second vertex. The datum axis is created along the two selected datum points or edge vertices. In Figure 4-36, the preview of the datum axis is shown with the two vertices highlighted.

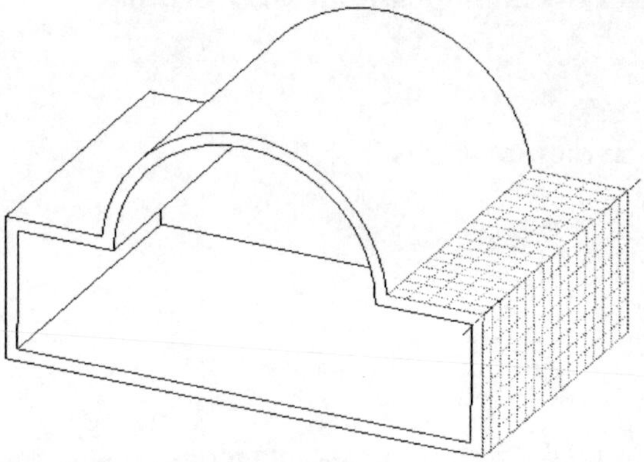

Figure 4-35 Datum axis created on the edge where the two selected planes meet

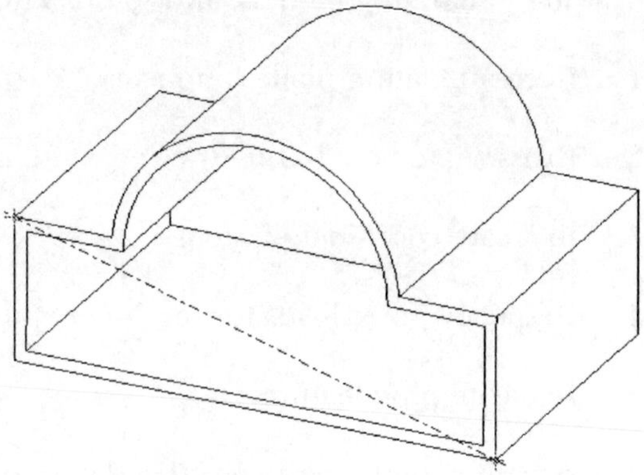

Figure 4-36 Datum axis created between the two selected vertices

Datum axis tangent to a curve and passing through its vertex

The **Tangent** and **Through** constraints create a datum axis tangent to a curve and passing through one of its vertex. Select the edge of the cylindrical surface as the first selection. Make sure you do not select the cylindrical surface. After you select a curve or an edge, use CTRL+left mouse button to select one vertex of the edge. The datum axis is created tangent to the curve and passes through its selected vertex. In Figure 4-37, the preview of the datum axis that is created using the **Tangent** and **Through** constraints is shown. The curve edge is also highlighted in this figure.

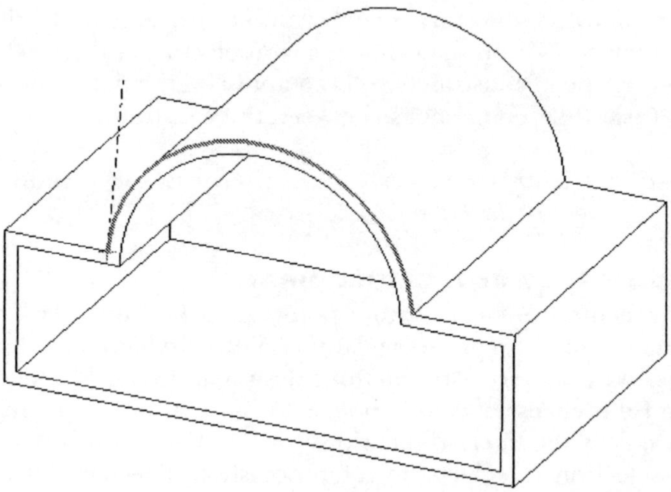

Figure 4-37 Datum axis created tangent to the selected curve

Datum Points

Datum points are imaginary points created in Pro/ENGINEER to aid in creating models, drawings, analyzing models, and so on. The uses of datum points are discussed next.

1. To create datum planes and axes.

2. To associate note in the drawings and attach datum targets.

3. To create coordinate system.

4. To specify point loads for mesh generation.

5. To create pipe features.

The default name associated with a datum point by Pro/ENGINEER is **PTN(Number)** where **Number** indicates the number of datum points created in a particular object. However, you can change the default name associated with the datum points.

When you choose **Insert > Model Datum > Point** from the menu bar or the **Datum Point Tool** button from the **Datum** toolbar, the **DATUM POINT** dialog box appears with various options to create datum points as shown in Figure 4-38. This dialog box can be used to create more than one datum point. The options in the **DATUM POINT** dialog box are explained next.

Placement tab

The **Placement** tab in the **DATUM POINT** dialog box has the options that change with the reference selected. However, the **References** display box shown in Figure 4-38 is always present. This area is used to select the reference where you need to place the datum point. These references are also known as collectors. The constraints available in the drop-down list in the **References** collector depends on the collectors you select from the model. This means that collectors narrow the type of constraints you can apply on the datum point you are creating. Thus, it becomes easier for you to choose between the constraints you can apply.

In the explanation that is discussed next, the various cases of creating datum points are discussed.

Datum point on a face or a datum plane

The **On** constraint is used to create datum points on a face or a datum plane. When you select a face or a datum plane to place the datum point, a yellow colored point is displayed at the selected point on the surface with the three drag handles as shown in Figure 4-39. Next, click in the **Offset References** collector to make it yellow in color. You are prompted to select two references to specify the linear dimensions for the placement of the datum point. Note that the second selection to select the reference should be made by using CTRL+left mouse button. After you select the two planes or edges for the placement dimension of the point, a default value is displayed in the **Offset References** collector. You can accept the default value or change it to the required value. After the datum point is located at the desired position on the face or datum plane, choose the **OK** button to exit the dialog box.

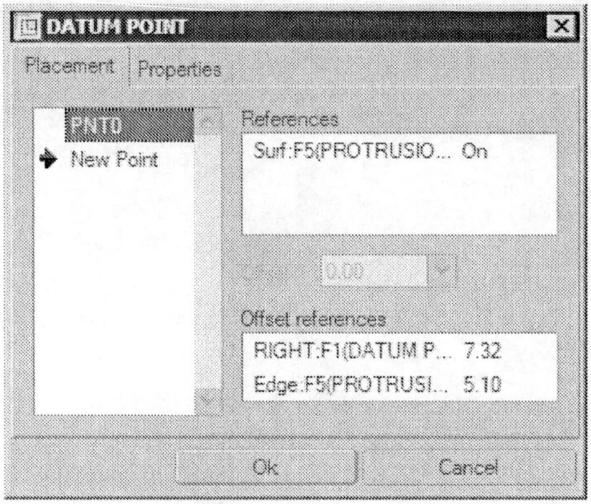

***Figure 4-38 DATUM POINT** dialog box with the **Placement** tab*

Tip: *While creating a datum feature, the collectors selected from the model help you to choose from the available constraints that you can apply to constraint the datum feature. Collectors reduce the range of available constraints, thus reducing the time taken in selecting the appropriate constraint. For example, the lesser the number of options available for a person, the lesser the time he needs to decide among them.*

Datum point offset to a face or a datum plane

The **Offset** constraint creates datum points at an offset distance from a specified face or a datum plane in a specified direction. Select a face or a datum plane from where the offset distance for the placement of the datum point will be measured. The **On** constraint is applied automatically. From the drop-down list in the **References** display box, change the constraint to **Offset**. The model and the **DATUM POINT** dialog box appear as shown in Figures 4-40 and 4-41.

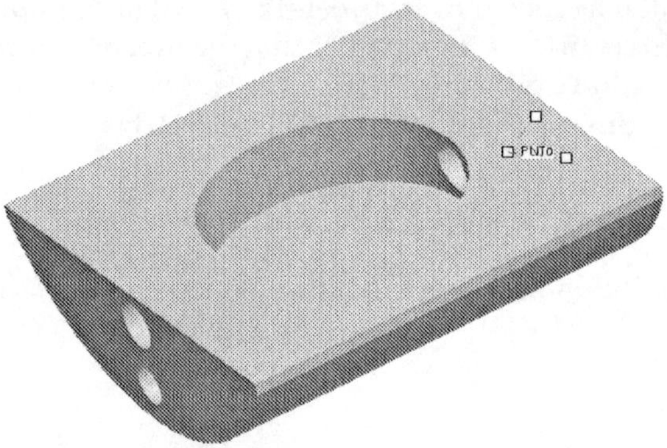

***Figure 4-39** Datum point on a face and the three drag handles*

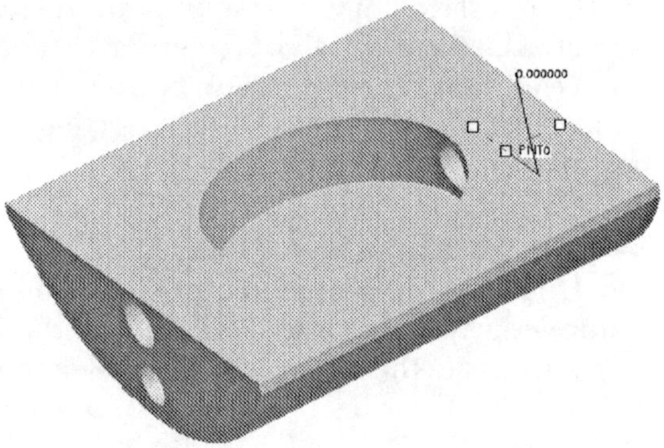

***Figure 4-40** Datum point on the top face with offset **0** and the three drag handles*

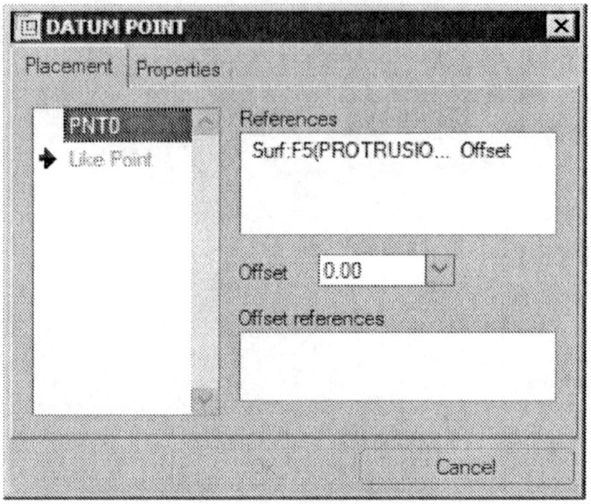

Figure 4-41 ***DATUM POINT*** *dialog box*

In the **Offset** edit box under the **References** collector, enter the offset value. You will be prompted to select the planes or the edges for dimensioning the point. Select two planes or edges for dimensioning and enter the distances from the highlighted references.

You can also specify the offset distance dynamically by using the middle drag handle. The two references for dimensioning the datum point can be specified by using the drag handles.

Note

When you use the ***Offset*** *constraint, the middle drag handle is used to specify the offset distance and when you use the* ***On*** *constraint, it is used to specify the location of the datum point on the selected face.*

Datum point at the intersection of three surfaces

The **On** constraint is used to create a datum point at the intersection of three surfaces. Select the three surfaces and the datum point is created as shown in Figure 4-42. The datum point is created on the vertex that is common to the three surfaces. Remember that the first reference is selected using the left mouse button. The second and third references are selected using CTRL+left mouse button. After selecting the three surfaces, the **DATUM POINT** dialog box appears as shown in Figure 4-43.

Datum point on a vertex

The **On** constraint is used to create a datum point on the vertex of a face, an edge, a surface feature edge, or a datum curve. Invoke the **DATUM POINT** dialog box and then select a vertex to create the datum point. Choose the **OK** button to exit the dialog box.

Note

You can create more than one datum point by using the ***DATUM POINT*** *dialog box.*

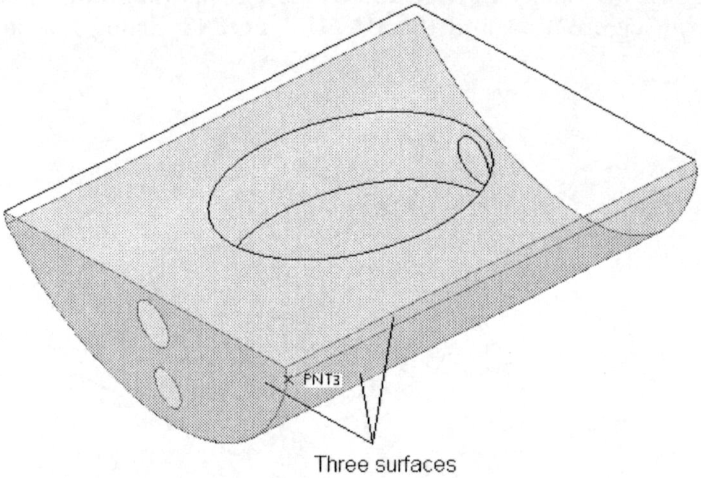

Figure 4-42 *The three highlighted surfaces and the datum point at the intersection of the three surfaces*

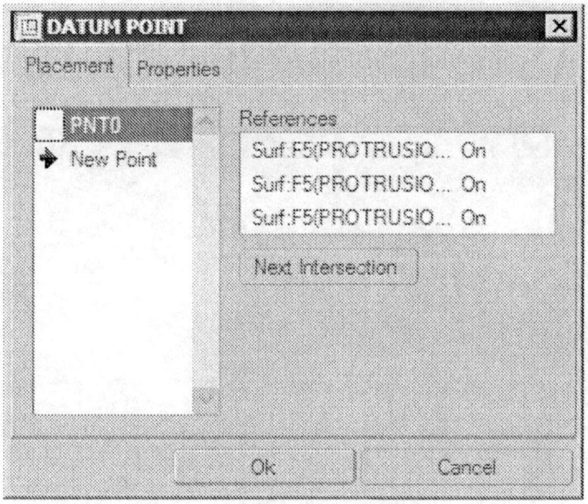

Figure 4-43 DATUM POINT *dialog box*

Datum point at the center of a curve edge

The **Center** constraint creates a datum point at the center of an arc or a curved edge. Invoke the **DATUM POINT** dialog box and select the curve edge. The **On** constraint is applied by default. To modify the constraint, select the constraint in the **References** collector; the drop-down list appears to the right of the constraint. Choose the **Center** constraint from the drop-down list. The datum point is created at the center of the curve edge.

Datum point on an edge or a curve

The **On** constraint is used to create a datum point on an edge or a curve. Invoke the **DATUM**

POINT dialog box, and select an edge or a curve. The datum point is placed on the selected edge as shown in Figure 4-44 and the **DATUM POINT** dialog box appears as shown in Figure 4-45.

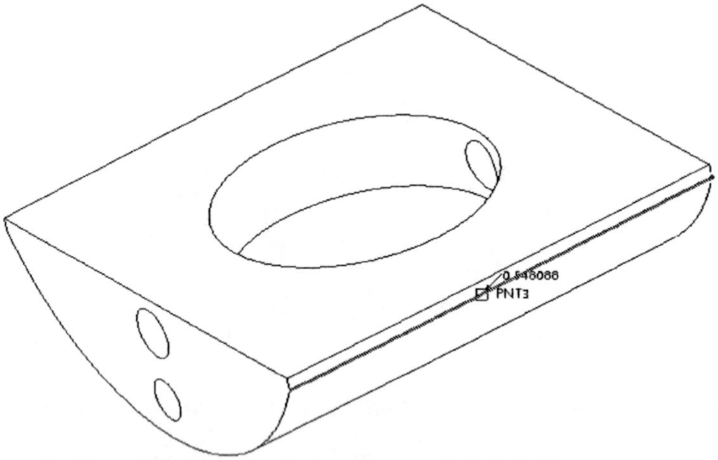

Figure 4-44 *The highlighted edge and the datum point at some offset distance from one end of edge*

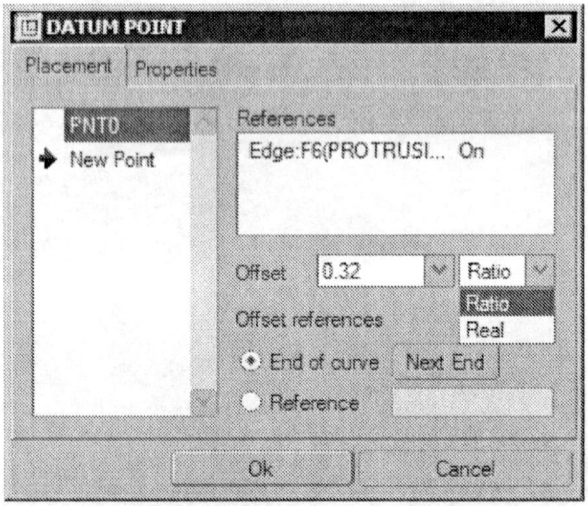

***Figure 4-45 DATUM POINT** dialog box*

In the **DATUM POINT** dialog box, after the point is placed, the dimension type can be defined. The **Offset** dimension box displays the offset distance of the datum point from one end of the edge. The offset distance is measured as a ratio or as the real value from the end of the edge.

In the **Offset References** collector, the **Next End** button is used to flip the end of the edge from where the offset distance is measured. This button is available only when the **End of**

curve radio button is selected. If you select the **Reference** radio button, you need to select a reference from which the datum point will be dimensioned.

Sketched Datum Point Tool button

When you choose the black arrow on the right of the **Datum Point Tool** button, a flyout is displayed. Choose the **Sketched Datum Point Tool** button from this flyout. The **Sketch Datum Point** dialog box is displayed. The function of this dialog box is same as the **Section** dialog box. Using the **Sketch Datum Point** dialog box, select the sketching plane and its orientation. When you choose the **Sketch** button, the system takes you to the sketcher environment. In the sketcher environment you can use the construction tools and the **Create Points** button to sketch the datum points. The datum points are dimensioned with the references. You can also modify these dimensions. Remember that you can exit the sketcher environment only if at least one point is drawn.

To create an array of datum points from a coordinate system

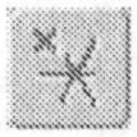

The **Offset Coordinate System Datum Point Tool** button is used to create an array of datum points at an offset distance from a coordinate system. You can change the array of the points by redefining the array. This button is available on the flyout.

When you choose this button, the **Offset Csys Datum Point** dialog box is displayed as shown in Figure 4-46 and you are prompted to select a coordinate system. After selecting a coordinate system, you can enter the values of the coordinates for the datum points. From the **Type** drop-down list, select the type of coordinate system: Cartesian, Cylindrical, or Spherical. In the figure, the dialog box is shown when the coordinates of three datum points are entered.

To create a Datum Point by clicking

When you choose the **Field Datum Point Tool** button from the flyout, the **Field Datum Point** dialog box is displayed as shown in Figure 4-47. Using this dialog box, you can create a datum point anywhere on the model. You just need to specify the location using the left mouse button to place a datum point.

CREATING CUTS

Cut is a material removal process and this option is available only when at least a base feature exists on the graphics screen. The tool that creates the cut is available on the dashboard of the feature creation tool. In this chapter, you will learn the extrude cut and the revolve cut. The procedure to create a cut on an existing feature is similar to that of adding material or protrusion.

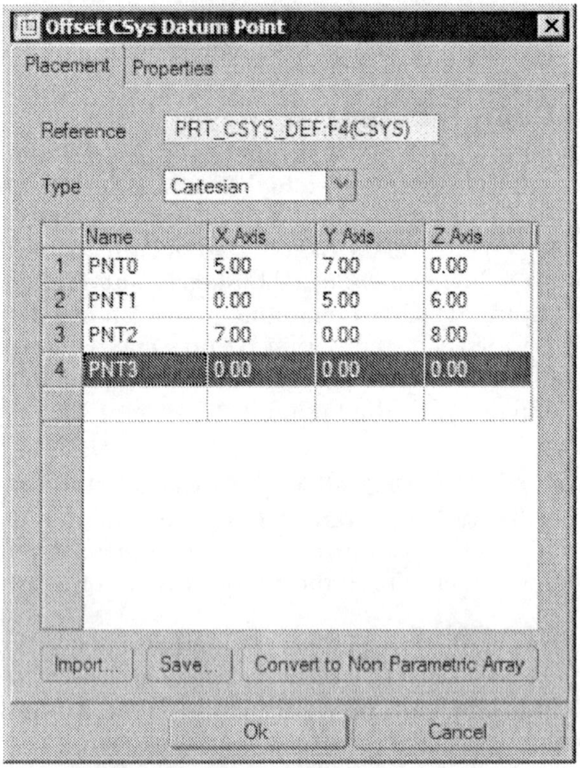

Figure 4-46 ***Offset CSys Datum Point*** *dialog box*

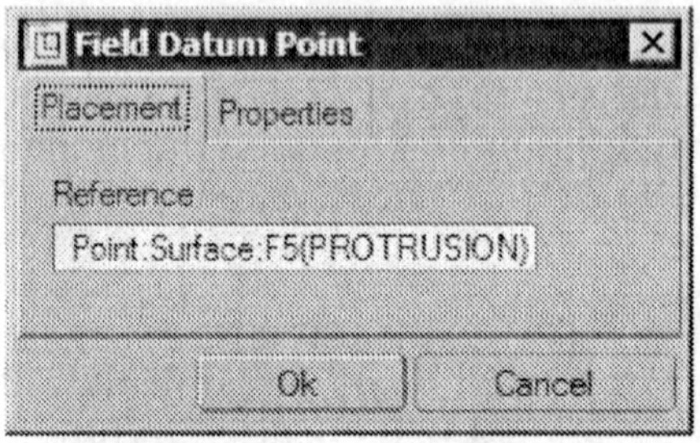

Figure 4-47 ***Field Datum Point*** *dialog box*

Removing Material by Extruding a Sketch

The **Extrude Cut** is an extruded feature that is created by removing material from an existing feature. The material that is removed is defined by the sketch you draw.

After drawing the sketch for the cut feature, you can specify the direction of material removal with respect to the sketch. For example, the yellow arrow in Figure 4-48 shows the direction of material removal. If the direction shown by the arrow is accepted then the cut feature will be created as shown in Figure 4-49.

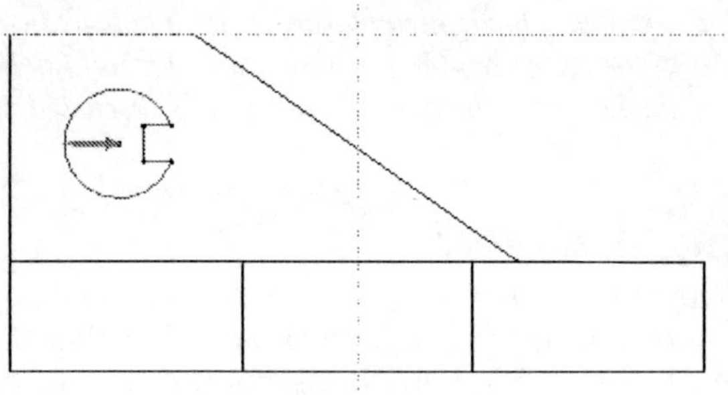

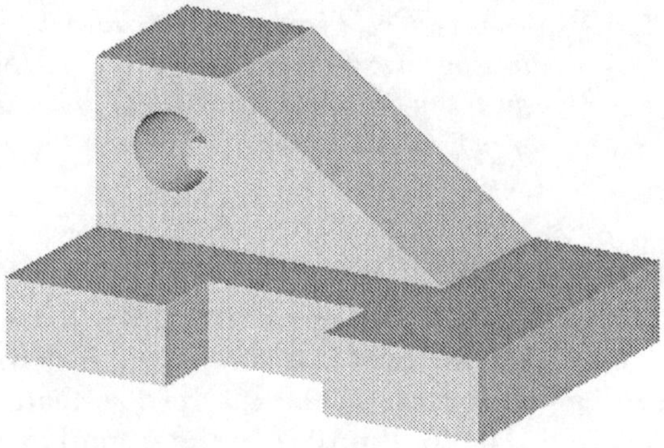

Figure 4-48 *Sketch for the extruded cut and arrow showing the direction of material removal* ***Figure 4-49*** *Cut feature created on the selected plane*

> **Tip**: *In the model shown in Figure 4-48, the sketching plane selected for creation of the extruded cut is not a datum plane but the front face of the base feature. You can also create a datum plane on the surface of an existing feature and select it as the sketching plane. But it is not recommended to create a datum plane in cases where a planar surface of the feature can be used as a sketching plane.*

However, if you choose the **Change material direction of extrude to other side of sketch** button from the Extrude dashboard, the arrow points in the direction shown in Figure 4-50. All the material on the plane selected for sketching will be removed leaving the extruded cut feature as shown in Figure 4-51.

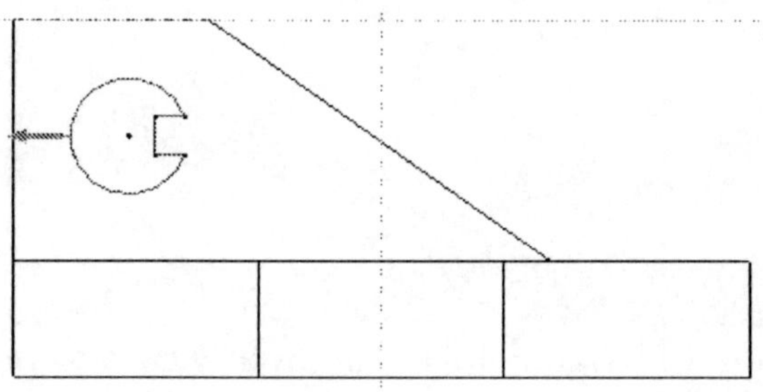

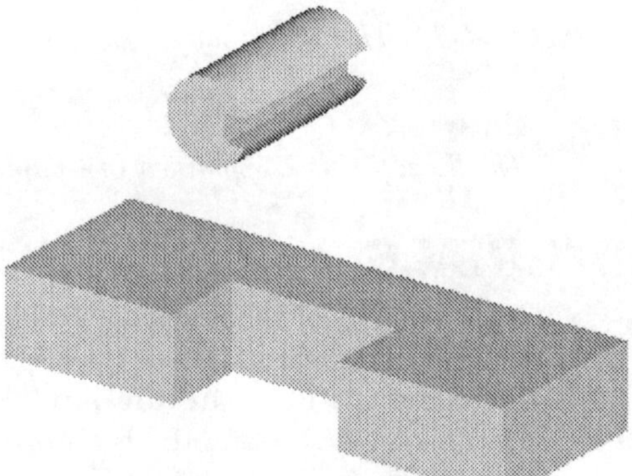

Figure 4-50 *Arrow showing the direction of material removal*

Figure 4-51 *Cut feature created in the direction shown in the adjacent figure*

Note

A straight hole can also be created by drawing its cross-section, that is, a circle, and then creating an extrude cut. But, Pro/ENGINEER provides predefined placement for a hole feature, which can be more desirable than dimensioning the cross-section of a cut feature. Straight holes do not require a sketch if you use the Hole dashboard. The Hole dashboard is discussed in Chapter 5.

Removing Material by Revolving a Sketch

The **Revolve Cut** is a revolved feature that is created by removing material from an existing feature. The material that is removed is defined by the sketch you draw. Remember that the center line is necessary in the revolve features. Figure 4-52 shows the section drawn to be revolved. The front surface of the second extruded feature is selected as the sketching plane. Figure 4-53 shows the revolve cut created on the selected surface.

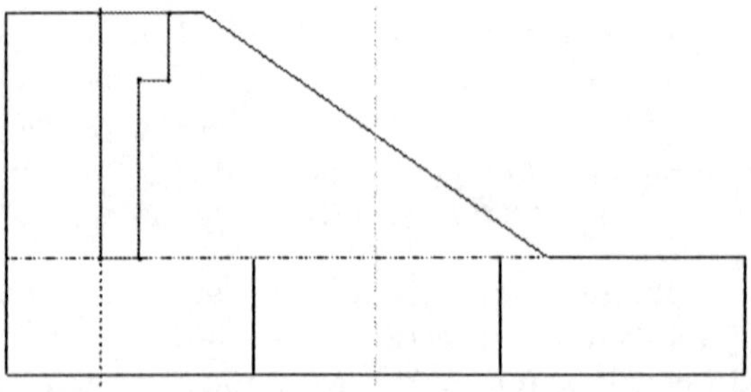

Figure 4-52 *The section for revolve cut*

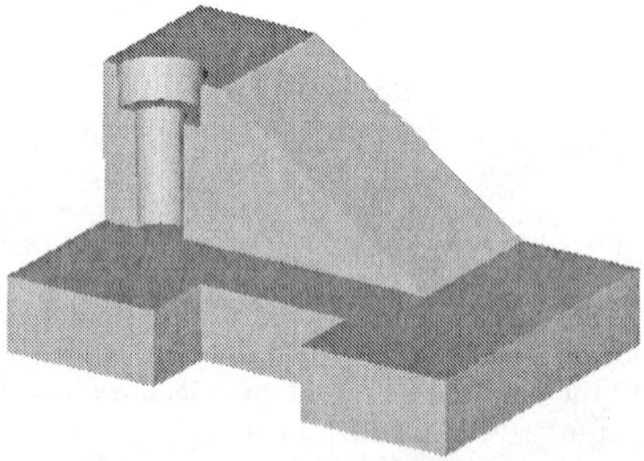

Figure 4-53 *Revolve cut created*

Note

The ***Sweep Cut*** *is explained in Chapter 7.*

TUTORIALS

Tutorial 1

In this tutorial you will create the model shown in Figure 4-54. The front view and the right-side view of the solid model is shown in Figure 4-55.

(Expected time: 25 min)

The following steps outline the procedure for creating this model:

a. First, examine the model and then determine the number of features in it, see Figure 4-54. The model is composed of four features: two at the top, one at the bottom, and one hole on the right face. Also, from the model it is evident that the two features at the top of the model can be created on the same plane.

b. Select the sketching plane for the base feature, draw the sketch using the sketching tools,

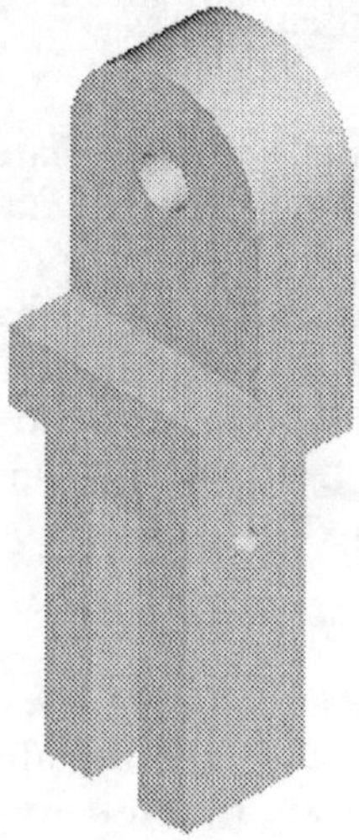

Figure 4-54 *Model for Tutorial 1*

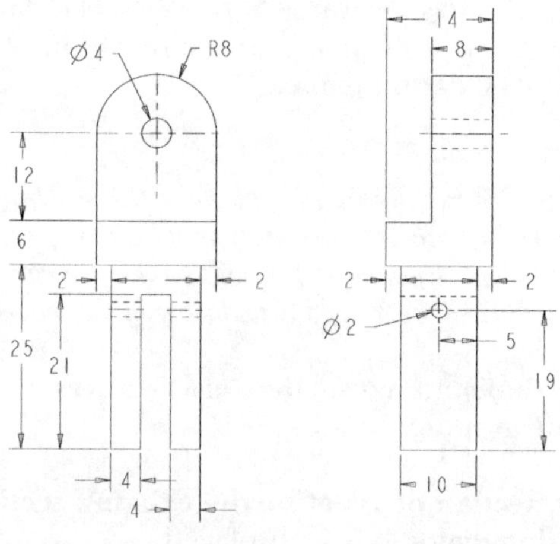

Figure 4-55 *Front and side views of the model*

apply the dimensions and constraints, and then extrude the sketch to the given distance, see Figures 4-56 and 4-57.

c. Select the sketching plane for the feature that is at the bottom of the base feature. This feature will be sketched on the same plane that was used by the base feature. Draw the sketch using the sketching tools, apply the dimensions and constraints, and then extrude the sketch to the given distance, see Figure 4-61.

d. The third feature that is at the bottom of the second feature will be created on a datum plane that is at an offset distance of 2 from the front planar surface of the second feature.

Draw the sketch, apply the dimensions and constraints, and then extrude it to the given distance, see Figure 4-67.

e. Similarly, select a sketching plane for the cut feature. The cut has a circular section. Draw the sketch for this feature and create the cut feature as shown in Figure 4-71 .

After understanding the procedure for creating the model, you are now ready to create it. When Pro/ENGINEER session is started, the first task is to set the working directory. Since this is the first tutorial of this chapter, you need to create a folder named *c04*, if it does not exist. Choose the **New Directory** button in the **Select Working Directory** dialog box and create a directory named *c04* at *C:\ProE-WF*.

Creating New Object File

1. Open a new part file and name it as **c04tut1**. The three default datum planes are displayed on the graphics window. The **Model Tree** also appears on the left in the Navigator. Close the **Model Tree** by clicking on the sash on its right edge.

Selecting the Sketching Plane for the Base Feature

To create the sketch for the base feature, you need to first select the sketching plane. In this model, you need to draw the base feature on the **FRONT** datum plane because from the isometric view of this model, it is evident that the direction of extrusion for this feature is perpendicular to the **FRONT** datum plane.

Note

The model can be created by selecting any plane as the sketching plane for the base feature. But when the base feature is created, the orientation of the base feature will not be proper. Hence, the final model will be oriented wrongly. You will have to be careful while defining the sketching plane for the base feature. The desired orientation of the model is shown in Figure 4-54.

1. Choose the **Extrude Tool** button from the **Base Features** toolbar. The Extrude dashboard is displayed below the graphics window.

2. Choose the **Create a section or redefine the existing section** button from the Extrude dashboard. The **Section** dialog box is displayed.

3. Select the **FRONT** datum plane as the sketching plane.

   A yellow arrow is displayed on the **FRONT** datum plane, pointing in the direction of viewing the sketch.

4. Select the **TOP** datum plane from the graphics window and then select the **Top** option from the **Orientation** drop-down list.

   The **TOP** datum plane is selected in order to orient the sketching plane.

5. Choose the **Sketch** button from the **Section** dialog box. Now you enter the sketcher environment.

Specifying References

In the sketcher environment, the **References** dialog box is displayed at the top right corner of the screen. The status displayed in the **Reference status** area is **Fully Placed**. Close the **References** dialog box by choosing the **Close** button from the dialog box.

Creating and Dimensioning the Sketch for the Base Feature

The base feature can be created by drawing the sketch and then extruding it to the given distance.

1. Draw the section sketch using various sketcher tools and add the required constraints and dimensions shown in Figure 4-56. The sketch is dimensioned automatically and some weak dimensions are assigned to it.

> **Tip**: *It is recommended that you use the* ***Modify the values of dimensions, geometry of splines, or text entities*** *button to modify the weak dimensions. In the* ***Modify Dimensions*** *dialog box that appears, clear the* ***Regenerate*** *check box and then modify the dimensions using the thumbwheel or the dimension edit box. This way the sketch will not regenerate as you edit dimensions.*

2. Modify the dimension values to the values shown in Figure 4-56.

3. After the sketch is completed, choose the **Continue with the current section** button. Now, you have exit the sketcher environment.

4. Choose the **Default View** option from the **Saved view list** button of the **View** toolbar.

   The model is displayed in the default view but it does not fit in the graphics window. The yellow colored arrow is also displayed on the model, indicating the direction of extrusion.

5. In the dimension box present on the Extrude dashboard, enter a value of **8** and press ENTER. The model appears as shown in Figure 4-57.

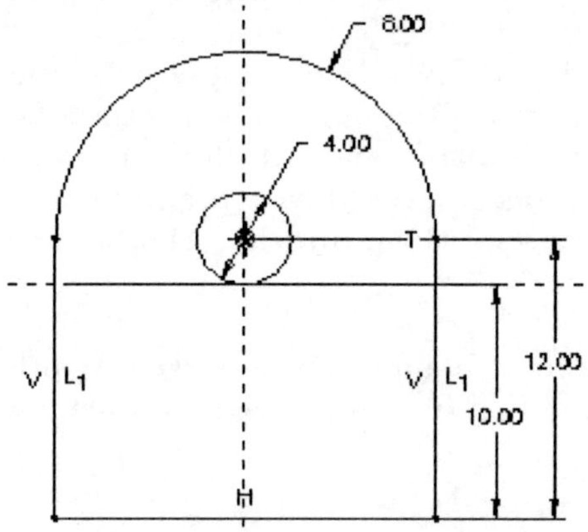

Figure 4-56 *Sketch for the base feature with dimensions and constraints*

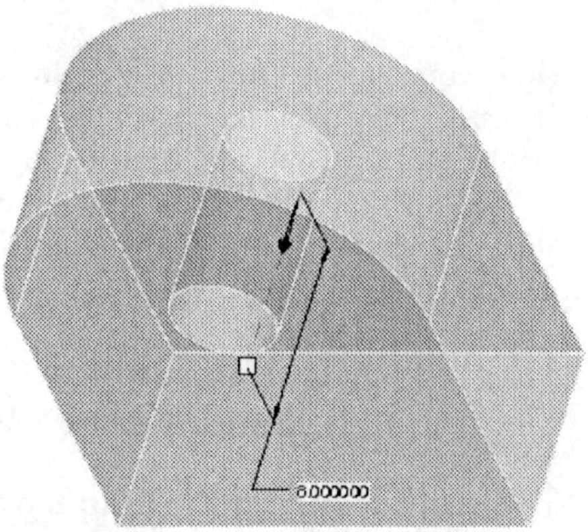

Figure 5-57 *Default orientation of the model and the arrow showing the direction of feature creation*

6. Choose the **Build feature** button from the Extrude dashboard.

   The base feature is completed. You can use the middle mouse button to spin the model in order to view it from various directions.

Tip: *It is recommended to check the orientation of the base feature of a model when it is completed. To check whether the plane you specified for sketching was correct or not, choose the* ***Saved view list*** *button from the* ***View*** *toolbar. Choose the* ***FRONT*** *option from the drop-down list; the base feature will reorient on the graphics window such that you can view the front view of the base feature, similar to that shown in Figure 4-55.*

Note

When you choose the ***Default View*** *option from the* ***Saved view list*** *button, the orientation of the model is trimetric and not isometric. If you want the isometric view of the model to be displayed whenever you choose the* ***Default*** *option, then you have to use the* ***Environment*** *dialog box. The* ***Environment*** *dialog box is displayed when you choose* ***Tools > Environment*** *from the menu bar. From the dialog box, in the* ***Standard Orient*** *drop-down list, choose the* ***Isometric*** *option. Now, the default orientation will be set to isometric.*

Selecting the Sketching Plane for the Second Feature

The second feature is an extrude feature and will be created on the previous plane that was used to create the base feature.

1. Choose the **Extrude Tool** button from the **Base Features** toolbar. The Extrude dashboard is displayed below the graphics window.

2. Choose the **Create a section or redefine the existing section** button from the Extrude dashboard. The **Section** dialog box is displayed.

3. Choose the **Use Previous** button from the **Sketch Plane** area in the **Section** dialog box. The yellow arrow is displayed on the graphics window as shown in Figure 4-58.

   When you choose the **Use Previous** button, the system selects the previous sketching plane that was used to create the base feature. This option is selected because the base feature and the second feature are on the same plane but have different depths of extrusion. If they had the same depth of extrusion, you could have drawn them on the same plane as a single feature. The **TOP** datum plane and its orientation is set automatically.

   In Figure 4-58, the model is oriented in its default orientation. The model can be oriented in its default position by choosing the **Default** option from the **Saved view list** button.

4. Choose the **Sketch** button from the **Section** dialog box.

   The system takes you to the sketcher environment.

Drawing the Sketch for the Second Feature

The second feature has a rectangular section that will be extruded to a depth of 14. To improve the clarity of the edges of the base feature, choose the **No Hidden** button from the **Model Display** toolbar before you start sketching the second feature.

1. Draw the sketch as shown in Figure 4-59.

> **Tip**: *You can use the **Create an entity from an edge** button from the **Sketcher Tools** toolbar to use the bottom edge of the base feature. The edge of the base feature is required to complete the sketch for the second feature. Or else, draw an aligned line on the edge.*

2. The sketch is automatically dimensioned and some weak dimensions are assigned to it. Add the required constraints and modify the weak dimensions as shown in Figure 4-59.

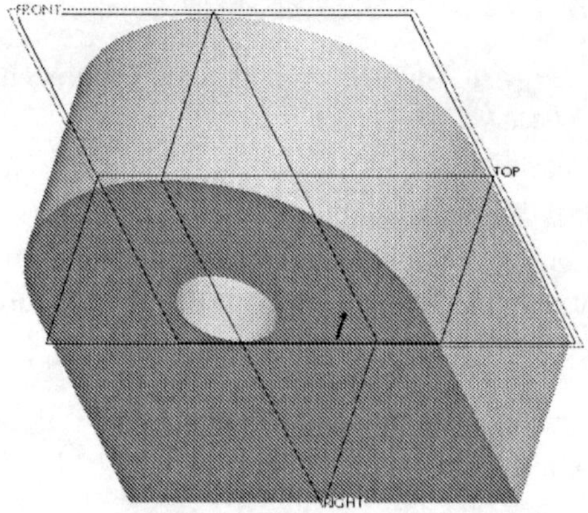

Figure 4-58 *Arrow showing the direction of viewing the sketching plane*

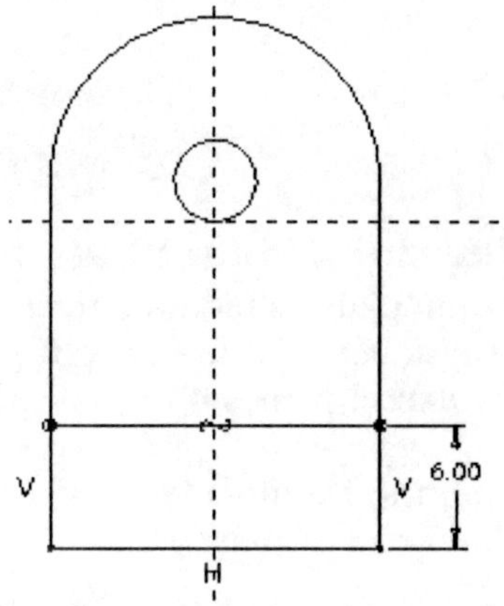

Figure 4-59 *Sketch for the second feature*

3. Choose the **Continue with the current section** button to exit the sketcher environment. The Extrude dashboard is enabled below the graphics window. Choose the **Shading** button from the **Model Display** toolbar to view the shaded model.

4. Use the middle mouse button to orient the model as shown in Figure 4-60. This orientation of the model gives you a better view of the sketch in the three-dimensional (3D) space. The yellow colored arrow is also displayed on the model, indicating the direction of extrusion.

5. Enter a value of **14** in the dimension box present on the Extrude dashboard and press ENTER. The second extruded feature is completed and its preview is displayed on the graphics window.

6. Choose the **Saved view list** button from the **View** toolbar. From the drop-down list choose the **Default View** option.

7. Now, choose the **Build feature** button from the Extrude dashboard to confirm the feature creation. The model orients on the screen as shown in Figure 4-61. You can also use the middle mouse button to spin the model.

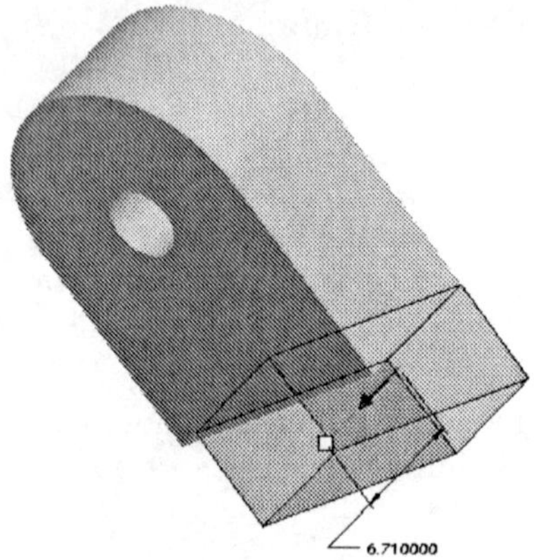

Figure 4-60 *Arrow showing the direction of feature depth*

Figure 4-61 *Second extruded feature with the base feature*

Creating the Datum Plane for the Third Feature

A new datum plane is required to create the next feature. The datum plane will be created at an offset distance of 2 from the front face of the second feature. You need to turn on the display of datum planes if their display is off.

1. Choose the **Datum Plane Tool** button from the **Datum** toolbar. The **DATUM PLANE** dialog box is displayed.

   Make sure that before opening the dialog box, there are no selections made. This can be confirmed from the **Status** bar that is below the graphics window. If there is a selection, you can remove it by left-clicking in the graphics window.

2. Using the left mouse button, select the front face of the second feature highlighted in Figure 4-62. The boundary of the selected front face is highlighted red in color.

   In the **DATUM PLANE** dialog box, under the **References** collector, the **Offset** constraint is displayed. This means that by the smart selection property of Pro/ENGINEER Wildfire, the **Offset** constraint is applied automatically to the datum plane you are going to create.

   Now, you need to specify the offset distance. If you enter a positive value, the datum plane will be created in the direction shown by the arrow and if you enter a negative value then the datum plane will be created in the direction opposite to that shown by the arrow.

3. In the **Translation** dimension box, under the **Offset** area of **DATUM PLANE** dialog box, enter a value of **-2** and press ENTER.

   The negative value is entered because the datum plane has to be created in the direction opposite to that shown by the yellow arrow. The datum plane named **DTM1** is created as shown in Figure 4-63.

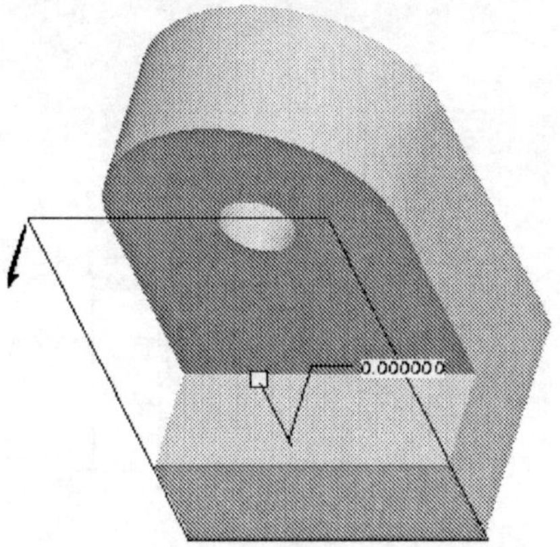

Figure 4-62 *The selected face and the arrow showing the positive direction of datum plane creation*

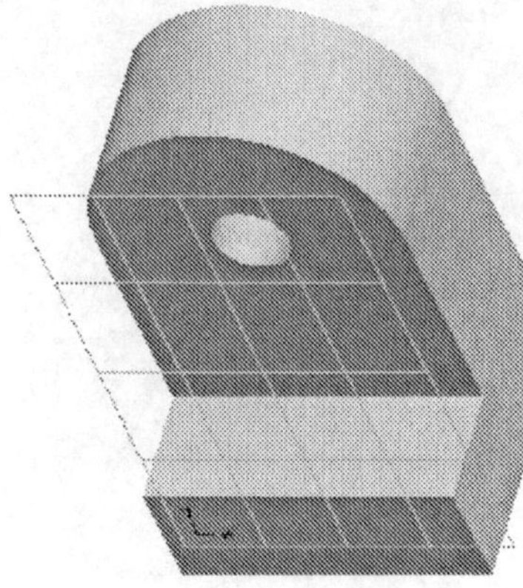

Figure 4-63 *The highlighted datum plane*

Note
Throughout the book, at some instances the datum planes are shown by a mesh plane as evident from Figure 4-63. This view of the datum plane is only for explanation. In Pro/ENGINEER, when you create a datum plane, they do not appear in the form of mesh.

4. Choose the **OK** button.

Creating the Third Feature on DTM1

The datum plane **DTM1** is created and can be seen in the **Model Tree** as well as on the graphics window. The section sketch of the next feature that will be extruded has to be created on the datum plane **DTM1**.

1. Choose the **Extrude Tool** button from the **Base Features** toolbar. The Extrude dashboard is displayed below the graphics window.

2. Choose the **Create a section or redefine the existing section** button from the Extrude dashboard. The **Section** dialog box is displayed.

3. Select **DTM1** as the sketching plane for the third feature. A yellow arrow is displayed on the selected datum plane as shown in Figure 4-64. This arrow shows the direction of viewing the sketching plane.

   The **TOP** datum plane and its orientation is selected by default.

4. Choose the **Sketch** button from the **Section** dialog box. The system takes you to the sketcher environment. Choose the **No Hidden** button from the **Model Display** toolbar.

5. Sketch the section for the third feature of the model and add constraints and dimensions to the sketch as shown in Figure 4-65.

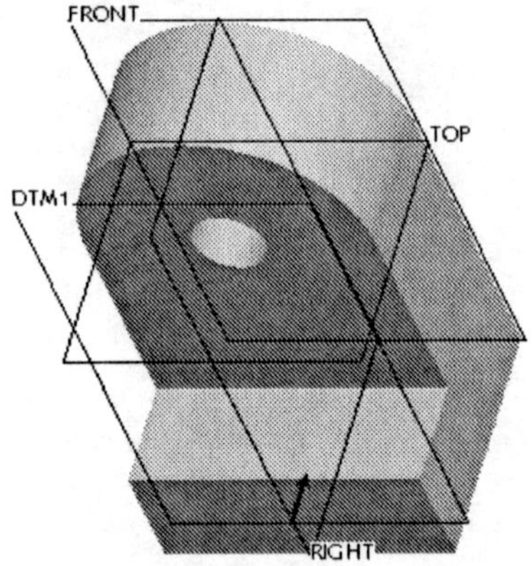

*Figure 4-64 Arrow on **DTM1** showing the direction of viewing the sketching plane*

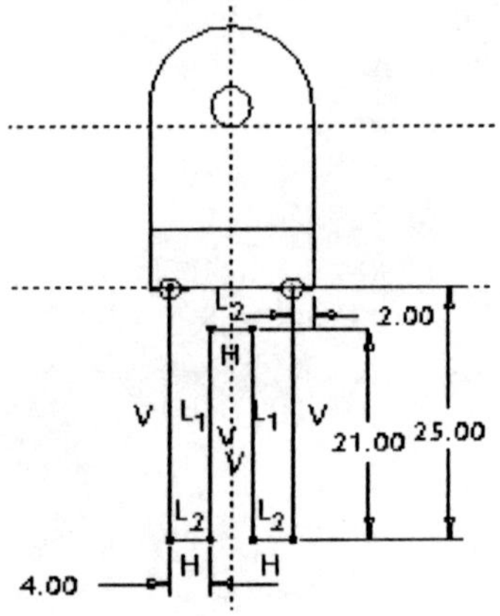

***Figure 4-65** Sketch with dimensions and constraints of the third feature*

6. Choose the **Continue with current section** button to exit the sketcher environment.

   The Extrude dashboard is enabled below the graphics window. Turn the model display to **Shading**. Use the middle mouse button to orient the model as shown in Figure 4-66. This orientation gives a better view of the model.

   Notice that the yellow arrow is pointing toward the direction of feature creation. But, you need to extrude the sketch in the opposite direction.

7. Choose the **Change depth direction of extrude to other side of sketch** button from the Extrude dashboard to change the direction of arrow. 

8. Enter a value of **10** in the dimension box present on the Extrude dashboard and press ENTER.

   You can see the preview of the model after creating the third extruded feature.

9. Now, choose the **Build feature** button from the Extrude dashboard to confirm the feature creation.

   The default view that is displayed when you choose the **Default View** option from the **Saved view list** button is shown in Figure 4-67.

Selecting the Sketching Plane for the Cut Feature

A through cut will be created on the outer right face of the third feature. The circular section for the cut will be sketched and using the **Remove Material** button in the Extrude dashboard,

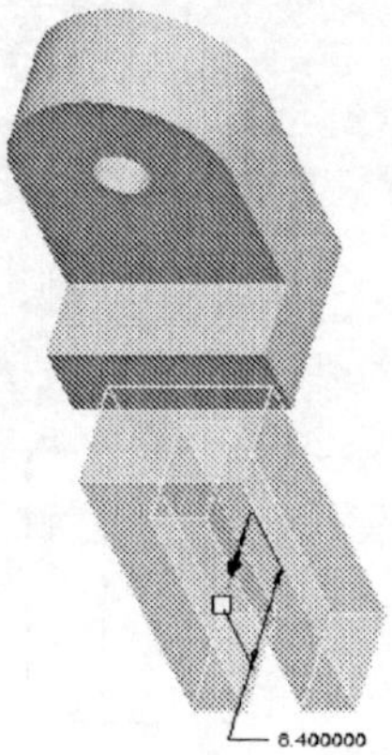

Figure 4-66 *Arrow showing the direction of material addition*

Figure 4-67 *Model after creating the third*

the circular cut will be created. The sketching plane for the cut feature is shown in Figure 4-68.

Note
The circular cut feature can also be created using the Hole dashboard, which will be discussed in Chapter 5.

1. Choose the **Extrude Tool** button from the **Base Features** toolbar. The Extrude dashboard is displayed below the graphics window.

2. Choose the **Remove Material** button from the Extrude dashboard.

3. Choose the **Create a section or redefine the existing section** button from the Extrude dashboard. The **Section** dialog box is displayed.

4. Select the face shown in Figure 4-68 for sketching.

5. Using the left mouse button select the **TOP** default datum plane and then select the **Top** option from the **Orientation** drop-down list.

6. Choose the **Sketch** button. The system takes you to the sketcher environment.

Sketching the Cut Feature

1. Turn the model display to **No Hidden**. Draw the sketch of the cut feature and add dimensions to it as shown in Figure 4-69. The hole dimension shown in Figure 4-69 appears as a radial dimension but it is a diameter dimension.

2. Choose the **Continue with current section** button to exit the sketcher environment. Turn the model display to **Shading**.

3. Choose the **Saved view list** button from the **View** toolbar. From the drop-down list choose

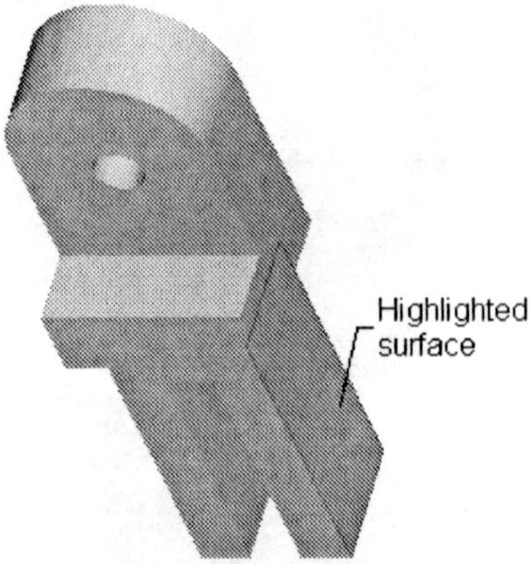

Figure 4-68 *Sketching plane for hole feature*

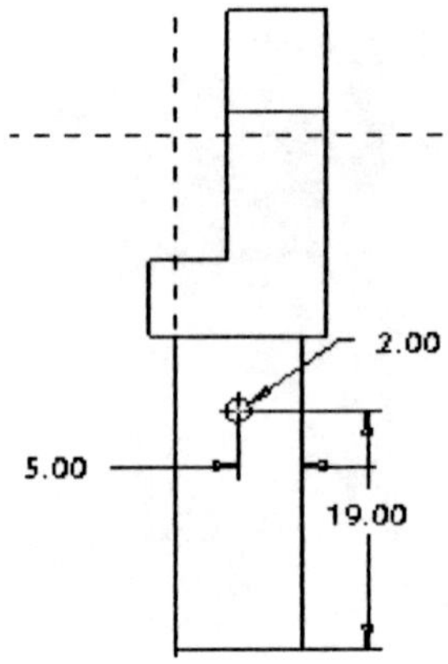

Figure 4-69 *Sketch and dimensions for cut*

the **Default View** option. The model is displayed as shown in Figure 4-70. Two yellow arrows also appear on the model. One arrow shows the direction of feature creation and the other arrow shows the direction with respect to the sketch from where the material will be removed. The arrows should point as shown in Figure 4-70.

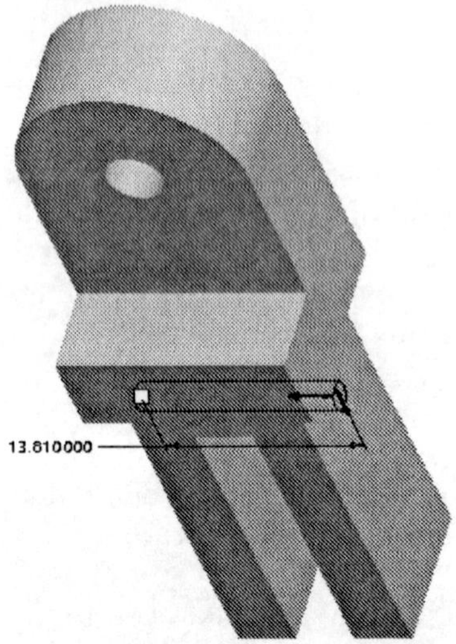

Figure 4-70 *The two arrows on the cut feature*

3. Choose the **Options** tab in the Extrude dashboard. The **Depth** slide-up panel is displayed.

4. From the **Side 1** drop-down list, choose the **Through All** option.

   The cut feature is completed and can now be previewed on the graphics window.

5. Choose the **Build feature** button from the Extrude dashboard. Turn the model display to shaded by choosing the **Shading** button from the **Model Display** toolbar.

The default view, when you choose the **Default View** option from the **Saved view list** button from the **View** toolbar, is shown in Figure 4-71.

Saving the Model

1. Choose the **Save the active object** button from the **File** toolbar and save the model.

   The order of feature creation can be seen from the **Model Tree** shown in Figure 4-72.

Figure 4-71 *Completed model for Tutorial 1*

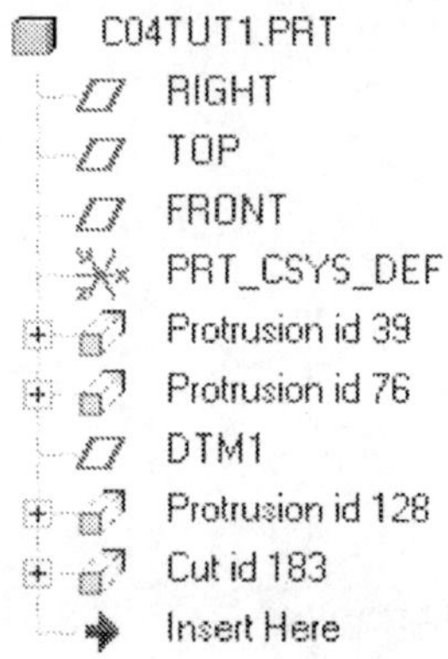

Figure 4-72 ***Model Tree*** *for Tutorial 1*

Note
The feature id numbers that are suffixed to the feature name in the ***Model Tree*** *may be different when you create the features.*

Tutorial 2

In this tutorial you will create the model shown in Figure 4-73. This figure also shows the front view, the top view, and the right-side view of the solid model.

(Expected time: 45 min)

The following steps outline the procedure for creating this model:

a. First, examine the model and then determine the number of features in it, see Figure 4-73. The model is composed of four features: one at the bottom (base feature), one on the left, and two at the back. Also, from the model it is evident that the two features on the back face of the model can be created on the same plane.

b. Select the sketch plane for the base feature, draw the sketch using the sketching tools, apply the dimensions and constraints, and then extrude the sketch to the given depth, see Figure 4-75.

c. Select the sketch plane for the second feature on the left face of the base feature. Draw the sketch using the sketching tools, apply dimensions and constraints, and then extrude the sketch to the given depth, see Figure 4-80.

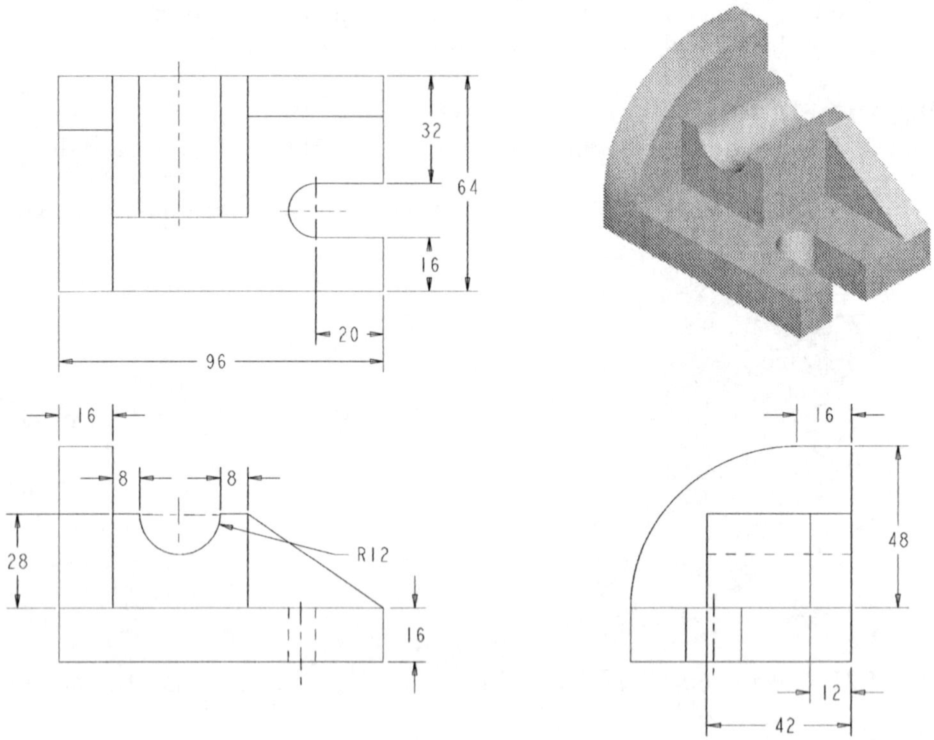

Figure 4-73 Top, front, right-side, and isometric views of the model

d. From the model it is evident that the third and fourth features are created on the same plane, but their depths of extrusion are different and hence they will be created as two separate extruded features. See Figures 4-85 and 4-89.

After understanding the procedure for creating the model, you are now ready to create it. The working directory has already been selected in Tutorial 1 and therefore you do not need to select it again. However, if you want to select the working directory, select it from the Navigator.

Creating New Object File

1. Open a new part file and name it as **c04tut2**. The three default datum planes are displayed on the graphics screen if the **Datum planes on/off** button is turned on. If the **Model Tree** is open, close it by selecting the sash.

Selecting the Sketching Plane for the Base Feature

To create the sketch for the base feature, you need to first select the sketching plane for the base feature. In this model, you need to draw the base feature on the **TOP** datum plane. This is because the direction of extrusion of the base feature is perpendicular to the **TOP** datum plane.

1. Choose the **Extrude Tool** button from the **Base Features** toolbar. The Extrude dashboard is displayed below the graphics window.

2. Choose the **Create a section or redefine the existing section** button from the Extrude dashboard. The **Section** dialog box is displayed.

3. Select the **TOP** datum plane as the sketching plane.

   A yellow arrow is displayed on the **TOP** datum plane, pointing in the direction of viewing the sketch. The **RIGHT** datum plane and its orientation is selected automatically.

4. Choose the **Sketch** button in the **Section** dialog box to enter the sketcher environment.

Specifying References

In the sketcher environment, the **References** dialog box is displayed at the top right corner of the screen. The status displayed in the **Reference status** area is **Fully Placed**. Exit the **References** dialog box by choosing the **Close** button from the dialog box.

Creating and Dimensioning the Sketch for the Base Feature

The sketch for the base feature consists of a rectangular shape with a slot as shown in Figure 4-74. When this sketch is extruded, it will create the base feature with the slot as shown in Figure 4-75.

1. Draw the section sketch using various sketcher tools and add the required constraints and dimensions to it. Modify these dimensions as shown in Figure 4-74.

2. Choose the **Continue with the current section** button to exit the sketcher environment.

3. Choose the **Default View** option from the **Saved view list** button of the **View** toolbar.

   The default view of the model is displayed, but it does not fit on the screen. You may zoom out to display the model on the graphics window. This gives you a better view of the sketch in the 3D space. A yellow arrow is also displayed on the model, indicating the direction of extrusion.

4. Enter a value of **16** in the dimension box present on the Extrude dashboard and press ENTER. The preview of the base feature is displayed in the graphics window.

5. Choose the **Build feature** button from the Extrude dashboard.

   The base feature is completed and is shown in Figure 4-75. You can use the middle mouse button to spin the object to view it from various directions.

Selecting the Sketching Plane for the Second Feature

The next feature is an extruded feature and will be created on the left face of the base feature. Therefore, you need to select the left face of the base feature as the sketching plane and then

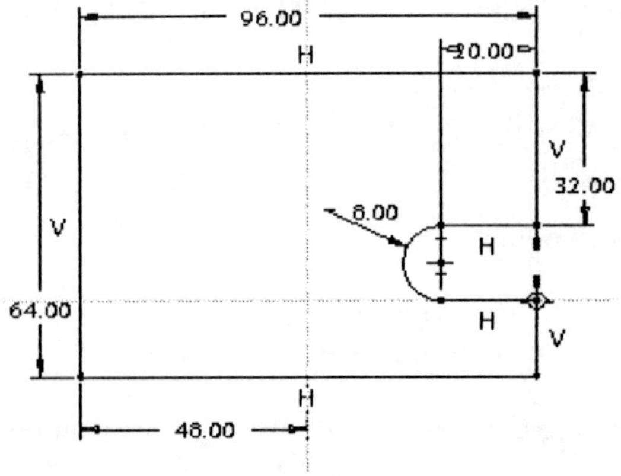

Figure 4-74 Sketch with dimensions and constraints for the base feature

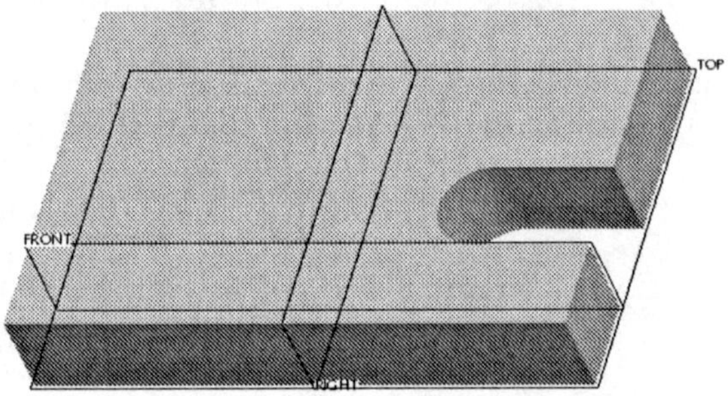

Figure 4-75 Base feature of the model

draw the sketch.

1. Choose the **Extrude Tool** button from the **Base Features** toolbar. The Extrude dashboard is displayed below the graphics window.

2. Choose the **Create a section or redefine the existing section** button from the Extrude dashboard. The **Section** dialog box is displayed.

3. Using the middle mouse button, spin the model as shown in Figure 4-76.

4. Now, select the left face of the base feature as the sketching plane.

   A yellow arrow that points in the direction of viewing the sketching plane appears on the left face of the base feature.

5. Choose the **Flip** button to flip the yellow arrow and to change the direction of viewing the sketching plane as shown in Figure 4-76.

Note

The arrow that appears on the sketching plane points in the direction of viewing the sketching plane. It should be remembered that the direction of this arrow should be set keeping in mind that the protrusion will always take place outside the screen.

6. Select the top face of the base feature shown in Figure 4-77 and choose the **Top** option from the **Orientation** drop-down list.

   By selecting the top surface of the base feature, the model will be oriented in such a way that the highlighted planar surface will be at the top while sketching. After you select the surface, the system takes you to the sketcher environment.

7. Choose the **Sketch** button from the **Section** dialog box to enter the sketcher environment.

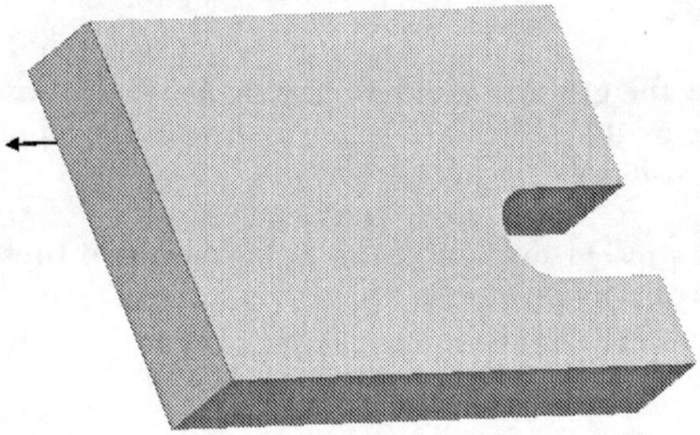

Figure 4-76 Arrow showing the direction of viewing the model

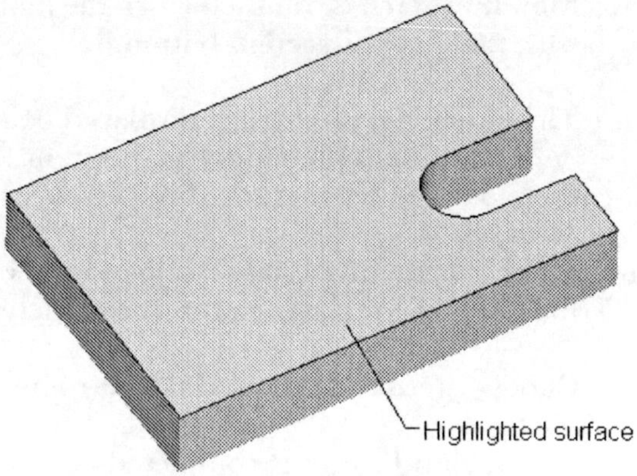

Figure 4-77 Surface selected to be at the top

Creating and Dimensioning the Sketch for the Second Feature

The sketch for the second feature consists of two lines and an arc. The bottom edge of the sketch coincides with the top edge of the base feature. This sketch is extruded to a depth of 16. Before drawing the sketch, turn the model display to **No Hidden**.

1. Close the **References** dialog box that is displayed on the top right corner of the screen.

2. Draw the sketch using various sketcher tools as shown in Figure 4-78. The sketch is dimensioned automatically and some weak dimensions are assigned to it.

3. Apply the constraints and modify the weak dimensions to the dimensions shown in Figure 4-78.

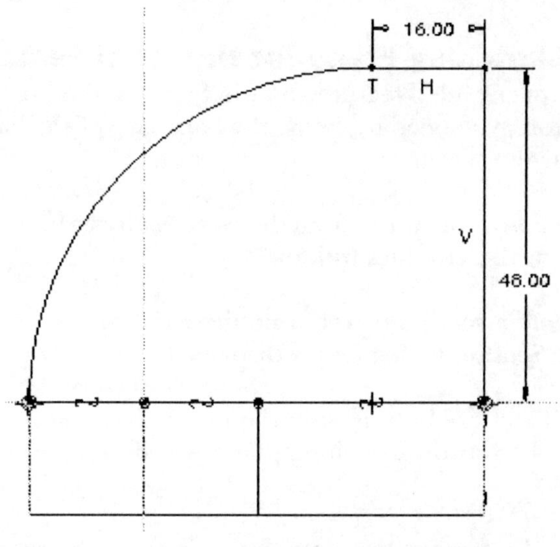

Figure 4-78 Sketch with dimensions and constraints for the second feature

4. After the sketch is complete, turn the model display to **Shading** and choose the **Continue with the current section** button.

   The Extrude dashboard is displayed below the graphics window. Use the middle mouse button to orient the model as shown in Figure 4-79. A yellow arrow is also displayed on the model, indicating the direction of extrusion.

5. Enter a value of **16** in the dimension box present in the Extrude dashboard and press ENTER. The second feature is completed and its preview is displayed on the screen.

6. Choose the **Build feature** button from the Extrude dashboard.

   The second feature is completed and is shown in Figure 4-80. You can use the middle mouse button to spin the model to view it from various directions.

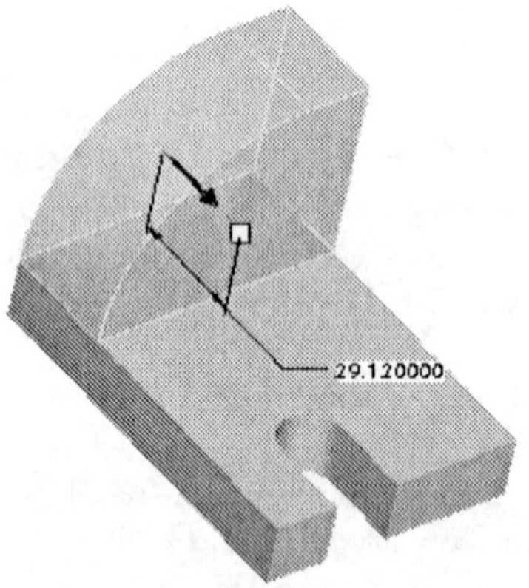

Figure 4-79 Arrow showing the direction of material addition

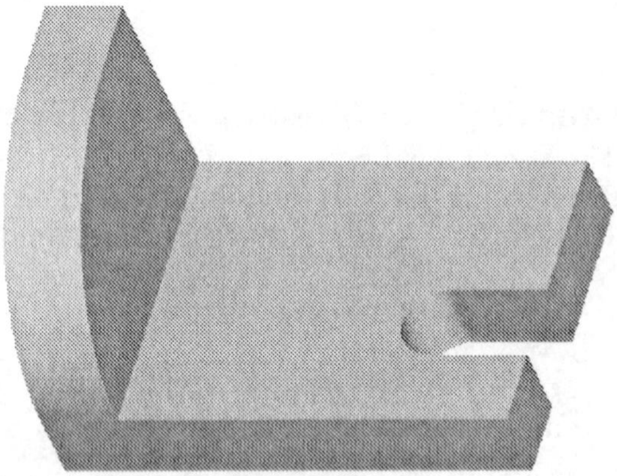

Figure 4-80 Model with the second extruded feature

Selecting the Sketching Plane for the Third Feature

The third feature is an extruded feature and will be created on the back planar surface of the base feature. Therefore, you need to define the back face of the base feature as the sketching plane and then draw the sketch.

1. Choose the **Extrude Tool** button from the **Base Features** toolbar. The Extrude dashboard is displayed below the graphics window.

2. Choose the **Create a section or redefine the existing section** button from the Extrude dashboard. The **Section** dialog box is displayed.

3. Using the middle mouse button, spin the model and select the face of the base feature shown in Figure 4-81 as the sketching plane.A yellow arrow is displayed on the selected face.

4. Choose the **Flip** button to flip the yellow arrow as shown in Figure 4-81. The arrow

points in the direction of viewing the sketching plane.

5. Select the top face shown in Figure 4-82 and then select the **Top** option from the **Orientation** drop-down list.

   The top face of the base feature is selected to be at the top while drawing the sketch.

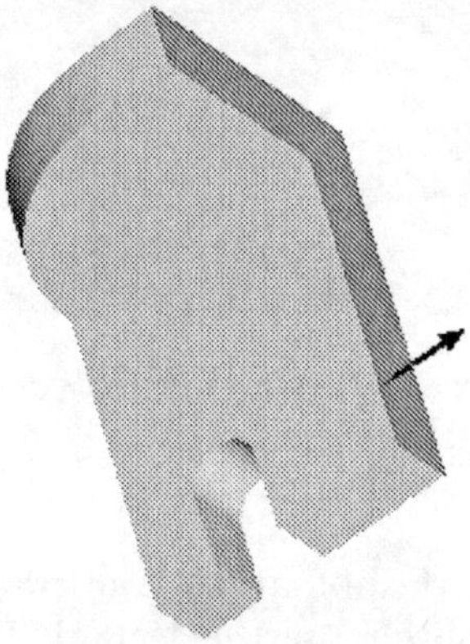

Figure 4-81 *Arrow showing the direction of viewing the model*

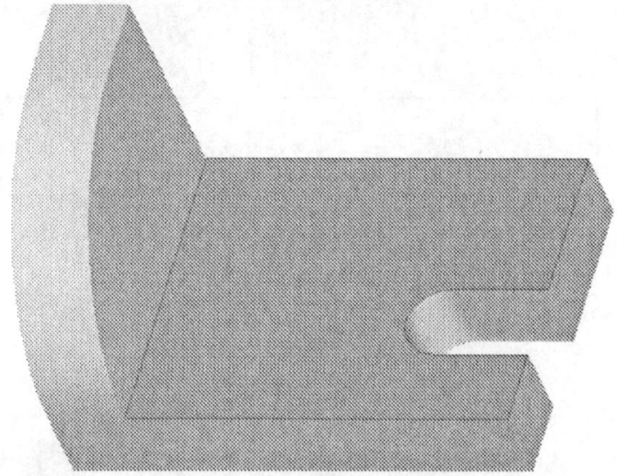

Figure 4-82 *Face selected to be at the top*

6. Choose the **Sketch** button to enter the sketcher environment.

Creating and Dimensioning the Sketch for the Third Feature

The sketch for the base feature consists of a rectangular section with a semicircular cut at the top. When this section is extruded, a feature with a semicircular slot will be created. Turn the display to **No Hidden**.

1. Draw the section sketch using various sketcher tools as shown in Figure 4-83.

2. The sketch is dimensioned automatically and some weak dimensions are assigned to it. Add the required constraints and modify the weak dimensions to the dimensions shown in Figure 4-83.

3. Choose the **Continue with the current section** button to exit the sketcher environment. The Extrude dashboard is enabled below the graphics window. Turn the display to **Shading.**

4. Use the middle mouse button to orient the model as shown in Figure 4-84. This orientation of the model gives a better view of the sketch in 3D space. A yellow arrow is displayed on the model, indicating the direction of extrusion.

5. Enter a value of **42** in the dimension box present in the Extrude dashboard and press ENTER. This feature is completed and its preview can be seen in the graphics window.

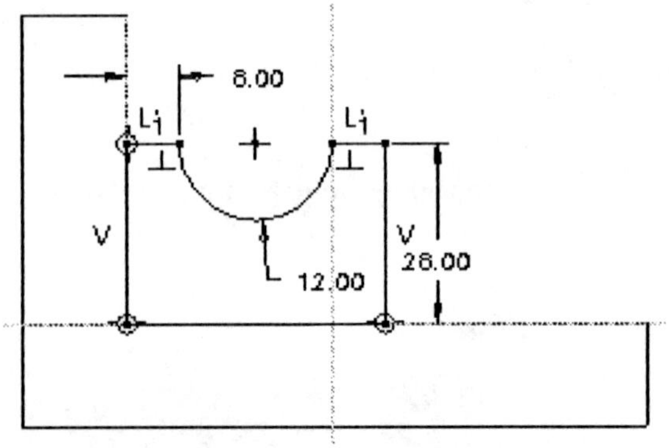

Figure 4-83 Sketch with dimensions and constraints

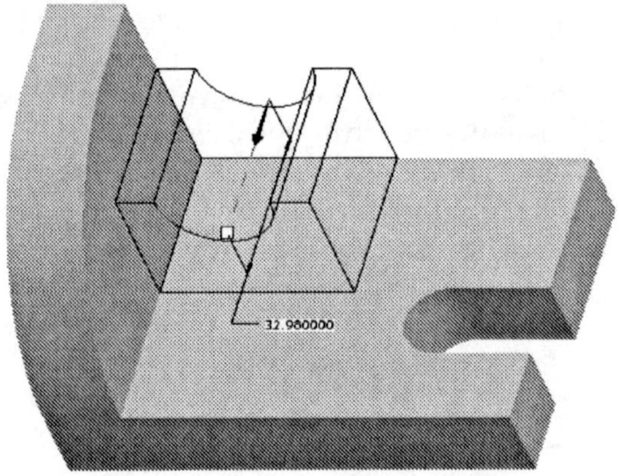

Figure 4-84 Arrow showing the direction of material addition

6. Choose the **Build feature** button from the Extrude dashboard to confirm the feature creation. The default view that is displayed when you choose the **Default View** option from the **Saved view list** button is shown in Figure 4-85.

Selecting the Sketching Plane for the Last Feature

As mentioned earlier, this feature and the third feature are created on the same plane. But the depth of extrusion is different for both of them. This is the reason they are considered as separate features. Therefore, for sketching this feature, you can use the sketching plane that was used for creating the third feature.

1. Choose the **Extrude Tool** button from the **Base Features** toolbar. The Extrude dashboard is displayed below the graphics window.

2. Choose the **Create a section or redefine the existing section** button from the Extrude dashboard. The **Section** dialog box is displayed.

3. Choose the **Use Previous** button from the **Sketch Plane** area in the **Section** dialog box. The yellow arrow is displayed on the graphics screen as shown in Figure 4-86.

   By specifying the **Use Previous** button, you are using the previous sketching plane that was defined for the previous feature. A yellow arrow is displayed on the face and points in the direction of viewing the sketch as shown in Figure 4-86.

4. Choose the **Sketch** button to enter the sketcher environment.

Creating and Dimensioning the Sketch for the Last Feature

The sketch of this feature consists of three lines. The bottom edge of the sketch is aligned with the top edge of the base feature and the left edge is aligned with the right edge of the third feature. When this sketch is extruded, it creates a rib shape.

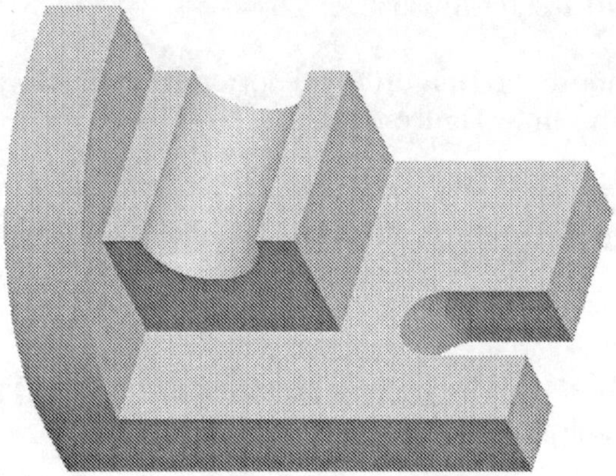

Figure 4-85 *Model with the second extruded feature*

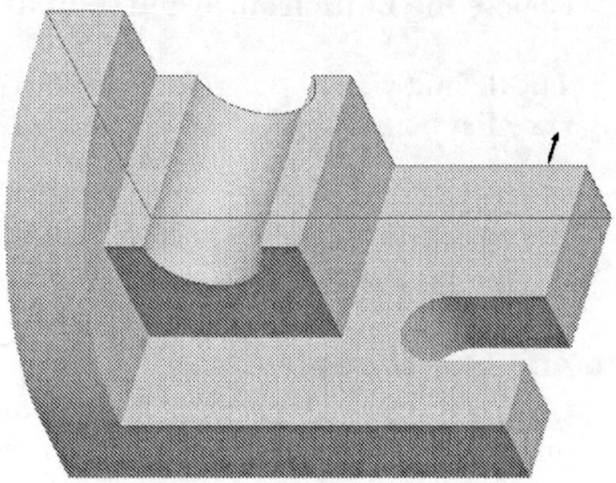

Figure 4-86 *Arrow pointing in the direction of viewing the sketch plane*

1. Draw the section sketch using various sketcher tools as shown in Figure 4-87.

   The sketch is dimensioned automatically and some weak dimensions are assigned to it. You can add the required constraints to align the lines and points in the sketch with the other features as shown in Figure 4-87. In the figure, the constraint symbol displayed on the line indicates that the edges of the adjacent features are used to close the section.

2. After the sketch is complete, choose the **Continue with the current section** button. The Extrude dashboard is enabled. Using the middle mouse button orient the model as shown in Figure 4-88.

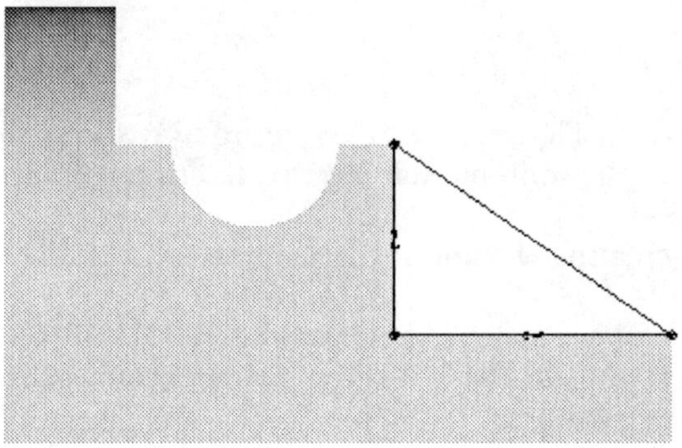

Figure 4-87 *Sketch and constraints for the last feature*

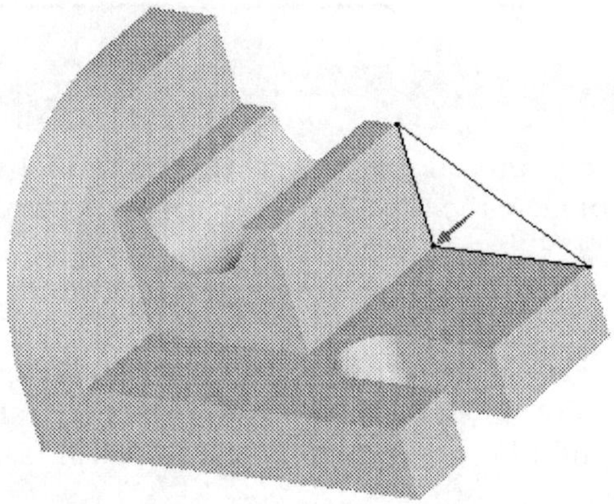

Figure 4-88 *Arrow showing the direction of material addition*

3. Enter a value of **12** in the dimension box present in the Extrude dashboard and press ENTER. The feature is completed and its preview can be seen on the graphics window.

4. Choose the **Build feature** button in the Extrude dashboard.

   The default view of the model, when you choose the **Default View** option from the **Saved view list** button from the **View** toolbar, is shown in Figure 4-89.

Note

*The last feature of this tutorial can also be created using the **Rib** feature creation tool. This tool will be discussed in Chapter 5.*

Saving the Model

1. Choose the **Save the active object** button from the **File** toolbar and save the model. The order of feature creation can be seen from the **Model Tree** shown in Figure 4-90.

Figure 4-89 Completed model for Tutorial 2

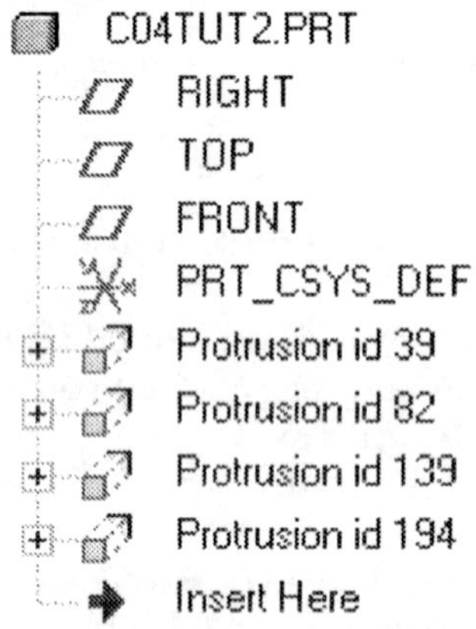

***Figure 4-90 Model Tree** for Tutorial 2*

Tutorial 3

In this tutorial you will create the model shown in Figure 4-91. This figure also shows the front view, the top view, and the right-side view of the solid model. **(Expected time: 45 min)**

The following steps outline the procedure for creating this model:

a. First, examine the model and then determine the number of features in it, see Figure 4-91. The model is composed of three features: one on the left (base feature), one in the middle, and one hollow cylindrical feature with a hole.

b. Select the sketch plane for the base feature, draw the sketch using the sketching tools, apply the dimensions and constraints, and then extrude the sketch to the given depth, see Figure 4-93.

c. Select the sketch plane for the middle feature, draw the sketch using the sketching tools, apply the dimensions and constraints, and then extrude the sketch to the given depth, see Figure 4-98.

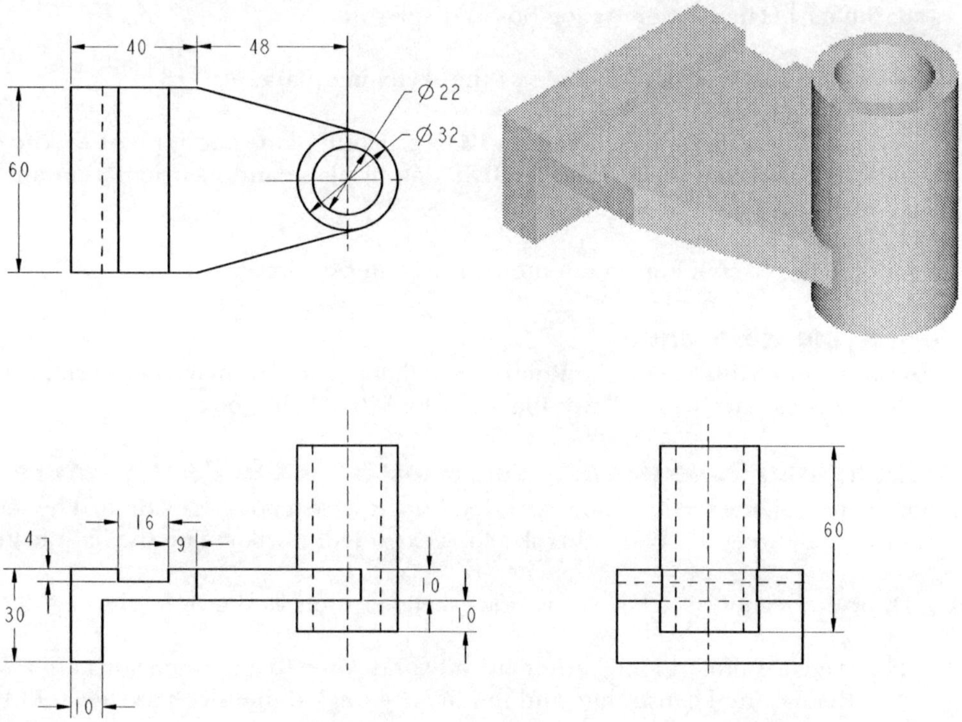

Figure 4-91 Top, front, right-side, and isometric views of the model

d. To create the hollow cylindrical feature, create a datum plane. Then draw the sketch, apply the dimensions and constraints, and then extrude the sketch to the given depth, see Figure 4-103.

The working directory has already been selected in Tutorial 2 and therefore you do not need to select it again. However, if you want to select the *c04* directory, select it from the Navigator.

Creating New Object File

1. Open a new part file and name it as **c04tut3**. The three default datum planes are displayed on the graphics window if the **Datum planes on/off** button is turned on. The **Model Tree** is not displayed on the graphics window as its display was turned off in the first tutorial.

Selecting the Sketching Plane for the Base Feature

To create the sketch for the base feature, you first need to select the sketching plane. In this model, you need to draw the base feature on the **FRONT** datum plane because the direction of extrusion is perpendicular to the **FRONT** datum plane.

1. Choose the **Extrude Tool** button from the **Base Features** toolbar. The Extrude dashboard is displayed below the graphics window.

2. Choose the **Create a section or redefine the existing section** button from the Extrude dashboard. The **Section** dialog box is displayed.

3. Select the **FRONT** datum plane as the sketching plane.

   A yellow arrow is displayed on the **FRONT** datum plane and it points in the direction of viewing the sketch plane. The **RIGHT** datum plane and its orientation are selected by automatically.

4. Choose the **Sketch** button to enter the sketcher environment.

Specifying References

In the sketcher environment, the **References** dialog box is displayed. The status displayed in the **Reference status** area is **Fully Placed**. Close this dialog box.

Creating and Dimensioning the Sketch for the Base Feature

From the model, the section to be extruded for the base feature is evident. The section sketch is shown in Figure 4-92. When this sketch is extruded, it will create the base feature.

1. Draw the section sketch using various sketcher tools as shown in Figure 4-92.

2. The sketch is dimensioned automatically and some weak dimensions are assigned to it. Add the required constraints and modify the weak dimensions as shown in Figure 4-92.

3. Choose the **Continue with the current section** button. The Extrude dashboard is displayed.

4. Choose the **Default View** option from the **Saved view list** button of the **View** toolbar.

   The default view is displayed but the model does not fit in the graphics window. A yellow arrow is also displayed on the model, indicating the direction of extrusion.

5. Enter a value of **60** in the dimension box that is present on the Extrude dashboard and press ENTER.

6. Choose the **Build feature** button from the Extrude dashboard.

   The base feature is completed as shown in Figure 4-93. You can use the middle mouse button to spin the model to view it from various directions.

Selecting the Sketching Plane for the Second Feature

The next feature is an extruded feature. The sketching plane for this feature is the top face of the base feature.

1. Choose the **Extrude Tool** button from the **Base Features** toolbar. The Extrude dashboard is displayed below the graphics window.

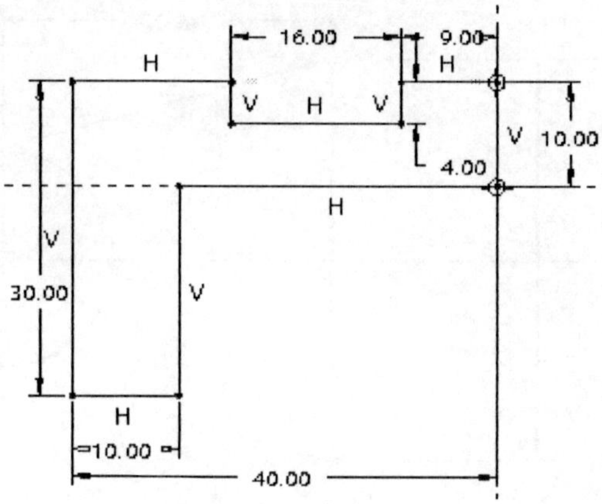

Figure 4-92 Sketch with dimensions and constraints for the base feature

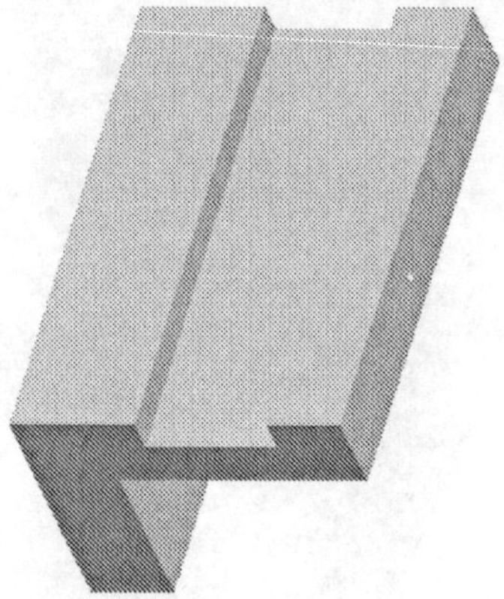

Figure 4-93 Base feature of the model

2. Choose the **Create a section or redefine the existing section** button from the Extrude dashboard. The **Section** dialog box is displayed.

3. Select the top face of the base feature shown in Figure 4-94 as the sketching plane. A yellow arrow that points in the direction of viewing the sketch is displayed on the top face.

4. Chose the **Flip** button to flip the yellow arrow to point in the direction shown in Figure 4-94.

5. Select the **RIGHT** datum plane and then choose the **Right** option from the **Orientation** drop-down list.

   By selecting the **RIGHT** datum plane, the model will be oriented in such a way that the highlighted planar surface will be at the right while drawing the sketch.

6. Choose the **Sketch** button to enter the sketcher environment.

Creating and Dimensioning the Sketch for the Second Feature

The next feature to be created is an extrude feature. The section for the extrude feature is shown in Figure 4-95. Before drawing the sketch, turn the model display to **No Hidden**.

1. Close the **References** dialog box that is displayed on the top right corner of the screen.

2. Draw the section sketch using various sketcher tools. In the sketch, draw a center line passing through the center of the arc as shown in Figure 4-95. This center line helps in dimensioning the sketch. Add the required constraints and dimensions to the sketch as shown in Figure 4-95.

   Before exiting the sketcher environment, turn the model display to **Shading**.

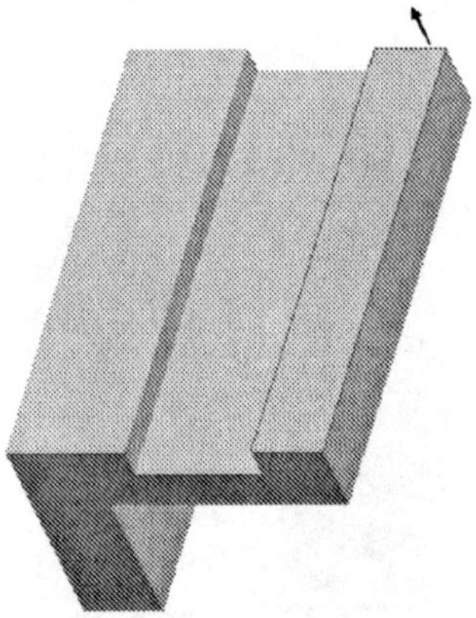

Figure 4-94 Arrow pointing from the sketching plane in the direction of feature creation

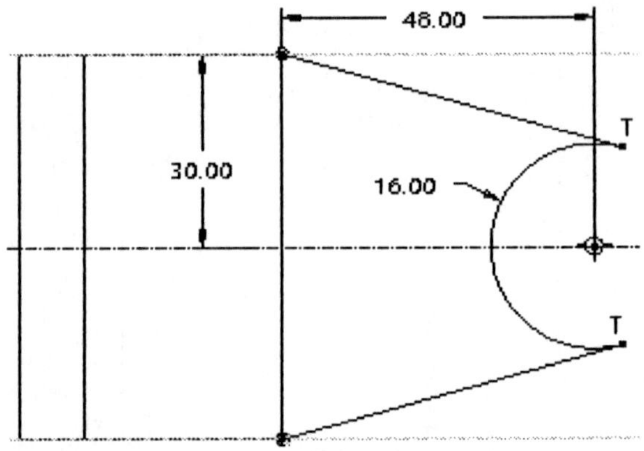

Figure 4-95 Sketch with dimensions and constraints

3. Choose the **Continue with the current section** button. The Extrude dashboard is enabled. Use the middle mouse button to orient the model as shown in Figure 4-96.

4. Enter **10** in the dimension box present on the Extrude dashboard and press ENTER.

5. Now, choose the **Build feature** button to confirm the feature creation. The default trimetric view of the extruded feature is shown in Figure 4-97.

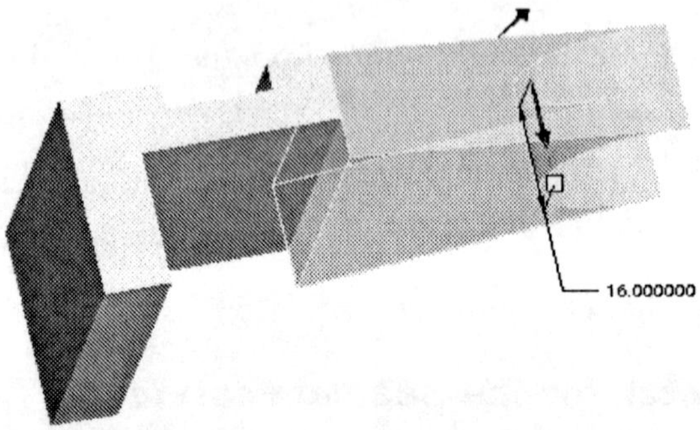

Figure 4-96 Arrow showing the direction of material addition

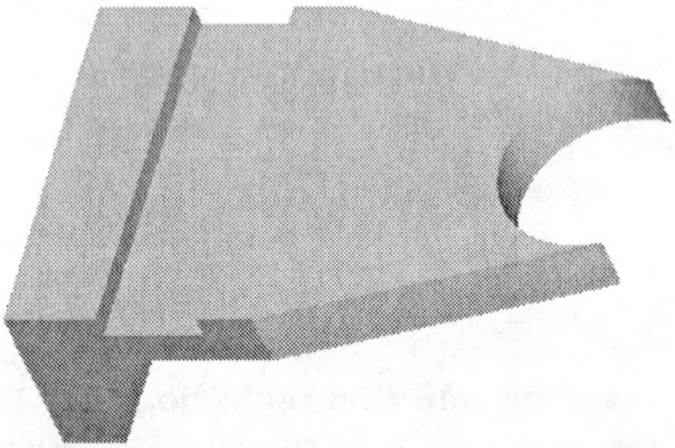

Figure 4-97 Model with the second extruded feature

Creating a Datum Plane for the Last Feature

To create the hollow cylindrical feature, you require a datum plane. This datum plane will be created at an offset distance of 10 from the bottom face of the second feature shown in Figure 4-98.

Note

The other method to create this feature is to select the top planar surface of the second feature as the sketching plane and extrude the sketch on both sides of the sketching plane. The depth of extrusion will be different at both the sides. If you use this method to create this cylindrical feature, you do not need to create a datum plane.

1. Choose the **Datum Plane Tool** button from the **Datum** toolbar. The **DATUM PLANE** dialog box is displayed.

2. Spin the model using the middle mouse button and then using the left mouse button, select the face shown in Figure 4-98.

   As you select the face of the second feature, the **Offset** constraint is displayed in the **References** collector of the dialog box.

3. In the **Translation** dimension box, enter a value of **10** and press ENTER.

4. Choose the **OK** button from the **DATUM PLANE** dialog box.

   Datum plane **DTM1** is created as shown in Figure 4-99 and will be selected as the sketching plane for creating the sketch.

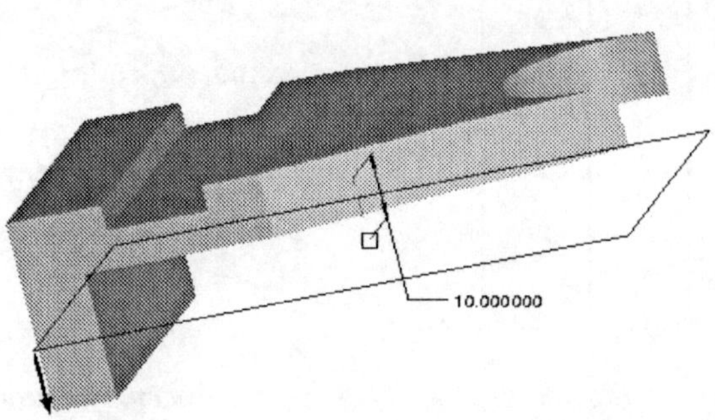

Figure 4-98 *Creating the datum plane*

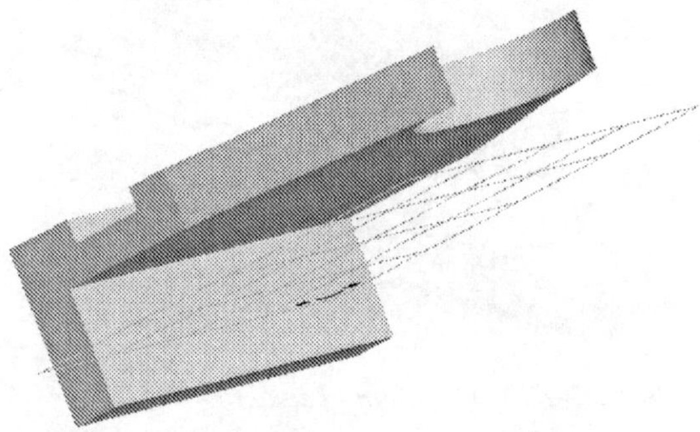

Figure 4-99 *Model after creating the datum plane*

Selecting the Sketching Plane for the Last Feature

The plane **DTM1** will be selected as the sketching plane and the depth of extrusion will be given from this plane.

1. Choose the **Extrude Tool** button from the **Base Features** toolbar. The Extrude dashboard is displayed below the graphics window.

2. Choose the **Create a section or redefine the existing section** button from the Extrude dashboard. The **Section** dialog box is displayed.

3. Select **DTM1** as the sketching plane. The yellow arrow appears on the datum plane.

4. Choose the **Flip** button to reverse the direction of viewing the sketch. Now, the arrow points in the direction shown in Figure 4-100.

5. Select the **FRONT** datum plane and choose the **Bottom** option from the **Orientation** drop-down list.

6. Choose the **Sketch** button to enter the sketcher environment.

Creating and Dimensioning the Sketch for the Last Feature

The sketch for the hollow cylindrical feature will be drawn on the datum plane **DTM1**. The sketch for the hollow cylindrical feature consists of two concentric circles. Before drawing the sketch, turn the model display to **No Hidden**.

1. Close the **References** dialog box that is displayed on the top right corner of the screen.

2. Draw the section sketch using various sketcher options and add the required constraints and dimensions to it as shown in Figure 4-101.

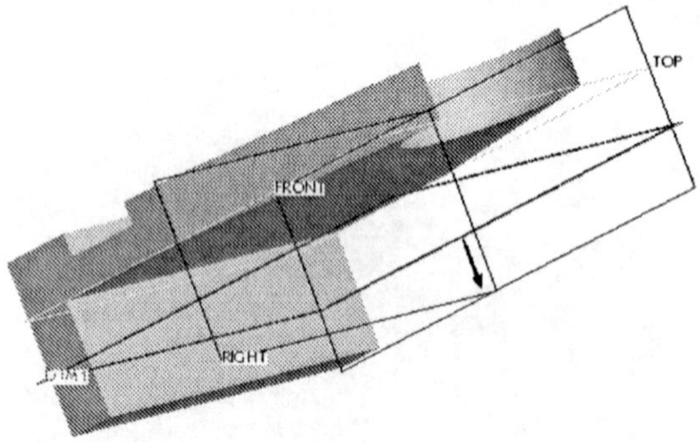

***Figure 4-100** Direction of viewing the sketching plane*

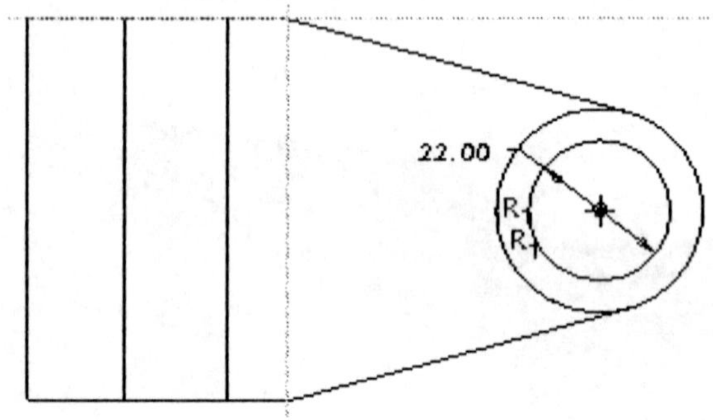

***Figure 4-101** Sketch with dimensions and constraints*

3. Choose the **Continue with the current section** button. The Extrude dashboard is displayed. Now, turn the model display to **Shading**.

   Use the middle mouse button to orient the model as shown in Figure 4-102. This view gives you a better view of the sketch in the 3D space.

4. Enter a value of **60** in the dimension box present on the Extrude dashboard and press ENTER.

5. Choose the **Build feature** button to confirm the feature creation. The default trimetric view of the complete model is shown in Figure 4-103.

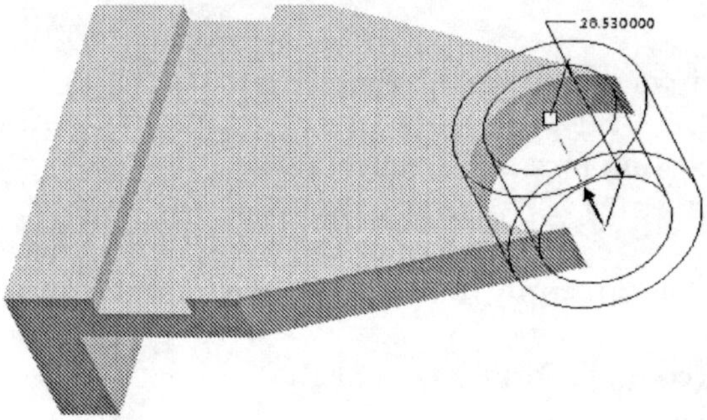

Figure 4-102 Preview of the feature

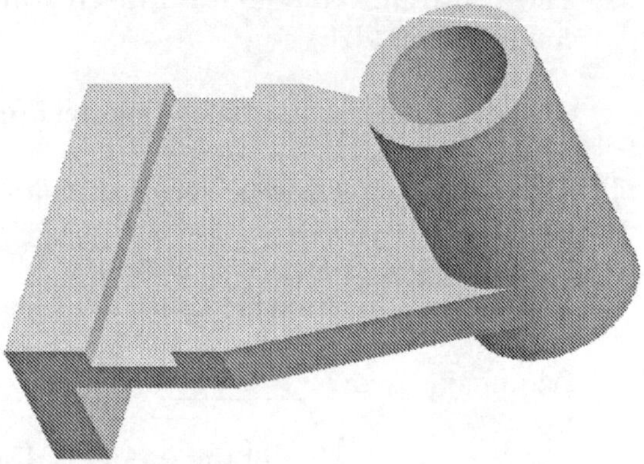

Figure 4-103 Final model of Tutorial 3

Saving the Model

1. Choose the **Save the active object** button from the **File** toolbar and save the model.

   The order of feature creation can be seen from the **Model Tree** shown in Figure 4-104. The feature id numbers that are displayed in the **Model Tree** may be different when you create the features.

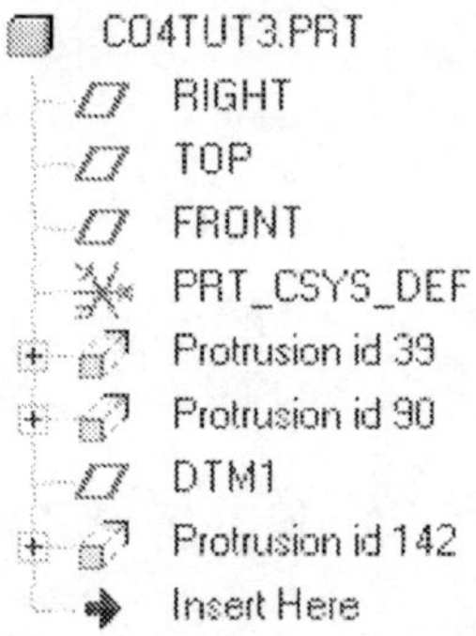

***Figure 4-104* Model Tree** *for Tutorial 3*

Self-Evaluation Test

Answer the following questions and then compare your answers to the answers given at the end of this chapter.

1. You can change the default names assigned to the datum planes. (T/F)

2. Datum points are also used to associate note in the drawings and attach datum targets. (T/F)

3. The constraint combination of creating an offset plane through cylinder is a valid combination. (T/F)

4. Generally all the features in a model are created on a single sketching plane. (T/F)

5. A sketching plane can be selected on an existing face of a feature. (T/F)

6. While creating a datum feature, the __________ selected from the model help you to choose from the available constraints that you can apply to constraint the datum feature.

7. Datum axis is an __________ axis that is created in Pro/ENGINEER.

8. The __________ view of the model is displayed when you choose the __________ option from the **Saved view list** button.

9. When you create a new object file, __________ default datum planes are displayed on the graphics screen.

10. The default datum planes in the Part mode are named as __________, __________, and __________.

Review Questions

Answer the following questions:

1. Which one of the following is not a type of datums available in Pro/ENGINEER?

   (a) Axis (b) Plane
   (c) Circle (d) Curve

2. How many methods are available in Pro/ENGINEER to create a datum plane?

   (a) One (b) Two
   (c) Three (d) Four

3. Which one of the following dialog boxes is displayed when you choose the **Datum Point Tool** button from the **Datum** toolbar while extruding a section sketch?

   (a) **DATUM POINT** dialog box (b) **DATUM PLANE** dialog box
   (c) **DATUM AXIS** dialog box (d) None

4. Which one of the following combinations of keys and mouse buttons can be used to spin a model on the graphics window?

   (a) CTRL+ALT (b) middle mouse button
   (c) left mouse button (d) CTRL+right mouse button

5. Which one of the following menus in the menu bar is used to set the default orientation of the model to isometric?

   (a) **File** (b) **Insert**
   (c) **Tool** (d) None

6. Datum planes are considered as feature geometry and have mass and volume. (T/F)

7. To set the default orientation of the model to isometric, you have to use the **Environment** dialog box. (T/F)

8. Generally, the sketching plane for the base feature of any model is decided after viewing the isometric view or the drawing views of the model. (T/F)

9. Datum planes are used as a reference for mirroring features, copying features, for creating a cross-section, as well as for orientation of references. (T/F)

10. Unlike the datum planes constraint options, all the datum axes constraint options are standalone. (T/F)

Exercises

Exercise 1

Create the model shown in Figure 4-105. The dimensions, front view, and right-side view of the model are shown in Figure 4-106. **(Expected time: 45 min)**

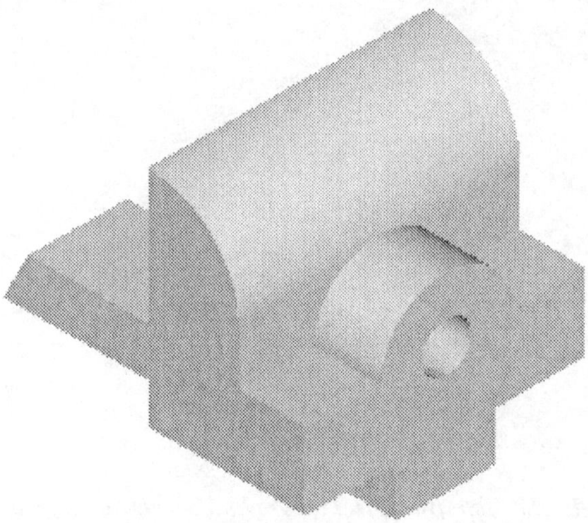

Figure 4-105 *Isometric view of the model*

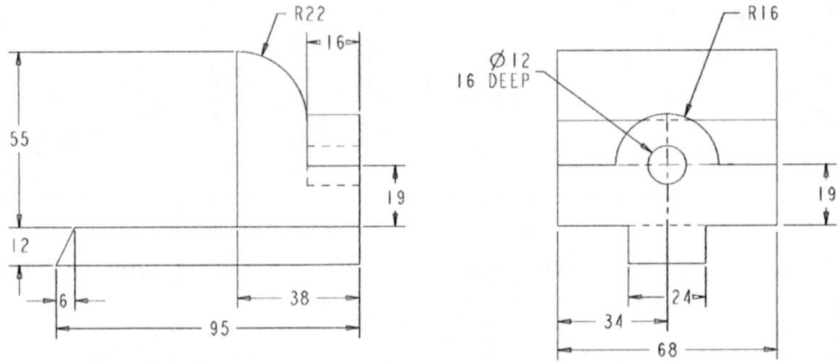

Figure 4-106 Front view and right-side view of the model

Exercise 2

Create the model shown in Figure 4-107. The dimensions, and the front, top, right-side, and isometric views of the model are also shown in the figure. **(Expected time: 45 min)**

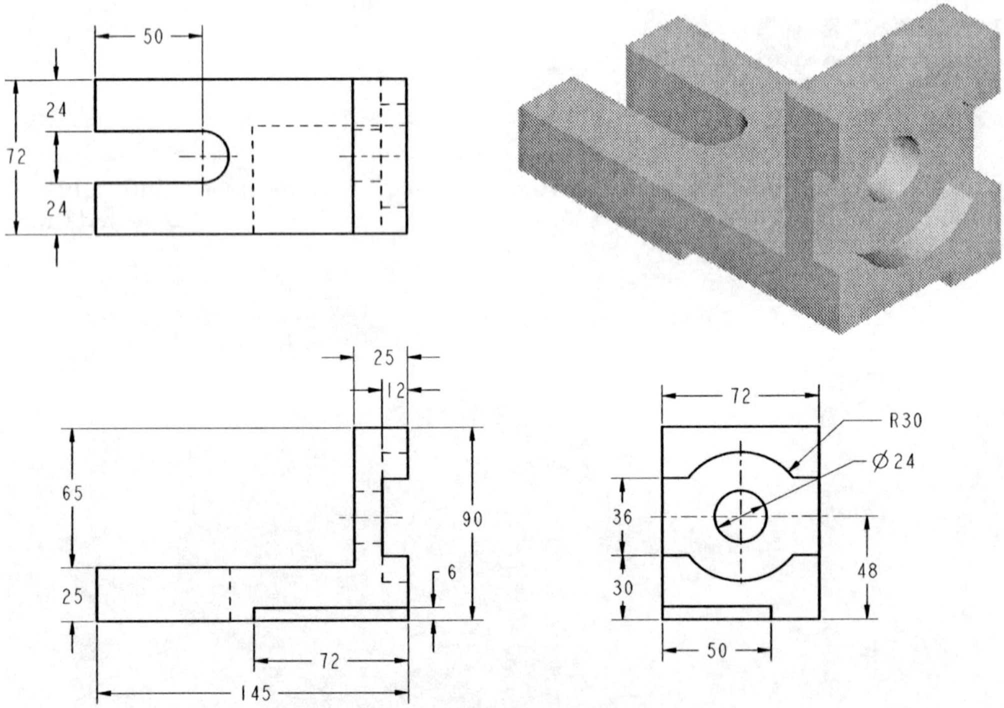

Figure 4-107 Top, front, right-side, and isometric views of the model

Exercise 3

Create the model shown in Figure 4-108. The dimensions, the top view, front section view, and the two auxiliary views of the model are shown in Figure 4-109. **(Estimated time: 1 hr)**

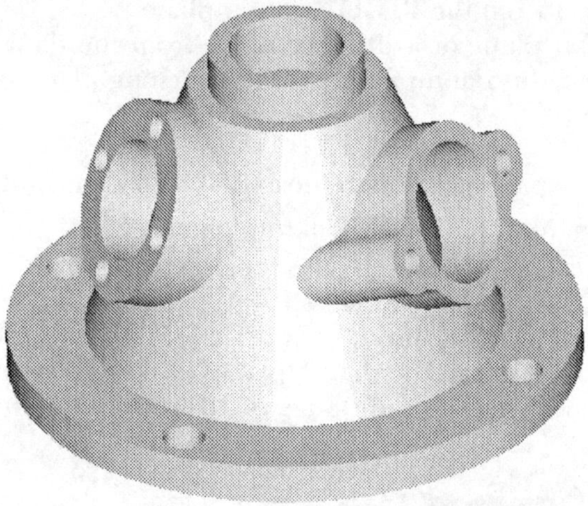

Figure 4-108 The 3D view of the model

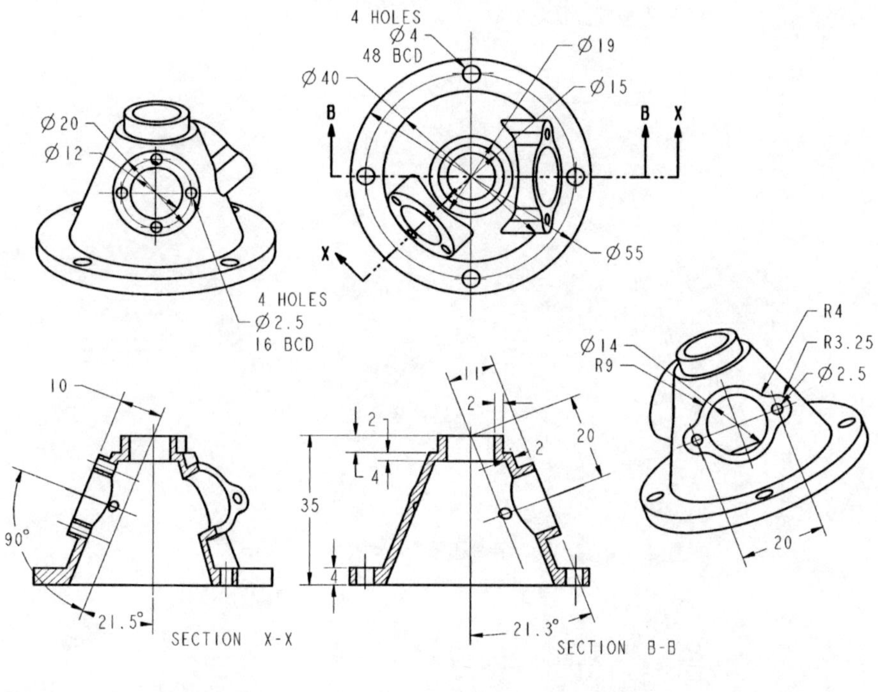

Figure 4-109 Top view, front section view, and the two auxiliary views of the model

Hint to create the two side features in Exercise 3:

Feature on the right

1. Create an axis passing through the top face of the base feature and through the **RIGHT** datum plane.
2. Create a datum plane passing through the axis (created in the previous step) and at an angle of 21.5 degrees from the **RIGHT** datum plane.
3. Now, create an offset plane at a distance of 11 from the datum plane (created in the previous step). Select this datum plane as the sketching plane.

Feature on the left

1. Create a datum plane passing through the axis of revolution of the base feature and at an angle of 45 degrees from the **FRONT** datum plane.
2. Create a datum axis passing through the top face of the base feature and through the datum plane (created in the previous step).
3. Create a datum plane passing through datum axis (created in the previous step) and at an angle of 21.5 degrees from the datum plane (created in step1).
4. Create a datum plane offset to a distance of 10 from the datum plane (created in the previous step).

Exercise 4

Create the model shown in Figure 4-110. The dimensions, the left-side view, auxiliary view, front view, and isometric view of the model are also shown in the figure.

(Expected time: 45 min)

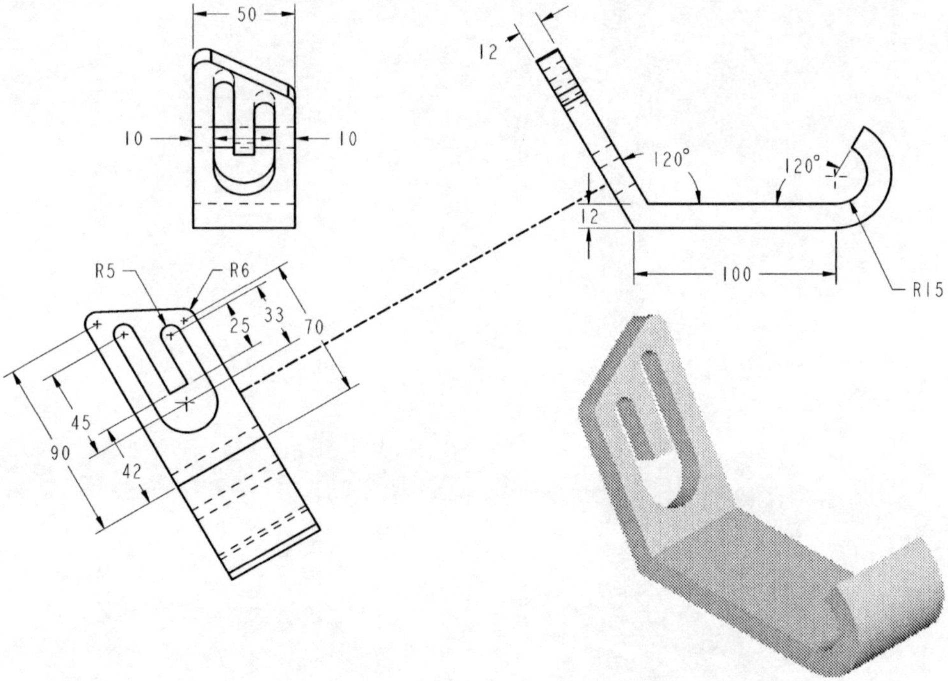

Figure 4-110 *Left-side view, auxiliary view, front view, and isometric view of the model*

Answers to the Self-Evaluation Test

1 - T, **2** - T, **3** - F, **4** - F, **5** - T, **6** - geometry, **7** - imaginary, **8** - trimetric, **Default View**, **9** - three, **10** - **TOP**, **FRONT**, **RIGHT**.

Chapter 5

Options Aiding Construction of Parts-I

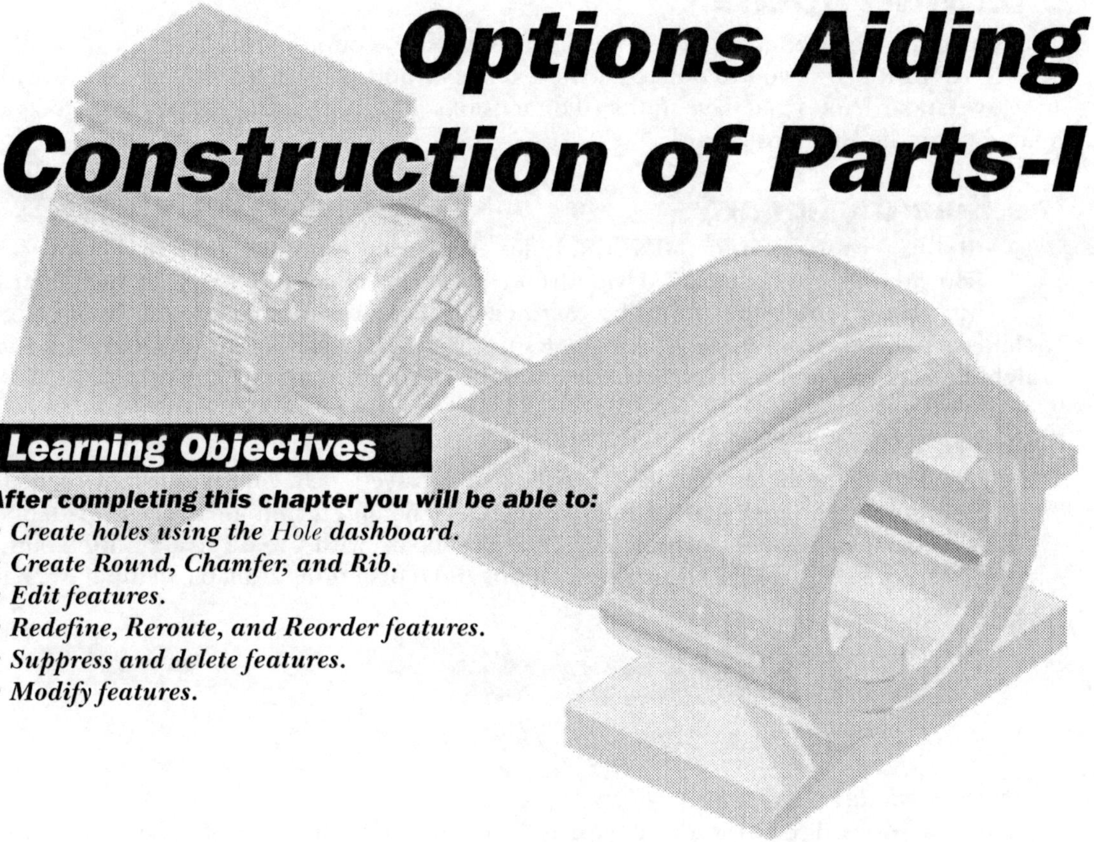

Learning Objectives

After completing this chapter you will be able to:

- ***Create holes using the** Hole **dashboard.***
- ***Create Round, Chamfer, and Rib.***
- ***Edit features.***
- ***Redefine, Reroute, and Reorder features.***
- ***Suppress and delete features.***
- ***Modify features.***

OPTIONS AIDING CONSTRUCTION OF PARTS

This chapter explains the feature creation tools provided in Pro/ENGINEER that help in creating a model and editing it once the model is completed. In this chapter you will learn to create various types of holes that are required in most of the engineering designs. In previous chapters you learnt to create holes using extrude cut, but in this chapter you will create holes using the Hole dashboard. Using the Hole dashboard, it becomes easy to create holes as well as to modify them.

In this chapter you will also learn to create rounds, chamfers, and ribs. All the options that are discussed in this chapter increase the efficiency in creating a design using Pro/ENGINEER.

CREATING HOLES

In engineering components, holes can be counterbore, countersink, tapered, or drilled. Pro/ENGINEER allows you to create all these types of holes. Pro/ENGINEER also provides industry standard holes that have standard dimensions. In Pro/ENGINEER Wildfire, holes are created using the Hole dashboard.

The Hole Dashboard

In this release of Pro/ENGINEER, holes are created using the Hole dashboard. The Hole dashboard is displayed when you choose **Insert > Hole** from the menu bar or the **Hole Tool** button from the **Engineering Features** toolbar present in the **Right Toolchest**. You can create three types of holes using the Hole dashboard. The first type is a straight hole, the second is a sketched hole, and the third is a standard hole.

Creating Straight holes

Straight holes are the holes that have a circular cross-section having a constant diameter throughout the depth. They start at the placement plane and terminate at the user-defined depth or at the specified end surface. The Hole dashboard with **Create straight hole** button selected is shown in Figure 5-1. The various options and tool buttons available in this dashboard are discussed next.

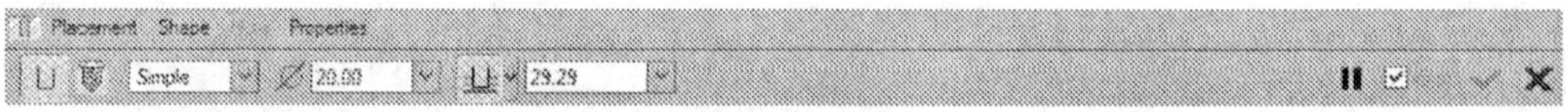

***Figure 5-1** Hole dashboard*

Drop-down list

The drop-down list in the dashboard is used to specify the type of hole. The **Simple** option is selected by default. The other option available in this drop-down list is **Sketched**. The **Simple** option is used to draw straight holes. The **Sketched** option of creating holes is discussed later in the chapter.

Dimension boxes

The first dimension box in the Hole dashboard is used to specify the diameter of the hole.

The second dimension box in the Hole dashboard is used to specify the depth of the hole. The options to specify the depth of the hole are same as that were available in the Extrude dashboard.

Placement tab

When you choose the **Placement** tab, the slide-up panel is displayed as shown in Figure 5-2. This slide-up panel is displayed with the parameters that are specified to define a linear hole. The **Primary** collector in the slide-up panel is displayed in yellow color. This indicates that Pro/ENGINEER is prompting you to select the placement plane (primary reference). You can select the placement plane only when this collector is in yellow color. Once you have selected the primary reference, its name appears in the box. The primary reference can be a plane, an axis, or a point.

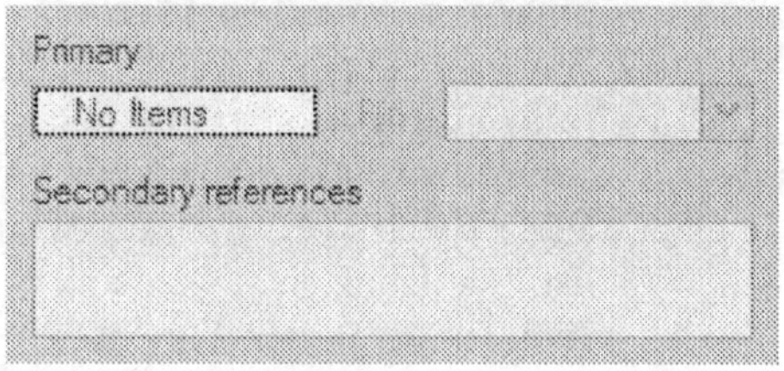

Figure 5-2 ***Placement*** *tab slide-up panel*

The **Flip** button is used to change the direction of hole creation with reference to the primary reference. When you choose this button, the direction of hole creation is reversed and you can view its preview on the graphics window. Remember that the hole will be created only if it is possible to create the hole in the specified direction.

The drop-down list to the right of the **Flip** button contains the options that determine the type of placement for the hole. Use the options in this drop-down list after selecting the placement plane or primary reference. These options are discussed next.

Linear. When you select this option, you need to specify the distances from two linear references. Generally, these linear references are the edges of the planar surface on the model, any two planar surfaces or axes, or a combination of any of these. Figure 5-3 shows a hole on the top face of the block. The procedure to create this hole using the Hole dashboard is discussed next.

1. Invoke the Hole dashboard.
2. Choose the **Placement** tab to open the slide-up panel.
3. Select the top face of the model to place the hole. The preview of the hole is displayed.
4. Click in the **Secondary References** collector to make the secondary selections.
5. Select the left face or the top left edge of the model.
6. Use CTRL+left mouse button to select the front face or the top front edge of the model. Now, you have selected the two references that were needed to dimension the hole.
7. Specify the diameter of the hole in the dimension box present on the Hole dashboard.
8. From the depth flyout, select the **Drill to intersect with all surfaces** button.
9. Choose the **Build feature** button to exit the Hole dashboard.

Note

When you have to select the secondary references, you need to click in the ***Secondary References*** *collector and then select the references on the model.*

Radial. This option is used to create a hole that can be referenced to an axis. When you select this option, you need to select an axial reference and an angular reference to place the hole. The distance from the axis and the angle from a plane are entered in the boxes under the **Secondary References** collector. Figure 5-4 shows a radial hole on a curved surface.

1. Invoke the Hole dashboard.
2. Choose the **Placement** tab to open the slide-up panel.
3. Select the curved surface of the model to place the hole. The preview of the hole is displayed.
4. Click in the **Secondary References** collector to make the secondary selections.
5. Select the datum plane that is normal to curved surface to specify the angular value.
6. Use CTRL+left mouse button to select the top face or the top edge of the model. Now, you have selected the two references that were needed to dimension the hole.
7. Specify the diameter of the hole in the dimension box present on the Hole dashboard.
8. From the depth flyout, select the **Drill to intersect with all surfaces** button.
9. Choose the **Build feature** button to exit the Hole dashboard.

Tip: *Note that since you have selected the curved surface as the placement plane for the hole, therefore, you do not need to select the axis to specify the radial distance. However, if the placement plane is a plane surface, then you do need to select an axis to create a radial hole at some radius.*

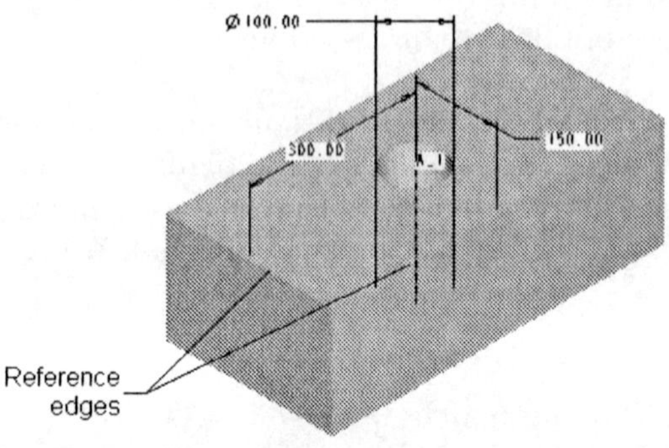

***Figure 5-3** Linear dimensioning of hole*

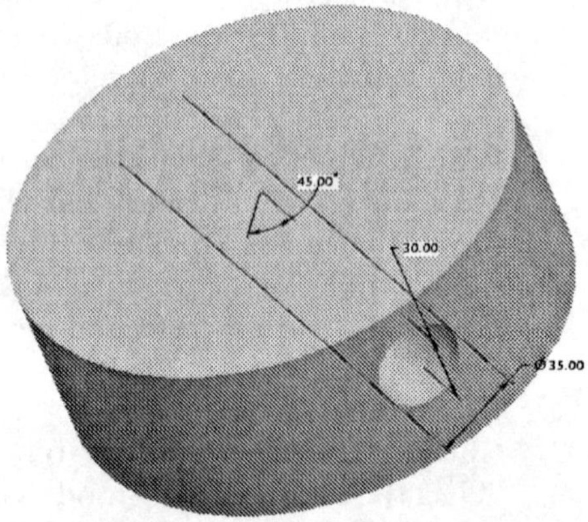

***Figure 5-4** Radial dimensioning of hole*

Diameter. This option creates a diametrically placed hole. When you select this option, you need to select an axial reference and an angular reference to place the hole. Figure 5-5 shows a diameter hole on a plane. The following steps explain the procedure to create this hole.

1. Invoke the Hole dashboard.
2. Choose the **Placement** tab to open the slide-up panel.
3. Select the top face of the model to place the hole. The preview of the hole is displayed.
4. Click in the **Secondary References** collector to make the secondary selections.

5. Select the edge shown in Figure 5-5 to specify the angular value.
6. Use CTRL+left mouse button to select the axis of the hole that is on the top face. Now, you have selected the two references that were needed to dimension the hole.
7. Specify the diameter of the hole in the dimension box present on the Hole dashboard.
8. From the depth flyout, select the **Drill to intersect with all surfaces** button.
9. Choose the **Build feature** button to exit the Hole dashboard.

Coaxial. This option creates a hole coaxially. When you select this option, you need to select an axis. No dimensions are required to place a coaxial hole. Figure 5-6 shows a coaxial hole on a plane. The procedure to create this hole is explained next.

1. Invoke the Hole dashboard.
2. Choose the **Placement** tab to open the slide-up panel.
3. Select the top surface of the model to place the hole. The preview of the hole is displayed.
4. From the drop-down list, select the **Coaxial** option.
5. Click in the **Secondary References** collector to make the secondary selections.
6. Select the axis of revolution of the base feature.
7. Specify the diameter of the hole in the dimension box present on the Hole dashboard.
8. From the depth flyout, select the **Drill to intersect with all surfaces** button.
9. Choose the **Build feature** button to exit the Hole dashboard.

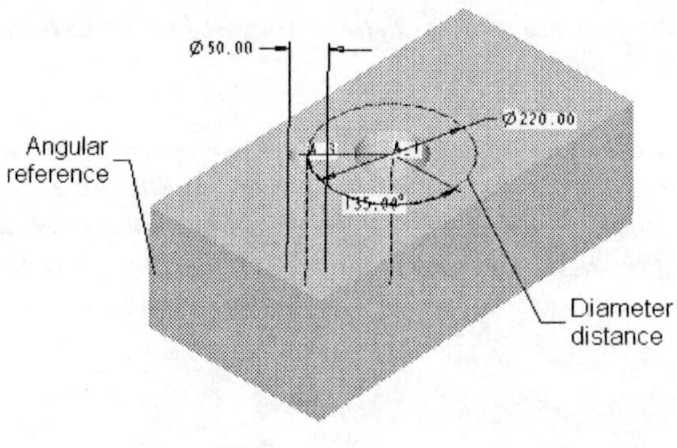

Figure 5-5 *Diameter dimensioning of hole*

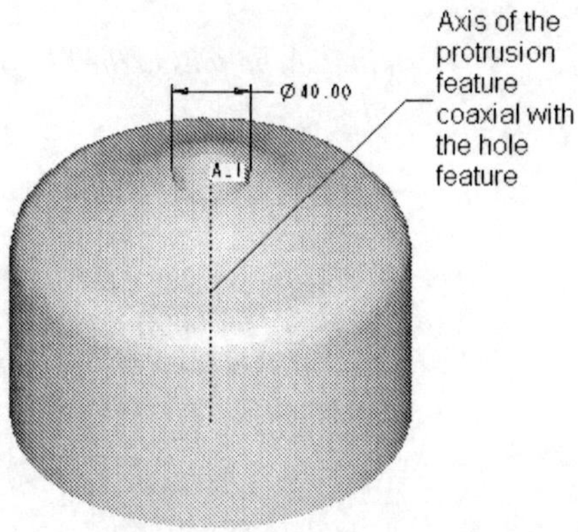

Figure 5-6 *Coaxial hole*

Shape tab

When you choose the **Shape** tab, the slide-up panel is displayed as shown in Figure 5-7. This slide-up panel can be used to specify the depth and the diameter of the hole. The depth of the hole can be specified in two directions with respect to the placement plane. The options to specify the depth of the hole are similar to those that were available in the Extrude dashboard. In the slide-up panel, notice the sketch for the hole, which gives an idea of the shape of the hole and its dimensions. The sketch that is shown in the slide-up panel is showing the depth of the hole in the Side 1 direction.

The **Side 2** drop-down list is used to specify the depth of the hole in the direction

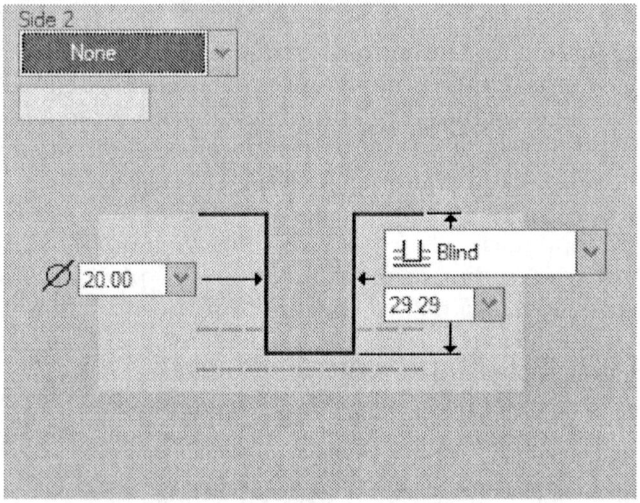

***Figure 5-7 Shape** slide-up panel*

opposite to Side 1. This means the hole can be created on both sides of the sketching or placement plane. By default, the **None** option is selected in this drop-down list.

Note

*The termination options in the **Shape** slide-up panel are similar to those discussed in the Extrude dashboard.*

When you specify the placement plane for a hole and then specify the two references in case of linear hole, you can notice the four drag handles on the hole. These four drag handles have four different functions to place the hole dynamically on the placement plane. Using these handles you can set the diameter of the hole, depth of the hole, and specify the two references to place the hole. Remember that these drag handles appear only when you have selected the primary reference or in other words the placement plane.

Tip: *The diameter of the hole and its depth in the Side 1 direction can be specified without using the slide-up panels. The options to specify diameter and depth in Side 1 direction are available on the Hole dashboard. Only when you need to specify the depth in the **Side 2** direction, you open the slide-up panel by choosing the **Shape** tab from the Hole dashboard.*

Creating Sketched hole

The **Sketched** option allows you to sketch the cross-section for the hole that is revolved about a center axis. This option is used to draw custom shapes for the hole. When you choose the **Sketched** option from the first drop-down list in the Hole dashboard, the Hole dashboard appears as shown in Figure 5-8. The tools and options available in the dashboard are discussed next.

Figure 5-8 Hole dashboard

Tip: *Remember that while placing any hole using the Hole dashboard, you need to define two steps. First is the placement plane on which the hole feature will be created and the second is the dimensional references for all holes other than the Co-axial hole. For co-axial hole, you need to select an axis.*

Activates Sketcher to create section button

Choose the **Activates Sketcher to create section** button to open a new window with the sketcher environment. The cross-section for the hole is sketched using the normal sketcher tools available. While drawing the sketch, a center line must be drawn that acts as the axis of revolution for the section of hole. The sketched holes can be blind or through all, depending on the dimensions of the section sketch.

When you complete the sketch for the hole and choose the **Continue with the current section** button to exit the sketcher environment, the Hole dashboard below the graphics window is enabled. Now, you need to specify the placement options to place the hole. The placement options are the same as discussed earlier.

Note

The placement options to place the hole can be specified after drawing the sketch for the hole or before drawing the sketch. The only difference is that when you want to specify the placement options after drawing the sketch, the preview of the hole is not available on the graphics window. This is because the primary reference is not selected yet. Therefore, it is recommended to specify the placement options before drawing the sketch for the hole.

Opens an existing sketched profile button

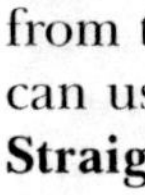

On the Hole dashboard, the **Opens an existing sketched profile** button is used to open an existing sketch that is saved in *.sec* file format. When you choose this button, the **Open Section** dialog box is displayed. This dialog box is used to select the *.sec* file to define the shape of the hole.

Placement tab

The options in the slide-up panel that is displayed when you choose the **Placement** tab from the Hole dashboard are the same that were available in case of **Straight** hole. You can use the same types of placement options that were available when you created the **Straight** hole. The types of placement options were discussed earlier.

Shape tab

The options in the slide-up panel that is displayed when you choose the **Shape** tab from the Hole dashboard are different than that were available in case of **Straight** holes. When you choose the **Shape** tab, the slide-up panel displays the preview of the sketch that you have drawn in the sketcher environment. You can pan and zoom the sketch in the slide-up panel.

Note

*If you choose the **Shape** tab before drawing the sketch for the hole, then the slide-up panel is blank and does not show a sketch.*

Standard Hole

The standard holes can be created by choosing the **Create standard hole** button from the Hole dashboard. The holes created using this button are based on industry standard fastener tables.

Note

Since the standard holes are based on industry standards and their dimensions are in millimeters (mm), therefore you need to select the template that measures the dimensions in mm. This can be done when you create a new file.

When you choose the **Create standard hole** button from the Hole dashboard, the dashboard appears as shown in Figure 5-9. The tools and options available in the dashboard are discussed next.

Figure 5-9 *Hole dashboard*

Drop-down lists

The first drop-down list from the left in the dashboard is used to specify the type of thread on the hole. The **UNC** option is selected by default. The other options available in this drop-down list are **ISO** and **UNF**.

The second drop-down list in the Hole dashboard is used to specify the size of the screw that will go in the hole.

The third drop-down list in the Hole dashboard is used to specify the depth of the hole. The options in this drop-down list are similar to those available in the Extrude dashboard.

Taps the drilled hole button

This button is not enabled if the depth option is selected to variable depth. As the name suggests, this button is used to tap a hole. Tapping is the process of threading of holes. In Pro/ENGINEER, these threads are cosmetic threads. This means that the threads created will not be visible in the hole. In the slide-up panel of the **Shape** tab, the preview of the cross-section of the tapped hole is displayed as shown in Figure 5-10. In the preview of the section, the dimensions can be edited as required. The dotted lines represent the threading outside the hole. If you exit the **Taps the drilled hole** button, these dotted lines will not appear.

Adds countersink drilling button

As the name suggests, this button is used to create a countersink hole. A countersink hole has two diameters and the transition between the bigger

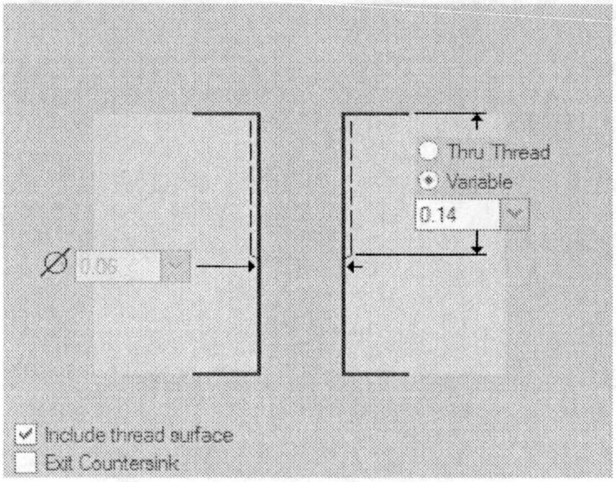

*Figure 5-10 **Shape** slide-up panel*

diameter and the smaller diameter is in the form of a cone. In the slide-up panel of the **Shape** tab, the preview of the cross-section of the countersink hole is displayed as shown in Figure 5-11. In the preview of the section, the dimensions can be edited as required.

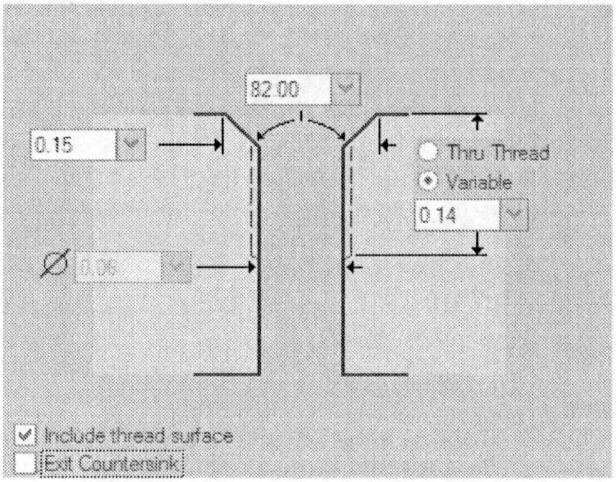

*Figure 5-11 **Shape** slide-up panel*

Adds counterbore drilling button

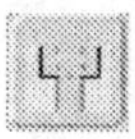

As the name suggests, this button is used to create a counterbore hole. A counterbore hole is a stepped hole and has two diameters, a larger one and a smaller one. The larger diameter is called counter diameter and the smaller diameter is called drill diameter. In the slide-up panel of the **Shape** tab, the preview of the cross-section of the counterbore hole is displayed as shown in Figure 5-12. In the preview of the section, the dimensions can be edited as required.

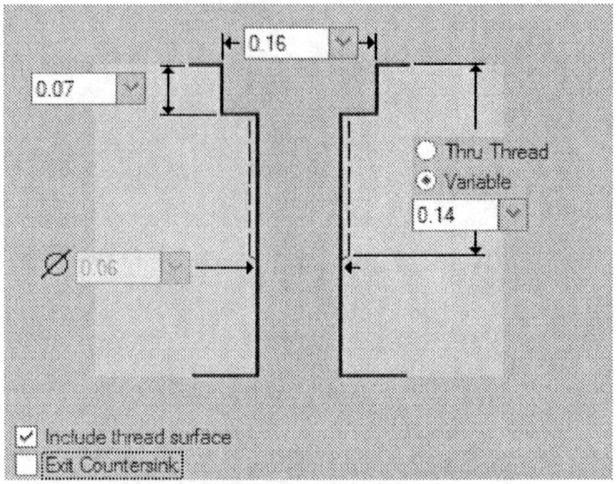

*Figure 5-12 **Shape** slide-up panel*

Placement tab

The options in the slide-up panel that is displayed when you choose the **Placement** option from the Hole dashboard are the same as those available in case of the **Straight** hole. You can use the same types of placement options that were available when you create the **Straight** hole.

Shape tab

The options in the slide-up panel that is displayed when you choose the **Shape** tab from the Hole dashboard are different than that were available in case of **Straight** holes. When you choose the **Shape** tab, the slide-up panel displays the preview of the hole. You can specify the dimensions for the hole in the dimension box.

Note tab

This tab displays the specifications of the hole in the text format. This option is available only when you are creating a **Standard** hole.

Properties option

This option displays the properties of the hole that you are creating.

Preview the hole

The **Feature Preview** button is used to preview the hole created before confirming its creation. Changes and modifications in the hole parameters can be made easily once the hole is previewed.

Note

*The holes created using the Hole dashboard are parametric in nature and hence can be modified at anytime using the **Model Tree**. The method of modification using the **Model Tree** is discussed later in this chapter.*

Important Points to Remember While Creating Hole

The following points should be remembered while creating a hole:

1. While drawing the sketch of a hole, the sketch should have axis of revolution and at least one entity normal to it. Now, when you place this hole on the primary reference, then this normal entity is aligned with the plane (primary reference). If there are two entities normal to the axis, then the normal entity at the top of the sketch is aligned with the placement plane.

2. While creating a hole, the primary reference for placement of hole can be selected without choosing any option from the Hole dashboard. However, when you need to specify the secondary references, then you have to choose the **Placement** tab from the Hole dashboard. From the slide-up panel that is displayed, click in the **Secondary References** collector and when the **Primary** collector appears white in color, you can select the references.

   Remember when the first reference is selected by using the left mouse button, then the second reference is selected by using CTRL+left mouse button.

3. It is recommended that if you are creating a **Standard** hole then the units of the model should be in mm. This can be achieved by using the appropriate template at the time of creating a new file or by setting the system of units.

4. An application of **Sketched** hole is a tapered hole. **Sketched** holes are always blind and have depth in one direction only.

5. You cannot select a convex or concave surface as the placement plane. To create hole on such a surface, you need to create a datum plane passing through the surface and then select this datum plane as the placement plane.

CREATING ROUNDS

The **Round** option in the **Insert** menu creates a fillet or smooth rounded transition with either a circular or a conic profile between two adjacent surfaces. This feature creation tool can be invoked from the menu bar or from the **Engineering Features** toolbar in the **Right Toolchest**. The **Round** option can either add or remove material, depending on the edge references selected. With the new interface of Pro/ENGINEER Wildfire, you can either select the edges or surfaces and create a round in a single set or define more than one set to create rounds.

Figures 5-13 and 5-14 show some examples of rounds. From the figures, it is evident that the geometry of the rounds are tangent to the references selected. These references can be edges or surfaces. In Pro/ENGINEER, rounds are created using the Round dashboard.

Note

The round created after you select references on the model has circular cross-section and rolling shapes. In Pro/ENGINEER, there are more than one shape that a round can have.

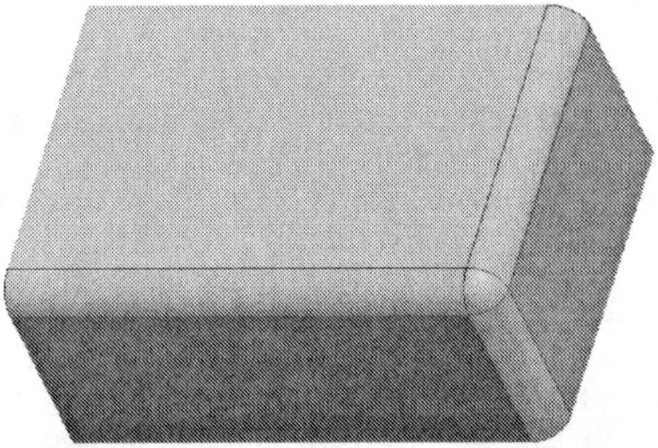

Figure 5-13 Round created on the edges

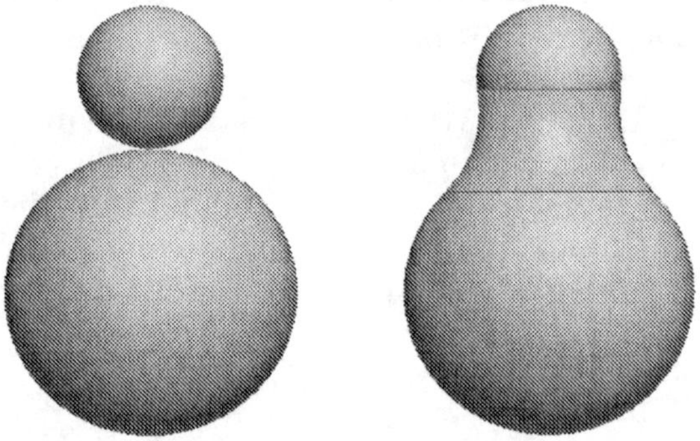

Figure 5-14 Round created on surfaces

The Round Dashboard

When you choose the **Round Tool** button from the **Engineering Features** toolbar or choose **Insert > Round** from the menu bar, the Round dashboard is displayed as shown in Figure 5-15 and you are prompted to select an edge or surface to create the round set.

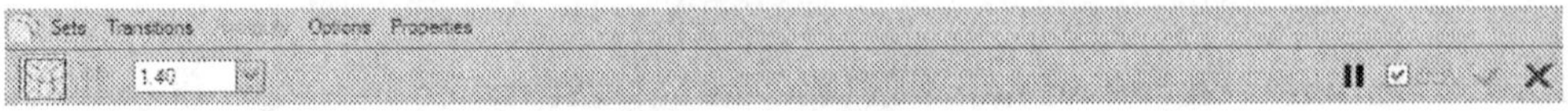

Figure 5-15 Round dashboard

After you select an edge or surface, you are prompted to select with the CTRL key an edge or a surface to fully specify the round set. Use CTRL+left mouse button to select another edge or a surface to include in the same set. The preview of the round with a default radius value and two drag handles appears on the model as shown in Figure 5-16. Note that since the radius of all the rounds in a set is equal, the drag handles are available on only one edge. You can either drag the handles to dynamically specify the radius of the round or specify the radius in the Round dashboard. The various options available in the Round dashboard are discussed next.

Switch to Set mode button

When you invoke the Round dashboard, the **Switch to Set mode** button is selected by default. This button is used to create rounds by defining sets. When this button is chosen, the dimension box to the right of this button in the Round dashboard is also displayed. This dimension box is used to specify the radius of round.

Tip: *You can create round on an edge by selecting the **Round Edges** option from the shortcut menu. Choose the **Geometry** option from the **Filter** drop-down list and then select an edge to turn it red in color. Hold down the right mouse button to invoke the shortcut menu. From the shortcut menu choose the **Round Edges** option. The Round dashboard is invoked and the preview of the round appears on the edge selected.*

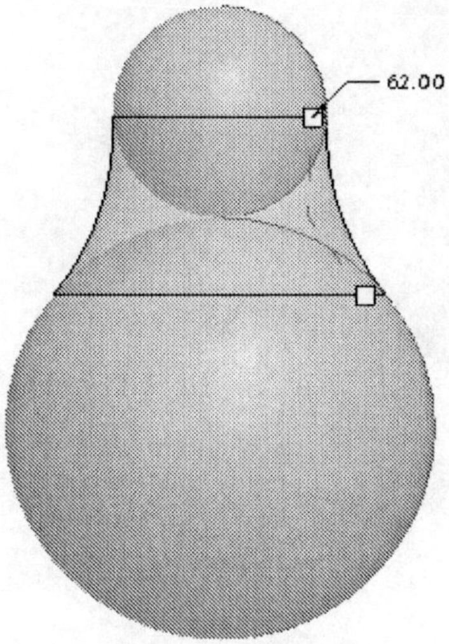

Figure 5-16 *Preview of the round feature along with the radius of round and drag handles*

Tip: *The need for creating rounds by defining more than one set arises when you want to create rounds that have different radii. The rounds created by using this method will be a single feature and appear as a single feature in the* ***Model Tree****.*

You can select more than one edge to create rounds.

Switch to Transition mode button

The **Switch to Transition mode** button is not available until you have defined at least one set of round. After you specify placement references and the radius of the round, the system generates the default round geometry by using some default attributes like the round shape, cross section, and so on. Now, the **Switch to Transition mode** button is active and can be chosen. This button allows you to select a reference to stop the round on an edge that is pre-selected. This means the round will be created on the selected edge up to the selected reference only. Figure 5-17 shows a simple example of transition round. In this case, the reference plane, which is highlighted by a mesh in the figure, is selected as the stopping reference. The stopping reference is the reference that stops the round on an edge.

Sets tab

When you choose the **Switch to set mode** button and then choose the **Sets** tab in the Round dashboard, the slide-up panel is displayed as shown in Figure 5-18. All the references and the geometry of the round are specified in this slide-up panel. The options in this slide-up panel are discussed next.

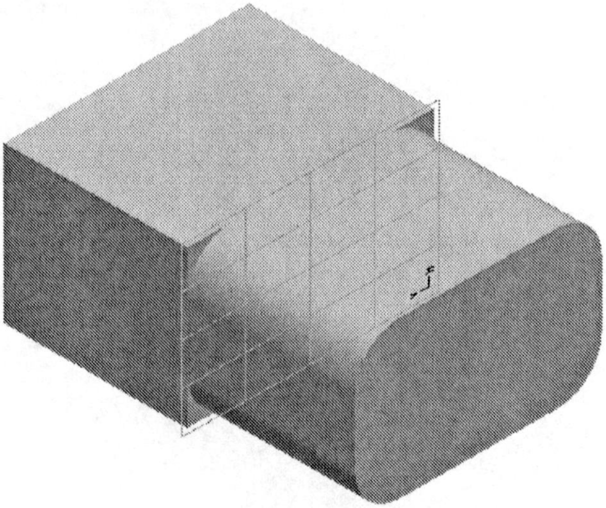

Figure 5-17 *Transition rounds created on four edges and stopped by the highlighted reference plane*

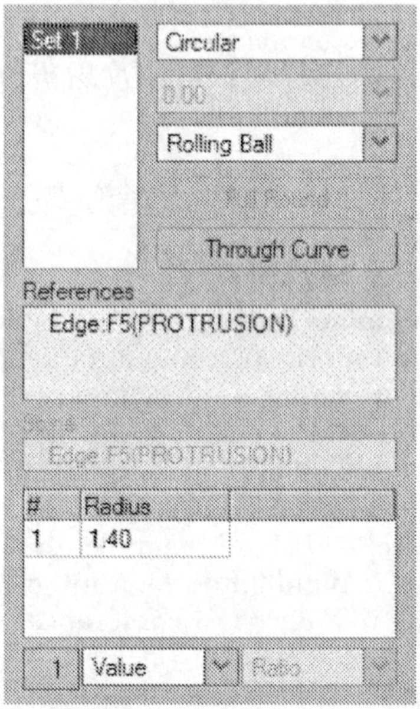

Figure 5-18 *Slide-up panel of the **Sets** tab*

First drop-down list

The first drop-down list that appears in the slide-up panel is used to specify the geometry of the round you need to create. This drop-down list contains the options as shown in Figure 5-19. These options are discussed next:

Circular. When you choose this option from the drop-down list, the round that is

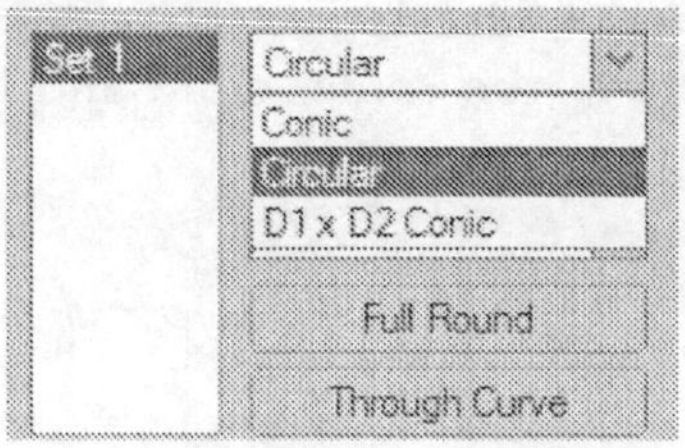

Figure 5-19 Drop-down list

created has a circular cross-section. This is the most widely used geometry type of rounds. The geometry is controlled by a radius value.

Conic. This type of round when created has a geometry of an ellipse. When you choose this option, the **Conic Parameter** drop-down list appears on the Round dashboard. The geometry of a conic round is controlled by two dimensions. The conic shape is controlled by a value that can lie between 0.5 and 0.95.

In Figure 5-20, the round on the model that is on the left is created using a circular geometry and the round on the model that is on the right is created using a conic geometry.

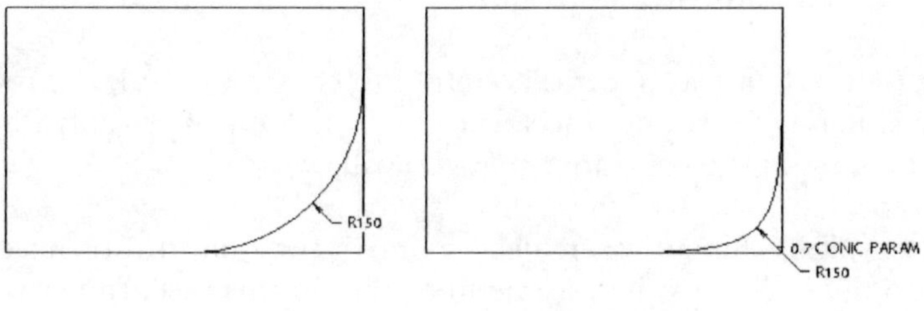

Figure 5-20 *Preview of the round feature along with the radius of round and drag handles*

D1xD2 Conic. The round created using this option has a geometry that can be controlled by three dimensions. Figure 5-21 shows the round that is created using the **D1xD2 Conic** option. This figure shows the two end legs dimensions and the middle dimension is the conic parameter. When you choose this option to create a round, the three edit boxes are displayed on the Round dashboard. The **Reverse**

direction of conic legs button is used to reverse the conic legs dimensions. Figure 5-22 shows the geometry of the conic round and the dimensions of the two legs after choosing this button.

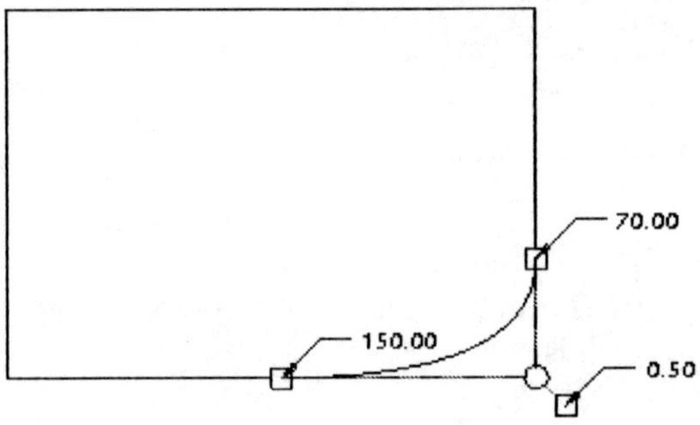

Figure 5-21 D1xD2 Conic round

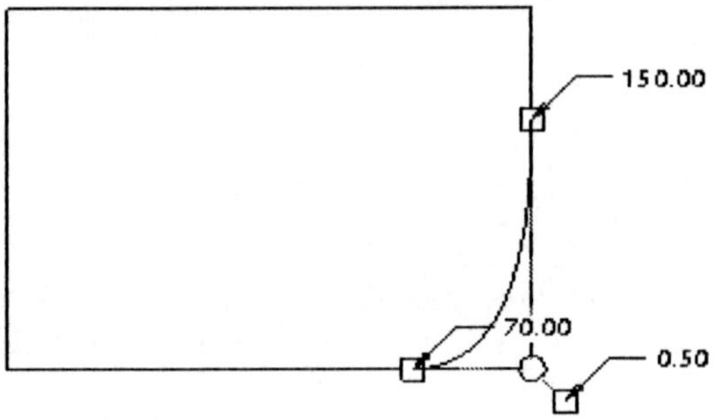

Figure 5-22 D1xD2 Conic round after reversing

Dimension box

The dimension box that is present in the slide-up panel is available only when you select the **Conic** or **D1xD2 Conic** option from the first drop-down list. This is because, this dimension box is used to enter the value for the sharpness of the conic round. The value lies between 0.5 and 0.95.

Second drop-down list

The second drop-down list that is present in the slide-up panel is used to specify the method that Pro/ENGINEER will use to create the round. The options in this drop-down list are **Rolling Ball** and **Normal to spine**. These options are discussed next.

Rolling Ball. When you select this option to create a round, the round is created by rolling a spherical ball along the surfaces to which it will possibly stay tangent. This is the most common type of round that you will create.

Normal to spine. This option creates a round by sweeping an arc or a conic cross-section normal to the selected spine. Remember that in this case, the cross-section remains normal to the spine.

The area left to the drop-down lists in the slide-up panel lists the number of sets formed for creating rounds. This means you can define more than one set to create the rounds. Each set may have different radius values and references. To add a set, hold down the cursor in this area until a shortcut menu appears. Choose the **Add** option from this shortcut menu. Whichever set is selected in this area, that set of round is active. It is recommended to define one set of round completely and then define the other set.

Full Round button

The **Full Round** button creates a complete round between two selected edges or two planar or non-planar surfaces. This button is not available when you choose the **Conic** or

D1xD2 Conic option. The following steps explain the procedure to create a full round by selecting the two edges.

1. Invoke the Round dashboard.
2. Select the first edge and then use CTRL+left mouse button to select the second edge. Note that the edges selected should have a surface between them that can be converted into a round.
3. Choose the **Full Round** button from the slide-up panel that is displayed when you choose the **Sets** tab from the Round dashboard. The round is created.

The following steps explain the procedure to create a full round by selecting the two surfaces.

1. To create a full round, invoke the Round dashboard.
2. Select the first surface and then use CTRL+left mouse button to select the second surface.
3. Now, choose the **Full Round** button from the slide-up panel that is displayed when you choose the **Sets** tab from the Round dashboard. You are prompted to select a surface to replace with a full round.
   Notice that in the slide-up panel, the **Driving surface** display box turns yellow in color. This indicates that now you can select the driving surface (surface to replace with a full round).

Figure 5-23 shows the round created using the **Full Round** option between the edge on the upper and the lower faces of the base of the model.

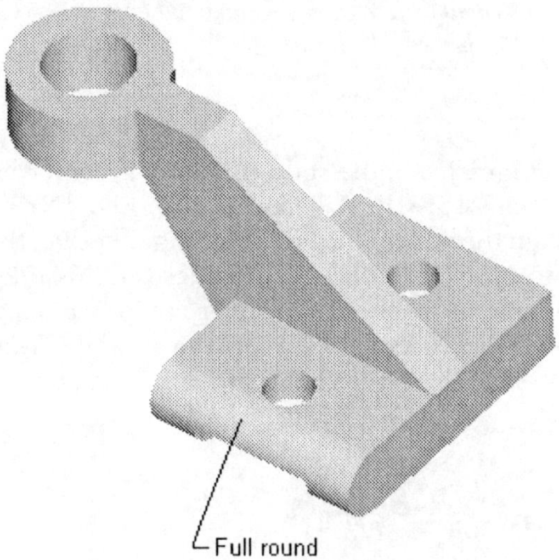

Figure 5-23 *Full round using* ***Edge Pair*** *option*

Through Curve button

The **Through Curve** button creates a round whose radius is specified by an existing curve.

References collector

The **References** collector displays the references selected on the model. These references are also known as collectors. The references that are displayed in this collector can be deleted by selecting them in the collector and then holding down the right mouse button to display a shortcut menu. Choose the **Remove** or **Remove All** option from this shortcut menu.

Spine collector

The **Spine** collector displays the spine selected on the model. This collector is active only when you choose the **Normal to spine** option from the third drop-down list. The **Spine** collector changes to **Driving surface** collector when you select two surfaces and then choose the **Full Round** button.

The area under the **Spine** collector is used to define the dimension values of the rounds that control their geometry.

Transitions tab

When you choose the **Transition** tab, the slide-up panel is displayed as shown in Figure 5-24. This slide-up panel displays the transitions if they are present in a round. For example, in Figure 5-17, there are four transitions (each at the four edges).

Tip: *When you choose the **Switch to Transition mode** button, the pieces of round are displayed on the selected edge. Now, if a corner is formed at the intersection of the three edges then you can specify the type of transition at the corner. To specify the type of transition, choose the options from the drop-down list that is available as soon as you select the corner piece from the model. These options can also be selected from the shortcut menu that can be invoked after the corner piece is selected.*

Ambiguity tab

When you choose the **Ambiguity** tab, the slide-up panel is displayed as shown in Figure 5-25. The items in this slide-up panel are displayed when Pro/ENGINEER is not able to decide the location of the round. When the system finds a non-tangent edge, the system tries to terminate the round feature at that point. Therefore, it provides the possible locations of the round in the form of pieces. When you move the cursor over any possible piece under the **Ambiguity** area, that piece is highlighted in cyan color on the model. From the pieces listed in the **Ambiguity** area, you can select the pieces to include or exclude by choosing the options from

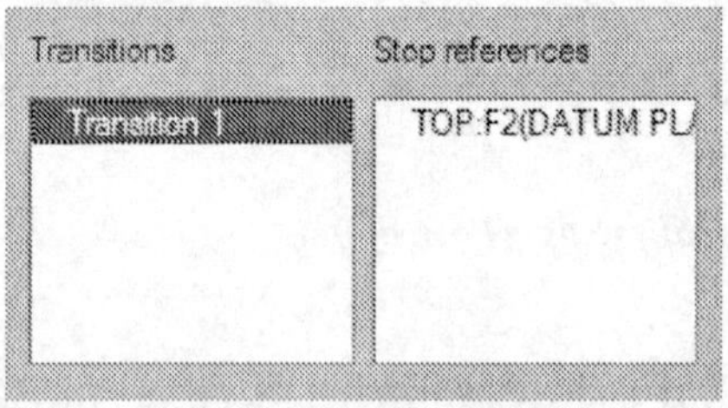

***Figure 5-24 Transitions** slide-up panel*

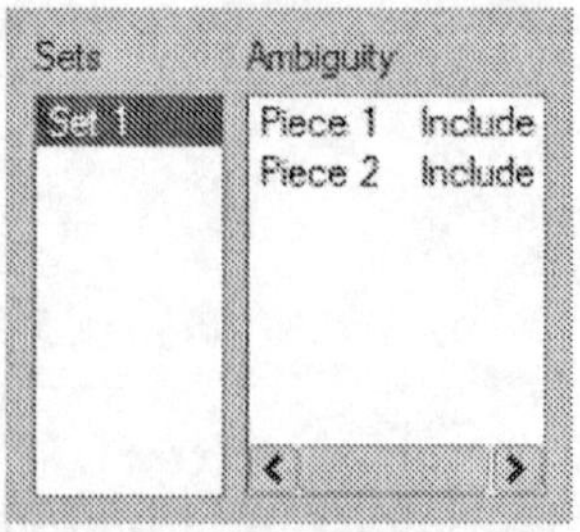

***Figure 5-25 Ambiguity** slide-up panel*

the shortcut menu that is displayed when you hold down the right mouse button in this area. It should be noted that to invoke this slide-up panel you should be in the set mode of creating rounds.

Options tab

When you choose the **Options** tab, the slide-up panel is displayed as shown in Figure 5-26. There are two radio buttons in this slide-up panel. These radio buttons allow you to make the round as a surface or as a solid.

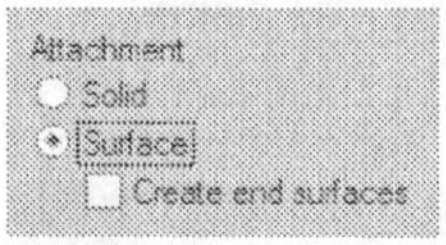

Figure 5-26* *Options* *tab

Properties tab

When you choose this option, the slide-up panel is displayed. This slide-up panel displays the feature identity. Click on the i button to read the information about the feature.

Creating a constant radius round on a chain of edge

This is one of the most simplest type of rounds that can be created using the Round dashboard. When the Round dashboard is invoked, select the two edges that you want to round and enter a radius value for the round. You can create a round by assigning the same radius value to every selected edge or surface. Figure 5-27 shows the constant radius round.

Creating a variable radius round between two surfaces

To create a variable round as shown in Figure 5-28, select the edge that will be converted into a round. When the preview of the round appears on the edge, hold down the right mouse button to display a shortcut menu. Choose the **Make Variable** option from this shortcut menu. Now, in the slide-up panel that is displayed when you choose the **Sets** tab, the two

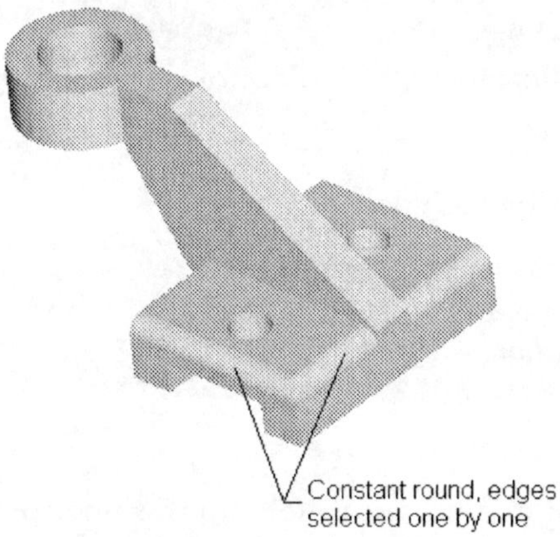

Figure 5-27 *Constant radius round by selecting the edges*

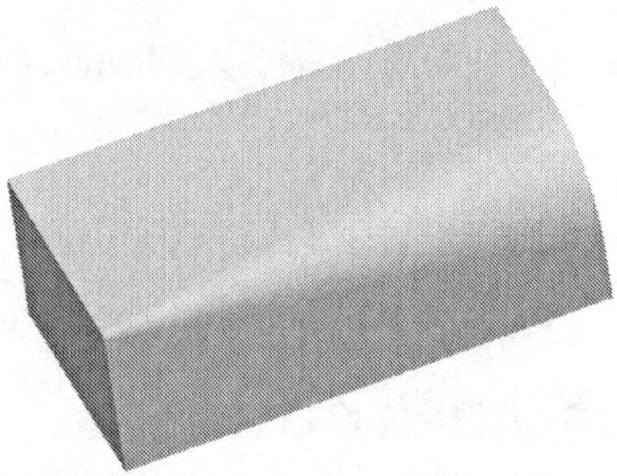

Figure 5-28 *Variable radius round by selecting the edge*

default values of the radius are available. You can modify the default values of radius of the variable round.

Note

You can also create a variable radius round by selecting the two surfaces. In this case, the edge that is common to the two surfaces is converted into a round.

The **Make Variable** option in the shortcut menu allows you to specify different radii at the end points of the selected edge or optionally at additional points along the selected edge. Additional points can be defined along the edge.

To define additional points, choose the **Sets** tab to invoke the slide-up panel. Bring the cursor to the area in the slide-up panel where the radius value and the location of the points are displayed and hold down the right mouse button to invoke the shortcut menu. Choose the **Add radius** option; one additional point appears in the same area of the slide-up panel and on the edge. This point by default is placed on the selected edge by a ratio value with respect to the previous point. You can select the **Reference** option from the drop-down list present under this area and select a reference on the edge. This reference can be used to specify the distance of the point.

Note

Rounds can be created on one edge, chain of edges, between two surfaces, an edge and surface, and between two edges.

Points to Remember While Creating Rounds

The following points should be remembered while creating rounds:

1. To avoid parent-child relationship, other features should not be referenced to the edges created by rounds or to the round surface.

2. Rounds should be added at the last. This means, the rounds should be the last features in any solid model.

3. The more the number of round features a solid model has, the more time the system takes to regenerate them.

4. Use the shortcut menus extensively while creating rounds.

Tip: *Rounds and chamfers are used in components to reduce the stress concentration at the sharp corners. Hence, they reduce the chances of failure of a component under a specified load condition.*

CREATING CHAMFERS

Chamfers are used to bevel the selected edges and corners with the help of some specified parameters. Pro/ENGINEER creates two type of chamfers. The first is the **Corner** chamfer and the second is the **Edge** chamfer. To create chamfer on a corner, choose **Insert > Chamfer**

> **Corner** from the menu bar. To create a chamfer on an edge, choose **Insert > Chamfer > Edge** from the menu bar. You can also choose the **Chamfer Tool** button from the **Engineering Features** toolbar to create the edge chamfers. Figure 5-29 shows the two types of chamfers.

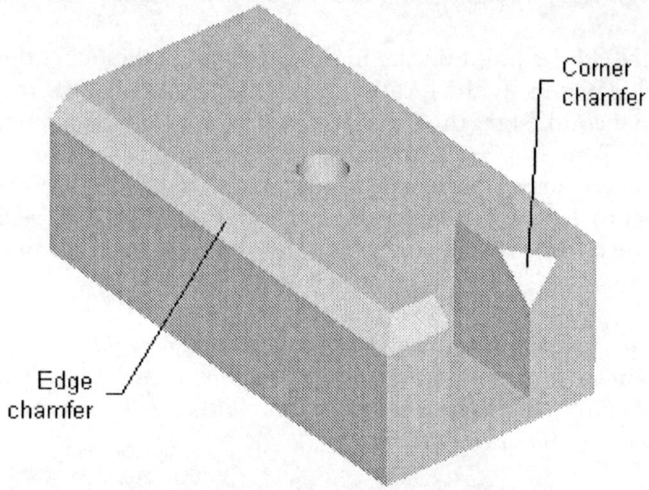

Figure 5-29 *Different types of chamfers*

Corner Chamfer

A **Corner** chamfer creates a beveled surface at the intersection of three edges. When you choose the **Corner Chamfer** option from the cascaded menu, the **SELECT** dialog box and the **CHAMFER (CORNER)** dialog box is displayed (Figure 5-30) and you are prompted to select a corner to chamfer. When you select the corner, the **PICK/ENTER** menu is displayed as shown in Figure 5-31. Use the options in the **PICK/ENTER** menu to specify the chamfer distance. The options in this menu are discussed next.

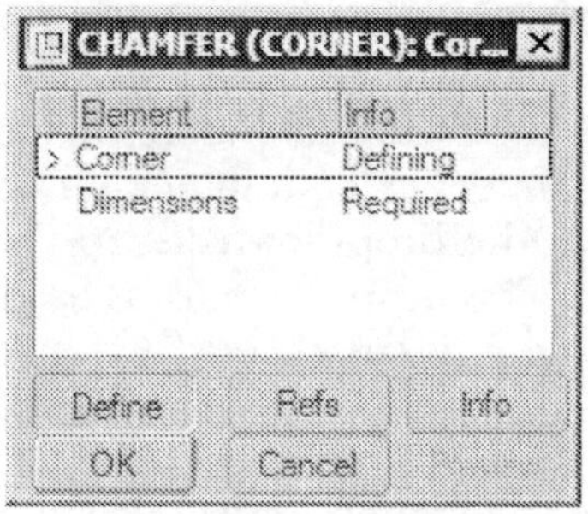

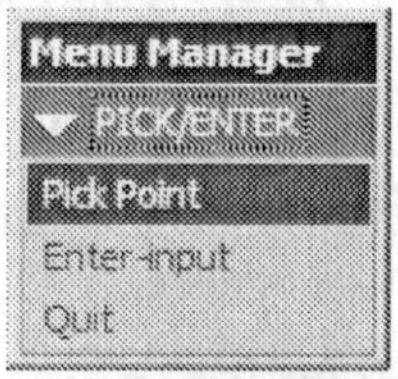

Figure 5-30 CHAMFER (CORNER) dialog box ***Figure 5-31 PICK/ENTER*** *menu*

Pick Point

The **Pick Point** option is used to select a point on the edge that is highlighted in blue. The point selected on the edge denotes the chamfer distance from the corner.

Enter-input

When you use the **Enter-input** option, the **Message Input Window** is displayed and you are prompted to enter the length dimension along the highlighted edge. Enter the chamfer distance for the highlighted edge in the **Message Input Window**.

After you have specified the point on the highlighted edge, the next edge will be highlighted in blue. Enter a value or specify the location using the **PICK/ENTER** menu. After you set the parameters for the second edge, the third edge is highlighted. Similarly, specify the location on the third edge. The **PICK/ENTER** menu disappears, indicating that the required information to create a corner chamfer is completed. Choose **Preview** from the **CHAMFER (CORNER)** dialog box to preview the corner chamfer and then choose the **OK** button. After you enter the distance value for all the three edges, the corner chamfer is created.

Note

*When you are creating the corner chamfer, notice that only the **Edge** filter is available in the **Filter** drop-down list in the **Status** bar. As mentioned earlier, the filters available with this version of Pro/ENGINEER narrow the entities available for selection. Thus, you can easily select the reference entities on the model.*

Edge Chamfer

An **Edge** chamfer creates a beveled surface along the selected edge. When you choose **Insert > Chamfer > Edge** from the menu bar or choose the **Chamfer Tool** button from the **Engineering Features** toolbar, the Chamfer dashboard is displayed as shown in Figure 5-32 and you are prompted to select any number of edges to create chamfer. The options in this menu are discussed next.

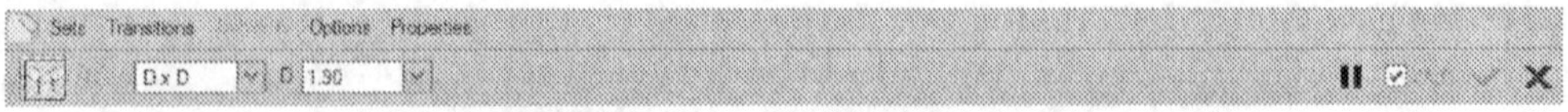

***Figure 5-32** Chamfer dashboard*

Switch to Set mode button

When you invoke the Chamfer dashboard, the **Switch to Set mode** button is chosen by default. This button is used to create chamfers by defining either one set or more than one set. When this button is chosen, the drop-down list to the right of this button in the Chamfer dashboard is also displayed. This drop-down list is used to specify the type of chamfer. The options in the drop-down list are discussed next.

D x D

The **D x D** option creates a chamfer such that the chamfer is created at an equal distance from the selected edge on both the faces connected by the edge. In Figure 5-33, the dimension value **10** is the value of **D**. From the figure it is evident that the distance **D** is measured from the corner.

D1 x D2

The **D1 x D2** option creates a chamfer at two user-defined distances from the selected edge. In Figure 5-34, the dimension value **10** is the value of **D1** and the dimension value **16** is the value of **D2**. The values for **D1** and **D2** can be entered on the Chamfer dashboard.

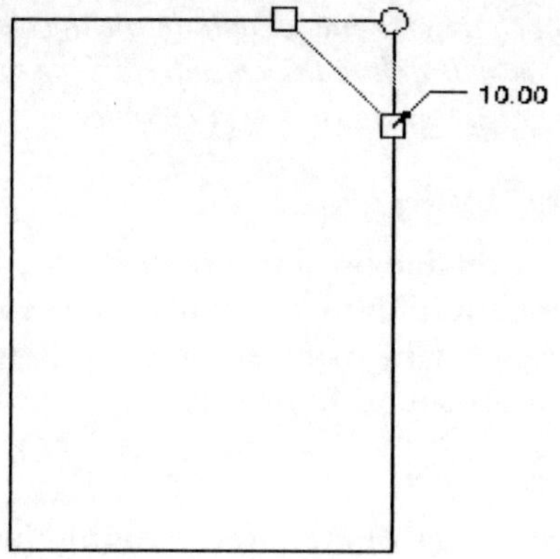

Figure 5-33 D x D option

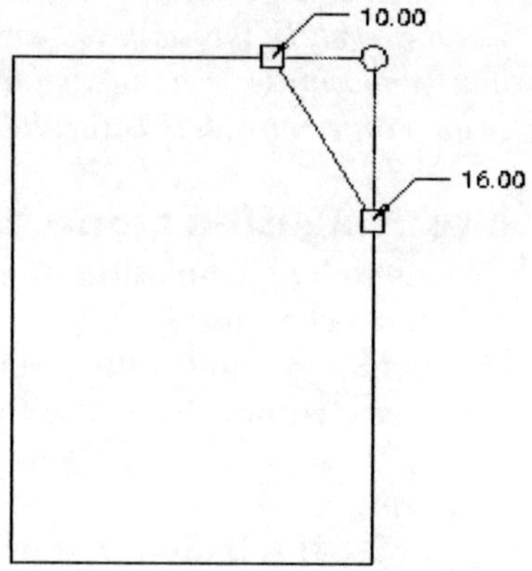

Figure 5-34 D1 x D2 option

Angle x D

The **Angle x D** option creates a chamfer at a user-defined distance from the selected edge and at a user-defined angle measured from a specified surface. In Figure 5-35, the angle is **60** and the value of **D** is **15**.

45 x D

The **45 x D** option creates a chamfer at the intersection of two perpendicular surfaces. The chamfer created is at an angle of 45-degree from both the surfaces and at a distance **D** from the edge along each surface. Once the chamfer is created, the distance **D** can be modified. Figure 5-36 shows this type of chamfer where the value of **D** is **25**.

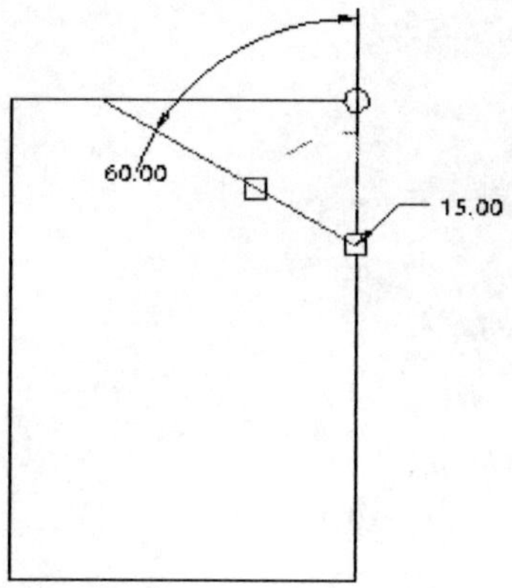

Figure 5-35 Angle x D option

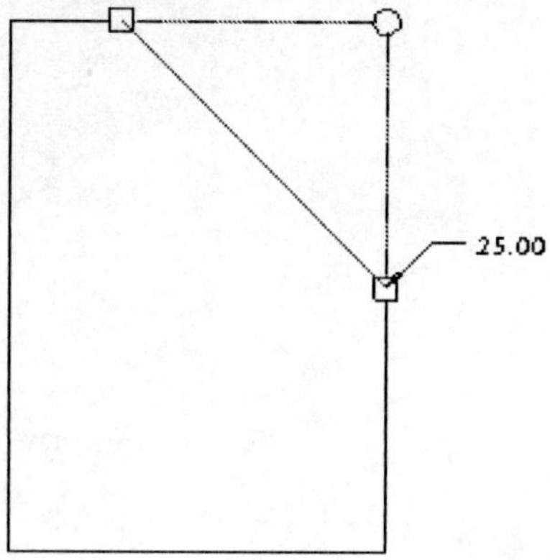

Figure 5-36 45 x D option

Note

The rounds and chamfers are defined in sets when there is a variation in their dimensions. For example, if you want to create chamfer on two edges and the dimensions for both the chamfers are different. In this case, you can define two sets of chamfers; one set will define the chamfer on one edge with some radius value and the second set will define the chamfer on the other edge with a different radius value. This is the basic use of sets in rounds and chamfers.

Switch to Transition mode button

The **Switch to Transition mode** button is not available until you have defined at least one set of chamfer. After you specify the placement reference and the type of chamfer, the system generates the default chamfer geometry by using some default attributes like the type of chamfer. Now, the **Switch to Transition mode** button is active and can be chosen.

When you choose this button, the preview of the two pieces of chamfer is highlighted. The highlighted piece that you select is not chamfered up to the selected reference. The reference to stop the transition can be selected after you select the **Stop at Reference** option from the drop-down list present on the Chamfer dashboard. This means the chamfer will be created on the selected edge starting from the opposite end and up to the selected reference only. Figure 5-37 shows the example of transition chamfer. The reference plane is highlighted by a mesh in the figure.

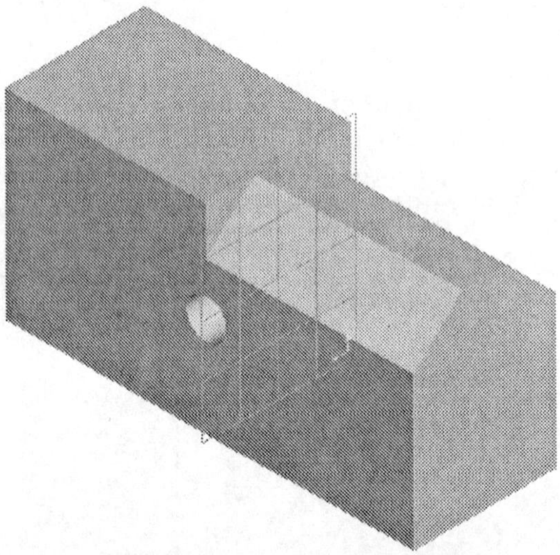

Figure 5-37 *Transition chamfer*

Sets tab

When you choose the **Sets** tab in the Chamfer dashboard, the slide-up panel is displayed as shown in Figure 5-38. All the references and the dimensions of the chamfer can be specified in this slide-up panel. The options in this slide-up panel are discussed next:

Set 1 display area

This display area displays the number of sets created. To add a set, right-click to invoke

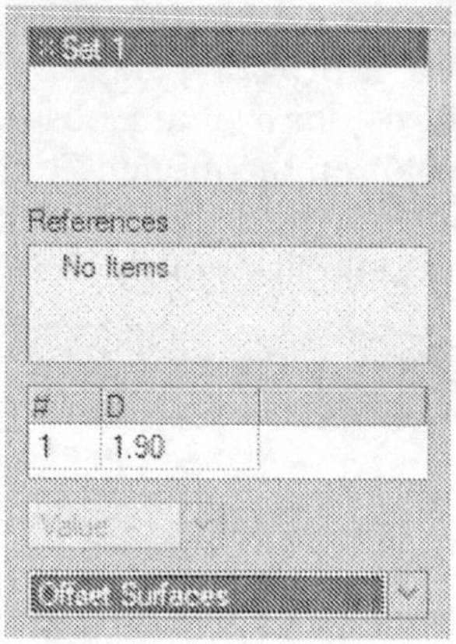

*Figure 5-38 Slide-up panel of the **Sets** tab*

a shortcut menu and choose the **Add** option. It is recommended that you define an additional set only when you have defined the first set. However, you can switch among the sets by selecting them from this area. The set that is highlighted in this area can be previewed in the graphics window.

References collector

The **References** collector displays the edge selected to chamfer. To remove an edge from this box, right-click to invoke a shortcut menu. Choose the **Remove** option from this shortcut menu. Now, you can select an edge from the model.

The area below the **References** collector is used to specify the dimension values. The dimension values can also be specified on the dashboard.

Transitions tab

When you choose the **Transition** tab, the slide-up panel is displayed as shown in Figure 5-39. This slide-up panel displays the transitions if they are present in a chamfer. For example, in Figure 5-37, there are two transitions (each at the two edges).

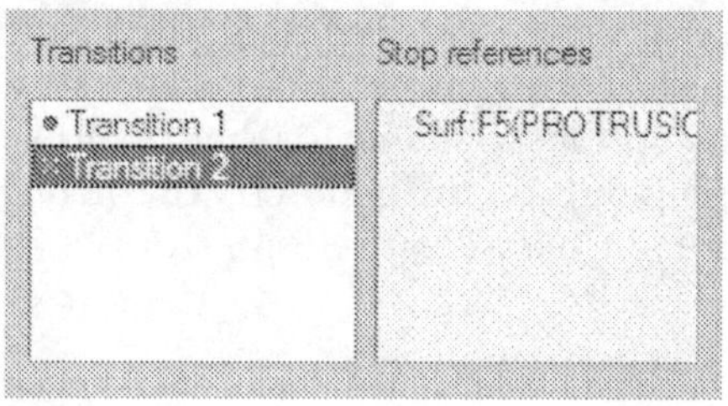

*Figure 5-39 **Transitions** slide-up panel*

Ambiguity tab

When you choose the **Ambiguity** tab, the slide-up panel is displayed as shown in Figure 5-40.

The items in this slide-up panel are displayed when Pro/ENGINEER is not able to decide the location of the chamfer. Therefore, it provides the possible locations of the chamfer in the form of pieces. When you move the cursor over any possible piece under the **Ambiguity** area, that piece is highlighted in cyan color on the model. From the pieces listed in the **Ambiguity** area, you can select the pieces to include or exclude by choosing the options from the shortcut menu that is displayed when you right-click in this area.

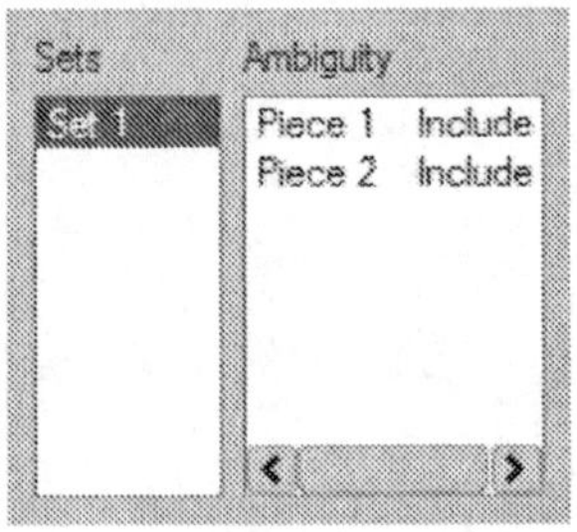

***Figure 5-39 Ambiguity** slide-up panel*

Options tab

When you choose the **Options** tab, the slide-up panel is displayed as shown in Figure 5-41. There are two radio buttons in this slide-up panel. These radio buttons allow you to make the chamfer as a surface or as a solid.

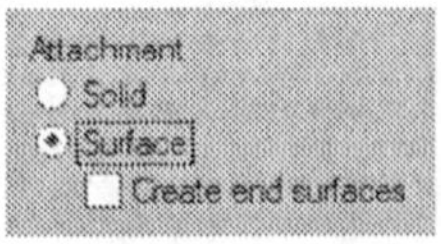

***Figure 5-41 Options** slide-up panel*

Properties option

When you choose this option, the slide-up panel is displayed. This slide-up panel displays the feature identity.

UNDERSTANDING RIBS

Ribs are defined as thin wall-like structures used to bind the joints together so that they do not fail under an increased load. In Pro/ENGINEER, the section for the rib is sketched as an open section and can be extruded equally in both the directions of the sketch plane or on either sides. The procedure of creating a rib is similar to that of creating a protrusion.

In Pro/ENGINEER, you can create two types of ribs: Rotational ribs and Straight ribs. Rotational ribs are constructed on cylindrical parts and straight ribs are created on planar faces. There are no separate options available for the creation of these ribs in Pro/ENGINEER. The creation of these ribs depends on the geometry of the feature on which the rib is created. Figure 5-42 shows a rotational rib and Figure 5-43 shows a straight rib.

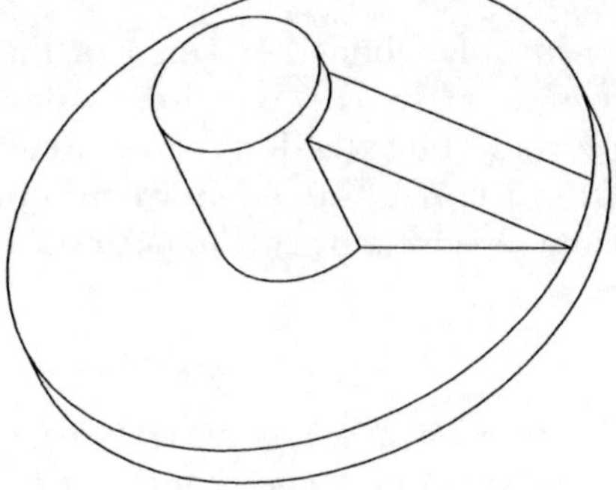

Figure 5-41 Rotational *Rib feature*

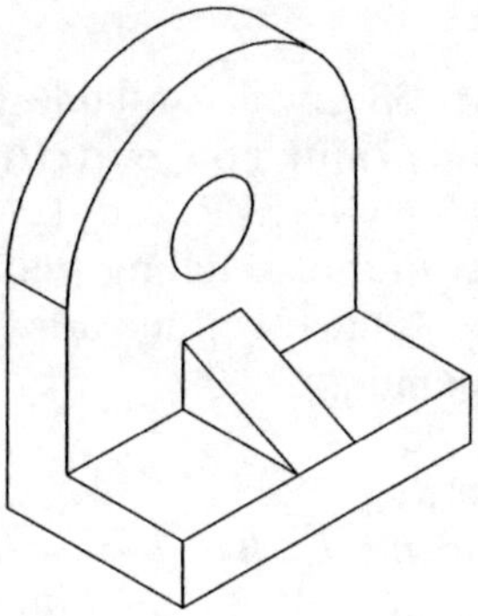

Figure 5-42 *Straight Rib feature*

Creating Ribs

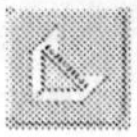

Ribs are created using the Rib dashboard shown in Figure 5-44. The Rib dashboard is displayed when you choose **Insert > Rib** from the menu bar or choose the **Rib Tool** button from the **Engineering Features** toolbar.

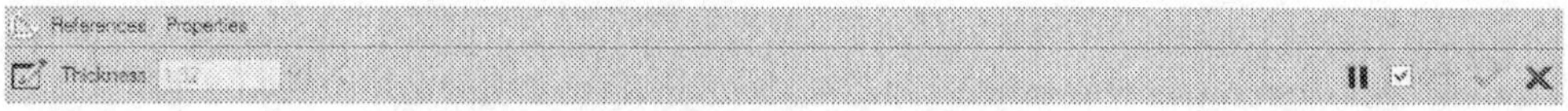

Figure 5-44 *Rib dashboard*

To create a rib, you first need to select a sketching plane and then draw the geometry for the rib feature in the sketcher environment. The sketch for the rib feature should be an open loop. The procedure to create a rib is the same as that of creating a protrusion.

The options and the tools available in the Rib dashboard are discussed next.

References tab

When you choose the **References** tab, the slide-up panel is displayed as shown in Figure 5-45. In the slide-up panel, the **Sketch** collector shows **No Items**. This indicates that the sketch for the rib feature is not drawn. The **Create a section or redefine the existing section** button is used to invoke the **Section** dialog box. Using the **Section** dialog box, you select the sketch plane and the reference plane.

Figure 5-44 References *slide-up panel*

The **Add material to the other side of the section** button is used to flip the side where the material should be added. This button will be available only when you draw the sketch and exit the sketcher environment.

Change the thickness option between both sides, side 1, and side 2 button

This button is available on the Rib dashboard only when the sketch of the rib is drawn and you exit the sketcher environment. In the previous releases of Pro/ENGINEER, the sketch of the rib feature could be extruded only symmetrically with respect to the sketching plane. But in Pro/ENGINEER Wildfire, you have an option to extrude the sketch on either sides of the sketching plane or symmetrically to both sides of the sketching plane.

Note

Since the sketch of the rib can be extruded on both sides of the sketching plane, therefore, while creating a rib you must always select an appropriate sketching plane such that it lies at the center of rib creation.

EDITING THE FEATURES OF A MODEL

Editing is one of the most important aspects of product designing. Most of the designs require editing either during their creation or after they are created. As mentioned earlier, Pro/ENGINEER is a parametric and feature-based solid modeling tool. Hence, the features that together constitute a model can be individually edited. For example, Figure 5-46 shows a cylindrical part that has six counterbore holes created at some bolt circle diameter (BCD).

Now, in case you have to edit the features such that the six holes are to be converted into eight holes and the counterbore holes are to be converted into countersink holes, as shown in Figure 5-47, all you need to do is to use two operations. The first operation will convert the six holes into eight holes by using the **Edit** option in the shortcut menu that is displayed when you select the hole and hold down the right mouse button. The second operation will open the Hole dashboard and convert the counterbore holes into countersink holes.

Figure 5-46 *Six counterbore holes*

Figure 5-47 *Eight countersink holes*

Similarly, you can also edit the datums and the features referenced to these datums. Since there exists a parent-child relationship between the two, therefore, the child feature is also modified when the parent feature is modified. For example, if you have created a feature using a datum plane that is at some offset distance, the feature will be automatically repositioned when the offset value of the datum plane is changed. The following methods explain how to edit the features in Pro/ENGINEER.

Editing Definition or Redefining Features

Editing the definition of features allow you to make changes in the parameters that were used to create a feature. You can also modify the sketches of the sketched features by editing their definition. A feature is redefined only when it is completed. To invoke the **Edit Definition** option, there are three methods available in Pro/ENGINEER Wildfire. These methods are listed next.

1. You can select the feature that has to be redefined from the graphics window. The selected feature is highlighted in red. Now, hold down the right mouse button. A shortcut menu is displayed; choose **Edit Definition** from this menu.

   Sometimes, it becomes difficult to select the desired feature from the model to redefine. Therefore, you should use the filters available in the **Filter** drop-down list on the **Status** bar. By default, the **Smart** filter is selected. The second filter in the **Filter** drop-down list is **Features**. This filter is used to select the features on the model. To redefine a feature, select this option from the drop-down list. The third filter is **Geometry**. This filter allows you to select only the edges, vertices, and surfaces. The fourth filter is **Datums**. This filter allows you to select the datum features. The fifth filter is **Quilts**. This filter allows you to select surfaces. The sixth filter is **Annotation**. This filter is used to select the notes from the graphics window.

   > **Tip**: *Suppose you have created a cube and the display is selected to be **Shaded**. Now, you need to select the top face of the cube. By default, in the **Filter** drop-down list the **Smart** option is selected. Select the cube on the top face; the whole cube turns red in color. Again click at the same location; the top face of the cube is selected. If you again click, the top face is highlighted in red color.*

2. You can also use the **Model Tree** to redefine a feature. Select the feature that has to be redefined from the **Model Tree**. The selected feature is highlighted in red on the graphics window. Now, right-click the feature listed in the **Model Tree**. A shortcut menu is displayed; choose **Edit Definition** from this menu.

3. Select the feature to redefine from the graphics window or from the **Model Tree**. Choose **Edit > Definition** from the menu bar. Remember that the **Definition** option in the **Edit** menu is not available if the feature is not selected first.

Reordering Features

Reordering the features is defined as the process of changing the order of features in a model. Sometimes, after creating a model it may be required to change the order in which the features of the model were created. A feature can be placed before or after another feature. For this purpose, the **Model Tree** or the **Menu Manager** is used.

If you use the **Model Tree**, you just have to drag and drop the feature you want to reorder below or above the destination feature.

The **Menu Manager** can be invoked by choosing **Edit > Feature Operations** from the menu bar. The **Menu Manager** was used extensively in the prior releases of Pro/ENGINEER. But in this release, the **Menu Manager** is used mainly to perform operations on features. The **Menu Manager** is shown in Figure 5-48. If you use the **Menu Manager**, the **Reorder** option is available in **FEAT** menu.

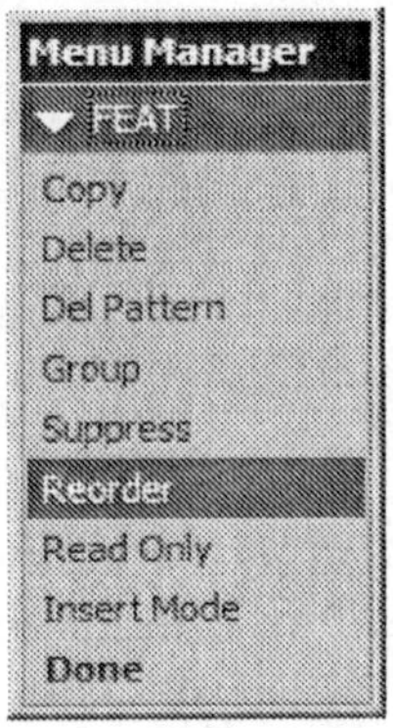

Figure 5-48 Menu Manager

When you choose the **Reorder** option from the **FEAT** menu, the **SELECT FEAT** submenu is displayed as shown in Figure 5-49 and you are prompted to select the features that you want to reorder. Select the feature to be reordered either from the **Model Tree** or from the graphics window and then choose the **Done** option from the **SELECT FEAT** submenu or use the middle mouse button to confirm the selection. In the **Message** area, the system will display the possible positions where the selected feature can be inserted. The features are numbered according to their occurrence in the **Model Tree**. Choose **Confirm** from the **CONFIRMATION** menu to confirm the reordering of the features.

***Figure 5-49 SELECT FEAT** submenu*

The example that is discussed here will explain why and when the need for reordering of features in a model arise. Consider the model shown in Figure 5-50. It consists of a rectangular pattern of nine columns and four rows. Now, a shell feature is created on this model, which removes the top and the front face as shown in Figure 5-51.

Figure 5-51 shows that the material equal to the wall thickness of the shell feature is added around the hole features. This was not desired, and here the need for reordering the feature arises. You will reorder the features such that the shell feature is inserted before the hole

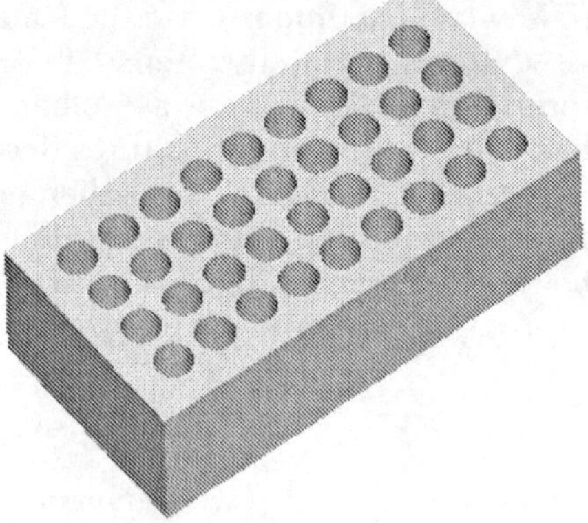

***Figure 5-50** Model with pattern*

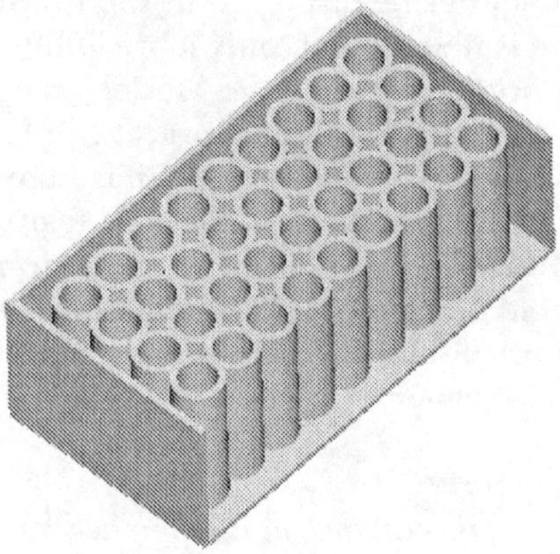

***Figure 5-51** Model after creating the shell feature*

feature and its pattern. When you reorder the features, all the features will be automatically adjusted in the new order as shown in Figure 5-52.

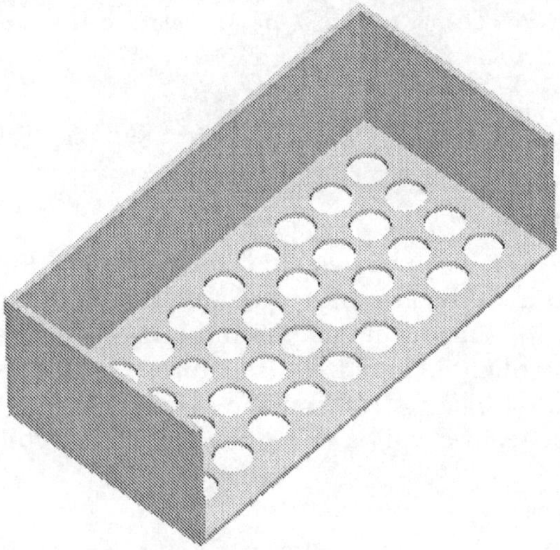

***Figure 5-52** Model after reordering the features*

Rerouting Features

The **References** option available in the **Edit** menu is used to modify the references of a feature and in turn break the parent-child relation that exists between the selected feature and the other features. From this release of Pro/ENGINEER, the rerouting of features is done using the **References** option.

Suppressing Features

When you do not want a feature to be displayed on the graphics window or to show up in the drawing views of a model, then that feature can be suppressed. Once the feature is suppressed, it will neither be displayed in the graphics window nor in the drawing views. Note that the feature is not deleted, only its visibility is turned off. You can anytime resume the feature by unsuppressing it using the **Model Tree** or using **Edit > Resume** from the menu bar. As soon as you unsuppress the feature, it will be displayed on the graphics window as well as in the drawing views. When a model has many features then the suppressing of features decreases the regeneration time of the new feature created. To suppress a feature, right-click on it in the **Model Tree** or select it on the graphics screen using the left mouse button and then press and hold down the right mouse button to display the shortcut menu. Choose **Suppress** from the shortcut menu to suppress the selected feature. You can also suppress a feature using the **Menu Manager**.

Note

If the feature that is suppressed has some child features then they will also be suppressed. When you select such a feature, the system prompts you to confirm the suppression of the highlighted features on the graphics window.

Tip: *By default, the suppressed features are not displayed in the* ***Model Tree****. To display the suppressed features, choose* ***Settings > Tree Filters*** *from the* ***Model Tree****. The* ***Model Tree Items*** *dialog box is displayed. Select the* ***Suppressed Objects*** *check box and choose* ***OK*** *from the* ***Model Tree Items*** *dialog box. The suppressed features will now be displayed in the* ***Model Tree****.*

The suppressed features can also be deleted by using the ***Model Tree****.*

Deleting a feature

The feature that is not required can be deleted from the model. Right-click on the feature in the **Model Tree** or select it on the graphics window and then press and hold down the right mouse button to display the shortcut menu. From this menu, choose the **Delete** option. The feature to be deleted is highlighted in red. If the feature to be deleted has some child features, they will also be highlighted. The system confirms before deleting the selected feature. If you confirm the deletion, the feature will be deleted along with its child features.

Modifying a feature

Once a feature is created, you can still modify the feature by modifying its dimensions. This editing operation reflects the parametric nature of Pro/ENGINEER. Select the feature on the graphics window using the left mouse button. The selected feature is highlighted in red, hold down the right mouse button to invoke the shortcut menu. Choose the **Edit** option from this shortcut menu. The selected feature is highlighted in red and the dimensions appear on the feature. These dimensions are the same that were defined while sketching or creating the feature.

In some cases, for example, rounds, holes, and so on, the dimensions of these features were defined as parameters. Hence, these dimensions are also displayed when you select them to modify. To modify the dimensions, double-click the dimension on the graphics window. The dimension is converted into an edit box. Enter the required value for the selected dimension and press ENTER. Once you modify a dimension, you also need to regenerate the feature. Choose the **Regenerates Model** button from the **Top Toolchest**.

If you want to modify a pattern, it is recommended that you select the instance other than the original one. This is because if you select the original feature, the incremental dimensions are not displayed and so you cannot modify the incremental values of the pattern. Patterns are discussed in Chapter 6.

Tip: *To modify a feature by modifying its dimensions, double-click on the feature. The dimensions of the feature are displayed. Now, you can modify the dimensions.*

Dynamically Modifying the Sketch of a Feature

If you want to modify the sketch of a feature without entering the sketcher environment, then you need to modify it dynamically. To dynamically modify a feature, follow these steps:

1. Select the feature preferably by using the **Features** filter from the **Filter** drop-down list.

2. When the feature turns red in color, hold down the right mouse button to invoke the shortcut menu.

3. From the shortcut menu, choose the **Edit** option. The dimensions of the sketch appear on the feature.

4. Select the sketch and hold down the right mouse button to invoke the shortcut menu. Choose the **Edit** option from this shortcut menu.

5. Now, bring the cursor close to the sketch. You will notice that the cursor changes to a hand. The hand symbol represents that you can select an entity from the sketch and drag it to modify the shape of the sketch.

6. Modify the shape of the sketch by dragging its entities.

7. Choose the **Regenerates Model** button from the **Top Toolchest** to apply the modifications.

TUTORIALS

Tutorial 1

Create the model shown in Figure 5-53. The dimensions, and the front, top, and left-side views of the model are also shown in the figure. **(Expected time: 45 min)**

The following steps outline the procedure for creating this model:

a. First examine the model and then determine the number of features in it. The model

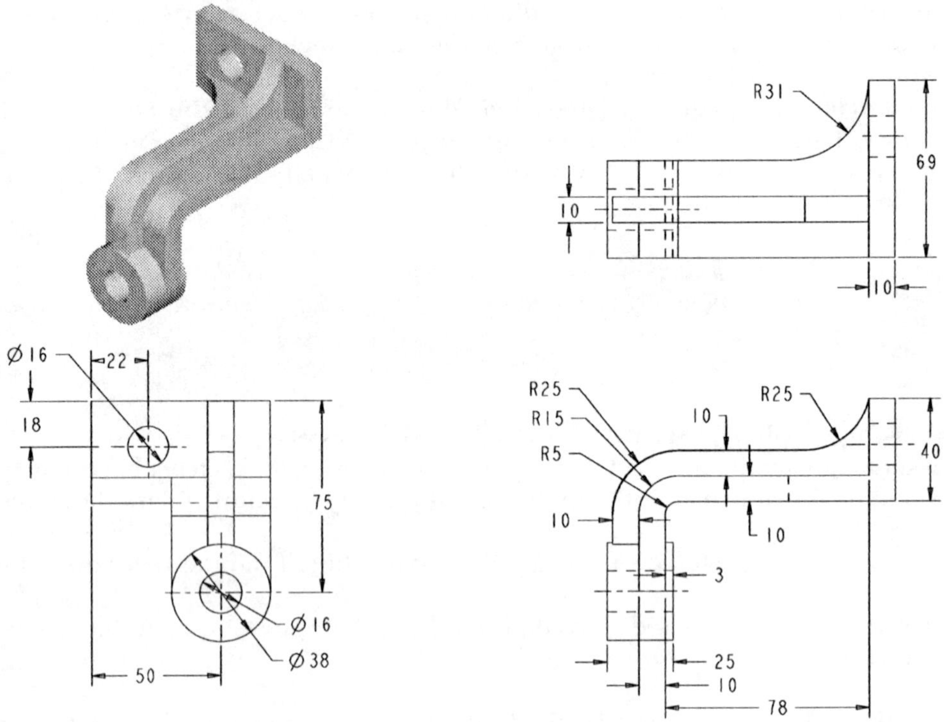

***Figure 5-53** Isometric shaded, left-side, front, and top views of the model*

is composed of seven features, see Figure 5-53.

b. The first four features are protrusion features, see Figure 5-62. First the sketch of base feature will be created on the **FRONT** datum plane, see Figure 5-54, and then it will be extruded to a depth of 10.

c. The sketch of the second feature will be created on the **TOP** datum plane and will be extruded on one side of the sketching plane. The depth of extrusion is 10, see Figure 5-57.

d. The sketch of the third feature will be created on the front planar surface of the second feature and will be extruded on one side of the sketching plane. The depth of extrusion is 10, see Figure 5-60.

e. The sketch of the cylindrical feature will be drawn on the front face of the third feature and extrusion will be on both sides of the plane. The depth of extrusion on one side is 12 and on the other side is 13, see Figure 5-62.

f. The next feature is a hole that is coaxial to the cylindrical feature, see Figure 5-63. This hole will be created using the Hole dashboard.

g. The next features that will be created are rounds. The two rounds have different radii. Hence, they will be defined in two sets.

h. The last feature is a rib. The sketch for this feature will be drawn on the **RIGHT** datum plane, see Figure 5-66.

After understanding the procedure for creating the model, you are now ready to create it. When Pro/ENGINEER session is started, the first task is to set the working directory. Since this is the first tutorial of this chapter, you need to create a folder named *c05*, if it does not exist. Choose the **New Directory** button in the **Select Working Directory** dialog box and create a directory named *c05* at *C:\ProE-WF*.

Creating New Object File

1. Open a new part file and name it *c05tut1*. The three default datum planes are displayed on the graphics window. The **Model Tree** is also displayed on the graphics screen. Close the **Model Tree** by clicking on the sash.

Creating the Base Feature

To create the sketch for the base feature, you need to first select the sketching plane for the base feature. In this model, you need to draw the base feature on the **FRONT** datum plane because the direction of extrusion of this feature is perpendicular to this plane.

1. Choose the **Extrude Tool** button from the **Base Features** toolbar. The Extrude dashboard is displayed.

2. Choose the **Create a section or redefine the existing section** button. The **Section** dialog box is displayed.

3. Select the **FRONT** datum plane as the sketching plane. The **RIGHT** datum plane and its orientation are selected by default.

4. Select the **TOP** datum plane and then choose the **Top** option from the **Orientation** drop-down list.

5. Choose the **Sketch** button from the **Section** dialog box to enter the sketcher environment.

6. Once you enter the sketcher environment, create the sketch of the base feature and apply constraints and dimensions as shown in Figure 5-54. Note that in the sketch, the bottom line of the rectangular section coincides with the **TOP** datum plane.

   As evident from the sketch of the base feature shown in Figure 5-54, the **RIGHT** datum plane is located at a dimension of 50 from the left edge because later in the tutorial, the rib feature will be created on this plane. The distance of 50 can be calculated from Figure 5-53. As mentioned earlier, the sketch for a rib is by default extruded to both the sides of the sketching plane. Therefore, when you create the sketch for the rib feature on the **RIGHT** datum plane, it will be extruded on both the sides of the sketching plane.

7. After the sketch is complete, choose the **Continue with the current section** button to exit the sketcher environment.

   The Extrude dashboard is enabled and appears below the graphics window. All the model attributes that are selected by default are accepted to create the model.

8. Enter **10** as the depth in the dimension box on the Extrude dashboard.

9. Choose the **Default View** option from the **Saved view** drop-down list and then choose the **Build feature** button from the Extrude dashboard.

   The base feature is completed and now the second feature will be created.

Creating the Second Feature

The second feature is also an extruded feature and will be created on the **TOP** datum plane. Therefore, you need to define the **TOP** datum plane as the sketching plane.

1. Choose the **Extrude Tool** button from the **Base Features** toolbar. The Extrude dashboard is displayed.

2. Choose the **Create a section or redefine the existing section** button. The **Section** dialog box is displayed.

3. Select the **TOP** datum plane as the sketching plane.

4. Select the front face of the base feature shown in Figure 5-55 and then select the **Bottom** option from the **Orientation** drop-down list.

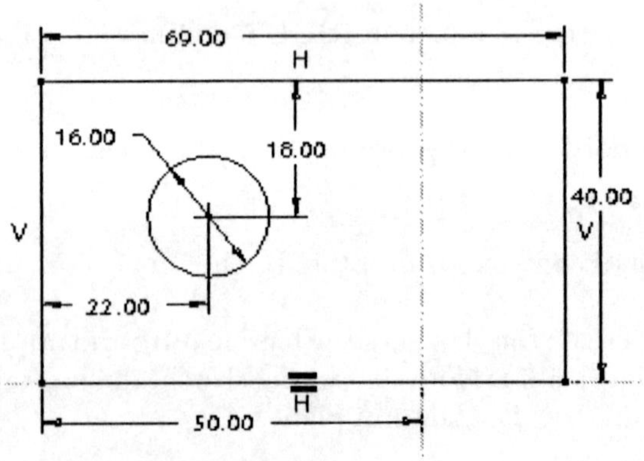

Figure 5-54 Sketch with dimensions and constraints for the base feature

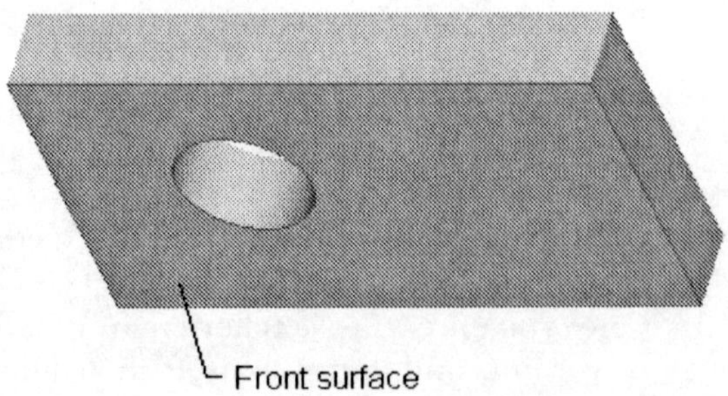

Figure 5-55 Front face of the base feature selected to be at the bottom while sketching

5. Choose the **Sketch** button from the **Section** dialog box to enter the sketcher environment. After entering the sketcher environment, turn the model display to **No Hidden**.

Note

Before you start sketching a section for any feature, it is recommended that you turn the model display to ***No Hidden*** *and turn off the datum plane display using the* ***Datum planes on/off*** *button. This is done to improve the clarity of the graphics screen while sketching.*

When you are working on these tutorials, make extensive use of buttons available in the ***Datum Display*** *toolbar.*

6. Create the sketch for the second feature and apply the constraints and dimensions as shown in Figure 5-56.

   You can use the **Create an entity from an edge** button from the **Sketcher Tools** toolbar to use the edge of the base feature. The edge of the base feature is required to close the section for the second feature. Note that if you use the **Create an entity from an edge** button, the 38 dimension in Figure 5-56 will not be required. However if you do not use the edge, you will have to draw a line to close the sketch. You will then have to align it with the bottom edge and add the constraint and dimensions as shown in Figure 5-56.

Note

If you do not close the section loop by drawing a line or using the edge of the base feature, the yellow arrow is displayed when you exit the sketcher environment. Using this arrow you can specify the direction in which the material will be added.

7. After completing the sketch, turn the model display to **Shading** and choose the **Continue with the current section** button. The Extrude dashboard is enabled below the graphics window.

8. Enter a value of **10** in the dimension box present on the Extrude dashboard and press ENTER.

9. Choose the **Build feature** button from the Extrude dashboard. The second feature is completed and the shaded default trimetric view is shown in Figure 5-57.

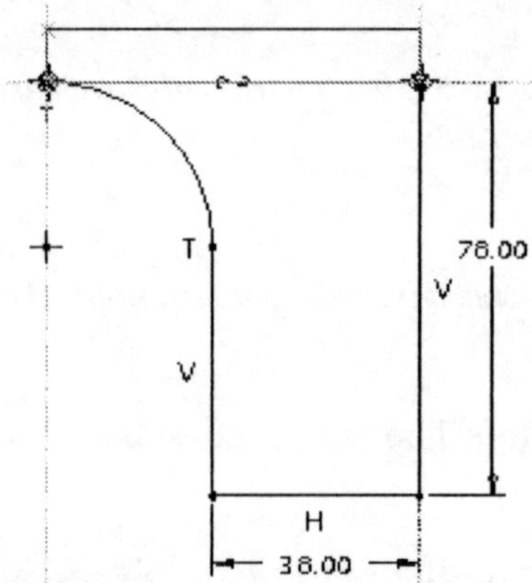

Figure 5-56 *Sketch with dimensions and constraints for the second feature*

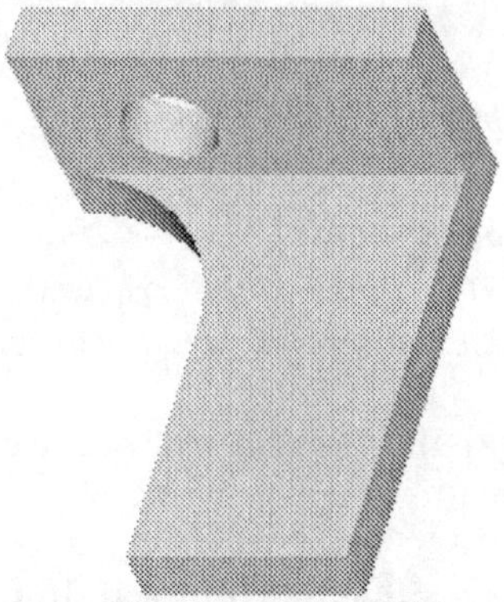

Figure 5-57 *The default trimetric view of the completed second feature and the base feature*

Creating the Third Feature

The sketch of the third feature will be drawn on the front planar surface of the second feature and will be extruded to the given depth.

1. Choose the **Extrude Tool** button from the **Base Features** toolbar. The Extrude dashboard is displayed.

2. Choose the **Create a section or redefine the existing section** button. The **Section** dialog box is displayed.

3. Select the face of the second feature shown in Figure 5-58 as the sketching plane.

4. Select the top face of the second feature and then select the **Top** option from the **Orientation** drop-down list.

5. Choose the **Sketch** button from the **Section** dialog box to enter the sketcher environment.

6. Once you enter the sketcher environment, turn the model display to **No Hidden**. Create the sketch for the third feature and apply constraints and dimensions as shown in Figure 5-59.

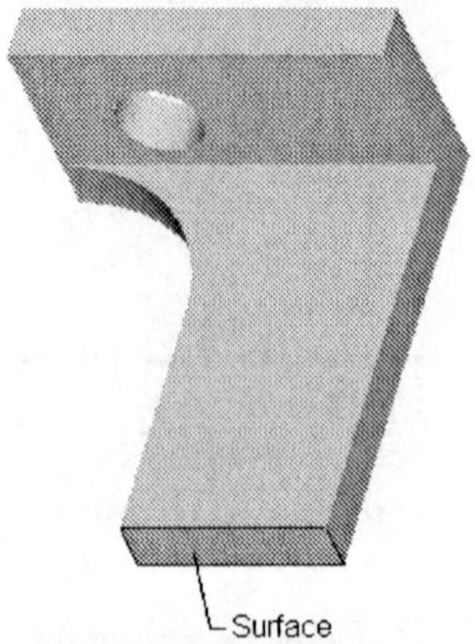

Figure 5-58 *Planar surface selected as the sketching plane for the third feature*

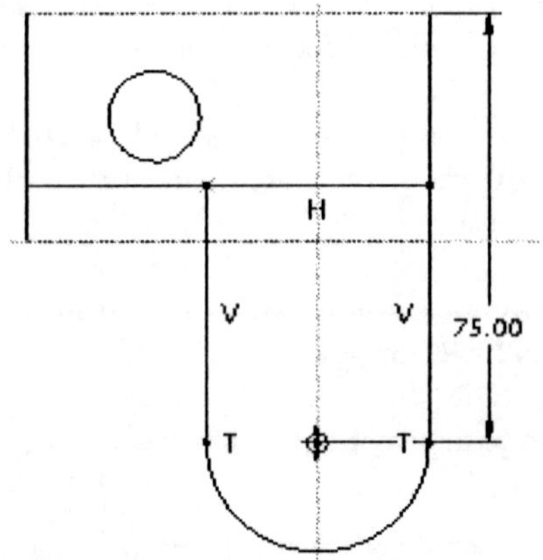

Figure 5-59 *Sketch with dimensions and constraints*

Note
While drawing the sketch for any feature, it is recommended that you first apply the required constraints and then dimension the sketch.

7. Choose the **Continue with the current section** button. The Extrude dashboard is enabled below the graphics window.

8. Enter a value of **10** in the dimension box present on the Extrude dashboard and press ENTER.

9. Choose the **Build feature** button from the Extrude dashboard. Turn the model display to **Shading**. The third feature is completed. You can use the middle mouse button to spin the model to view it from various directions.

Creating the Fourth Feature

The fourth feature of the model is an extruded feature and the sketch of this feature is drawn on the front planar surface of the third feature.

1. Choose the **Extrude Tool** button from the **Base Features** toolbar. The Extrude dashboard is displayed.

2. Choose the **Create a section or redefine the existing section** button. The **Section** dialog box is displayed.

3. Select the front face shown in Figure 5-60 as the sketch plane. The extrusion of the cylindrical feature will be on both sides of the front face.

   The attributes of the feature, like extruding to both sides, and the depth of extrusion will be specified after the sketch of the feature is drawn.

4. Select the top face of the second feature and then select the **Top** option from the **Orientation** drop-down list.

5. Choose the **Sketch** button from the **Section** dialog box to enter the sketcher environment.

6. Once you enter the sketcher environment, turn the model display to **No Hidden**.

7. Draw the circular section for the fourth feature as shown in Figure 5-61. Choose the **Create concentric circle** button from the **Sketcher Tools** toolbar to draw the sketch for this feature. Select the arc using the left mouse button. As you move the mouse, the yellow rubber-band circle changes its size. Move the mouse cursor close to the arc, the cursor snaps to the arc. Use the left mouse button and select a point on the arc. You will notice that the equal radius constraint is applied to the sketch. Now, abort this option.

8. After the sketch is completed, choose the **Continue with the current section** button. The Extrude dashboard is enabled below the graphics window.

9. Choose the **Options** option from the Extrude dashboard. From the slide-up panel, choose the **Blind** option from the **Side 2** drop-down list.

10. In the dimension box that is on the right of the **Side 1** drop-down list, enter a value of **12**.

11. In the dimension box that is on the right of the **Side 2** drop-down list, enter a value of **13**.

12. Choose the **Build feature** button from the Extrude dashboard.

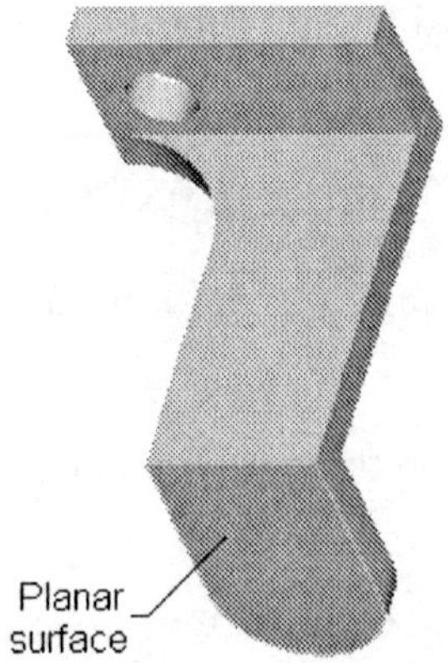

Figure 5-60 *Front face*

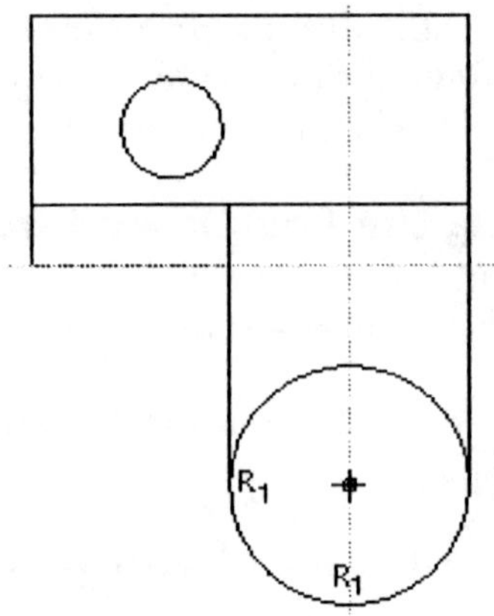

Figure 5-61 *Sketch for the fourth feature*

Creating the Hole Feature

The hole feature will be created using the Hole dashboard. The hole will be placed coaxial to the fourth cylindrical feature.

Note

Since the hole will be created coaxially, you will be required to select the axis of the circular feature. You may need to turn on the axis display if it is turned off. Choose the ***Datum axes on/off*** *button from the* ***Datum Display*** *toolbar to turn on the datum axis display.*

1. Choose the **Hole Tool** button from the **Engineering Features** toolbar. The Hole dashboard is displayed. In the Hole dashboard, the **Create straight hole** button is selected by default.

2. Choose the **Placement** tab from the Hole dashboard. The slide-up panel is displayed.

3. Select the front face of the cylindrical feature shown in Figure 5-62 to place the hole.

   As you select the front face of the cylindrical feature, the preview of the hole is displayed on the graphics window. Now, you need to specify the reference for the hole placement.

4. From the drop-down list present in the slide-up panel, select the **Coaxial** option. Now, you have specified the type of hole you need to create.

5. Select the axis of the cylindrical feature from the graphics window.

6. Click in the **Secondary References** collector so that the **Primary** collector turns white in color.

7. Select the front face of the cylindrical feature. The coaxial hole is created.

8. In the diameter dimension box on the Hole dashboard, enter **16** and press ENTER.

9. From the depth flyout in the Hole dashboard, select the **Drill to intersect with all surfaces** button.

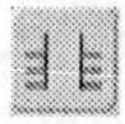

10. Choose the **Build feature** button from the Hole dashboard. The hole is created and the trimetric view of the shaded model with the hole is shown in Figure 5-63.

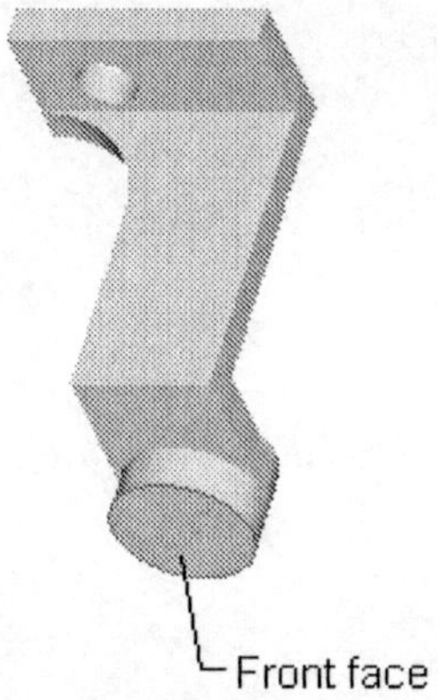

Figure 5-62 *Planar face*

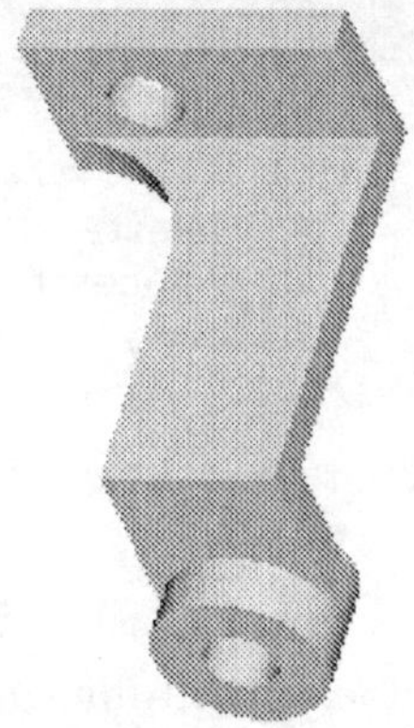

Figure 5-63 *The default trimetric view*

Creating the Two Round Features

Now, the two round features will be created. Both the rounds have different radii and will be created by defining two sets.

1. Choose the **Round Tool** button from the **Engineering Features** toolbar. The Round dashboard is displayed.

2. Choose the **Sets** tab to display the slide-up panel. Let this slide-up panel remain open so that you can view the selections you make on the model.

3. Spin the model and select one of the faces shown in Figure 5-64. To select the second face, you need to press the CTRL key and then select the face.

   The preview of the round is created on the edge that connects the two selected faces. A default value of the radius is also displayed.

4. Double-click on the default radius value that is displayed on the preview of the round. In the edit box that appears, enter a value of **5** and press ENTER. The first round is created.

5. Spin the model as shown in Figure 5-65 to select the two faces.

6. After spinning the model, select one of the faces shown in Figure 5-65 and then use CTRL+left mouse button to select the second face. You will notice that in the slide-up panel, under **Set 1**, **Set 2** appears. This indicates that second set is defined.

   The preview of the round is created on the edge that connects the two selected faces. A

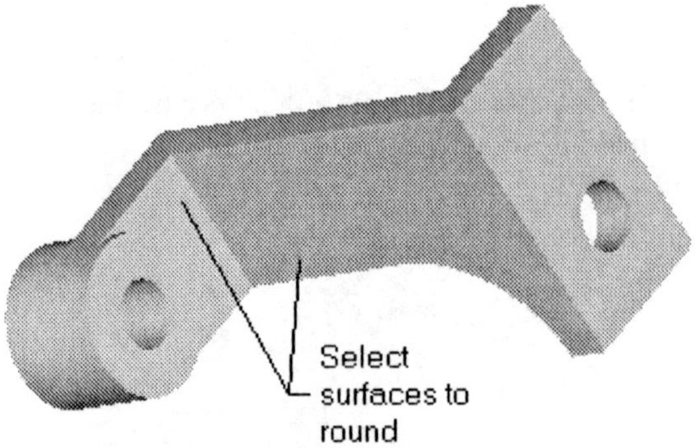

Figure 5-64 Faces for creating round of radius 5

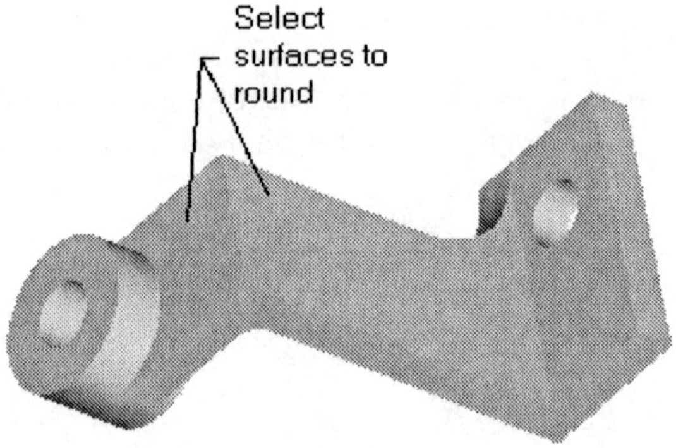

Figure 5-65 Faces for creating round of radius 15

default value of the radius is also displayed.

7. Double-click on the default radius value that is displayed on the preview of the round. In the edit box that appears, enter a value of **15** and press ENTER.

8. Choose the **Build feature** button from the Round dashboard.

   The second round is created. In the **Model Tree** the two rounds created appear as one feature.

Creating the Rib Feature

Some advance planning that was done while drawing the sketch for the base feature will help you to draw the rib feature now. The location for the **RIGHT** datum plane was calculated while sketching the base feature. The section for the rib feature will be drawn on the **RIGHT** datum plane.

1. Choose the **Rib Tool** button from the **Engineering Features** toolbar. The Rib dashboard is displayed. 

   As mentioned earlier, a rib feature is always sketched from the side view. Now, turn on the display of datum planes from the **Datum Display** toolbar.

2. Choose the **Create a section or redefine the existing section** button from the Rib dashboard. The **Section** dialog box is displayed.

3. Select the **RIGHT** datum plane as the sketching plane from the graphics window.

4. Select the **TOP** datum plane and then select the **Top** option from the **Orientation** drop-down list.

5. Choose the **Sketch** button to enter the sketcher environment.

   When you enter the sketcher environment, the **References** dialog box is displayed and the status displayed is **Fully Placed**. Close the **References** dialog box.

   You will need to turn the model display to **No Hidden** before drawing the sketch for the rib feature.

6. Draw the open section sketch for the rib feature and apply the required constraints and dimensions as shown in Figure 5-66.

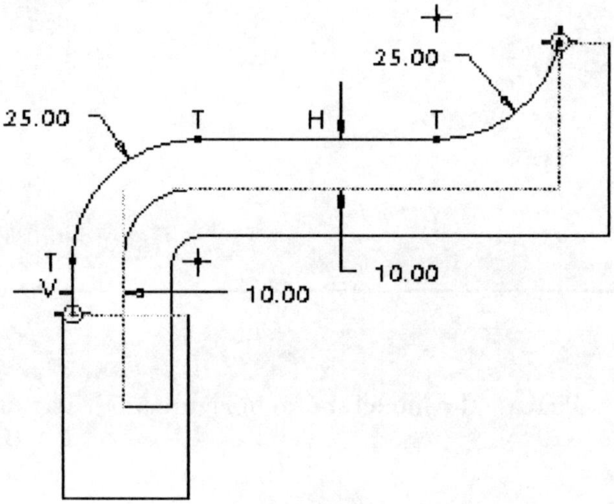

***Figure 5-66** Open sketch of the rib feature with dimensions and constraints*

7. After the sketch is completed, choose the **Continue with the current section** button.

   A yellow arrow is displayed on the sketch pointing in the direction of material addition. Since, the section for the rib feature is open, therefore Pro/ENGINEER allows you to specify the direction where the material should be added.

8. Enter **10** in the dimension box present on the Rib dashboard.

9. Choose the **Build feature** button from the Rib dashboard to exit this feature creation tool.

   All the features in the model have been created and the model is now complete. The trimetric shaded view of the completed model is shown in Figure 5-67.

Saving the Model

1. Choose the **Save the active object** button from the **File** toolbar and save the model.

The order of feature creation can be seen from the **Model Tree** shown in Figure 5-68. The feature id numbers displayed in the **Model Tree** may be different when you create the features.

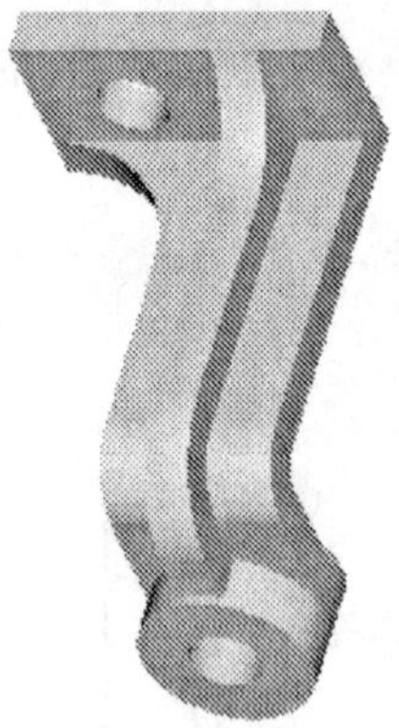

Figure 5-67 *The default trimetric view of the model*

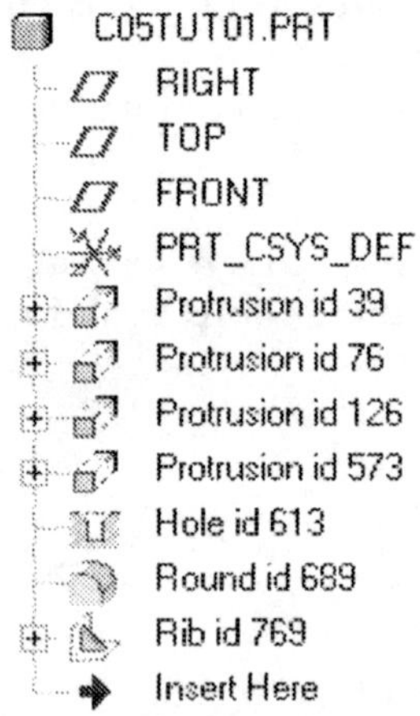

***Figure 5-68 Model Tree** for Tutorial 1*

Tutorial 2

In this tutorial you will create the model shown in Figure 5-69. The dimensions of the model are shown in Figure 5-70. **(Expected time: 30 min)**

Figure 5-69 *Isometric view of the solid model*

The following steps outline the procedure for creating this model:

a. First examine the model and then determine the number of features in it. The model

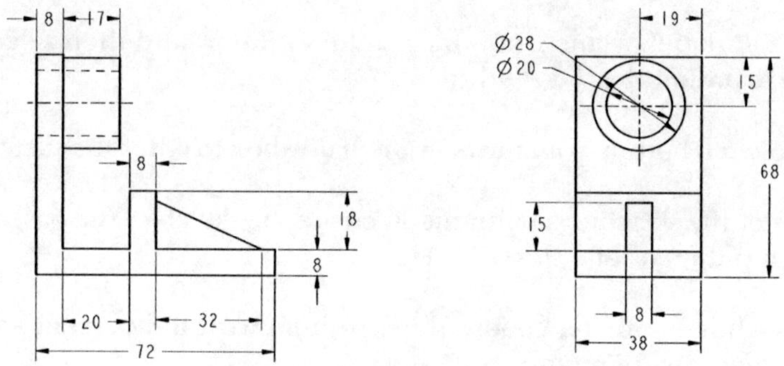

Figure 5-70 *Front view and the right-side view of the model for Tutorial 2*

is composed of four features; one at the bottom (base feature), one cylindrical feature, one hole feature on the cylindrical feature, and one rib feature.

b. Select the sketching plane for the base feature, draw the sketch using the sketcher tools, apply the constraints and dimensions, and then extrude the sketch to the given depth, see Figure 5-72.

c. Select the sketching plane for the cylindrical feature, draw the sketch using the sketcher tools, apply the dimensions and constraints, and then extrude the sketch to the given depth, see Figure 5-75.

d. Create the hole on the cylindrical feature using the Hole dashboard, see Figure 5-76.

e. Create a sketching plane for the rib feature, draw the sketch using the sketcher tools, apply the dimensions and constraints, and then give the thickness to the sketch, see Figure 5-78.

After understanding the procedure for creating the model, you are now ready to draw it. The working directory has already been selected in Tutorial 1 and therefore you do not need to select it again. However, if you want to change the working directory, choose **File > Set Working Directory** and then select *c05* in the **Select Working Directory** dialog box.

Creating New Object File

1. Open a new part file and name it as *c05tut2*. The three default datum planes and the **Model Tree** appears on the graphics window. Exit the **Model Tree**. The datum planes and **Model Tree** will not appear if they were previously turned off.

Creating the Base Feature

1. Choose the **Extrude Tool** button from the **Base Features** toolbar. The Extrude dashboard is displayed.

2. Choose the **Create a section or redefine the existing section** button from the Extrude dashboard. The **Section** dialog box is displayed.

3. Select the **FRONT** datum plane for drawing the sketch of the base feature.

4. Select the **TOP** datum plane from the graphics window and then select the **Top** option from the **Orientation** drop-down list.

5. Choose the **Sketch** button from the **Section** dialog box to enter the sketcher environment.

6. Once you enter the sketcher environment, create the sketch of the base feature and apply constraints and dimensions shown in Figure 5-71.

7. After the sketch is complete, choose the **Continue with the current section** button and exit the sketcher environment.

   The Extrude dashboard is enabled and appears below the graphics window.

8. Enter a depth of **38** in the dimension box that is present on the Extrude dashboard.

   The default trimetric view of the base feature is shown in Figure 5-72.

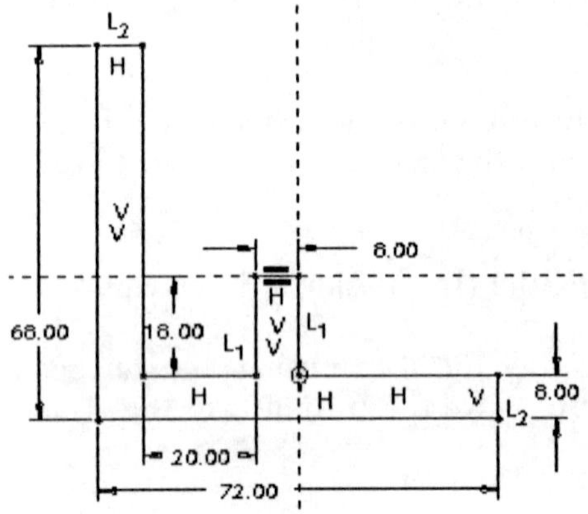

Figure 5-71 *Sketch with dimensions and constraints for the base feature*

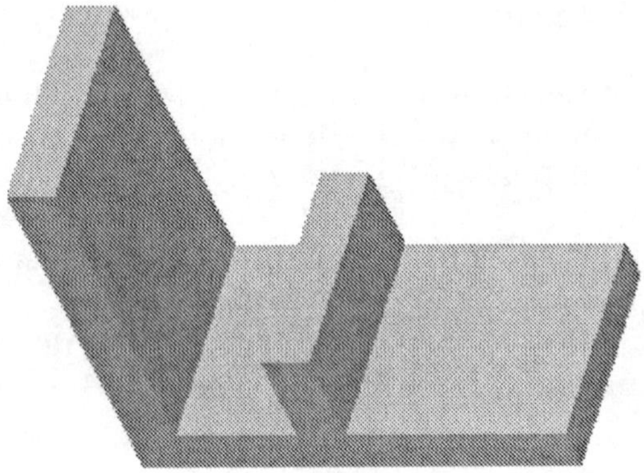

Figure 5-72 *The default trimetric view of the base feature*

9. Choose the **Build feature** button from the Extrude dashboard to exit the feature creation tool.

Note

*In this tutorial the base feature can also be created by extruding it on both the sides of the sketching plane, that is the **FRONT** datum plane. This would reduce the model creation by one feature. This is because, as evident form the model, later in the tutorial you will have to create a datum plane to create the rib feature. However, to make you familiar with creating datum planes on-the-fly, the base feature is extruded on one side of the sketching plane.*

Creating the Second Feature

The second feature is a cylindrical feature that is sketched on the planar surface of the base feature that is highlighted in Figure 5-73.

1. Choose the **Extrude Tool** button from the **Base Features** toolbar. The Extrude dashboard is displayed.

2. Choose the **Create a section or redefine the existing section** button from the Extrude dashboard. The **Section** dialog box is displayed.

3. Select the face of the base feature shown in Figure 5-73 as the sketching plane.

4. Select the **TOP** datum plane from the graphics window and then select the **Top** option from the **Orientation** drop-down list.

5. Choose the **Sketch** button from the **Section** dialog box to enter the sketcher environment.

   After entering the sketcher environment, turn the model display to **No Hidden**.

6. Draw the sketch for the second feature. Apply and modify the dimensions as shown in Figure 5-74.

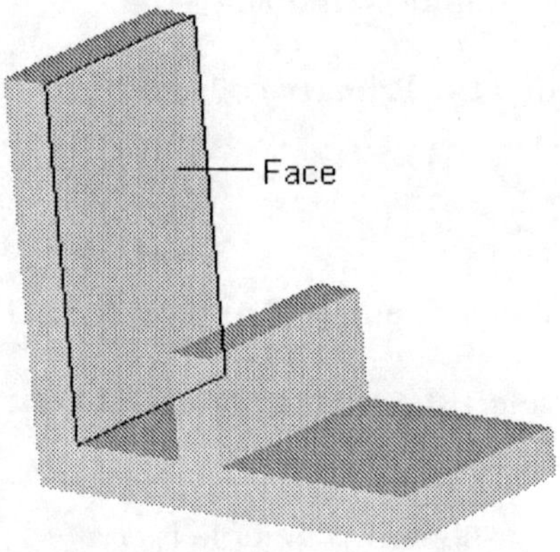

Figure 5-73 *The face of the base feature selected as the sketching plane*

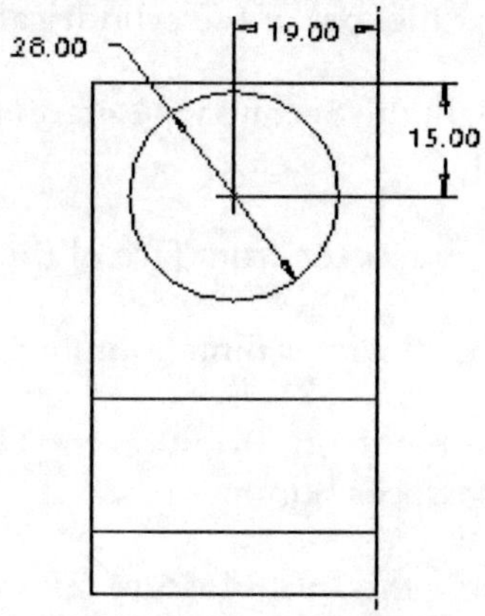

Figure 5-74 *Sketch and dimensions for the second feature*

7. Turn the model display to **Shading**. Exit the sketcher environment by choosing the **Continue with the current section** button. The Extrude dashboard is enabled and appears below the graphics window.

8. Enter a value of **17** in the dimension box present on the Extrude dashboard and press ENTER.

9. Choose the **Build feature** button from the Extrude dashboard to exit the feature creation tool. The trimetric view of the model with the second feature is shown in Figure 5-75.

Creating the Hole Feature

The hole feature will be created using the Hole dashboard. The coaxial hole will be created on the cylindrical feature. The axis of the cylindrical feature will be used as axial reference to create the coaxial hole.

1. Choose the **Hole Tool** from the menu bar. The Hole dashboard is displayed.

   The **Create straight hole** button in the Hole dashboard is selected by default.

2. Choose the **Placement** tab from the Hole dashboard. The slide-up panel is displayed.

3. Select the front face of the cylindrical feature to place the hole.

   As you select the front face of the cylindrical feature, the preview of the hole is displayed on the graphics window. Now, you need to specify the reference for the hole placement.

4. From the drop-down list present in the slide-up panel, select the **Coaxial** option. You have specified the type of hole you need to create.

5. Select the axis of the cylindrical feature from the graphics window.

6. Click in the **Secondary References** collector so that the **Primary** collector turns white in color.

7. Now, select the front face of the cylindrical feature.

8. In the diameter dimension box on the Hole dashboard, enter **20** and press ENTER.

9. From the depth flyout on the Hole dashboard, select the **Drill to intersect with all surfaces** button.

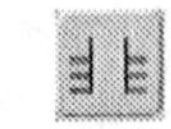

10. Choose the **Build feature** button from the Hole dashboard. The hole is created and the trimetric view of the shaded model with the hole is shown in Figure 5-76.

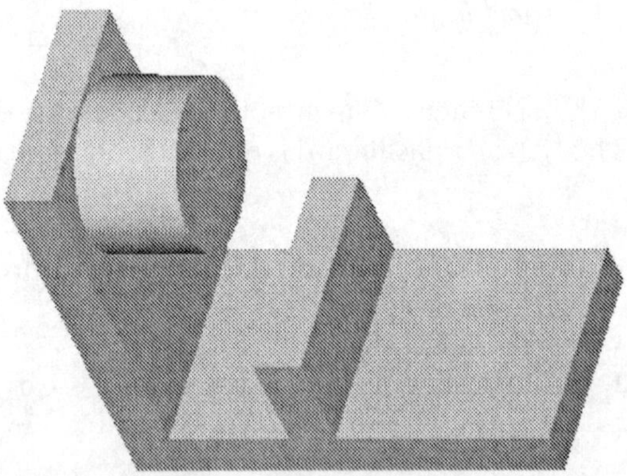

Figure 5-75 *Model with the second feature*

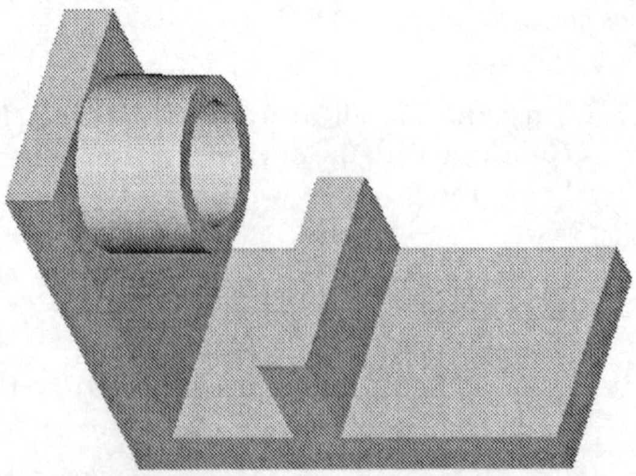

Figure 5-76 *Model with the hole feature*

Creating the Rib Feature

To create the rib feature, a datum plane is created on-the-fly. As mentioned earlier, rib features are always drawn from the side view.

1. Choose the **Rib Tool** button from the **Engineering Features** toolbar. The Rib dashboard is displayed.

2. Choose the **Create a section or redefine the existing section** button from the Rib dashboard.

3. Choose the **Datum Plane Tool** button from the **Datum** toolbar. The **DATUM PLANE** dialog box is displayed.

4. Select the axis of the hole and then press the CTRL key and select the **FRONT** datum plane. You can view your selections in the **References** collector.

5. Click the **FRONT** reference in the **References** collector. A drop-down list appears to the right of the reference. Select the **Parallel** option from this drop-down list.

6. Choose **OK** from the **DATUM PLANE** dialog box to exit it.

   A datum plane that passes through the selected axis and is parallel to the **FRONT** datum plane will be created. This datum plane is selected automatically as the sketching plane. Now, you need to select the reference plane.

7. Select the **TOP** datum plane and then select the **Top** option from the **Orientation** drop-down list.

8. Choose the **Sketch** button; the system takes you to the sketcher environment.

9. Specify the references and close the reference dialog box. Draw the open sketch of the rib feature and apply the dimensions as shown in Figure 5-77. Exit the sketcher environment by choosing the **Continue with the current section** button.

   The Rib dashboard is enabled and appears below the graphics window. Turn the model display to **Shading**.

10. Choose the **Add material to the other side of section** button on the **References** slide-up panel to flip the arrow. The preview of the rib is displayed on the model.

11. Enter a value of **8** in the edit box present on the Rib dashboard and press ENTER. This value is the thickness of the rib. Choose the **Build feature** button.

    The trimetric view of the complete model with the rib feature is shown in Figure 5-78.

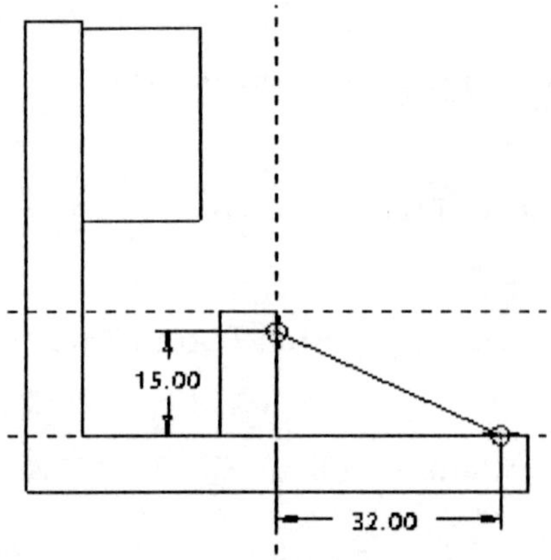

Figure 5-77 *Sketch for the rib feature with the model display set to* ***No Hidden***

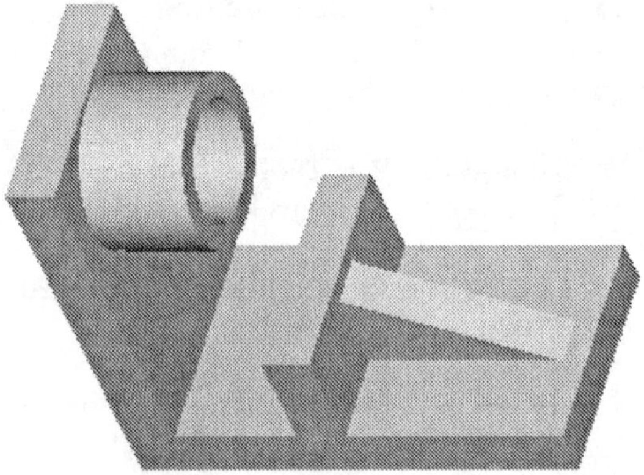

Figure 5-78 *The default trimetric view of the final model*

Saving the Model

1. Choose the **Save the active object** button from the **File** toolbar and save the model.

   The order of feature creation is evident from the **Model Tree** shown in Figure 5-79. The feature id numbers displayed in the **Model Tree** may be different when you create the features.

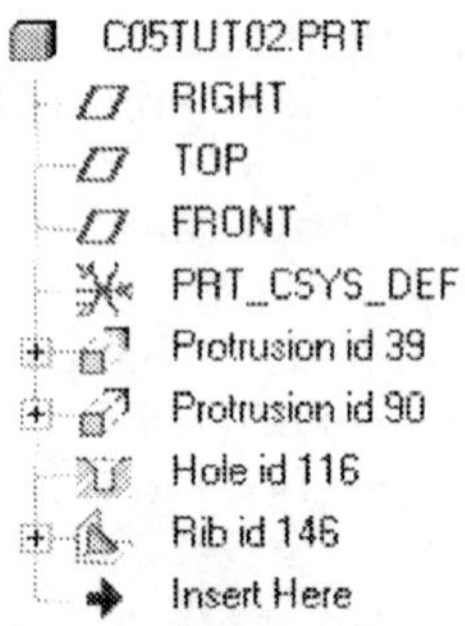

Figure 5-79 *Model Tree for Tutorial 2*

Tutorial 3

In this tutorial you will create the model shown in Figure 5-80. The solid model, dimensions, and the front, and the right-side views are also shown in this figure. **(Expected time: 45 min)**

The following steps outline the procedure for creating this model:

a. First examine the model and then determine the number of features in it. The model is composed of four features; one base feature, one cut feature, one counterbore hole on the front planar surface of the base feature, and one round feature, see Figure 5-80.

b. Select the sketching plane for the base feature, draw the sketch using the sketcher tools, apply the dimensions and constraints, and then extrude the sketch to the given depth, see Figure 5-82.

c. Create a sketching plane for the cut feature, draw the sketch using the sketcher tools, apply the dimensions, and then extrude the sketch to both sides of the plane, see Figure 5-85.

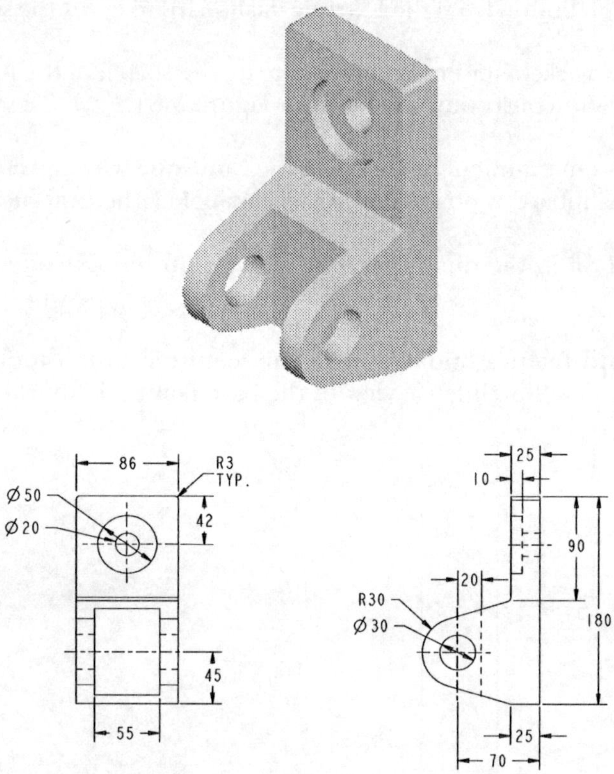

Figure 5-80 Isometric view, front view, and the right-side view of the model

d. Create a counterbore hole as shown in Figure 5-88

e. Create the round as shown in Figure 5-90.

Creating New Object File

1. Open a new part file and name it as **c05tut3**. The three default datum planes and the **Model Tree** appear on the graphics window if they were not turned off previously. Close the **Model Tree**.

Creating the Base Feature

1. Choose the **Extrude Tool** button from the **Base Features** toolbar. The Extrude dashboard is displayed.

2. Choose the **Create a section or redefine the existing section** button from the Extrude dashboard. The **Section** dialog box is displayed.

3. Select the **RIGHT** datum plane as the sketching plane.

4. Select the **TOP** datum plane from the graphics window and then select the **Top** option from the **Orientation** drop-down list.

5. Choose the **Sketch** button from the Extrude dashboard to enter the sketcher environment.

6. Once you enter the sketcher environment, create the sketch of the base feature and apply the dimensions and constraints as shown in Figure 5-81.

7. Exit the sketcher environment by choosing the **Continue with the current section** button. The Extrude dashboard is enabled and appears below the graphics window.

8. Enter a value of **86** in the dimension box present on the Extrude dashboard and press ENTER.

9. Choose the **Build feature** button to exit the feature creation tool. The base feature is completed. The default trimetric view of the base feature is shown in Figure 5-82.

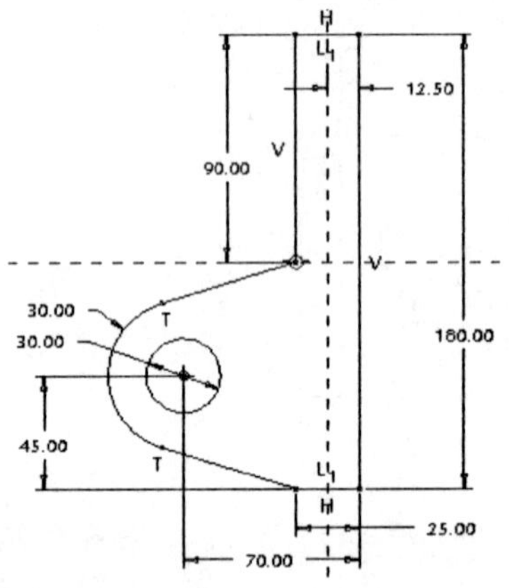

***Figure 5-81** Sketch with dimensions and constraints for the base feature*

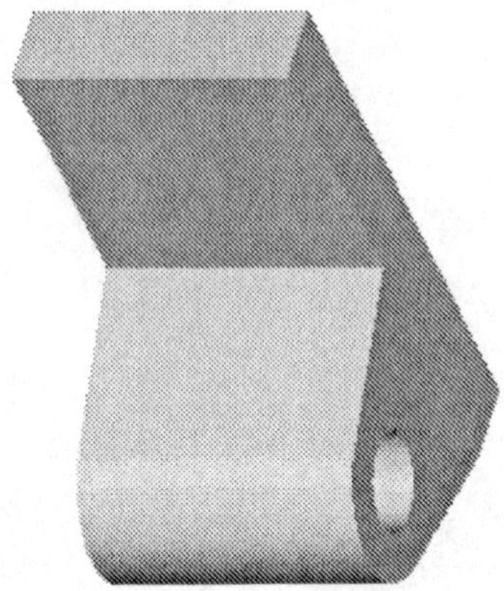

***Figure 5-82** The default trimetric view of the base feature*

Creating the Second Feature

The second feature is an extruded cut. This cut is created on a datum plane that passes through the center of the base feature.

1. Choose the **Extrude Tool** button from the **Base Features** toolbar. The Extrude dashboard is displayed.

2. Choose the **Remove material** button.

3. From the depth flyout, choose the **Extrude on both sides of sketch plane by half the specified depth value in each** button. 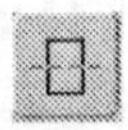

4. Choose the **Create a section or redefine the existing section** button from the Extrude dashboard. The **Section** dialog box is displayed.

5. Choose the **Datum Plane Tool** button from the **Datum** toolbar. The **DATUM PLANE** dialog box is displayed.

6. Select the **RIGHT** datum plane. In the **Translation** edit box that appears in the dialog box, enter a value of **43**. This value is half of the width of the base feature.

7. Choose **OK** to exit the **DATUM PLANE** dialog box.

8. Select the **FRONT** datum plane and then select the **Left** option from the **Orientation** drop-down list.

9. Choose the **Sketch** button. The system takes you to the sketcher environment.

10. Draw the sketch for the cut feature and apply and modify the dimensions as shown in Figure 5-83.

    In Figure 5-83, some dimensions appear light in color. These dimensions are weak dimensions and it is not important to convert them into strong dimensions. These dimensions are not important for the creation of this feature. However, the geometry for the cut should be similar to that shown in Figure 5-83.

11. After the sketch is completed, exit the sketcher environment. The Extrude dashboard is enabled and appears below the graphics window.

12. Enter a depth of **55** in the dimension box present on the Extrude dashboard and press ENTER. The system will accept this depth symmetrical to the sketching plane.

13. Choose the **Build feature** button to exit the feature creation tool. The cut feature is completed now and the default trimetric view of the cut feature along with the base feature is shown in Figure 5-84.

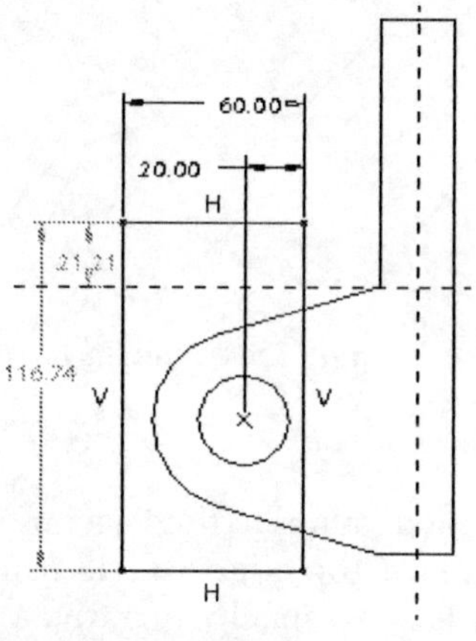

Figure 5-83 *Sketch with dimensions and constraints for the cut feature*

Figure 5-84 *The default trimetric view of the model*

Creating the Hole Feature

The hole is created using the Hole dashboard.

1. Choose the **Hole Tool** button from the **Engineering Features** toolbar. The Hole dashboard is displayed.

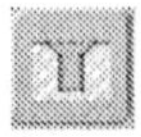

2. From the drop-down list present on the Hole dashboard, select the **Sketched** option.

3. Choose the **Activates Sketcher to create section** button to enter the sketcher environment.

4. In the sketcher environment, draw the sketch of the counterbore hole as shown in Figure 5-85. After drawing the sketch, exit the sketcher environment.

5. Select the front face of the base feature to place the hole. The preview of the hole appears on the selected face.

6. Choose the **Placement** tab from the Hole dashboard. The slide-up panel is displayed.

7. Click in the **Secondary References** collector so that the **Primary** collector turns white in color.

8. Select the two references as shown in Figure 5-86 for dimensioning. Note that the second reference should be selected using CTRL+left mouse button.

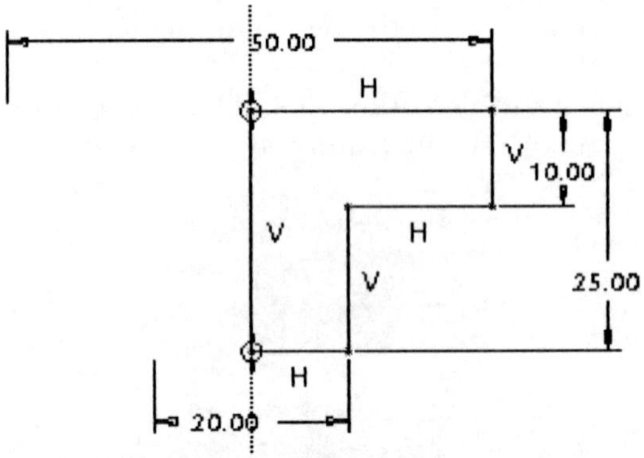

Figure 5-85 *Cross-section of the hole with dimensions and constraints*

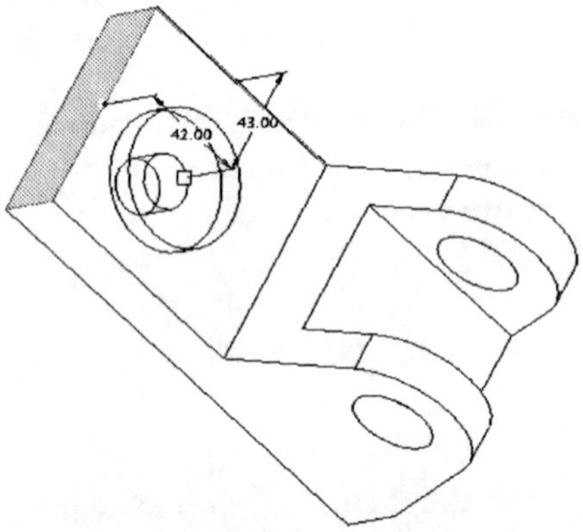

Figure 5-86 *Edges to be selected for placing hole*

9. Modify the dimensions of the references by entering their values in the **Placement** slide-up panel. The center of hole is at the distance of **42** from the top edge and **43** from the right edge. After viewing the preview of the hole you may have to specify negative value of any one of the distances.

10. Choose the **Build feature** button to exit this feature creation tool. The hole will be created as shown in Figure 5-87.

Note

When you select a feature, entity, or item from the graphics window and right-click to invoke the shortcut menu, the right mouse button should be held down until the shortcut menu is displayed. Once the shortcut menu is displayed, the right mouse button can be released.

Creating the Round Feature

The round is created using the **Simple** option.

1. Choose the **Round Tool** button from the **Engineering Features** toolbar. The Round dashboard is displayed.

2. Using the CTRL key, select the edges that are shown in Figure 5-88. The selected edges turns red in color.

Figure 5-87 *The default trimetric view*

Figure 5-88 *Edges to be selected to round*

3. Enter **3** in the dimension box that is present on the Round dashboard. The round is created and its preview appears on the model.

Note

The value for the radius of round can also be specified by double-clicking on the default value displayed on the preview of the round. When the edit box appears in the preview, enter the radius value and press ENTER.

4. Choose the **Build feature** button to exit the round feature creation tool. The trimetric view of the final model with round feature created is shown in Figure 5-89.

Saving the Model

1. Choose the **Save the active object** button from the **File** toolbar and save the model.

The order of feature creation can be seen from the **Model Tree** shown in Figure 5-90.

Figure 5-89 *Final model for Tutorial 3*

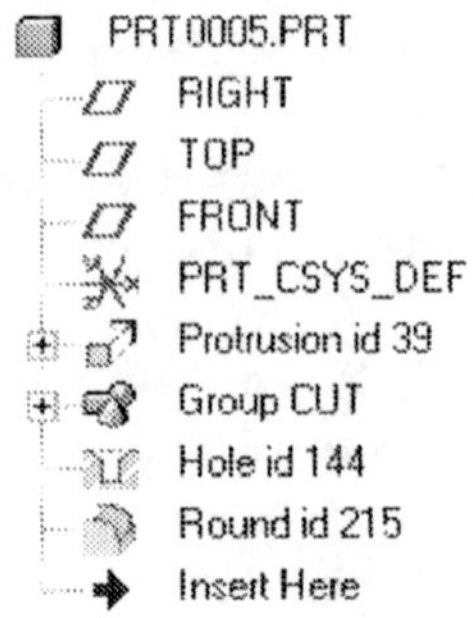

Figure 5-90 *Model Tree for Tutorial 3*

Tutorial 4

Create the model shown in Figure 5-91. The dimensions, the top view, the front section view, and the right section view of the model are shown in Figure 5-92. **(Expected time: 1 hr)**

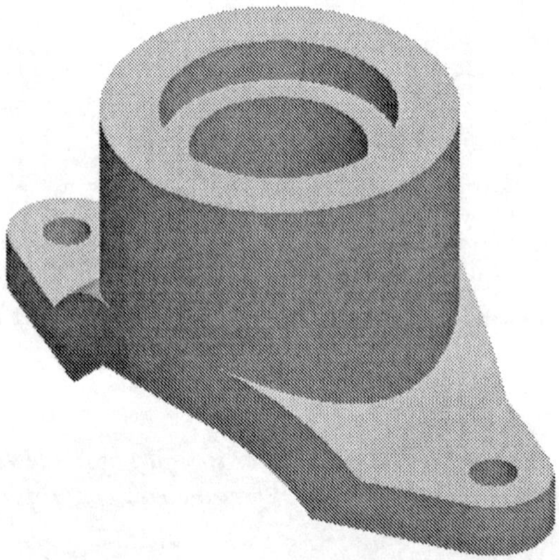

Figure 5-91 *Isometric view of the model*

The following steps outline the procedure for creating this model:

a. First examine the model and then determine the number of features in it, see Figure 5-91. The model is composed of five features: one at the bottom (base feature), one cut feature on the base feature, one protrusion on the top of the base feature, and two cut features that will create one hole at the top and two holes on the base feature.

b. Select the sketch plane for the base feature, draw the sketch using the sketching tools,

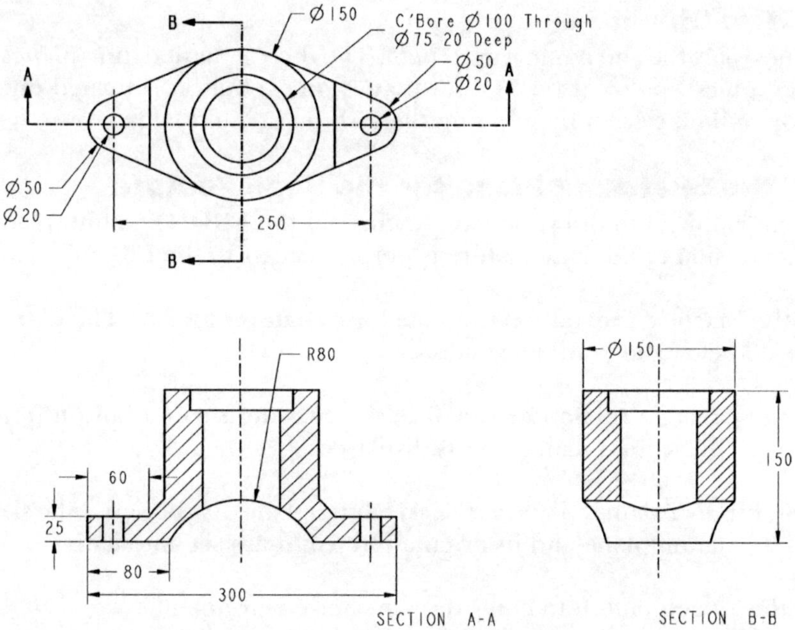

Figure 5-92 *Top view, front section view, and right section view of the model*

apply the dimensions and constraints, and then extrude the sketch to the given depth, see Figures 5-93 and 5-94.

c. To create the cut feature, select the bottom face of the base feature as the sketching plane, draw the sketch, apply dimensions and constraints, and then extrude the sketch, see Figure 5-97.

d. To create the protrusion feature, you need to create a datum plane offset to the top face of the base feature, draw the sketch using the sketching tools, apply dimensions and constraints, and then extrude the sketch, see Figure 5-100.

e. The next feature is a counterbore hole. You will sketch the section of the counterbore hole and then select the top face of the cylindrical feature as the placement plane. This hole is coaxial with the cylindrical feature, see Figure 5-102.

f. The last feature is the set of two straight holes that are created on the top face of the base feature.

After understanding the procedure for creating the model, you are now ready to create it. The working directory has already been selected in Tutorial 1 and therefore you do not need to select it again. However, if you want to select the working directory, select it from the Navigator.

Creating New Object File

1. Open a new part file and name it as *c05tut5*. The three default datum planes are displayed on the graphics window if the **Datum planes on/off** button is turned on. If the **Model Tree** is open then close it by selecting the sash on its right edge.

Selecting the Sketching Plane for the Base Feature

In this model, you need to draw the base feature on the **FRONT** datum plane because the direction of extrusion of the base feature is perpendicular to the **FRONT** datum plane.

1. Choose the **Extrude Tool** button from the **Base Features** toolbar. The Extrude dashboard is displayed below the graphics window.

2. Choose the **Create a section or redefine the existing section** button from the extrude dashboard. The **Section** dialog box is displayed.

3. Select the **FRONT** datum plane as the sketching plane. As you select the sketching plane, the **RIGHT** datum plane and its orientation **Right** are set automatically.

4. Choose the **Sketch** button to enter the sketcher environment.

Specifying References

In the sketcher environment, the **References** dialog box is displayed. The status displayed in the **Reference status** area is **Fully Placed**. Close this dialog box.

Creating and Dimensioning the Sketch for the Base Feature

From the model, the section to be extruded for the base feature is not evident. Therefore, you need to visualize the sketch for the base feature. The section sketch is shown in Figure 5-93. When this sketch is extruded, it will create the base feature.

1. Draw the section sketch using various sketcher tools as shown in Figure 5-93.

2. The sketch is dimensioned automatically and some weak dimensions are assigned to it. Add the required constraints and modify the weak dimensions as shown in Figure 5-93.

3. Choose the **Continue with the current section** button. The Extrude dashboard is displayed.

4. Choose the **Default View** option from the **Saved view list** button of the **View** toolbar.

   The default view is displayed which gives you a better view of the sketch in the 3D space. A yellow arrow is also displayed on the model, indicating the direction of extrusion.

5. Enter a value of **150** in the dimension box that is present on the Extrude dashboard and press ENTER.

6. Choose the **Build feature** button from the Extrude dashboard.

   The base feature is completed as shown in Figure 5-94. You can use the middle mouse button to spin the model to view it from various directions.

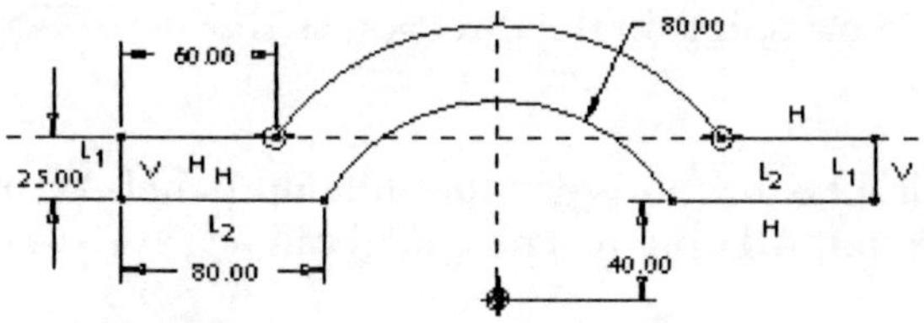

Figure 5-93 Sketch with dimensions and constraints

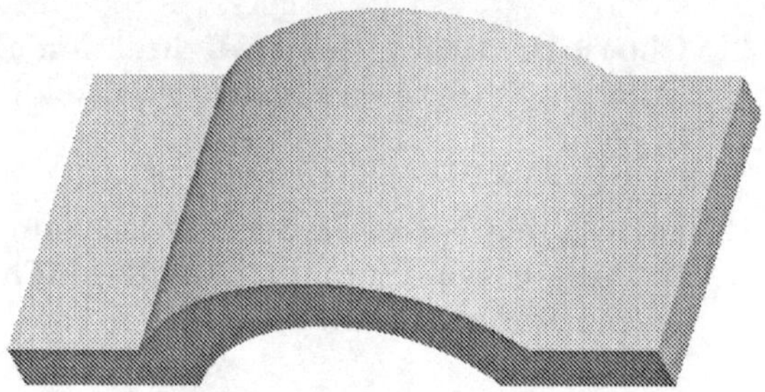

Figure 5-94 Default view of the base feature

Creating the Second Feature

The next feature is an extruded cut feature. The sketching plane for this feature is the bottom face of the base feature. To get the required shape that is at the base of the model, you need to cut the base feature in such a way such that you get the required shape. This sketch and its dimensions can be referred to from the top view of the model shown in Figure 5-92. Before drawing the sketch, change the model display to **No Hidden**.

1. Choose the **Extrude Tool** button from the **Base Features** toolbar. The Extrude dashboard is displayed below the graphics window.

2. Choose the **Remove Material** button from the Extrude dashboard.

3. Choose the **Create a section or redefine the existing section** button from the Extrude dashboard. The **Section** dialog box is displayed.

4. Select the bottom face of the base feature as the sketching plane.

5. Select the **RIGHT** datum plane and from the **Orientation** drop-down list, choose the **Right** option.

6. Choose the **Sketch** button to enter the sketcher environment.

7. Draw the sketch of the cut feature, and apply constraints and dimensions as shown in Figure 4-95.

8. Choose the **Continue with the current section** button to exit the sketcher environment.

The Extrude dashboard is displayed below the graphics window.

9. Choose the **Shading** button from the **Model Display** toolbar.

10. Choose the **Default View** option from the **Saved view list** drop-down list. The model orients in its default orientation.

11. Choose the **Change material direction of extrude to other side of sketch** button from the Extrude dashboard. The yellow arrow points in the direction as shown in Figure 4-96.

12. Choose the **Options** tab in the Extrude dashboard to display the slide-up panel. From the **Side 1** drop-down list, choose the **Through All** option. The model can be previewed on the graphics window.

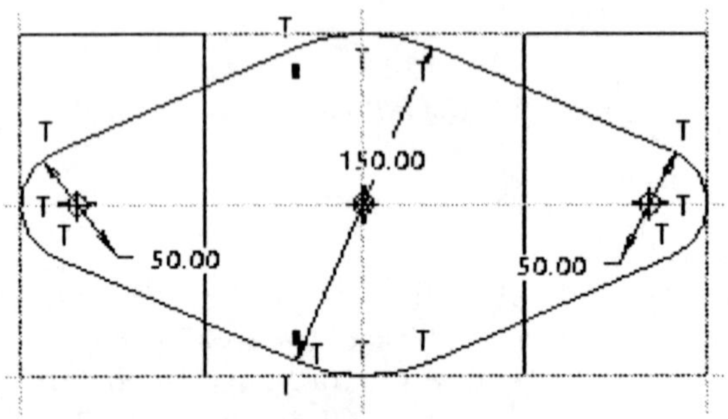

Figure 5-95 *Sketch with dimensions and constraints for the cut feature*

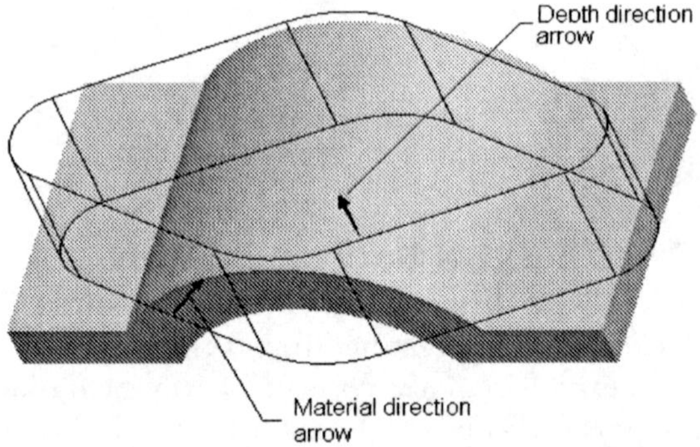

Figure 5-96 *Two arrows and the preview of the cut feature*

13. Choose the **Build feature** button from the Extrude dashboard to complete the feature creation. The model after creating the cut feature is shown in Figure 5-97.

Creating the Sketching Plane for the Third Feature

You need to create a datum plane to create the third feature. The datum plane will be at a distance of 150 from the bottom face of the model.

1. Choose the **Datum Plane Tool** button from the **Datum** toolbar. The **DATUM PLANE** dialog box is displayed.

2. Spin the model using the middle mouse button and then using the left mouse button, select the bottom face of the model.

   When you select the bottom face of the model, the **Offset** constraint is displayed in the **References** collector of the dialog box. Note that the yellow arrow is pointing in the opposite direction. Therefore, you need to enter a negative value of the offset distance.

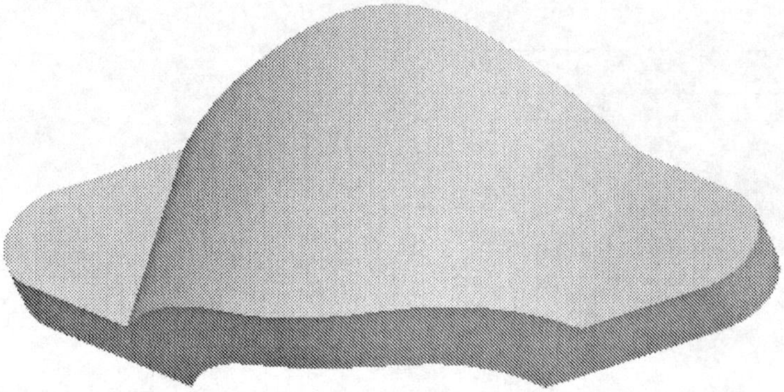

Figure 5-97 *Model after creating the cut feature*

3. In the **Translation** dimension box, enter a value of **-150** and press ENTER.

4. Choose the **OK** button from the **DATUM PLANE** dialog box.

   Datum plane **DTM1** is created as shown in Figure 5-98 and will be selected as the sketching plane for creating the sketch.

Creating the Third Feature

The plane **DTM1** will be selected as the sketching plane and the depth of extrusion will be given from this plane. The third feature is cylindrical in shape and its outer edge is tangent to the edge of the base feature.

1. Choose the **Extrude Tool** button from the **Base Features** toolbar. The Extrude dashboard is displayed below the graphics window.

2. Choose the **Create a section or redefine the existing section** button from the Extrude dashboard. The **Section** dialog box is displayed.

3. Select **DTM1** as the sketching plane. The yellow arrow appears on the datum plane.

4. Select the **FRONT** datum plane and then from the **Orientation** drop-down list, choose the **Top** option.

5. Choose the **Sketch** button to enter the sketcher environment.

6. Draw the sketch of the cylindrical feature by drawing a concentric circle as shown in Figure 5-99.

7. Choose the **Continue with the current section** button to exit the sketcher environment.

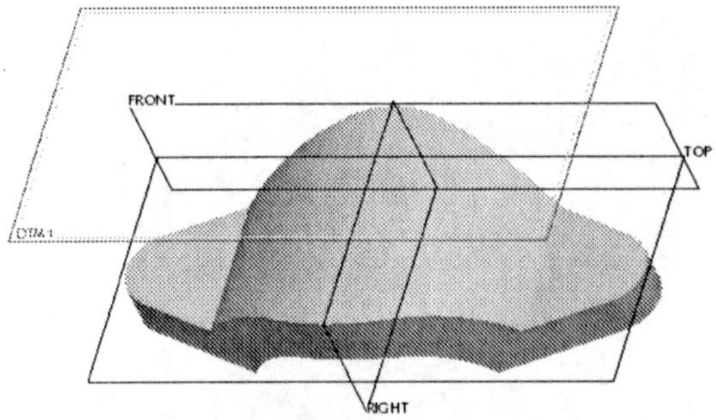

Figure 5-98 ***DTM1*** *created at an offset distance*

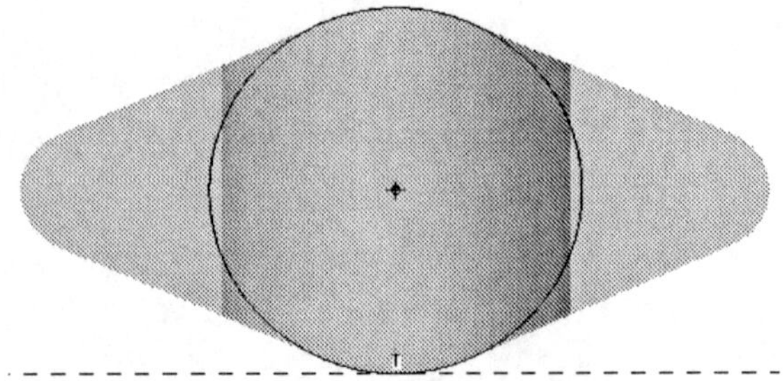

Figure 5-99 *Sketch for the cylindrical feature*

8. Choose the **Default View** option from the **Saved view list** drop-down list. The model orients in its default orientation.

9. Choose the **Options** tab in the Extrude dashboard to display the slide-up panel. From the **Side 1** drop-down list, choose the **To Next** option. The model can be previewed on the graphics window.

10. Choose the **Build feature** button from the extrude dashboard to complete the feature creation. The model after creating the cut feature is shown in Figure 5-100.

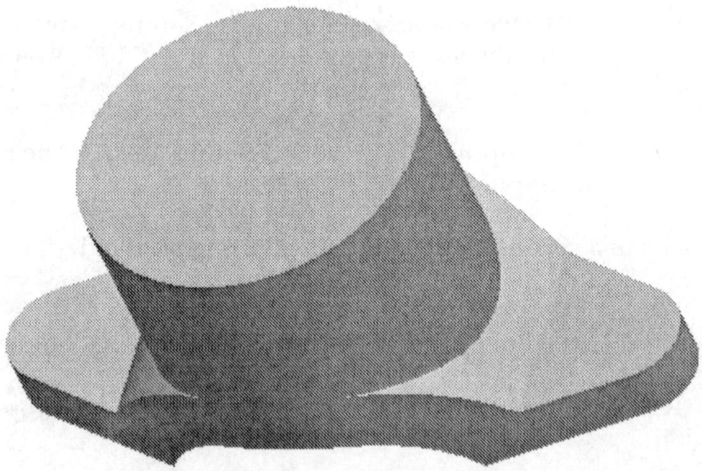

Figure 5-100 *Model after creating the cylindrical feature*

Creating the Counterbore Hole

The fourth feature is a counterbore hole that can be easily created using the **Hole** feature creation tool.

1. Choose the **Hole Tool** button from the **Engineering Features** toolbar. The Hole dashboard is displayed.

2. From the drop-down list present on the Hole dashboard, select the **Sketched** option.

3. Choose the **Activates Sketcher to create section** button to enter the sketcher environment.

4. In the sketcher environment, draw the sketch of the counterbore hole as shown in Figure 5-101. After drawing the sketch, exit the sketcher environment.

5. Select the top face of the cylindrical feature to place the hole. The preview of the hole appears on the selected face.

6. Choose the **Placement** tab from the Hole dashboard. The slide-up panel is displayed.

7. Select the **Coaxial** option from the drop-down list in the slide-up panel.

8. Click in the **Secondary References** collector so that the **Primary** collector turns white in color.

9. Select the axis of the cylindrical feature to which the hole will be coaxial. The preview of the hole can be viewed on the model.

10. Choose the **Build feature** button to exit the Hole dashboard. The counterbore hole is shown in Figure 5-102.

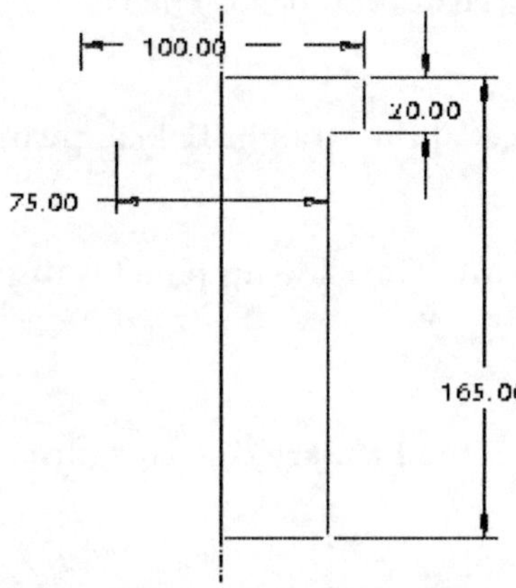

Figure 5-101 *Sketch with dimensions*

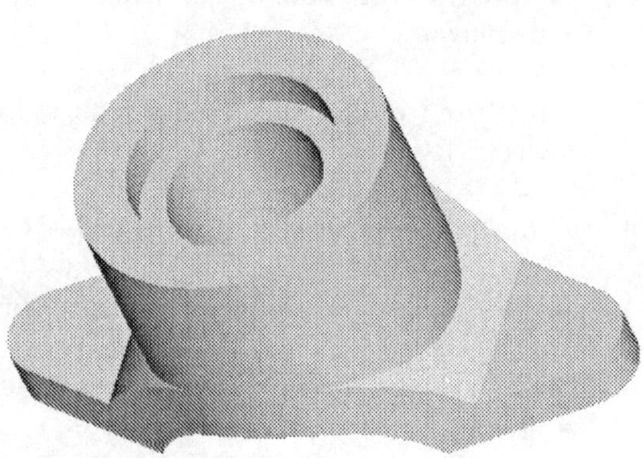

Figure 5-102 *Model with fourth feature*

Creating the Datum Axis for the Hole features

The last feature is a pair of two holes that are on the top face of the base feature. The two holes will be placed coaxially with the two axes that you need to create.

1. Choose the **Datum Axis Tool** button from the **Datum** toolbar. The **DATUM AXIS** dialog box is displayed.

2. Select the curved surface shown in Figure 5-103 to create the datum axis.

3. Similarly, create a datum axis on the left of the model. Figure 5-104 shows the model after creating the two datum axes.

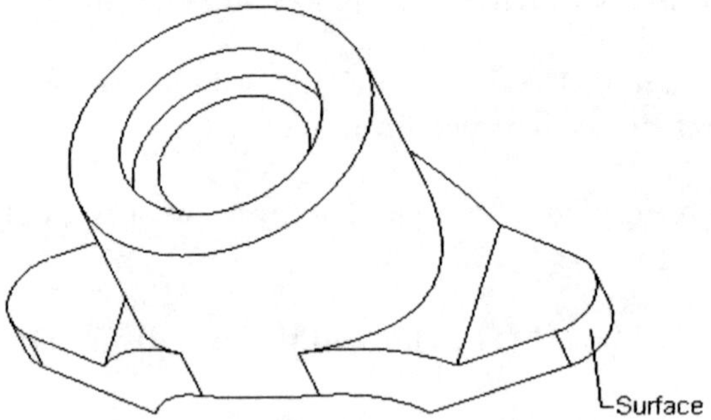

Figure 5-103 Surface to select for creating the datum axis

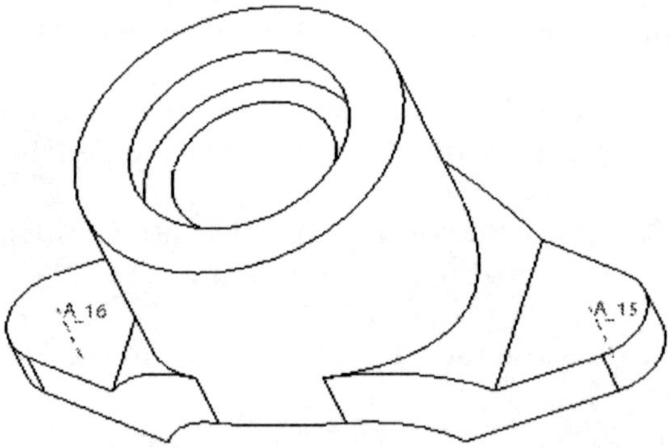

Figure 5-104 The two datum axis

Creating Holes

To create the two coaxial holes, you need to invoke the Hole dashboard twice. This is because only one hole can be created at a time by using the Hole dashboard.

1. Choose the **Hole Tool** button from the **Engineering Features** toolbar. The Hole dashboard is displayed.

2. Select the top face of the base feature. The preview of the straight hole appears on the selected face.

3. Choose the **Placement** tab from the Hole dashboard. The slide-up panel is displayed.

4. Select the **Coaxial** option from the drop-down list.

5. Click in the **Secondary References** collector so that the **Primary** collector turns white in color.

6. Select the datum axis to which the hole will be coaxial.

7. Enter the diameter of hole as **20** in the dimension box present on the Hole dashboard.

8. Choose the **Build feature** button to exit the Hole dashboard.

9. Again, invoke the Hole dashboard and similar to the first hole create the second hole.

   The two holes on the base feature are shown in Figure 5-105.

Saving the Model

1. Choose the **Save the active object** button from the **File** toolbar and save the model.

The order of feature creation can be seen from the **Model Tree** shown in Figure 5-106. The feature id numbers that are displayed in the **Model Tree** may be different when you create the features.

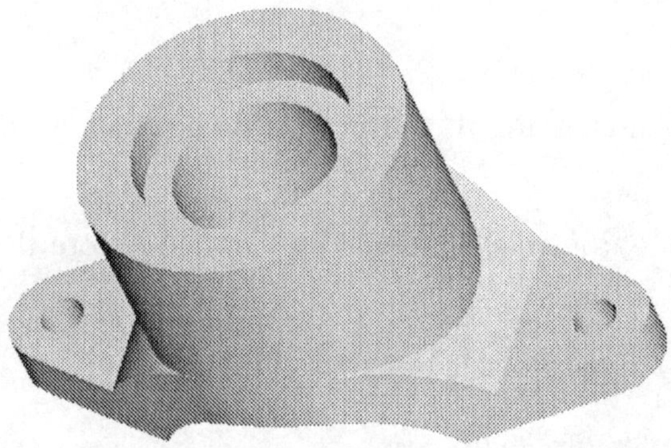

Figure 5-105 *Model after creating the two holes*

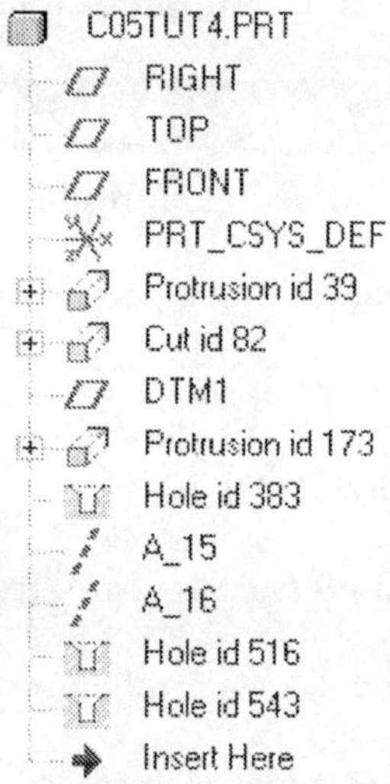

Figure 5-106 *Model Tree* ***of Tutorial 4***

Self-Evaluation Test

Answer the following questions and then compare your answers to the answers given at the end of this chapter.

1. A hole created using the Hole dashboard is parametric in nature. (T/F)
2. A hole cannot be created on both sides of the sketching plane or the placement plane. (T/F)
3. In Pro/ENGINEER, holes can also be sketched. (T/F)
4. The **Full Round** button in the **Sets** slide-up panel allows you to enter the radius of round. (T/F)
5. The **Model Tree** is used extensively in Pro/ENGINEER for editing a model. (T/F)
6. The rib feature is always created from the __________ view.
7. A __________ hole is a stepped hole and has two diameters, a larger one and a smaller one.

8. Straight holes are the holes that have circular cross-section having __________ diameter throughout the depth.

9. __________ are defined as thin wall-like structures used to bind the joints together so that they do not fail under an increased load.

10. By default, the suppressed features are not displayed in the __________.

Review Questions

Answer the following questions:

1. If the feature that is suppressed has some child features then they will also be suppressed. (T/F)

2. While in the sketcher mode, the constraints to a sketch should be applied before the dimensions. (T/F)

3. The chamfers created in Pro/ENGINEER are parametric in nature. (T/F)

4. The **Model Tree** can be used to redefine a feature. (T/F)

5. When you redefine a rib feature, the __________ dashboard is displayed.

6. The sketch of a rib feature can be extruded to __________ side(s) of the sketching plane.

   (a) one (b) both
   (c) either (d) All of the above

7. __________ is a stepped hole with two diameters.

   (a) Counterbore (b) Countersink
   (c) Straight (d) None

8. Which of the following options of the **PART** menu can modify the dimensions of a feature after it is completed?

   (a) **Feature** (b) **Modify**
   (c) **Regenerate** (d) None

9. At the intersection of how many edges does a **Corner** chamfer create a beveled surface?

   (a) One (b) Two
   (c) Three (d) None

10. Rounds and chamfers are used in engineering components to reduce the __________ on the corners.

(a) Stress concentration
(b) Tension
(c) Conduction
(d) None

Exercises

Exercise 1

Create the model shown in Figure 5-107. The dimensions and front and top views of the model are shown in Figure 5-108. **(Expected time: 45 min)**

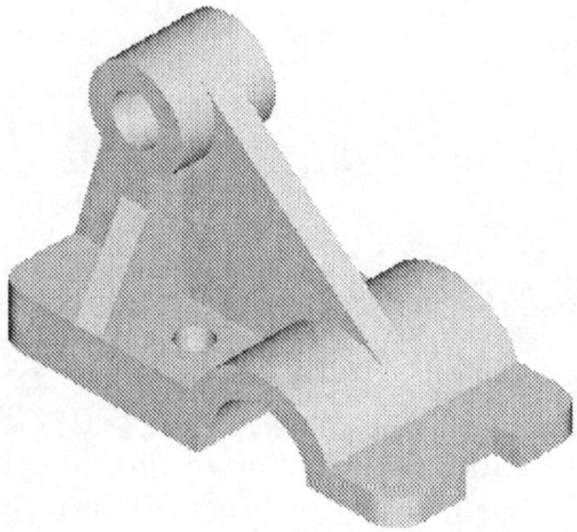

Figure 5-107 *Isometric view of the model*

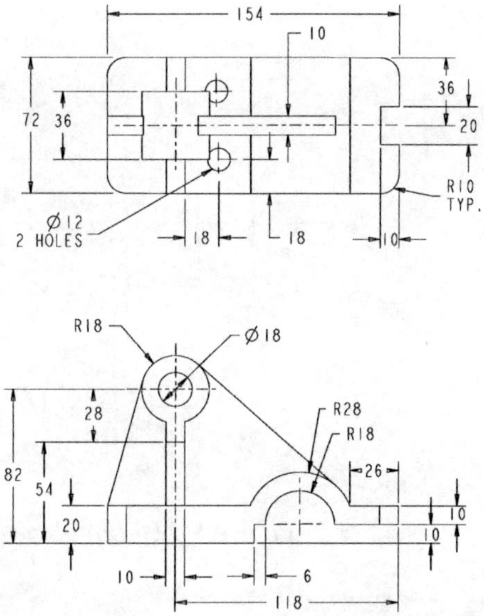

Figure 5-108 *Top and front views of the model; hidden lines are suppressed for clarity*

Exercise 2

Create the model shown in Figure 5-109. The dimensions and front and right-side views of the model are shown in Figure 5-110. **(Expected time: 30 min)**

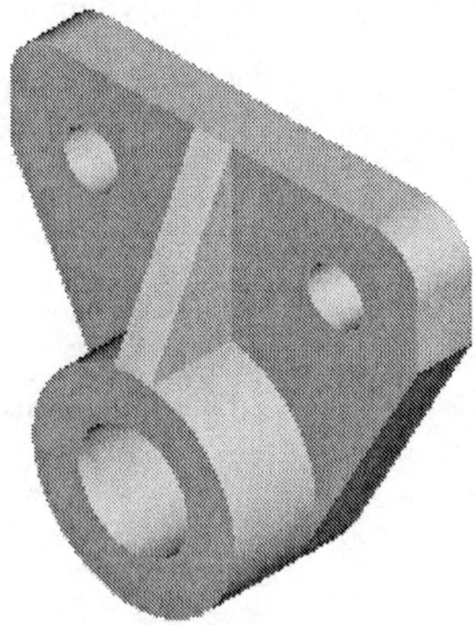

Figure 5-109 *Isometric view of the model for Exercise 2*

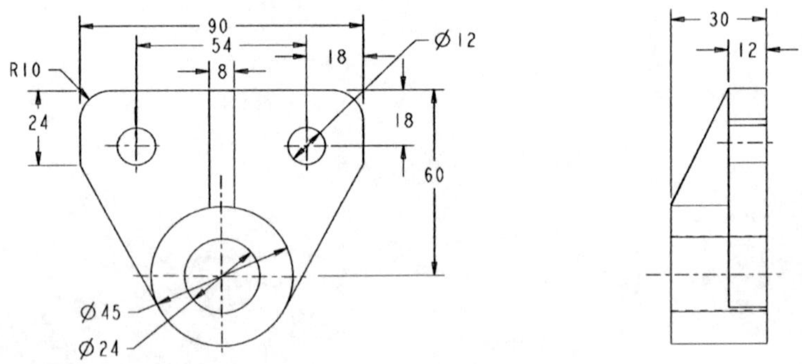

Figure 5-110 *Front view and the right-side view of the model*

Exercise 3

Create the model shown in Figure 5-111. The dimensions and front and right-side views of the model are shown in Figure 5-112. **(Expected time: 30 min)**

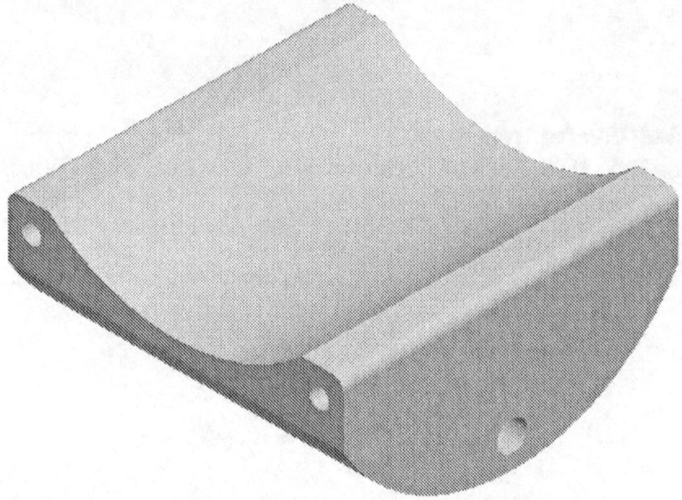

Figure 5-111 *Isometric view of the model for Exercise 3*

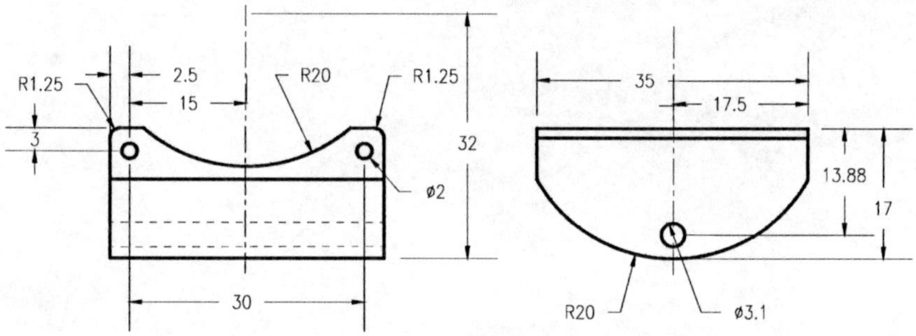

Figure 5-112 *Front view and the right-side view of the model*

Answers to the Self-Evaluation Test

1 - T, **2** - F, **3** - T, **4** - F, **5** - T, **6** - side, **7** - counterbore, **8** - constant, **9** - Ribs, **10** - **Model Tree**.

Chapter 6

Options Aiding Construction of Parts-II

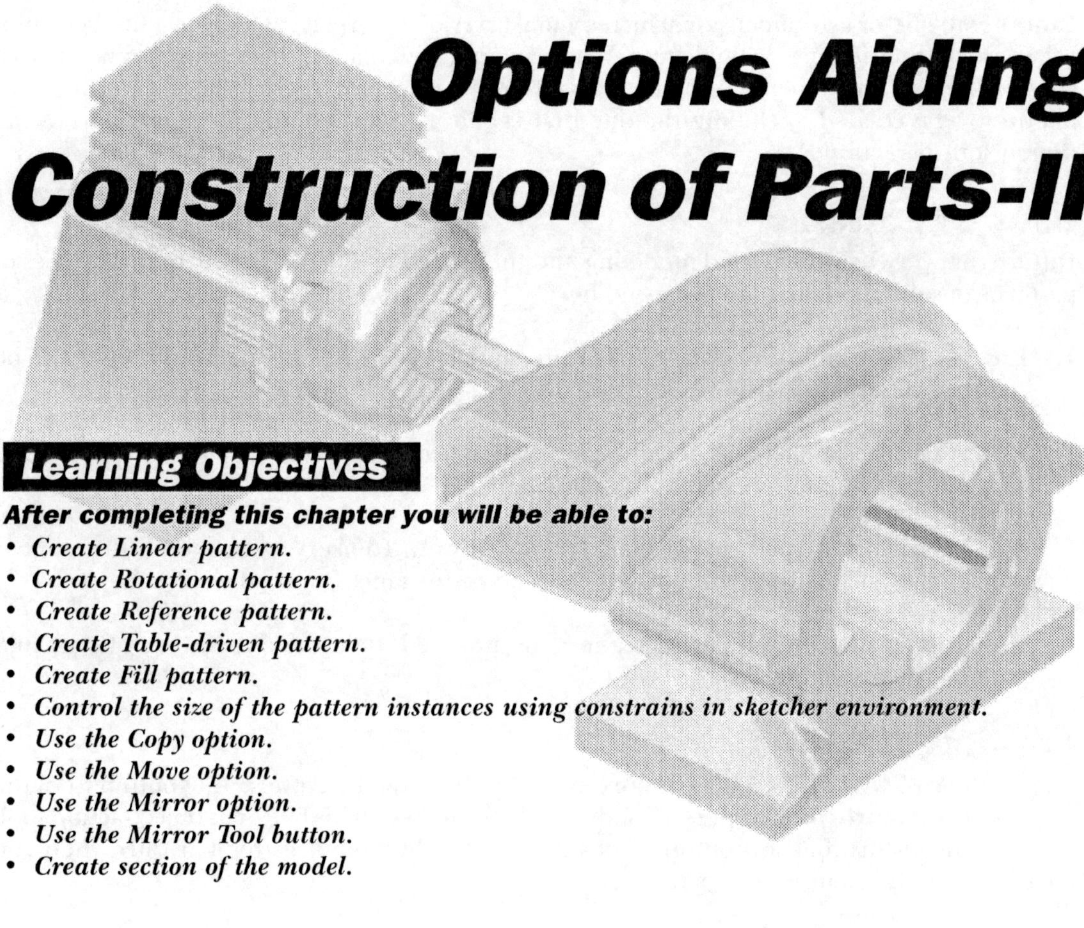

Learning Objectives

After completing this chapter you will be able to:

- *Create Linear pattern.*
- *Create Rotational pattern.*
- *Create Reference pattern.*
- *Create Table-driven pattern.*
- *Create Fill pattern.*
- *Control the size of the pattern instances using constrains in sketcher environment.*
- *Use the Copy option.*
- *Use the Move option.*
- *Use the Mirror option.*
- *Use the Mirror Tool button.*
- *Create section of the model.*

In this chapter you will learn about the various methods of duplicating the existing features. In Pro/ENGINEER, you can duplicate a feature by using the following methods:

- **Pattern**
- **Copy**
- **Mirror**

You will also learn to create sections in solid models. A solid model is sectioned for presenting a model to display the section and for creating section drawing views.

PATTERNING THE FEATURES

Patterns are one-or two-dimensional incremental array of features created from a single feature called the parent feature or the leader. When a pattern is created, the leader also becomes a part of the pattern. When you pattern a feature, you need to specify the total number of features to be created, including the one that is being patterned and the increment in the dimensions if required.

Uses of Patterns

Patterns are very helpful in solid modeling and they speed up the model creation. The uses of patterns in solid modeling are discussed next.

1. Patterns create multiple copies of a feature, and hence save time that would otherwise be spent in creating the features individually.

2. All the instances in a pattern, including the parent feature, act as a single feature. Therefore, they can easily be suppressed or mirrored.

3. All the instances in a pattern are related parametrically. Hence, you can modify the number of instances in a pattern, the spacing between the instances, and other parameters.

4. If the dimensions of the parent feature are modified, then the dimensions of the child features are also modified.

Creating Patterns

In Pro/ENGINEER, patterns are created by choosing the **Pattern Tool** button from the **Edit Features** toolbar present in the **Right Toolchest**. This button is object-action tool. This means that this button is enabled in the **Edit Features** toolbar only when you have selected the feature to be patterned.

To invoke the pattern tool from the menu bar, choose **Edit > Pattern**. The Pattern dashboard is displayed as shown in Figure 6-1. The options and the tools available on the dashboard depend on the type of pattern you are creating. There are four types of patterns that can be created in Pro/ENGINEER. The options and tools available in the Pattern dashboard are discussed next.

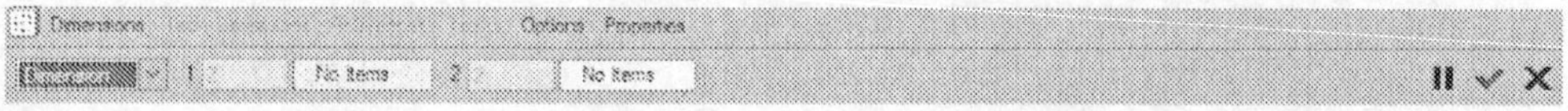

Figure 6-1 *Pattern dashboard*

Drop-down list

In the drop-down list present on the Pattern dashboard, the **Dimension** option is selected by default. The other options in this drop-down list are **Table**, **Reference**, and **Fill**. When you select the various options from this drop-down list, the options and tools available in the Pattern dashboard change accordingly. The types of patterns that can be created in Pro/ENGINEER are discussed next:

Dimension Patterns

In the dimension patterns, the existing dimensions of the parent feature are used to create a pattern. This pattern can be created in either one direction or in two directions. When you select the option to create the pattern in the second direction, all instances that were created in the first direction can also be created in the second direction. You need to specify the increment value for the instances. You can enter a positive or a negative value of the increment. Once you have specified the increment value in a direction, the system creates the specified number of instances (including the parent feature) in that direction. Figure 6-2 shows a hole that is patterned. The patterned hole is shown in Figure 6-3.

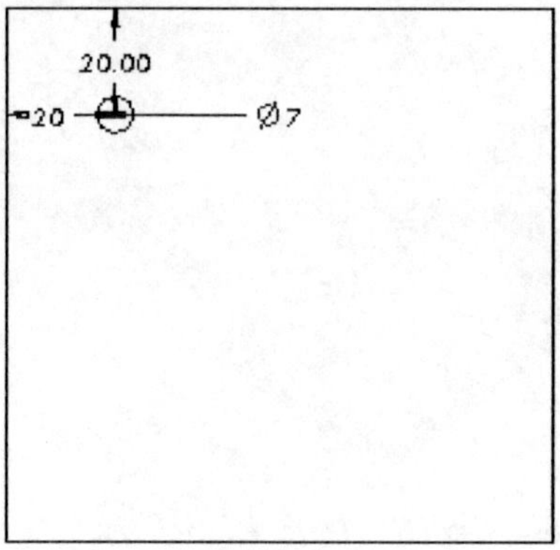

Figure 6-2 *Hole on a surface*

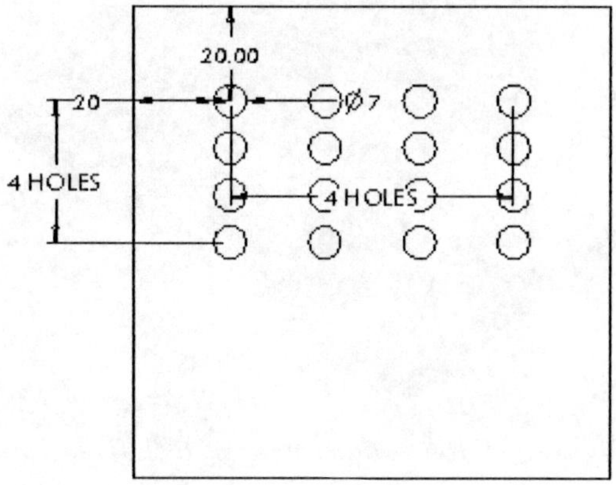

Figure 6-3 *Hole patterned in two directions*

To create a dimension pattern, follow the following steps:

1. Click the dimension in the first direction. The edit box is displayed; enter the increment in the dimension where you need to place the second instance of the pattern.
2. In the **1** edit box present on the dashboard, enter the number of instances.
   The instances in the first direction are specified and now you need to specify the number of instances and the increment in dimension in the second direction. Note that the dimension in the second direction should be selected only if you need to

create instances in the second direction.

3. Click on **No Items** in the **2** collector. It turns yellow. Now, click the dimension in the second direction; the edit box appears. Enter the dimension increment in the edit box.
4. In the **2** edit box, enter the number of instances needed in the second direction. The pattern is created but you cannot preview the pattern.
5. Choose the **Build feature** button to exit the feature creation tool and to view the pattern.

Reference Patterns

In the reference patterns, an existing pattern is referenced to create a new pattern. In this type of pattern, the parent feature of the new pattern should be referenced to the parent feature of the existing pattern.

Figure 6-4 shows a rib feature that is referenced to the parent hole feature. **In this figure, the parent rib feature is created on a plane that was created on-the-fly while creating the rib feature. This plane passes through the axis of the parent hole feature**. Hence, a relationship is built between the parent rib feature and the parent hole feature. Now, when you select the rib feature to pattern, the pattern shown in Figure 6-5 is created without specifying the increment in dimensions.

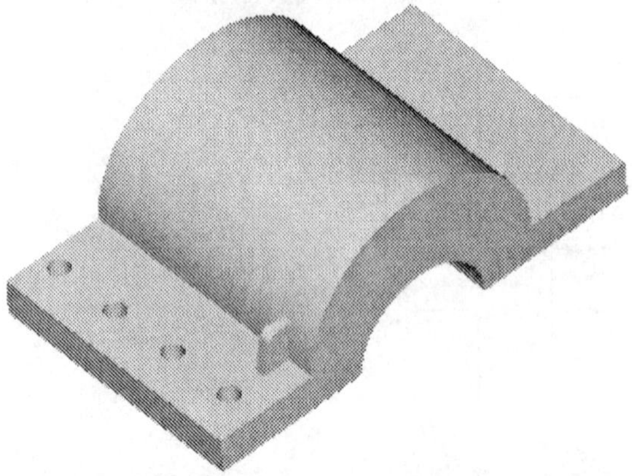

Figure 6-4 *Rib feature referenced to the hole feature*

Figure 6-5 *Rib feature is reference patterned*

Tip: *To select the reference feature for creating the reference pattern, you may need to open the **Model Tree**. Select the pattern that will be used as reference from the **Model Tree**. Since, you cannot preview the pattern, you need to choose the **Build feature** button.*

Note

If you create the datum plane passing through the axis of the hole using the ***Datum Plane Tool*** *button and then create the rib feature on it, the rib feature will not be patterned with reference to the holes. This is because the rib does not have any direct relation with the hole. The rib has relation with the datum plane and the datum plane has a relation with the hole. However, since the datum plane has relation with the hole, you can pattern the datum planes with reference to the holes using the* ***Reference*** *option.*

Table Patterns

Table-driven patterns are created by defining a table. In this table you need to specify the dimensions of the instances from the edge or faces from where the leader of the pattern is referenced (leader of a pattern is the feature that is selected to create a pattern). When you select the **Table** option from the drop-down list, the dashboard appears as shown in Figure 6-6.

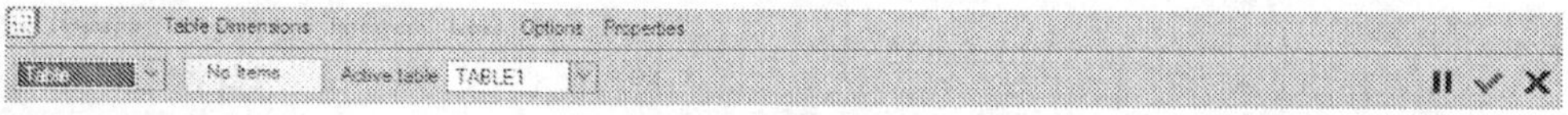

Figure 6-6 *Pattern dashboard*

In Figure 6-6, the **Active Table** collector shows the table that is active. You can create more than one table to create a pattern. You can choose the **Edit** button to display the **Pro/TABLE** shown in Figure 6-7. This button will be available only after you have

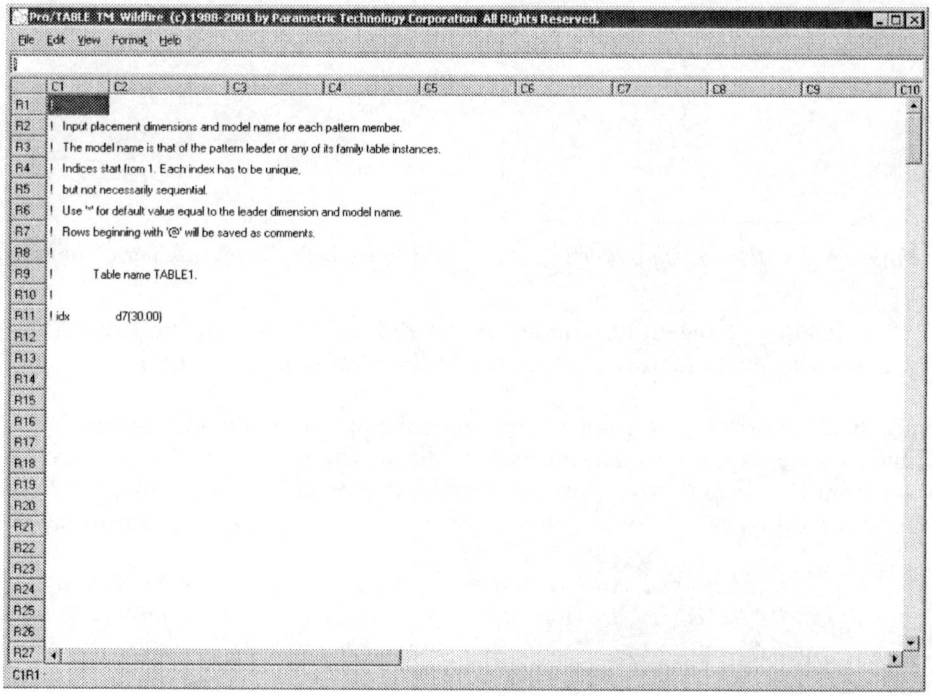

Figure 6-7 Pro/TABLE

selected at least one dimension.

To create the pattern shown in Figure 6-8, follow the steps given next:

1. Create a solid protrusion of dimension 120x100x15.
2. Create a hole of diameter **5** on the top face of the base feature at a distance of 20 from the left edge and 20 from the bottom edge.
3. Select the hole feature and choose the **Pattern Tool** button from the **Edit Features** toolbar. The Pattern dashboard is displayed and the dimensions of the hole feature appear on the hole.
4. Select the **Table** option from the drop-down list present on the dashboard.
5. Select the dimension **20** that is from the left edge and then use CTRL+left mouse button to select the dimension **20** that is from the bottom edge.
6. Choose the **Edit** button to display the **Pro/TABLE.**
7. In column **C1** enter 1 under **idx (index** number). The number 1 signifies the first instance. To enter the value 1, click under **idx**. The cell is highlighted, enter 1.
8. Similarly, under the rows that are below **idx,** enter the dimensions of other instances of the holes as shown in Figure 6-9. Remember that distance of each hole is measured from where the leader is dimensioned.

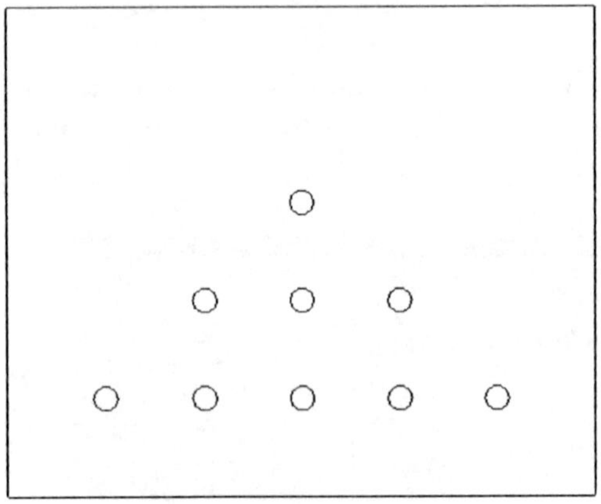

Figure 6-8 *Table-driven pattern*

! idx	d5(20.00)	d6(20.00)
1	40.00	40.00
2	40.00	20.00
3	60.00	20.00
4	60.00	40.00
5	60.00	60.00
6	80.00	20.00
7	80.00	40.00
8	100.00	20.00

Figure 6-9 *Coordinate values of instances*

9. Choose **File > Save** and then choose the **Exit** option to exit the **Pro/TABLE.**
10. Choose the **Build feature** button to exit the feature creation tool.

Generally, table-driven patterns are used where the pattern is already created. For example, you have created a dimensional pattern. Redefine the pattern and now select the **Table** option from the drop-down list in the Pattern dashboard. Now, invoke the **Pro/TABLE** and edit the dimension values of the instances of pattern to get the desired pattern.

Tip: *Pattern tables can be created from scratch by picking the dimensions and filling in all the values. In most cases, it is easier and less time-consuming to create a dimensional pattern that is similar to what you need and then convert it to a table.*

Fill Patterns

Fill patterns are used to fill the sketched area by a selected feature. This is the easiest and fastest method to create pattern in the sketched area. When you select the **Fill** option from the drop-down list, the Pattern dashboard appears as shown in Figure 6-10.

Figure 6-10 *Pattern dashboard*

The options and tools available in this dashboard are discussed next:

Create a section or redefine the existing section button. This button is used to sketch the surface that will be filled by the pattern instances. When you choose this button, the **Section** dialog box is displayed. Select the sketching plane and its orientation and then enter the sketcher environment to draw the sketch.

Drop-down list. The drop-down list that is on the right of the **Create a section or redefine the existing section** button is used to specify the shape of the pattern. The options that are available in this drop-down list are **Square**, **Diamond**, **Triangle**, **Circle**, **Curve**, and **Spiral**. Generally, it is advised that the sketch should be similar to the shape of the fill pattern you need. Figures 6-11 through 6-16 show the patterns of various shapes.

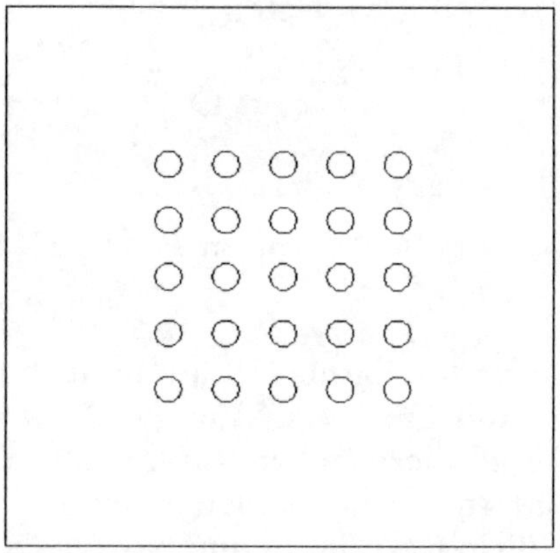

Figure 6-11 *Fill pattern in* ***Square*** *shape on the square sketched area*

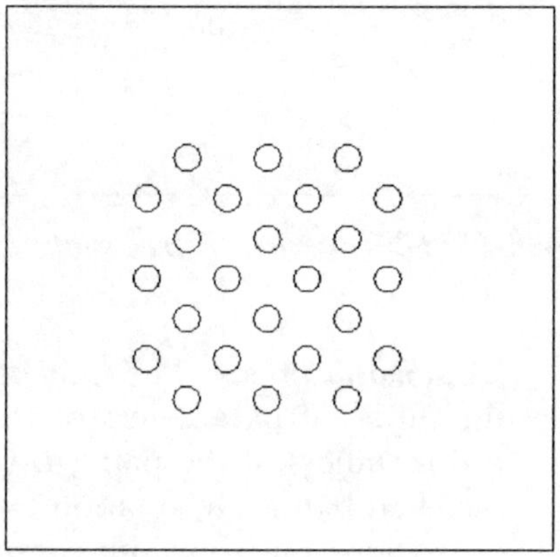

Figure 6-12 *Fill pattern in* ***Diamond*** *shape on the square sketched area*

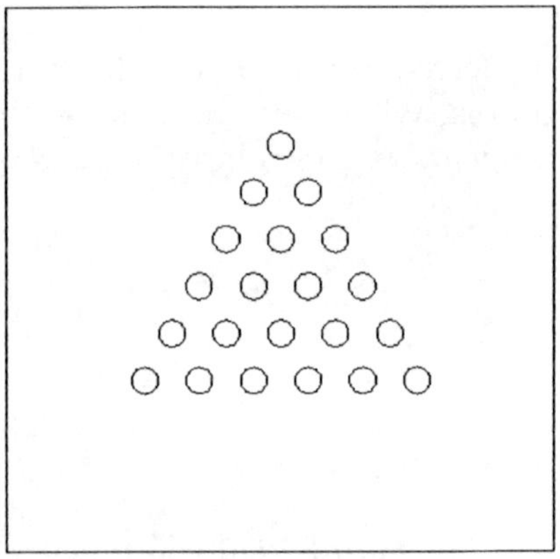

Figure 6-13 *Fill pattern in* ***Triangle*** *shape on the sketched triangle*

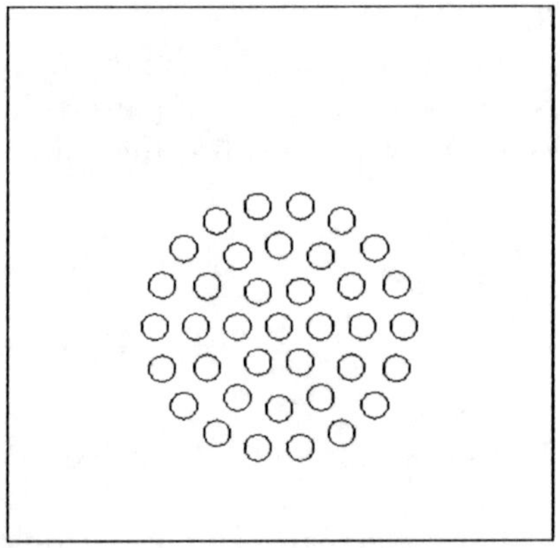

Figure 6-14 *Fill pattern in* ***Circle*** *shape on the sketched circle*

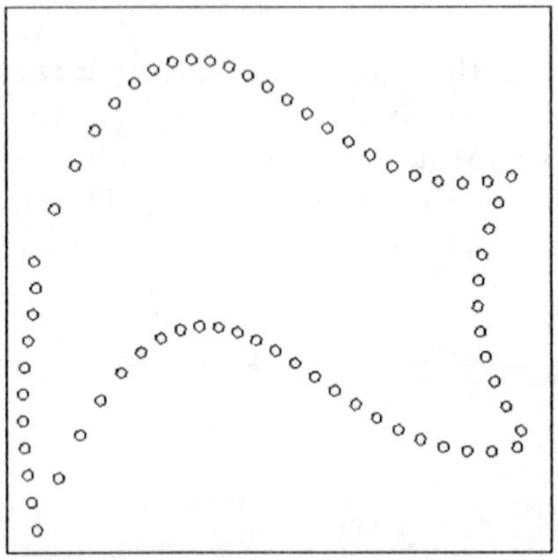

Figure 6-15 *Fill pattern in* ***Curve*** *shape on the sketched curve*

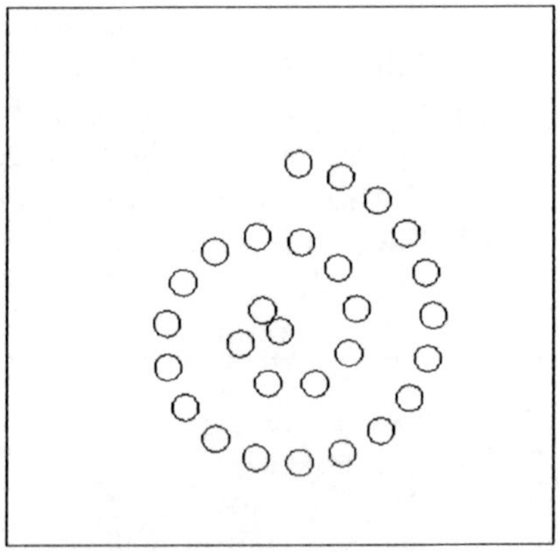

Figure 6-16 *Fill pattern in* ***Spiral*** *shape on the sketched circle*

Dimension boxes. The edit boxes available on the dashboard are used to specify the different parameters of the fill pattern. You can specify the gap between the two members of the pattern, the distance of all the members (instances) from the sketched boundary, rotation angle of the pattern in case of circular shapes, and so on. When you bring the cursor on these edit boxes, the **Status** bar on the main window of Pro/ENGINEER displays the function of that edit box.

Tip: *When you create a Fill pattern, you can select the member of the pattern that you want to remove from the pattern. This can be done before exiting the Pattern dashboard. To remove the member of a pattern, select it in the preview of the Fill pattern. The selected member turns white, suggesting that it is removed. To resume the removed member of the pattern, select it once again.*

Options tab

When you choose the **Options** tab, the slide-up panel is displayed as shown in Figure 6-17. The various options in this slide-up panel are discussed next:

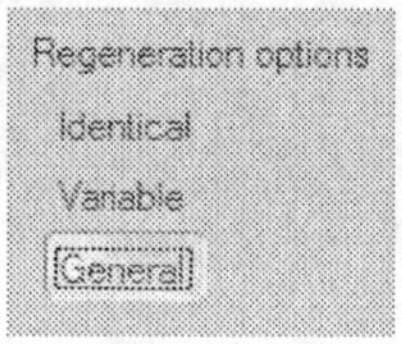

Figure 6-17 ***Options*** *slide-up panel*

Identical Pattern

The **Identical** option is used to create an identical pattern. You need to select at least one incremental dimension to pattern the feature. Depending on the incremental dimension selected, the resultant pattern will be linear or rotational. A linear pattern is created when the driving dimension is linear and a rotational pattern is created when the driving dimension is angular. You can enter a positive or a negative value as the increment in a pattern dimension. All the instances of a pattern that are created using this option are identical in size and geometry. This is the reason the patterns created using this option are known as the **Identical** patterns. Figure 6-18 shows a hole feature on the base feature and Figure 6-19 shows holes patterned linearly.

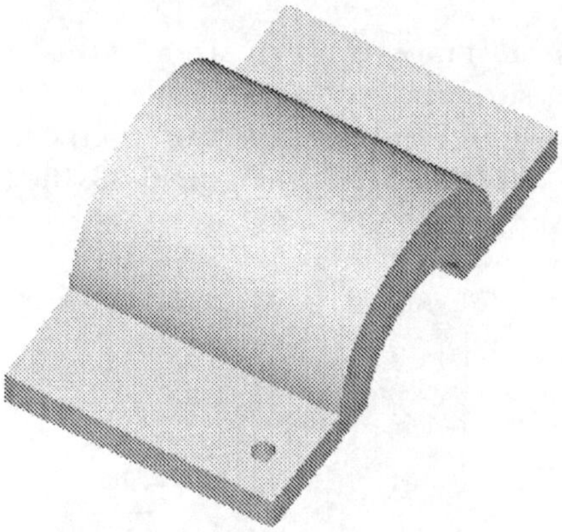

Figure 6-18 *Hole on the base feature*

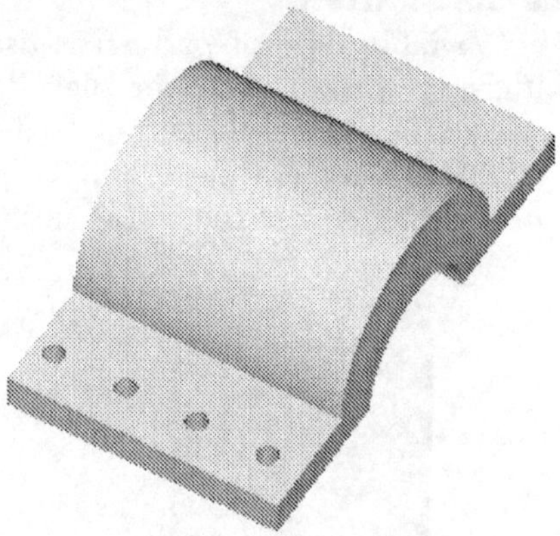

Figure 6-19 *Hole patterned on the base feature*

Similarly, Figure 6-20 shows a hole feature on the base feature and Figure 6-21 shows the rotational pattern of the hole feature.

Figure 6-20 Hole on the base feature

Figure 6-21 Rotational pattern of hole feature

As evident from Figures 6-19 and 6-21, all the instances in the identical patterns are placed on the same placement surface and no feature intersects the edges of the placement surface, any other instance, or any other feature other than the placement surface. Note that it is not possible to pattern the hole feature shown in Figure 6-18 on the right flap using the identical patterns. However, you can use the **General** option to pattern the hole on the right flap of the model shown in Figure 6-18.

Variable Pattern

The **Variable** type of pattern is used when the instances vary in size. In this type of pattern the instances can be placed on different surfaces and can also intersect with the edges of the placement surface. The feature shown in Figure 6-22 is patterned using the **Variable** option. The patterned feature is shown in Figure 6-23. In Figure 6-23, the length and the diameter of the rod varies in all the instances.

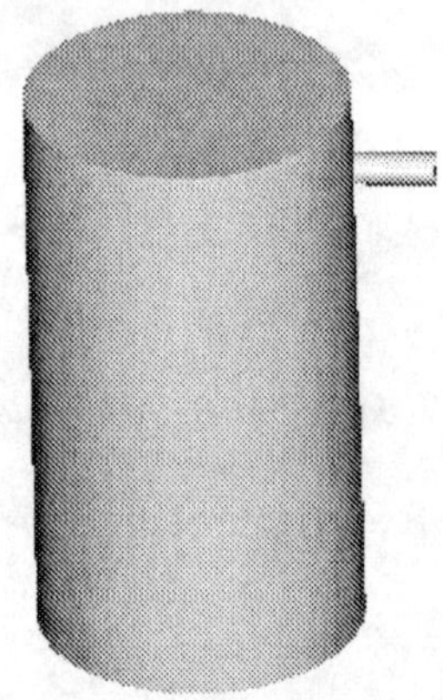

Figure 6-22 A rod on the base feature

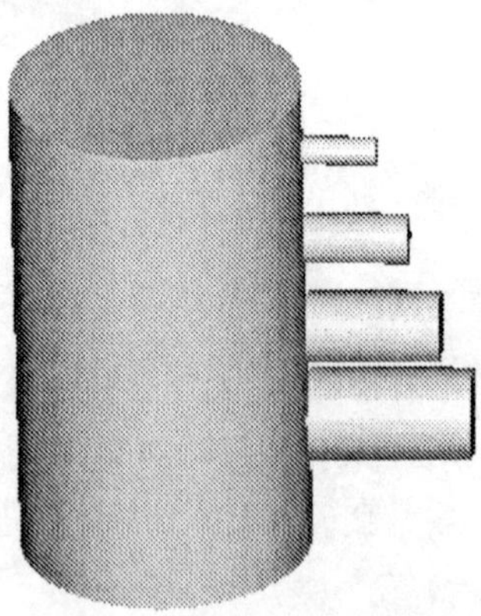

Figure 6-23 Varying pattern of the rod

General Pattern

The most complex patterns can be created using the **General** pattern. This option is used to create patterns in which the instances touch each other and intersect with other instances or the edges of the surface. This option of creating patterns is also used when instances intersect with the base feature and the intersection is not visible. The hole shown in Figure 6-24 is patterned using the **General** option. The pattern is shown in Figure 6-25.

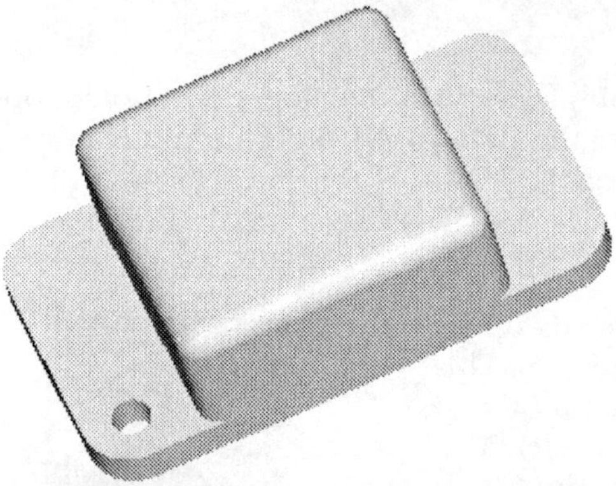

Figure 6-24 *Hole on the base feature*

Figure 6-25 *General pattern*

Note

*If the feature that you need to pattern has formed a group and is listed in the **Model Tree** as a group feature, then you need to select the group from the **Model Tree** to pattern the desired feature.*

*A feature forms a group when it has used other features to form itself. For example, if you create datum on-the-fly and use it as a sketch plane then after the feature is created, it is listed in the **Model Tree** as a group feature. This group comprises of the feature, the datum plane, and the sketch of the feature.*

Creating a Rotational Pattern

To create a rotational pattern in Pro/ENGINEER, you need to remember three important points. These points are discussed next.

1. There should be an angular value that is directly related to the feature to be patterned. This is achieved by creating a datum plane on-the-fly. This plane can be either used as the sketch plane or the reference plane by the feature that will be patterned.

2. When you enter the sketcher environment to sketch the feature to be patterned, the sketch of the feature should be referenced to the axis around which the feature will be patterned. This can be done by selecting the axis in the **References** dialog box. Remember that the sketch of the feature should not be constrained with any datum plane.

3. When the feature to be patterned is created, in the **Model Tree** it appears as a grouped feature. This group contains the datum plane created on-the-fly and the feature. For creating the rotational pattern, the entire group should be selected.

Deleting a Pattern

You can delete a pattern by selecting it from the **Model Tree**. Select the pattern feature from the **Model Tree** and right-click to invoke the shortcut menu. Choose the **Delete Pattern** option from the shortcut menu. However, note that the parent feature (leader) is not deleted when you delete the pattern by using the **Delete Pattern** option even if you selected the parent feature for deleting the pattern.

If you want to delete the pattern along with the parent feature, choose the **Delete** option from the shortcut menu. The system highlights the pattern on the graphics window and confirms the deletion of the pattern from the user.

COPYING FEATURES

The **Copy** option facilitates the model and speeds up its creation by copying and mirroring the selected features. This option reduces the time required in the model creation and also helps in maintaining the symmetry of the model. The **Copy** option is available in the **FEAT** menu in the **Menu Manager**. To invoke the **Menu Manager**, choose **Edit > Feature Operations** from the menu bar.

When you choose the **Copy** option from the **FEAT** menu, the **COPY FEATURE** submenu appears with various options as shown in Figure 6-26. The various options of this submenu are explained next.

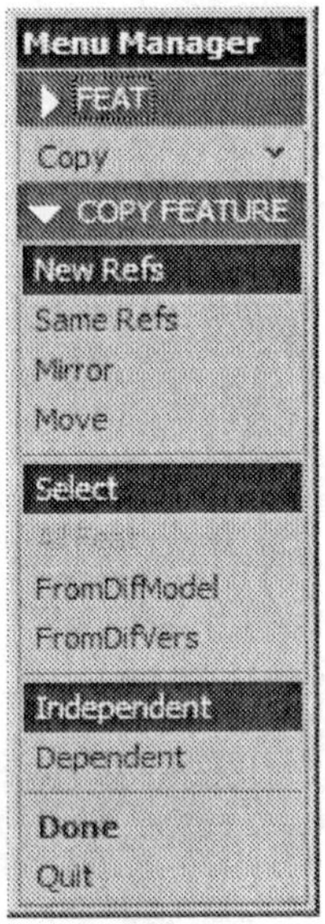

***Figure 6-26** COPY FEATURE submenu*

New Refs

The **New Refs** option is used to copy a feature by varying the dimension values and by selecting new references. The references can be edges, axes, or placement planes. When you select the feature to be copied, the **GP VAR DIMS** menu is displayed as shown in Figure 6-27. Using this menu, you can specify the dimensions that are to be varied while copying a feature with new references.

You can copy a feature using the **New Refs** option by two methods. You can keep the same dimensional or placement reference for the copied feature as that of the original feature by using the **Same** option in the **WHICH REF** menu shown in Figure 6-28. This means that you can use the same edge or surface as the references for the copied part. The other possibility is that you can use new references for the copied feature. This can be achieved by using the **Alternate** option in the **WHICH REF** menu. This provides you with a greater flexibility to copy the features.

Same Refs

When you use the **Same Refs** option, you have the flexibility to vary the dimensions of the copied features. You need to select the dimensions you want to vary while copying a feature.

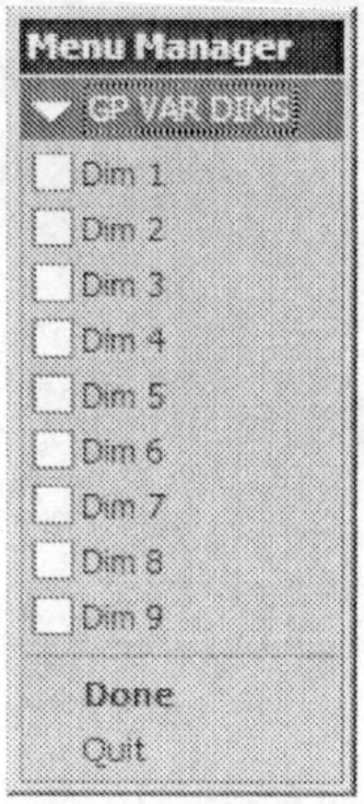

***Figure 6-27 GP VAR DIMS** menu*

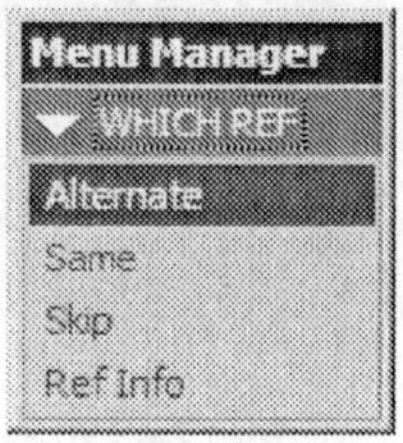

***Figure 6-28 WHICH REF** menu*

The dimensional and placement references of the copied feature are the same as for the source feature.

Mirror

The **Mirror** option is used to copy a feature by mirroring it about a specified datum plane or a face. When you invoke this option, you are prompted to select the features to be mirrored. Once you select the features to be mirrored, you are prompted to select a plane or create a datum about which the features will be mirrored. As soon as you select a datum plane or a planar surface, the selected feature will be mirrored. As mentioned earlier, this option not only reduces the model completion time, it also helps in maintaining the symmetry in the features of a model. Figure 6-29 shows a rib feature that is to be mirrored and the datum plane about which the rib feature will be mirrored. Figure 6-30 shows the model after mirroring the rib feature and turning off the visibility of the model.

Tip: *To mirror a feature at an angle of 90 degrees to the parent feature, create a datum plane at an angle of 45 degrees to the parent feature and then use this datum plane to mirror the feature.*

Move

The **Move** option is used to copy features by translating or rotating them. When you invoke this option, you will be prompted to select the features to be translated. To invoke the **Move** option, choose **Move > Select> Independent > Done** from the **COPY FEATURE** menu. The **SELECT FEAT** menu is displayed and you are prompted to select the features to copy. After selecting the features and choosing the **Done** option from the **SELECT FEAT** menu,

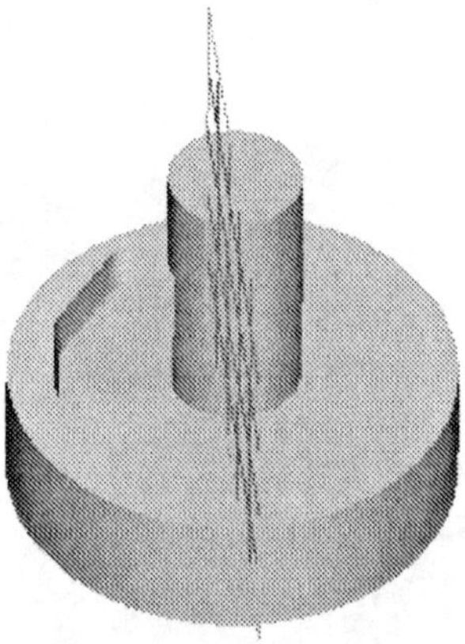

Figure 6-29 Rib feature and the mirror plane

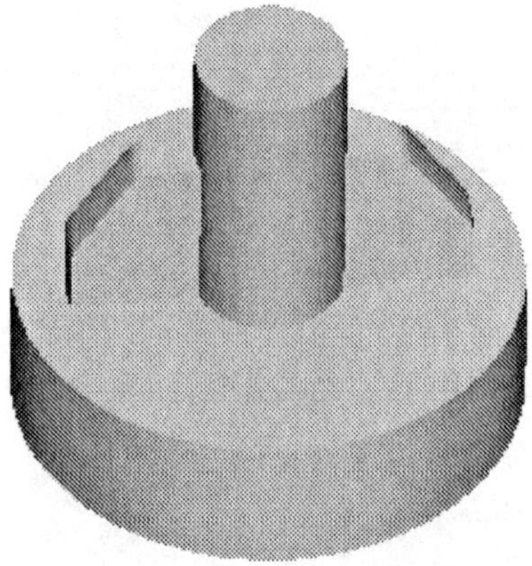

Figure 6-30 Mirrored rib feature

you are prompted to define the movements by combination of translation and rotation. The **MOVE FEATURE** submenu is displayed as shown in Figure 6-31. The selected feature can be translated or rotated using the options in this submenu.

Figure 6-31 MOVE FEATURE submenu

Translating the features

You can select a feature from the graphics screen or from the **Model Tree** to copy it by translating it. You need to specify a perpendicular plane to define the direction in which the feature will be copied and the offset distance. You can also select multiple features to copy.

Rotating the features

You can select a feature from the graphics screen or from the **Model Tree** and then rotate it about an axis, edge, datum curve, or coordinate system. Using this option you can select multiple features to copy.

Select

The **Select** option provides the flexibility of choosing the features to be copied. When you use this option, you need to select the features either from the graphics screen or from the **Model Tree**. You can select any number of features to copy.

All Feat

The **All Feat** option is available only when you copy a feature using the **Mirror** or the **Move** option. When you use this option, all the features created are copied. Remember that you need to specify a coordinate system while using this option.

FromDifModel Option

The **FromDifModel** (From different model) option is used to copy a feature from a different model. This option is available only when you are using the **New Refs** option of the **COPY FEATURE** submenu. This is due to change in the references required to copy from one model to another. Therefore, all the references will be new.

FromDifVers Option

You can copy features from a different version of the current model using the **FromDifVers** (From different versions) option. This option is available only when you are using the **New Refs** and the **Same Refs** options of the **COPY FEATURE** submenu.

Independent Option

The **Independent** option specifies that the dimensions of the copied features are independent of the dimensions of the parent feature. The features that you copy from a different model or different versions are independent by default. The dimensions of such copied features have no relation with those of the original feature. This is the reason the **Dependent** option is not available while using the **FromDifModel** or the **FromDifVers** options.

Dependent Option

The **Dependent** option specifies that the dimensions of the copied features are dependent on the dimensions of the parent feature. Therefore, if you make any modification in the section of the parent feature, the changes are automatically reflected in the copied feature.

MIRRORING A GEOMETRY

Choose the **Mirror Tool** button from the **Edit Features** toolbar. The Mirror dashboard is displayed and you are prompted to select a datum plane about which the model will be mirrored. This button is available only when the name of the model is selected from the **Model Tree**. The name of the model is at the top in the **Model Tree**.

Pro/ENGINEER allows you to mirror an existing model about a datum plane. All the features in the model are mirrored about a datum plane. This option is different from the **Copy > Mirror** option. When you use the **Copy > Mirror** option, you can select features to be copied. But, while using the **Mirror Tool** button, the whole model is mirrored about a specified datum plane and the mirrored portion is automatically merged with the original portion. All the features in the mirrored model are related to the parent model. Any modification in the parent model reflects that modification in the mirrored model. Figure 6-32 shows a model and a datum plane for mirroring the model. Figure 6-33 shows the resultant mirrored model. Notice the merging in the resultant model.

CREATING A SECTION OF THE SOLID MODEL

The complex geometry of a model sometimes needs sectioning so that you can understand the inner sections of various features that are not visible from outside. These sections are mainly used to display the inner sections of features of a model and for showing these sections in the drawing views.

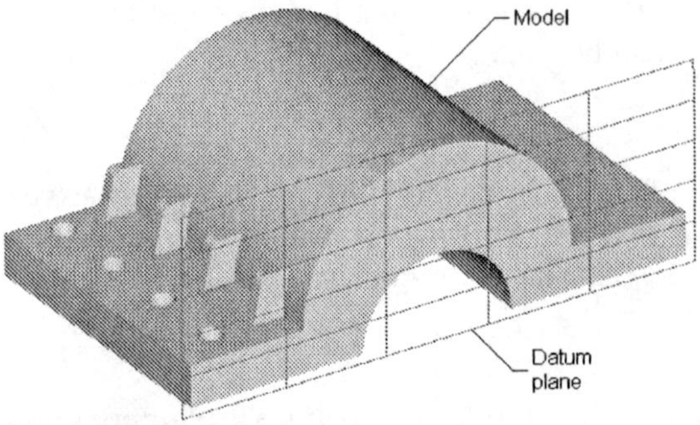

Figure 6-32 *Model and a plane*

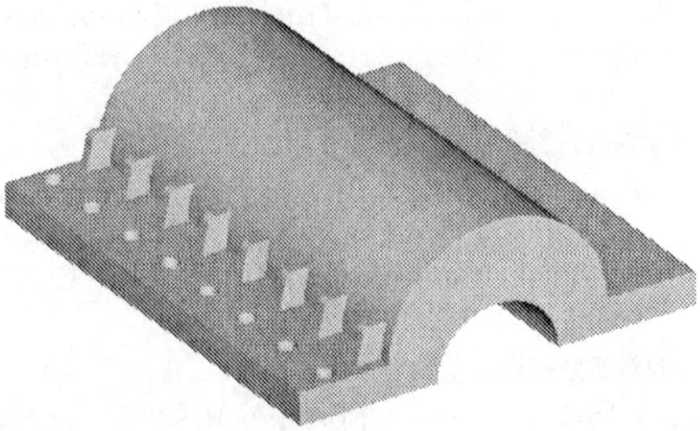

Figure 6-33 *Resultant mirrored model*

Choose **View > View Manager** from the menu bar. The **View Manager** dialog box is displayed as shown in Figure 6-34.

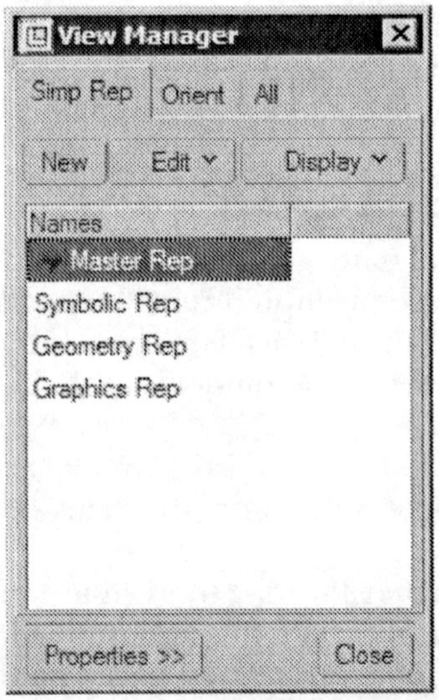

Figure 6-34 View Manager *dialog box*

Choose the **New** button from the dialog box. The **Rep0001** is displayed in the **Names** display box. Press ENTER. The dialog box disappears and then appears again. Select **Rep0001** and then right-click. From the shortcut menu choose the **Redefine** option. The **EDIT METHOD** menu is displayed as shown in Figure 6-35.

In this menu the methods of representing a model are listed. Here, you will learn the **Work Region** method of representing the model.

*Figure 6-35 **EDIT METHOD** menu*

Work Region Method

This method of representing a model is very similar to removing material (creating cut) from a model. The following steps explain the procedure to create the section of the model shown in Figure 6-36.

***Figure 6-36** Solid model*

1. Choose the **Work Region** option from the **EDIT METHOD** menu. The **SOLID OPTS** submenu is displayed as shown in Figure 6-37.

2. Choose **Extrude** > **Solid** > **Done** from the **SOLID OPTS** submenu. The Extrude dashboard is displayed.

3. Choose the **Create a section or redefine the existing section** button to invoke the **Section** dialog box.

4. Select the top face of the plate as the sketch plane and select the **RIGHT** datum plane to be at the right while drawing the sketch.

5. When you enter the sketcher environment, draw the section lines as shown in Figure 6-38.

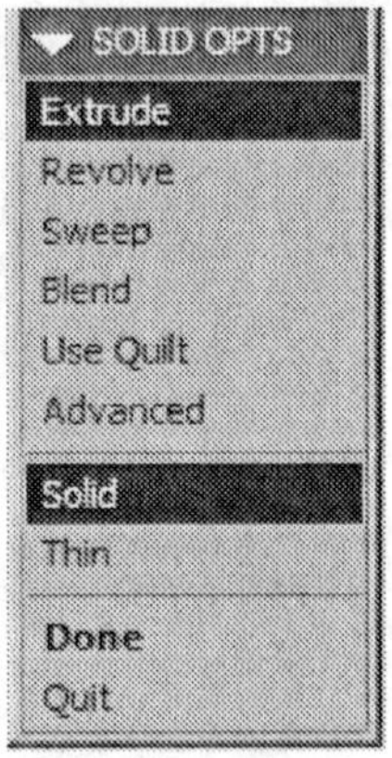

***Figure 6-37 SOLID OPTS** submenu*

This section is an open loop. The dimensions are not of importance to create this section.

6. Exit the sketcher environment and select the **Extrude to intersect with all surface** button from the depth flyout. The yellow arrow is pointing in the reverse direction.

7. Choose the **Change material direction of extrude to other side of sketch** button to reverse the direction of arrow.

8. Choose the **Build feature** button from the Extrude dashboard. The sectioned model is shown in Figure 6-39.

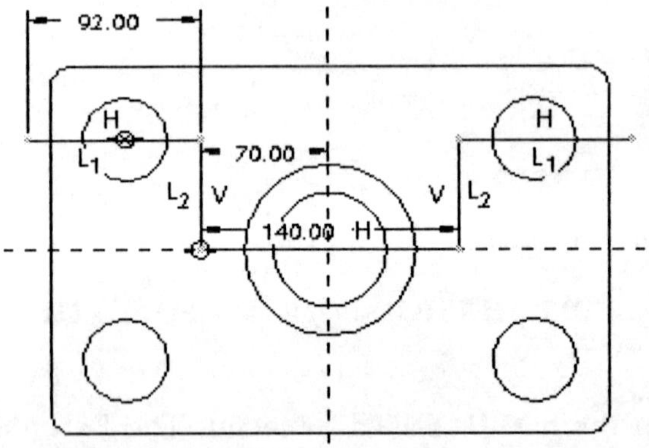

Figure 6-38 *Sketch of the section lines*

Figure 6-39 *Sectioned model*

TUTORIALS

Tutorial 1

In this tutorial you will create the model shown in Figure 6-40. This figure also shows the top view, the front view, and the right-side view of the model. **(Expected time: 30 min)**

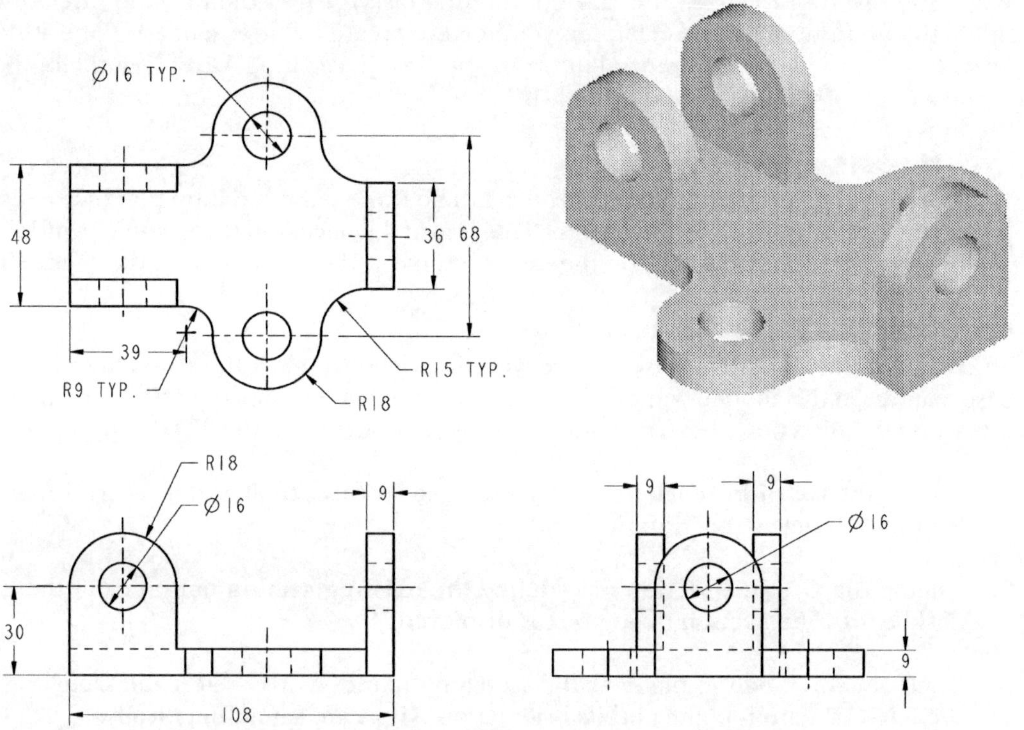

Figure 6-40 *Top, front, isometric, and right-side views of the model*

The following steps outline the procedure for creating this model:

a. First examine the model and then determine the number of features in it. The model is composed of four features, see Figure 6-40.

b. The base feature is an extruded feature, see Figure 6-42. First the sketch of the base feature will be created on the **TOP** datum plane, see Figure 6-41, and then it will be extruded to a depth of 9.

c. The second feature is also an extruded feature, see Figure 6-45. The sketch of the second feature will be created on the right face of the base feature, see Figure 6-44, and then it will be extruded to a depth of 9. You need to precisely follow the steps given to create this feature. Else, it may not get copied later.

d. The third feature is the same as the second feature, and therefore, a copy of it will be created by defining new references, see Figure 6-47.

e. The fourth feature is the same as the third feature and therefore, a mirror copy of it will be created as shown in Figure 6-48.

After understanding the procedure for creating the model, you are now ready to create it.

When Pro/ENGINEER session is started, the first task is to set the working directory. Since this is the first tutorial of this chapter, you need to create a folder named *c06*, if it does not exist. Choose the **New Directory** button in the **Select Working Directory** dialog box and create a directory named *c06* at *C:\ProE-WF*.

Creating New Object File

1. Open a new part file and name it *c06tut1*. The three default datum planes are displayed on the graphics window. The **Model Tree** is also displayed on the graphics window. Close the **Model Tree** by clicking on the sash present on the right edge of the **Model Tree**.

Creating the Base Feature

To create the sketch for the base feature, you first need to select the sketching plane for the base feature. In this model, you need to draw the base feature on the **TOP** datum plane. This is because the direction of extrusion of this feature is normal to the **TOP** datum plane.

1. Choose the **Extrude Tool** button from the **Base Features** toolbar. The Extrude dashboard is displayed below the graphics window.

2. Choose the **Create a section or redefine the existing section** button from the Extrude dashboard. The **Section** dialog box is displayed.

3. Select the **TOP** datum plane as the sketching plane. As you select the sketching plane, the **RIGHT** datum plane and its orientation **Right** are set automatically.

4. Choose the **Sketch** button to enter the sketcher environment.

5. Once you enter the sketcher environment, create the sketch of the base feature and apply constraints and dimensions as shown in Figure 6-41.

   Note that in the sketch, the bottom half of the sketch is mirrored to create the top half of the sketch. This is evident from the constraints of symmetry applied to the sketch in Figure 6-40. Also, as evident from the sketch of the base feature shown in Figure 6-41, the **RIGHT** datum plane is located at a dimension of 18 from the left edge because later in the tutorial, this datum plane will be used as a reference to copy a feature.

6. After the sketch is complete, choose the **Continue with the current section** button to exit the sketcher environment.

   The Extrude dashboard is enabled and is displayed below the graphics window.

7. Enter **9** as the value of the depth in the dimension box present on the Extrude dashboard and press ENTER.

8. Choose the **Build feature** button from the Extrude dashboard.

   The base feature is completed and now the second feature will be created. The default trimetric view of the base feature is shown in Figure 6-42.

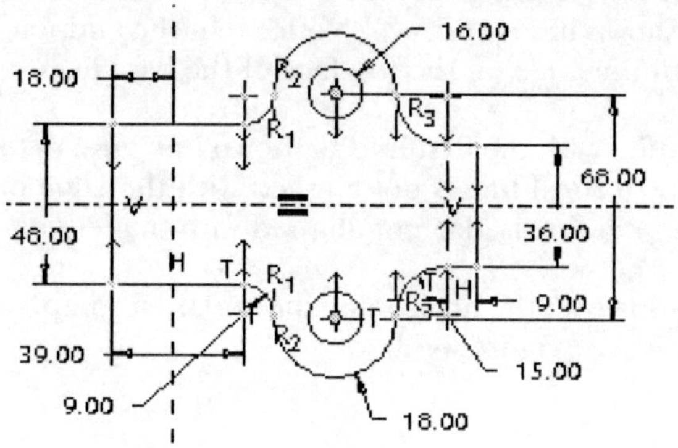

Figure 6-41 *Sketch of the base feature*

Figure 6-42 *The default trimetric view of the base feature*

Note

The two holes are integrated in the base feature. These holes are sketched while drawing the sketch for the base feature. Hence, the base feature is created as one single feature that includes the two holes. The other method is to create the two holes separately on the base feature using the Hole dashboard. In the method to create the features separately, the total features created will be three.

Creating the Second Feature

The second feature is also an extruded feature and is created on the right planar surface of the base feature. Therefore, you need to select the right planar surface as the sketching plane. **Note that the sketch of this feature has to be created keeping in mind some important steps. The feature may not get copied later on if you do not use the following steps in creating the sketch of this feature.**

1. Choose the **Extrude Tool** button from the **Base Features** toolbar. The Extrude dashboard is displayed below the graphics window.

2. Choose the **Create a section or redefine the existing section** button from the Extrude dashboard. The **Section** dialog box is displayed.

3. Select the right face of the base feature as the sketching plane.

4. Choose the **Flip** button to reverse the direction of the yellow arrow.

5. Select the **TOP** datum plane and then select the **Top** option from the **Orientation** drop-down list.

6. Choose the **Sketch** button to enter the sketcher environment.

After entering the sketcher environment, turn the model display to **No Hidden**.

7. Choose the **Create 2 point lines** button from the **Sketcher Tools** toolbar and draw the right vertical line starting from the point shown in Figure 6-43. Notice that the endpoint of the right vertical line is not aligned with the edge on the top face of the base feature.

8. Next, draw the horizontal line in continuation with the first line as shown in Figure 6-43. Notice that since the endpoint of the right vertical line is not aligned with the edge on the top face of the base feature, the horizontal line is also not aligned with that edge.

9. Next, draw the left vertical line in continuation with the horizontal line and then complete the sketch by drawing the arc and the circle, see Figure 6-43.

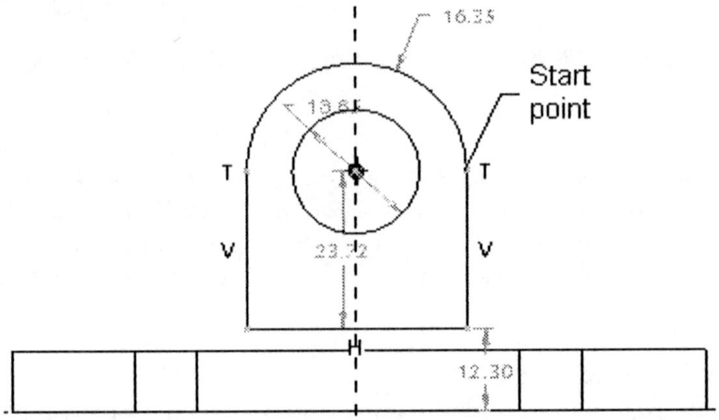

Figure 6-43 *Sketch of the second feature with weak dimensions*

10. Choose the **Impose sketcher constraints on the section** button to display the **Constraints** dialog box. Choose the **Create same points, points on entity or collinear constraint** button and select the bottom horizontal line of the sketch. Now, select the edge on the top face of the base feature as the second entity to apply the constraint. Notice that the sketch will extend such that the horizontal line is now aligned with the top face.

11. If the center points of the arc and the circle are not aligned with the **FRONT** datum plane, align them.

12. Add the dimensions and then modify them as shown in Figure 6-44. After completing the sketch, turn the model display to **Shading** and choose the **Continue with the current section** button. The Extrude dashboard is enabled.

13. Enter the depth of extrusion in the dimension box present on the Extrude dashboard and press ENTER.

14. Choose **Build feature** from the Extrude dashboard. The default shaded trimetric view of the model after creating the second feature is shown in Figure 6-45.

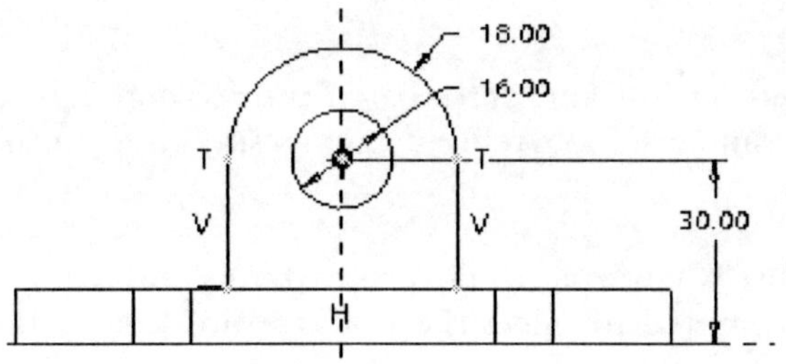

Figure 6-44 *Sketch of the second feature*

Figure 6-45 *Model after creating the second feature*

Creating the Third Feature

You can create the third feature using two methods. The first method is to draw the sketch of the third feature by defining a sketching plane and then extruding it to the given depth. The second method is to copy the second feature by defining new references. This is because the third feature is similar in geometry and dimensions to the second feature of the model. In this tutorial you will use the second method to create the third feature.

1. Invoke the **Menu Manager** by choosing **Edit > Feature Operations** from the menu bar.

2. Choose **FEAT > Copy** from the **Menu Manager**. The **COPY FEATURE** submenu is displayed.

3. Choose **New Refs > Select > Dependent** from the **COPY FEATURE** submenu and choose **Done**. The **SELECT FEAT** submenu is displayed and you are prompted to select the feature to be copied.

4. Select the second feature from the graphics window. The selected feature is highlighted with red boundary.

5. Choose **Done** from the **SELECT FEAT** submenu. The **GP VAR DIMS** menu is displayed. Since you do not need to vary any dimension of the source feature, you can proceed further without selecting any dimension.

6. Choose **Done** from the **GP VAR DIMS** menu. The **WHICH REF** menu is displayed and you are prompted to select a sketching plane reference corresponding to the highlighted surface.

7. Select the face shown in Figure 6-46 from the graphics window as the surface on which the copied feature will be placed. The **TOP** datum plane is highlighted and you are prompted to select a horizontal sketcher reference corresponding to the highlighted surface.

8. Choose the **Same** option from the **WHICH REF** menu. The **FRONT** datum plane is highlighted and you are prompted to select section dimensioning reference corresponding to the highlighted surface.

9. Select the **RIGHT** datum plane. The top left vertical edge of the base feature is highlighted and you are prompted to select section dimensioning reference corresponding to the highlighted surface.

10. Select the top front edge of the base feature where you need to copy the second feature. A red arrow is displayed and you are prompted to select the upward direction of the horizontal plane for protrusion.

11. Choose **Okay** from the **DIRECTION** submenu and then choose **Done** from the **GRP PLACE** menu. The feature is copied at the required location on the model as shown in Figure 6-47.

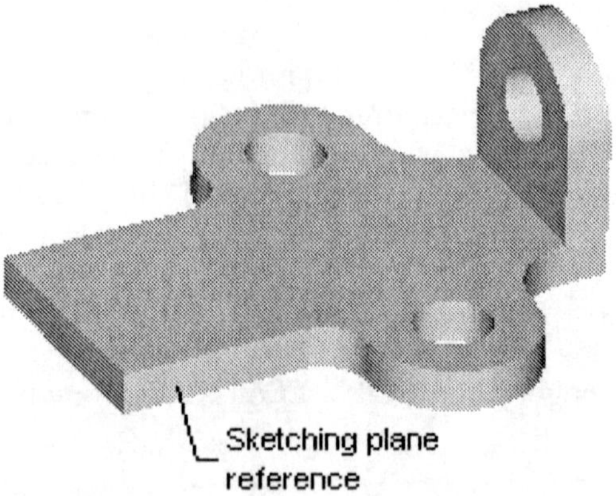

Figure 6-46 *Sketching plane reference for third feature*

Figure 6-47 *The model after copying the feature with new references*

Creating the Fourth Feature

The fourth feature can be created by sketching and extruding it to the given depth. You can also create this feature by placing a mirrored copy of the third feature at the required location. In this tutorial you will use the second method because it consumes less time.

1. Choose **FEAT** > **Copy** from the **Menu Manager**. The **COPY FEATURE** submenu is displayed.

2. Choose **Mirror** > **Select** > **Dependent** > **Done** from the **COPY FEATURE** submenu. The **SELECT FEAT** submenu is displayed and you are prompted to select the features to be mirrored.

3. Select the third feature from the graphics screen and choose **Done** from the **SELECT FEAT** submenu. You are prompted to select a plane or create a datum to mirror about.

4. Select the **FRONT** datum plane as the mirror plane. The third feature is mirrored about the **FRONT** datum plane. The trimetric view of the final model is shown in Figure 6-48.

Figure 6-48 *Default trimetric view of the model*

5. Choose the **Save the active object** button from the **File** toolbar and save the model. The order of feature creation can be seen from the **Model Tree** shown in Figure 6-49. Note that the feature id numbers in your model may be different from the ones shown in this figure.

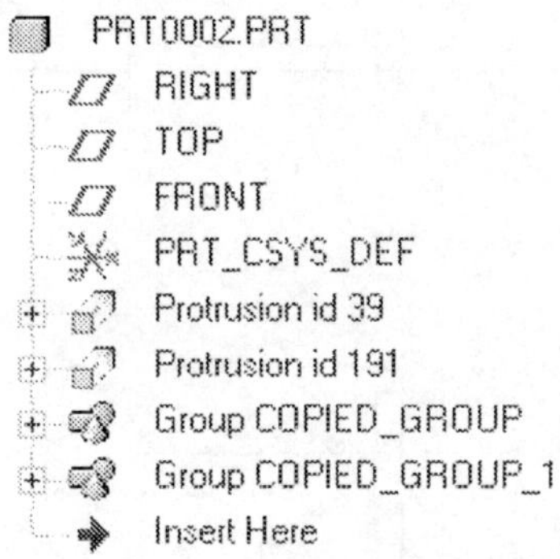

Figure 6-49 Model Tree *for Tutorial 1*

Tutorial 2

In this tutorial you will create the model shown in Figure 6-50. The dimensions of the model are given in the top view and the front view of the model shown in Figure 6-51.

(Expected time: 45 min)

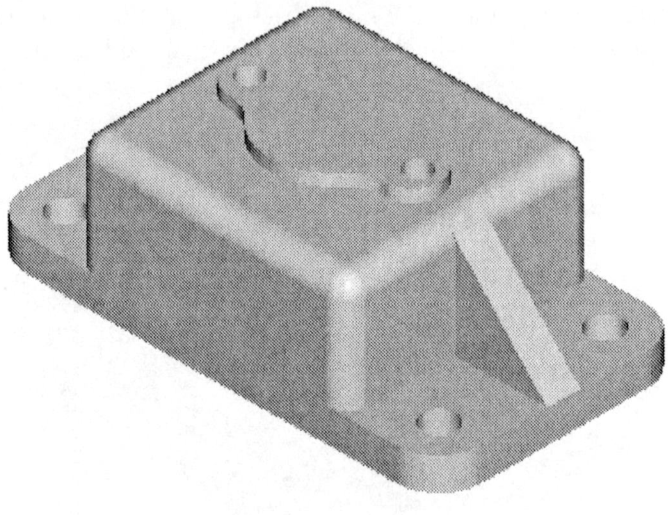

Figure 6-50 *Solid model for Tutorial 2*

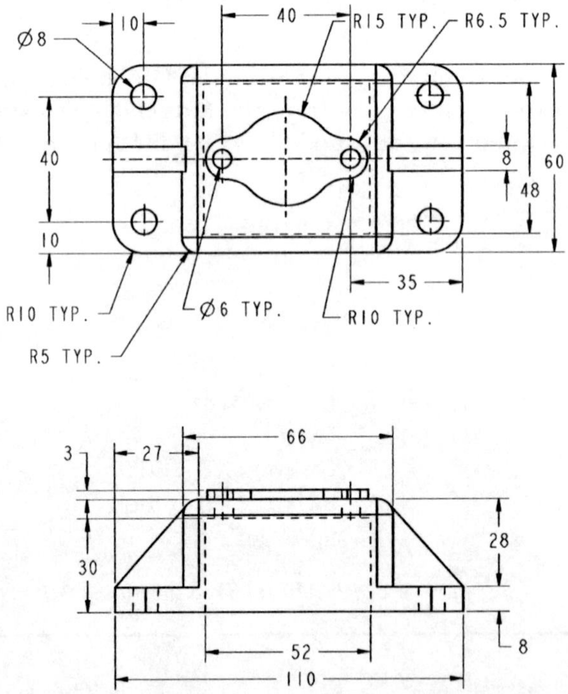

Figure 6-51 *Top view and front view of the model*

The following steps outline the procedure for creating this model:

a. First examine the model and then determine the number of features in it. The model is composed of nine features, see Figure 6-50.

b. The base feature is an extruded feature, see Figure 6-53. First the sketch of the base feature will be created on the **TOP** datum plane, see Figure 6-52, and then it will be extruded to a depth of 9.

c. The second feature is a set of rounds of radius 5 on the top portion of the base feature and of radius 10 on the bottom portion of the base feature, see Figure 6-55.

d. The third feature is a hole feature, see Figure 6-56, and then this hole will be patterned, see Figure 6-57. After you pattern the hole feature, the hole and the pattern features will be combined in a single feature.

e. The fourth feature is a rib feature, see Figure 6-59. The sketch of the rib feature will be created on the **FRONT** datum plane, see Figure 6-58, and then the given thickness will be applied to it.

f. The fifth feature is the mirror copy of the rib feature, see Figure 6-60.

g. The sixth feature is an extruded feature, see Figure 6-62. The sketch for this feature will be created on the top planar surface of the base feature and then it will be extruded to the given distance, see Figure 6-61.

h. The seventh feature is a cut feature, see Figure 6-64. The sketch for this feature will be created on the bottom planar surface of the base feature and then it will be extruded to the given distance, see Figure 6-63.

i. The eighth feature is a hole feature created on the sixth feature, see Figure 6-65.

j. The ninth feature is a copy of the hole feature, see Figure 6-66.

After understanding the procedure for creating the model, you are now ready to draw it. The working directory has already been selected in Tutorial 1 and therefore you do not need to select it again. However, if you want to change the working directory, select the working directory from the Navigator.

Creating New Object File

1. Open a new part file and name it *c06tut2*. The three default datum planes are displayed on the graphics window.

Creating the Base Feature

To create the sketch for the base feature, you first need to select the sketching plane for the base feature. In this model, you need to draw the base feature on the **FRONT** datum plane because the direction of extrusion of this feature is perpendicular to the **FRONT** datum plane. The base feature will also be created symmetric to the **FRONT** datum plane, and therefore the sketch will be created equally on both the sides of the **FRONT** datum plane.

1. Choose the **Extrude Tool** button from the **Base Features** toolbar. The Extrude dashboard is displayed below the graphics window.

2. Choose the **Create a section or redefine the existing section** button from the Extrude dashboard. The **Section** dialog box is displayed.

3. Select the **FRONT** datum plane as the sketch plane.

4. Select the **TOP** datum plane and then select the **Top** option from the **Orientation** drop-down list.

5. Choose the **Sketch** button to enter the sketcher environment.

6. Once you enter the sketcher environment, create the sketch of the base feature and apply constraints and dimensions as shown in Figure 6-52.

   In the sketch, you should note that since the base feature is symmetrical, therefore, a vertical center line is drawn and then the right half of the sketch is mirrored to create the left half. This is evident from the constraints of symmetry applied to the sketch in Figure 6-52. These constraints appear as arrow symbols in the sketch.

   As evident from the sketch of the base feature shown in Figure 6-52, the **TOP** datum plane is aligned with the bottom line segment.

7. After the sketch is complete, choose the **Continue with the current section** button to exit the sketcher environment.

   The Extrude dashboard is enabled and appears below the graphics window. Now, you need to extrude the sketch symmetrically to both sides of the sketching plane in order to create the base feature symmetrical to the **FRONT** datum plane. This is because, later in the tutorial, you will need to select the default datum planes as mirror planes for mirroring features. Hence, you do not need to create datum planes.

8. Choose the **Extrude on both sides of sketch plane by half the specified depth value in each** button from the depth flyout present on the Extrude dashboard. 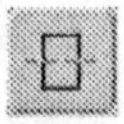

9. Enter a depth of **60** in the dimension box present on the Extrude dashboard and press ENTER.

10. Choose the **Build feature** button from the Extrude dashboard to exit the feature creation tool.

    The base feature is completed and now the second feature will be created. The default trimetric view of the base feature is shown in Figure 6-53.

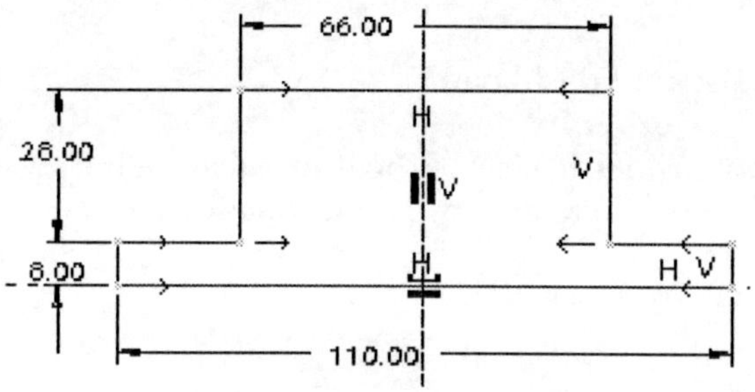

Figure 6-52 Sketch with dimensions and constraints for the base feature

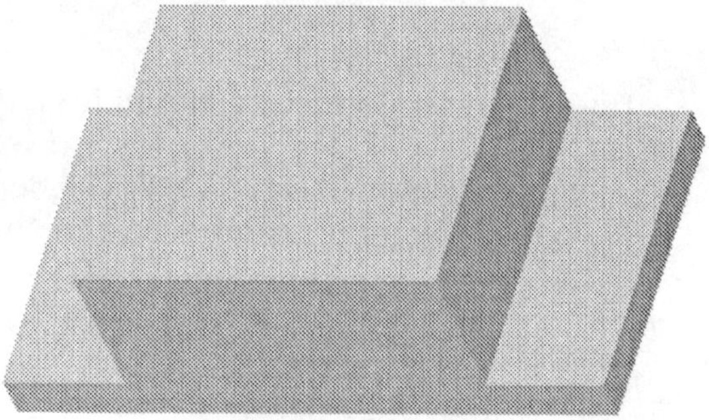

Figure 6-53 Default trimetric view of the base feature

Creating the Second Feature

The second feature is a round feature of radius 5.

1. Choose the **Round Tool** button from the **Engineering Features** toolbar. The Round dashboard is displayed.

2. Choose the **Sets** tab to display the slide-up panel. Let this slide-up panel remain open so that you can view the selections you make on the model.

3. Select one of the edges shown in Figure 6-54. To select the second edge and the subsequent edges you need to press the CTRL key and then select the edge.

   The preview of the round is created on the edges that you have selected. A default value of the radius is also displayed.

4. Double-click on the default radius value that is displayed on the preview of the round. In the edit box that appears, enter a value of **5** and press ENTER. The first set of rounds is created.

   Now, you need to create the second set of rounds.

5. After spinning the model, select the four vertical edges of the bottom portion of the base feature to round. Remember to use the CTRL+left mouse button to select the second and subsequent edges.

   You will notice that in the slide-up panel, **Set 1**, **Set 2** appears. This indicates that second set is defined. The preview of the round is created on the edges. A default value of the radius is also displayed.

6. Double-click on the default radius value that is displayed on the preview of the round. In the edit box that appears, enter a value of **10** and press ENTER.

   The second round is created.

7. Choose the **Build feature** button from the Round dashboard.

   The round feature is completed. The default trimetric view of the round feature is shown in Figure 6-55. In the **Model Tree** the two rounds created appear as one feature.

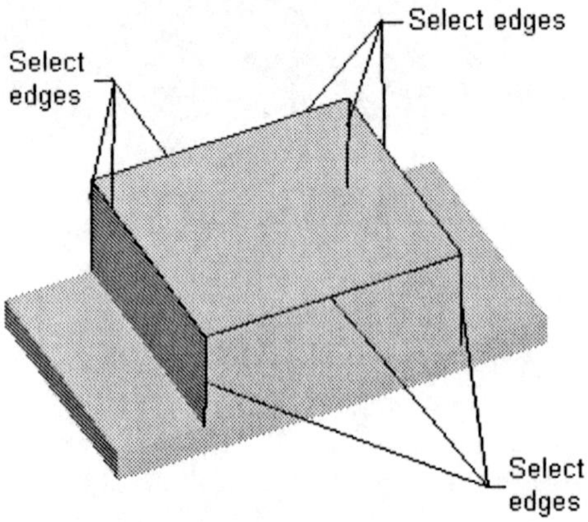

Figure 6-54 *Edges to be selected to round*

Figure 6-55 *The default trimetric view of the base feature with two sets of round*

Creating the Third Feature

The third feature is a through hole and will be created using the Hole dashboard.

1. Choose the **Hole Tool** button from the **Engineering Features** toolbar. The Hole dashboard is displayed. The **Create straight hole** button is selected by default.

2. Create the hole as shown in Figure 6-56 by specifying the placement parameters. The placement parameters can be referred from Figure 6-51.

Creating a Pattern of the Hole Feature

As evident from Figure 6-50, you need to create four instances of the hole. The first instance is created by using the Hole dashboard and you can use the **Pattern Tool** button to create the remaining three instances. You will create a rectangular pattern of the hole feature. You can also create all the holes using the Hole dashboard and specify the placement parameters for each of them. But to save time, it is recommended that you create a pattern of the hole.

1. Select the hole feature and then choose the **Pattern Tool** button from the **Edit Features** toolbar.

   The Pattern dashboard is displayed. You are prompted to select the dimensions to vary in the first direction.

2. Select the **General** option from the **Options** slide-up panel.

   You cannot use the **Identical** option to create the rectangular pattern of the hole feature. This is because when you use the **Identical** option to pattern, the pattern cannot intersect the base feature on which the hole is created. If the top portion of the base feature is created as a separate feature, then the hole can be patterned using the **Identical** option.

3. Select the dimension value **10** from the graphics window. Since the two dimensions that are displayed on the graphics screen are both 10, therefore, select the dimension 10 that is along the shorter side of the base feature. After you select the dimension in the first direction, the edit box is displayed.

4. Enter a value of **40** in the edit box and press ENTER.

5. Hold down the right mouse button to display a shortcut menu. Select the **Direction 2 Dimensions** option from the shortcut menu.

6. Select the dimension value **10** that is along the longer side of the base feature. After you select the dimension in the second direction, the edit box is displayed.

   Note that in the edit boxes present on the dashboard, the number of instances, **2**, is specified by default.

7. Enter a value of **90** in the edit box and press ENTER.

8. Choose the **Build feature** button from the Pattern dashboard.

   The rectangular pattern of holes is displayed as shown in Figure 6-57. You can use the middle mouse button to spin the model and to display the model as shown in Figure 6-57.

Figure 6-56 *The hole feature on the base feature*

Figure 6-57 *Model after creating the hole pattern*

Note

If you have multiple holes to be created on a model, it is recommended that you create their pattern, if possible. This is because when you assemble bolts in these holes in the Assembly mode, it becomes very easy to assemble them using the reference pattern.

Creating the Rib Feature

The sketch of the rib feature will be created on the **FRONT** datum plane and the given thickness will be applied to it. This is the fourth feature of the model.

1. Choose the **Rib Tool** from the **Engineering Features** toolbar. The Rib dashboard is displayed.

2. Choose the **Create a section or redefine the existing section** button from the Rib dashboard. The **Section** dialog box is displayed.

3. Select the **FRONT** datum plane as the sketching plane.

4. Select the **TOP** datum plane from the graphics window and then select the **Top** option from the **Orientation** drop-down list.

5. Choose the **Sketch** button to proceed to the sketcher environment.

6. Draw the sketch of the rib feature as shown in Figure 6-58.

Note

As evident from Figure 6-58, the top end of the inclined line in the sketch is aligned with the curve and the tangent constraint is applied to the line and the curve. Similarly, the bottom end of the inclined line is also aligned with the two edges. This is the reason there are no dimensions in the sketch and the sketch for the rib feature is fully constrained.

7. Exit the sketcher environment.

8. From the **References** slide-up panel, choose the **Add material to the other side of the section** button.

9. Specify the rib thickness as 8 in the dimension box present on the Rib dashboard.

   The rib feature is shown in Figure 6-59. You can use middle mouse button to spin the model in the orientation as shown in Figure 6-59.

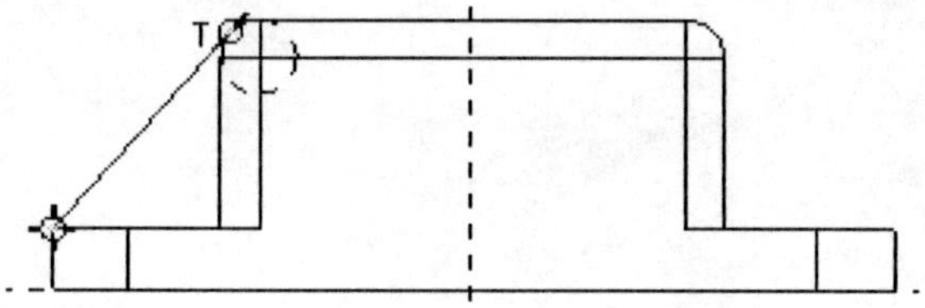

Figure 6-58 Fully constrained sketch for the rib

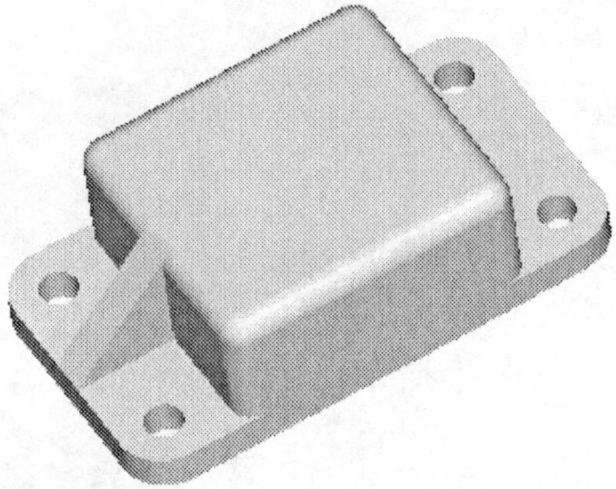

Figure 6-59 Model after creating the rib feature

Creating a Copy of the Rib Feature

A copy of the rib feature will be created as shown in Figure 6-60. This copy feature is the fifth feature of the model. The other method to create the same is to create the sketch of the rib feature on a sketching plane. But this method will consume a lot of time. Therefore, the rib feature will be copied about the **RIGHT** datum plane.

1. Choose **Edit > Feature Operations** from the menu bar. The **Menu Manager** is displayed.

2. Choose **FEAT > Copy** from the **Menu Manager**. The **COPY FEATURE** submenu is displayed.

3. Choose **Mirror > Select > Independent > Done** from the **COPY FEATURE** submenu. The **SELECT FEAT** submenu is displayed and you are prompted to select the features to be mirrored.

4. Select the rib feature from the graphics screen and choose **Done** from the **SELECT FEAT** submenu. You are prompted to select a plane or create a datum plane to mirror about.

5. Select the **RIGHT** datum plane as the mirror plane. The selected feature is mirrored about the **RIGHT** datum plane as shown in Figure 6-60.

6. Choose **Done** from the **FEAT** menu to exit the **Menu Manager**.

Creating the Protrusion Feature

The sixth feature is an extruded feature that will be created on the top face of the base feature.

1. Choose the **Extrude Tool** button from the **Base Features** toolbar. The Extrude dashboard is displayed below the graphics window.

2. Choose the **Create a section or redefine the existing section** button from the Extrude

Figure 6-60 *Model after mirroring the rib*

dashboard. The **Section** dialog box is displayed.

3. Select the top face of the base feature as the sketching plane.

4. Select the **RIGHT** datum plane and then select the **Right** option from the **Orientation** drop-down list.

5. Choose the **Sketch** button to enter the sketcher environment.

6. After you enter the sketcher environment, draw the sketch of the extruded feature as shown in Figure 6-48.

   Here in the sketch, you should note that the tangent and equal radii constraints are applied. Also the center of the top arc and the bottom arc coincides with the intersection of the two datum planes.

7. After the sketch is complete, choose the **Continue with the current section** button to exit the sketcher environment.

   The Extrude dashboard is enabled and appears below the graphics window.

8. Enter a depth of **3** in the dimension box present on the Extrude dashboard. Choose the **Build feature** button from the Extrude dashboard.

   The extruded feature is completed and the default trimetric view is shown in Figure 6-62.

Creating the Cut Feature

You need to create an extruded cut on the bottom planar surface of the base feature and this is the seventh feature of the model.

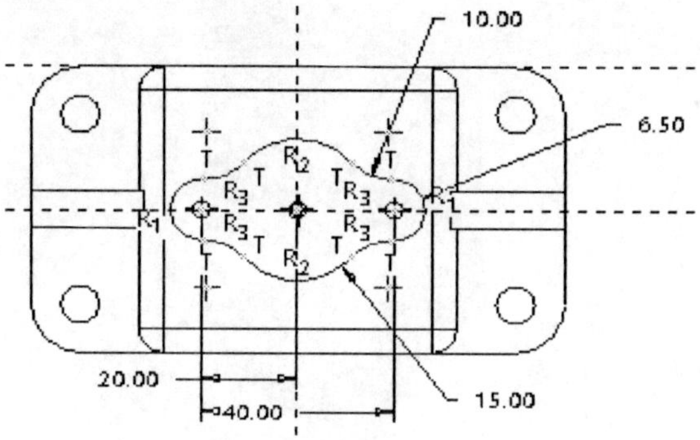

Figure 6-61 *Sketch with dimensions and constraints for the extruded feature*

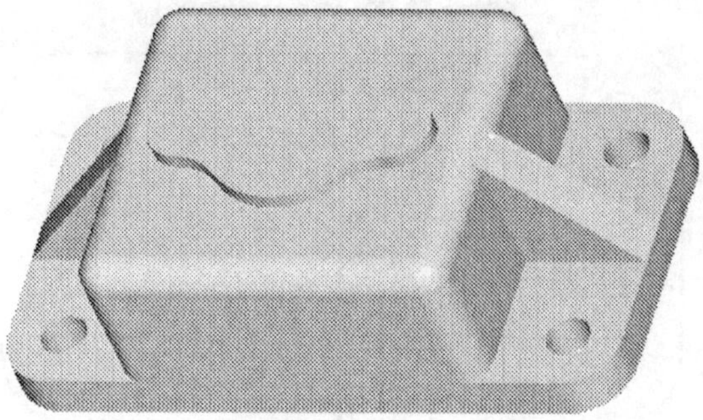

Figure 6-62 *The default trimetric view after creating the extruded feature*

1. Choose the **Extrude Tool** button form the **Base Features** toolbar. The Extrude dashboard is displayed below the graphics window.

2. Select the **Remove Material** button from the Extrude dashboard.

3. Choose the **Create a section or redefine the existing section** button from the Extrude dashboard. The **Section** dialog box is displayed.

4. Select the bottom face of the base feature as the sketching plane.

5. Select the **RIGHT** datum plane and then select the **Right** option from the **Orientation** drop-down list.

6. Choose the **Sketch** button to enter the sketcher environment.

7. After you enter the sketcher environment, draw the sketch and dimension it as shown in Figure 6-63.

8. After completing the sketch, choose the **Continue with the current section** button. The Extrude dashboard is enabled and appears below the graphics window.

9. Enter a value of **30** in the dimension box present on the Extrude dashboard and press ENTER.

10. Choose the **Build feature** button. You can spin the model using the middle mouse button. The cut feature is shown in Figure 6-64.

Creating the Hole

Next, a through hole will be created on the top extruded feature using the Hole dashboard.

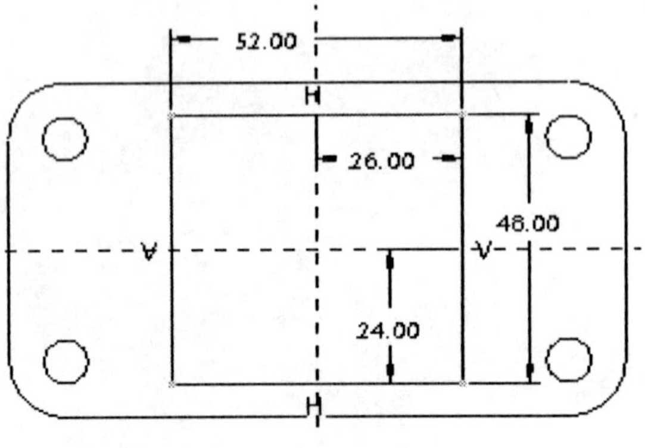

Figure 6-63 Sketch for the cut feature with dimensions

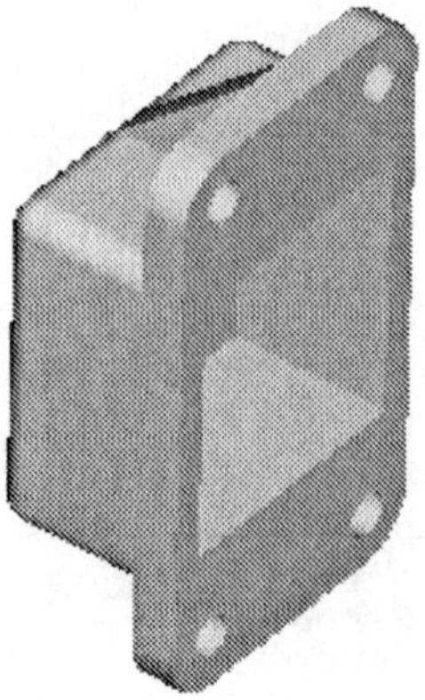

Figure 6-64 Model after creating the cut feature

1. Choose **Hole Tool** button from the **Engineering Features** toolbar. The Hole dashboard is displayed. The **Create straight hole** button is selected by default. Specify the placement parameters of the hole as given in Figure 6-50 and create the hole. The model after creating the hole is shown in Figure 6-65.

Creating a Copy of the Hole

You need to create a mirror copy of the hole as shown in Figure 6-66 and this will be the tenth feature of the model. The hole feature will be mirrored about the **RIGHT** datum plane.

1. Invoke the **Menu Manager** using the **Edit** menu. Choose **FEAT** > **Copy** from the **Menu Manager**. The **COPY FEATURE** submenu is displayed.

2. Choose **Mirror** > **Select** > **Independent** > **Done** from the **COPY FEATURE** submenu. The **SELECT FEAT** submenu is displayed and you are prompted to select the features to be mirrored.

3. Select the previous hole feature from the graphics screen and choose **Done** from the **SELECT FEAT** submenu. You are prompted to select a plane or create a datum plane to mirror about.

4. Select the **RIGHT** datum plane as the mirror plane and choose **Done**. The hole feature is mirrored about the **RIGHT** datum plane as shown in Figure 6-66.

Note

The default trimetric view of the model is displayed when you choose the ***Default*** *option from the* ***Saved view list*** *button. If you want to change the default view to isometric, then you need to use the* ***Environment*** *dialog box. The* ***Environment*** *dialog box is displayed when you choose the* ***Environment*** *option from the* ***Tools*** *menu in the menu bar.*

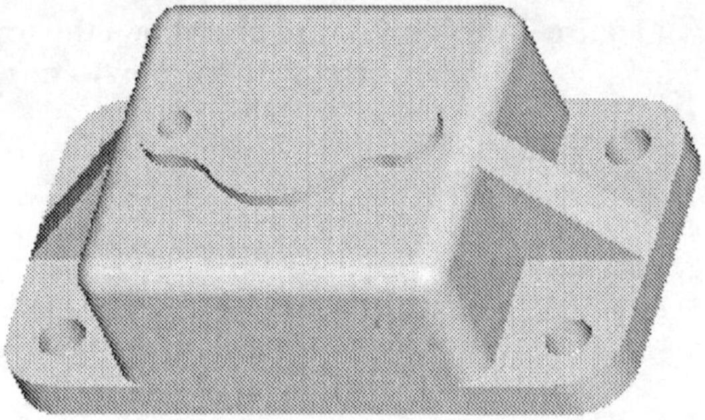

Figure 6-65 Hole feature on the extruded feature

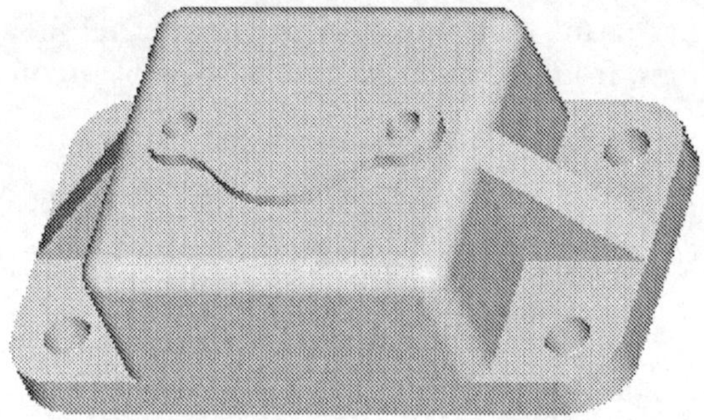

Figure 6-66 The copied hole feature

Saving the Model

You have to save the model because you may need it later.

1. Choose the **Save the active object** button from the **File** toolbar and save the model.

   The order of feature creation can be seen from the **Model Tree** shown in Figure 6-67.

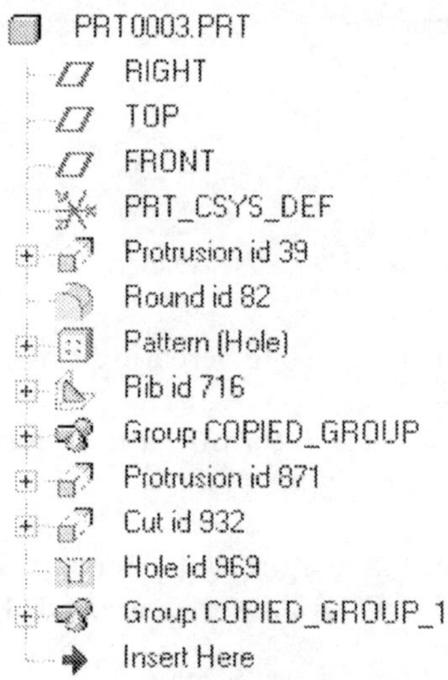

***Figure 6-67 Model Tree** for Tutorial 2*

   The feature id numbers displayed in the **Model Tree** may be different when you create the features.

Tutorial 3

In this tutorial you will create the model shown in Figure 6-68. This figure also shows the top view, front view, and isometric view of the model. **(Expected time: 30 min)**

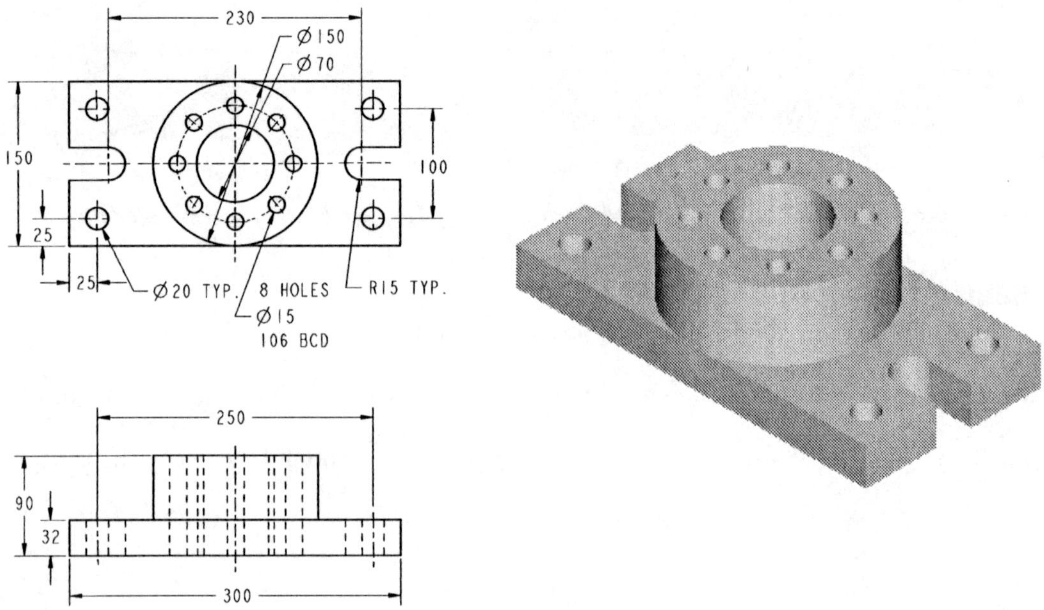

Figure 6-68 *Top, front, and isometric views of the model*

The following steps outline the procedure for creating this model:

a. First examine the model and then determine the number of features in it. The model is composed of five features, see Figure 6-68.

b. The base feature is an extruded feature, see Figure 6-70. First the sketch of the base feature will be created on the **TOP** datum plane, see Figure 6-56, and then it will be extruded to a depth of 32.

c. The second feature is a cylindrical extruded feature, see Figure 6-72. The sketch of the second feature will be created on the top planar surface of the base feature, see Figure 6-71, and then it will be extruded to a depth of 58.

d. The third feature is a hole feature and this hole is coaxial with the cylindrical feature, see Figure 6-73.

e. The fourth feature is a hole feature and will be created on the top planar surface of the base feature, see Figure 6-74. A rectangular pattern of this hole will be created, see Figure 6-75.

f. The fifth feature is also a hole feature and will be created on the top planar surface of the cylindrical feature, see Figure 6-76. A rotational pattern of this hole will be created, see Figure 6-77.

After understanding the procedure for creating the model, you are now ready to draw it. The working directory is already selected in Tutorial 1 and therefore you do not need to select it again. However, if you want to change the working directory, choose **File > Set Working Directory** and then select *c06* in the **Select Working Directory** dialog box.

Creating New Object File

1. Open a new part file and name it *c06tut3*. The three default datum planes are displayed on the graphics window.

Creating the Base Feature

To create the sketch for the base feature, you first need to select the sketching plane for the base feature. In this model, you need to draw the base feature on the **TOP** datum plane. This is because the direction of extrusion of this feature is perpendicular to the **TOP** datum plane.

1. Choose the **Extrude Tool** button from the **Base Features** toolbar. The Extrude dashboard is displayed below the graphics window.

2. Choose the **Create a section or redefine the existing section** button from the Extrude dashboard. The **Section** dialog box is displayed.

3. Select the **TOP** datum plane as the sketching plane.

4. Select the **RIGHT** datum plane and then select the **Right** option from the **Orientation** drop-down list, if it is not selected by default.

5. Choose the **Sketch** button to enter the sketcher environment.

6. Once you enter the sketcher environment, create the sketch of the base feature and apply constraints and dimensions as shown in Figure 6-69.

7. After the sketch is complete, choose the **Continue with the current section** button to exit the sketcher environment.

   The Extrude dashboard is enabled and appears below the graphics window.

8. Enter a depth of **32** in the dimension box that is present on the Extrude dashboard and press ENTER. Choose **Build feature** from the dashboard.

The base feature is completed and now the second feature will be created. The default trimetric view of the base feature is shown in Figure 6-70.

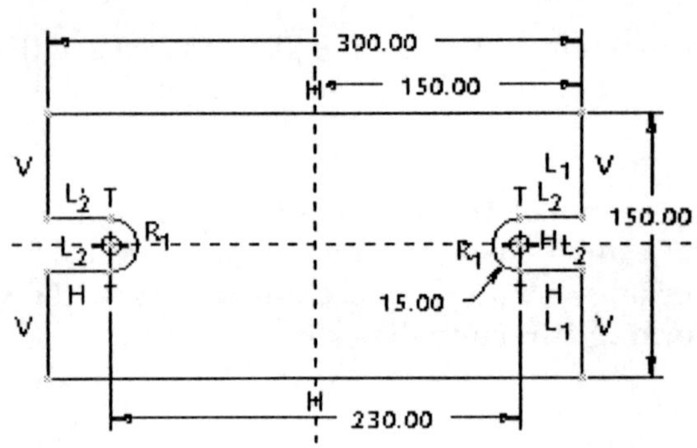

Figure 6-69 *Sketch with dimensions and constraints for the base feature*

Figure 6-70 *Default trimetric view of the base feature*

Creating the Second Feature

The second feature is also an extruded feature and will be created on the top face of the base feature. Therefore, you need to define the top face as the sketching plane for the second feature.

1. Choose the **Extrude Tool** button from the **Base Features** toolbar. The Extrude dashboard is displayed below the graphics window.

2. Choose the **Create a section or redefine the existing section** button from the Extrude dashboard. The **Section** dialog box is displayed.

3. Select the top face of the base feature as the sketching plane.

4. Select the **RIGHT** datum plane and then select the **Right** option from the **Orientation** drop-down list.

5. Choose the **Sketch** button to enter the sketcher environment.

6. Create the sketch for the second feature and dimension it as shown in Figure 6-71.

7. After creating the sketch, choose the **Continue with the current section** button. The Extrude dashboard is enabled.

8. Enter a value of **58** in the dimension box present on the Extrude dashboard and press ENTER.

9. Choose the **Build feature** button. The second feature is completed and the default trimetric view is shown in Figure 6-72.

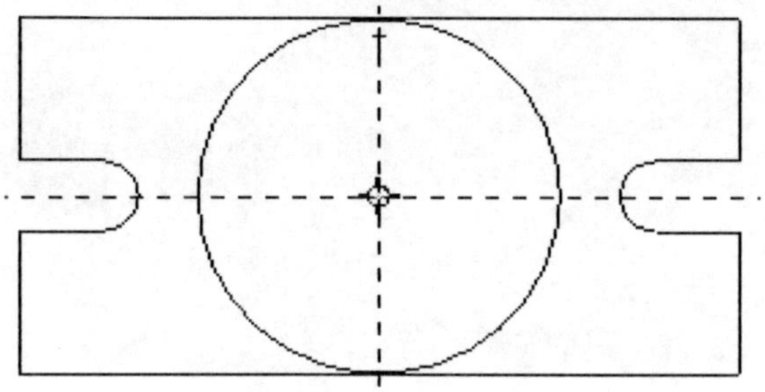

Figure 6-71 Sketch of the cylindrical feature with diameter of the cylinder

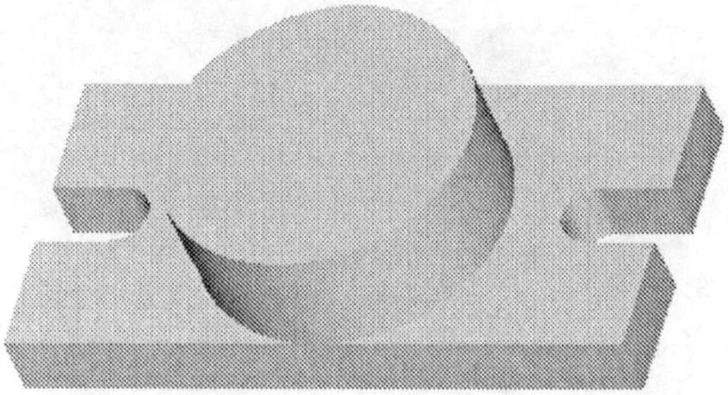

Figure 6-72 Default trimetric view of the cylindrical feature

Creating the Third Feature

The third feature is a through hole that is coaxial to the cylindrical feature. The hole feature will be created using the Hole dashboard.

1. Choose the **Hole Tool** button from the **Engineering Features** toolbar. The Hole dashboard is displayed. The **Create straight hole** button in the Hole dashboard is chosen by default.

2. Create a hole of diameter **70** as shown in Figure 6-73 by specifying the placement parameters.

Creating the Fourth Feature

The fourth feature is a through hole and will be created on the top planar surface of the base feature. The hole feature will be created using the Hole dashboard.

1. Choose the **Hole Tool** button from the **Engineering Features** toolbar. The Hole dashboard is displayed. The **Create straight hole** button in the Hole dashboard is selected by default.

2. Create the hole as shown in Figure 6-74 by specifying the placement parameters.

Patterning the Hole Feature

Next, you need to create a rectangular pattern of the hole feature that is created on the base feature. You can also create the individual holes, but this will consume a lot of time and the number of features in the **Model Tree** will increase. Hence, it is recommended that you create a rectangular pattern of the hole feature.

1. Select the hole feature and then choose the **Pattern Tool** button from the **Edit Features** toolbar.

   The Pattern dashboard is displayed. You are prompted to select the dimensions to vary in the first direction.

Figure 6-73 *Coaxial hole on the cylindrical feature*

Figure 6-74 *Hole on the base feature*

2. Select the **Identical** option from the **Options** slide-up panel.

   Here, you can use the **Identical** option because the feature on which the pattern is created is not intersecting the pattern.

3. Select the dimension value **25** from the graphics window. Since the two dimensions that are displayed on the graphics screen are both 25, hence select the dimension 25 that is along the shorter side of the base feature. After you select the dimension in the first direction, the edit box is displayed.

4. Enter a value of **100** in the edit box and press ENTER.

5. Hold down the right mouse button to display the shortcut menu. Choose the **Direction 2 Dimensions** option from the shortcut menu.

6. Select the dimension value **25** that is along the longer side of the base feature. After you select the dimension in the second direction, the edit box is displayed.

7. Enter a value of **250** in the edit box and press ENTER. Note that in the edit boxes present on the dashboard, the number of instances, **2**, is specified by default.

8. Choose the **Build feature** button from the Pattern dashboard.

   The rectangular pattern of holes will be displayed as shown in Figure 6-75.

Creating a Hole on the Cylindrical Feature

The hole on the cylindrical feature will created diametrically using the Hole dashboard.

1. Choose the **Hole Tool** button from the **Engineering Features** toolbar. The Hole dashboard is displayed. The **Create straight hole** button in the Hole dashboard is chosen by default.

2. Choose the **Placement** tab from the Hole dashboard to display the slide-up panel.

3. Select the top face of the cylindrical feature as the placement plane.

4. From the drop-down list in the slide-up panel, select the **Diameter** option.

5. Click in the **Secondary References** collector and now select the axis of the cylindrical feature.

6. Enter a value of **106** in the second dimension box present on the right of the **Angular** option in the **Secondary References** display box and press ENTER.

7. Use CTRL+left mouse button and select the **FRONT** datum plane from the graphics window. Enter a value of **90** in the dimension box present in the **Secondary References** collector.

8. Enter a value of **20** as the diameter of the hole in the diameter dimension box present on the Hole dashboard.

9. Choose the **Build feature** button from the Hole dashboard. The hole is created as shown in Figure 6-76.

Figure 6-75 *Rectangular pattern of the hole feature*

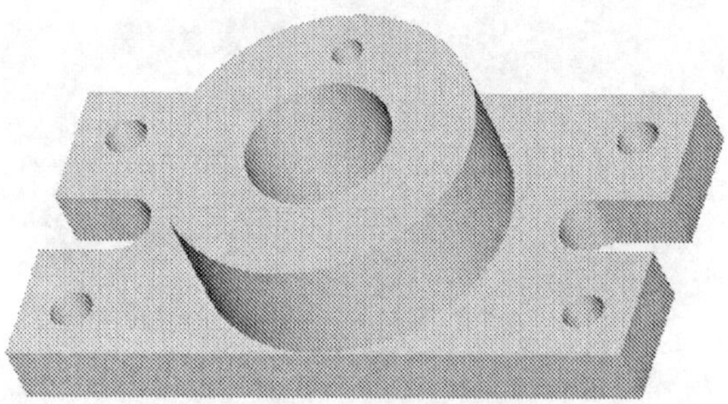

Figure 6-76 *Diametrical hole on the cylindrical feature*

Creating the Rotational Pattern of the Hole Feature

Creating the remaining holes individually on the cylindrical feature will consume a lot of time. Therefore, you need to create a rotational pattern of the hole feature.

1. Select the hole feature and then choose the **Pattern Tool** button from the **Edit Features** toolbar.

   The Pattern dashboard is displayed. You are prompted to select the dimensions to vary in the first direction.

2. Select the **Identical** option from the **Options** slide-up panel.

3. The dimensions of the hole feature are displayed on the graphics window. Select the angular dimension **90**. The edit box is displayed.

4. Enter a value of **45** in the edit box and press ENTER. Now you need to specify the number of instances of the hole feature in the pattern.

5. Enter **8** in the **1** edit box and press ENTER. Choose the **Build feature** button. The rotational pattern is created as shown in Figure 6-77.

Saving the Model

You have to save the model because you may need it later.

1. Choose the **Save the active object** button from the **File** toolbar and save the model.

   The order of feature creation can be seen from the **Model Tree** shown in Figure 6-78. The feature id numbers displayed in the **Model Tree** may be different when you create the features.

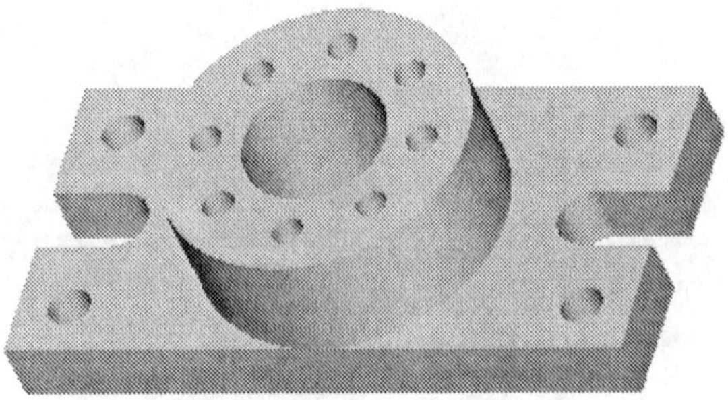

Figure 6-77 The complete model

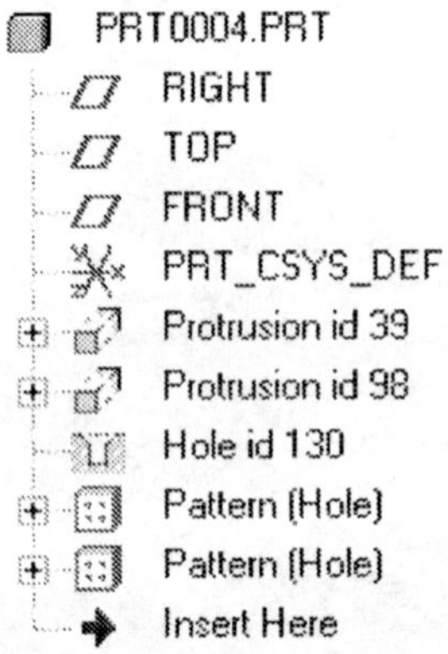

***Figure 6-78 Model Tree** for Tutorial 3*

Tutorial 4

In this tutorial you will create the model of the cylinder head shown in Figure 6-79. Figure 6-80 shows the top view and the front section view of the model. **(Expected time: 45 min)**

The following steps outline the procedure for creating this model:

a. First examine the model and then determine the number of features in it. The model is composed of twelve features, see Figure 6-79.

b. The base feature is a protrusion feature, see Figure 6-82. First the sketch of the base feature will be created on the **TOP** datum plane, see Figure 6-81, and then it will be

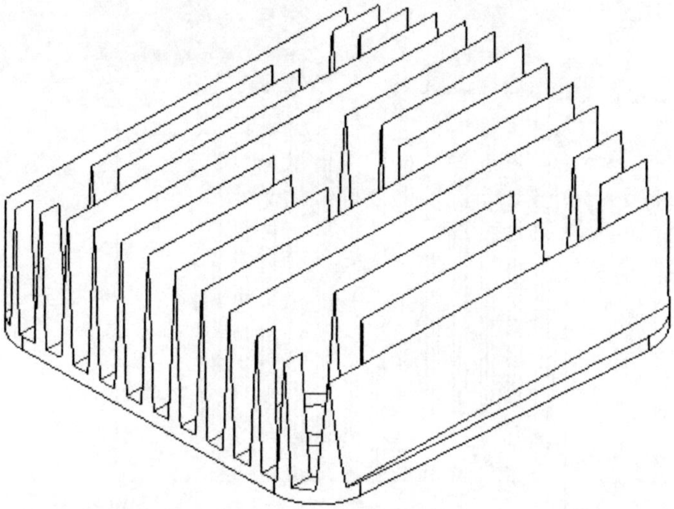

Figure 6-79 Isometric view of the model

extruded to a depth of 4.

c. The second feature is a set of rounds created on the edges of the four corners, see Figure 6-83.

d. The third feature is a cylindrical feature and is created on the bottom face of the base feature, see Figure 6-84.

e. The fourth feature is a revolve cut and will be created on the plane passing through the center of the cylindrical feature, see Figure 6-86.

f. The fifth feature is a fin that later will be patterned. The cross-section of the fin (Figure 6-87) will be sketched on the plane passing through the center of the cylindrical feature and then the sketch will be extruded on both the sides of the plane, see Figure 6-88. A pattern of this fin will be created by constraining the sketch, see Figure 6-89.

g. The sixth feature is a cut feature that will remove the projections of the fins that are projecting out of the base feature. This cut feature will be created on the bottom face of the base feature, see Figure 6-90.

h. The seventh feature is a circular cut that will be created on the top face of the base feature. This cut feature cuts the fins in the circular shape, see Figure 6-92.

i. The eighth feature is a cylindrical feature that will be created on the top face of the base feature. This sketch of this feature is drawn by using the edge of the circular cut feature, see Figure 6-93.

j. The ninth feature is a cut feature, see Figure 6-95. This feature is reference patterned

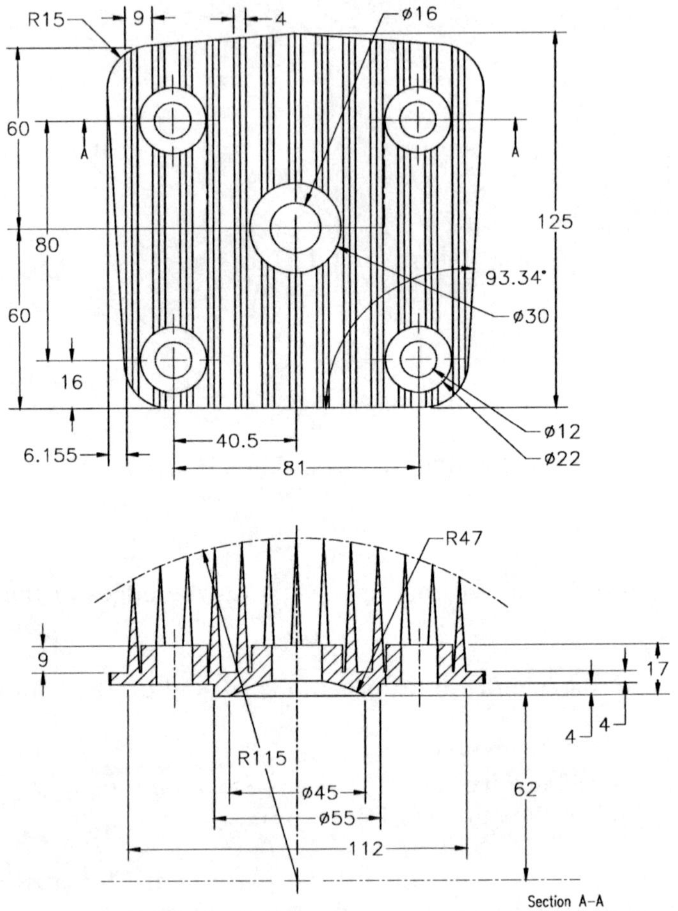

Figure 6-67 Top view and the front section view of the model

after creation to create its other instances, see Figure 6-96.

k. The tenth feature is a cylindrical feature and is created on the top face of the base feature, see Figure 6-97. Later, this feature is referenced patterned, see Figure 6-98.

l. The eleventh feature is a coaxial hole. After creating the feature, create a reference pattern of this hole, see Figure 6-99.

m. The last feature is the coaxial hole that will be created on the eighth feature, see Figure 6-100.

After understanding the procedure for creating the model, you are now ready to draw it. The working directory has already been selected in Tutorial 1 and therefore you do not need to select it again. However, if you want to change the working directory, choose **File > Set Working Directory** and then select *c06* in the **Select Working Directory** dialog box.

Creating New Object File

1. Open a new part file and name it *c06tut4*. The three default datum planes are displayed on the graphics window.

Creating the Base Feature

In this model, you need to draw the sketch of the base feature on the **TOP** datum plane. The sketch is a polygon. The right half of the polygon will be drawn first and then it will be mirrored about the centerline to create the complete polygon.

1. Choose the **Extrude Tool** button from the **Base Features** toolbar. The Extrude dashboard is displayed below the graphics window.

2. Choose the **Create a section or redefine the existing section** button from the Extrude dashboard. The **Section** dialog box is displayed.

3. Select the **TOP** datum plane as the sketching plane.

4. Select the **RIGHT** datum plane and then select the **Right** option from the **Orientation** drop-down list, if it is not selected automatically.

5. Choose the **Sketch** button to enter the sketcher environment.

6. Once you enter the sketcher environment, create the sketch of the base feature and apply constraints and dimensions as shown in Figure 6-81.

7. After the sketch is complete, choose the **Continue with the current section** button to exit the sketcher environment.

   The Extrude dashboard is enabled and appears below the graphics window.

8. Enter a depth of **4** in the dimension box that is present on the Extrude dashboard and press ENTER. Choose **Build feature** from the dashboard.

   The base feature is completed and now the second feature will be created. The default trimetric view of the base feature is shown in Figure 6-82.

Creating the Second Feature

The second feature is a set of rounds of radius 15.

1. Choose the **Round Tool** button from the **Engineering Features** toolbar. The Round dashboard is displayed.

2. Choose the **Sets** tab to display the slide-up panel. Let this slide-up panel remain open so

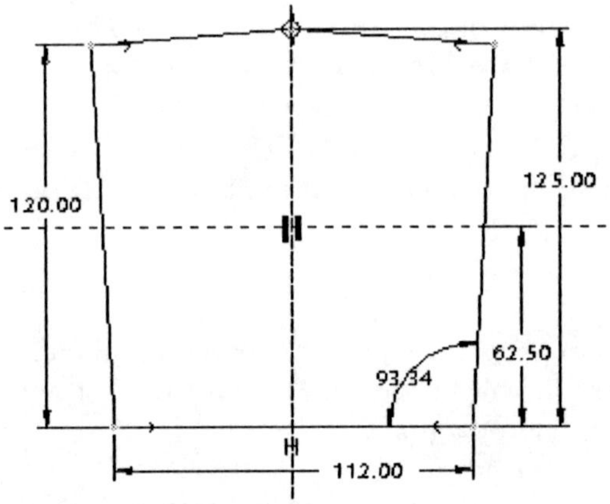

Figure 6-81 Sketch of the base feature with dimensions and constraints

Figure 6-82 Top view of the base feature

that you can view the selections you make on the model.

3. Select one of the four vertical edges on the corners. To select the second edge and the subsequent edges you need to press the CTRL key and then select the edge.

   The preview of the round is created on the edges that you have selected. A default value of the radius is also displayed.

4. Double-click on the default radius value that is displayed on the preview of the round. In the edit box that appears, enter a value of **15** and press ENTER. The set of rounds is created.

5. Choose the **Build feature** button.

   The default trimetric view of the model after creating the rounds is shown in Figure 6-83.

Creating the Third Feature

The third feature is a cylindrical feature and will be created on the bottom face of the base feature.

1. Invoke the Extrude dashboard and using the **Section** dialog box select the bottom face of the base feature as the sketch plane.

2. Enter the sketcher environment and draw the sketch of the cylindrical feature.

3. Exit the sketcher environment and enter **4** as the depth of extrusion.

4. Exit the Extrude dashboard.

   The cylindrical feature is completed and is shown in Figure 6-84.

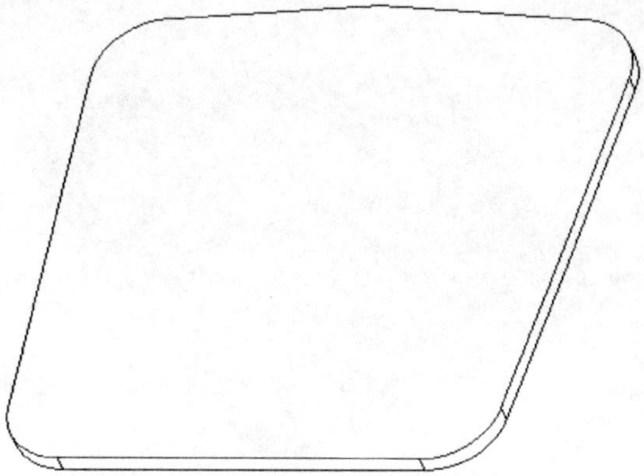

Figure 6-83 Model with the rounds

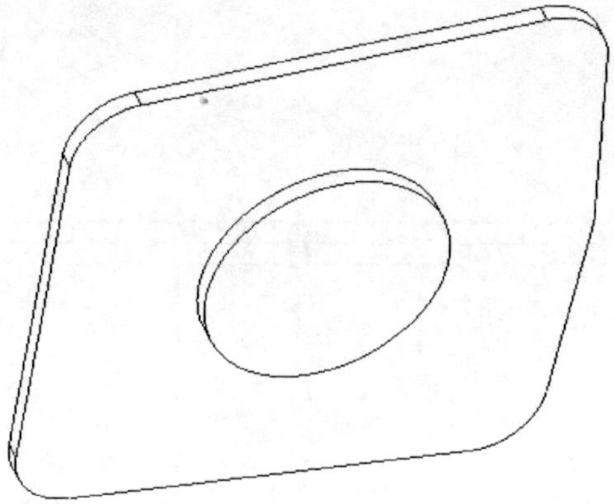

Figure 6-84 Model after creating the cylindrical feature at the bottom face

Creating the Fourth Feature

The fourth feature is a revolve cut that will be created on the datum plane passing through the center of the cylindrical feature.

1. Invoke the Revolve dashboard and select the **Remove Material** button.

2. Select the **FRONT** datum plane as the sketch plane.

3. Select the **RIGHT** datum plane and then select the **Right** option from the **Orientation** drop-down list.

4. Once you enter the sketcher environment, create the sketch of the revolve cut feature and apply constraints and dimensions as shown in Figure 6-85.

5. After the sketch is complete, exit the sketcher environment.

   The Extrude dashboard is enabled and appears below the graphics window. The value for angle of revolution, **360**, is selected by default.

6. Choose **Build feature** from the dashboard.

   The revolve cut feature is created and is shown in Figure 6-86.

Creating the Fifth Feature

The fifth feature is the fin and its pattern. The sketch of the fin will be created on the **FRONT** datum plane. The pattern of the fin feature will be created by controlling the dimensions of the sketch and constraining it.

This type of pattern can only be created if dimensions of the feature that is to be patterned is proper. In the sketch of the feature, a construction entity must exist. This construction entity

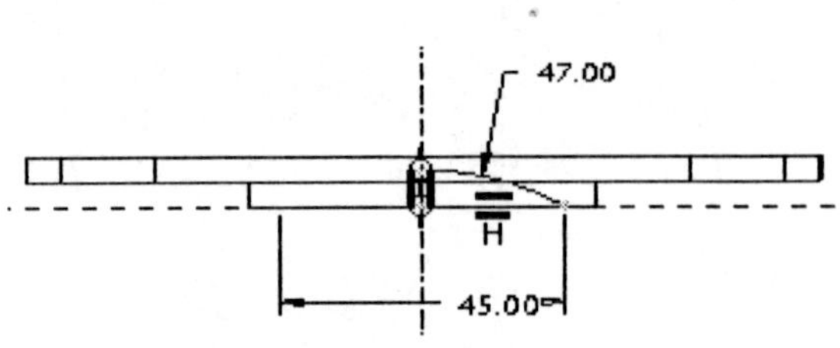

Figure 6-85 Sketch of the revolve feature with dimensions and constraints

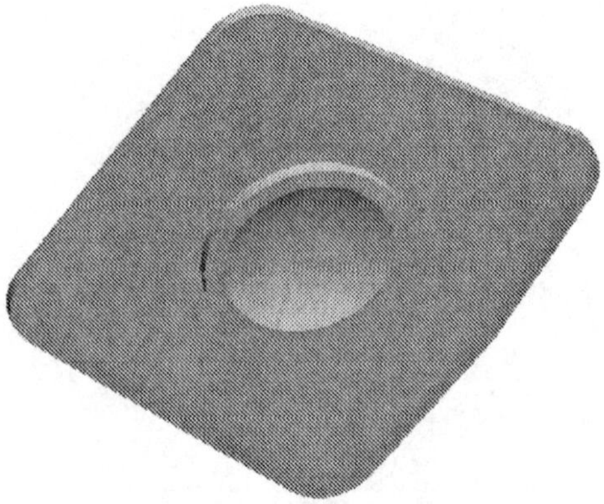

Figure 6-86 Revolve cut feature

can be a circle, an arc, or a line. The construction entity is used as the driving entity for the pattern that will be created later. The pattern instances are created in such a way that they follow the geometry of the construction entity.

1. Invoke the Extrude dashboard and using the **Section** dialog box select the **FRONT** datum plane as the sketch plane.

2. Select the **RIGHT** datum plane and then select the **Right** option from the **Orientation** drop-down list.

3. Choose the **Sketch** button to enter the sketcher environment.

4. Once you enter the sketcher environment, draw the arc shown in Figure 6-87 by using the **Create an arc by picking its center and endpoints** button.

Note

In the sketch of the fin feature, you need to draw the sketch such that later you can control the size of the instances of the pattern. You will use the dimension 6.155 to create the pattern. Also, the vertical height of the fin feature is not specified because in that case the fin feature will not slide along the construction arc and will be locked with the top construction arc.

5. Select the arc and then hold down the right mouse button to invoke the shortcut menu. From the shortcut menu, choose the **Construction** option. The arc is converted to a construction arc. Create the sketch of the fin feature and apply constraints and dimensions as shown in Figure 6-87.

6. After the sketch is complete, exit the sketcher environment.

7. Select the **Extrude on both sides of sketch plane by half the specified depth value in each** button from the depth flyout that is present on the Extrude dashboard.

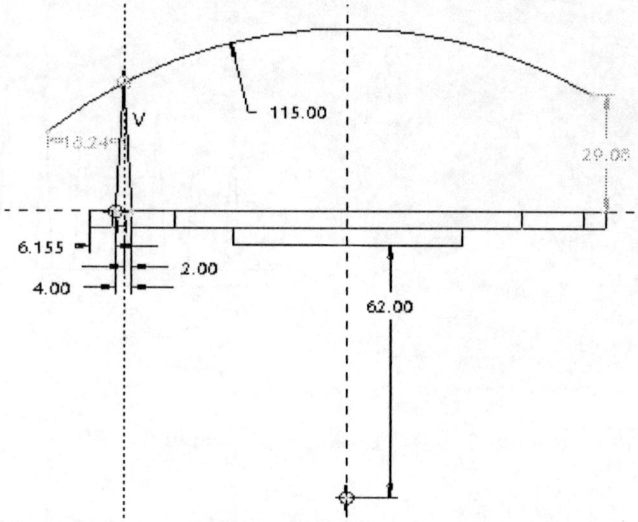

Figure 6-87 *Sketch for the fin feature with two weak dimensions*

8. Enter a value of **135** in the dimension box present on the dashboard.

9. Choose **Build feature** from the dashboard.

   The default trimetric view of the fin feature is shown in Figure 6-88. Now, you need to pattern the fin feature to create the remaining 12 fins.

10. Select the fin and invoke the Pattern dashboard. The dimensions of the fin feature are displayed on the graphics window.

11. Select the value **6.155**; the edit box appears. Enter a value of **9** in this edit box.

12. Enter a value of **13** in the dimension box present on the dashboard. These are the number of instances of the fin feature.

13. Choose the **Build feature** button to exit the pattern dashboard. The model after patterning the fin feature is shown in Figure 6-89.

Creating the Sixth Feature

The sixth feature is an extrude cut that will be created on the bottom face of the base feature. This cut will remove the fins that are projecting out of the base feature.

1. Invoke the Extrude dashboard and select the **Remove Material** button.

2. Select the bottom face of the base feature as the sketch plane.

3. Select the **RIGHT** datum plane and then select the **Right** option from the **Orientation** drop-down list.

Figure 6-88 *Fin created on the base feature*

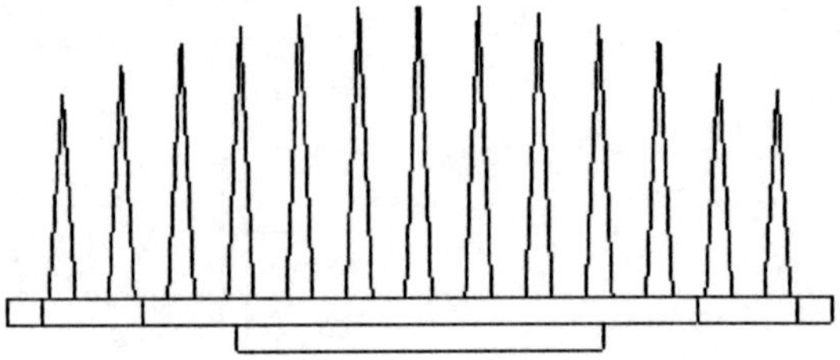

Figure 6-89 *Pattern of the fin feature*

4. Once you enter the sketcher environment, use the edge of the base feature to create the sketch that will be later extruded to create the cut.

5. After the sketch is complete, exit the sketcher environment.

   The Extrude dashboard is enabled and appears below the graphics window.

6. Select the **Extrude to intersect with all surface** button from the depth flyout.

7. Choose the **Change material direction of extrude to other side of the sketch** button to reverse the direction of yellow arrow. This yellow arrow points in the direction from where the material will be removed.

8. Choose the **Build feature** button from the dashboard. The default trimetric view of the model after the cut feature is created is shown in Figure 6-90.

Creating the Seventh Feature

The seventh feature is a is an extrude cut that will be created on the top face of the base feature.

1. Invoke the Extrude dashboard and select the **Remove Material** button.
2. Select the top face of the base feature as the sketch plane.

3. Select the **RIGHT** datum plane and then select the **Right** option from the **Orientation** drop-down list.

4. Once you enter the sketcher environment, draw the sketch of the circular cut feature as shown in Figure 6-91.

5. After the sketch is complete, exit the sketcher environment.

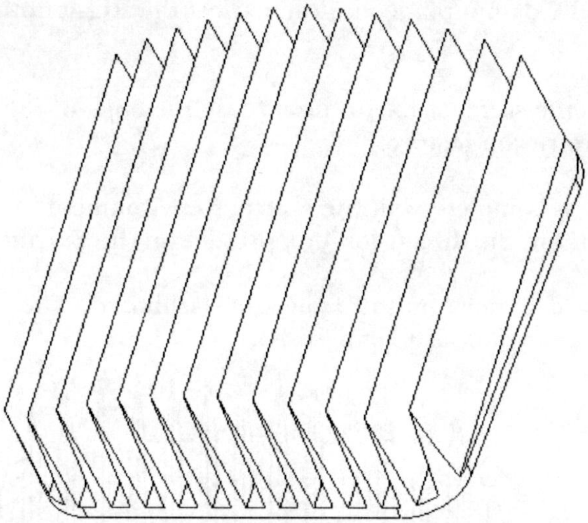

Figure 6-90 *Model after trimming the fins by creating the cut*

The Extrude dashboard is enabled and appears below the graphics window.

6. Select the **Extrude to intersect with all surface** button from the depth flyout.

7. Choose the **Build feature** button from the dashboard. The top view of the model after creating the circular cut is shown in Figure 6-92.

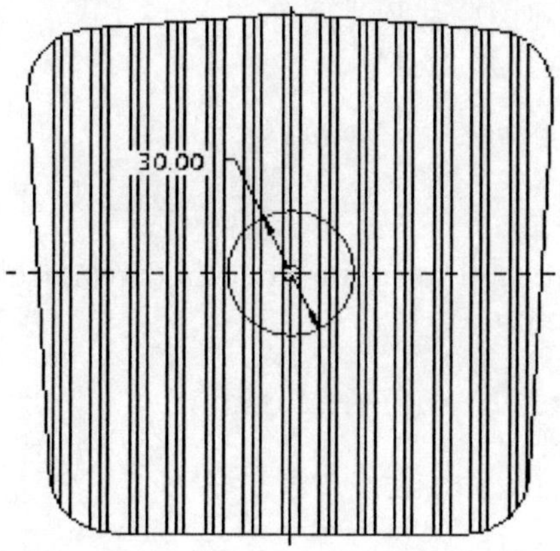

Figure 6-91 *Sketch of the cut feature*

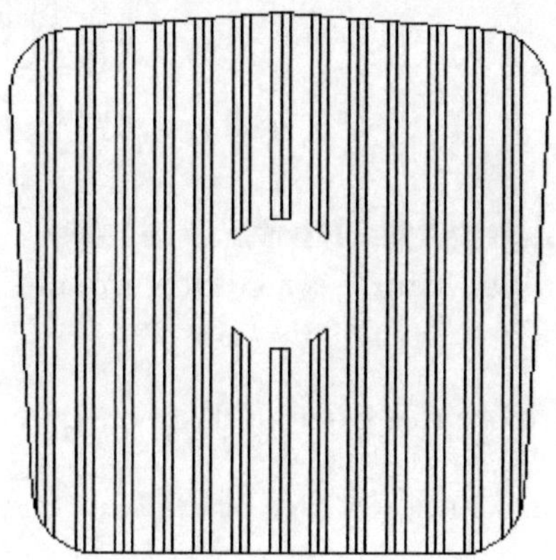

Figure 6-92 *Model after creating cut*

Creating the Eighth Feature

The eighth feature is a protrusion feature that will be created on the top face of the base feature.

1. Invoke the Extrude dashboard and select the top face of the base feature as the sketch plane.

2. Select the **RIGHT** datum plane and then select the **Right** option from the **Orientation** drop-down list.

3. Once you enter the sketcher environment, use the edge of the cut feature to create the sketch of the protrusion feature.

4. After the sketch is complete, exit the sketcher environment.
5. Enter a value of **9** in the dimension box present on the Extrude dashboard.

6. Choose the **Build feature** button from the dashboard. The model after creating the cylindrical feature is shown in Figure 6-93.

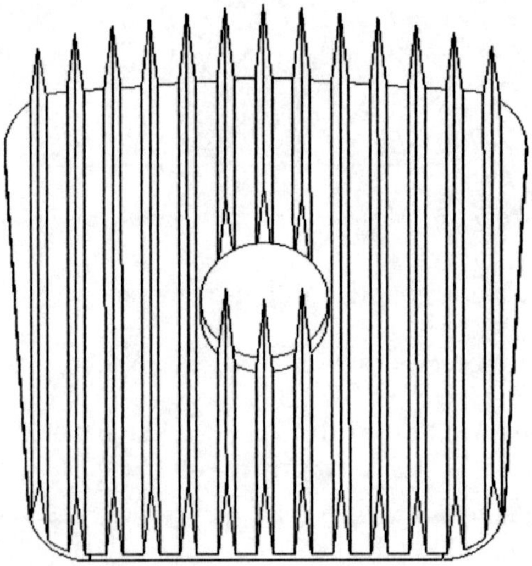

Figure 6-93 Model after creating the cylindrical feature

Creating the Ninth Feature

The ninth feature is a cut feature and it will be patterned to create the other instances. The cut feature is created on the top face of the base feature.

1. Invoke the Extrude dashboard and select the **Remove Material** button.

2. Select the top face of the base feature as the sketch plane.

3. Select the **RIGHT** datum plane and then select the **Right** option from the **Orientation** drop-down list.

4. Once you enter the sketcher environment, draw the sketch of the circular cut feature as shown in Figure 6-94.

5. After the sketch is complete, exit the sketcher environment.

   The Extrude dashboard is enabled and appears below the graphics window.

6. Select the **Extrude to intersect with all surface** button from the depth flyout.

7. Choose the **Build feature** button from the dashboard. The model after creating the circular cut is shown in Figure 6-95.

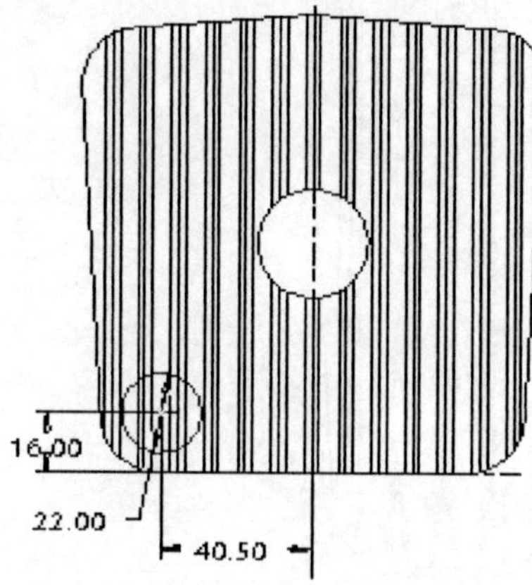

Figure 6-94 *Sketch of the cut feature*

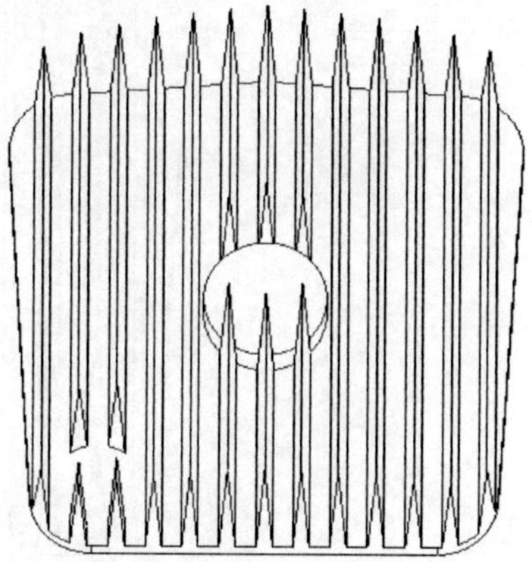

Figure 6-95 *Model after creating cut*

Now, the cut feature will be patterned to create its other instances.

8. Select the cut feature from the **Model Tree** and then choose the **Pattern Tool** button from the **Edit Features** toolbar. Select the **Dimension** option if it is not selected by default in the drop-down list.

   The Pattern dashboard is displayed. You are prompted to select the dimensions to vary in the first direction.

9. Select the **General** option from the **Options** slide-up panel if it is not selected.

10. Select the dimension value **40.5** from the graphics window. After you select the dimension in the first direction, the edit box is displayed.

11. Enter a value of **-81** in the edit box and press ENTER.

12. Hold down the right mouse button to display a shortcut menu. Choose the **Direction 2 Dimensions** option from the shortcut menu.

13. Select the dimension value **16**. After you select the dimension in the second direction, the edit box is displayed.

14. Enter a value of **80** in the edit box and press ENTER. Note that in the edit boxes present on the dashboard, the number of instances, **2**, is specified by default.

15. Choose the **Build feature** button from the Pattern dashboard.

The rectangular pattern of cut features is displayed as shown in Figure 6-96.

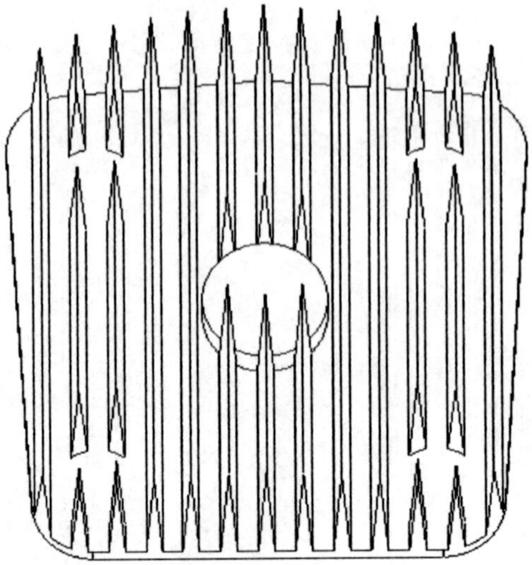

Figure 6-96 Model after patterning the cut feature

Creating the Tenth Feature

The tenth feature is a protrusion feature and it will be patterned to create the other instances. The protrusion feature is created on the top face of the base feature.

1. Invoke the Extrude dashboard and select the **Remove Material** button.

2. Select the top face of the base feature as the sketch plane.

3. Select the **RIGHT** datum plane and then select the **Right** option from the **Orientation** drop-down list.

4. Once you enter the sketcher environment, draw the sketch of the protrusion feature using the edges of the cut feature.

5. After the sketch is complete, exit the sketcher environment.

6. Enter a value of **9** in the dimension box present on the dashboard.

7. Choose the **Build feature** button from the dashboard. The model after creating the protrusion feature is shown in Figure 6-97.

   Now, a reference pattern of the protrusion feature will be created in order to create its other instances.

8. Select the protrusion feature from the **Model Tree** and then choose the **Pattern Tool** button from the **Edit Features** toolbar.

The **Reference** option is selected by default in the drop-down list present on the Pattern dashboard.

9. Select the **General** option from the **Options** slide-up panel.

10. Select the pattern of the cut feature from the **Model Tree** to create the reference pattern.

11. Choose **Build feature** from the Pattern dashboard. The model after creating the pattern is shown in Figure 6-98.

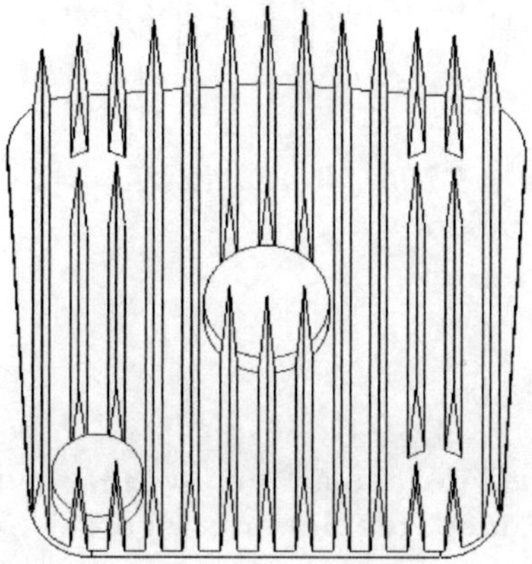

Figure 6-97 *Model after creating the cylindrical feature*

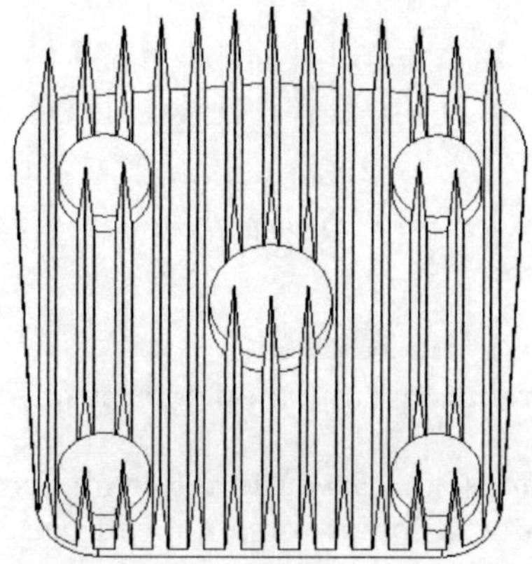

Figure 6-98 *Model after creating the reference pattern of the cylindrical feature*

Creating the Eleventh Feature

The eleventh feature is a hole that is created on the top face of the tenth feature and later it is reference patterned to create its other instances.

1. Create a hole of diameter **12** on the top face of the protrusion feature created on the lower left corner.

Note

Make sure that the primary reference while creating the hole is the top face of the protrusion feature and the secondary reference is the axis. If you take the axis as the primary reference, the pattern of holes may not be what is desired.

2. Pattern the feature using the **Reference** option. The model after creating the pattern of the hole feature is shown in Figure 6-99.

Creating the Last Feature

The last feature is a hole that is created on the top face of the eighth feature. This hole is coaxial with the cylindrical feature on which the hole is created. The model after creating the hole feature is shown in Figure 6-100.

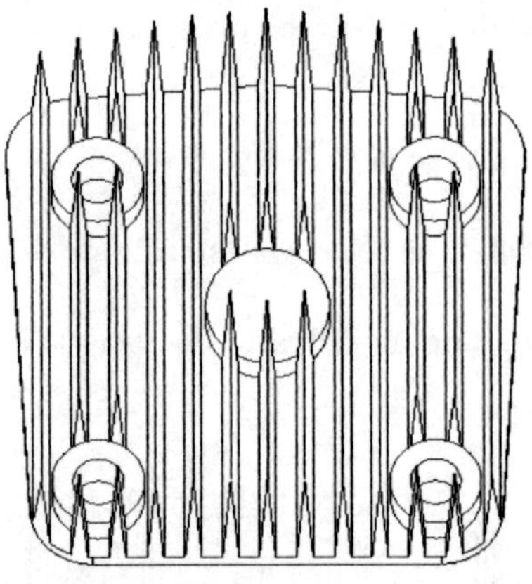

Figure 6-99 Model after creating the reference pattern of the hole feature

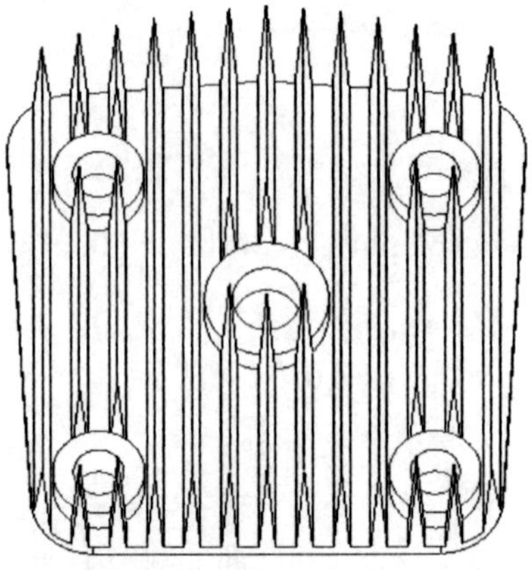

Figure 6-100 Model after creating the hole on the cylindrical feature

Saving the Model

You have to save the model because you may need it later.

1. Choose the **Save the active object** button from the **File** toolbar and save the model. The order of feature creation can be seen from the **Model Tree** shown in Figure 6-101.

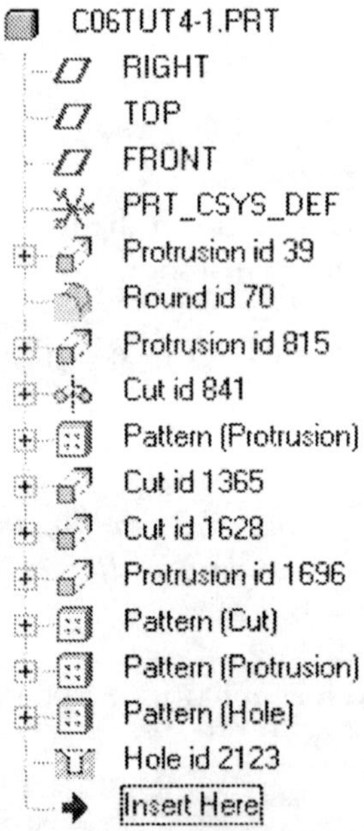

Figure 6-101 ***Model Tree*** *of Tutorial 4*

Creating a section view of the model

1. Choose **View > View Manager** from the menu bar. The **View Manager** dialog box is displayed.

2. Choose the **New** button from the dialog box. **Rep0001** is displayed in the **Names** display box.

3. Press ENTER. The dialog box disappears and then appears again.

4. Select **Rep0001** and then right-click. From the shortcut menu choose the **Redefine** option. The **EDIT METHOD** menu is displayed.

5. Choose the **Work Region** option from the **EDIT METHOD** menu. The **SOLID OPTS** submenu is displayed.

6. Choose **Extrude > Solid > Done** from the **SOLID OPTS** submenu. The Extrude dashboard is displayed.

7. Choose the **Create a section or redefine the existing section** button to invoke the **Section** dialog box.

8. Select the top face of the polygonal plate as the sketch plane and select the **RIGHT** datum plane to be at the right while drawing the sketch.

9. Choose the **Flip** button to reverse the direction of yellow arrow. You need to draw the sketch from the side of the sketch plane where the fins are facing. This is because when you draw the sketch, you have to make sure that the section lines are not cutting the fins.

10. When you enter the sketcher environment, draw the section lines as shown in Figure 6-102. This section is an open loop. The dimensions are not of importance to create this section.

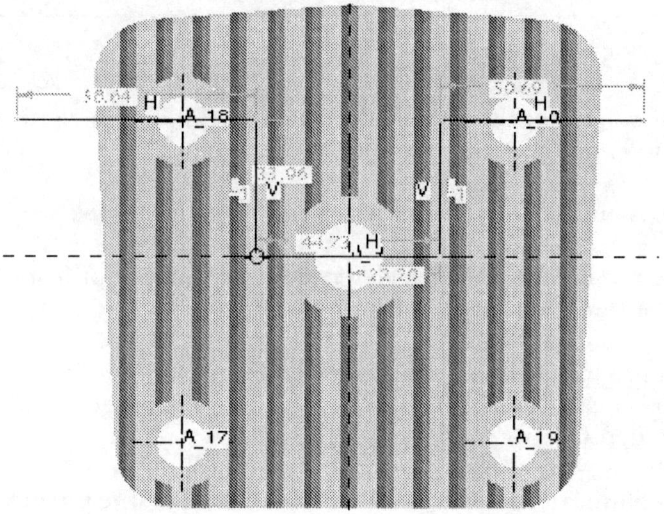

***Figure 6-102** Sketch of the section lines with weak dimensions*

11. Exit the sketcher environment and select the **Extrude to intersect with all surface**

button from the depth flyout. The yellow arrow is pointing in the reverse direction.

12. Choose the **Change material direction of extrude to other side of sketch** button to reverse the direction of arrow.

13. Choose the **Build feature** button from the Extrude dashboard. The sectioned model is shown in Figure 6-103.

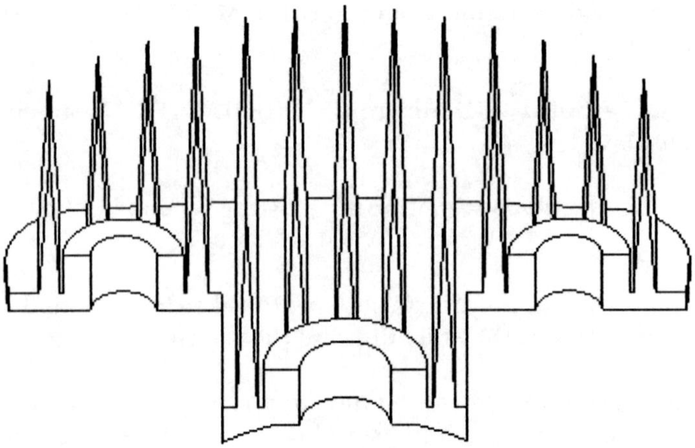

Figure 6-103 *Sectioned model*

14. Choose the **Done/Return** option from the **EDIT METHOD** menu.

Tutorial 5

In this tutorial you will create the model of ratchet wheel shown in Figure 6-104. Figure 6-105 shows the front view and the right-side view of the model. **(Expected time: 30 min)**

The following steps outline the procedure for creating this model:

a. First examine the model and then determine the number of features in it. The model is composed of three features, see Figure 6-104.

b. The base feature is a protrusion feature, see Figure 6-107. First the sketch of the base feature will be created on the **TOP** datum plane, see Figure 6-106, and then it will be extruded to a depth of 15.

c. The second feature is a cut feature that will be created on the top face of the base feature, see Figure 6-109. The datum plane on-the-fly will be created that will reference the sketch of the cut feature. Then the pattern of this cut feature will be created, see Figure 6-108.

d. The third feature is a cut that will create the keyway in the ratchet wheel, see Figure 6-112.

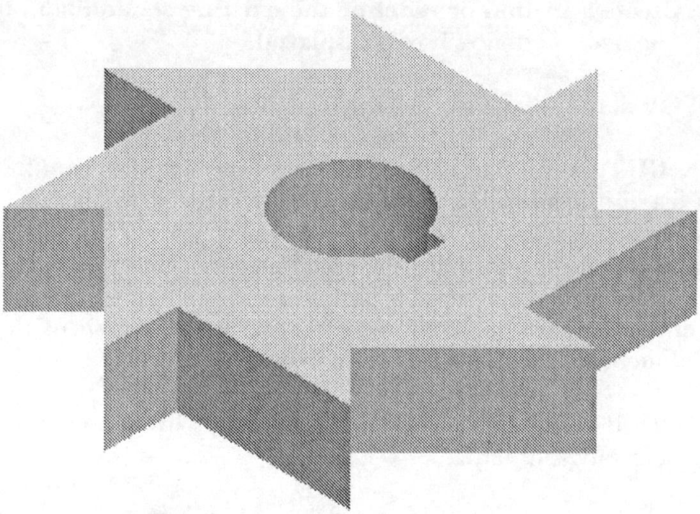

Figure 6-104 Solid model of ratchet wheel

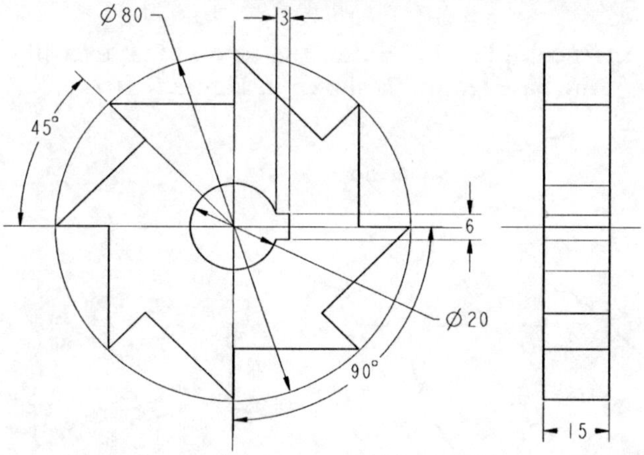

Figure 6-105 Front and right-side views of the ratchet wheel

Creating New Object File

1. Open a new part file and name it *c06tut5*. The three default datum planes are displayed on the graphics window.

Creating the Base Feature

In this model, you need to draw the sketch of the base feature on the **TOP** datum plane.

1. Choose the **Extrude Tool** button from the **Base Features** toolbar. The Extrude dashboard is displayed below the graphics window.

2. Choose the **Create a section or redefine the existing section** button from the Extrude dashboard. The **Section** dialog box is displayed.

3. Select the **TOP** datum plane as the sketch plane.

4. Select the **RIGHT** datum plane and then select the **Right** option from the **Orientation** drop-down list, if it is not selected automatically.

5. Choose the **Sketch** button to enter the sketcher environment.

6. Once you enter the sketcher environment, create the sketch of the base feature and modify the dimension as shown in Figure 6-106.

7. After the sketch is complete, choose the **Continue with the current section** button to exit the sketcher environment.

   The Extrude dashboard is enabled and appears below the graphics window.

8. Enter a depth of **15** in the dimension box that is present on the Extrude dashboard and press ENTER. Choose **Build feature** from the dashboard.

   The base feature is completed and now the second feature will be created. The default trimetric view of the base feature is shown in Figure 6-107.

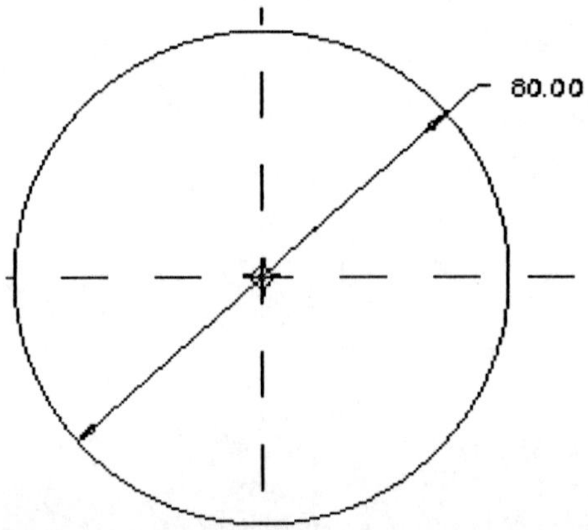

Figure 6-106 *Sketch of the base feature*

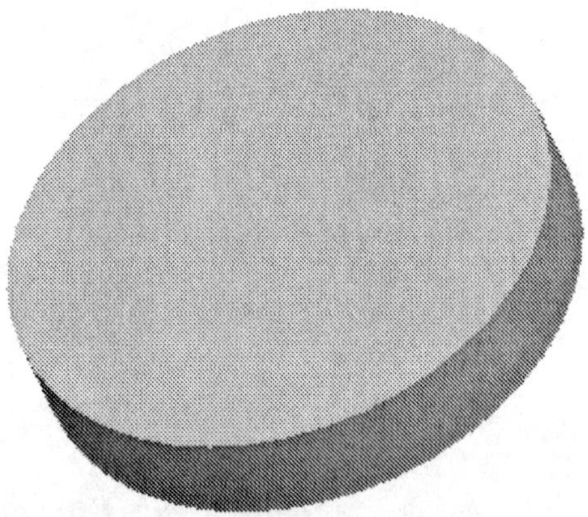

Figure 6-107 *Default view of the base feature*

Creating the Second Feature

The second feature is an extrude cut that will be created on the top face of the base feature. Then this feature will be patterned. As discussed earlier, an angular value is needed to create a rotational pattern. Therefore, a datum plane on-the-fly will be created that will be at an angle to the existing datum plane. But the sketch plane for the cut feature should be the top face of the base feature. Therefore, this datum plane that will be created on-the-fly will be selected as the reference plane for orienting the sketch plane.

1. Choose the **Extrude Tool** button from the **Base Features** toolbar.

2. Choose the **Remove Material** button from the dashboard.

3. Choose the **Create a section or redefine the existing section** button from the Extrude dashboard. The **Section** dialog box is displayed.

4. Select the top face of the base feature as the sketch plane. Drag the **Section** dialog box downwards if it is hiding the **Datum Plane Tool** button.

5. Choose the **Datum Plane Tool** button from the **Datum** toolbar. The **DATUM PLANE** dialog box.

6. Select the axis of the base feature and then use CTRL+left mouse button to select the **RIGHT** datum plane. When you select the **RIGHT** datum plane, the angular value is automatically applied to the new datum plane.

7. Modify this value to **90**.

8. Exit the dialog box. Notice, that in the **Reference** collector the **RIGHT** datum plane is automatically selected.

9. Right-click in the **Reference** collector to invoke the shortcut menu, choose the **Remove** option.

10. Select **DTM1** either from the graphics window or from the **Model Tree** to select as the reference for sketching. Choose the **Top** option from the **Orientation** drop-down list.

11. Choose the **Sketch** button from the **Section** dialog box to enter the sketcher environment.

12. As you enter the **References** dialog box is displayed. Select the axis of the base feature as the reference. Close the dialog box after selecting the axis.

13. Draw a vertical centerline passing through the center of the base feature and a centerline passing through the center at some angle as shown in Figure 6-108.

14. Draw the sketch of the cut feature as shown in Figure 6-108. Dimension the angle between the two centerlines. Align the endpoint of the horizontal line with the outer edge of the base feature and with the inclined centerline.

    In Figure 6-108, there are some weak dimensions that are of no use because the feature you are creating is a cut feature. This means the geometry of the part of the sketch, like the arc will lie outside and will not cut any material.

Note

*Before exiting the sketcher environment, invoke the **References** dialog box and check that no datum plane is selected as the reference. If any datum plane(s) are selected, delete them from the dialog box. This is because the feature will not be patterned if any datum plane is selected as the reference.*

15. After the sketch is complete, choose the **Continue with the current section** button to exit the sketcher environment.

    The Extrude dashboard is enabled and appears below the graphics window.

16. Choose the **Extrude to intersect with all surfaces** button from the depth flyout. Choose the **Build feature** button from the dashboard.

    The base feature is completed and now the second feature will be created. The default trimetric view of the base feature is shown in Figure 6-109.

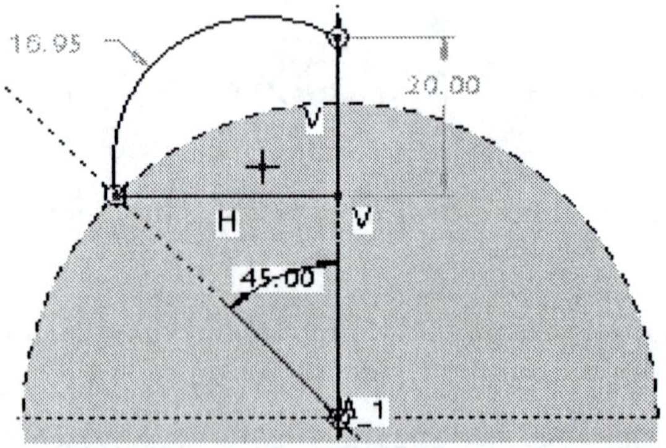

***Figure 6-108** Sketch of the cut feature*

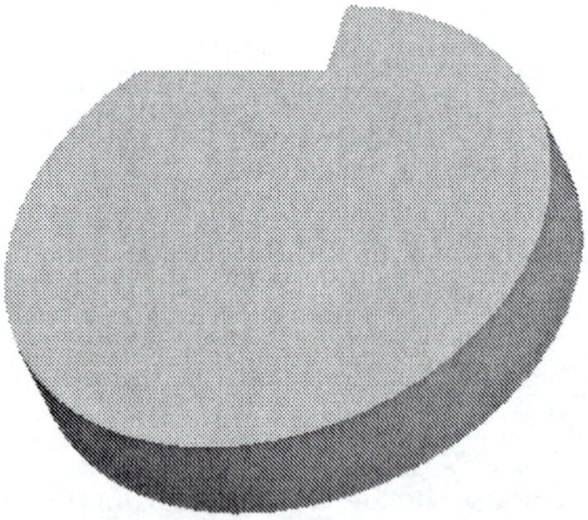

***Figure 6-109** Cut feature*

Creating the Pattern of the Cut Feature

To create the pattern of the cut feature, the grouped feature should be selected from the **Model Tree**.

1. Select the grouped feature from the **Model Tree**. Right-click to invoke the shortcut menu.

2. Choose the **Pattern** option from the shortcut menu. The Pattern dashboard is displayed. You are prompted to select the dimensions to vary in the first direction.

3. Select the dimension value **90** from the graphics window. After you select the dimension in the first direction, the edit box is displayed.

4. Enter a value of **45** in the edit box and press ENTER. Note that in the edit boxes present on the dashboard, the number of instances, **2**, is specified by default.

5. Enter **8** as the number of instances in the edit box present on the dashboard.

6. Choose the **Build feature** button from the Pattern dashboard.

   The rotational pattern of the cut feature is displayed as shown in Figure 6-110.

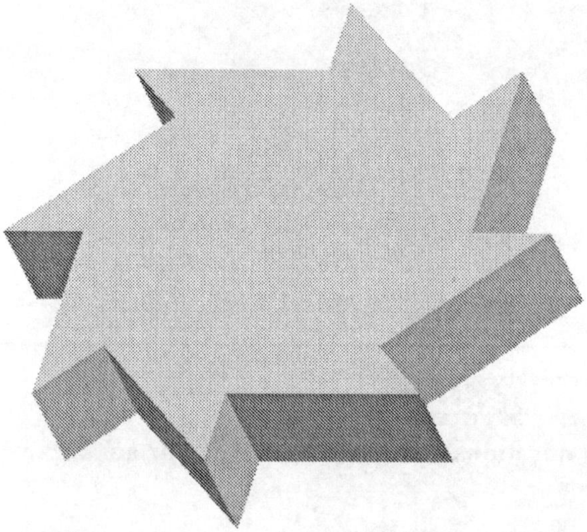

Figure 6-110 *Default view of the model after creating pattern*

Creating the Keyway

The keyway is the cut feature which will be created on the top face of the base feature.

1. Draw the sketch of the keyway on the top face of the base feature, apply dimensions and constraints as shown in Figure 6-111.

2. Extrude the sketch to create the keyway, see Figure 6-112.

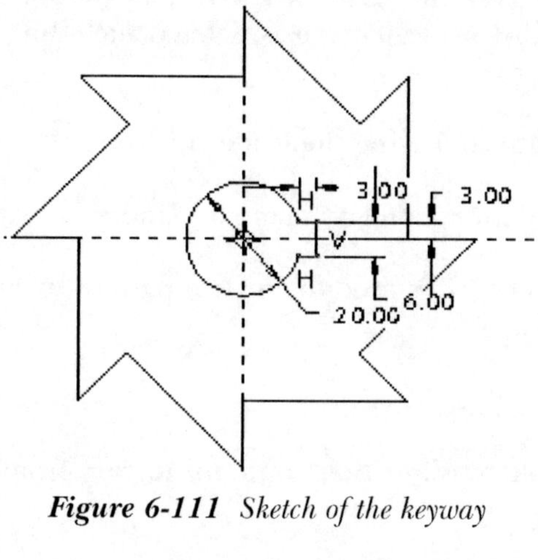

Figure 6-111 *Sketch of the keyway*

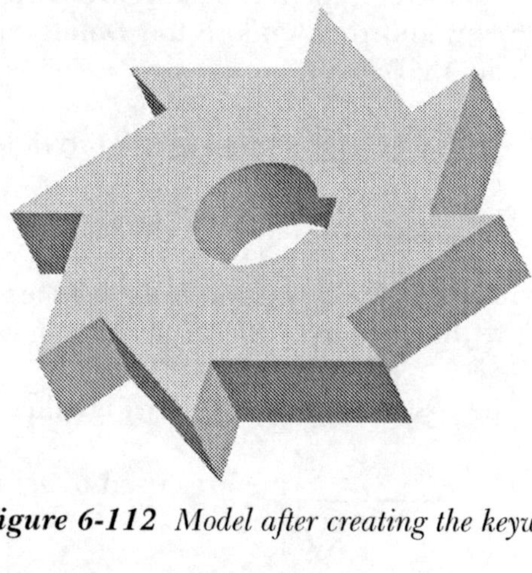

Figure 6-112 *Model after creating the keyway*

Saving the Model

You have to save the model because you may need it later.

1. Choose the **Save the active object** button from the **File** toolbar and save the model. The order of feature creation can be seen from the **Model Tree** shown in Figure 6-113.

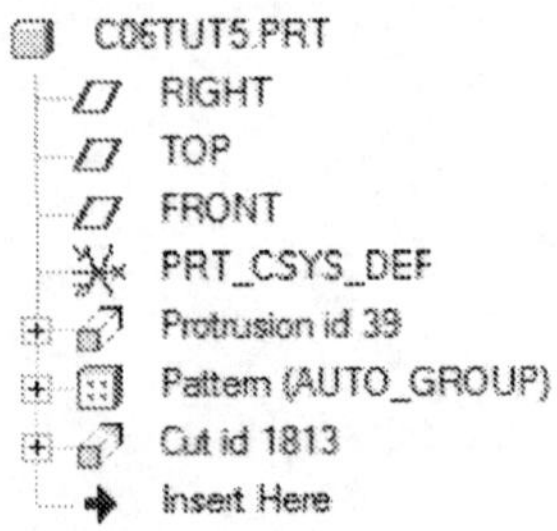

***Figure 6-113 Model Tree** of Tutorial 5*

Self-Evaluation Test

Answer the following questions and then compare your answers with the answers given at the end of this chapter.

1. When a pattern is created, the leader or the parent feature also becomes a part of the pattern. (T/F)

2. Once a pattern is created, all the instances in the pattern including the parent feature act as a single feature. (T/F)

3. In Reference pattern, an existing pattern is referenced to create a new pattern. (T/F)

4. Using the Pattern dashboard, you can create only linear patterns. (T/F)

5. If you select a pattern feature from the **Model Tree** and right-click to display the shortcut menu and then choose the **Delete** option, the whole pattern is deleted including the leader. (T/F)

6. The __________ option is used to delete a pattern leaving the leader feature.

7. The __________ option is used to mirror an entire geometry about a plane.

8. The __________ option from the Pattern dashboard is used to create a pattern in which the dimensions of the instances can be varied.

9. The first feature in a pattern is called __________.

10. The __________ menu is used to select the dimensions that you want to vary from the leader while copying.

Review Questions

Answer the following questions:

1. Which of the following options from the Pattern dashboard are used to create patterns in which the instances touch each other, and intersect with other instances or the edges of the surface?

   (a) **Identical** (b) **Variable**
   (c) **General** (d) None

2. Which of the following datums is required to create a rotational pattern?

   (a) Graph (b) Curve
   (c) Axis (d) Point

3. Which of the following options from the Pattern dashboard cannot be used to create a pattern that intersects an edge of a feature on which the pattern has to be created?

   (a) **Identical** (b) **Variable**
   (c) **General** (d) None

4. Which of the following options from the Pattern dashboard is used to create a pattern that has all instances of different size?

   (a) **Identical** (b) **Variable**
   (c) **General** (d) None

5. Which of the following options in the **COPY FEATURE** submenu creates a parent-child relationship between the copied feature and the source feature?

   (a) **FromDifModel** (b) **FromDifVers**
   (c) **Independent** (d) **Dependent**

6. You can mirror features using the datum planes or planar surfaces. (T/F)

7. To create a rotational pattern, you should specify an angular increment. (T/F)

8. The option **New Refs** is in the **COPY FEATURE** submenu. (T/F)

9. When you copy a feature using the **New Refs** option, then it is not related with the original feature. (T/F)

10. The features copied from different models or from different versions are always independent. (T/F)

Exercises

Exercise 1

Create the model shown in Figure 6-114. The top view, front view, right-side view, detailed, and sectioned views of the model are shown in Figure 6-115. **(Expected time: 45 min)**

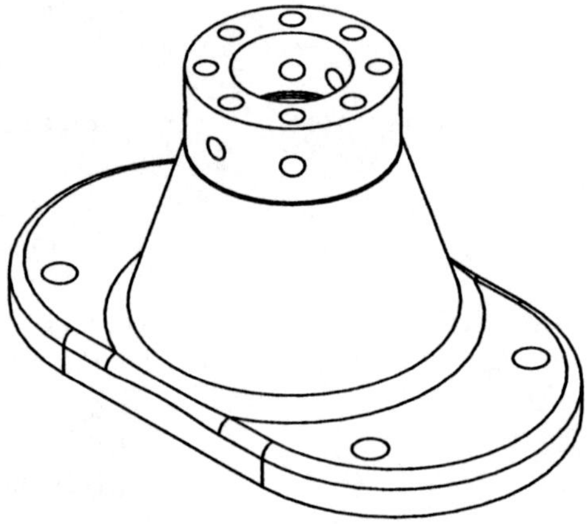

Figure 6-114 Solid model for Exercise 1

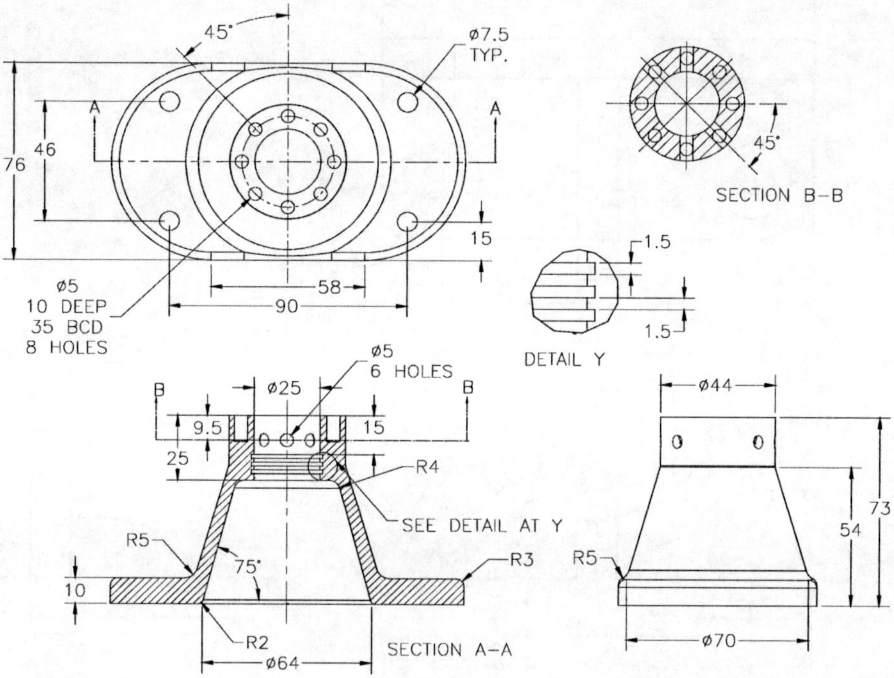

Figure 6-115 *Top, front, right-side, detailed, and sectioned views of the model*

Exercise 2

Create the model shown in Figure 6-116. The figure also shows the top view, front view, and the right-side view of the model. **(Expected time: 30 min)**

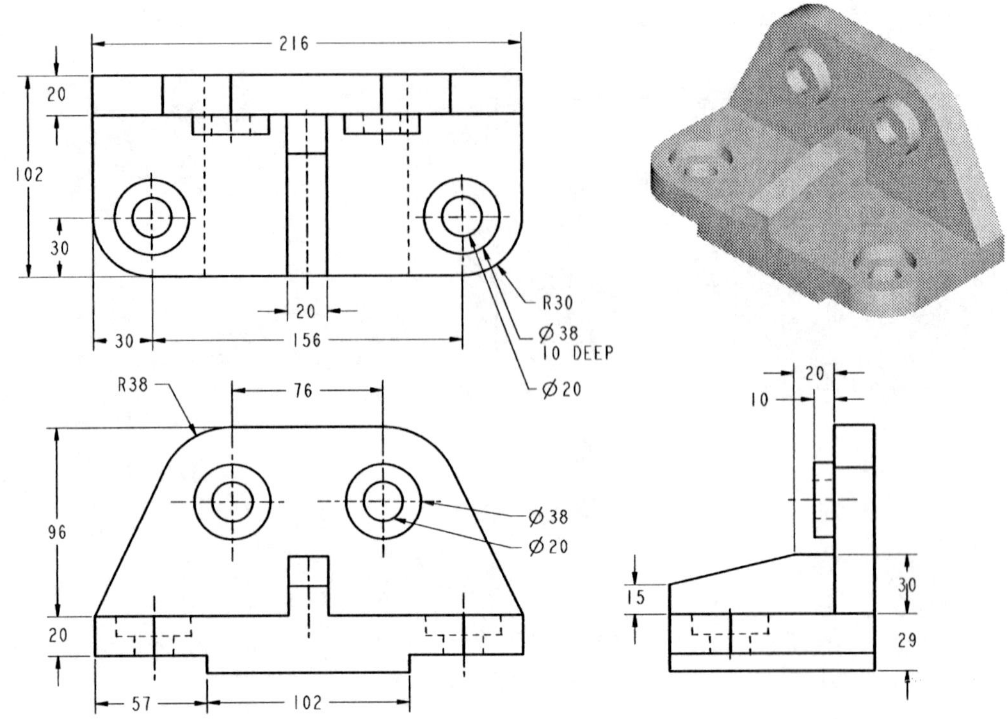

Figure 6-116 *Top, front, right-side, and isometric views of the*

Exercise 3

Create the model shown in Figure 6-117. The top view, front view, and the isometric view is shown in the figure. **(Expected time: 30 min)**

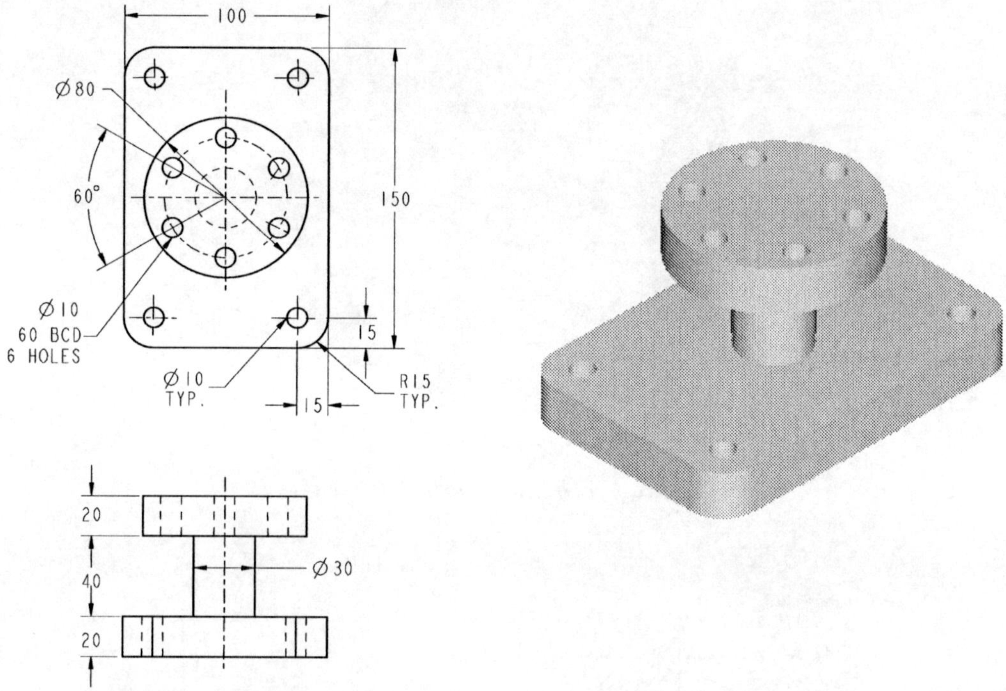

Figure 6-117 *Top view, front view, and the isometric view of the model*

Exercise 4

Create the model shown in Figure 6-118. The front and sectioned right-side views are shown in the Figure 6-119. **(Expected time: 1 hr)**

Note

To create the three protrusion features on the outer cylindrical surface, you will have to create a datum on-the-fly using the ***Datum Plane Tool*** *button to orient the sketch view. Also, when you enter the sketcher environment, you need to select the central axis of the model for defining the references using the* ***References*** *dialog box. As you select the central axis, the message in the* ***Reference status*** *area of the* ***References*** *dialog box will be* ***Fully Placed****. This suggests that you do not require any datum plane to define the references. Use this central axis to dimension the sketch. If you do not select the central axis to define the references using the* ***References*** *dialog box, the resultant feature may not get patterned.*

You will learn about the aligned section views and total aligned section views in Chapter 10.

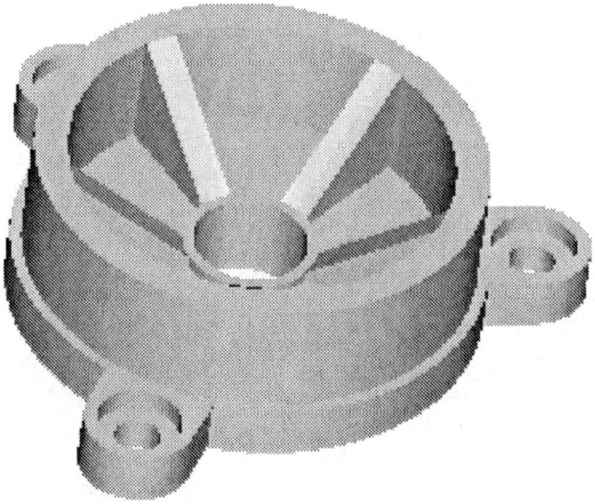

Figure 6-118 Solid model for Exercise 4

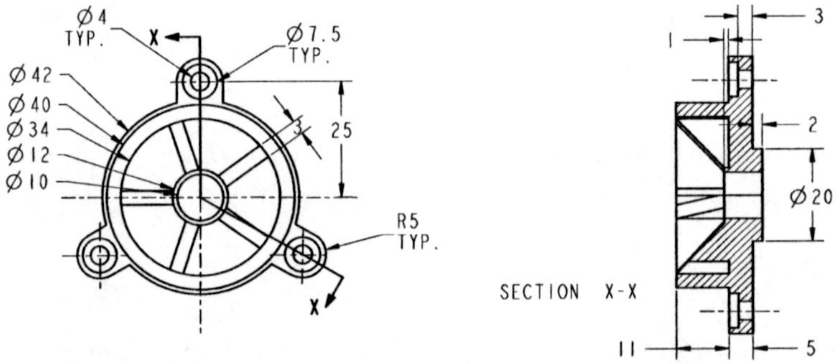

Figure 6-119 Top view and the total aligned sectioned view of the model

Answers to the Self-Evaluation Test

1 - T, **2** - T, **3** - T, **4** - F, **5** - T, **6** - **Delete Pattern**, **7**- **Mirror**, **8** - **Variable**, **9** - leader, **10** - **GP VAR DIMS**.

Chapter 7

Advanced Modeling Tools-I

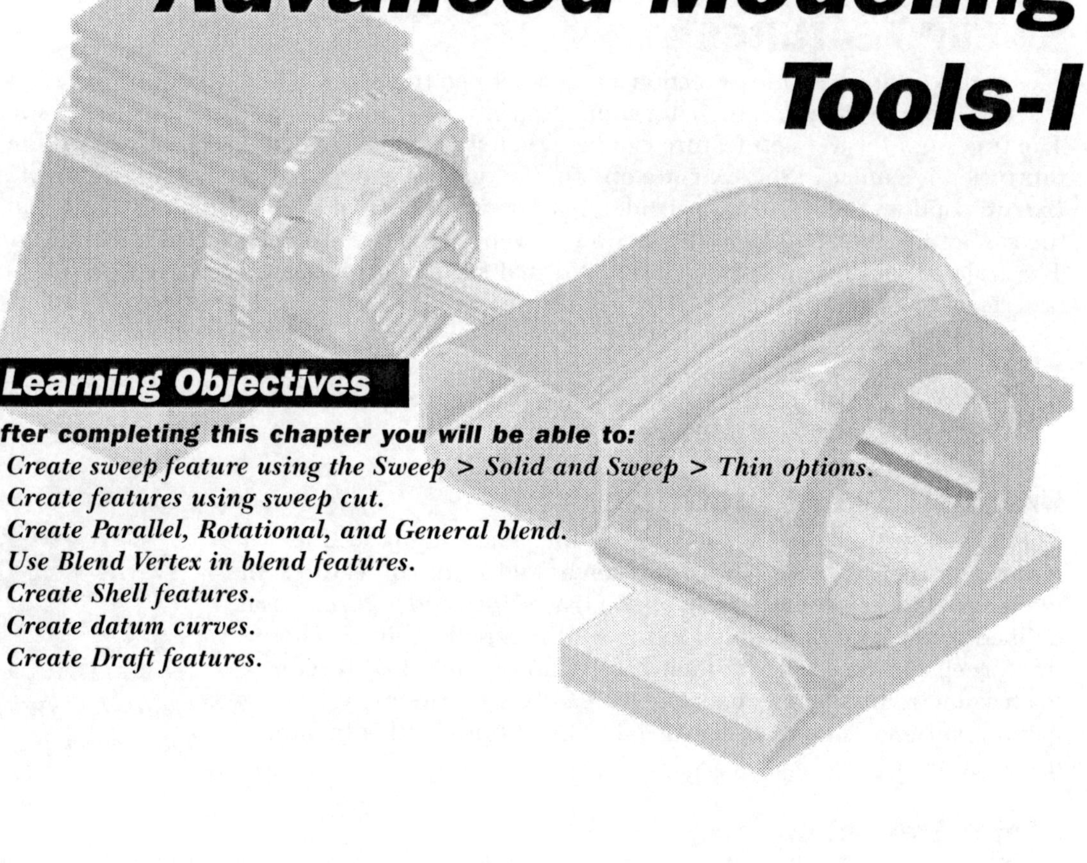

Learning Objectives

After completing this chapter you will be able to:

- *Create sweep feature using the Sweep > Solid and Sweep > Thin options.*
- *Create features using sweep cut.*
- *Create Parallel, Rotational, and General blend.*
- *Use Blend Vertex in blend features.*
- *Create Shell features.*
- *Create datum curves.*
- *Create Draft features.*

OTHER PROTRUSION OPTIONS

The **Extrude Tool** and **Revolve Tool** buttons available in the **Base Features** toolbar were discussed in Chapter 3, and the **Sweep** and **Blend** options are discussed in this chapter. As mentioned earlier, Protrusion and Cut are the two basic options available in Pro/ENGINEER that are used to create a feature.

Note

All the options that are available for creating a cut are similar to those that are available for creating a protrusion. Remember, that a cut is performed on an existing feature and therefore, the Cut option is available only when at least a base feature exists on the graphics window.

In this chapter you will also learn about the tools of solid modeling that make the creation of a complex model easy. In the next section you will learn about protrusion and cut sweeps.

SWEEP FEATURES

The **Sweep** option extrudes a section along a defined trajectory. The order of operation is to first create a trajectory and then a section. Trajectory is a path along which a section is swept. The trajectory for a sweep feature can be either sketched or selected. The **Sweep** option of protrusion is similar to the **Extrude** option, the only difference being that in the case of the **Extrude** option, the feature is extruded in a direction normal to the sketching plane, but in the case of the **Sweep** option, the section is swept along the sketched or selected trajectory. The trajectory can be open or closed. Normal sketching tools are used for sketching the trajectory. The cross section of the swept feature remains constant throughout the sweep.

Note

Some important points to remember while drawing a trajectory and a section for a sweep feature are discussed later in this chapter.

Creating Swept Protrusions

The **Sweep** option can be used for adding material as well as for removing material, that is, for protrusion as well as for cut features. In the **Insert > Sweep > Protrusion** option that is discussed here, material defined by the section is added in the specified path. You can choose the **Sweep** option from the menu bar. Choose **Insert > Sweep > Protrusion** from the menu bar. The **SWEEP TRAJ** (Sweep Trajectory) menu appears as shown in Figure 7-1. The options in this menu are discussed next.

***Figure 7-1** SWEEP TRAJ menu*

Sketch Traj option

The **Sketch Traj** (Sketch Trajectory) option is used when you want to sketch the trajectory for the sweep feature. This is the most commonly used option for defining the trajectory. As mentioned earlier, the trajectory can be open or closed. There are some limitations for using closed or open trajectory with closed or open section. These limitations are discussed in the next section. When you choose the **Sketch Traj** option, you are prompted to select a sketching plane. The sketching plane you select will be parallel to the screen when you sketch the trajectory. Figure 7-2 shows how the section is swept along the sketched trajectory, and Figure 7-3 shows the shaded image of the swept feature.

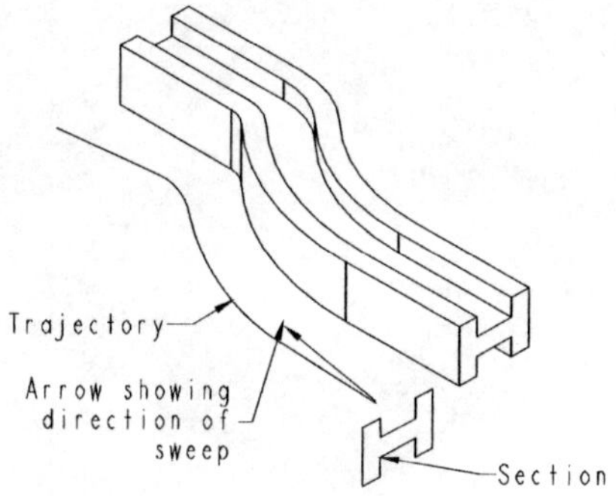

Figure 7-2 *Sweep along the sketched trajectory*

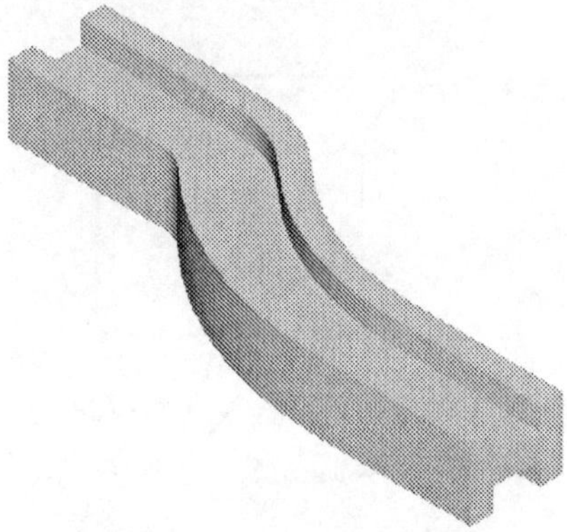

Figure 7-3 *Shaded image*

The following points specify the combinations of trajectories and sections that are possible/not possible to create.

1. Open section and open trajectory are not possible.
2. Closed section and open trajectory are possible.
3. If the sketched trajectory is a closed loop then after you exit the sketcher environment, the **ATTRIBUTES** menu is displayed as shown in Figure 7-4. There are two options that are available: **Add Inn Fcs** (Add inner faces) and **No Inn Fcs** (No inner faces).

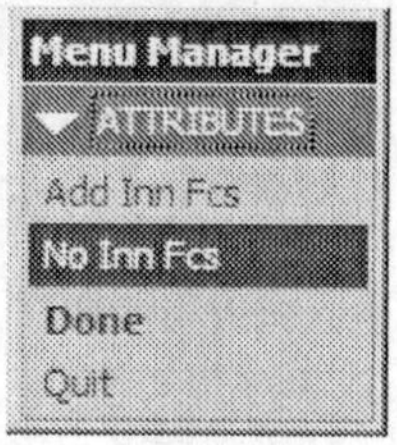

Figure 7-4 ATTRIBUTES menu

For **Add Inn Fcs**, only open sections are possible, as shown in Figure 7-5. The shaded image of the corresponding swept feature is shown in Figure 7-6. The two figures below explain the **Add Inn Fcs** option.

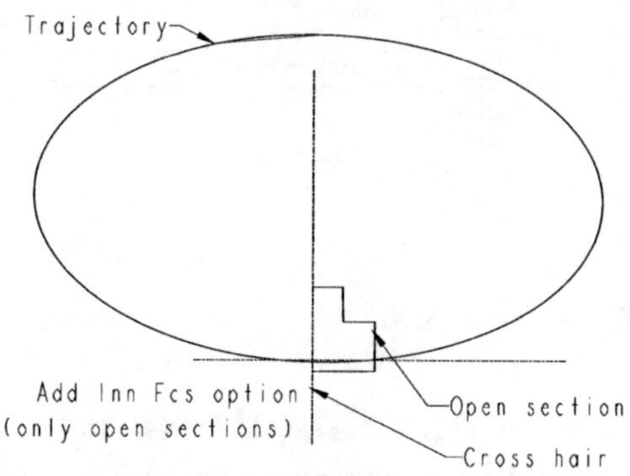

Figure 7-5 Add Inn Fcs *option*

Figure 7-6 *Shaded image*

For **No Inn Fcs**, only closed sections are possible, as shown in Figure 7-7. The shaded image of the corresponding swept feature is shown in Figure 7-8.

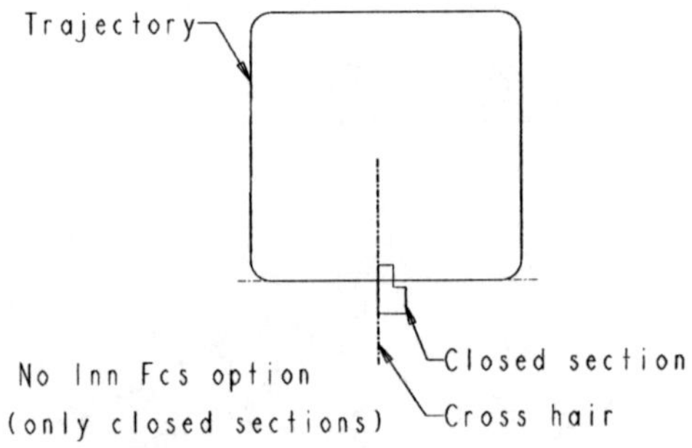

Figure 7-7 No Inn Fcs option

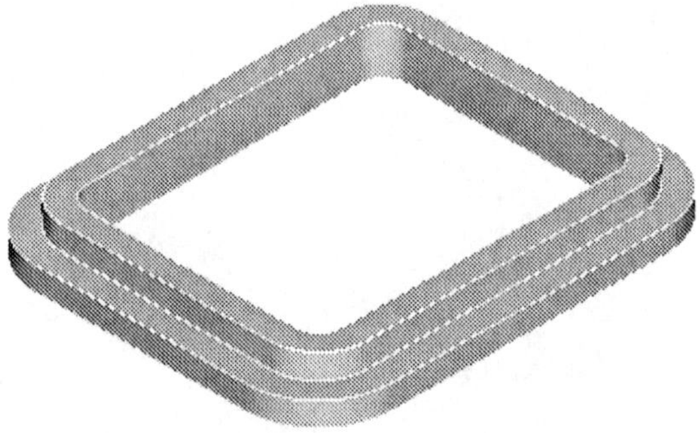

Figure 7-8 Swept feature

Select Traj option

The **Select Traj** (Select Trajectory) option allows you to select a trajectory on the graphics screen. The trajectory to be selected can be an existing edge or a datum curve. Creation of datum curves will be discussed later in the chapter. When you choose this option from the **SWEEP TRAJ** menu, the **CHAIN** menu is displayed. The **CHAIN** menu is also discussed later in this chapter.

Figures 7-9 and 7-10 show two examples of selecting the edges of the base feature and then using these as a trajectory to sweep and their corresponding swept features.

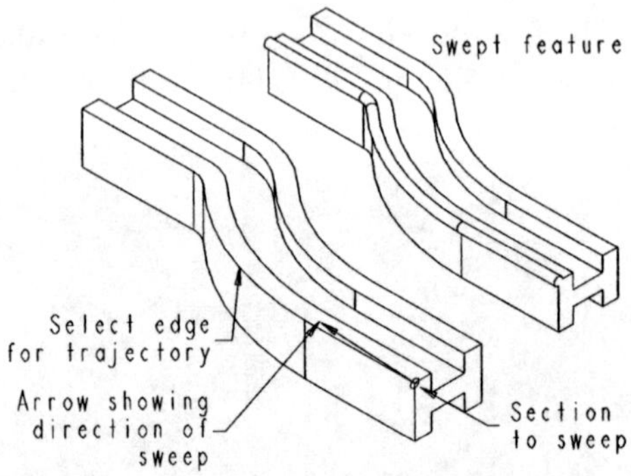

Figure 7-9 Sweep along the selected trajectory

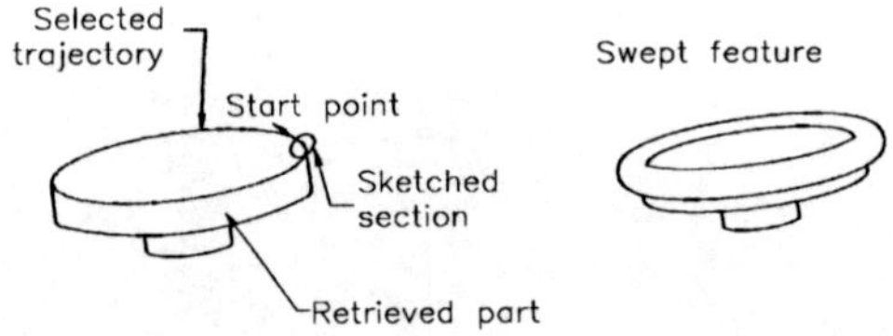

Figure 7-10 Sweep along the selected trajectory

Sketching a Trajectory Aligned to an Existing Geometry

When one end of the sketched trajectory is aligned to the adjacent geometry of the existing feature, Pro/ENGINEER provides two options. The first option is to merge the ends of the sweep feature with the adjacent geometry and the second option is to leave the ends of the sweep feature free. These options are available in the **ATTRIBUTES** menu that is displayed after you complete the sketch of the trajectory and choose the **Continue with the current section** button. The **ATTRIBUTES** menu is shown in Figure 7-11.

Figure 7-11 ATTRIBUTES menu

The **ATTRIBUTES** menu is displayed only when the trajectory is aligned to an edge or surface of the feature that already exists on the graphics screen. This means that the **ATTRIBUTES** menu does not appear if the sweep feature you are drawing is the base feature of a model or, in other words, if there is no adjacent geometry to which the trajectory can be merged. The options available in the **ATTRIBUTES** menu are discussed next.

Merge Ends

The **Merge Ends** option merges the end of a sweep feature to the surface to which the end of the trajectory is aligned. For this option the trajectory should be aligned to the adjacent geometry.

Free Ends

The **Free Ends** option leaves the sweep feature partially attached to the adjacent feature even if the end of the trajectory is aligned to the adjacent geometry.

Figure 7-12 show the **Merge Ends** and **Free Ends** options. In the figure shown below, the trajectory is aligned with the adjacent geometry in both the cases. Figure 7-13 shows the corresponding shaded image.

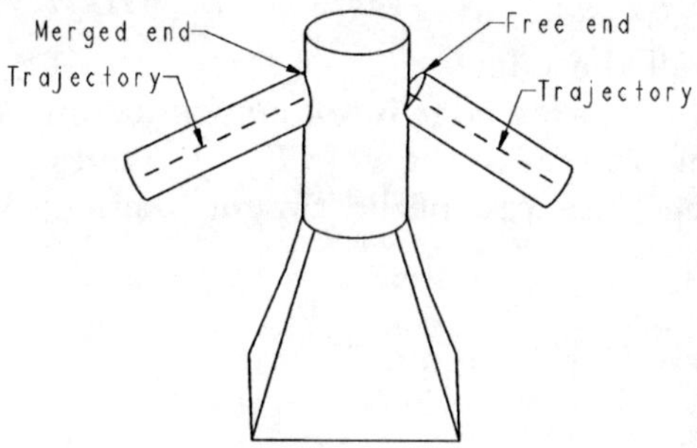

Figure 7-12 Merge Ends and Free Ends

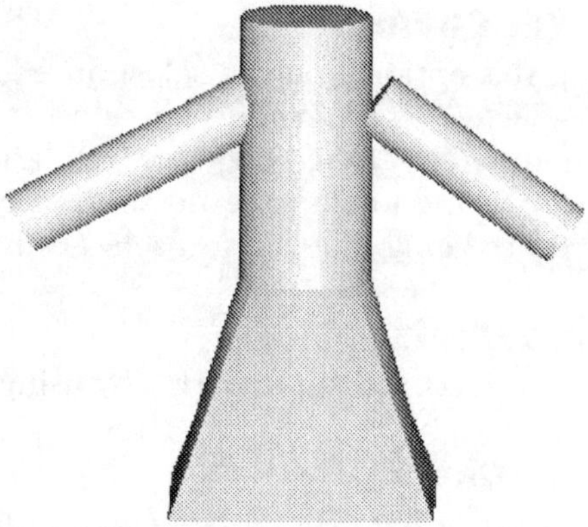

Figure 7-13 Shaded image

Creating Sweep Feature by Selecting a Trajectory

When you choose the **Select Traj** option from the **SWEEP TRAJ** menu, the **CHAIN** menu is displayed as shown in Figure 7-14. You can use the **CHAIN** menu only if you have created a feature that will be used to select the trajectory. The options in this menu are used to select a

Tip: *The following points should be remembered while creating a sweep feature:*

1. *Similar to other sketched features the trajectory of the sweep feature is also sketched after selecting a sketching plane.*

2. *The section for the sweep trajectory is sketched using the normal sketcher tools when the sketch trajectory option is selected.*

3. *At bends in a trajectory, the radius of the bend should be proportionate to the cross section to be swept to avoid overlapping. If the section size is large and the radius of the curve or bend is small, overlapping takes place and the sweep feature will not be created. Therefore, make sure that the ratio of the size of the section to the size of the trajectory is appropriate.*

trajectory. These options are discussed next.

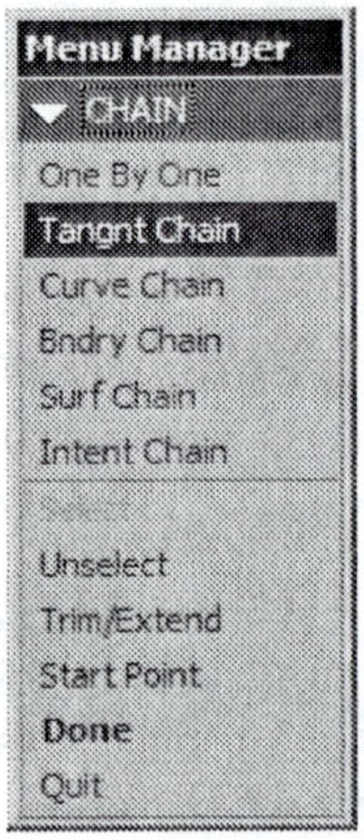

***Figure 7-14 CHAIN** menu*

One By One

The **One By One** option of the **CHAIN** menu is selected by default. Using this option you can select an edge or curve individually, one by one. When you select an edge using the left mouse button, the color of the selected edge changes to blue. Before selecting the edge, make sure that the **Select** option in the **CHAIN** menu is highlighted. The edge once selected and confirmed by choosing **Done Sel** from the **GET SELECT** menu or by using the middle mouse button can also be unselected by selecting the **Unselect** option from the **CHAIN** menu.

Tangnt Chain

Using this option, you can select an edge or edges tangent to the selected edge. When you select an edge, all the edges tangent to the selected edge are highlighted. If the selected edge is not tangent to any other edge then the function of this option is the same as that of the **One By One** option, the difference being that you can select only one edge in the case of the **Tangnt Chain** option.

Curve Chain

You can select a chain of curves by using the **Curve Chain** option.

Bndry Chain

The **Bndry Chain** (Boundary Chain) option is used only for surface features. You can define a chain by selecting a quilt and using its one-sided edges. If the quilt has more than one loop, select a specific loop to define the chain. When you select the edge of the quilt, it is highlighted in blue and the **CHOOSE** menu is displayed with the options as shown in Figure 7-15.

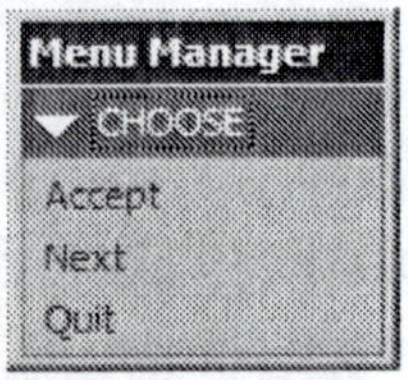

***Figure 7-15 CHOOSE** menu*

Surf Chain

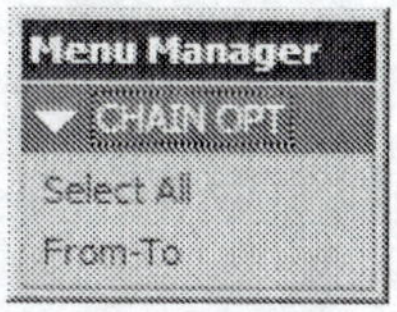

Figure 7-16 *CHAIN OPT menu*

Using the **Surf Chain** option, you can define a chain by selecting a surface and using its edges. If the surface has more than one loop, then you are prompted to specify a loop to define the chain. When a surface is selected, the **CHAIN OPT** menu is displayed as shown in Figure 7-16. Choose either **Select All** or **From-To** from the **CHAIN OPT** menu.

Intent Chain

The **Intent Chain** option is used to select multiple edges. When a section is extruded, the edges formed by the extrusion consists of intent chains. The intent chains can either be the edges of the section or the edges of the extruded surface. The edges selected should form a closed loop.

Creating thin Sweep Protrusion

The **Sweep > Thin Protrusion** option creates a thin sweep feature with a specified thickness. This option is similar to the **Extrude > Thin** option that was discussed in Chapter 3. In the case of thin features, a certain thickness has to be specified. The thickness is specified on one side of the section or symmetrically to both the sides of the section. The resultant sweep is similar to the solid sweep created with a section comprising of two closed loops at some offset distance. Figure 7-17 shows the sections that can be used to create the model shown in Figure 7-18.

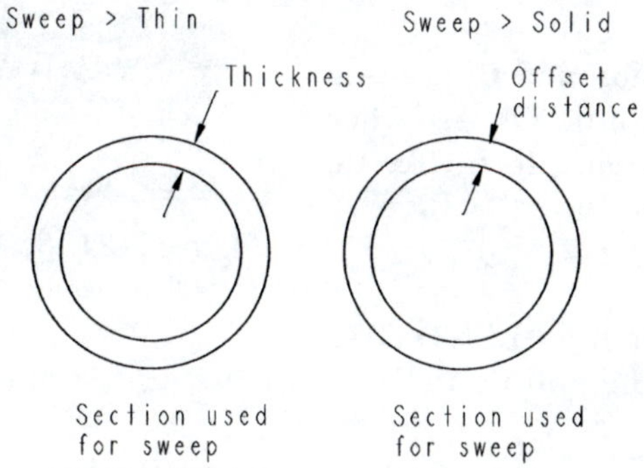

Figure 7-17 *Two possible options to create the same model*

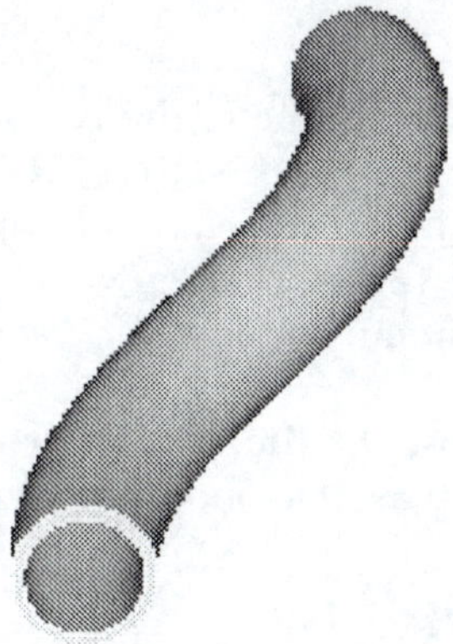

Figure 7-18 *Model created using the same sections*

Creating a Sweep Cut

To create a **Sweep Cut** feature, the procedure to be followed is the same as that in **Sweep Protrusion**. The only difference is that in case of cut features, the material is removed from an existing feature. The **Cut** option can be invoked by choosing **Insert > Sweep > Cut** from the menu bar. Cut can be a solid swept cut or thin swept cut. Figure 7-19 shows trajectories for the **Sweep Cut** feature. Figure 7-20 shows the shaded model of an open trajectory sweep cut and a closed trajectory sweep cut.

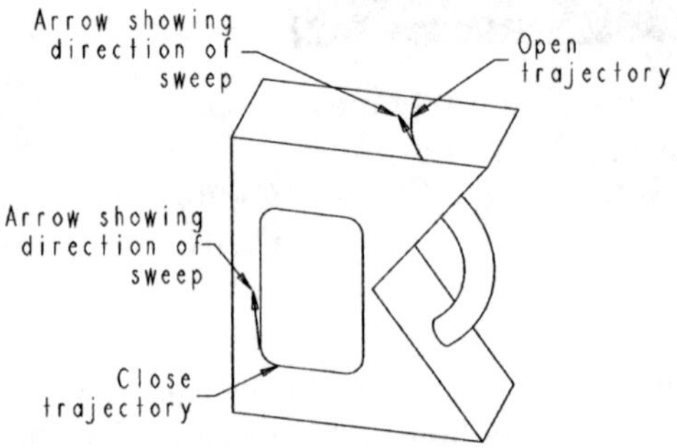

Figure 7-19 *Trajectories for* ***Sweep Cut***

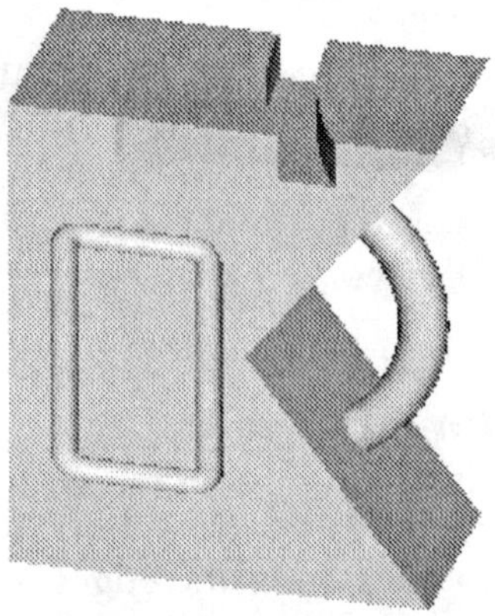

Figure 7-20 *Shaded model with sweep cuts*

BLEND FEATURES

Blend features are composed of two or more sections that are joined through transitional faces at their edges so as to form a continuous feature. The number of entities in each section that creates the blend feature should be the same. For example, you cannot blend a circle with a rectangle. This is because a rectangle is composed of four entities and a circle of one. It can be acheived only if the circle is divided into four entities.

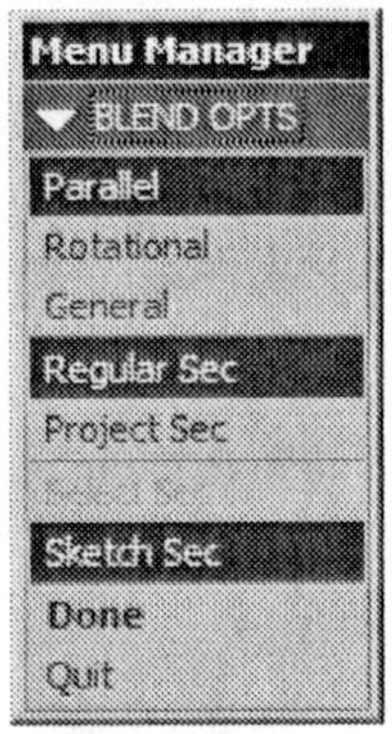

Figure 7-21 BLEND OPTS *submenu*

The **Blend** option is one of the types of **Protrusion** and **Cut** options that are available in Pro/ENGINEER. The **Blend** option is used where the feature to be created has varying cross sections. To invoke this option, choose **Insert > Blend > Protrusion** or **Insert > Blend > Cut** from the menu bar.

When you choose the **Blend > Protrusion** option, the **BLEND OPTS** submenu is displayed as shown in Figure 7-21. The options in this menu are discussed next.

Parallel Option

Parallel blends have sections that are drawn parallel to each other and a distance is defined between the parallel sections.

After choosing **Parallel > Regular Sec > Sketch Sec > Done** from the **BLEND OPTS** submenu, the **ATTRIBUTES** menu is displayed as shown in Figure 7-22. The options in this menu are discussed next.

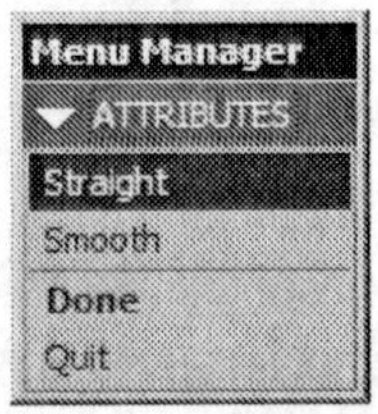

Figure 7-22 ATTRIBUTES *menu*

Straight option

The **Straight** option is used to connect the vertices of all sections in a blend feature with straight lines.

Smooth option

The **Smooth** option is used to connect the vertices of all sections in a blend feature with curves.

Figure 7-23 shows three sections that are used to create the blend feature. Figures 7-24 and 7-25 show the parallel blend features with straight edges and smooth edges respectively created using the sections shown in Figure 7-23.

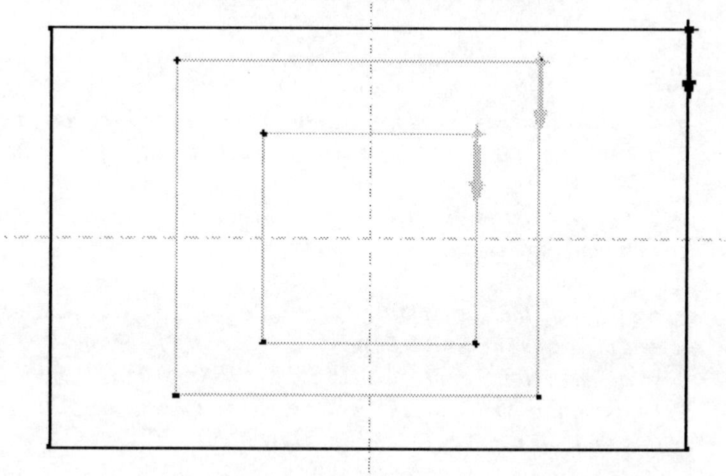

***Figure 7-23** Parallel sections*

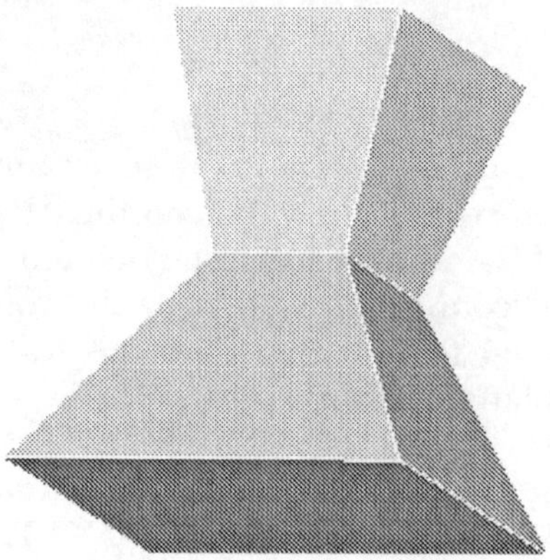

***Figure 7-24** Parallel blend with straight edges*

***Figure 7-25** Parallel blend with smooth edges*

Note

While drawing a section, the start point of all the sections should be in the same direction in order to avoid twisted blend features.

Tip: *The following points should be remembered while creating a* ***Parallel*** *blend feature:*

1. *After completing first section, choose* ***Sketch > Feature Tools > Toggle Section*** *to proceed for drawing the second section. You can also hold down the right mouse button and choose the* ***Toggle Section*** *option from the shortcut menu that appears. If you choose the* ***Continue with the current section*** *button before drawing the second section, the system prompts you to use the* ***Toggle Section*** *option to continue with the second section.*

2. *Active section appears in cyan color and the other sections appears in gray.*

3. *All the sections in a blend feature must have the same number of entities. However, you can blend a point with any section irrespective of the number of entities.*

4. *System prompts for depth between subsequent sections after completion of all the sections in the blend.*

5. *By default, the start point of any entity that is drawn to define a section is considered as the* ***Start Point*** *of the section. To change the* ***Start Point*** *of a section, select the point to be defined as the* ***Start Point*** *and hold down the right mouse button to display the shortcut menu. Choose the* ***Start Point*** *option from this shortcut menu to change the start point.*

6. *After defining the sections in a blend feature and before choosing the* ***Continue with the current section*** *button, the sections can be modified by using the* ***Toggle Section*** *option.*

Rotational Blend

Rotational blends have sections that are rotated about the Y-axis up to a maximum of 120-degree and the distance between two sections is measured from the coordinate system. Between each section, an angle called the **rotational blend angle** has to be defined. In this type of blend, each section has its own user-defined coordinate system. If the rotational blend angle entered between the two sections is equal to 0-degree then the **Rotational** blend option functions the same way as the **Parallel** blend option.

Note that all the nonparallel blends can be open or close. Therefore, after choosing **Rotational > Regular Sec > Sketch Sec > Done** from the **BLEND OPTS** submenu, the **ATTRIBUTES** menu is displayed as shown in Figure 7-26. The options in this menu are discussed next.

Open option

The **Open** option is used when the blend feature to be created has to be kept open.

Closed option

The **Closed** option is used to create a closed blend feature. In this type of blend feature,

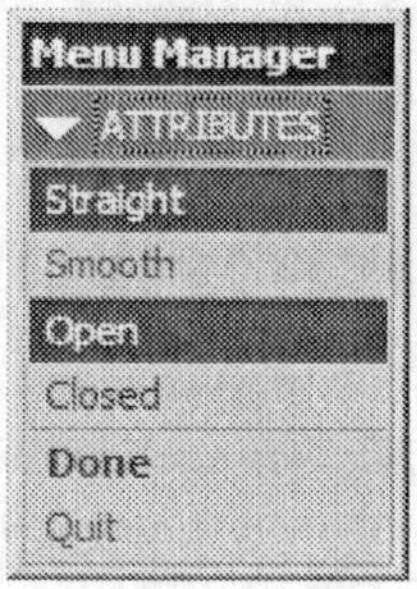

Figure 7-26 *ATTRIBUTES* ***menu***

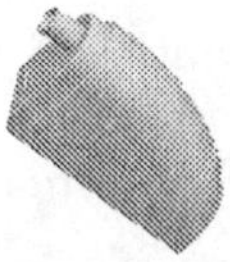

Tip: *It is recommended that the closed blend should have at least three sections.*

Pro/ENGINEER closes the feature by automatically blending the last section with the first section. Figure 7-27 shows the sections used to create a rotational smooth blend feature. The three default datum planes can also be seen. From Figure 7-27, it is evident that two sections are used to create the blend feature and that these sections are at an angle of 45-degree. It is also evident from the figure that the second section is dimensioned from the coordinate system that was defined in the first section. Figure 7-28 shows the shaded model of the same blend feature.

The following steps explain in brief the procedure to create the blend feature shown in Figure 7-28.

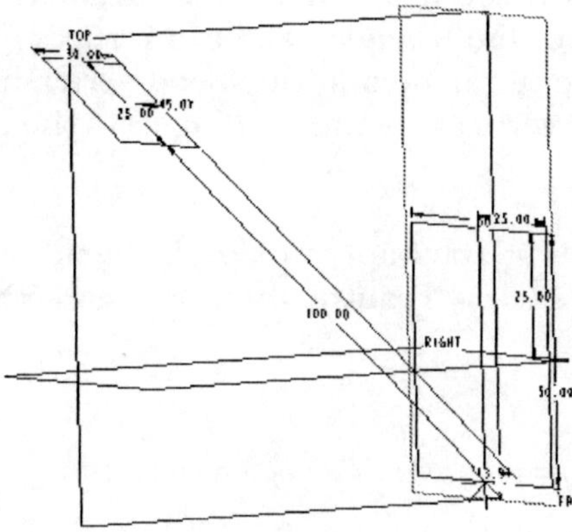

Figure 7-27 *The two sections with dimensions and the default datum planes used to create the blend feature shown in the adjacent figure*

Figure 7-28 *Shaded model of rotational open blend feature*

1. Invoke the **Blend** option and from the **BLEND OPTS** menu choose **Rotational** > **Regular Sec** > **Sketch Sec** > **Done**. The **ATTRIBUTES** menu is displayed.

2. Choose **Smooth** > **Open** > **Done** from the **ATTRIBUTES** menu. You are prompted to select a plane to sketch.
3. Select the **RIGHT** datum plane as the sketch plane and orient the **FRONT** datum plane to be on top.
4. After selecting the sketch plane and its orientation, draw the first section that is a square of side 50. The start point on this square should be at the lower right corner.
5. Place the coordinate system at the lower left corner of the square.
6. Choose the **Continue with the current section** button to proceed for drawing the next section. Accept the default Y-axis rotation angle of 45-degrees for section 2.
7. Draw the second section that is a square of size 25.
8. Place the coordinate system for this section at a distance of 100 toward the left of the lower left corner of the square. Remember that the start point of this section should also lie at the lower right corner of the rectangle.
9. Exit the sketcher environment. The **Message Input Window** appears and you are prompted that do you want to continue to draw the next section.
10. Choose the **No** button. Now, you can see the preview of the feature that you have created and then exit the feature creation tool.

General Option

Using the **General** option, sections are translated and rotated about the x, y, and z axes. The sections are aligned using the user-defined coordinate system. The coordinate system has to be manually placed in every section sketch that constitutes the blend feature.

USING BLEND VERTEX

As mentioned earlier, each section of the blend feature must have an equal number of entities. However, you can use the **Blend Vertex** option if the number of entities in all the sections are not the same. For example, to create a blending between a rectangle and a triangle, add blend vertex on a point other than the start point of the triangle. The two vertices of the triangle will be blended with the two vertices of the rectangle and the blend vertex in the triangle will be blended with the remaining two vertices in the rectangle as shown in Figure 7-29.

To add a blend vertex in a sketch, select the point where you want to place the blend vertex. The selected point is highlighted in red. Choose **Sketch** > **Feature Tools** > **Blend Vertex**. The blend vertex is placed at the selected point.

Note

*The **Blend Vertex** can be used only in either the first or last section of a blend feature.*

SHELL OPTION

The **Shell** option scoops out the material from the model and at the same time removes the selected faces, leaving behind a thin model with some specified wall thickness. The **Shell** option can be invoked by choosing the **Shell Tool** from the **Engineering Features** toolbar. You can also choose this option from the menu bar by choosing **Insert** > **Shell**. The Shell dashboard is displayed as shown in Figure 7-30.

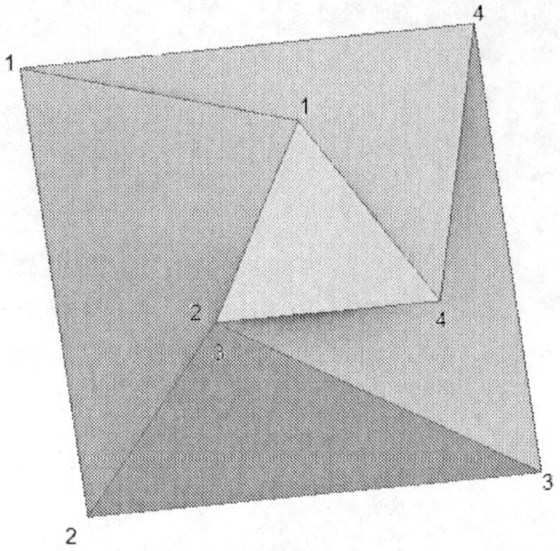

***Figure 7-29** Blending a square with a triangle using the blend vertex*

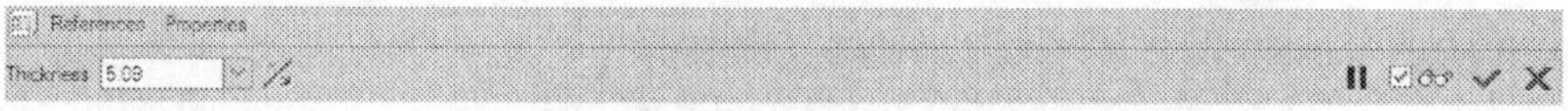

***Figure 7-30** Shell dashboard*

Note

*The **Shell** option is used on existing models and hence this option is available only when a model exists on the graphics screen.*

Using the Shell dashboard you can create two types of shell.

1. Constant thickness shell
2. Variable thickness shell

Creating Constant Thickness Shell

The constant thickness shell is a shell that has a uniform thickness on all the four faces of the model. The following steps explain the procedure to create a constant thickness shell.

1. Invoke the Shell dashboard.
2. Select the top face of the model to remove as shown in Figure 7-31. The selected face will be removed from the model, leaving the specified thickness from the boundary of the selected face.
3. Enter the thickness value of the shell in the dimension box present on the dashboard.
4. Choose the **Feature Preview** button from the dashboard to preview the shell feature.
5. Choose the **Build feature** button from the Shell dashboard to exit it.

The shell is created on the selected face as shown in Figure 7-32. The thickness of the shell is uniform on all the faces of the model.

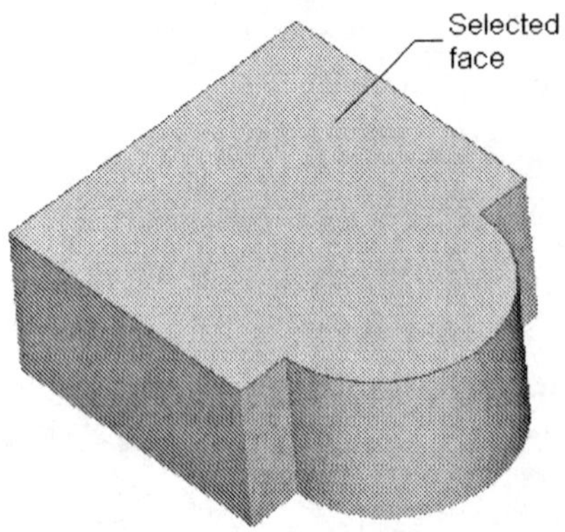

Figure 7-31 Highlighted surface to shell

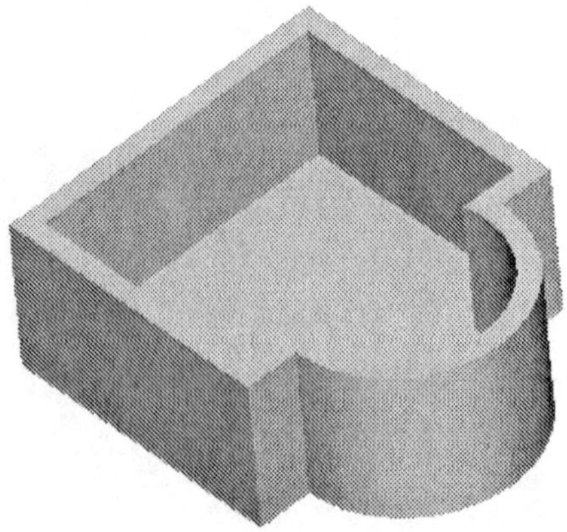

Figure 7-32 Shell created on the selected surface

Note
*When you invoke the Shell dashboard, the **Face** filter is selected by default.*

Creating Variable Thickness Shell

The variable thickness shell is a shell that has different thickness values with the adjacent faces. The following steps explain the procedure to create a variable thickness shell.

1. Invoke the Shell dashboard.
2. Select the top face of the model to remove as shown in Figure 7-33.
3. Choose the **References** tab to invoke the slide-up panel. In the slide-up panel, there are two collectors. The **Removed Surfaces** collector shows the surface id of the face that you have selected to remove. The **Non Default Thickness** collector shows the surfaces that you will select for creating the variable thickness shell.
4. Click in the **Non Default Thickness** collector and now select the adjacent faces that are shown in Figure 7-33. To select the second face and the successive faces you need to use CTRL+left mouse button.
   Now you need to specify the different thickness values with reference to the selected face.
5. Enter the thickness values in the dimension boxes that are present in front of the surfaces in the **Non Default Thickness** collector.

The shell is created on the selected face and with the different thickness values with respect to the adjacent selected faces, see Figure 7-34.

Note
The thickness value entered can be positive or negative. If the value entered is positive, the material is removed, leaving the shell thickness inside the boundary of the face selected. But when the value entered is negative, the shell thickness is added outside the boundary of the face selected.

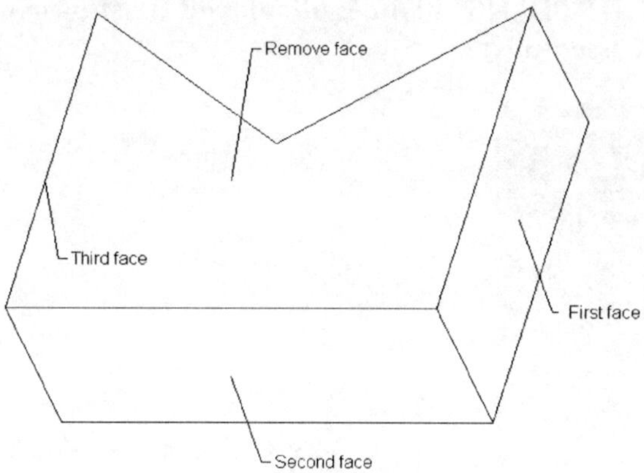

Figure 7-33 *Faces of the model to select for variable thickness and the top face to remove*

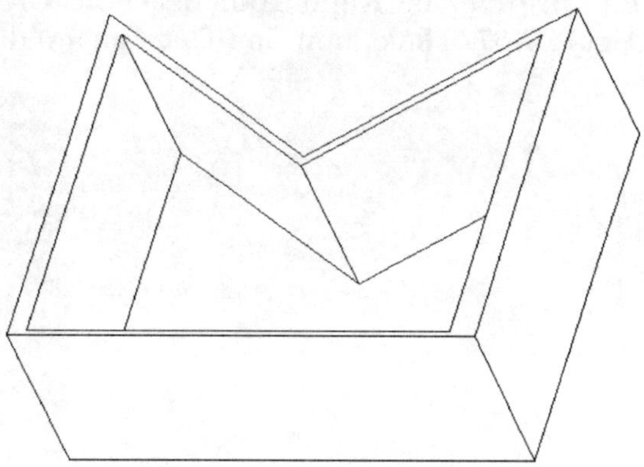

Figure 7-34 *Variable thickness shell*

Figure 7-35 shows the model whose top surface is selected to remove in order to create a shell. Figure 7-36 shows the model after shelling. Notice that the shell thickness is left on the selected face and the remaining material is removed.

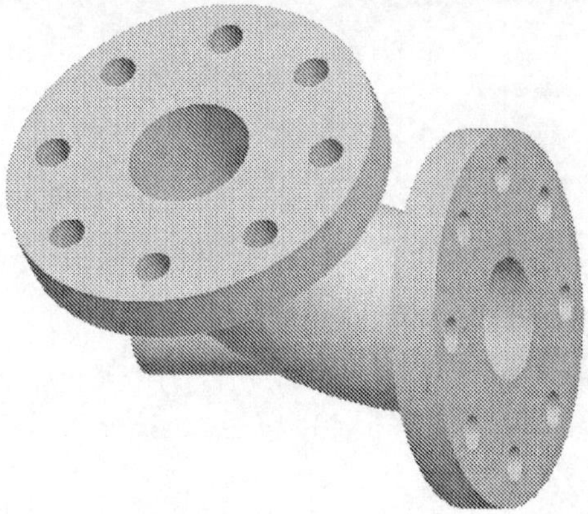

Figure 7-35 *Solid model*

Figure 7-36 *Model after shelling the top face*

DATUM CURVES

Datum curves are useful in creation of advanced solid and surface features such as the sweep trajectories to create a sweep feature. A datum curve is considered as a feature and is displayed in the **Model Tree**. There are various tools and options to create a datum curve. You can use the **Insert a datum curve** button and **Sketched Datum Curve Tool** button from the **Datum** toolbar to create a datum curve. You can also choose **Intersect**, **Project**, and **Wrap** options in the **Edit** menu from the menu bar to create a datum curve.

Creating Datum Curve Using the Insert a datum curve button

The **Insert a datum curve** button is used to create datum curves. When you choose this button from the **Right Toolchest**, the **CRV OPTIONS** menu is displayed as shown in Figure 7-37. The options in this menu are discussed next.

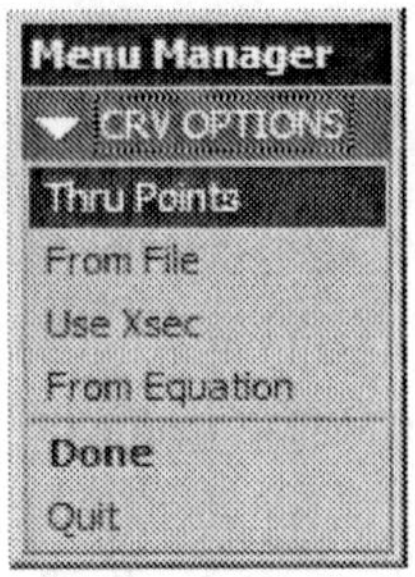

*Figure 7-37 **CRV OPTIONS** menu*

Thru Points

The **Thru Points** option creates a datum curve by selecting the existing datum points or vertices. The resulting datum curve may be a spline curve or can have a user-defined radii.

When you choose **Done** from this menu, the **CONNECT TYPE** menu is displayed as shown in Figure 7-38. The options in this menu are discussed next.

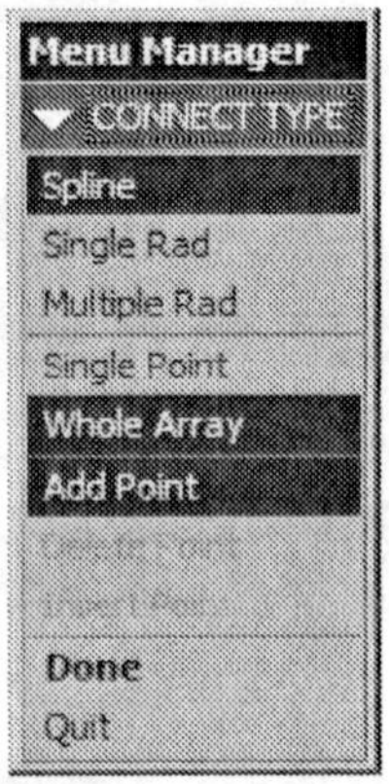

*Figure 7-38 **CONNECT TYPE** menu*

Spline

The **Spline** option creates a datum curve in the form of a spline and passing through the selected datum points or vertices.

Single Rad

The **Single Rad** option creates a datum curve that has a constant radius at the bends. This option is called single radius because the system prompts you to specify a radius value at the bend. The bend is the location on the datum curve that lies between two datum points or vertices.

Multiple Rad

As the name suggests, the **Multiple Rad** option allows you to specify radius at every bend that occurs on a datum curve. Using this option you can either specify the same radius that was specified at the previous bend or a new radius value.

Single Point

The **Single Point** option is used to select the datum points individually. This option is provided because if the datum points are created such that they are a single feature then they all may be selected.

Whole Array

The **Whole Array** option is used to select all the datum points that act as a single feature.

From File

The **From File** option is used to import a datum curve from IGES, VDA, *.ibl file formats. You can save a model that is in any of the above-mentioned formats and export the geometry in the form of a curve. When you open the exported file to create a datum curve then the geometry of the model is converted to a datum curve.

Use Xsec

The **Use Xsec** option is used to create a datum curve that has the geometry of an existing cross section. The creation of sections is discussed in Chapter 6.

From Equation

The **From Equation** option is used to create datum curves by defining equations using the coordinate systems. When you choose this option, you are prompted to select a coordinate system. Select the default coordinate system from the graphics window. The **SET CSYS TYP** menu is displayed as shown in Figure 7-39.

Figure 7-39 SET CSYS TYP menu

This menu provides the types of coordinate system that you can choose from. Choose the coordinate system depending on the equation you want to use to create a datum curve. If you choose the **Cartesian** option from the menu, the **rel.ptd - Notepad** window is displayed as shown in Figure 7-40.

Using this notepad you can define the equations. The notepad shows the instructions that should be followed while writing an equation in the notepad. These instructions varies and depend on the type of coordinate system you have selected. After you have entered the equations in the notepad, save the file and then exit the notepad.

Creating a datum curve in spiral shape

The following steps explain the procedure to create a spiral-shaped datum curve.

1. Choose the **Cylindrical** option from the **SET CSYS TYP** menu. The notepad is displayed.
   In the notepad the parametric equations to create a circle are given as follows:

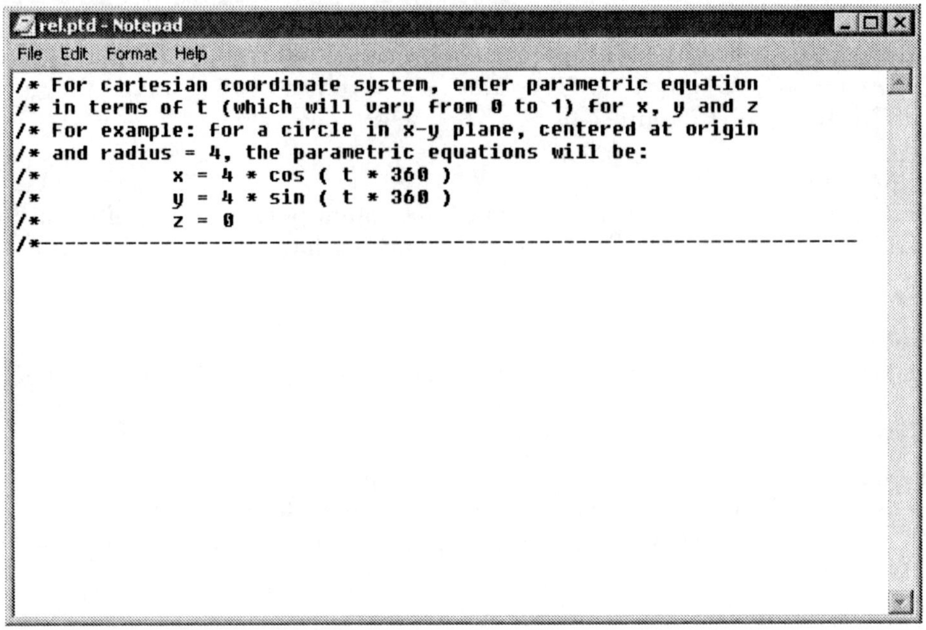

Figure 7-40 *Notepad window*

r = 4
theta = t * 360
z = 0
In the above equations, the variable t varies from 0 to 1, r is the radius of the circle, and z is the third equation that is set equal to 0. Now, to understand the given equation of the circle, notice that the radius of the circle is given. The only value that is unknown is theta. The value of theta depends on the variable t. Therefore,
when t = 0, theta = 0
when t = 1/2, theta = 180 (semicircle)
and when t = 1, theta = 360 (circle)

2. Enter the following parametric equation of the spiral curve below the dashed line in the notepad:
   IR = 8
   OR = 80
   TURNS = 10
   r = IR + t * (OR - IR)
   theta = t * 360 * TURNS
   z = 0
   In the above equations, the value of r is selected to vary because in the spiral curve, the value of r always increases from center (at center of spiral, r = 0). The internal radius of the spiral IR = 8, outer radius OR = 80, and number of turns TURNS = 10 are given.
3. Choose **File > Save** from the notepad and then exit the notepad.
4. Choose the **Preview** button from the **CURVE: From Equation** dialog box to preview the datum curve.

5. Choose **OK** from the dialog box to exit. The datum curve appears on the graphics window as shown in Figure 7-41.

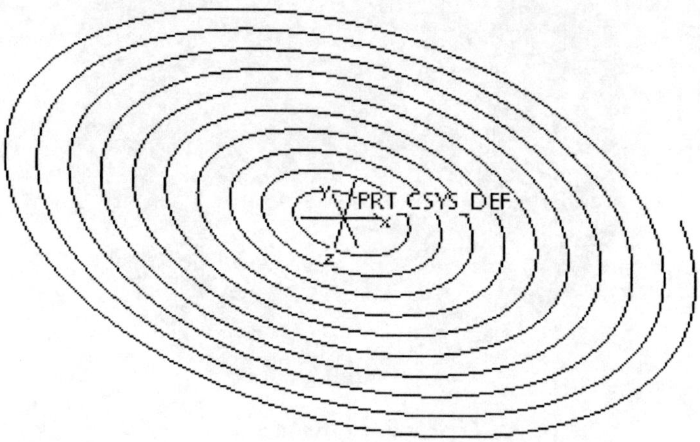

Figure 7-41 *Spiral-shaped datum curve*

Creating Datum Curve by Sketching

The **Sketched Datum Curve Tool** button is used to sketch a datum curve using the sketcher tools. This is the most commonly used option to create a datum curve. The sketch can be open or closed and is drawn in the sketcher environment.

Intersect Option

The **Intersect** option can be invoked by choosing **Edit > Intersect** from the menu bar. This option is object action tool; this means that you need to select a datum plane and then invoke this option. The **Intersect** option creates a datum curve at the intersection of a face of the model and a datum plane, intersection of a face of the model and a quilt surface, intersection of a quilt surface and a datum plane. Note that you cannot create a datum curve at the intersection of two datum planes, two quilts, or two model faces using this option.

When you choose this option, the Intersect dashboard appears as shown in Figure 7-42. You may need to choose the **Intersect** option twice to invoke the Intersect dashboard.

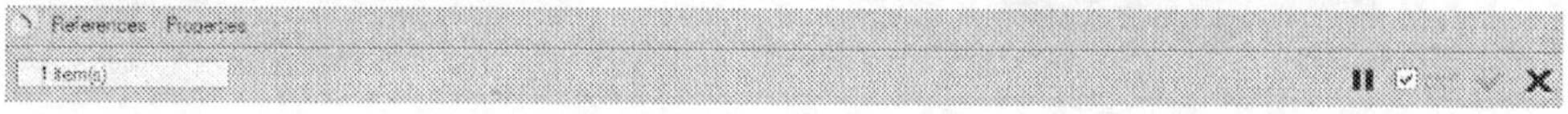

Figure 7-42 *Intersect dashboard*

To create the ring on the circumference of the cylindrical feature as shown in Figure 7-43 you need to create a datum curve that is on the circumferential surface of the cylindrical feature. To create the datum curve that lies on the outer surface, follow these steps:

1. Select the datum plane that is intersecting with the circumferential surface of the cylinder.

Figure 7-43 *Solid model*

2. Choose **Edit > Intersect** from the menu bar. The Intersect dashboard is displayed.
3. Choose the **References** tab to open the slide-up panel. In the **Surfaces** collector, the selected datum plane is displayed.
4. Use CTRL+left mouse button and select the top and the bottom circumferential surface of the cylindrical feature as shown in Figure 7-44.
5. Choose the **Build feature** button from the dashboard to exit the feature creation tool. The datum curve is created on the circumferential surface of the cylinder as shown in Figure 7-45.

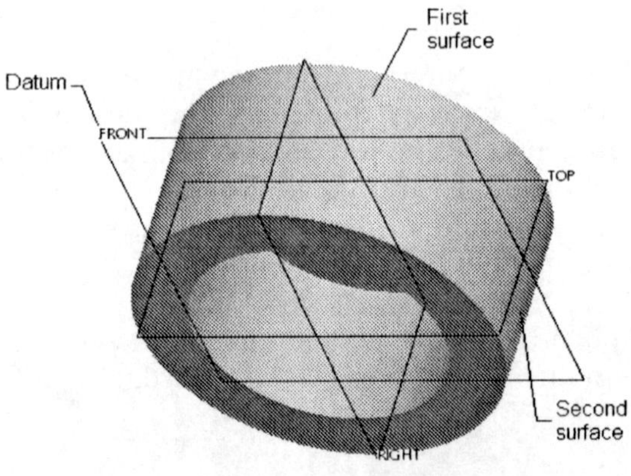

Figure 7-44 *Selections to make on the model*

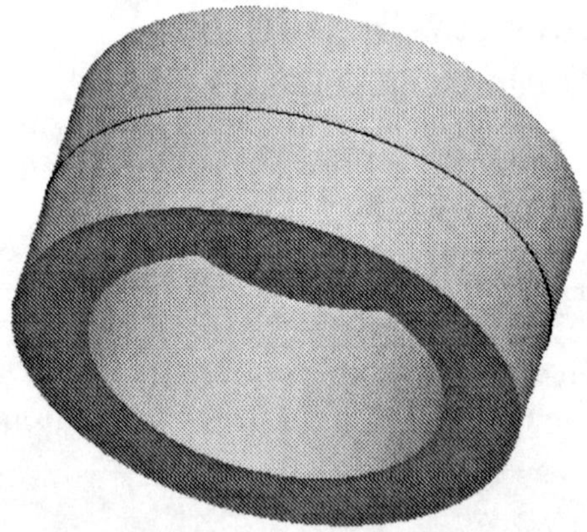

Figure 7-45 *Datum curve created on the circumferential surface of the cylindrical feature*

Project Option

The **Project** option projects a selected or sketched entity on one or more planar or non-planar surfaces or datum planes. The projected datum curve forms a true projection of the selected or sketched entity on the specified surfaces. The dimensions of the original entity may distort while projecting.

Choose **Edit > Project** from the menu bar. The Project dashboard is displayed as shown in Figure 7-46. The tools and options available in this dashboard are discussed next.

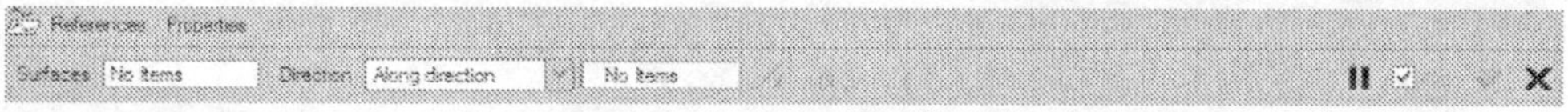

Figure 7-46 *Project dashboard*

Surfaces Collector

The **Surfaces** collector is used to select the surface on which you need to project the sketched or selected datum curve.

Direction drop-down list

The **Direction** drop-down list is used to specify the method of projection of the datum curve on the receiving surface or plane. The two options available in this drop-down list are:

1. Along direction
2. Normal to surface

Along direction

This option projects the datum curve in a direction shown by the yellow arrow. To specify the direction of projection you can select the default coordinate system, datum plane, edge, or surface. Figure 7-47 shows the datum curves that are overlapping. The datum curve that is selected to project on the receiving surface and the datum curve after projection are of same geometry. This is because, using the **Along direction** option, the true geometry is obtained after projection. Figure 7-48 shows the sketched datum curve after projecting on the receiving surface. The curve is sketched on a datum plane that is parallel to the bottom face of the model.

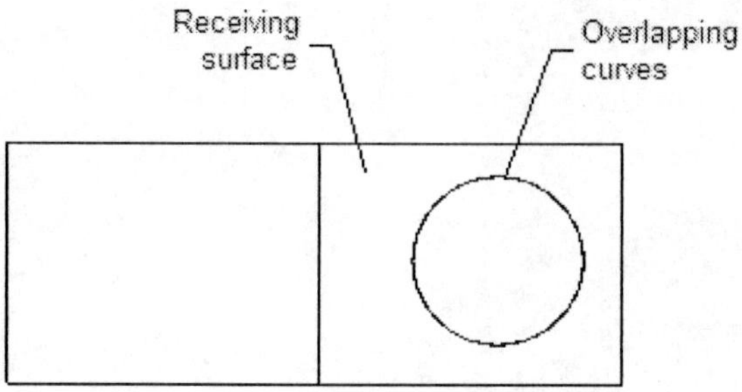

Figure 7-47 *Top view of projected datum curve*

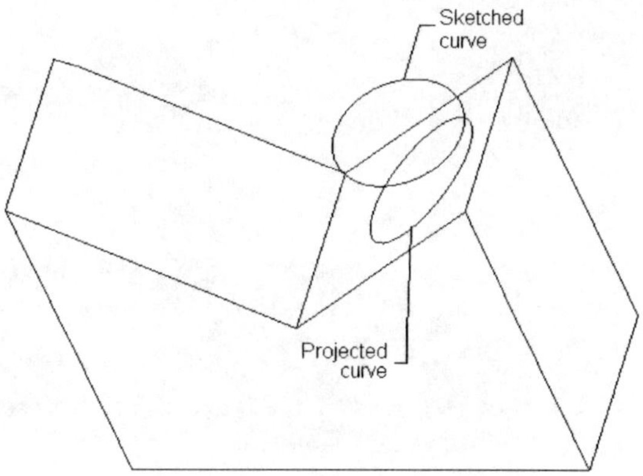

Figure 7-48 *Projecting a datum curve*

Normal to Surface

This option of projecting a datum curve projects the datum curve normal to the receiving surface. Figure 7-49 shows the datum curve that is selected to project on the receiving surface and the datum curve after projection. Figure 7-50 shows the sketched datum curve after projecting on the receiving surface.

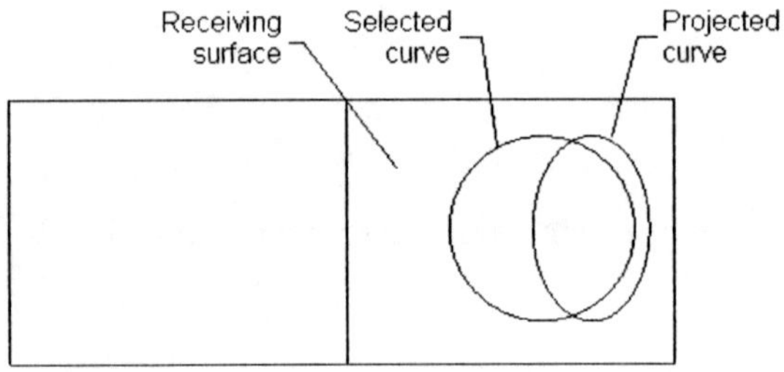

Figure 7-49 Top view of projected datum curve

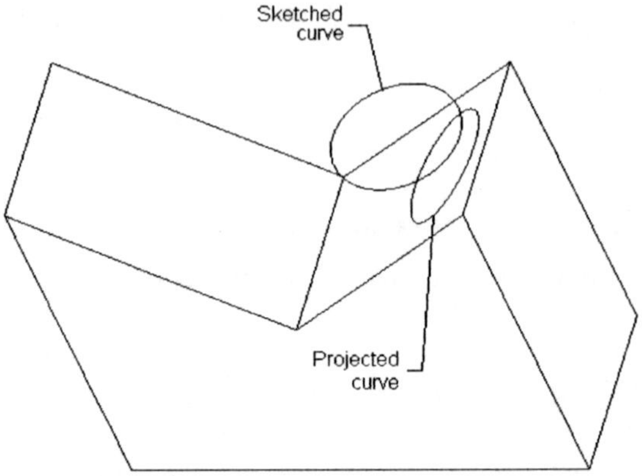

Figure 7-50 Projecting a datum curve

References tab

When you choose the **References** tab, the slide-up panel is displayed as shown in Figure 7-51. Using this slide-up panel you can either select an existing datum curve to project or sketch a datum curve to project. The drop-down list in the slide-up panel lists two options, **Project Curve** and **Project Sketch**. These two options are discussed next.

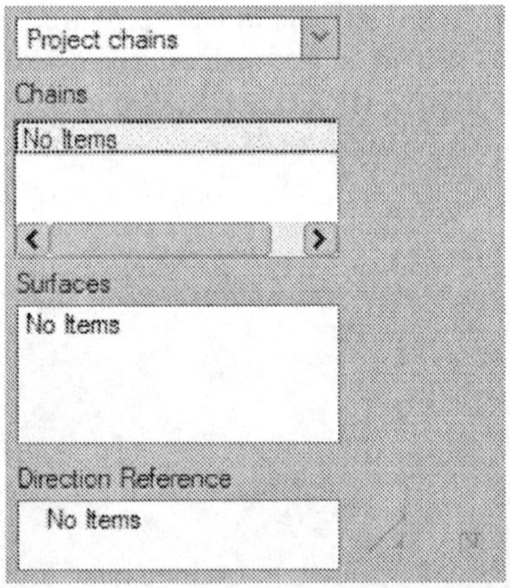

Figure 7-51 ***References*** *tab slide-up panel*

Project chains

The **Project chains** option is used when the datum curve to project exists. Click in the **Chains** collector to make it yellow in color and then select the datum curve that you need to project. After selecting the datum curve, click in the **Surfaces** collector and select the receiving surface (plane or surface to project on to). Then you need to specify the direction of projection. Remember that if you are using the **Normal to surface** option then you do not need to specify the direction of projection. Also in this case, the **Direction Reference** collector displayed in Figure 7-51 does not appear.

Project a sketch

When you select the **Project a sketch** option from the drop-down list in the slide-up panel, the slide-up panel appears as shown in Figure 7-52.

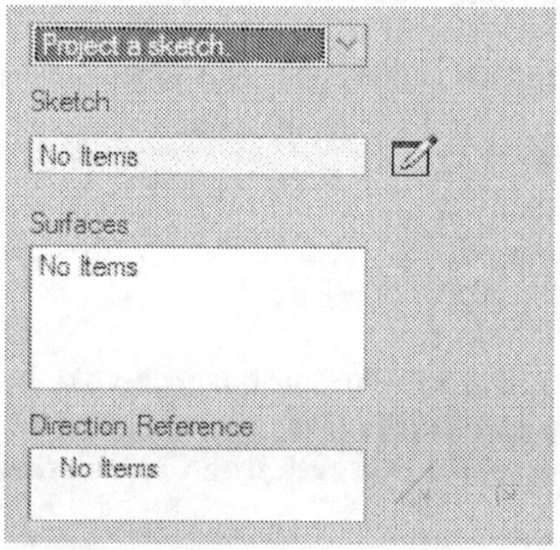

***Figure 7-52** **References** tab slide-up panel*

Choose the **Create a section or redefine the existing section** button to invoke the **Section** dialog box and to select the sketch plane for the curve to be projected. After sketching, exit the sketcher environment and select the surface or plane on which the curve will be projected. If you are using the **Along direction** option to project the curve then you need to click in the **Direction Reference** collector to specify the direction of projection.

Wrap Option

This option is used to create a datum curve by wrapping a sketched entity around a solid or a quilt. Choose **Edit > Wrap** from the menu bar, the Wrap dashboard is displayed as shown in Figure 7-53. The tools and options in this dashboard are discussed next.

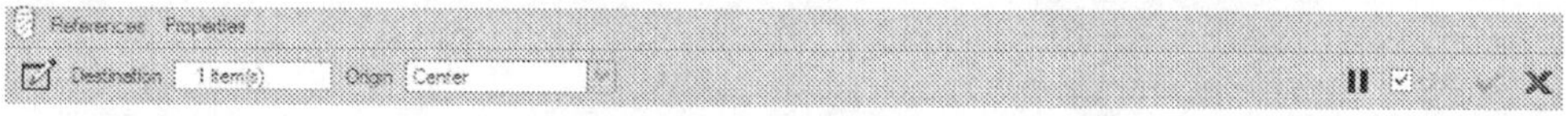

***Figure 7-53** Wrap dashboard*

Destination collector

The **Destination** collector is used to select the object on which you need to wrap the curve. Generally, if there is a single feature on the graphics window then you do not need to select the object to wrap on. Pro/ENGINEER automatically wraps the selected curve or the sketched curve on the object. However, if you want to select a different object to wrap on then you can click in this collector to make it yellow in color and then select the object.

Origin drop-down list

The **Origin** collector is used to specify the origin of the sketched or the selected curve.

References tab

When you choose the **References** tab, the slide-up panel is displayed as shown in Figure 7-54. The **Create a section or redefine the existing section** button is used to invoke the **Section** dialog box using which you can select the sketch plane and enter the sketcher environment to draw the curve.

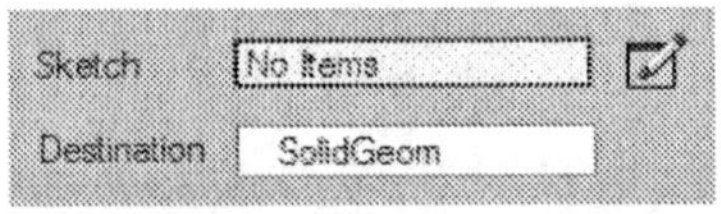

Figure 7-54 ***References** tab slide-up panel*

The **Destination** collector automatically selects the object on which the curve will be wrapped.

The curve is sketched and wrapped on the cylinder as shown in Figure 7-55. To wrap a curve on the cylinder as shown in Figure 7-55, follow the steps given next. It is assumed that a cylinder of diameter 20 and height 50 exists.

1. Choose **Edit > Wrap** from the menu bar to invoke the Wrap dashboard.
2. Choose the **References** tab to invoke the slide-up panel.
3. Choose the **Create a section or redefine the existing section** button to invoke the **Section** dialog box.
4. Choose the datum plane that is passing through the center of the cylinder as the sketch plane and then choose a reference for orienting the sketch plane.
5. After entering the sketcher environment, draw a line that starts from the bottom of the cylinder. Its start point is aligned with the center axis of the cylinder and with the bottom edge of the cylinder. The length of the line is **1000** and is at a distance of **50** from the bottom of the cylinder, see Figure 7-56.
6. Place a user-defined coordinate system at the start point of the line.
7. Exit the sketcher environment. The sketched curve is automatically wrapped on the circumference of the cylinder.
8. Choose the **Build feature** button to exit the Wrap dashboard.

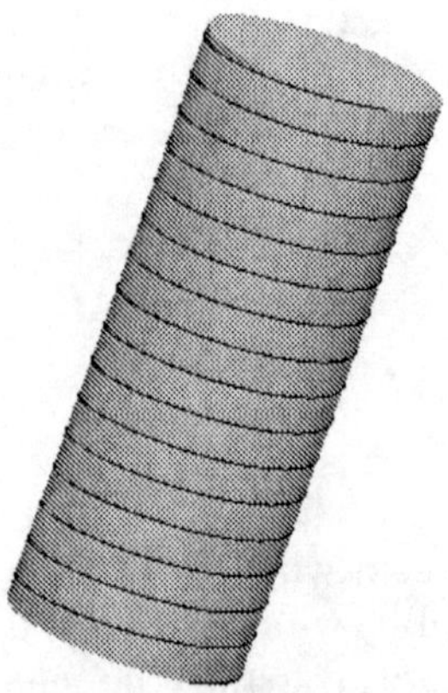

Figure 7-55 *Curve wrapped on the cylinder*

Figure 7-56 *Sketch of the curve to wrap*

CREATING DRAFT FEATURES

In Pro/ENGINEER, drafts are created on existing surfaces. Drafts are created by rotating the selected surface by a certain angle. One of the applications of draft features is in moulds and castings where a taper is required to separate the casting from the mould or vice versa.

Choose the **Draft Tool** button from the **Engineering Features** toolbar. The Draft dashboard is displayed as shown in Figure 7-57. The options in this dashboard are discussed next.

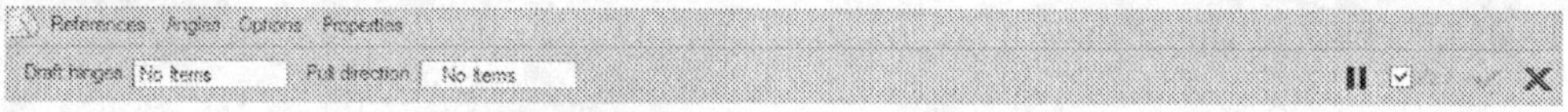

***Figure 7-57** Draft dashboard*

References Tab

When you choose the **References** tab, the slide-up panel is displayed as shown in Figure 7-58. The options in this slide-up panel are discussed next.

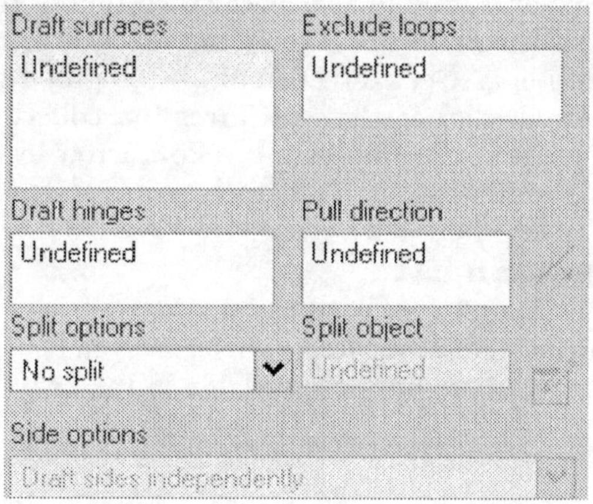

***Figure 7-58 References** tab slide-up panel*

Draft surfaces collector

When you invoke the Draft dashboard, this collector is selected by default. If this collector is not selected by default, click in this collector and then select the surface to which you need to add a draft angle. The maximum draft angle that can be added to a surface is 30-degrees.

Draft hinges collector

The draft hinges collector is used to select the hinge of the draft surface. The hinge that you select can be an edge, a surface, an axis, or a datum plane. The draft surface is pivoted on the hinge that you select. In other words, the draft surface is rotated about the hinge. The hinge that you select need not intersect the draft surface. Figure 7-59 shows the draft surface and the hinge selected to create the draft is shown in Figure 7-60.

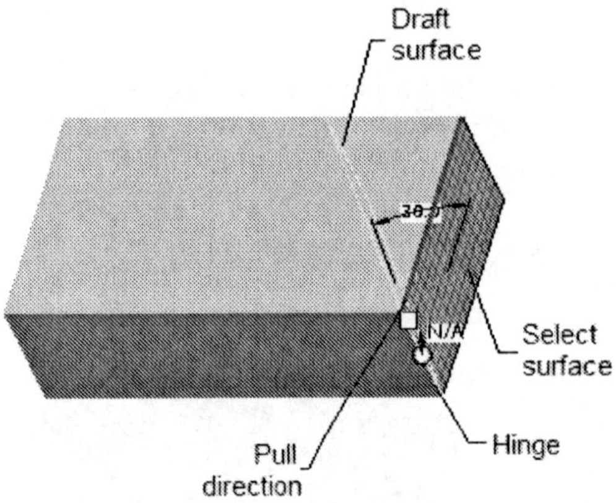

Figure 7-59 *Parameters needed to create a basic draft surface*

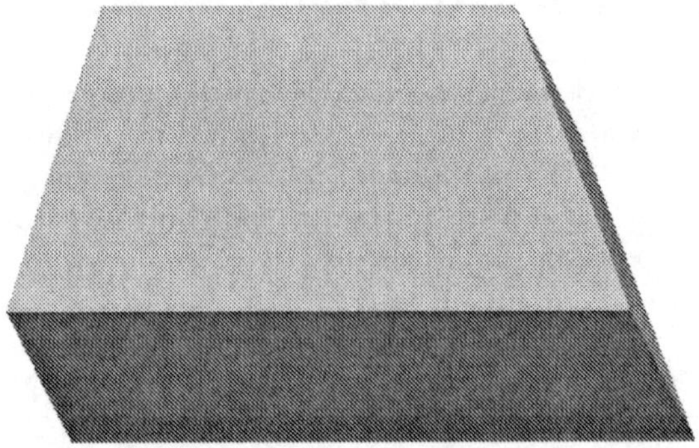

Figure 7-60 *Resultant draft surface*

Pull direction collector

The **Pull direction** collector is used to specify the direction of rotation of the draft surface. The direction of pull is shown by the direction of the yellow arrow. Generally, when you select the hinge, the pull direction is selected by default. If the pull direction shown by the yellow arrow is not that is desired, click in the **Pull direction** collector and then select the pull direction. You can change the direction of the yellow arrow by choosing the **Reverse pull direction** button.

Split options drop-down list

The options in this drop-down list are used to split the surface selected to draft. Using these options the selected surface gets split into two surfaces and different draft angles can be applied to both the surfaces. The options available in the **Split options** drop-down list are discussed next:

No Split

This option is used when you do not want to split the surface selected to give the draft angle.

Split by draft hinge

This option is available only when you have selected the hinge for the draft surface. When you select this option, the selected surface is divided into two surfaces at the location on the surface where the hinge intersects it. Figure 7-61 shows the surface that is split into two surfaces at the hinge.

Split by split object

This option when chosen activates the **Split object** collector and the **Create and edit sketch for split** button. This option is used to split the surface selected to draft by drawing a sketch. The surface gets split into two surfaces and the sketch defines the profile of the

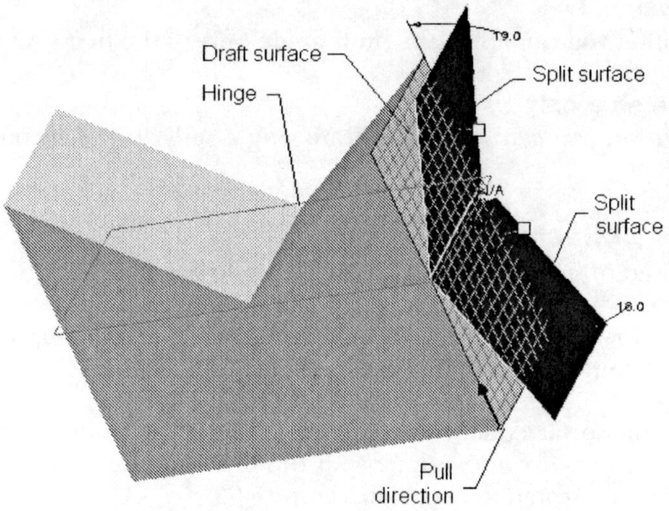

Figure 7-61 Split at draft hinge

split. Figure 7-62 shows the different parameters that you need to define to create a draft split by sketching. Figure 7-63 shows the draft created on the cylindrical surface.

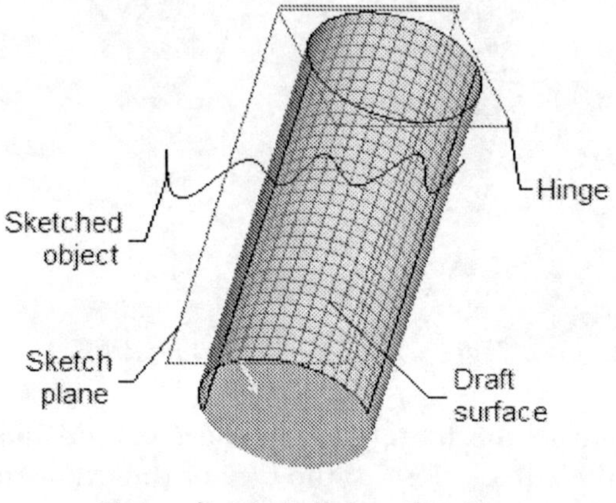

Figure 7-62 Split by split object

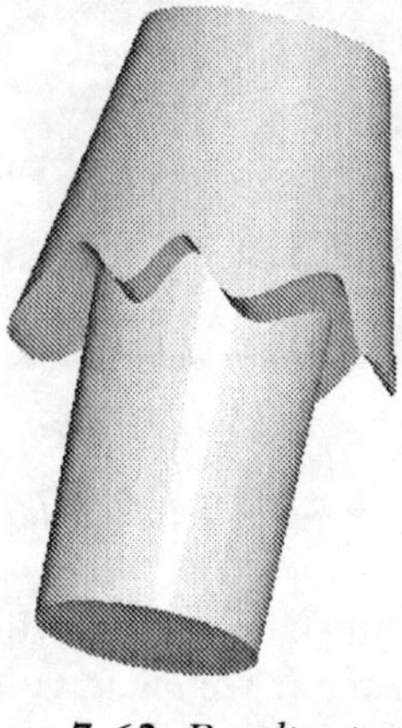

Figure 7-63 Resultant model

Side options drop-down list

The options in this drop-down list are used to specify how to apply the draft once the surface has been split. The options in this drop-down list are discussed next.

Draft sides independently

The two sides of the draft surface that are formed when the surface is split can be given different draft angles. When you choose this option, the edit boxes appear on the Draft dashboard. You can enter the values of the draft angle in these edit boxes.

Draft sides dependently

The two sides of the split surface are given the same draft angle.

Draft first side only

Using this option you can apply the draft angle only to the first side of the split surface.

Draft second side only

Using this option you can apply the draft angle only to the second side of the split surface.

Exclude loops collector

This collector is used to select a surface to which you do not want to add a draft angle. When you select a surface to add a draft angle, all the loops that are on the surface are applied the same draft angle. However, using this collector you can select the loop to which you do not want to add a draft angle.

Figure 7-64 shows the surface that when selected to add a draft angle also selects the surface of the cylindrical feature. If you continue with the draft feature creation and exit the Draft dashboard, the draft is created as shown in Figure 7-65.

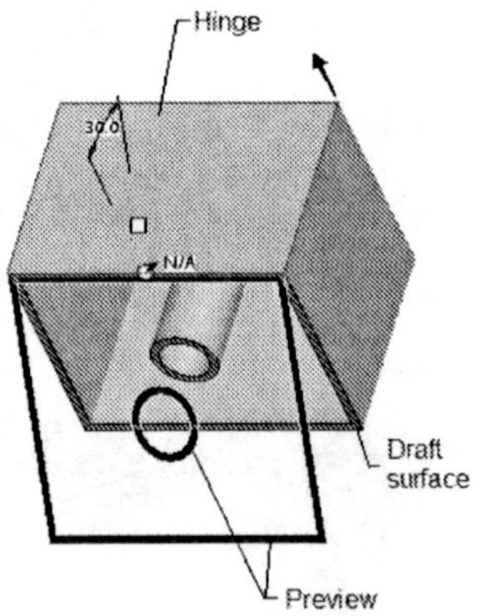

Figure 7-64 Surfaces for draft

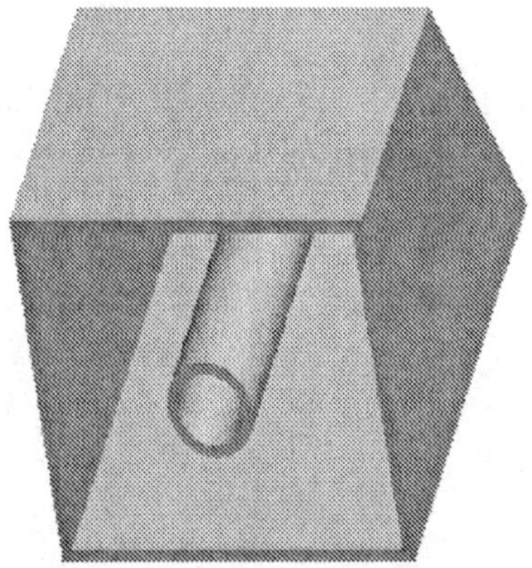

Figure 7-65 After creating draft

To exclude the face of the cylindrical feature from the loop, click in the **Exclude loops** collector to turn it yellow in color. Now, bring the cursor close to the face of the cylindrical feature. The boundary of the face is highlighted in cyan. Select the face to exclude it from the loop. Figure 7-66 shows the draft surface after excluding the face of the cylindrical feature from the loop.

Angles Tab

Choose this tab from the Draft dashboard after you have specified some angle for draft surface. When you choose this option, the slide-up panel is displayed as shown in Figure 7-67. In Pro/ENGINEER, you can create constant and as well as variable angle draft.

Constant Angle Drafts

By default, Pro/ENGINEER creates a constant angle draft. This means, the same draft angle is applied to the surface that you have selected.

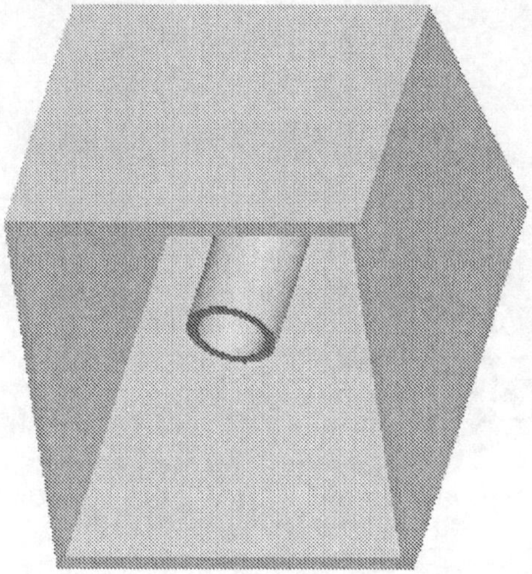

Figure 7-66 Exclude loop draft

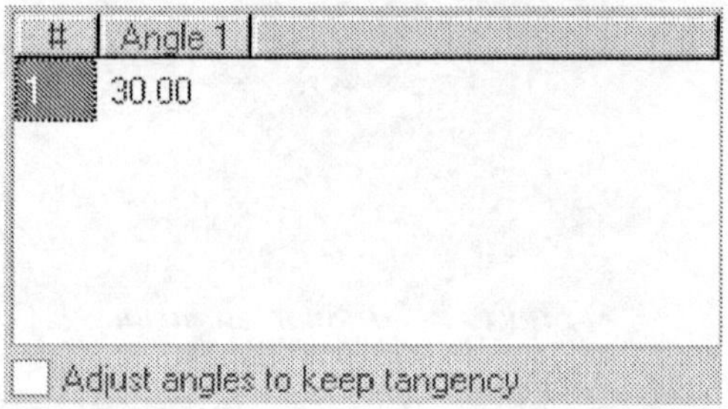

Figure 7-67 ***Angles*** *option slide-up panel*

Variable Angle Drafts

When you apply more than one value of the draft angle to the selected surface, it is called variable angle draft. Apply the variable angle draft after you have selected the hinge, the pull direction, and the draft surface.

To create the draft surface shown in Figure 7-68, follow the steps given next. It is assumed that you have the base feature of the model as shown in Figure 7-69.

1. Invoke the Draft dashboard and click in the **Draft hinges** collector to make it yellow in color.
2. Select the top face of the base feature. Notice that in the **Pull direction** collector the direction of pull is selected automatically. Reverse the pull direction by choosing the **Reverse pull direction** button available on the right of the **Pull direction** collector.

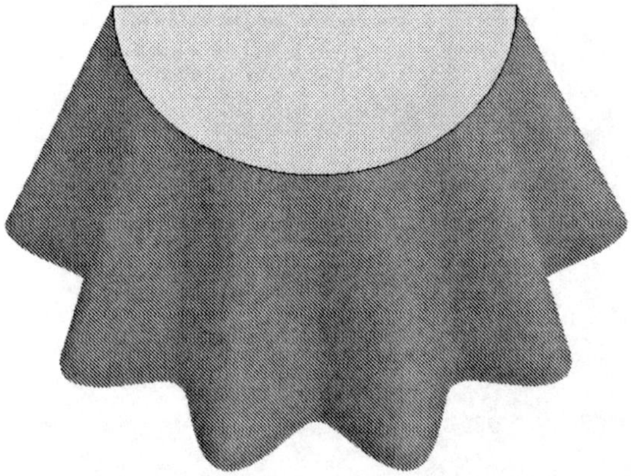

Figure 7-68 Draft feature on the model

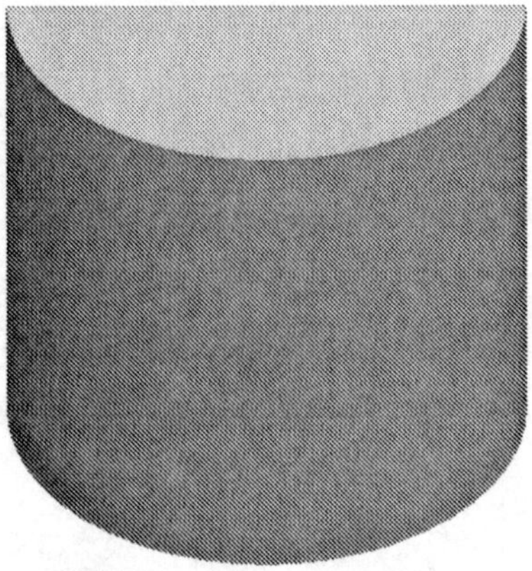

Figure 7-69 Base feature

3. Hold down the right mouse button to invoke the shortcut menu shown in Figure 7-70.

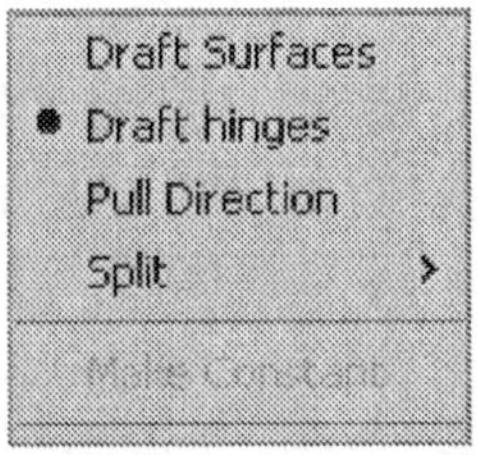

Figure 7-70 Shortcut menu

4. Choose the **Draft Surfaces** option from the shortcut menu and select the cylindrical surface of the base feature to add the draft angle. The surface that you select is highlighted by a mesh. The drag handle appears on the base feature; you can use it to vary the angle of the draft. The draft angle value of **1** also appears on the graphics window. Also notice the white ball that appears on the edge of the top face (face selected as hinge).
5. Bring the cursor on the white ball and hold down the right mouse button to invoke the shortcut menu shown in Figure 7-71.

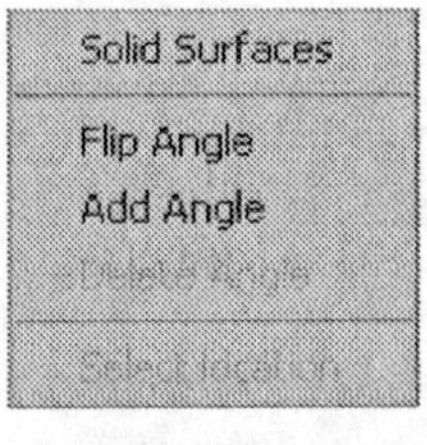

Figure 7-71 Shortcut menu

6. Choose the **Add Angle** option from the shortcut menu. Notice that now there are two white balls on the edge of the top face. On the first ball the value of **0.5** appears. This value varies from one end of the face to the other end. This value represents the location

of the point that you need to use for reference in order to apply the draft angles.

7. Again bring the cursor to any one of the two white balls and invoke the shortcut menu.
8. Choose the **Add Angle** option from the shortcut menu. Another point is added with a location value on the edge.
9. Add eleven such points for locations varying from **0** to **1**.
10. After adding eleven points, double-click on any of the location value that is present on the graphics window. The edit box appears, enter the value of **0** in this edit box. Similarly, locate all the remaining ten points with an increment of 0.1. Figure 7-72 shows the model after locating all the eleven points on the edge.
11. Double-click on the value of the angle that corresponds to the location **0**. The edit box appears, enter a value of **15**.
12. Double-click on the value of the angle that corresponds to the location **0.1** (these values appear above the model and have a default value of 1.0). The edit box appears; enter a value of **5**. Similarly, vary the angle at other locations. Every alternate location should have an angle of **15**. Figure 7-73 shows the preview of the model after modifying the values of the angle at all the eleven locations.

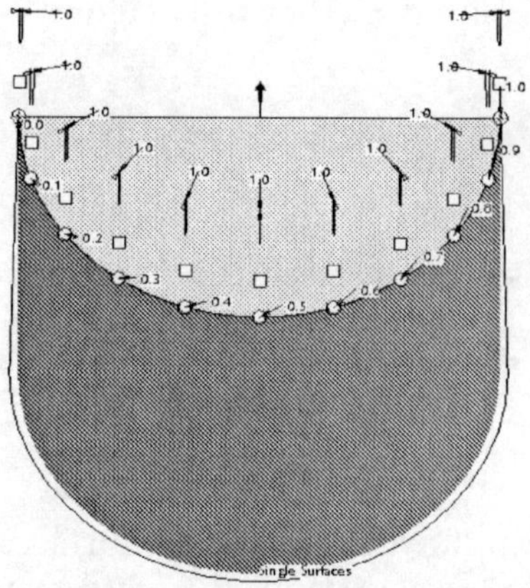

***Figure 7-72** Location points on the edge*

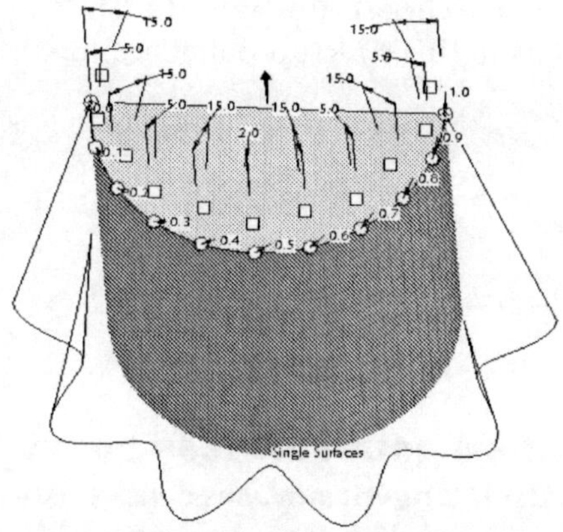

***Figure 7-73** Location points with modified angle values*

13. After modifying the values of the angle at all the locations, choose the **Feature Preview** button from the Draft dashboard. The model appears as shown in Figure 7-74.
14. The model after mirroring the geometry and then shelling the top and bottom faces is shown in Figure 7-75.

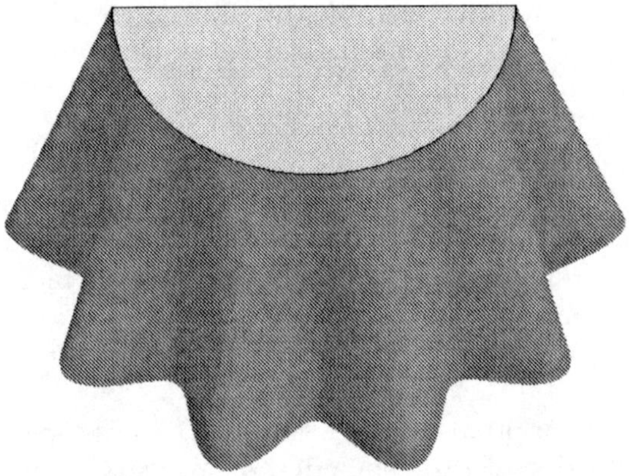

Figure 7-74 Model with the draft feature

Figure 7-75 Model of the lamp shade

Options Tab

When you choose this tab, the slide-up panel is displayed as shown in Figure 7-76. The options in this slide-up panel are discussed next.

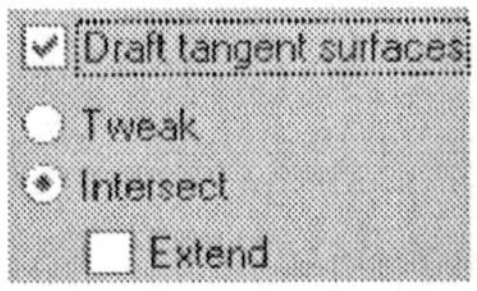

*Figure 7-76 **Options** tab slide-up panel*

Draft tangent surfaces check box

The **Draft tangent surfaces** check box when selected applies the draft to the surfaces tangent to the selected surface. In Figure 7-77, the surface shown is selected to add draft angle. Because the **Draft tangent surfaces** check box is selected, all the surfaces that are tangent to the selected surface and to the other surfaces are automatically selected. Figure 7-78 shows the resultant model with the draft feature.

Tweak radio button

The **Tweak** radio button is used to create a draft when the draft surface does not intersect with any other surface or edge. Generally, this radio button can be selected when you are creating the draft on a base feature and the draft is the second feature on the model.

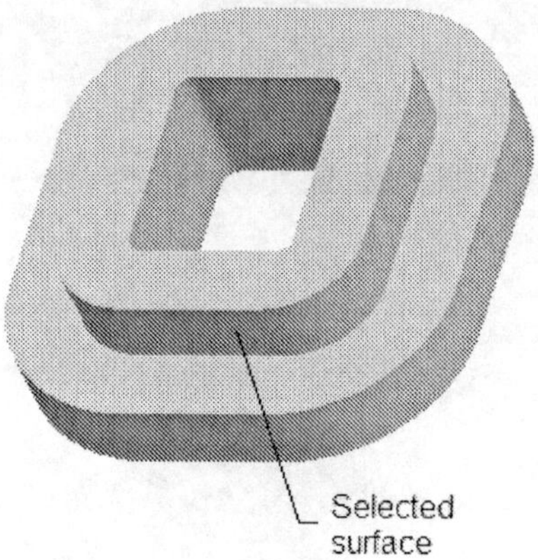

Figure 7-77 *Single surface selected to draft*

Figure 7-78 *Draft applied to all the tangent surfaces*

Intersect radio button

This radio button is selected by default. The **Intersect** radio button is used when the draft surface you are creating can intersect with other surfaces or edges. Figure 7-79 shows the example of the draft that intersects with the adjacent feature and projects outside the edge of the feature at the bottom.

Extend check box

The **Extend** check box is available only when the **Intersect** radio button is selected. When this check box is selected, the draft surface extends in the direction with the feature it intersects. Figure 7-80 shows that the feature when intersected, gets extended.

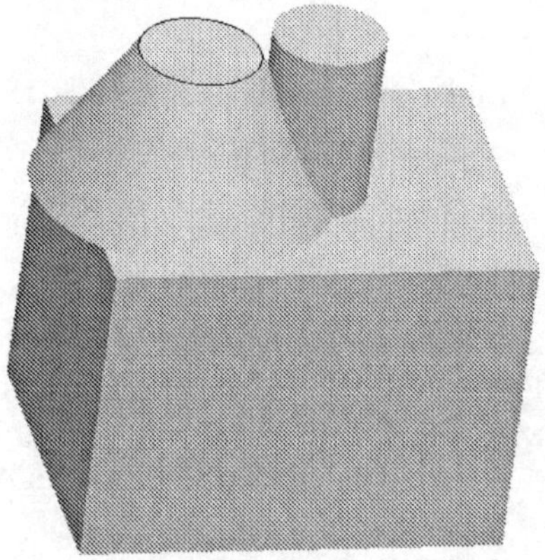

Figure 7-79 *Intersect*

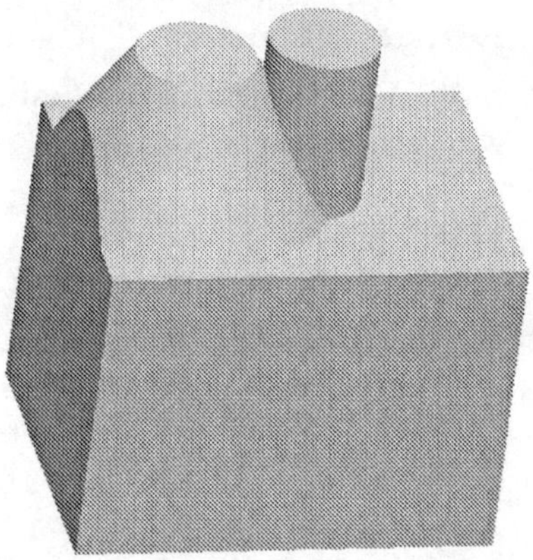

Figure 7-80 *Extend intersect*

Figure 7-81 shows another example of extended draft feature. The following steps explain in brief the procedure to create the pencil-shaped model. It is assumed that you have created the model shown in Figure 7-82. The initial model shown is an octagon on the top of which a cylinder is created.

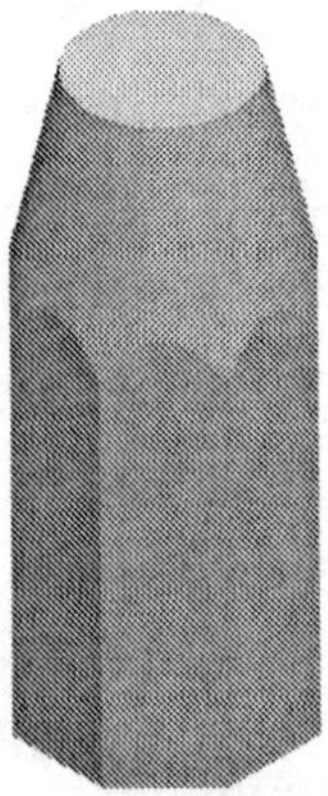

Figure 7-80 Example of extend intersect in a draft surface

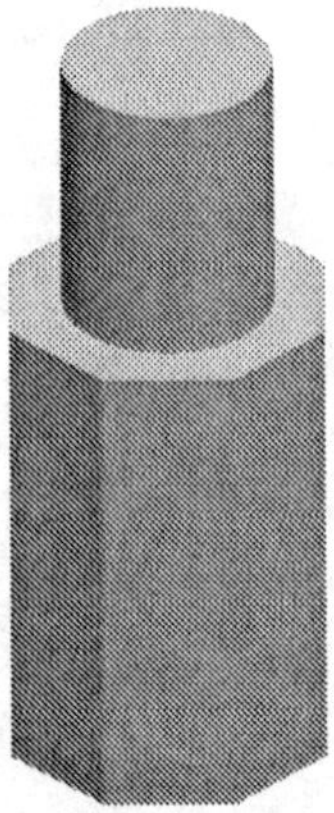

Figure 7-82 Model initially created

1. Invoke the Draft dashboard and select the cylindrical surface to draft.
2. Select the top face of the cylindrical feature as the hinge. The yellow arrow points in the upward direction.
3. Choose the **Options** tab and from the slide-up panel, select the **Extend** check box.
4. Use the drag handle and increase the draft angle upto the vertex of the octagon. This can be easily done by viewing the preview of the draft surface as you increase the angle. You may even have to use the edit box present on the dashboard to enter the small increase in angle value. Remember that you will get the desired shape of the intersect only when the draft surface intersects the vertices of the octagonal feature.

Properties Tab

When you choose the **Properties** tab, the slide-up panel that shows the feature id of the feature you are creating is displayed.

TUTORIALS

Tutorial 1

In this tutorial you will create the model shown in Figure 7-83. This figure also shows the sectioned front view, top view, and the right-side view of the solid model with dimensions.

(Expected time: 40 min)

The following steps outline the procedure for creating this model:

a. First, examine the model and determine the number of features in it. The model is composed of six features, see Figure 7-83.

b. The base feature is a sweep feature, see Figure 7-86. Select the sketching plane for the base feature, draw the sketch using the sketcher tools, and apply constraints and dimensions.

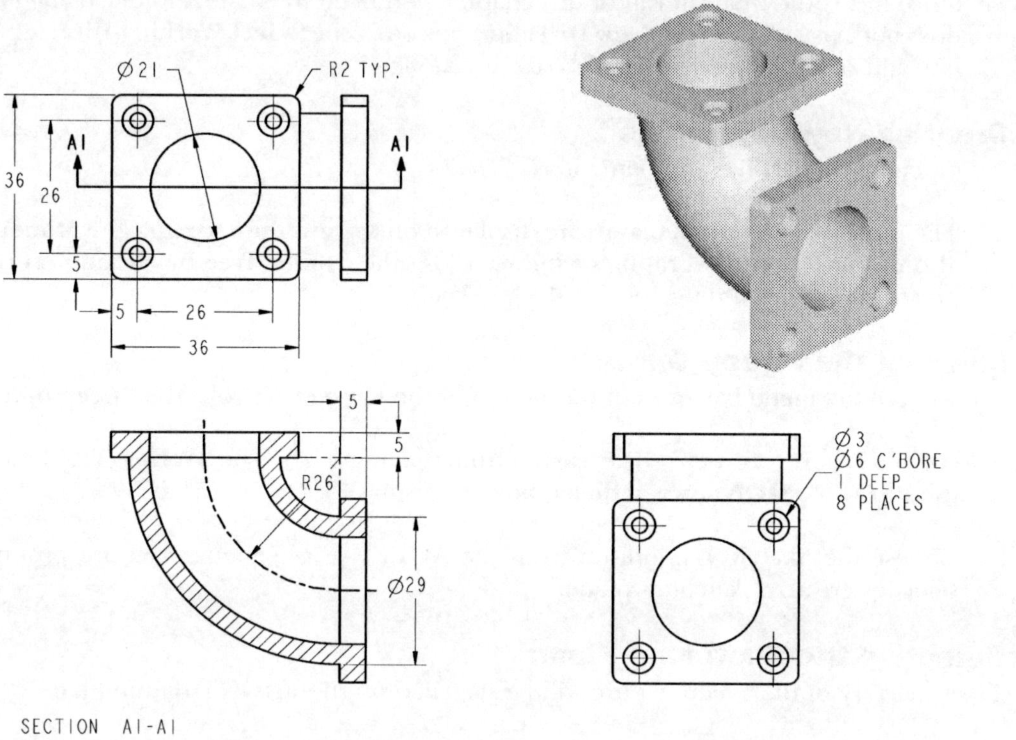

Figure 7-83 *Top, front, right-side, and isometric views of the model. The hidden lines are suppressed for clarity*

c. The second feature is a shell of given thickness. Both the end surfaces of the sweep feature will be removed during shelling, see Figure 7-87.

d. The third and the fourth features are protrusion features and will be created on the two ends of the swept feature respectively. These features have different sketching planes and hence will be created as separate features. Select the sketching plane, draw the sketch using the sketcher tools, apply constraints and dimensions, and then extrude the sketch to the given distance, see Figure 7-89.

e. The fifth feature is a counterbore hole that will be created on the third feature, see Figure 7-91. This hole will be created using the Hole dashboard.

f. The sixth feature is a counterbore hole that will be created on the fourth feature, see Figure 7-89. This hole will be created using the Hole dashboard. Then, the two counterbore holes will be patterned on the third and the fourth feature, see Figure 7-92.

After understanding the procedure for creating the model, you are now ready to create it. When the Pro/ENGINEER session is started, the first task is to set the working directory.

Since this is the first tutorial of this chapter, you need to create a folder named *c07* if it does not exist. Choose the **New Directory** button in the **Select Working Directory** dialog box and create a directory named *c07* at *C:\ProE*.

Creating New Object File

1. Open a new part file and name it as *c07tut1*.

   The three default datum planes are displayed on the graphics screen. The **Model Tree** is also displayed on the graphics window. Close the **Model Tree** by clicking on the sash present on the right edge of the **Model Tree**.

Invoking the Sweep Option

You will use the menu bar present on the top of the screen to invoke the **Sweep** option.

1. Choose **Insert > Sweep > Protrusion** from the menu bar. The **SWEEP TRAJ** menu and the **PROTRUSION: Sweep** dialog box are displayed.

2. Choose the **Sketch Traj** option from the **SWEEP TRAJ** menu. You are prompted to select or create a sketching plane.

Selecting the Sketching Plane

The trajectory of the sweep feature will be sketched on the **FRONT** datum plane.

1. Select the **FRONT** datum plane as the sketching plane. A red arrow points in the direction of viewing the sketch plane.

2. Choose **Okay** from the **DIRECTION** submenu. The **SKET VIEW** submenu is displayed.

3. Choose **Top** from this submenu and select the **TOP** datum plane.

   After you select the planes for orientation, the system takes you to the sketcher environment.

Drawing the Trajectory

From the model it is evident that the trajectory for the sweep feature is a quarter circle.

1. Exit the **References** dialog box by choosing the **Close** button from the dialog box.

2. Choose the **Create an arc by picking its center and endpoints** button from the **Sketcher Tools** toolbar. This button is available on the flyout that is displayed when you choose the black arrow on the right of the **Create an arc by 3 points or tangent to an entity at its endpoint** button.

3. Draw an arc such that the center of the arc lies at the intersection of the **TOP** and **RIGHT** datum planes as shown in Figure 7-84. As you specify the center of the arc, the cursor snaps to the point of intersection of the two planes. Now, draw the arc and exit this tool.

The endpoints of the arc are automatically aligned to the **TOP** and **RIGHT** datum planes. You will notice that an arrow is attached at the start point of the trajectory. This arrow points in the direction of sweep.

Tip: *To change the start point on the trajectory, select the point where you want the start point. When the point is highlighted in red color, hold down the right mouse button to invoke a shortcut menu. Choose the* ***Start Point*** *option.*

Modifying the Dimensions of the Trajectory

When you were drawing the arc, the arc was dimensioned automatically and a weak radial dimension was assigned to it. You need to modify the dimension as per your requirement.

1. Double-click on the dimension and modify the radial dimension to 26 as shown in Figure 7-85. You will notice the sketch refits on the screen.

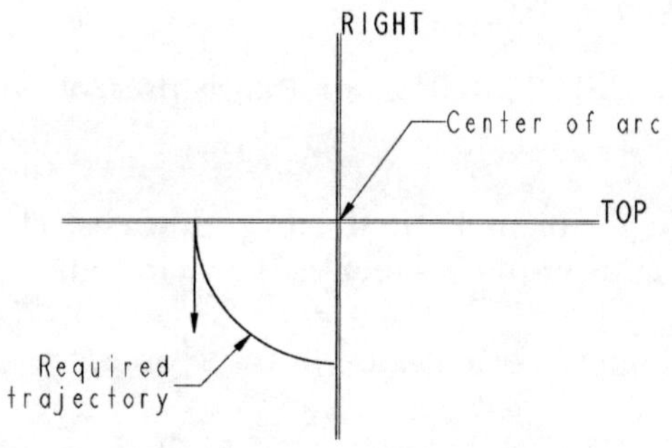

Figure 7-84 *The sketch of required trajectory*

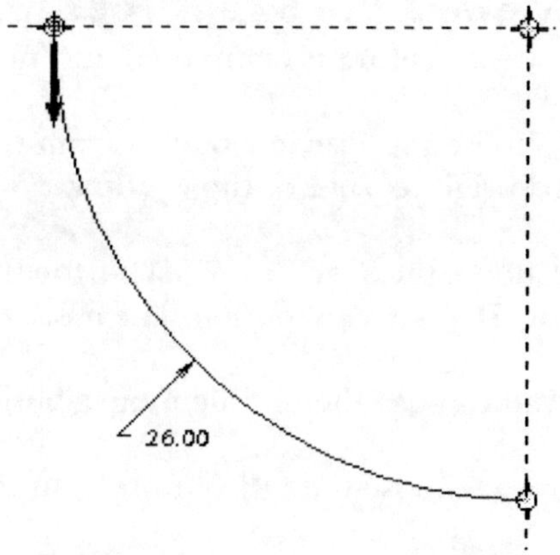

Figure 7-85 *Dimension for the arc*

2. Choose the **Continue with the current section** button.

Drawing the Section for Sweep

After choosing the **Continue with the current section** button, the direction of viewing is modified such that the start point of the trajectory becomes normal to the screen. A yellow cross of infinite length appears on the screen. This cross consists of two perpendicular lines of infinite length. The intersection point of these lines is the start point of the trajectory. The center of the circle should lie at the intersection of these lines. The **References** dialog box is also displayed. The status displayed in the **Reference status** area of the **References** dialog box is **Fully Placed**. Close the **References** dialog box. You are also prompted to draw the cross-section for the sweep.

Tip: *When you enter the sketcher environment to define a section for sweep trajectory, it is often difficult to understand the orientation of the view. For this purpose, a yellow-colored cross that determines the orientation of the section with respect to the trajectory is available.*

1. Choose the **Create circle by picking the center and a point on the circle** button and create a circle such that the center of circle lies at the intersection of the two infinite perpendicular lines.

   When you draw the circle, the cursor snaps to the intersection point of the cross.

Modifying the Dimensions of the Section

1. Choose the **Select items** button and using the left mouse button double-click on the dimension and modify the diameter dimension to 29.

   The sketch accepts the value and refits on the screen.

2. Choose the **Continue with the current section** button.

Previewing the Swept Feature

The sweep feature is completed and now it can be previewed.

1. Choose the **Preview** button from the **PROTRUSION** dialog box that is present on the top right corner of the window.

2. Choose the **Saved view list** button from the **View** toolbar. From the drop-down list, choose the **Default View** option. The model orients on the graphics screen as shown in Figure 7-86.

   You can use the middle mouse button to change the orientation of the model.

3. Now, choose the **OK** button from the **PROTRUSION** dialog box to exit it.

Creating the Shell Feature

The sweep feature is completed and now you can create the next feature. The next feature is shell.

> **Note**
> *Instead of using the* ***Shell*** *option, two concentric circles can be drawn while drawing the section for the sweep feature in order to obtain the desired hollow feature. Also, the* ***Sweep > Thin Protrusion*** *option can be used to obtain the same hollow feature. However, in this tutorial you will use the* ***Shell*** *option.*

1. Choose the **Shell Tool** button from the **Engineering Features** toolbar. The Shell dashboard is displayed and you are prompted to select the surfaces to be removed.

2. Select the one end surface of the swept feature and then using CTRL+left mouse button, select the other end surface. The two surfaces selected are highlighted in red.

3. Enter the thickness value of the shell as **4** in the dimension box present on the Shell dashboard and press ENTER.

4. Choose the **Build feature** button to exit the Shell dashboard.

   The default trimetric view of the shell feature is shown in Figure 7-87. You can use the middle mouse button to view the model from various directions.

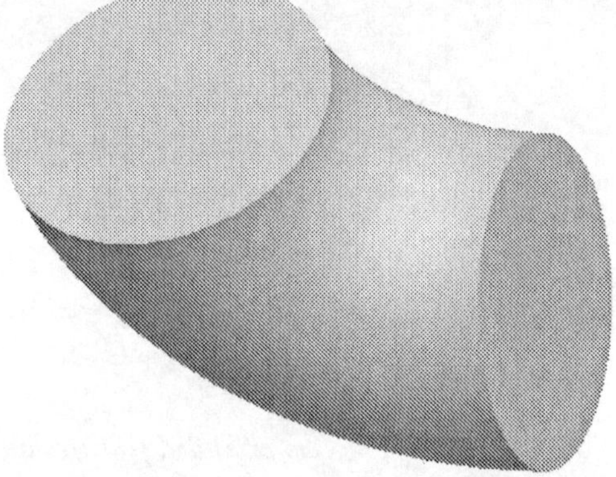

Figure 7-86 Sweep feature without datum planes

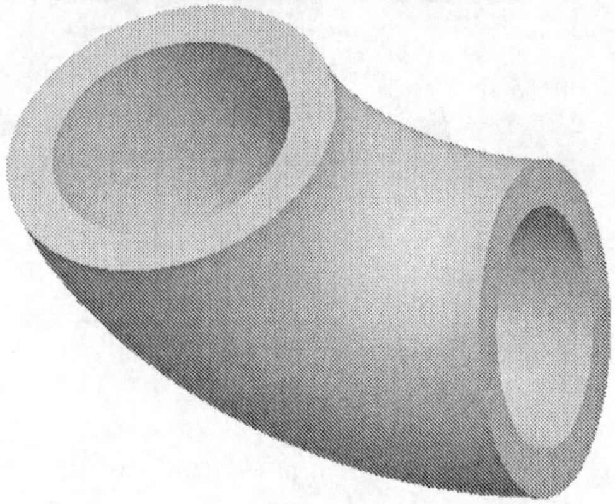

Figure 7-87 Shell feature without datum

Creating the Extrude Features

The next feature is a protrusion feature with a depth of extrusion of 5 and is created at both the ends of the swept feature. While drawing the circle of the sketch for the extrude feature, remember to use the edge of the shell in order to create a hole in the extruded feature also.

1. Choose the **Extrude Tool** button from the **Base Features** toolbar. The Extrude dashboard is displayed.

2. Choose the **Create a section or redefine the existing section** button from the Extrude dashboard. Select the top face of the swept feature as the sketching plane.

3. Select the **RIGHT** datum plane and then choose the **Right** option from the **Orientation** drop-down list.

4. The **References** dialog box is displayed with the status **Not Placed**. Select the **FRONT** and **RIGHT** datum planes. Now, note that the status displayed under the **Reference Status** area in the **References** dialog box is **Fully Placed**.

5. Draw the sketch of the extrude feature and apply constraints and dimensions as shown in Figure 7-88.

6. Create the extruded feature having a depth of extrusion of 5. Similarly, create the next extruded feature at the other end of the swept feature. Select the sketching plane, draw the sketch similar to the first extruded section, apply the same dimensions and constraints, and extrude the sketch to the given distance.

The protrusion features created at both the ends of the swept feature are shown in Figure 7-89.

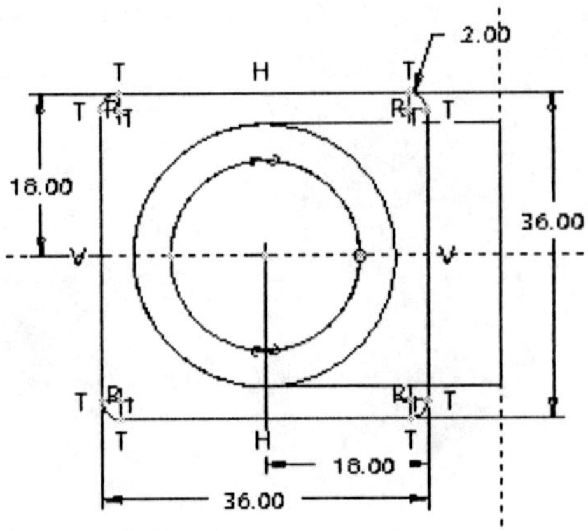

Figure 7-88 Sketch with dimensions and constraints

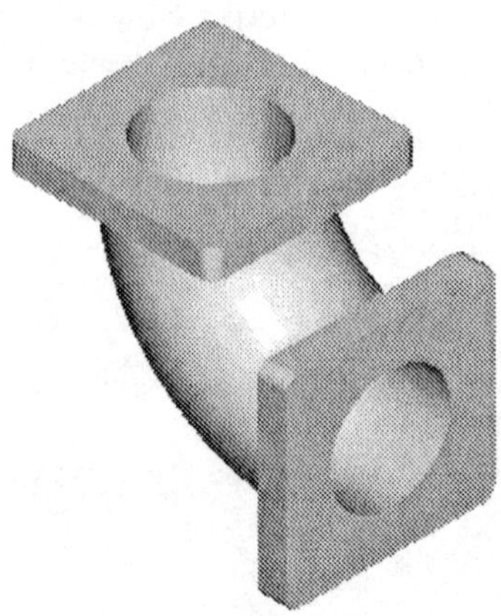

Figure 7-89 The two extruded features on the ends of the sweep feature

Creating the Hole Feature

After creating extruded features at both the ends of the swept feature, counterbore holes will be created. One hole is to be created on each extruded surface and then they are to be patterned on individual planes separately to create the remaining three instances.

1. Choose the **Hole Tool** button from the **Engineering Features** toolbar. The Hole dashboard is displayed with the **Create straight hole** button and the **Simple** option is selected by default.

2. Choose the **Sketched** option from the drop-down list in the Hole dashboard.

3. Choose the **Activates Sketcher to create section** button. The system takes you to the sketcher environment; sketch the section for the counterbore hole as shown in Figure 7-90. Sketch a center line, apply constraints, and diametrically dimension the entities.

4. After completing the sketch, choose the **Continue with the current section** button to exit the sketcher environment.

   The system exits the sketcher environment and the Hole dashboard is redisplayed. Now, you are prompted to select a surface, axis, or point to place the hole.

5. Select the top face of the third extruded feature for the placement of hole. The preview of the hole appears on the graphics window.

   Now, you need to specify the placement parameters for the hole. For this purpose refer to Figure 7-83 and look for the dimensions that can help in placing the hole. The two edges are used to dimension the hole.

6. Choose the **Placement** tab and click in the **Secondary References** collector to turn it yellow.

7. Select the front edge and then use CTRL+left mouse button to select the left edge of the third feature for specifying the linear references. The hole is at a distance of 5 from both the edges. Default dimensions appear on the hole.

   You can also drag the handle and place them at the two edges.

8. Double click on the dimensions and modify the linear distances to **5**.

9. Choose the **Build feature** button from the Hole dashboard. The hole is created on the selected face.

10. Create another hole on the fourth feature using the same procedure as discussed above. The default trimetric view of the model that is completed until now is shown in Figure 7-91.

> **Tip**: *To draw the sketch of the second hole when you enter the sketcher environment do not again draw the same sketch that was drawn for the first hole. Instead, import the sketch for the hole from the* ***In Session*** *folder. The* ***In Session*** *folder contains all the files that you have created in the current session.*
>
> *To import the sketch of the hole in the sketcher environment, choose* ***Sketch > Data from File****. The* ***Open*** *dialog box appears. Choose the* ***In Session*** *folder to display the files created in the current session. Select the third .sec file and open it. The sketch file for the hole will be the third file because you have created three skecthes since this tutorial was started.*

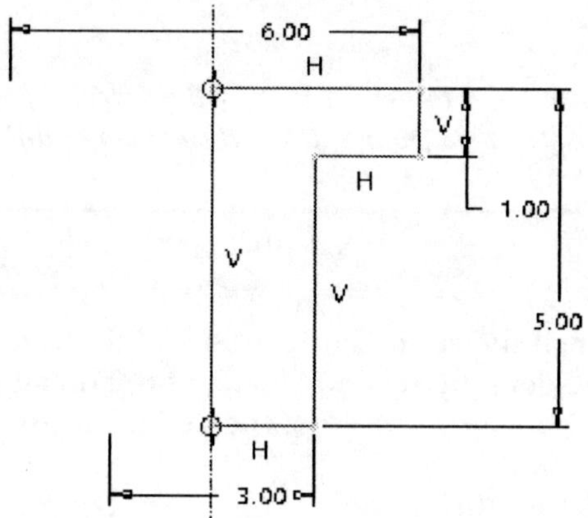

Figure 7-90 *Sketch with dimensions and constraints for the counterbore hole*

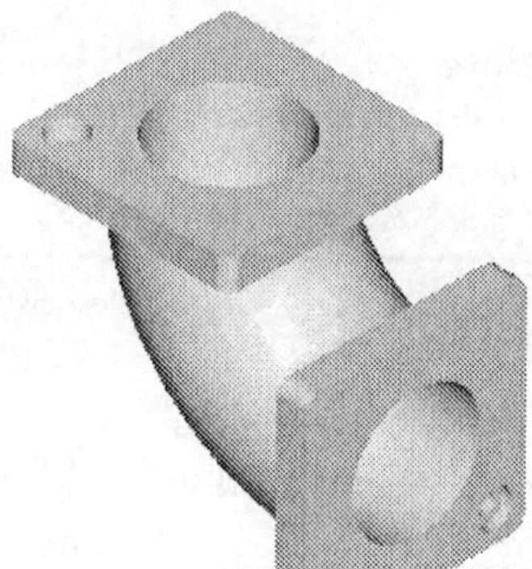

Figure 7-91 *One hole each on the two extruded features*

Creating the Pattern of the Holes

After one hole is placed on each of the two faces, they are patterned.

1. Select the hole feature from the **Model Tree** or from the graphics window and hold down the right mouse button to invoke the shortcut menu. Choose the **Pattern** option from the shortcut menu. The Pattern dashboard is displayed.

2. Create the pattern of the hole.

   Similarly, create pattern of the hole on the other extruded feature also. The default trimetric view of the complete model is shown in Figure 7-92.

Saving the Model

1. Choose the **Save the active object** button from the **File** toolbar and save the model. The order of feature creation can be seen from the **Model Tree** shown in Figure 7-93.

Figure 7-92 *The default trimetric view of the model*

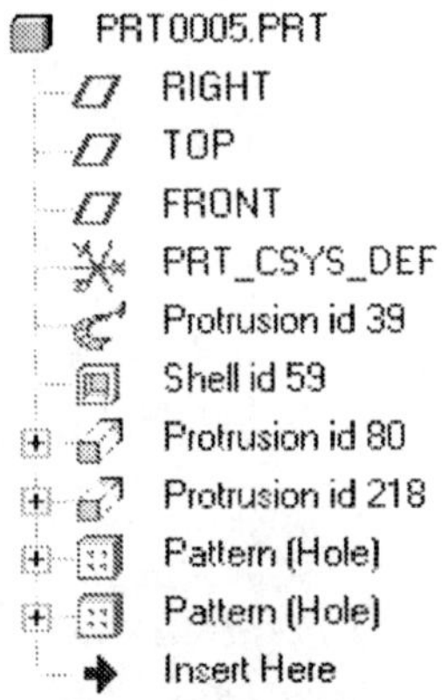

Figure 7-93 Model Tree *for Tutorial 1*

Note

*As discussed in earlier chapters, the model tree is used to get an idea of the order of feature creation. In the **Model Tree**, the id numbers displayed in front of the features may be different when you create the features.*

Tutorial 2

In this tutorial you will create the model of the pencil shown in Figure 7-94 and then change the color of the pencil. Figure 7-95 shows the front view, the section view, and the detail view of the model. **(Expected time: 30 min)**

The following steps outline the procedure for creating this model:

a. First, examine the model and determine the number of features in it. The model is composed of six features, see Figure 7-94.

b. The base feature is a protrusion feature in which the octagon is extruded to a depth of extrusion, see Figure 7-97.

Figure 7-94 Solid model of the pencil

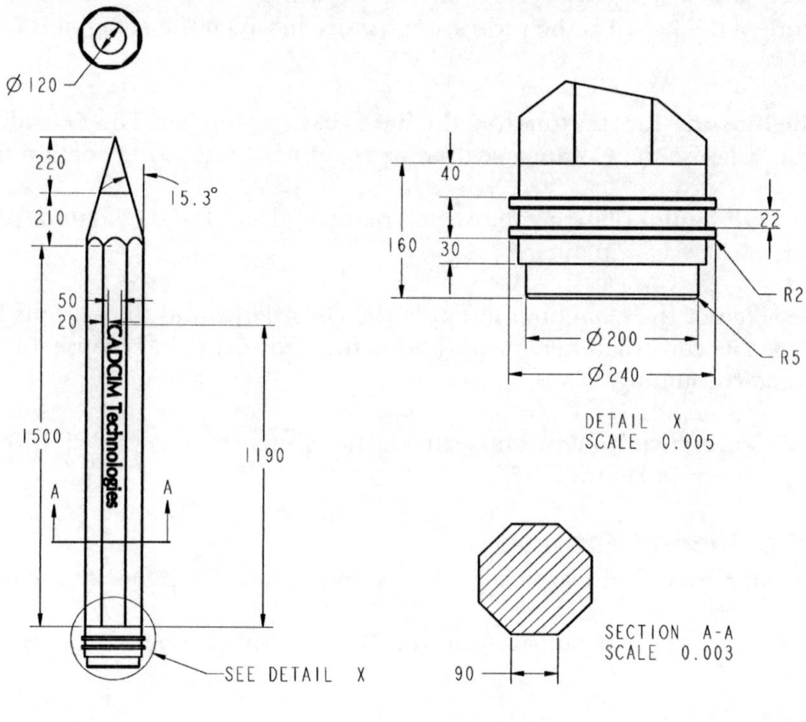

Figure 7-95 Top, front, sectioned, and detail views of the model

c. The second feature is an cylinder on the top face of the base feature, see Figure 7-98.

d. The third feature is a draft feature, see Figure 7-99.

e. The fourth feature is a revolve feature that will create the tip of the pencil, see Figure 7-101.

f. The fifth feature is a revolve feature that will create the tail-end of the pencil, see Figure 7-103.

g. The sixth feature is the text written on the pencil. The text will be written by using the sketched datum curves, see Figure 7-105.

After understanding the procedure for creating the model, you are now ready to create it. The working directory has already been selected in Tutorial 1. Make sure that *c07* is the current working directory.

Creating New Object File

1. Open a new part file and name it as *c07tut2*.

   The three default datum planes are displayed on the graphics window.

Creating the Base Feature

The base feature of the pencil is the protrusion feature in which the octagon is extruded to a depth of 1500.

1. Choose the **Extrude Tool** button from the **Base Features** toolbar. The Extrude dashboard is displayed. Choose the **Create a section or redefine the existing section** button.

2. Select the **TOP** datum plane as the sketch plane and the **RIGHT** datum plane for the orientation of the sketch plane.

3. Draw the sketch of the octagon and apply the constraints and dimensions as shown in Figure 7-96. The construction circle is used to draw the octagon because it becomes easy and less time-consuming.

4. Exit the sketcher environment and extrude the sketch to a depth of 1500. The base feature is as shown in Figure 7-97.

Creating the Second Feature

The second feature is a cylindrcial feature that is drawn on the top face of the base feature.

1. Choose the **Extrude Tool** button from the **Base Features** toolbar. The **Section** dialog box is displayed.

2. Enter the sketcher environment and draw the sketch for the cylinder. The diameter of the circle is 120.

3. After exiting the sketcher environment, extrude the circle to a depth of 210. The model

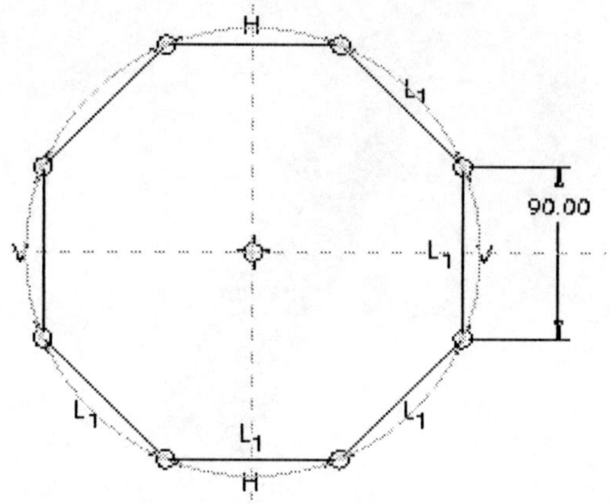

Figure 7-96 Sketch of the octagonal feature

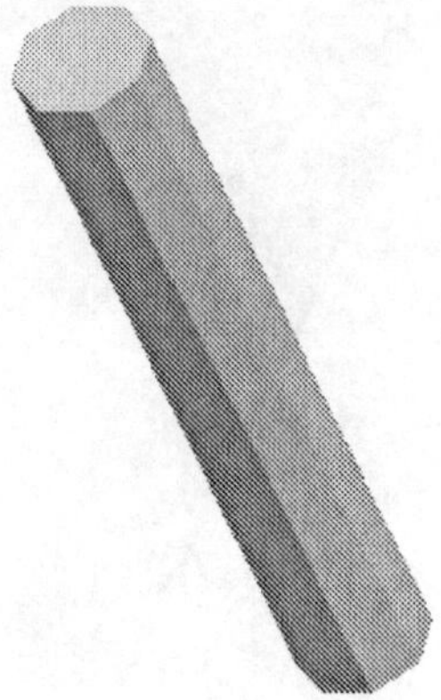

Figure 7-97 Octagonal sketch after extruding

after creating the second feature is shown in Figure 7-98.

Creating the Draft Feature

The draft feature on the cylindrical surface will be created and the draft surface will be allowed to intersect the base feature. This is because when the draft surface intersects the base feature and is extended, the required shape is obtained at the intersecting edge.

1. Choose the **Draft Tool** button from the **Engineering Features** toolbar. The Draft toolbar is displayed.

2. Select the cylindrical surface of the cylinder to draft.

3. Choose the **References** tab to invoke the slide-up panel. Click the **Draft hinges** collector and then select the top face of the cylindrical feature as the hinge. The yellow arrow points in the upward direction.

4. Choose the **Reverse pull direction** button present on the right of the **Pull direction** collector to change the direction of the arrow so that it points downward.

5. Choose the **Options** tab to invoke the slide-up panel; select the **Extend** check box.

6. Enter the angle value of **15.3** in the dimension box present on the Draft dashboard.

7. Choose the **Build feature** button from the Draft dashboard to complete the draft feature as shown in Figure 7-99.

Creating the Fourth Feature

The fourth feature is a revolve feature that will be used to create the tip of the pencil.

1. Choose the **Revolve Tool** button from the **Base Features** toolbar. The Revolve dashboard is displayed.

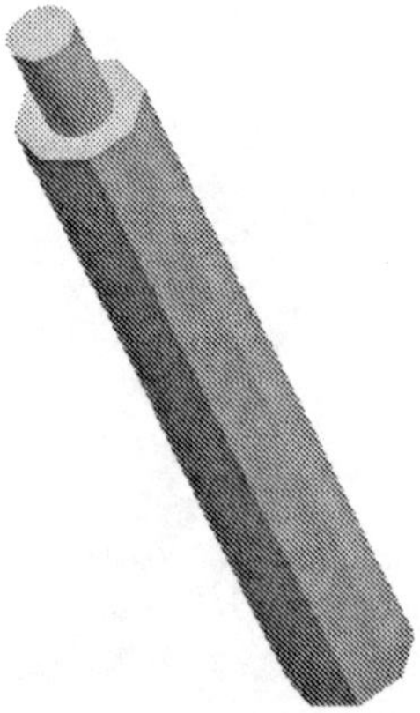

***Figure 7-98** Model after creating the cylindrical feature*

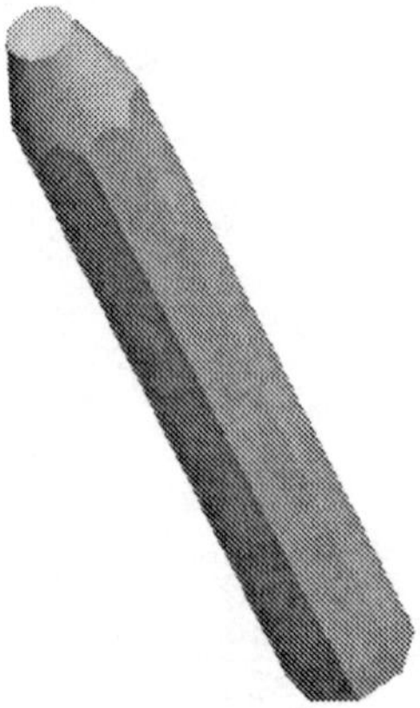

***Figure 7-99** Model after creating the draft feature*

2. Invoke the **Section** dialog box and select the **FRONT** datum plane as the sketch plane.

3. After entering the sketcher environment, draw the sketch, apply constraints, and dimension it as shown in Figure 7-100.

4. After exiting the sketcher environment, the angle of revolution of 360-degrees is selected by default.

5. Choose the **Build feature** button to complete the feature. The default trimetric view of the model after creating the fourth feature is shown in Figure 7-101.

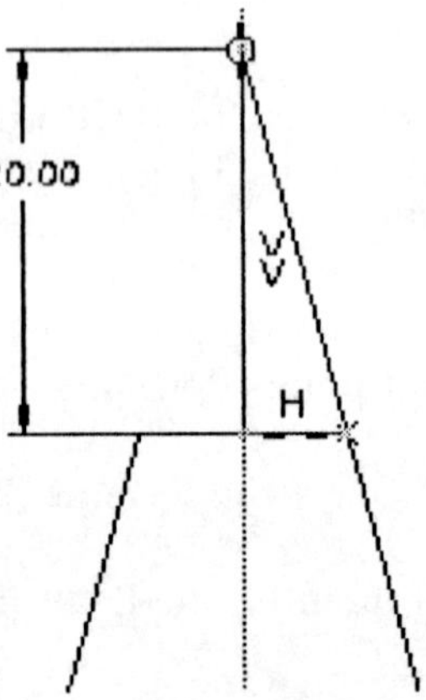

***Figure 7-100** Sketch for the tip of the pencil*

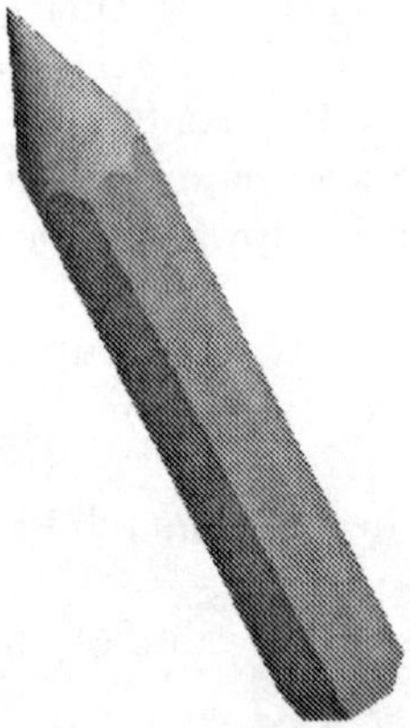

***Figure 7-101** Model after creating the tip*

Creating the Fifth Feature

The fifth feature is a revolve feature that is used to create the top portion of the pencil.

1. Choose the **Revolve Tool** button from the **Base Features** toolbar. The Revolve dashboard is displayed.

2. Invoke the **Section** dialog box and select the **FRONT** datum plane as the sketch plane.

3. After entering the sketcher environment, draw the sketch, apply constraints, and dimension it as shown in Figure 7-102.

4. After exiting the sketcher environment, the angle of revolution of 360-degrees is selected by default.

5. Choose the **Build feature** button to complete the feature. The default trimetric view of the model after creating the fourth feature is shown in Figure 7-103.

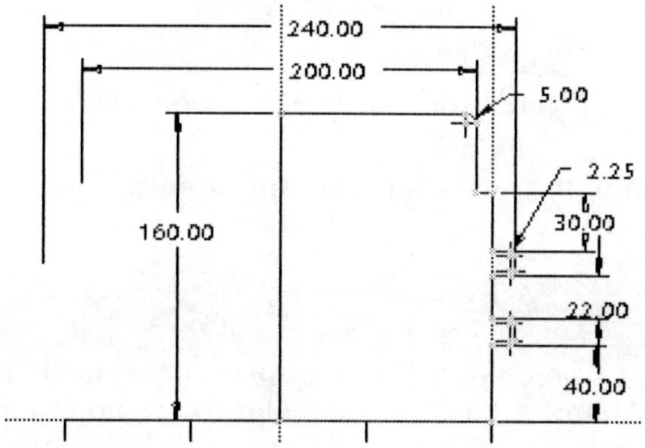

***Figure 7-102** Sketch with dimensions and constraints*

***Figure 7-103** Model after creating the top of the pencil*

Writing the Text

Generally, the text on a model should be written using the datun curves.

1. Choose the **Sketched Datum Curve Tool** button from the **Right Toolchest**. The **Sketch** dialog box is displayed.

2. Select the front face of the base feature as the sketch plane.

3. After entering the sketcher environment, create the text using the **Create text as a part of a section** button. Figure 7-104 shows the text written, with the dimensions.

4. Choose the **Continue with the current section** button to exit the sketcher environment.

   The model after creating the text is shown in Figure 7-105.

Changing the Colors

The colors in Pro/ENGINEER can be applied on selected surfaces. Remember that the colors you apply are saved with the model and remain on the model until you clear them.

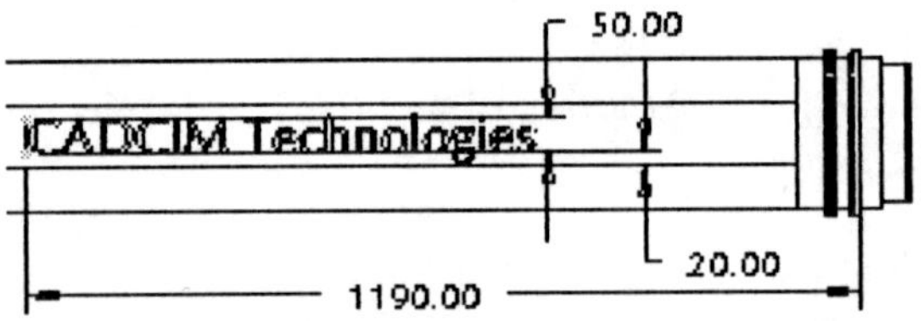

Figure 7-104 Text with dimensions

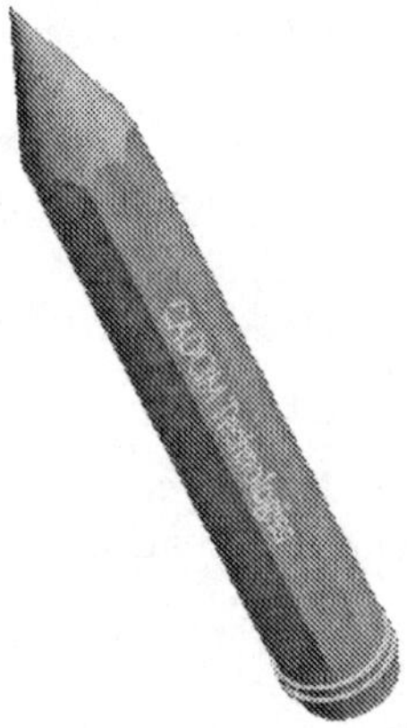

Figure 7-105 Model after creating the text

1. Choose **View > Color and Appearance** from the menu bar. The **Appearance Editor** is displayed.

2. Select the +sign. The new color is added.

3. In the **Color** area, choose the **Color** button. The **Color Editor** is displayed. Set the RGB values to 0. The color obtained is black. You need to apply this color to the tip of the pencil.

4. Choose the **Close** button from the **Color Editor**.

5. Choose the **Surfaces** option from the drop-down list present in the **Assignment** rollout. The **SELECT** dialog box is displayed and you are prompted to select the surfaces to alter colors.

6. Select the tip of the pencil and press the middle mouse button. Then choose the **Apply** button to apply the color to the selected surface. The tip of the pencil is changed to black.

7. To add another color to any of the surfaces of the pencil, select the +sign and modify the color to the required color. Follow steps 3 to 6 to add colors to other surfaces of the pencil.

8. After modifying the colors, choose the **Properties** rollout and then choose the **Close** button.

Saving the Model

1. Choose the **Save the active object** button from the **File** toolbar and save the model.

The order of feature creation can be seen from the **Model Tree** shown in Figure 7-106.

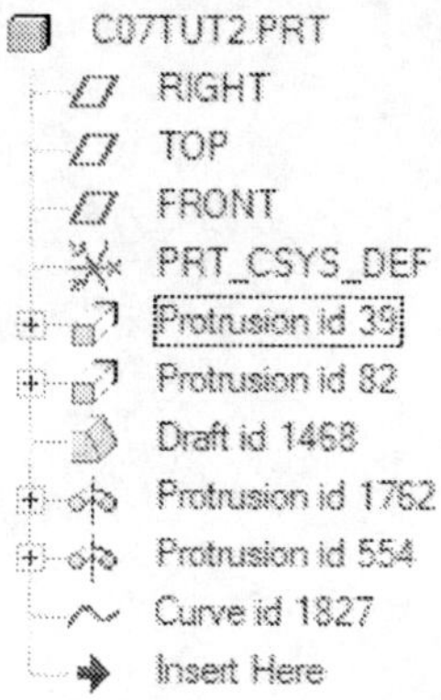

***Figure 7-106 Model Tree** for Tutorial 2*

Note

*Colors are not a feature of the model and therefore they will not appear in the **Model Tree**.*

Tutorial 3

In this tutorial you will create a blend feature shown in Figure 7-107. The two views of the blend feature are shown in Figure 7-108 with dimensions. After creating the model, you will redefine it such that the straight blending is changed into a smooth blending.

(Expected time: 45 min)

The following steps outline the procedure for creating this model:

a. First, examine the blend feature and determine the number of sections in this feature. The blend consists of four sections, see Figure 7-108.

b. Select the sketching plane for the blend feature. Draw the first section, apply dimensions and constraints, see Figure 7-109, and then toggle the section to draw the sketch for the second section.

c. Draw the second section, apply the dimensions and constraints, see Figure 7-110, and then toggle the section to draw the third section.

d. Draw the third section, apply the dimensions and constraints, see Figure 7-111, and then toggle the section to draw the fourth section.

e. Draw the fourth section, apply the dimensions and constraints, see Figure 7-111, and then give the depth between section numbers 1 and 2, 2 and 3, and 3 and 4.

f. Redefine the model to change the straight blending into a smooth blending.

After understanding the procedure for creating the model, you are now ready to create it.

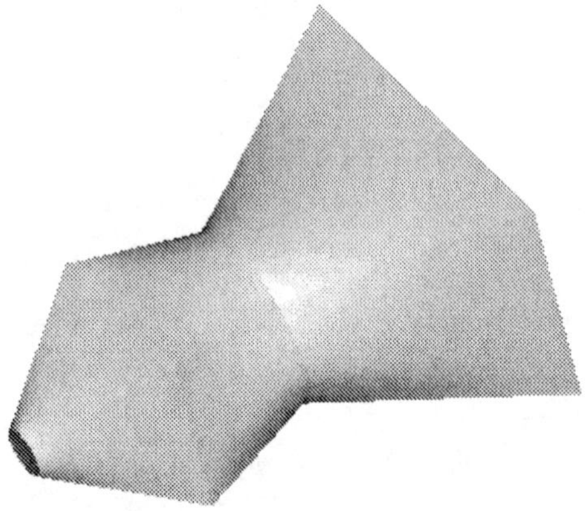

Figure 7-107 Isometric view of the model

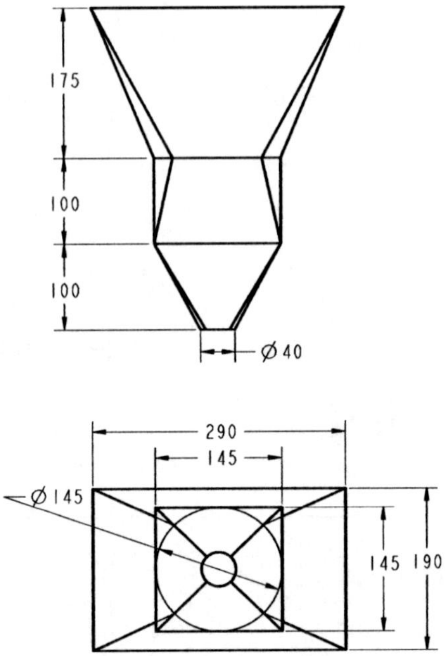

Figure 7-108 Top and front views of the model

Creating New Object File

1. Open a new part file and name it as *c07tut3*.

The three default datum planes are displayed on the graphics screen.

Invoking the Blend Option

1. Choose **Insert > Blend > Protrusion** from the menu bar. The **BLEND OPTS** submenu is displayed.

2. Choose **Parallel > Regular Sec > Sketch Sec > Done** options from the **BLEND OPTS** submenu. The **ATTRIBUTES** menu is displayed and you will be prompted to choose the **Straight** or **Smooth** option from this menu.

3. Choose **Straight > Done** option from the **ATTRIBUTES** menu.

A **Smooth** blend will be created during the redefining of the model using the same sections that will be used to create the given model .

Selecting the Sketch Plane

1. Select the **FRONT** datum plane as the sketching plane. A red arrow points in the direction of feature creation and you are prompted to specify the direction of feature creation.

2. Choose **Okay** from the **DIRECTION** submenu. The **SKET VIEW** submenu is displayed.

3. Choose **Top** from this menu and using the left mouse button select the **TOP** datum plane present on the graphics screen.

Drawing the First Section

The first section is a rectangle of 290x190 units.

1. Draw the sketch of the rectangular section and then add constraints and dimensions to it as shown in Figure 7-109.

   After drawing the rectangular section, you need to toggle the section and draw the next section.

Note

While drawing the sections for the blend feature, the start point is very important. The start point should be similar to those shown in the figures. If the start point is not at the desired point then select the point where you need the start point. Hold down the right mouse button to invoke a shortcut menu and choose the ***Start Point*** *option.*

Toggling the Section

Toggling of a section is required in order to sketch the next section. Since in this tutorial four sections are used to create the required model, therefore, whenever you finish drawing one section you need to toggle to the next section.

1. Choose **Sketch > Feature Tools > Toggle Section** from the menu bar. You can also hold down the right mouse button to display the shortcut menu and choose the **Toggle Section** option from the shortcut menu.

   When you choose **Toggle Section**, the previous section becomes inactive and appears gray in color.

Tip: *To toggle the section, simply right-click when you are in the selection mode. The selection mode in the sketcher environment can be invoked by choosing the* ***Select items*** *button.*

Drawing the Next Section

The next section is a circle.

1. Draw the sketch of the circular section, refer to Figure 7-110. Modify the diameter of the circle to 145.

   As discussed earlier, the number of entities per section must be equal in a blend feature. Since, a circle is a single entity, it should be divided at four points.

Dividing the Circular Section

The circular section should be divided at four points because the rectangle and square have four entities. When you divide a circle at four points, the number of entities becomes four.

1. Choose the black arrow on the right of the **Dynamically trim section entities** button and then choose the **Divide an entity at the point of selection** button.

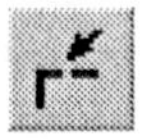

2. Select the circle at four points as shown in Figure 7-110.

   As you select points on the circle to divide it, some weak dimensions appear on the circle. Next, you need to apply constraints on the four points that were selected to divide the circle.

Applying Constraints on the Four Points

1. Choose the **Impose sketcher constraints on the section** button. The **Constraints** dialog box is displayed.

2. Choose the **Make line or two vertices vertical** constraint button from the **Constraints** dialog box and select the two division points on the left side of the circle to lie in a vertical line. Similarly, select the two points on the right to apply the constraint.

3. Now, choose the **Make line or two vertices horizontal** constraint button and select the two division points on the upper half and the lower half to lie in a horizontal line.

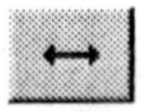

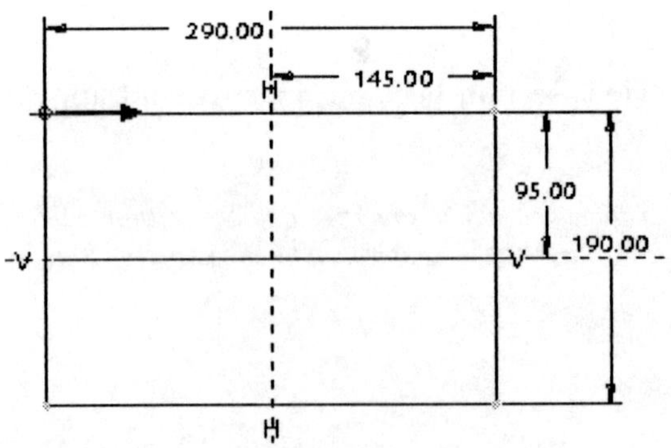

Figure 7-109 First rectangular section with dimensions

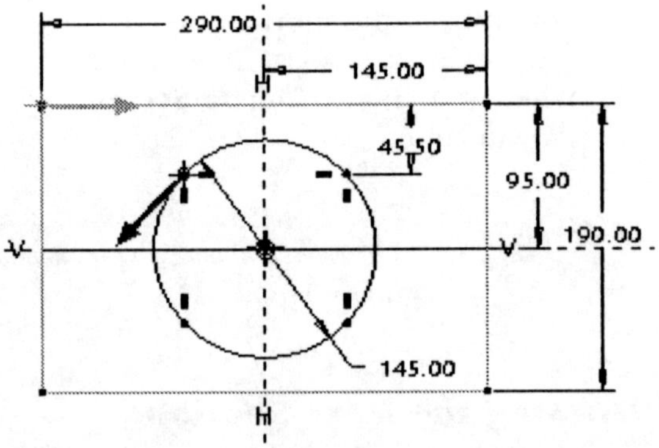

Figure 7-110 The two completed sections

4. Modify the vertical dimension of the upper left division point as shown in Figure 7-110. After the circular section is completed, the two sections with dimensions should look similar to the sections shown in Figure 7-110.

5. Now, toggle the section and create the next section. The next section to be drawn is a square. After drawing the square section, draw the circular section. Divide the circular section into four entities similar to section 2 and then constrain and dimension it. Figure 7-111 shows all the sections completed with dimensions.

Note

In Figure 7-111, note the direction of the start points shown by arrows. These are important to avoid a twisted feature.

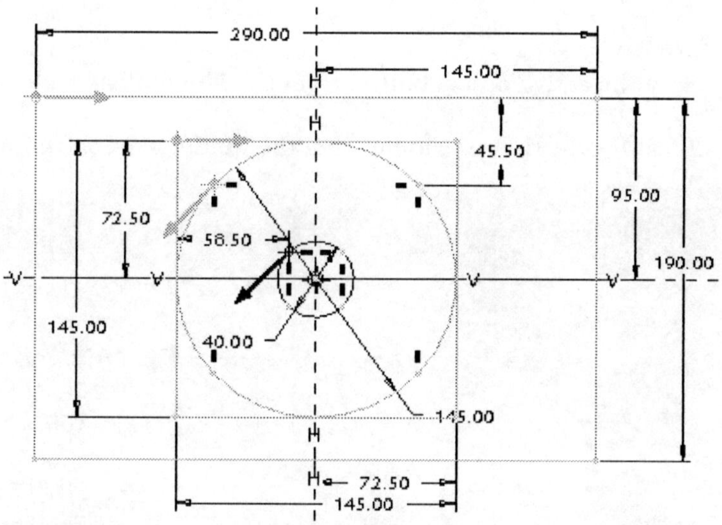

Figure 7-111 *All four completed sections before giving depth*

Giving Depth to the Sections

After the sketches of all the sections are completed, you need to specify the depth between each section. The dimensions for depth between each section can be referred to from Figure 7-108.

1. Choose the **Continue with the current section** button.

   The **Message Input Window** appears and you are prompted to enter the depth for section 2.

2. Enter a value of **175** in the window and press ENTER.

   Similarly, the **Message Input Window** appears again and you are prompted to enter the depth for section 3. Enter 100 for section 3 and 100 for section 4.

Previewing the Blend Feature

The blend feature is completed and it can now be previewed.

1. Choose the **Preview** button from the **PROTRUSION** dialog box that is displayed on the top right corner of the window.

2. Choose the **Saved view list** button from the **View** toolbar. From the drop-down list choose the **Default View** option. The model orients on the graphics window as shown in Figure 7-112. You can use the middle mouse button to change the orientation of the model.

3. Now, choose the **OK** button that is present on the **PROTRUSION** dialog box.

Saving the Model

1. Choose the **Save the active object** button from the **File** toolbar and save the model.

   The order of feature creation is evident from the **Model Tree** shown in Figure 7-113.

Figure 7-112 Shaded model

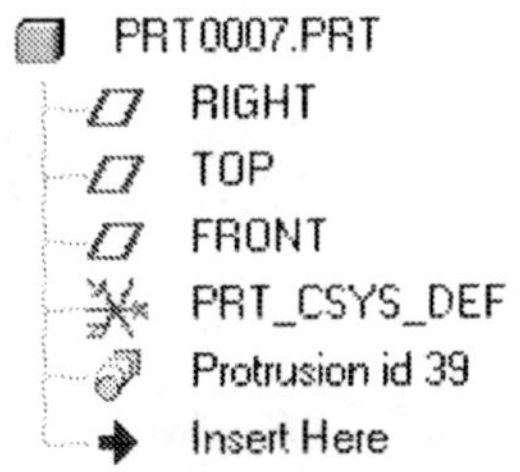

Figure 7-113 ***Model Tree*** *for Tutorial 3*

Redefining the Blend Feature

After saving the straight blend feature, you will redefine this feature so that it is converted into a smooth blend.

1. Select the model on the graphics window. The edges of the model turn red in color.

2. Press and hold down the right mouse button on the graphics screen until a shortcut menu appears.

3. Choose the **Redefine** option from the shortcut menu. The **PROTRUSION** dialog box is displayed.

4. Select the **Attributes** option under the **Elements** tab. The **Attributes** option is highlighted.

5. Choose the **Define** button from the **PROTRUSION** dialog box. The **ATTRIBUTES** menu is displayed.

6. Choose **Smooth** > **Done** from the **ATTRIBUTES** menu.

7. Choose the **OK** button from the **PROTRUSION** dialog box. The smooth blend is created as shown in Figure 7-114.

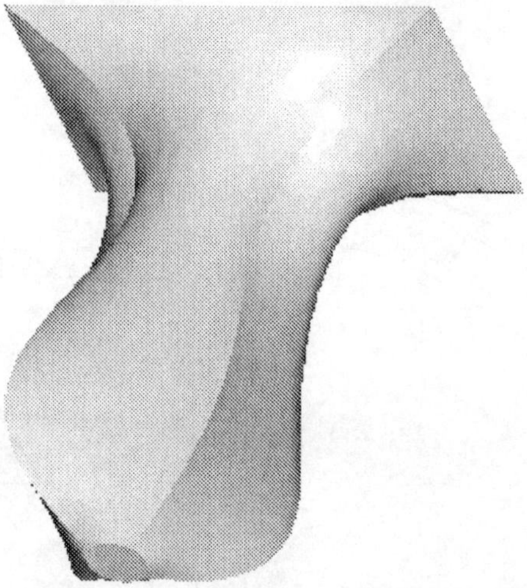

Figure 7-114 *Smooth blend feature*

Tutorial 4

In this tutorial you will create the model of a tap shown in Figure 7-115. All dimensions for the three sections are shown in the figure. **(Expected time: 30 min)**

This model is created using the general blend. In general blend, each section should have a coordinate system. The coordinate system helps in the alignment of the sections. Each section will be dimensioned with its coordinate system.

The following steps outline the procedure for creating this model:

a. First, examine the model and determine the number of sections in the blend feature. The model consists of three sections, see Figure 7-115. The first section is the bottom of tap, Figure 7-116. The second section, Figure 7-117, is tilted at 50-degree with respect to y-axis and the third section, Figure 7-118, is further located 30-degree with respect to y-axis.

b. Select the sketching plane for the blend feature, draw the first section, insert a coordinate system, apply dimensions and constraints, see Figure 7-116, and then continue to the second section.

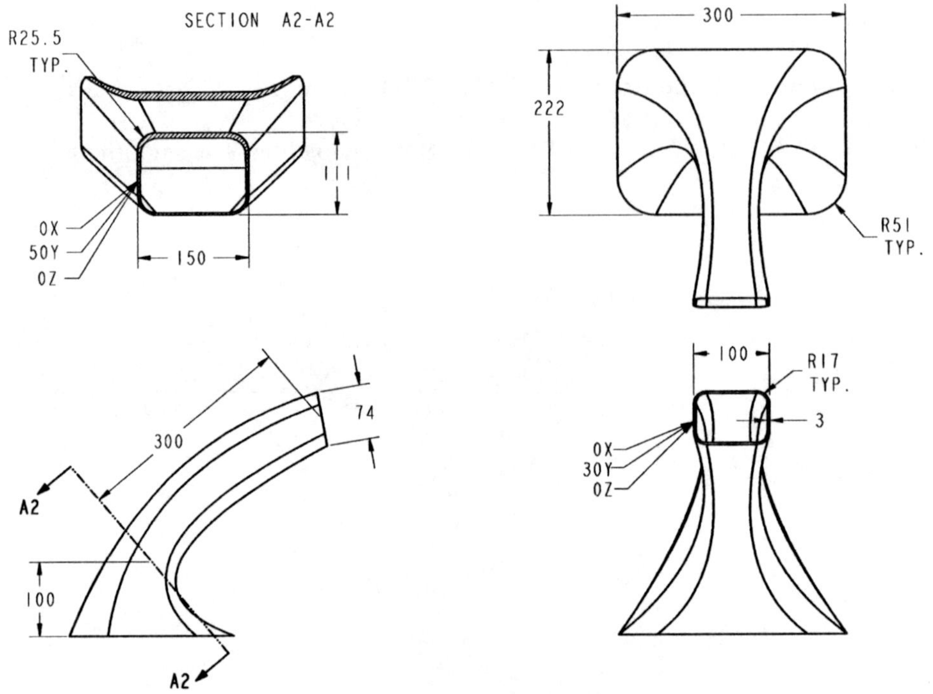

Figure 7-115 *Sectioned, left-side, front, and top views of the model*

c. Draw the second section, insert a coordinate system, apply dimensions and constraints, see Figure 7-117, and then continue to the third section.

d. Draw the third section, insert a coordinate system, apply dimensions and constraints, see Figure 7-118, and then specify the distance between the sections.

After understanding the procedure for creating the model, you are now ready to create it.

Creating New Object File

1. Open a new part file and name it as *c07tut4*. The three default datum planes appear on the graphics window if they were not turned off in the previous tutorial.

Invoking the Blend Option

The given model is created using the general type of blend.

1. Choose **Insert** > **Blend** > **Protrusion** from the menu bar. The **BLEND OPTS** submenu is displayed.

2. Choose **General** > **Regular Sec** > **Sketch Sec** > **Done** from this submenu. The **ATTRIBUTES** menu is displayed.

3. Choose **Smooth** > **Done**.

   You are prompted to select a sketching plane.

4. Select the **TOP** datum plane and choose **Okay** from the **DIRECTION** submenu. From the **SKET VIEW** submenu, choose the **Right** option and select the **RIGHT** datum plane. The system takes you to the sketcher environment.

Drawing the Sketch for the First Section of the Blend

The blend consists of three sections. The dimensions for the second section are half of the dimensions of the first section. Similarly, the dimensions for the third section are one-third of the dimensions of the first section.

1. Draw the sketch for the first section.

2. Choose **Sketch** > **Coordinate System** from the menu bar. A reference coordinate system is attached to the cursor. Place the coordinate system at the intersection of the two planes using the left mouse button. It will be automatically aligned with the two datum planes.

3. Apply constraints and modify the dimensions of the sketch as shown in Figure 7-116.

4. Choose the **Continue with the current section** button. The **Message Input Window** is displayed and you are prompted to enter the x-axis rotation angle for section 2.

5. Enter **0** as the value in this window and press ENTER. You are prompted to enter the y-axis rotation angle for section 2. Enter **50** as the value and press ENTER. Now, you are prompted to enter the z-axis rotation angle for section 2. Enter **0** as the value and press ENTER.

   The system takes you to the sketcher environment and allows you to draw the sketch for the second section. Here, you will notice that the datum planes are not displayed even if they are turned on. The reason for this is that the datum planes are not required now. The use of datum planes is to reference the sketch you draw. But, the second section that you will draw is already referenced to the first section. Hence, the datum planes are not displayed and you can draw the next section anywhere on the graphics screen.

Drawing the Sketch for the Second Section

1. Draw the sketch for the second section, insert a reference coordinate system, apply constraints and dimensions to the sketch, and modify the dimensions as shown in Figure 7-117.

2. After completing the sketch, choose the **Continue with the current section** button.

   The **Message Input Window** is displayed and you are prompted to specify if you want to continue to the next section.

3. Enter **Y** in this window and press ENTER. You have entered **Yes** in this window because you have to draw the third section of the blend feature.

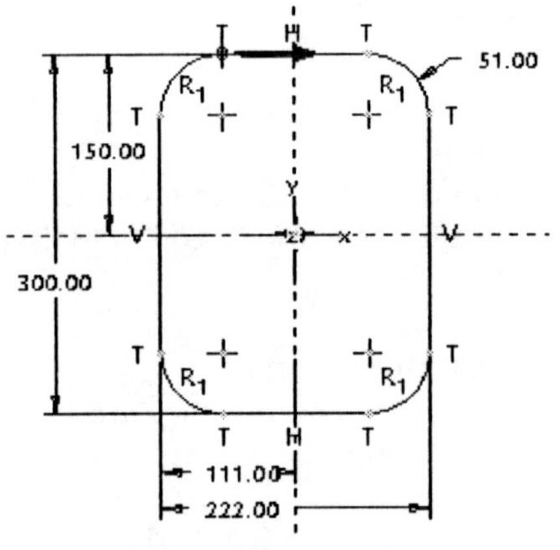

Figure 7-116 *Sketch with dimensions and constraints of section 1*

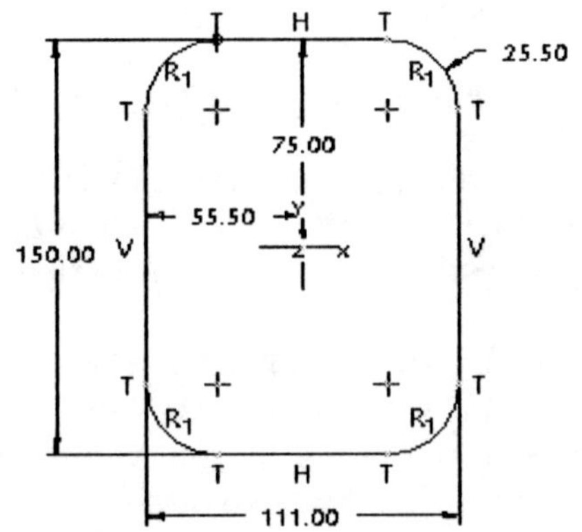

Figure 7-117 *Sketch with dimensions and constraints of section 2*

The **Message Input Window** is displayed and you are prompted to enter the x-axis rotation angle for section 3.

4. Enter **0** as the value in this window and press ENTER. You are prompted to enter the y-axis rotation angle for section 3. Enter **30** as the value and press ENTER. Now, you are prompted to enter the z-axis rotation angle for section 3. Enter **0** as the value and press ENTER.

   The system takes you to the sketcher environment and allows you to draw the sketch for the third section.

Drawing the Sketch for the Third Section

1. Draw the sketch for the third section, insert a reference coordinate system, apply constraints and dimension to the sketch, and modify the dimensions as shown in Figure 7-118.

2. After completing the sketch, choose the **Continue with the current section** button.

   The **Message Input Window** is displayed and you are prompted if you want to continue to the next section.

3. Enter **N** in this window and press ENTER. You have entered **No** in this window because all the sections that are needed to create the blend feature are completed.

   The **Message Input Window** is displayed and you are prompted to enter the depth for **section 2**.

Specifying the Depths between Sections

1. Enter a value of **100** in the **Message Input Window** and press ENTER. This depth is the distance between the first section and the second section.

Now, you are prompted to enter the depth for **section 3**.

2. Enter a value of **300** in this window and press ENTER. This depth is the distance between the second section and the third section.

   The blend feature is complete and you can now preview it.

3. Choose **Preview** from the **PROTRUSION** dialog box and choose **OK**. The default trimetric view of the blend feature is shown in Figure 7-119.

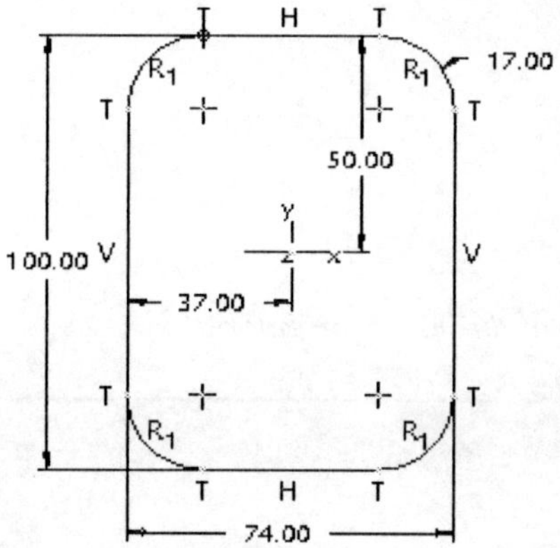

***Figure 7-118** Sketch with dimensions and constraints of section 3*

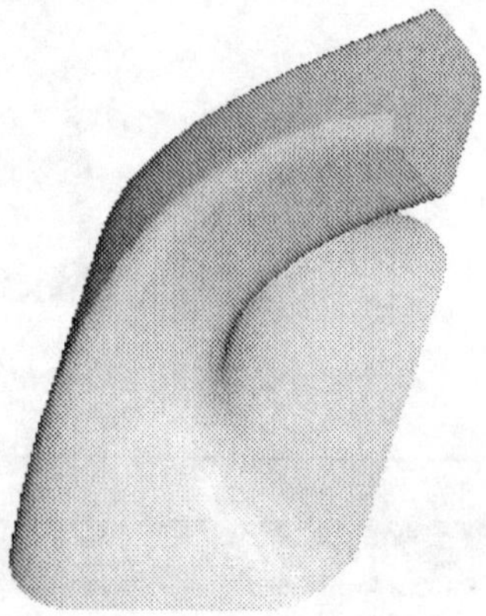

***Figure 7-119** The default trimetric view of the blend feature*

Creating Shell

1. Choose the **Shell Tool** button from the **Engineering Features** toolbar. The Shell dashboard is displayed and you are prompted to select the surfaces to be removed from the part.

2. Select the one end surface of the swept feature using the left mouse button and then using CTRL+left mouse button, select the other end surface. The two surfaces selected are highlighted red in color.

3. Enter the thickness value of the shell as **3** in the dimension box present on the Shell dashboard and press ENTER.

4. Choose the **Build feature** button to exit the Shell dashboard.

   The default trimetric view of the shell feature is shown in Figure 7-120. You can use the middle mouse button to change the orientation of the model.

Saving the Model

1. Choose the **Save the active object** button from the **File** toolbar and save the model.

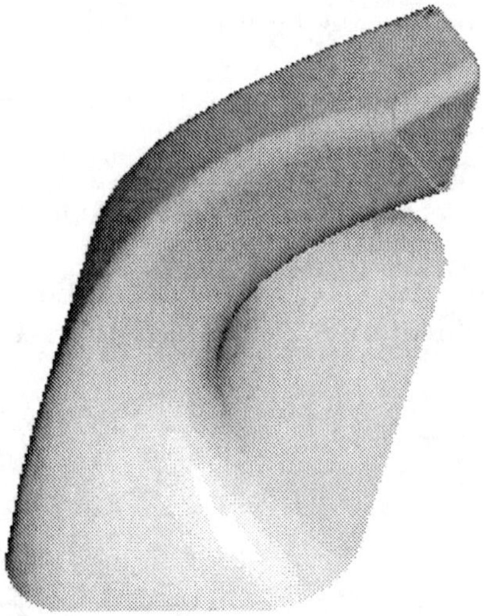

Figure 7-120 The default trimetric view of the model for Tutorial 4

Self-Evaluation Test

Answer the following questions and then compare your answers to the answers given at the end of this chapter.

1. The types of Protrusion available in Pro/ENGINEER are different from those that are available for Cut. (T/F)

2. **Cut** option is available only when at least a base feature exists. (T/F)

3. If the shell thickness value is negative then the shell thickness is added outside the boundary of the face selected for shelling. (T/F)

4. When trajectory is closed then there are two options that are available: **Add Inn Fcs** and **No Inn Fcs**. (T/F)

5. To create a Sweep Cut feature the procedure to follow is the same as in the case of Sweep Protrusion. (T/F)

6. The **Sweep** option extrudes a section along a _________.

7. The cross-section of the swept feature remains _________ throughout the sweep.

8. The sketching plane you select will be _________ to the screen when you draw the trajectory.

9. _________ section and open trajectory are not possible.

10. A Quilt is a __________ feature.

Review Questions

Answer the following questions:

1. What is the maximum permissible angle for the rotation of sections in a **Rotational** blend?

   (a) 120 (b) 90
   (c) 180 (d) 45

2. In which of the following numbered section of a blend feature, **Blend Vertex** option can be used?

   (a) second (b) fourth
   (c) last (d) None of the above

3. What is the minimum number of sections required for a blend feature?

   (a) one (b) two
   (c) three (d) None of the above

4. Can a trajectory of a sweep feature be modified independent of the geometry of the section?

   (a) No (b) Yes
   (c) In some cases (d) None of the above

5. In which one of the following types of blend, sections are translated and rotated about the x, y, and z-axes?

   (a) **Parallel** (b) **Rotational**
   (c) **General** (d) None of the above

6. You can create a **Cut** feature using the **Sweep** option. (T/F)

7. In creating a Rotational or a General blend you need to create a coordinate system. (T/F)

8. In **General** blend the section is rotated about the y-axis of the coordinate system. (T/F)

9. The **Rotational** blend option is same as the **Parallel** blend option if the rotational blend angle entered between the two sections equals 0-degree. (T/F)

10. There must be an equal number of vertices in each section for blending. (T/F)

Exercises

Exercise 1

Create the foundation bolt shown in Figure 7-121. The shaded model is shown in Figure 7-122.
(Expected time: 30 min)

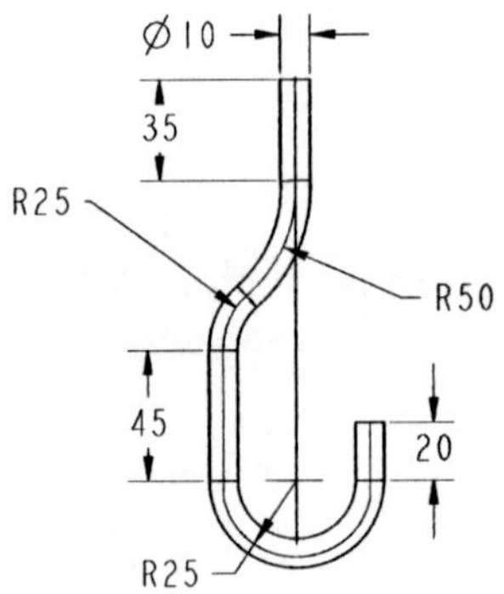

Figure 7-121 *Figure for Exercise 1*

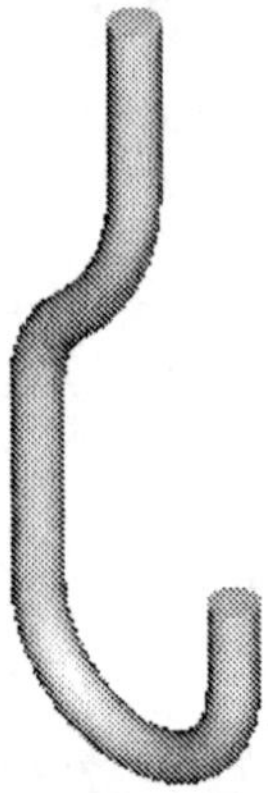

Figure 7-122 *Shaded model*

Exercise 2

In this exercise you will create the model shown in Figure 7-123. This figure also shows the sectioned front view, top view, and the right-side view of the solid model with dimensions.
(Expected time: 40 min)

Exercise 3

In this exercise you will create the model of a soap case shown in Figure 7-124. Figure 7-125 shows the sectioned front view, top view, right-side view, and the detail view with dimensions.
(Expected time: 50 min)

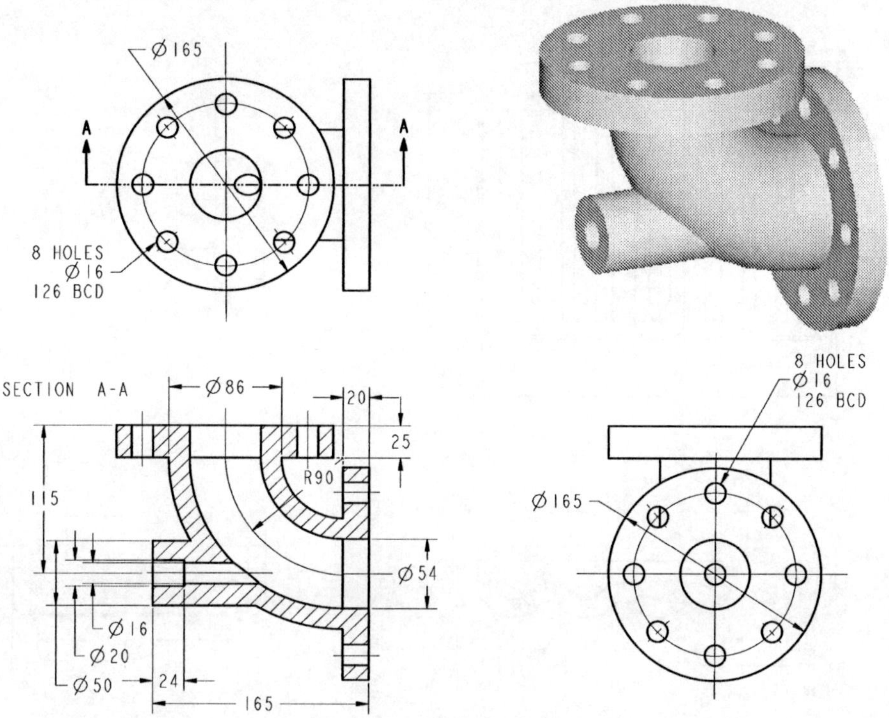

Figure 7-123 *Top, front, right-side, and isometric views of the model*

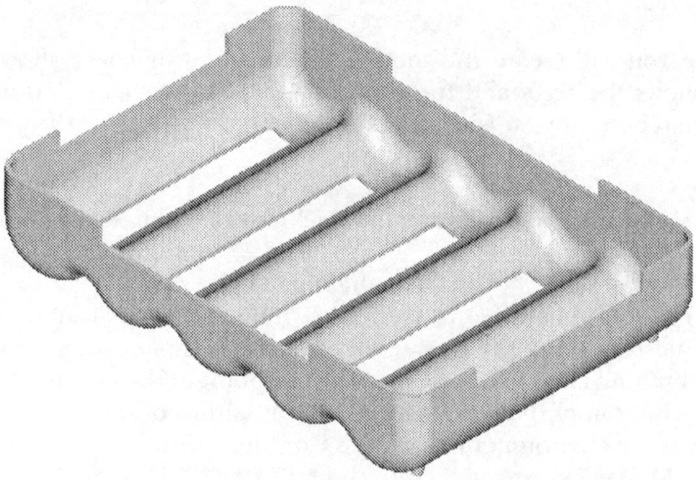

Figure 7-124 *Isometric view of the soap case*

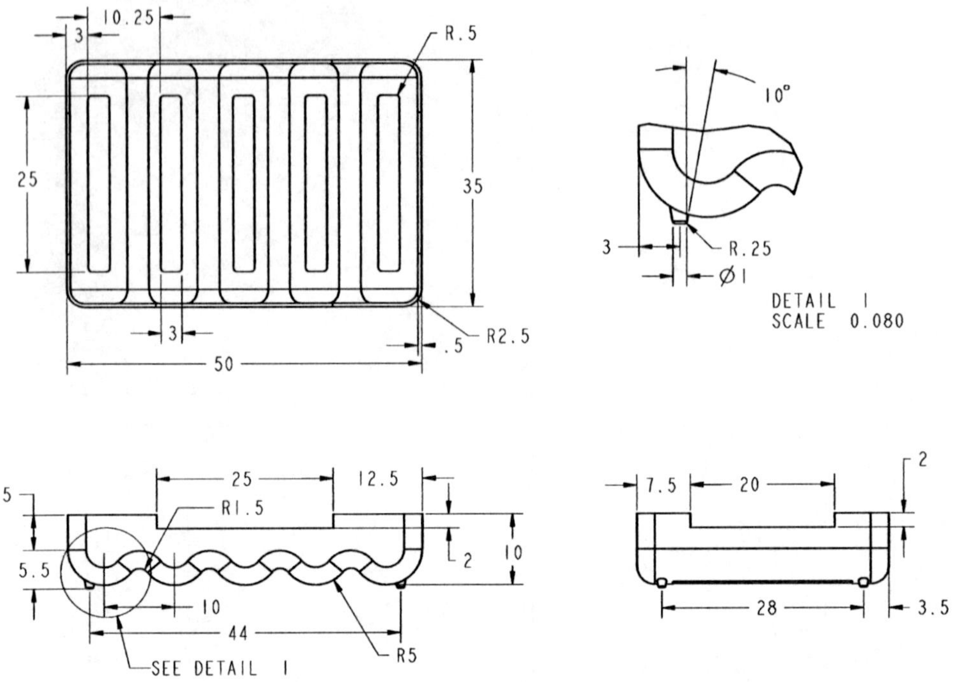

Figure 7-125 Top, sectioned front, right-side, and detail view of the soap case

Exercise 4

In this exercise you will create the model of a carburettor cover shown in Figure 7-126. Figure 7-127 shows the sectioned top view, sectioned front view, sectioned right-side view, and the sectioned bottom view with dimensions. **(Expected time: 50 min)**

Hint

1. Create the sketch of the base feature that includes a rectangle of 125x50 and then extrude it.
2. Choose **Insert > Sweep > Cut** from the menu bar.
3. Select the **FRONT** datum plane as the sketching plane for sketching a trajectory.
4. Create a trajectory using the **Create an arc by 3 points or tangent to an entity and its endpoints** button. The start point and the endpoint of the arc should be at a distance of 28 from the bottom of the base feature and the radius of the arc should be 100.
5. Exit the sketcher environment using the **Continue with the current section** button.
6. Choose the **Merge Ends** option from the **ATTRIBUTES** menu.
7. You will again enter the sketching environment to create the section for the sweep feature. Choose the **Create an arc by 3 points or tangent to an entity and its endpoints** button

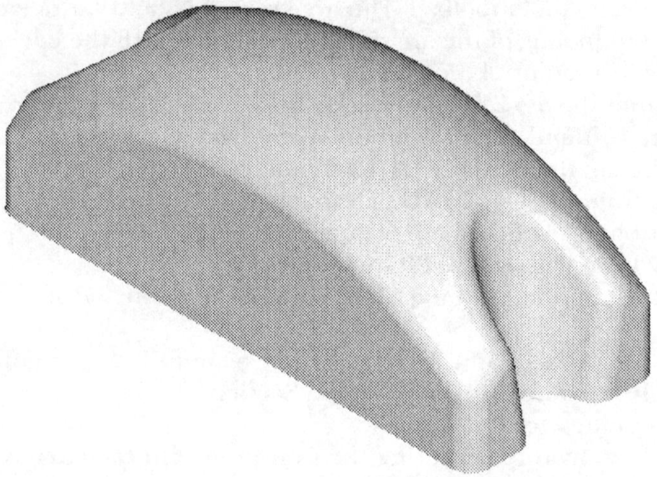

Figure 7-126 Isometric view of the carburettor cover

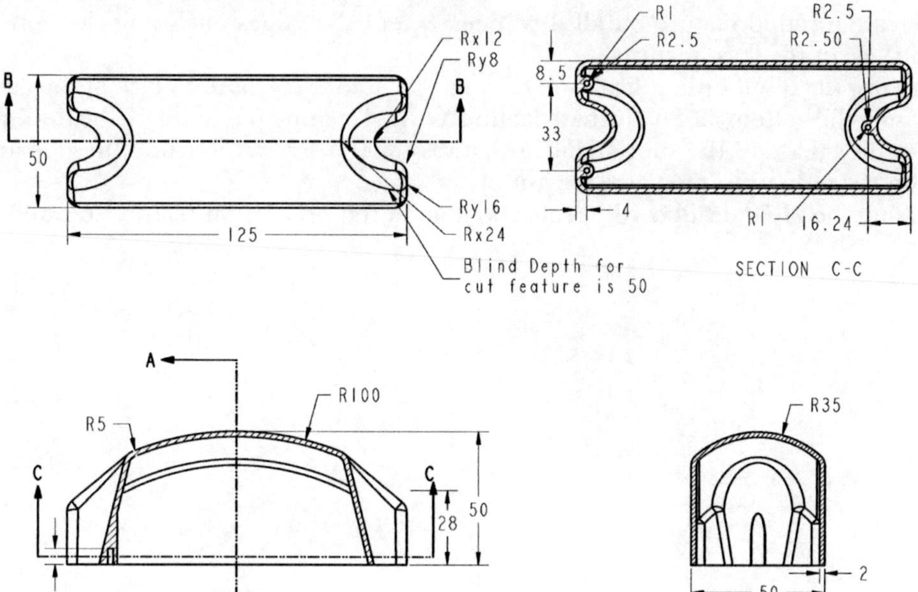

Figure 7-127 Top view, sectioned front view, sectioned right-side view, and sectioned bottom view of the carburettor cover

from the **Sketcher Tools** toolbar. The arc created should be tangent to the reference lines, and the endpoints of the arc should be aligned with the edges of the base feature and should have a radius of 35.

8. Choose **OK** from the **SWEEP: Cut** dialog box.
9. Choose **Insert** > **Blend** > **Cut** from the menu bar.
10. Choose **Done** from the **BLEND OPTS** submenu.
11. Choose **Done** from the **ATTRIBUTES** menu.
12. Select the bottom face of the base feature as the sketching plane.
13. Choose **Okay** from the **DIRECTION** submenu.
14. Choose the **TOP** option from the **SKET VIEW** submenu and select the **RIGHT** datum plane.
15. Create an ellipse of Rx12 and Ry8 using the **Create full ellipse** button.
16. Choose **Sketch** > **Feature Tools** > **Toggle Section**.
17. Create another ellipse of Rx24 and Ry16.
18. Exit the sketcher environment using the **Continue with the current section** button.
19. Choose **Okay** from the **DIRECTION** menu.
20. Choose **Done** from the **DEPTH** menu and enter the depth of cut in the **Message Input Window** and press ENTER.
21. Mirror the cut feature about the **RIGHT** datum plane.
22. Create a round feature on all the edges except the edges enclosing the bottom planar surface of the base feature.
23. Invoke the **Shell** option from the menu bar. Remove the bottom face of the base feature.
24. Using the bottom face of the base feature as the sketching plane create the three protrusion features that are the supporting structures for the screws. Extrude these features using the **Extrude upto next surface** button.
25. Using the Hole dashboard, create the hole in the protrusion features created earlier.

Answers to the Self-Evaluation Test

1 - F, **2** - T, **3** - T, **4** - T, **5** - T, **6** - trajectory, **7**- constant, **8** - parallel, **9** - open, **10** - surface.

Chapter 8

Advanced Modeling Tools-II

Learning Objectives

After completing this chapter you will be able to:

- *Use various Advanced options for creating complex models.*
- *Create features using Var Sec Swp option.*
- *Create features using Swept Blend option.*
- *Create features using Helical Sweep option.*
- *Create features using Sect to Srfs option.*
- *Create features using Srfs to Srfs option.*
- *Create features using From File option.*
- *Create features using Toroidal Bend option.*
- *Use Relations.*

ADVANCED FEATURE CREATION TOOLS

Pro/ENGINEER facilitates its users with some advanced options to create complex features with greater ease. All these options are available in the **Insert** menu present in the menu bar.

In this chapter you will learn about the following feature creation tools that are available in Pro/ENGINEER.
1. Variable Section Sweep
2. Swept Blend
3. Helical Sweep
4. Blend Section to Surfaces
5. Blend between Surfaces
6. Blend from File

Variable Section Sweep Option

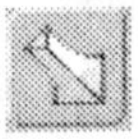

The **Variable Section Sweep** option can be invoked from the **Insert** menu in the menu bar or you can choose the **Variable Section Sweep Tool** button from the **Base Features** toolbar. This option is used to create a sweep feature in which the section of the sweep feature varies along the shape of the trajectories. This feature is created using more than one trajectory. The section is swept along the origin trajectory and the variation of the section is controlled by the X-trajectory and the other trajectories. You can sketch the sweep trajectory using various sketcher tools. You can also select a previously created entity as the trajectory for sweep. The **Variable Section Sweep** option is available for both protrusion and cut options.

The following points should be remembered while creating a variable section sweep feature.

1. You should define the origin trajectory to which the section is normal.

2. After creating the origin trajectory you have to create an X-trajectory that defines the horizontal vector of the section.

3. You can also define additional trajectories to facilitate the creation of a complex profile.

4. When you draw the section for the variable section sweep feature, the section should be aligned to the endpoints of the X-trajectory and the other trajectories. If the section is not aligned then the feature will be created without following the trajectory path and will be similar to a feature created using the **Sweep** option.

When you choose this option the Variable Section Sweep dashboard is displayed as shown in Figure 8-1. Remember that you should have the trajectories ready on the graphics window so that they can be selected for the creation of the variable section sweep feature. This is because you cannot sketch the trajectories by using the Variable Section Sweep dashboard. However, you can draw the trajectories, that is, datum-on-the-fly using the **Sketched Datum Curve Tool** button. As mentioned earlier, the process of creating a datum feature when a feature creation tool is active is called datum-on-the-fly. The options in the Variable Section Sweep dashboard are discussed next.

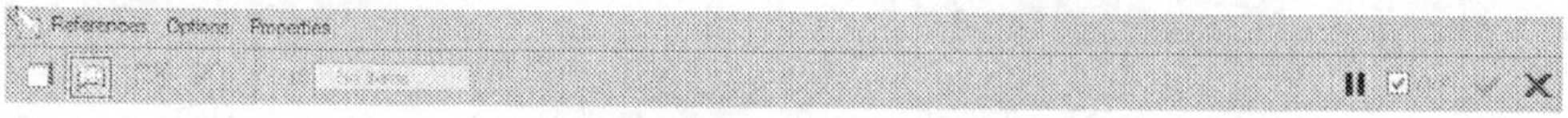

Figure 8-1 Variable Section Sweep dashboard

References tab

When you choose the **References** tab, the slide-up panel is displayed as shown in Figure 8-2. The options available in this slide-up panel are discussed next.

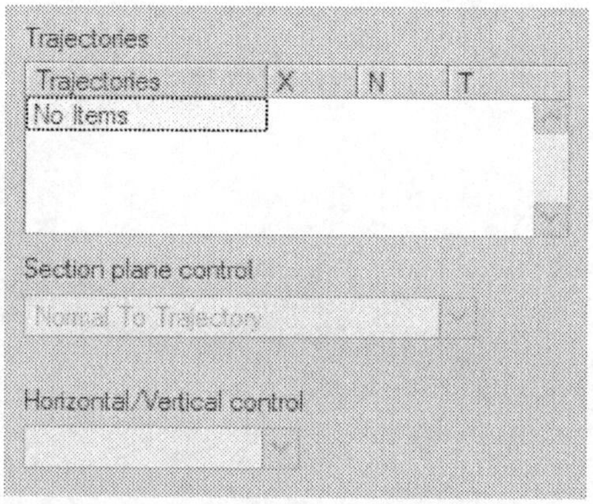

Figure 8-2 ***References*** *tab slide-up panel*

Trajectories collector

This collector is used while selecting the trajectories and after drawing the section. The trajectories that you select are listed in this collector. The first trajectory that you select is by default listed as the origin trajectory.

Note that you need to first select the normal and X trajectories and then draw the sketch of the section. This is because the sketch plane will be oriented only if you have defined the normal trajectory. Also, since the sketch needs to pass through the endpoints of the X-trajectory, it is better to define the X-trajectory before drawing the section. This way you will have the reference point from which the sketch need to pass.

This collector has three columns and the selected trajectories are listed in the rows. The first column is **X**. The check box in this column when selected converts the trajectory to the X trajectory. The second column is **N**. The check box in this column when selected converts the trajectory in that row to the normal trajectory. Remember that only one trajectory can be a normal trajectory and only one trajectory can be a X trajectory. The third column is **T**. There are two check boxes available in this column and in each row. All these check boxes are gray in color. When you select the check box, the name of the trajectory is filled in that cell and values T=0 appears on the trajectory in the graphics window. This value is displayed on both the ends of the trajectory. Initially, this value is 0. When you enter a value, the point of alignment of the section with the trajectory varies.

Section plane control drop-down list. The options in this list are used to specify the direction of the section in which it start sweeping.

Normal to trajectory option. When you use the Normal to trajectory option to create the variable sweep, the section is normal to the origin trajectory and is swept along the X trajectory. The section should be aligned with the X trajectory to sweep along it. If the section is not aligned with the X trajectory then this option works similar to the Sweep option and the section is swept along the origin trajectory only. Figure 8-3 shows the two trajectories and the section that are used to create the swept feature shown in Figure 8-4.

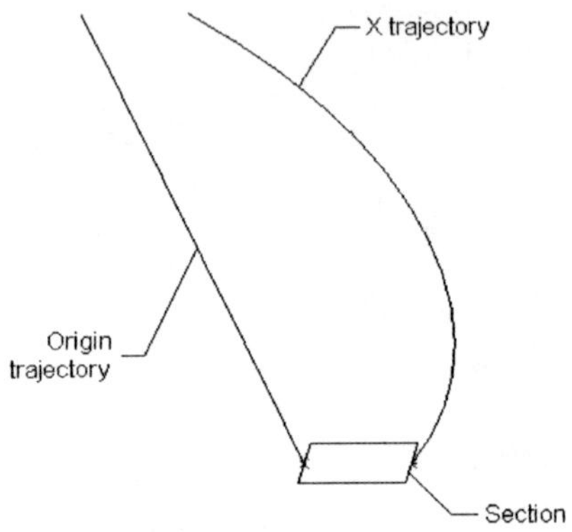

Figure 8-3 *The two trajectories*

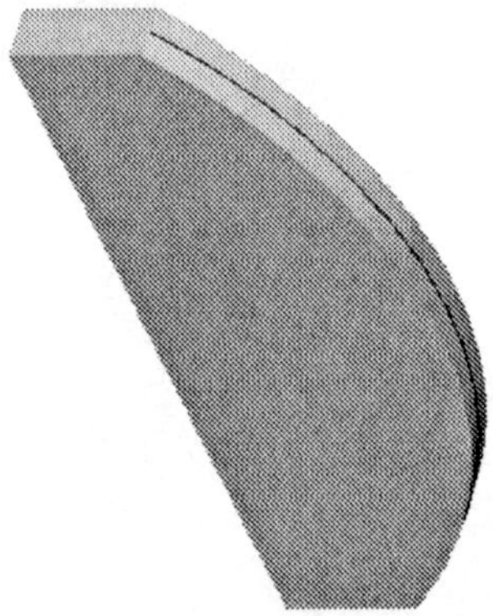

Figure 8-4 *Variable section sweep feature*

Now, in Figure 8-5, the X trajectory and the origin trajectory are interchanged and the sweep feature that is obtained is shown in Figure 8-6. Note the difference in the swept features. In Figure 8-5, the section is normal to the selected origin trajectory and when the curve is selected as the origin trajectory, the section is normal to it as shown in Figure 8-6.

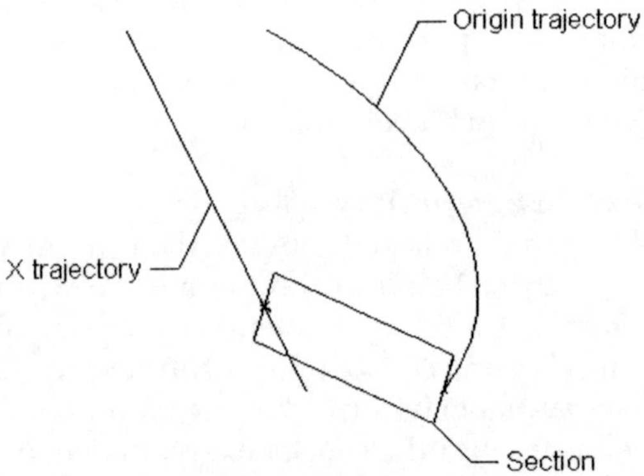

Figure 8-5 *The two trajectories and section*

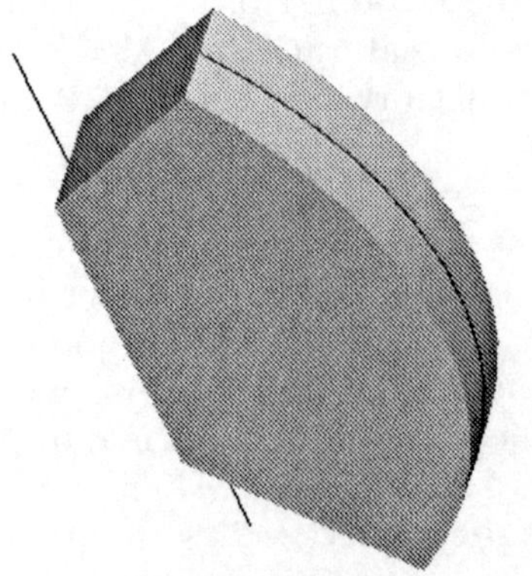

Figure 8-6 *Resultant variable section sweep feature*

Normal to projection option. The Normal to projection option is used to create a variable section sweep feature in which the section is normal to the two-dimensional (2D) projection of trajectories. The swept section is normal to the origin trajectory and the variation in the section is controlled by the X-trajectory and the other trajectories. When you use this option, you need to select or sketch an origin trajectory, an X-trajectory, other trajectories if required, and the direction of projection. The direction of projection can be specified by using the default coordinate system.

Figure 8-7 shows the origin trajectory and the X trajectory. The two trajectories are used to create the sweep feature shown in Figure 8-8.

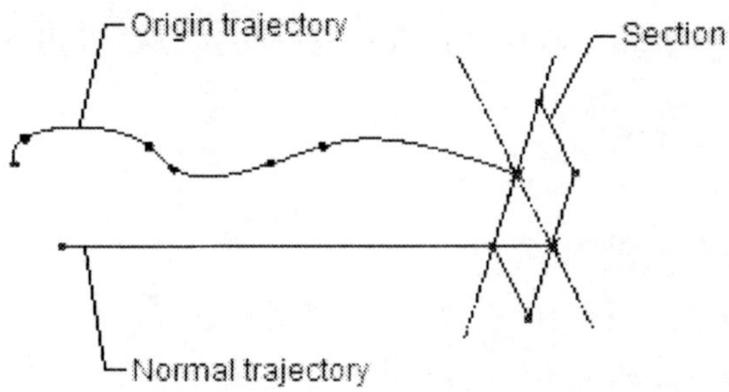

Figure 8-7 *Trajectories and section*

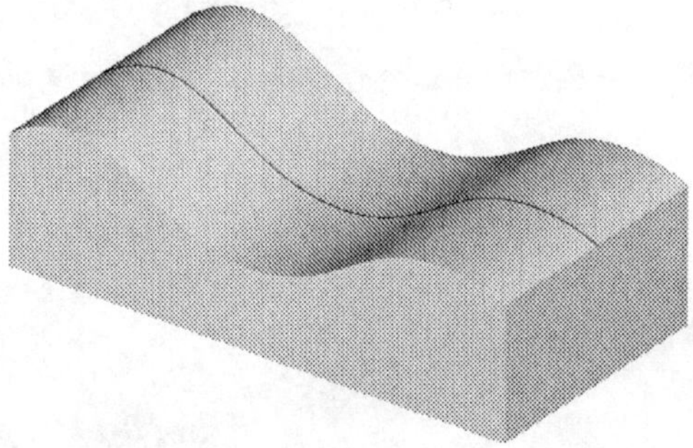

Figure 8-8 *Resultant swept feature*

Constant normal direction option. The **Constant normal direction** option is used to create a variable section sweep in which the section to be swept is normal to an edge, normal of the selected plane, or an axis.

When you select the origin trajectory and then select the **Constant normal direction** option, you are required to select or create a plane, an edge, a curve, an axis, or a coordinate system to define the constant normal direction. Select a datum plane and then sketch the section. The system takes you to the sketcher environment and automatically orients the sketch plane such that it is normal to the selected direction. Figure 8-9 shows the origin trajectory, the section, and the normal vector of the datum plane to which the section is perpendicular. Figure 8-10 shows the section that is normal to the origin trajectory. This figure explains the difference between the **Normal to trajectory** option and the **Constant normal direction** option.

Options tab

When you choose the **Options** tab, the slide-up panel is displayed as shown in Figure 8-11. The options in this slide-up panel are discussed next.

Variable section

This option is used to create a sweep feature whose section varies along the trajectory.

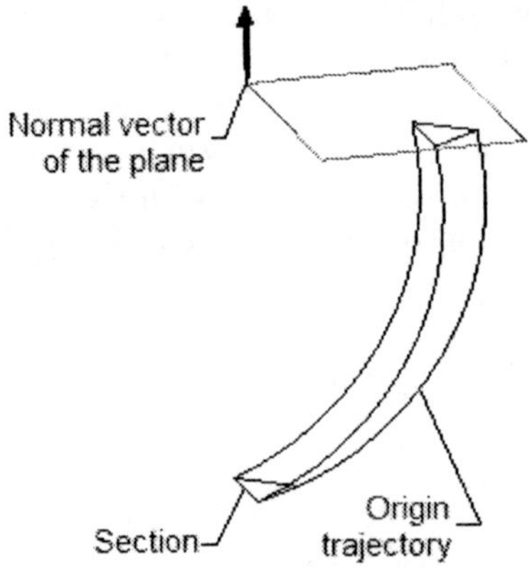

Figure 8-9 Origin trajectory and the pivot plane

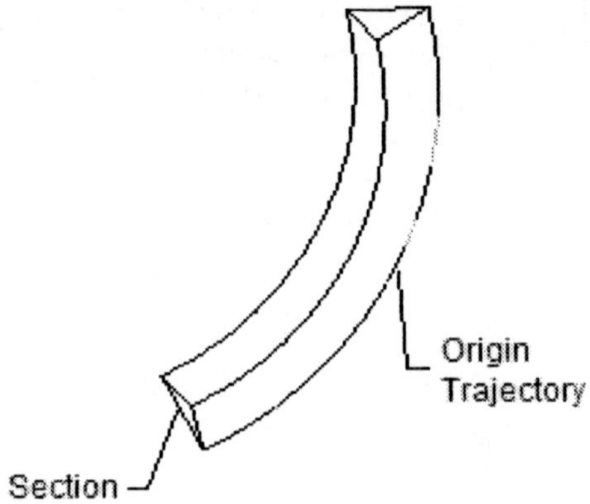

Figure 8-10 Swept feature on the base feature

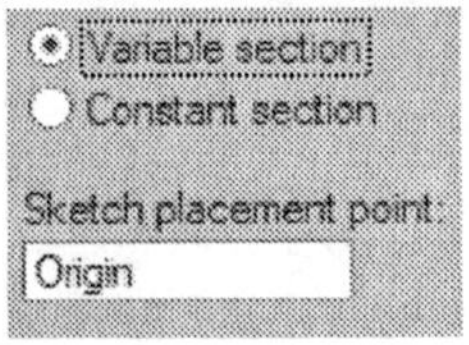

Figure 8-11 ***Options*** *option slide-up panel*

Constant section

This option is used to create a sweep feature whose section is constant throughout the trajectory. The feature created using this option is similar to the feature created using the **Sweep** option.

Sketch placement point collector

By default, **Origin** is selected as the start point of the section. This is the reason you draw the section at the start point of the trajectory. However, if you click in this collector to activate and then select a datum point that lies on the origin trajectory, the section will start from that point. This means that the selected datum point becomes the start point of the trajectory.

Note

The options available on the lower half of the dashboard are the same as those available on the Extrude dashboard. Using these options you can specify the type of feature you want to create. For example, solid, surface, or thin feature.

Swept Blend

The **Swept Blend** feature is a combination of sweep and blend. To create a sweep feature you need a trajectory along which the section is swept. You have an option of sketching or selecting a previously created feature as the trajectory of the sweep. To create a parallel blend you need to have more than one section to be blended. The combination of both these options is the **Swept Blend** option.

In this option you can blend two or more than two sections having same number of entities and the path of the transition of blending is defined by selecting or sketching a trajectory. You can invoke the **SWEPT BLEND** option from the **Insert** menu in the menu bar.

Choose **Insert > Swept Blend > Protrusion** from the menu bar; the **BLEND OPTS** menu is displayed as shown in Figure 8-12. The options in this menu are discussed next.

Figure 8-12 BLEND OPTS menu

Select Sec
The **Select Sec** (Select section) option is used to select existing sections that will be used to create a swept blend.

Sketch Sec
The **Sketch Sec** (Sketch section) option is used to sketch the section that will be used to create the swept blend.

NormalToOriginTraj
The **NormalToOriginTraj** option is used to keep the section normal to the origin throughout the trajectory. The sweep option behaves in this manner.

Pivot Dir
The **Pivot Dir** (Pivot direction) option is used to create a swept blend in which the section to be swept is normal to the origin trajectory. You need to select the pivot direction.

Norm To Traj
The **Norm To Traj** (Normal to trajectory) option is used to create a swept blend in which the section is perpendicular to the normal trajectory and is swept along the origin trajectory.

Figure 8-13 shows the three rectangular sections. The three sections are connected through a trajectory by connecting their vertices. The shaded view of the model is shown in Figure 8-14.

The model is also shelled. In Figure 8-15, the trajectory and the sections that are used to create the feature shown in Figure 8-16 are shown. It should be noted that the number of entities should be same in all the sections.

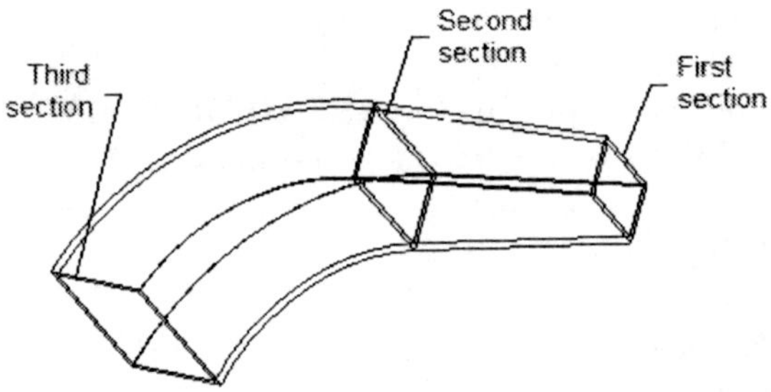

Figure 8-13 Three sections in swept blend

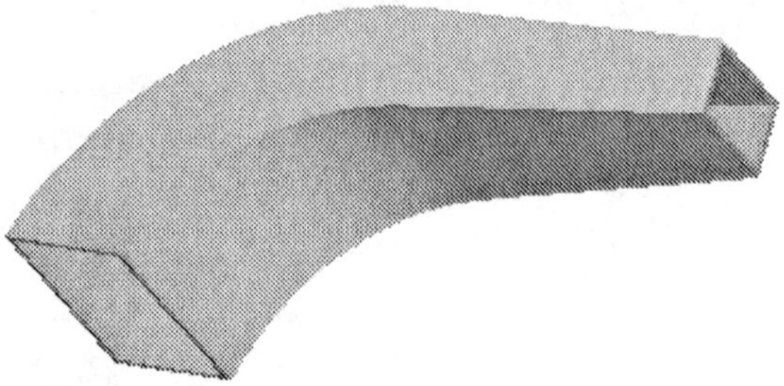

Figure 8-14 Shaded view of the feature

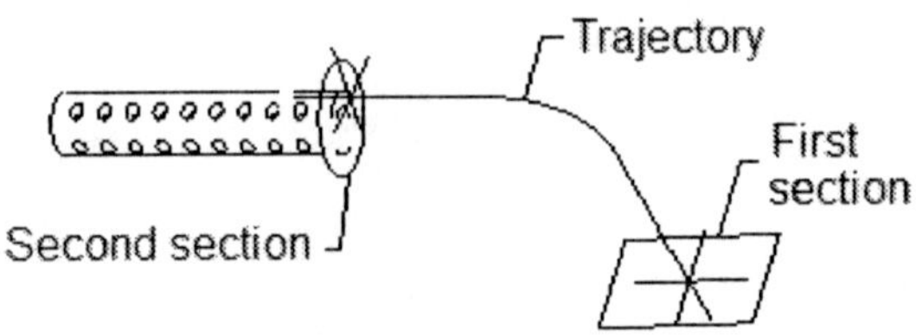

Figure 8-15 Two sections and a trajectory

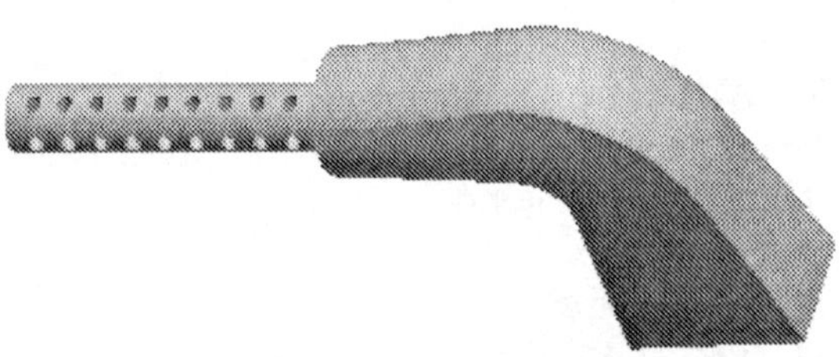

Figure 8-16 Resultant feature

Helical Sweep

The **Helical Sweep** option is used to create helical swept features. You have to define a trajectory that will specify the shape and height of the helix, a pitch value, and a cross-section to create a helical feature using this option. The distance of the trajectory from the center line defines the radius of the helical path and the length of the trajectory defines the length of the swept feature. The main use of this option is to create the helical springs and threads.

***Figure 8-17 ATTRIBUTES** menu*

When you choose the **Helical Sweep** option from the **Insert** menu in the menu bar, the **ATTRIBUTES** menu is displayed as shown in Figure 8-17. The menu options are discussed next.

Figure 8-18 shows the parameters to be defined for the helical sweep and Figure 8-19 shows the resultant helical sweep feature.

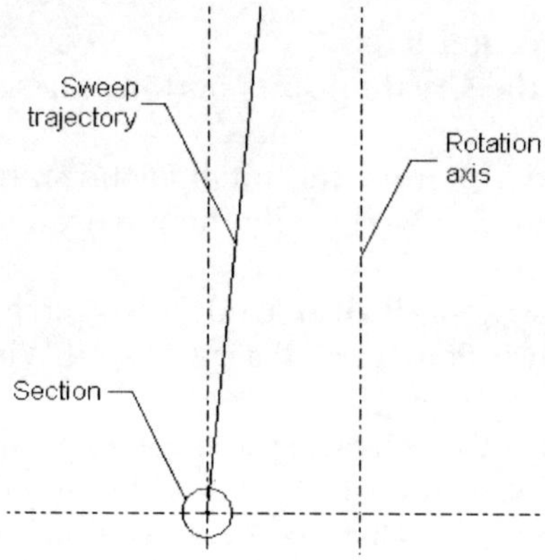

Figure 8-18 Sweep trajectory, sweep section, and axis of rotation to create a helical feature

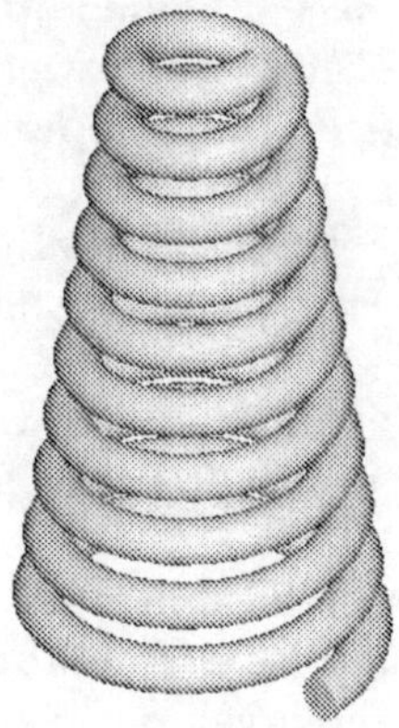

Figure 8-19 The resultant helical feature

Constant

The **Constant** option is used to create a helical feature with constant pitch as shown in Figure 8-20.

Variable

The **Variable** option is used to create a helical feature of varying pitch. While using this option you have to define the start pitch and the end pitch. You can also define a pitch between the start point and the endpoint of the helix. This pitch value between the endpoint and the start point is specified by creating points on the sweep trajectory. The helical sweep feature with variable pitch is shown in Figure 8-21.

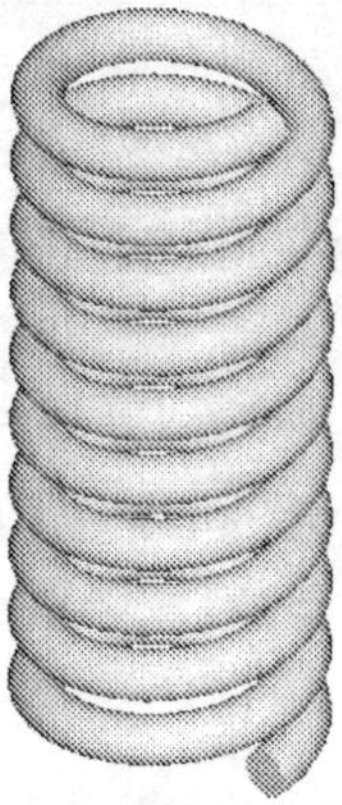

Figure 8-20 Helical sweep feature with constant pitch

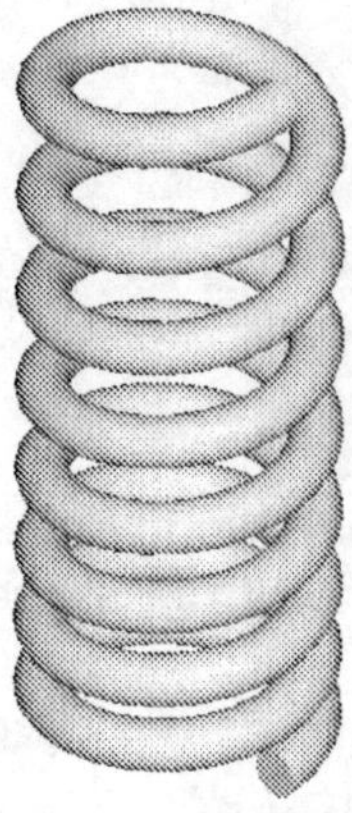

Figure 8-21 Helical sweep feature with variable pitch

The brief procedure to create a variable pitch spring is discussed next.

1. Choose **Variable** > **Thru Axis** > **Right Handed** > **Done** from the **ATTRIBUTES** menu.

2. Select the sketch plane and then orient it.
3. Draw the sketch of the sweep trajectory and the center axis.
4. Add points on the sweep trajectory by choosing the **Create points** button. These points will be later used to vary the pitch of the spring.
5. Exit the sketcher environment. You are prompted to enter the pitch at the start point. After entering the pitch value at start you are prompted to specify the pitch value at the end.
6. After specifying the pitch value at the end, a subwindow is displayed with a pitch graph.
7. Select the points on the trajectory from the main window. The **Message Input Window** is displayed.
8. Enter the pitch value at the selected point. Similarly, specify the pitch values at the other points that were defined on the trajectory while sketching it. It should be noted that a specified pitch value can be modified by selecting the **Change Value** option from the **DEFINE GRAPH** submenu.
   As you specify the pitch values to the points, the graph in the subwindow also changes.
9. After all the pitch values are entered, choose the **Done** option from the **GRAPH** menu. Now, you enter the sketcher environment to draw the section of the spring.
10. Draw the section of the spring at the intersection of the two infinite yellow dotted lines.
11. Exit the sketcher environment and choose **OK** from the **PROTRUSION** dialog box.

Note
You can use any geometric entity to create the sweep trajectory. For example spline, line, or arcs.

Thru Axis
The **Thru Axis** option is used to create a helical feature around an axis.

Norm To Traj
The **Norm To Traj** option is used to create a helical feature perpendicular to the sketched trajectory.

Right Handed
The **Right Handed** option is used to create a helical feature in which the section is swept in the counterclockwise direction from the start sketch.

Left Handed
The **Left Handed** option is used to create a helical feature in which the section is swept in the clockwise direction from the start sketch.

Blend Section To Surfaces
The **Blend Section to Surfaces** option is used to blend a selected set of tangential surfaces with a sketched contour. You can invoke this option by choosing **Insert > Advanced > Blend Section To Surfaces** from the menu bar. The contour sketched to create the feature using this option must be closed. This option can be used to create a surface or a solid feature. Figure 8-22 shows the section to be blended with the selected set of tangential surfaces and Figure 8-23 shows the resultant blended feature.

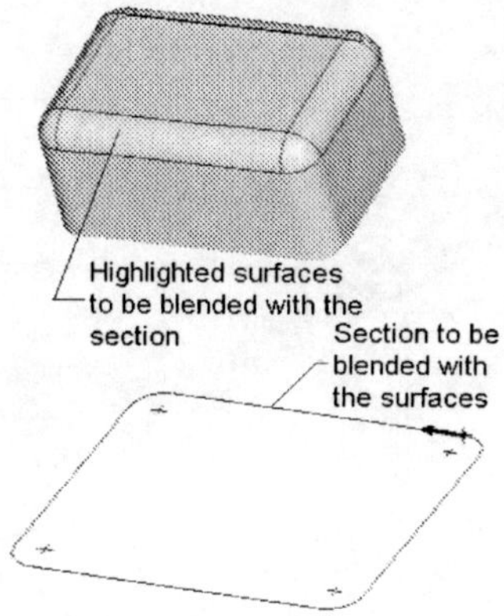

Figure 8-22 *Highlighted set of tangential surfaces and the contour sketch.*

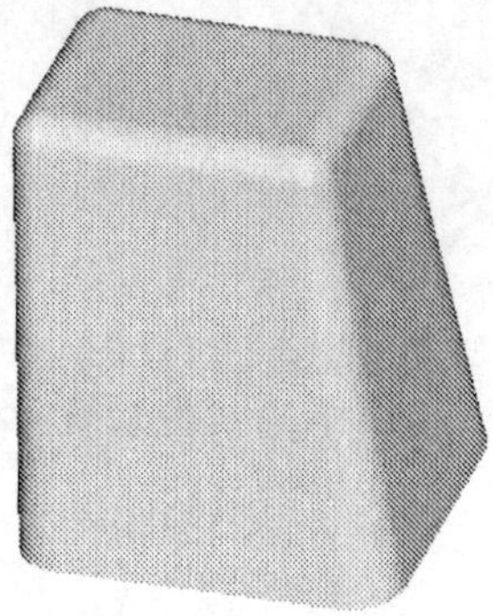

Figure 8-23 *Feature created using the* ***Sect to Srfs*** *option*

Blend Between Surfaces

The **Blend Between Surfaces** option is used to create a blending between two curved quilts, two curved solid surfaces, or a combination of both. The curved surfaces may be revolved features or spheres. You can invoke this option by choosing **Insert > Advanced > Blend Between Surfaces** from the menu bar. Figure 8-24 shows the two spheres to be blended and Figure 8-25 shows the resultant blended feature created using this option.

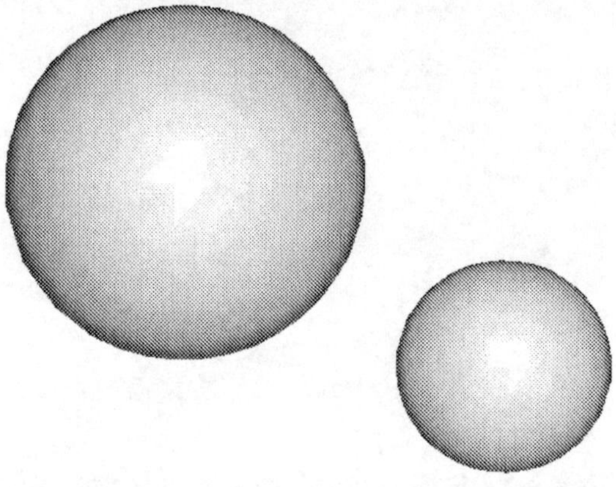

Figure 8-24 *Two spheres to be blended*

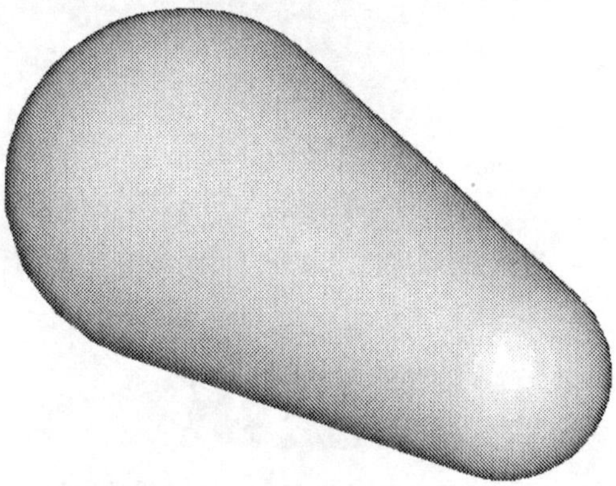

Figure 8-25 *Feature created using the* ***Srfs to Srfs*** *option*

Figure 8-26 shows the two curved features to be blended and Figure 8-27 shows the resultant blended feature created using this option.

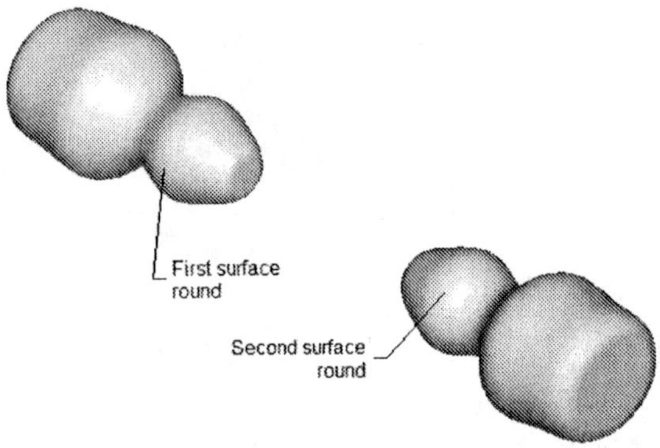

Figure 8-26 Two curved features to be blended

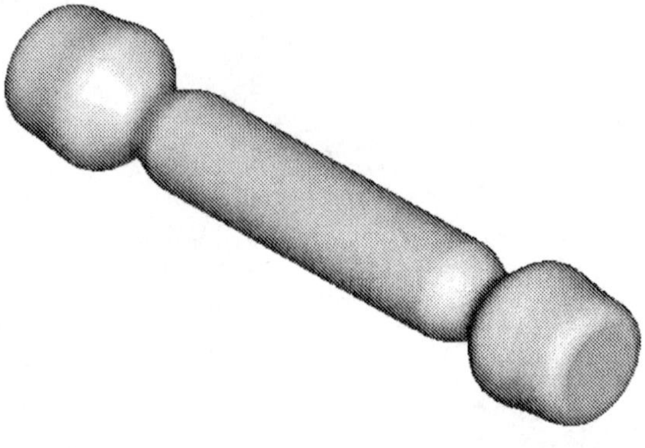

***Figure 8-27** Feature created using the **Srfs to Srfs** option*

Toroidal Bend

The **Toroidal Bend** option is used to create features with curved surfaces or the models that generally have a cut profile on the curved surface. For example, if you want to create a feature shown in Figure 8-28, you will first create a rectangular plate and cut the profile on it. Then you will bend the rectangular plate shown in Figure 8-29 to 360-degrees.

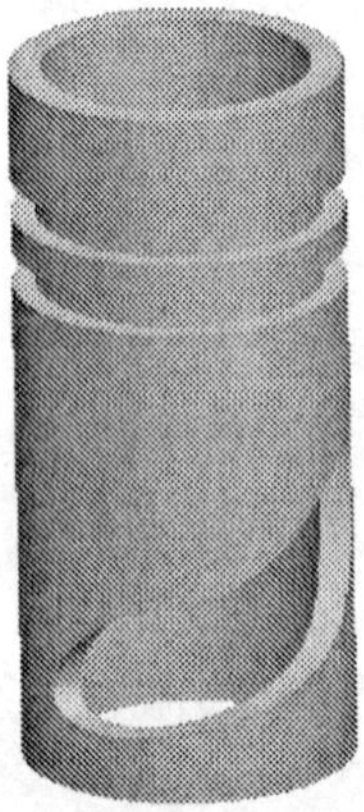

***Figure 8-28** Example of toroidal bend*

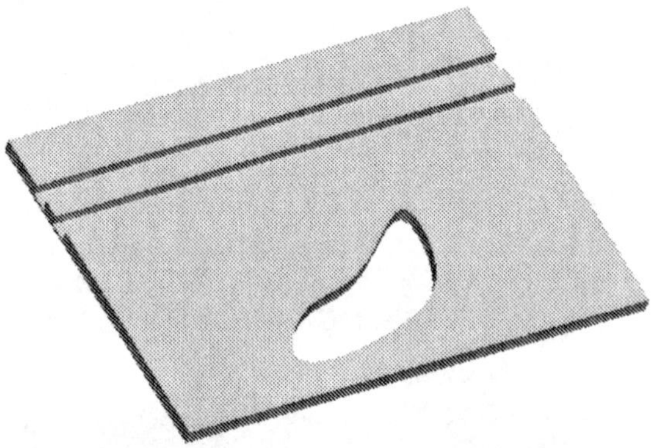

***Figure 8-29** Rectangular plate with a profile cut*

To invoke the **Toroidal Bend** option, choose **Insert > Advanced > Toroidal Bend** from the menu bar. The **OPTIONS** menu is displayed as shown in Figure 8-30. This menu is divided into three parts. The first part of the menu has options to specify the angle of rotation of the bend. The second part has options that specify whether you want to create the feature on one side of the sketch plane or on both sides. The third part of the menu is used to specify the method of creating the toroidal bend.

The following steps explain the procedure to create the model shown in Figure 8-28. It is assumed that you have created the rectangular plate shown in Figure 8-29.

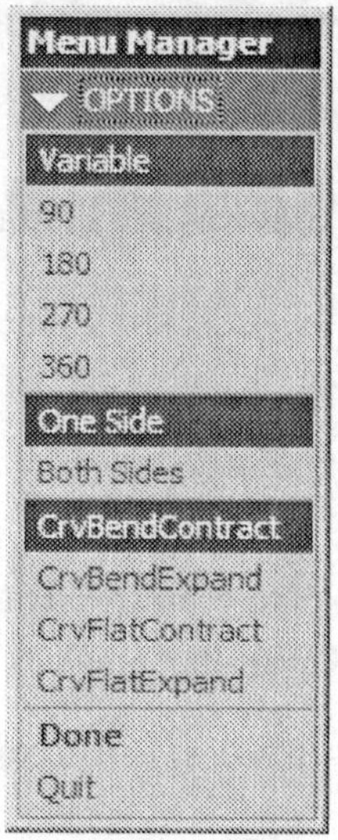

Figure 8-30 ***OPTIONS*** *menu*

1. Choose **Insert > Advanced > Toroidal Bend** to invoke the **OPTIONS** menu. Choose **360 > One Side > CrvBendContract > Done**. The **DEFINE BEND** menu is displayed. In this menu, the **Add** option is selected by default and you are prompted to select solids, quilts, or datum curves to bend.
2. Select the top face of the base feature and choose **Done** from the **DEFINE BEND** menu. You are prompted to select the sketch plane.
3. Select the right face of the base feature as the sketch plane and orient the sketch plane.
4. After entering the sketcher environment, draw the sketch as shown in Figure 8-31. The sketch comprises of a line that is drawn from the left edge to the right edge and is at a distance of 7 from the top edge. A user-defined coordinate system is placed at the midpoint of the line.

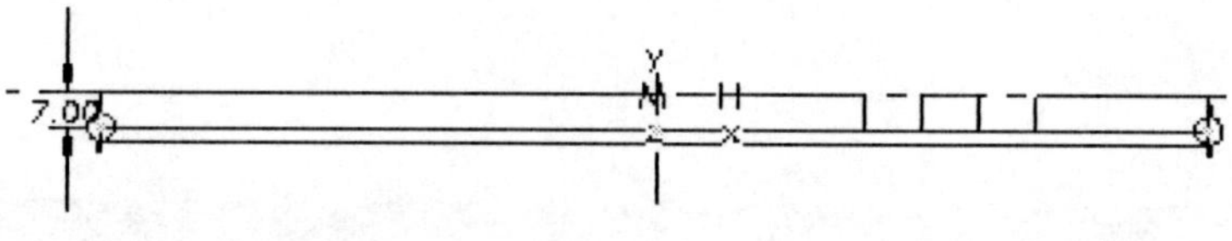

Figure 8-31 *Sketch along with user-defined coordinate system*

5. Exit the sketcher environment. You are prompted to select the two parallel planes to define the length of bend.
6. Select the face that was selected as the sketch plane and the face opposite to it (the left face of the base feature.)
   The face of the base feature that was selected to bend is bent at an angle of 360-degrees and the desired model is created.

USING RELATIONS

Relations are user-defined equations that relate the dimension symbols or values. This option is used to create relations between various dimensions of a model and to keep the design intent of a model. For example, if it is required that the centre of the circle should lie at a distance of 1/3 from the vertex of a triangle, then it can be achieved using the relations. Relations can also be used to control the effect of modification on a model.

When you choose **Tools** > **Relations** from the menu bar, the **Relations** dialog box is displayed as shown in Figure 8-32.

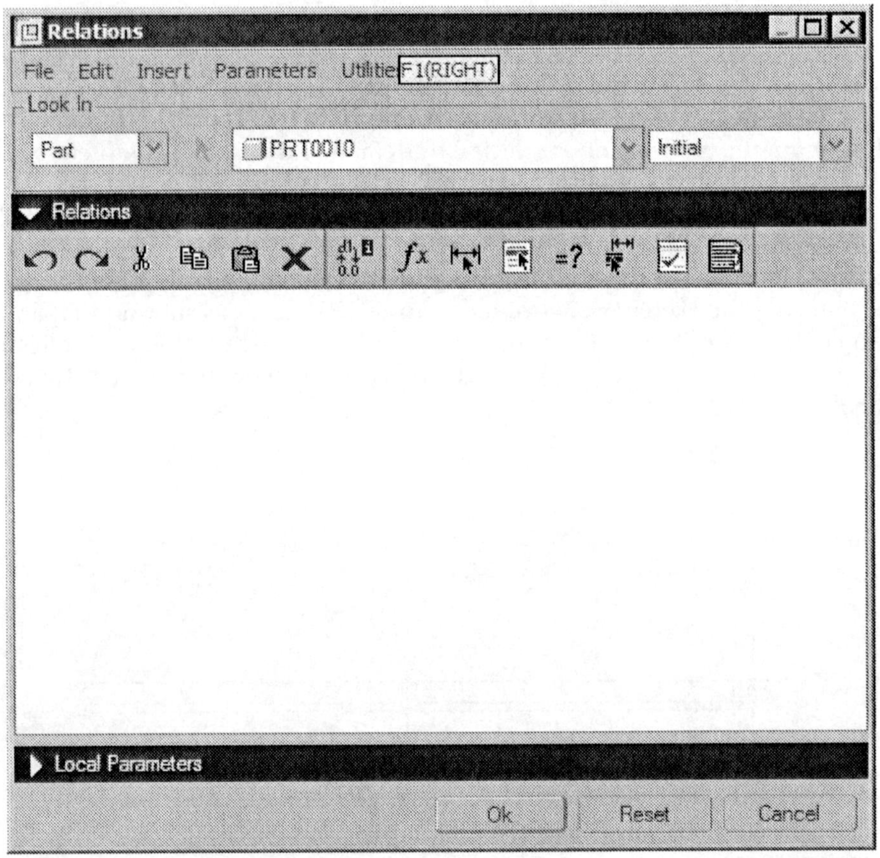

Figure 8-32 *Relations dialog box*

Relations applied to a sketch can be modified and used in the part or assembly mode also. In the Sketch mode, the **Relations** dialog box can be invoked by choosing **Sketch > Relations** from the menu bar. Relations are of two types and are discussed next.

Equality Relations

The Equality relations are used to assign a value to a parameter or to a symbolic dimension. The variable on left of the equation is called driven and the variable on right of the equation is called driver. Another way of representing this is

driven=driver

For example,

d1=25
d2=d1+10
d5=d3*sin(d4)

In the above equations, the symbol **d** represents the dimension created in the Part mode. Dimensions in the Sketch mode are represented by the symbol **sd**, where **s** represents the Sketch mode and **d** the dimension.

In all of the above equations the expression on the right-hand side of the equation determines the value of the variable on the left.

Comparison Relations

The Comparison relation compares the two sides of the equation. This type of relation is used in two ways: by giving conditional statements or by defining constrains.

For example,

As conditional statements

If sd1>=10
sd2=20
else
sd2=30

As a constraint

sd1+sd2>sd3+10
sd3>sd2/2

Note

The assignment operator '=' assigns a value to the driven that is on the left of the equation, whereas in case of the comparison operator the driven is compared with the driver and the value is returned as True or False. Hence, the assignment operator '=' (equal to) is different from the comparison operator '=' (equal to). The comparison operator '=' is always used with 'IF', '>', or '<'.

For example,   sd1>=10

If sd1 is greater than or equal to 10 then the value returned is True and if sd1 is less than or equal to 10 then the value returned is False.

To Add Relations

To add relations in the **Relations** dialog box, choose **Insert** > **From Screen**. The dimensions on the model are displayed in the form of symbols. Remember to use the appropriate dimension symbols while giving relations.

1. Choose **Relations** from the **Tools** menu in the menu bar. The dimensions are displayed in a symbolic form.

2. Choose **Insert** > **From Screen** from the **Relations** dialog box. You are prompted to select the feature.

3. The relations are added using the sketcher dimension symbols or the known dimension symbols. After adding all the relations, choose **Utilities** > **Verify** from the menu bar in the **Relations** dialog box. Regenerate the sketch or the model. The changes will appear in the sketch as per the relations added.

To Evaluate a Relation

The **Utilities** > **Evaluate** option gives you the value of any dimensional symbol used in a relation. When you choose this option in the **Relations** dialog box, the **Evaluate Expression** dialog box appears. Enter in this window the dimensional symbol whose value you want to evaluate.

To Sort Relations

In the **Relations** rollout of the **Relations** dialog box, choose the **Sort relations** button to sort the relations in an order. The relations are ordered so that a relation that depends on the value of another relation is evaluated after this relation.
For example,   consider relations

sd0=sd1+sd2*2
sd2=sd3 - 2*sd4

Now when you choose the **Sort relations** button, the relation will be ordered as shown below.

sd2=sd3 - 2*sd4

sd0=sd1+sd2*2

Note
*The conditional relations are not sorted using the **Sort relations** button.*

Toggling the Display of Dimensions

The **Toggle between dimensions values and names** button in the **Relations** dialog box toggles the display between the symbols and the value mode of the dimensions. You can choose this option to display the dimensions as values or as symbols.

User Prog

The **Utilities** > **User Prog** option in the **Relations** dialog box allows you to link a C program module with the sketch. When you choose this option the cascaded menu that has three options is displayed: **Edit**, **Link**, and **Run**. Pro/ENGINEER provides a template C file that helps you to read and use the C programming environment.

TUTORIALS

Tutorial 1

In this tutorial you will create the model shown in Figure 8-33. This figure also shows the sectioned top, front, right-side, and isometric views of the model. **(Expected time: 45 min)**

The following steps outline the procedure for creating this model:

a. First, examine the model and then determine the number of features in it. The model is composed of five features, see Figure 8-33.

b. The first feature is the set of datum curves that will be drawn on the **FRONT** datum plane. First, the datum curve for the origin trajectory will be sketched, see Figure 8-34, and then the datum curve for the X trajectory will be sketched, see Figure 8-35. One more datum curve that will sweep the section along their paths will be sketched, see Figure 8-36.

c. The second feature is the datum curve that will be drawn on the **RIGHT** datum plane, see Figure 8-37.

d. The third feature is the variable section sweep that will be created using the sketched datum curves. Then the section that will vary with the shape of the trajectories will be sketched, see Figure 8-39. The variable section sweep feature is created as shown in Figure 8-40.

e. The fourth feature is a round, see Figure 8-42.

f. The fifth feature is a shell of thickness 2, see Figure 8-44.

After understanding the procedure for creating the model, you are now ready to create it.

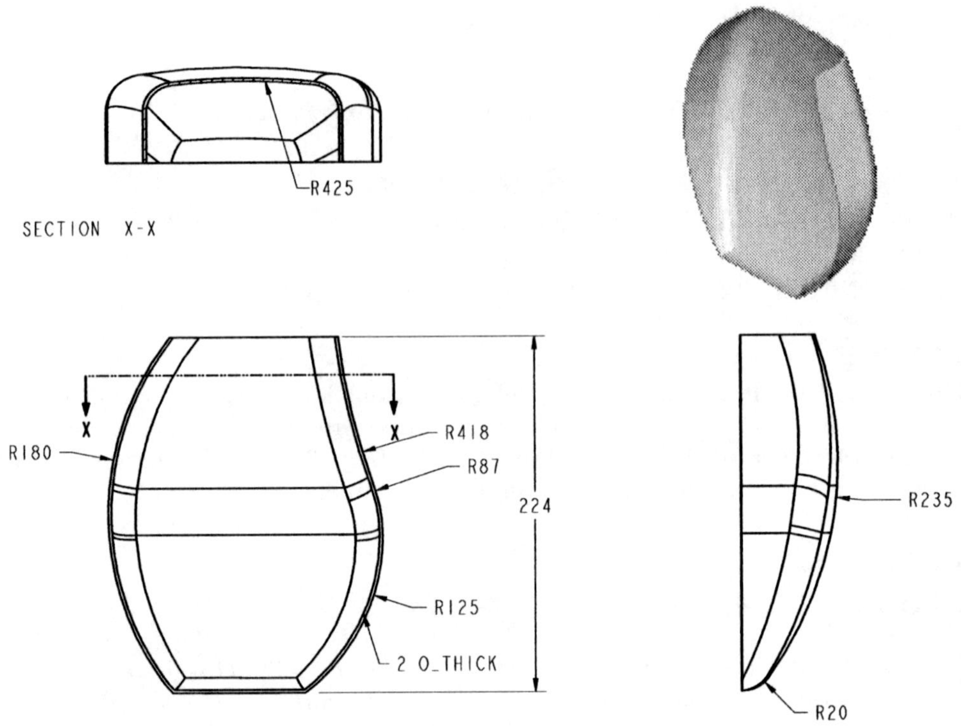

Figure 8-33 *Sectioned-top, front, right-side, and isometric views of the model*

When the Pro/ENGINEER session is started, the first task is to set the working directory. Since this is the first tutorial of this chapter, you need to create a directory named *c08*, if it does not exist. Choose the **New Directory** button in the **Select Working Directory** dialog box and create a directory named *c08* at *C:\ProE*.

Creating New Object File

1. Open a new part file and name it *c08tut1*. The three default datum planes are displayed on the graphics window.

Creating the Trajectories

You will draw the datum curves that will be selected as trajectories to create the variable section sweep feature. After selecting the sketching plane, draw the sketch for the origin trajectory and dimension it. Then you need to draw the sketch for the X trajectory on the same plane and dimension it. The X trajectory defines the horizontal vector of a section. If the section is aligned with the X trajectory, the section varies along the path of the X trajectory. Similarly, two additional trajectories will be drawn to guide the section throughout the sweep.

1. Choose the **Sketched Datum Curve Tool** button from the **Datum** toolbar. The **Sketched Datum Curve** dialog box is displayed.

2. Select the **FRONT** datum plane from the graphics window.

3. Select the **TOP** datum plane from the graphics window and then select the **Top** option from the **Orientation** drop-down list.

4. Once you enter the sketcher environment, close the **References** dialog box and create the sketch of the origin trajectory. The origin trajectory is a straight line segment aligned to the **RIGHT** datum plane and the start point is aligned to the **TOP** datum plane. The sketch of the origin trajectory is shown in Figure 8-34.

Note

In the sketcher environment you will draw all the three datum curves because they are on the same datum plane.

You are drawing the datum curves as individual features because you cannot sketch the trajectories using the tools available in the Variable Section Sweep dashboard.

5. After the sketch of the origin trajectory is complete, draw the sketch of the X trajectory. The X trajectory is drawn using three arcs. It is recommended that you start the sketch from the top arc. After drawing the first arc, dimension it and then modify its dimension.

6. Draw the second arc and then modify the dimension as shown in Figure 8-35.

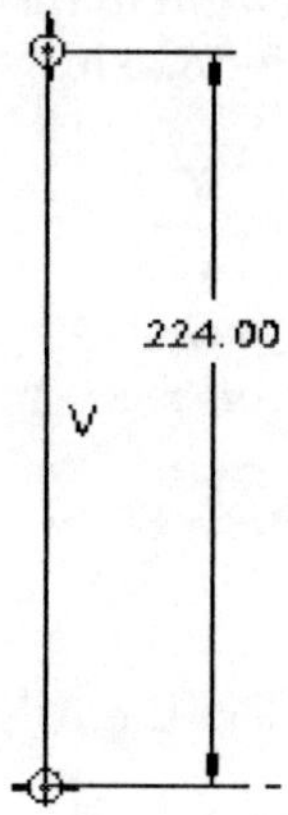

Figure 8-34 *Sketch with dimension of the origin trajectory*

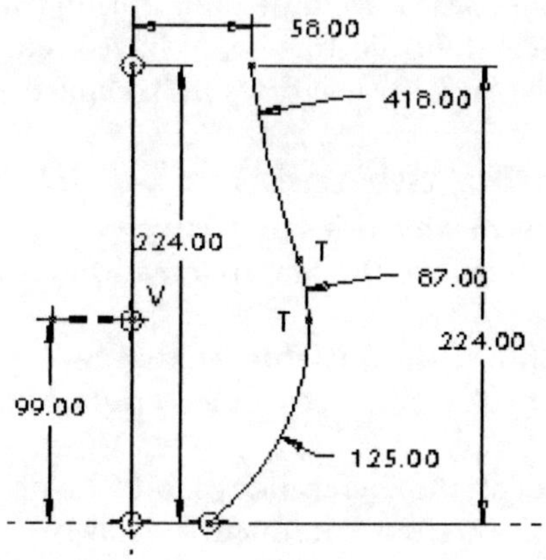

Figure 8-35 *Sketch with dimensions of the X trajectory*

7. Next, draw the third arc. As evident from the sketch of the X trajectory, the center points of the bottom and the second arcs are aligned horizontally.

8. Create the sketch of the third trajectory as shown in Figure 8-36. This trajectory is created using a single arc.

9. After the sketch is complete, choose the **Continue with the current section** button to exit the sketcher environment.

10. Similarly, create another trajectory on the **RIGHT** datum plane and keep the **TOP** datum plane at the left. The sketch of the trajectory is shown in Figure 8-37.

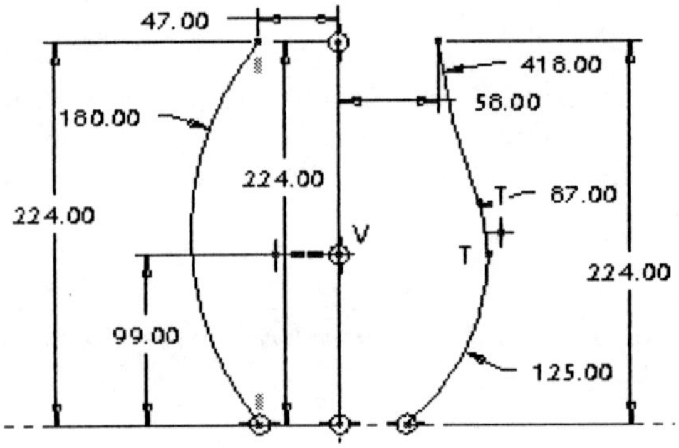

Figure 8-36 Sketch with dimensions of the additional trajectory

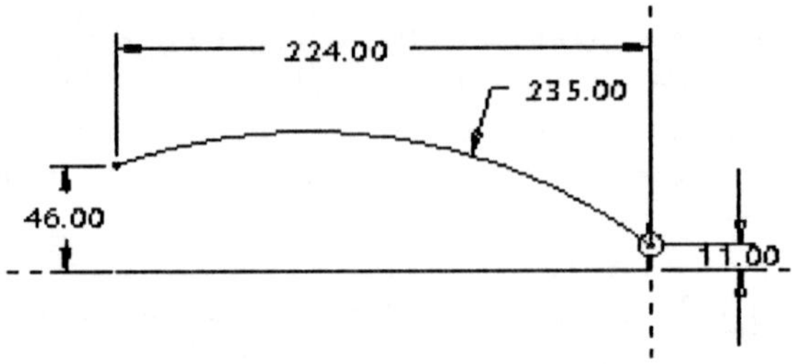

Figure 8-37 Sketch with dimensions of the additional trajectory

11. After the sketch is complete, choose the **Continue with the current section** button to exit the sketcher environment.

12. Choose the **Default View** option from the **Saved view list** button in the **Top Toolchest** to view the trajectories you have created, as shown in Figure 8-38. This view gives a better display of trajectories in 3D space.

Creating the Variable Section Sweep feature

The four trajectories that will create the variable section sweep have been created. Now, you need to invoke the feature creation tool and select these trajectories to create the feature.

1. Choose the **Variable Section Sweep Tool** button from the **Base Features** toolbar. The Variable Section Sweep dashboard is displayed.

2. Select the **References** tab to invoke the slide-up panel. Select the origin trajectory that was the first datum curve drawn.

3. Use CTRL+left mouse button to select the X trajectory (second datum curve). Note as you are selecting the trajectories, they are being listed in the **Trajectories** collector.

4. Use CTRL+left mouse button to select the remaining two trajectories. Select the check box under the X column for **Chain 1** row. This makes the trajectory that you select second as the X trajectory.

5. Now, choose the **Create or edit sweep section** button from the dashboard to enter the sketcher environment.

6. Draw the closed sketch of the section and apply the dimensions as shown in Figure 8-39.

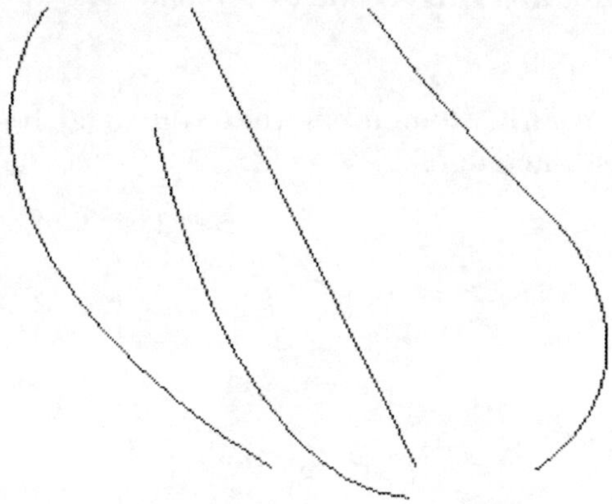

Figure 8-38 *Default view of the four trajectories*

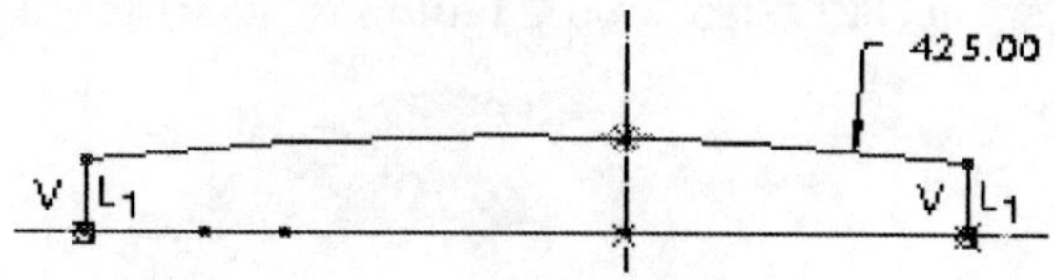

Figure 8-39 *Sketch of the section with dimensions*

Note that the section is aligned to all the start points of the trajectories that are displayed by crosses. The cursor will automatically snap to these points as you move it close to these points. In order to vary the section with the trajectories, it is necessary to align the section with the trajectories.

7. After the sketch is complete, choose the **Continue with the current section** button to exit the sketcher environment.

   The preview of the variable section sweep feature is displayed on the graphics window.

8. Choose the **Sweep as solid** button from the dashboard if the preview shows a surface feature.

9. Choose the **Build feature** button from the dashboard. Now you need to hide the datum curves.

10. Choose the **Set layers, layer items and display states** button from the **Top Toolchest** and open the **Model Tree**. In the Navigator, the layers of the current model are displayed. Right-click on the **PRT_ALL_CURVES** to invoke the shortcut menu.

11. From the shortcut menu, choose the **Blank layer** option. Regenerate the model, if needed. All the datum curves in the model will disappear from the display.

    The variable section sweep feature is completed and is shown in Figure 8-40. You can use middle mouse button to view the feature as shown in the figure.

Creating the Round Feature

You will create round on the edges of the base feature. The edges you need to round are shown in Figure 8-41.

1. Choose the **Round Tool** button from the **Engineering Features** toolbar. The Round dashboard is displayed.

2. Select the edges shown in Figure 8-41 to round. Remember that you need to use CTRL+left mouse button to select more than one edge.

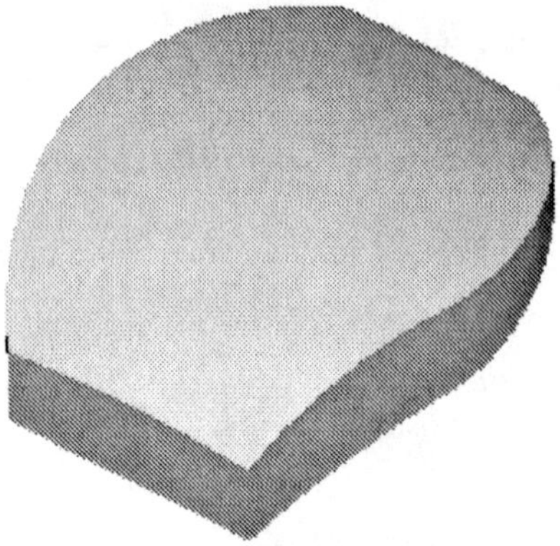

Figure 8-40 *Variable section swept feature*

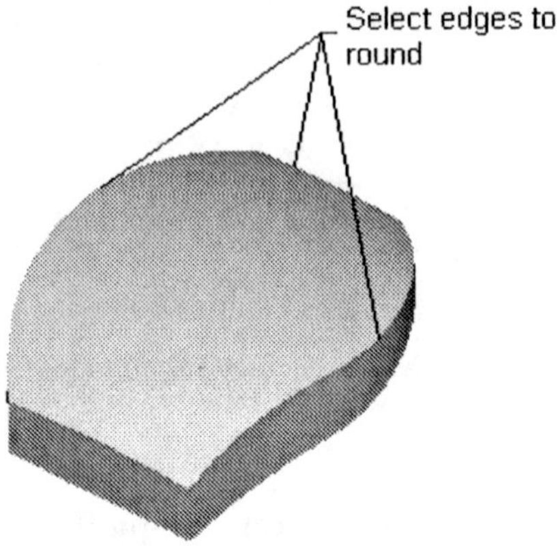

Figure 8-41 *Edges to be selected to round*

3. Enter **20** in the dimension box present on the dashboard. Choose **Build feature** from the dashboard. The base feature after creating the round is shown in Figure 8-42.

Creating the Shell

Now, you need to create the shell and remove the front and bottom faces of the base feature.

1. Choose the **Shell Tool** button from the **Engineering Features** toolbar. You are prompted to select the surfaces to be removed.

2. Select the surfaces shown in Figure 8-43 to remove. Remember that to select more than one face you need to use CTRL+left mouse button.

Figure 8-42 *Model after creating round*

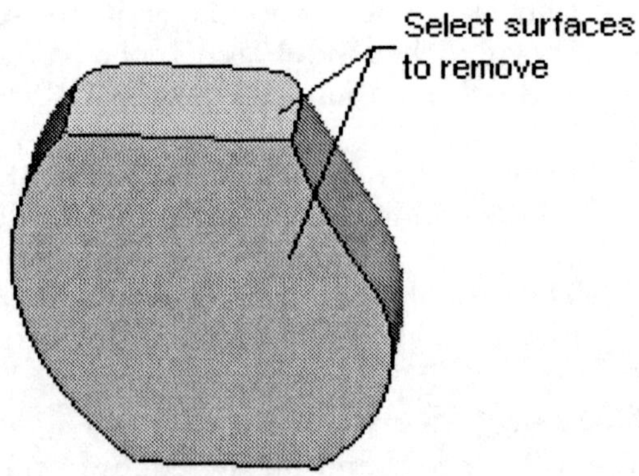

Figure 8-43 *Surfaces to be selected to remove*

3. Enter **2** in the dimension edit box. Choose the **Build feature** button from the dashboard. The model after creating the shell feature is shown in Figure 8-44.

Saving the Model

You have to save the model because you may need it later.

1. Choose the **Save the active object** button from the **File** toolbar and save the model.

   The order of feature creation can be seen from the **Model Tree** shown in Figure 8-45.

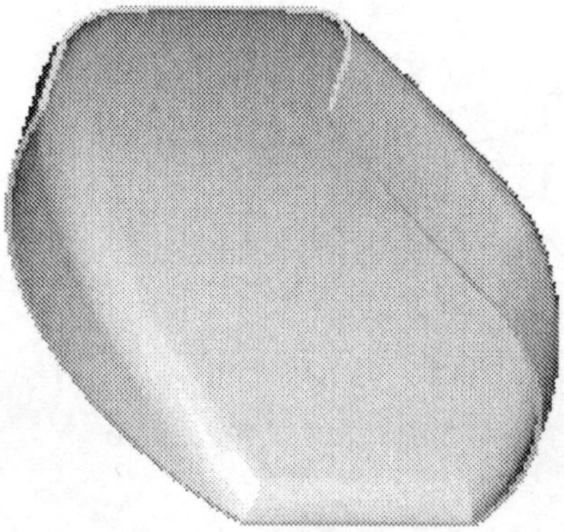

Figure 8-44 Final model after shelling

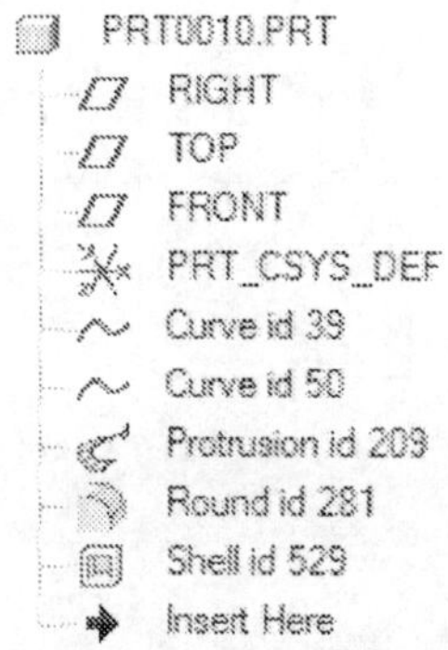

Figure 8-45 Model Tree for Tutorial 1

Tutorial 2

In this tutorial you will create the model shown in Figure 8-46. This model is a Upper Housing of a motor blower assembly. Figure 8-47 shows the left-side view of the top view, top view, front view, and the sectioned left-side view of the model. **(Expected time: 1 hr)**

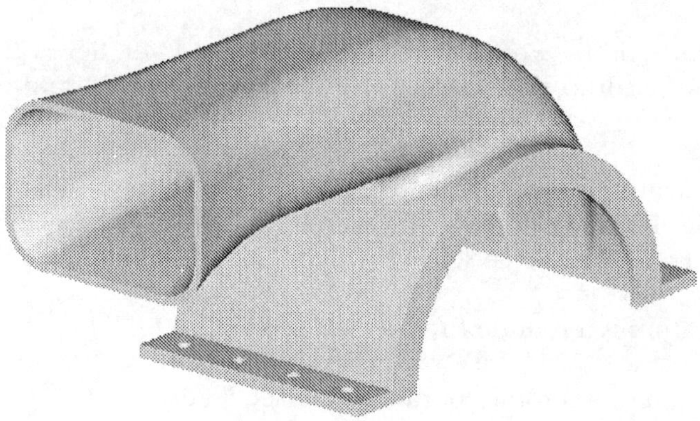

Figure 8-46 Solid model of the Upper Housing

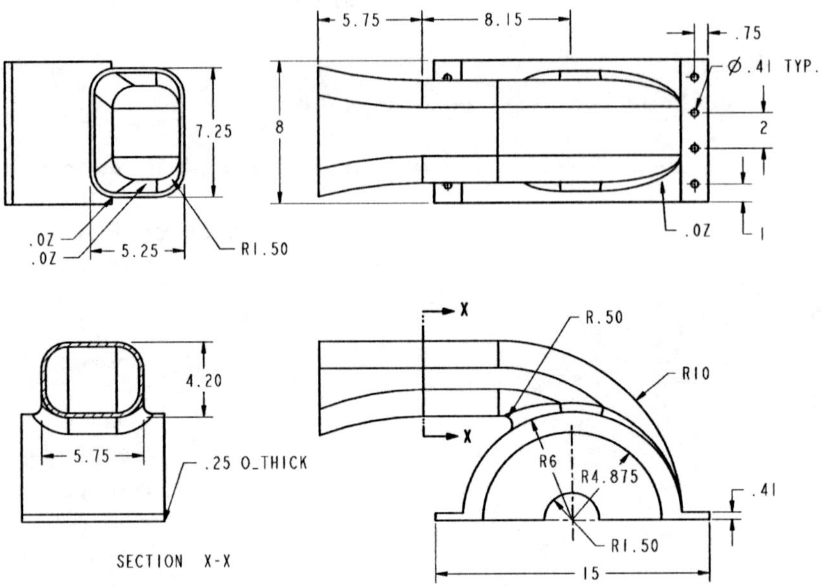

Figure 8-47 *Left-side view of the top view, top view, front view, and the sectioned left-side view of the model*

The following steps outline the procedure for creating this model:

a. First examine the model and then determine the number of features in it. The model is composed of ten features, see Figures 8-46 and 8-47.

b. The base feature is an extruded feature, see Figure 8-49. Select the sketching plane for the base feature, draw the sketch using the sketcher tools, and apply dimensions, see Figure 8-48.

c. The second feature is a swept blend feature in which the section is normal to the origin trajectory, see Figure 8-55. First the sketch for the origin trajectory will be drawn and dimensioned and then three sections will be defined along the origin trajectory.

d. The third feature is a round of radius 1.5, see Figure 8-57.

e. The fourth feature is a round of radius 0.5, see Figure 8-59.

f. The fifth feature is a shell of thickness 0.25 that will be created on the front face of the swept blend feature and on the bottom face of the base feature, see Figure 8-60.

g. The sixth feature is an extruded cut, see Figure 8-62. Select the sketching plane for the cut feature, draw the sketch using sketching tools, and apply the dimensions, see Figure 8-61.

h. The seventh feature is also an extruded cut, see Figure 8-64. Select the sketch plane for the cut feature, draw the sketch using sketching tools, and apply the dimensions, see Figure 8-63.

i. The eighth feature is an extruded feature, see Figure 8-66. Select the sketch plane for the extruded feature, draw the sketch, and apply constraints and dimensions, see Figure 8-65.

j. The ninth feature is a mirror copy of the eighth feature, see Figure 8-67.

k. The tenth feature is a hole, see Figure 8-68, and this hole will be patterned. After you pattern the hole, the hole and the pattern feature will be combined in a single feature.

After understanding the procedure for creating the model, you are now ready to create it. The working directory was already selected in Tutorial 1 and therefore you do not need to select it again. However, if you want to change the working directory, choose **File > Set Working Directory** and then select *c08* in the **Select Working Directory** dialog box.

Creating New Object File

1. Open a new part file and name it *c08tut2*. The three default datum planes and the **Model Tree** are displayed on the graphics screen. Exit the **Model Tree**. The **Model Tree** will not appear if it was previously closed.

Creating the Base Feature

To create the sketch for the base feature, you first need to select the sketching plane for the base feature. In this model, you need to draw the base feature on the **RIGHT** datum plane.

1. Choose the **Extrude Tool** button from the **Base Features** toolbar.

2. Select the **RIGHT** datum plane as the sketch plane.

3. Select the **TOP** datum plane from the graphics window and then select the **Top** option from the **Orientation** drop-down list.

4. Choose the **Sketch** button to enter the sketcher environment.

5. Once you enter the sketcher environment, create the sketch of the base feature and apply dimensions as shown in Figure 8-48.

6. After the sketch is complete, choose the **Continue with the current section** button to exit the sketcher environment.

   The Extrude dashboard reappears below the graphics window. The **Extrude from sketch plane by specified depth value** button is selected by default.

7. Choose the **Extrude on both sides of sketch plane by half the specified depth value in each** button from the depth flyout.

8. Enter a depth of **8** in the dimension box present in the Extrude dashboard. Choose the **Build feature** button from the Extrude dashboard.

   The base feature is completed and the default trimetric view is shown in Figure 8-49.

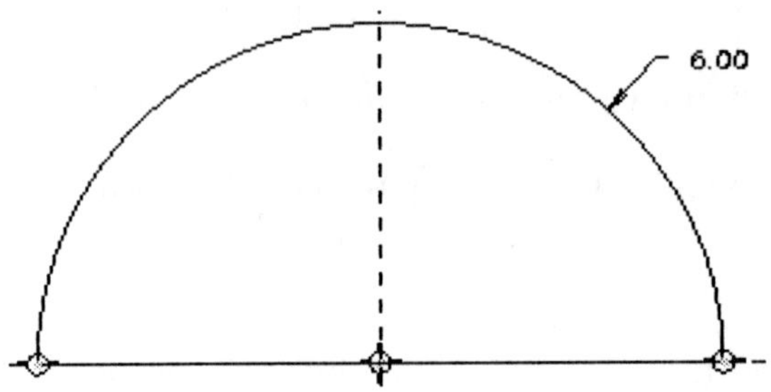

Figure 8-48 *Sketch for the base feature*

Figure 8-49 *Default trimetric view of the base feature*

Creating the Swept Blend

The second feature is a swept blend. You will sketch the trajectory for the blend on the **RIGHT** datum plane. Note that the trajectory is tangent to the curve of the base feature and the start point of the trajectory is aligned with the **TOP** datum plane.

1. Choose **Insert > Swept Blend > Protrusion** from the menu bar. The **BLEND OPTS** menu is displayed. The **Sketch Sec** and the **NrmToOriginTraj** options are selected by default.

2. Choose **Done** from the **BLEND OPTS** menu. The **SWEEP TRAJ** menu is displayed.

3. Choose the **Sketch Traj** option from the **SWEEP TRAJ** menu. You are prompted to select a sketching plane.

4. Select the **RIGHT** datum plane from the graphics screen. The **DIRECTION** menu is displayed. Choose **Okay**. The **SKET VIEW** submenu is displayed.

5. Select the **Top** option from the submenu and choose the **TOP** datum plane.

6. Once you enter the sketcher environment, draw the sketch of the trajectory and add constraints and dimensions as shown in Figure 8-50.

   As evident from Figure 8-50, the start point of the trajectory is aligned with the **TOP** datum plane and the trajectory is tangent to the curve of the base feature.

7. After the sketch is complete, choose the **Continue with the current section** button to exit the sketcher environment.

   The **CONFIRM SEL** menu is displayed. Note that on the origin trajectory one red, one white, and two green points are displayed. The green points indicate that these are fixed points for defining the sections. The location for red point can be changed by using the options in the **CONFIRM SEL** menu.

Note

The two green points are fixed at the start point and at the endpoint of the trajectory, indicating that it is necessary to draw sections at these points in order to complete the swept blend feature. This is because swept blend is created by blending the sections along the trajectory. There can be any number of sections that can be possible between the start point and the endpoint of the trajectory.

8. Choose **Next** from the **CONFIRM SEL** menu to shift the position of the red point. You need to shift the position of the red point because the section of the swept blend feature (SECTION X-X) that is given in Figure 8-50 is at the point where the red point is shifted now. Hence, now you will need to define three sections at three points.

9. Choose the **Accept** option from the **CONFIRM SEL** menu. The **Message Input Window** is displayed and you are prompted to enter the z-axis rotation angle for section 1. Section 1 is at the start point of the origin trajectory.

10. The value **0** is entered by default in the **Message Input Window**; press ENTER.

    The system takes you to the sketcher environment. Now, you need to sketch the first section of the blend at the start point of the trajectory. You will notice a coordinate system is placed at the start point of the trajectory. You need to align the sketch of the section to this coordinate system.

11. Use the middle mouse button to spin the view of the sketch as shown in Figure 8-51 to have a better understanding of the sketch in 3D space.

12. Choose the **Orient the sketching plane parallel to the screen** button from the **Top Toolchest**. The sketching plane reorients parallel to the graphics screen.

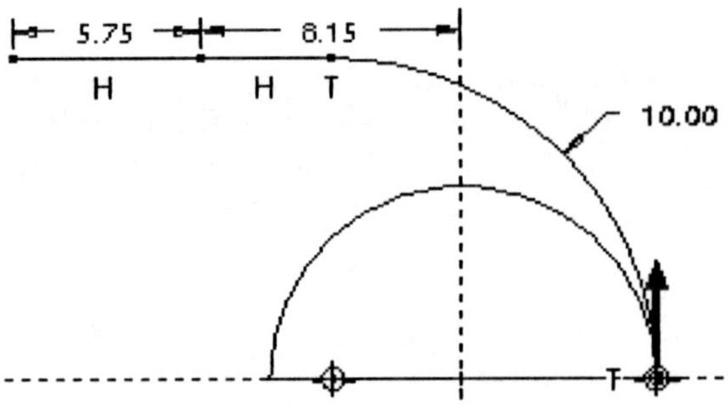

Figure 8-50 Sketch of the origin trajectory

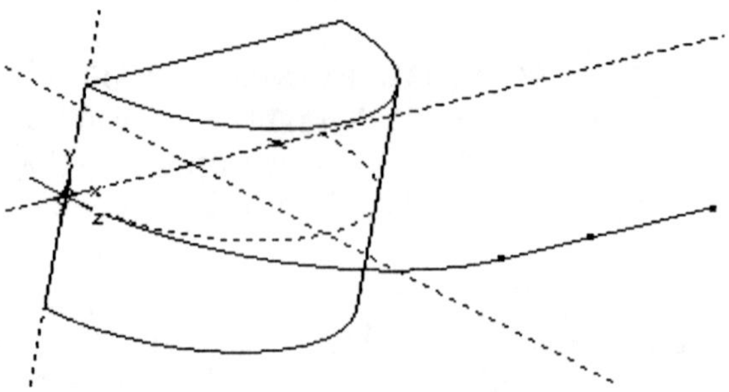

Figure 8-51 Spinned view of the sketch showing the origin trajectory, base feature, and the default coordinate system

13. Draw the sketch of the first section and add dimensions as shown in Figure 8-52.

14. After the sketch is complete, choose the **Continue with the current section** button to exit the sketcher environment. The **Message Input Window** is displayed and you are prompted to enter the z-axis rotation angle for section 2.

15. The value **0** is entered by default in the **Message Input Window**; press ENTER.

16. Draw the sketch of the second section and add dimensions as shown in Figure 8-53. This section is at the point between the first and the third sections and in Figure 8-47, this section is named as X-X.

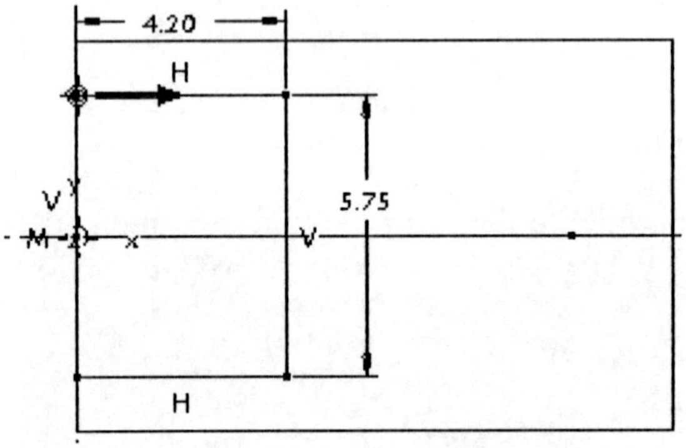

Figure 8-52 Sketch for first section with dimensions

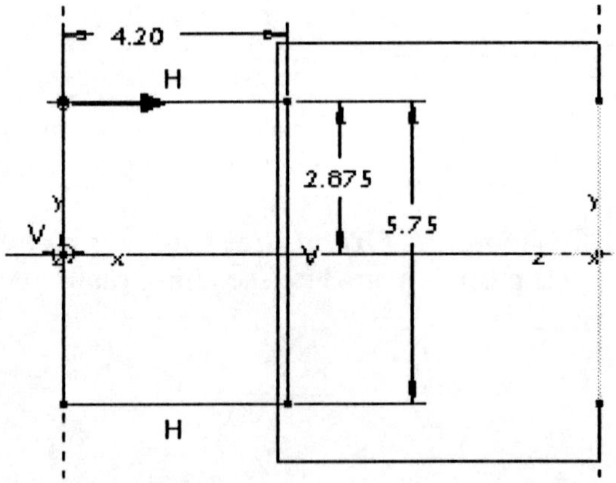

Figure 8-53 Sketch for second section with dimensions

17. After the sketch is complete, choose the **Continue with the current section** button to exit the sketcher environment. The **Message Input Window** is displayed and you are

prompted to enter the z-axis rotation angle for section 3.

18. The value **0** is entered by default in the **Message Input Window**; press ENTER.

19. Draw the sketch of the third section and add dimensions as shown in Figure 8-54. This is the last section on the origin trajectory. Remember to align the section to the coordinate system.

20. After the sketch is complete, choose the **Continue with the current section** button to exit the sketcher environment.

21. Choose **OK** from the **PROTRUSION** dialog box. The default trimetric view of the swept blend feature with the base feature is shown in Figure 8-55.

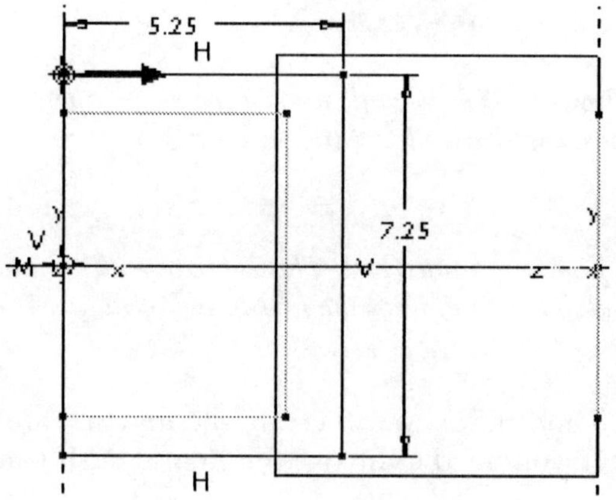

Figure 8-48 *Sketch of the third section with dimensions*

Figure 8-49 *Default view of the swept blend*

Creating the Round Features

The two round features are created before shelling because the rounds are also shelled along with the other faces of the model. The third feature in the model is a round of radius 1.5 and the fourth feature is also a round of radius 0.5.

1. Choose the **Round Tool** button from the **Engineering Features** toolbar. The Round dashboard is displayed.

2. Select all the four longer edges of the swept feature shown in Figure 8-56 to round them. Remember that the first edge is selected by using the left mouse button and the other remaining edges are selected by using CTRL+left mouse button.

   The preview of the round is displayed on the graphics window along with the default radius value.

3. Double-click on the radius value, enter **1.5** in the edit box that appears, and then press ENTER.

4. Choose **Build feature** from the Round dashboard. The default view of the model after creating the round is shown in Figure 8-57.

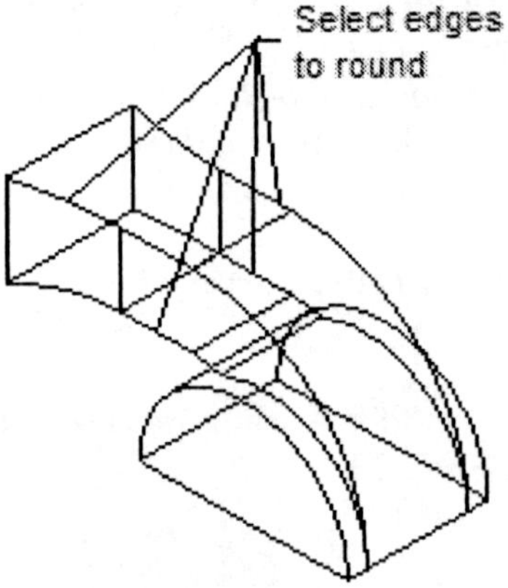

Figure 8-56 Edges to round

Figure 8-57 Default view of the model after creating a round feature of radius 1.5.

Note

The round of radius 0.5 cannot be created by adding a round set. This is because this round is created on the geometry of the previous round. Therefore, until the first round of radius 1.5 is created you cannot create the second round.

The third feature of the model is completed and now you will create the fourth feature, that is, a round of radius 0.5. For this purpose you need to invoke the Round dashboard.

5. Choose the **Round Tool** button from the **Engineering Features** toolbar. The Round dashboard is displayed.

6. Select the edge where the base feature and the swept feature joins as shown in Figure 8-58.

   The preview of the round is displayed on the graphics window along with the default radius value.

7. Double-click the radius value, enter **0.5** in the edit box that appears, and then press ENTER.

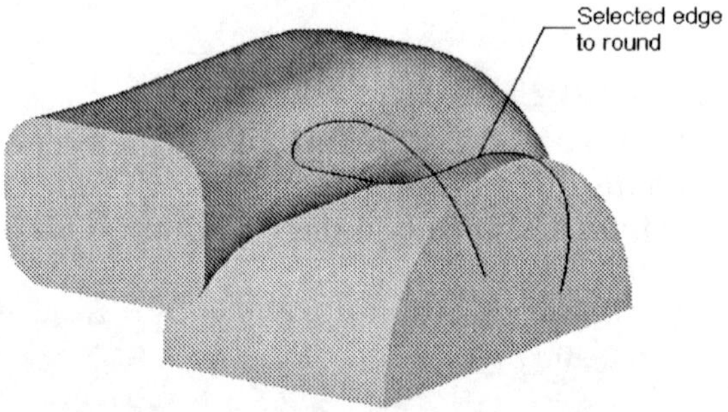

Figure 8-58 Figure showing a solid model

8. Choose **Build feature** from the Round dashboard. The default view of the model after creating the round is shown in Figure 8-59.

Creating the Shell

You need to create a shell of thickness 0.25 on the swept blend feature.

1. Choose the **Shell Tool** button from the **Engineering Features** toolbar. The Shell dashboard is displayed and you are prompted to select the surface to remove.

2. Using the left mouse button, select the front face of the swept blend feature and then use CTRL+left mouse button to select the bottom face of the base feature.

3. Enter a value of **0.25** in the dimension box available on the Shell dashboard and press ENTER.

4. Choose the **Build feature** button from the Shell dashboard. The shell feature is completed and is shown in Figure 8-60.

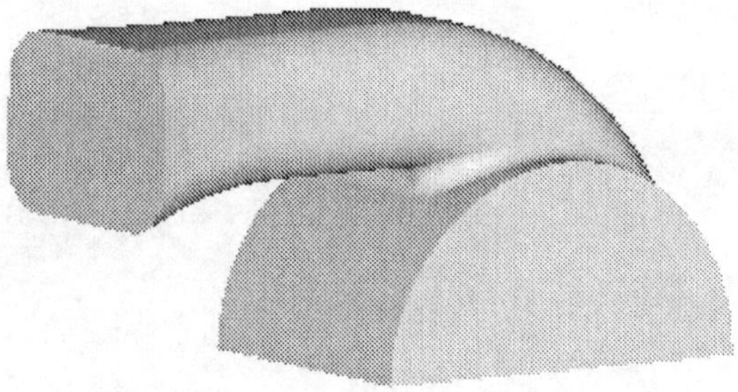

Figure 8-53 *Model after creating a round feature of radius 0.5*

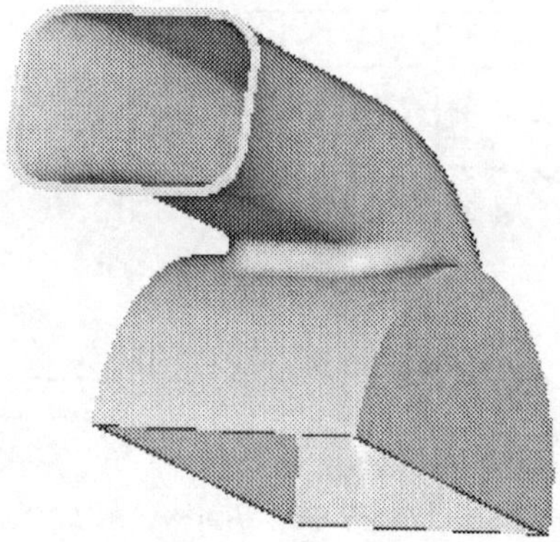

Figure 8-54 *Default trimetric view after shelling*

Creating the Extruded Cut

The extruded cut is the sixth feature of this model. You need to create the cut on the front face of the base feature.

1. Choose the **Extrude Tool** button from the **Base Features** toolbar. The Extrude dashboard is displayed.

2. Choose the **Remove Material** button from the dashboard.

3. Select the front face of the base feature as the sketch plane.

4. Select the **TOP** datum plane from the graphics window and then select the **Top** option from the **Orientation** drop-down list.

5. Choose the **Sketch** button to enter the sketcher environment.

6. Draw the sketch of the cut feature and apply dimension as shown in Figure 8-61. You can use the **Create concentric arc** button from the **Sketcher Tools** toolbar to draw the sketch.

7. After the sketch is complete, choose the **Continue with the current section** button to exit the sketcher environment.

   The Extrude dashboard reappears below the graphics window. The **Extrude from sketch plane by specified depth value** button is selected by default.

8. Choose the **Extrude upto next surface** button from the depth flyout.

9. Choose the **Build feature** button from the Extrude dashboard. The cut feature is created and is shown in Figure 8-62.

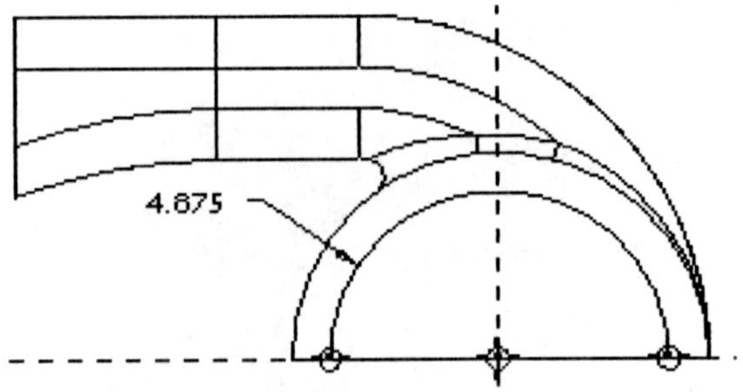

Figure 8-55 Sketch with dimension for cut feature

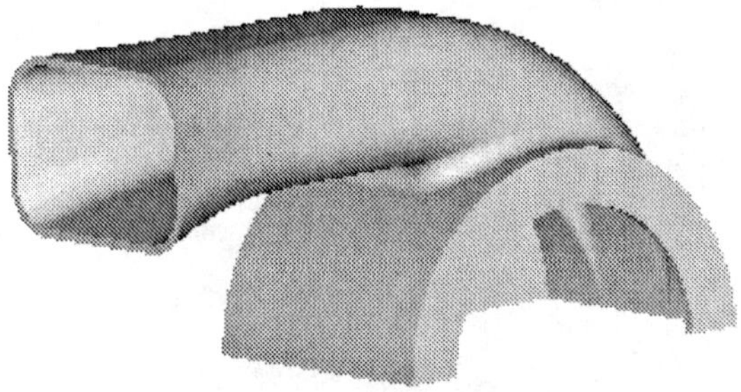

Figure 8-56 Model with cut feature

Similarly, create the next extruded cut feature at the back face of the base feature. Select the sketch plane, draw the sketch, apply dimension, and extrude the sketch using the **Extrude upto next surface** button. The sketch for the cut feature is shown in Figure 8-63.

The seventh feature of the model, that is, cut feature is shown in Figure 8-64.

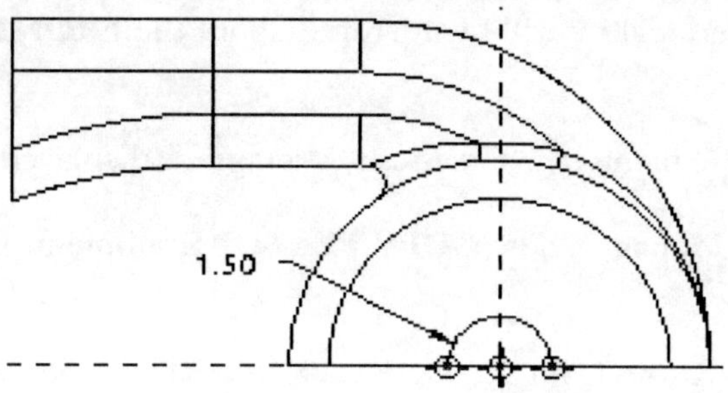

Figure 8-63 Sketch with dimension for cut

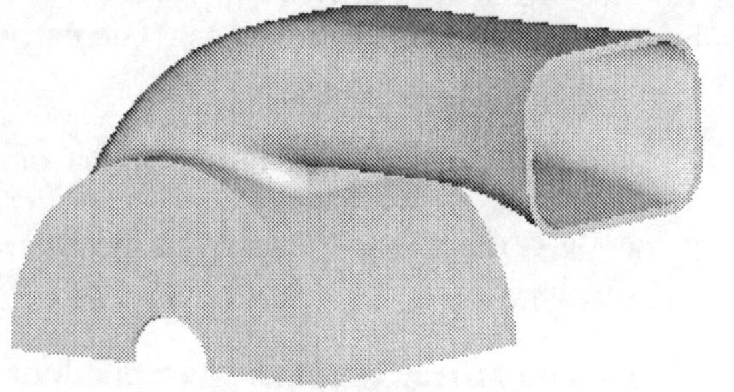

Figure 8-64 Model with the cut feature

Creating the Extruded Feature

The eighth feature of the model is an extruded feature and will be created on the front face of the base feature.

1. Select the front face of the base feature as the sketching plane for the extrude feature. Once you enter the sketcher environment, draw the sketch and apply constraints and dimensions as shown in Figure 8-65.

2. After the sketch is complete, choose the **Continue with the current section** button to exit the sketcher environment.

3. From the depth flyout, select the **Extrude to selected point, curve, plane or surface** button and select the back face of the base feature.

4. Choose the **Build feature** button from the Extrude dashboard. The extruded feature is completed and is shown in Figure 8-66.

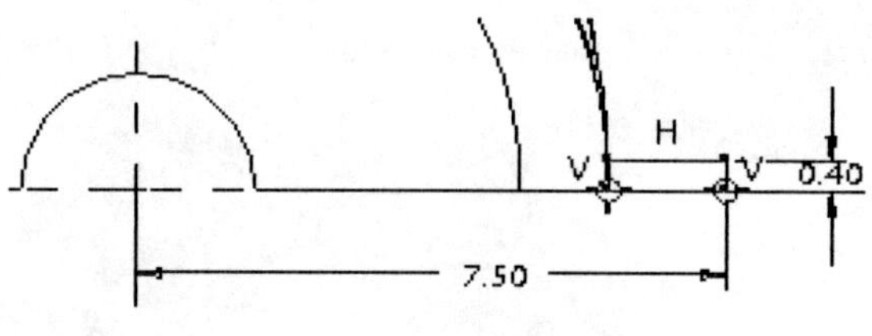

Figure 8-65 Sketch for the extruded feature

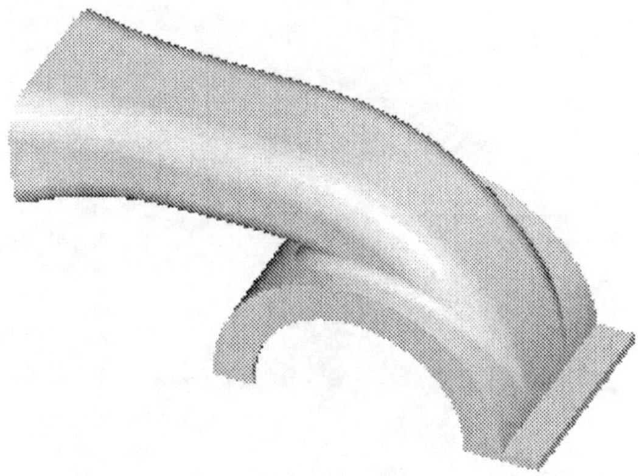

Figure 8-66 Model after creating the extrude feature

Creating a Copy of the Eighth Feature

You need to create a mirror copy of the extruded feature as shown in Figure 8-67 and this will be the ninth feature of the model. The extruded feature will be mirrored about the **FRONT** datum plane.

1. Choose **Edit > Feature Operations** from the menu bar. The **Menu Manager** is displayed.

2. Choose the **Copy** option from the **Menu Manager**. The **COPY FEATURE** submenu is displayed.

3. Choose **Mirror > Select > Dependent > Done** from the **COPY FEATURE** submenu. The **SELECT FEAT** submenu is displayed and you are prompted to select the features to be mirrored.

4. Select the previous extruded feature from the graphics window and choose **Done** from the **SELECT FEAT** submenu. You are prompted to select a plane or create a datum plane to mirror about.

5. Select the **FRONT** datum plane as the mirror plane. The protrusion feature is mirrored about the **FRONT** datum plane as shown in Figure 8-67.

6. Choose **Done** from the **Menu Manager**.

Creating the Tenth Feature

The tenth feature is a through hole and will be created using the Hole dashboard.

1. Choose the **Hole Tool** button from the **Engineering Features** toolbar. The Hole dashboard is displayed. The **Create straight hole** button is selected by default.

   Create the hole as shown in Figure 8-68 by specifying the placement parameters. The placement parameters can be referred from Figure 8-47.

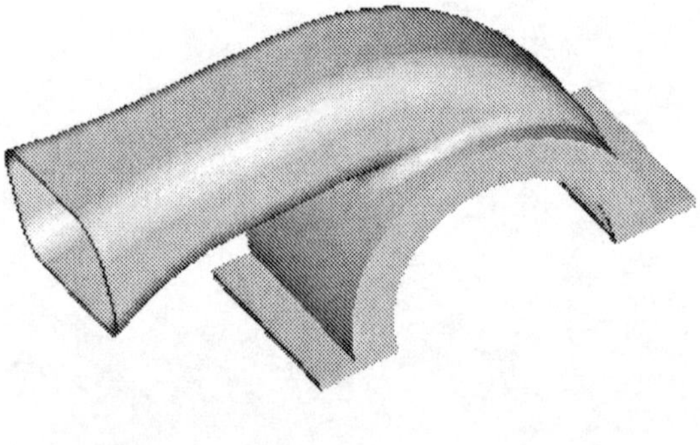

Figure 8-67 *Model after mirroring the extruded feature*

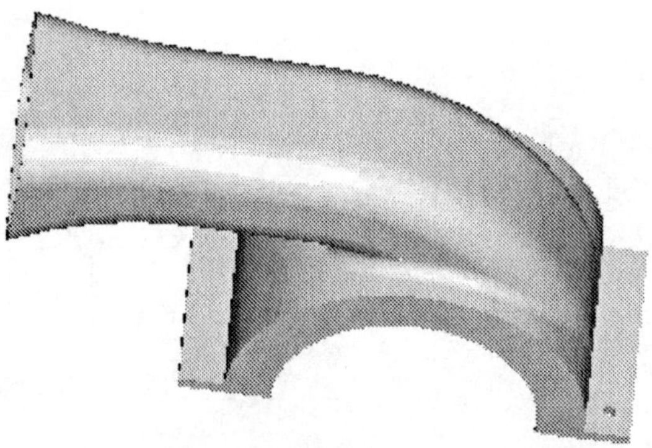

Figure 8-68 *Model after creating the hole*

Creating a Pattern of the Hole Feature

As evident from Figure 8-47, you need to create eight instances of the hole. The first instance is created by using the Hole dashboard and you can use the **Pattern** option to create the remaining seven instances. You will create a rectangular pattern of the hole feature. You can also create all the holes using the Hole dashboard and specifying the placement parameters for each of them. But to save time, you will create a pattern of the hole.

1. Select the hole feature from the **Model Tree** or from the model. Hold down the right mouse button to invoke the shortcut menu.
2. Choose the **Pattern** option from the shortcut menu. The Pattern dashboard is displayed and the dimensions are displayed on the hole feature.
3. Select the **General** option from the **Options** option slide-up panel.
4. Double-click the dimension value **1** from the graphics window. Enter a value of **2** in the edit box and press ENTER.
5. Enter **4** in the **1** edit box present on the dashboard and press ENTER. Now, you need to enter dimension increment in the second direction.
6. Click the collector that displays **No Items** on the right of the **2** edit box. Double-click the dimension value **0.75** from the graphics window. Enter a value of **13.5** in the edit box and press ENTER.
7. Enter **2** in the **2** dimension box present on the dashboard and press ENTER.
8. Choose the **Build feature** button from the dashboard. You can use the middle mouse button to spin the model and to display the model as shown in Figure 8-69.

Saving the Model

You have to save the model because you may need it later.

1. Choose the **Save the active object** button from the **File** toolbar and save the model. The order of feature creation can be seen from the **Model Tree** shown in Figure 8-67.

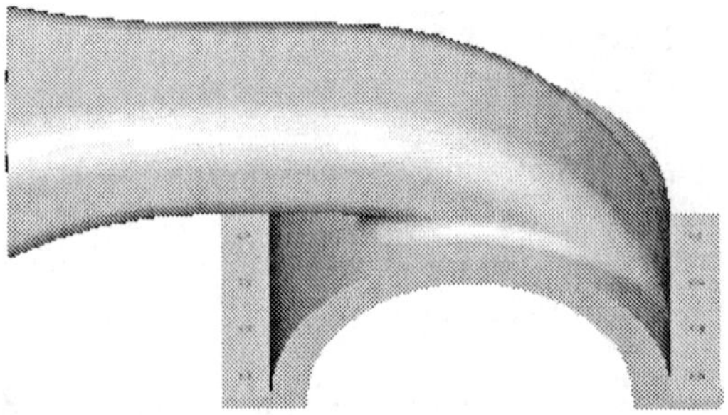

Figure 8-69 *Completed model*

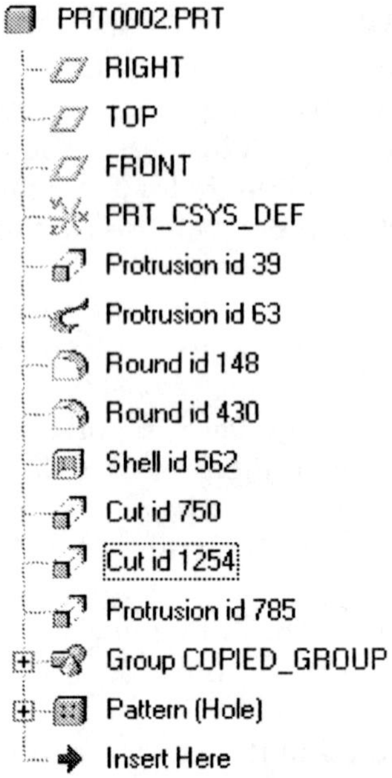

Figure 8-70 ***Model Tree*** *for Tutorial 2*

Tutorial 3

In this tutorial you will create the model shown in Figure 8-71. This model is a wheel of a car. Figure 8-72 shows the top view, sectioned front view, sectioned right view, detail view, and two blend sections with dimensions. **(Expected time: 45 min)**

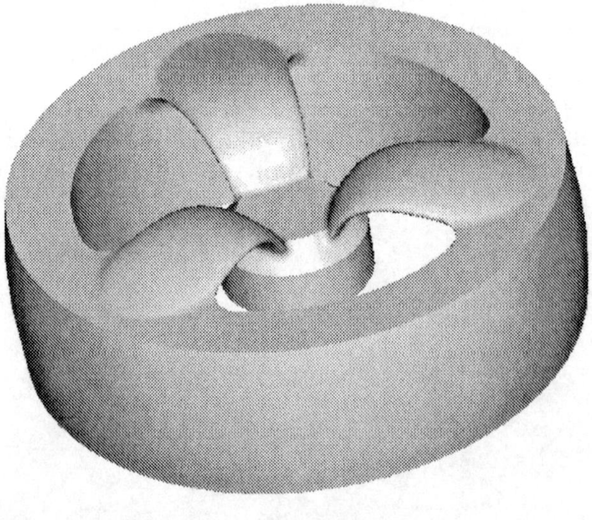

Figure 8-71 *Solid model of the wheel*

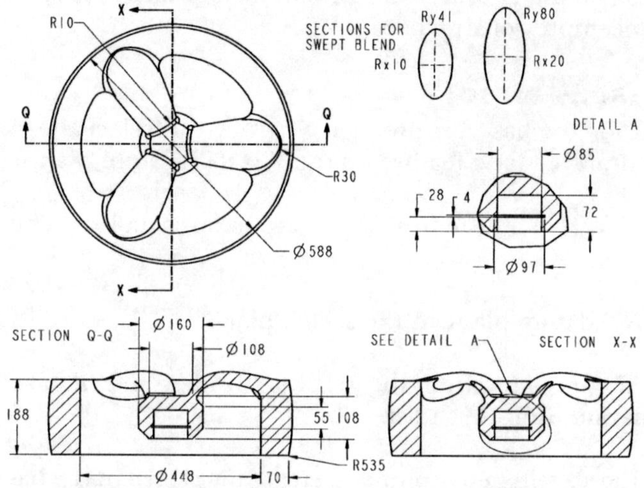

Figure 8-72 *Top view, sectioned front view, sectioned right view, detail view, and two blend sections with dimensions*

The following steps outline the procedure for creating this model:

a. First, examine the model and then determine the number of features in it. The model is composed of four features, see Figure 8-71.

b. The base feature is a revolved feature, see Figure 8-74. Select the sketching plane for the base feature, draw the sketch using the sketcher tools, and apply constraints and dimensions, see Figure 8-73.

c. The second feature is a swept blend feature in which the section is normal to the origin trajectory, see Figure 8-78. First the sketch for the origin trajectory will be drawn and dimensioned and then two sections at the start points and endpoints will be defined along the origin trajectory. Then this feature will be patterned, see Figure 8-79.

d. The third feature is a round of radius 30 and a round of radius 10, see Figures 8-81 and 8-82 respectively.

e. The fourth feature is a revolved cut, see Figure 8-84. First you will draw the sketch of the cut feature on the sketching plane, then apply constraint and dimensions, and finally revolve the sketch to 360-degree.

After understanding the procedure for creating the model, you are now ready to create it. The working directory was already selected in Tutorial 1 and therefore you do not need to select it again.

Creating New Object File

1. Open a new part file and name it *c08tut3*. The three default datum planes and the **Model Tree** are displayed on the graphics screen. Exit the **Model Tree**. The **Model Tree** will not appear if it has been previously closed.

Creating the Base Feature

To create the sketch for the base feature, you first need to select the sketch plane. In this model, you need to draw the base feature on the **FRONT** datum plane.

1. Choose the **Revolve Tool** button from the **Base Features** toolbar. The Revolve dashboard is displayed.

2. Select the **FRONT** datum plane as the sketch plane.

3. Select the **TOP** datum plane from the graphics window and then select the **Top** option from the **Orientation** drop-down list.

4. Once you enter the sketcher environment, create the sketch of the base feature and apply constraints and dimensions as shown in Figure 8-73. The sections must be closed.

5. After the sketch is complete, choose the **Continue with the current section** button to exit the sketcher environment. The Revolve dashboard is displayed.

6. Choose the **Build feature** button from the Revolve dashboard to exit it.

   The base feature is completed and the default trimetric view is shown in Figure 8-74.

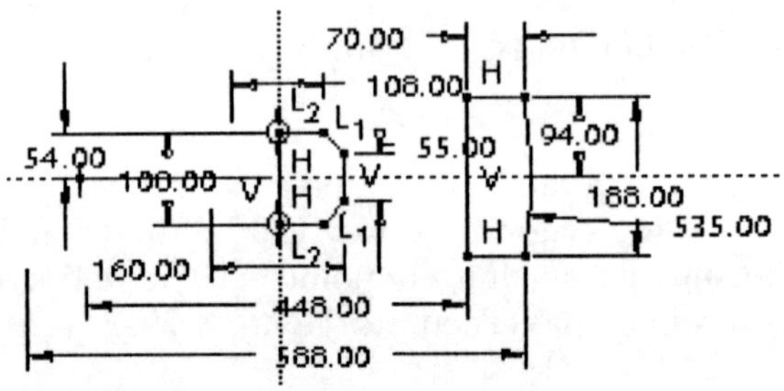

Figure 8-73 Sketch with dimensions for the base feature

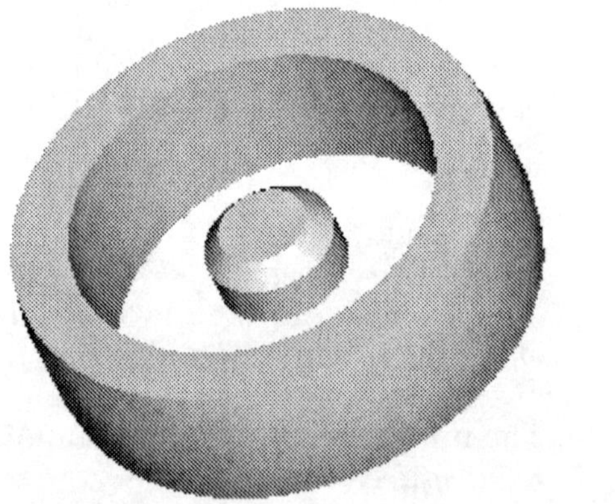

Figure 8-74 Default trimetric view of the base feature

Creating the Swept Blend

The second feature of the model is a swept blend. You need to create a datum plane that is at an angle of 90-degree from the **RIGHT** datum plane. This is because later in the tutorial, you will create a rotational pattern of the swept blend feature. The 90-degree angle value will be used for the dimensional increment for creating the rotational pattern.

1. Choose **Insert** > **Swept Blend** > **Protrusion** from the menu bar. The **BLEND OPTS** submenu is displayed. The **Sketch Sec** and the **NrmToOriginTraj** options are selected by default.

2. Choose **Done** from the **BLEND OPTS** submenu. The **SWEEP TRAJ** menu is displayed.

3. Choose the **Sketch Traj** option from the **SWEEP TRAJ** menu. You are prompted to select a sketching plane or create a datum plane.

4. Choose the **Make Datum** option from the **SETUP PLANE** submenu. **The DATUM PLANE** submenu is displayed.

5. Choose the **Through** option from the **DATUM PLANE** submenu and select the axis named **A_1** from the graphics screen. The **A_1** axis is the axis of revolution of the base feature.

6. Choose the **Angle** option from the **DATUM PLANE** submenu and select the **RIGHT** datum plane.

7. Choose **Done** from the **DATUM PLANE** submenu. The **OFFSET** submenu is displayed.

8. Choose the **Enter Value** option from the **OFFSET** submenu. The **Message Input Window** is displayed.

9. Enter **90** in the **Message Input Window** and press ENTER. The **DIRECTION** menu is displayed. Choose **Okay**. The **SKET VIEW** submenu is displayed.

10. Select the **Top** option from the submenu and choose the **TOP** datum plane.

11. Once you enter the sketcher environment, first you need to specify two references to dimension. Select the **TOP** and **RIGHT** datum planes.

12. Draw the sketch of the trajectory and add constraints and dimensions as shown in Figure 8-75. In Figure 8-75, the hidden lines are also shown because in this case the display of the sketch is more clear while drawing the sketch with display of hidden lines. The curve is drawn using the spline. The spline has only four control points: one start point, one endpoint, and two intermediate points. The dimensions shown in Figure 8-75 are enough to define the sketch.

Note

The shape of the spline will differ from user to user, and so the shape of the protrusion feature may vary.

13. After the sketch is complete, choose the **Continue with the current section** button to exit the sketcher environment.

The **CONFIRM SEL** menu is displayed. Note that two green points are displayed on the origin trajectory. The green points indicate that they are fixed points for defining the sections.

14. Choose the **Next** option twice from the **CONFIRM SEL** menu. The **Message Input Window** is displayed and you are prompted to enter the Z-axis rotation angle for section 1. Section 1 is at the start point of the origin trajectory.

15. The value **0** is entered by default in the **Message Input Window**; press ENTER.

    The system takes you to the sketcher environment. Now, you need to sketch the first section of the blend at the start point of the trajectory. You will notice a coordinate system is placed at the start point of the trajectory. You need to align the sketch of the section to this coordinate system.

16. Draw the sketch of the first section and add dimensions as shown in Figure 8-76.

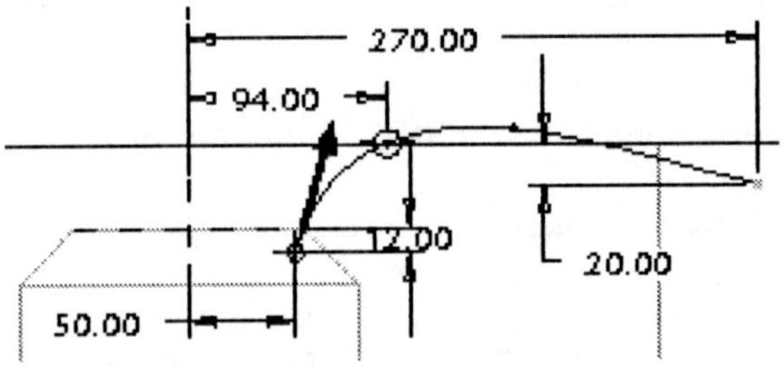

Figure 8-75 Sketch with dimensions for the origin trajectory

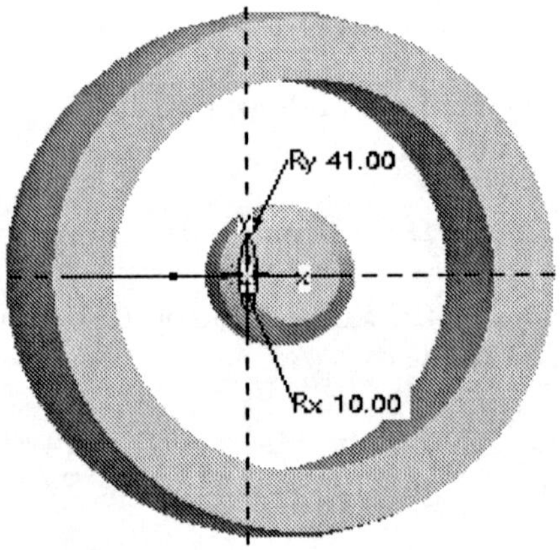

Figure 8-76 Sketch for section 1 with dimensions

17. After the sketch is complete, choose the **Continue with the current section** button to exit the sketcher environment. The **Message Input Window** is displayed and you are prompted to enter the Z-axis rotation angle for section 2.

18. The value **0** is entered by default in the **Message Input Window**; press ENTER.

19. Draw the sketch of the second section and add dimensions as shown in Figure 8-77. The second section is also elliptical in shape, and therefore an ellipse is drawn in the sketch. The dimensions for this section are shown in Figure 8-72.

20. After the sketch is complete, choose the **Continue with the current section** button to exit the sketcher environment.

21. Choose **OK** from the **PROTRUSION** dialog box. The default trimetric view of the swept

blend feature with the base feature is shown in Figure 8-78.

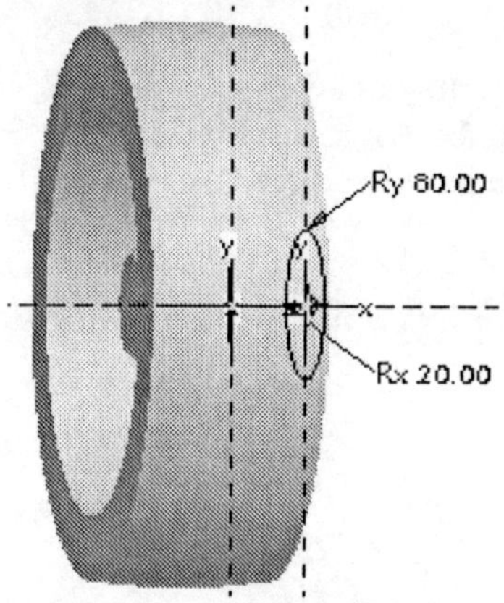

Figure 8-77 *Sketch for section 2 with dimensions*

Figure 8-78 *Model after creating the swept blend*

Creating the Rotational Pattern of the Swept Blend

Creating the swept blend features individually on the base feature will consume a lot of time. Therefore, you need to create a rotational pattern of the hole feature. You will use the angular dimension that was defined to create datum on the fly to pattern the feature.

1. Select the swept blend feature from the **Model Tree** or from the model. Hold down the right mouse button to invoke the shortcut menu.

2. Choose the **Pattern** option from the shortcut menu. The Pattern dashboard is displayed and the dimensions are displayed on the hole feature.

3. Select the **General** option from the **Options** option slide-up panel.

4. Double-click the dimension value **90** that is actually the angular dimension of the datum plane created on the fly. Enter a value of **120** in the edit box and press ENTER.

   You have entered a value of 120 because there are three instances of the swept blend feature on a circular feature. Y

Note

You may need to zoom in the display of the model to select the angular dimension value ***90***.

5. Enter **3** in the **1** edit box present on the dashboard and press ENTER.

6. Choose the **Build feature** button from the dashboard. The rotational pattern is created as shown in Figure 8-79.

Creating the Round Feature

The third feature is a round of radius 30.

1. Choose the **Round Tool** button from the **Engineering Features** toolbar. The dashboard is displayed.

2. Select the edges shown in Figure 8-80 to round. You may need to spin the model to select the inner edges of the swept blend feature. Remember that the first edge is selected by using the left mouse button and the other remaining edges are selected by using CTRL+left mouse button.

   The preview of the round is displayed on the graphics window along with the default radius value.

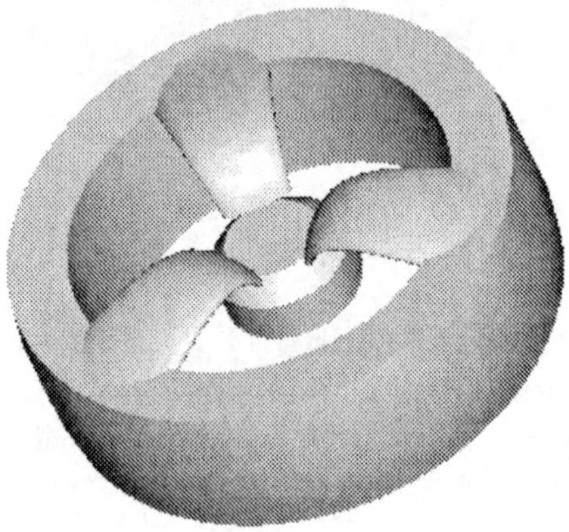

Figure 8-79 *Model after creating pattern*

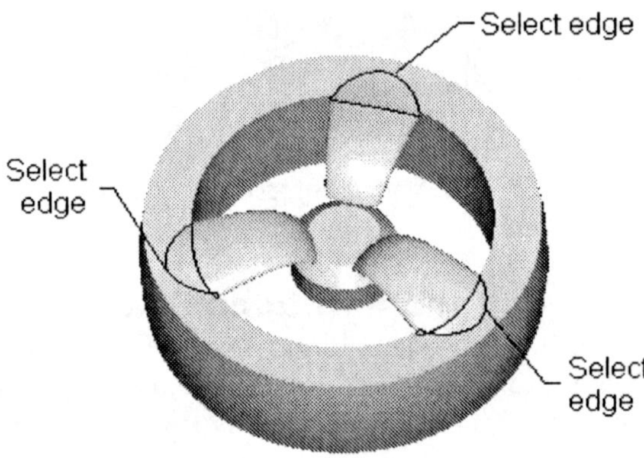

Figure 8-80 *Edges to round*

3. Double-click the radius value, enter **30** in the edit box that appears, and then press ENTER.

4. Choose **Build feature** from the Round dashboard.

   The round feature is completed. The default trimetric view of the round feature is shown in Figure 8-81. Add another set and using the same options create a round feature of radius 10 highlighted in Figure 8-82.

Creating the Revolve Cut Feature

You need to first select a sketching plane to draw the sketch of the cut feature and then specify the angle of revolution.

1. Choose the **Revolve Tool** button from the **Base Features** toolbar. The Revolve dashboard is displayed.

2. Select the **FRONT** datum plane as the sketch plane.

3. Select the **TOP** datum plane from the graphics window and then select the **Bottom** option from the **Orientation** drop-down list.

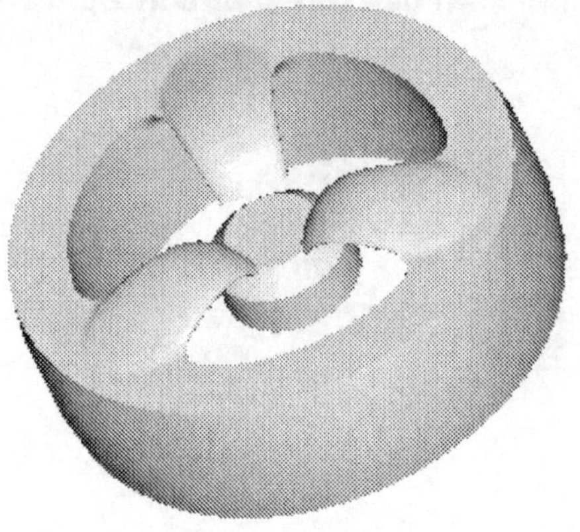

Figure 8-81 Model after creating a round of radius 30

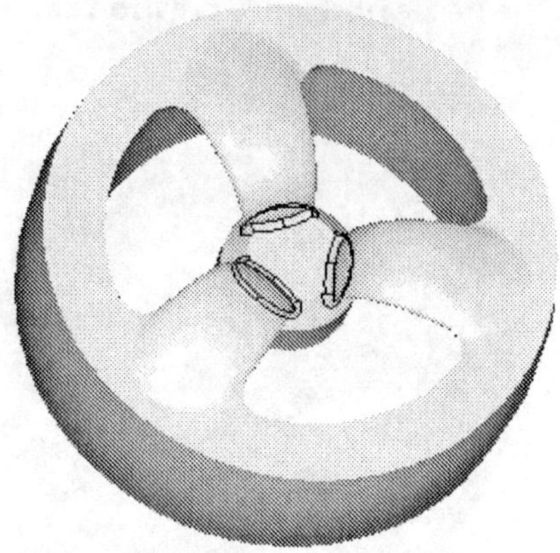

Figure 8-82 Model highlighting the round feature of radius 10

4. Once you enter the sketcher environment, draw a center line aligned to the **RIGHT** datum plane. Draw the sketch of the revolved cut feature and add dimensions and constraints as shown in Figure 8-83.

5. After the sketch is complete, choose the **Continue with the current section** button to exit the sketcher environment. The Revolve dashboard is displayed.

6. Choose the **Build feature** button from the Revolve dashboard to exit it. The revolve cut feature is shown in Figure 8-84.

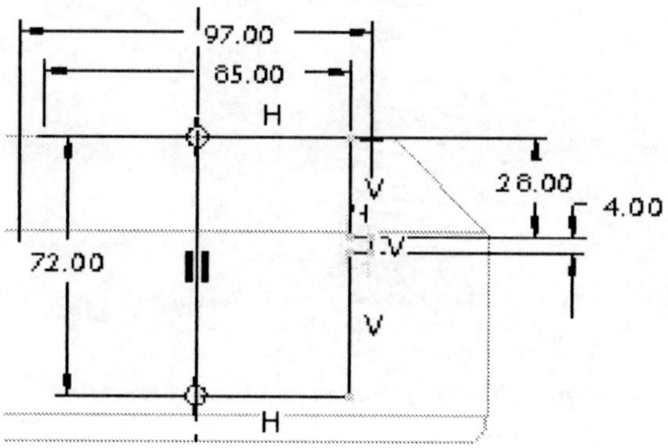

Figure 8-83 *Sketch for the revolve cut feature*

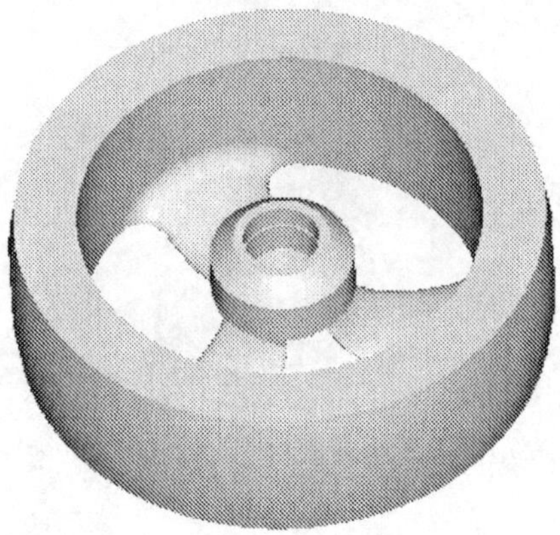

Figure 8-84 *Model after creating the revolve cut feature*

The model is completed and the isometric view of the model is shown in Figure 8-85.

Saving the Model

You have to save the model because you may need it later.

1. Choose the **Save the active object** button from the **File** toolbar and save the model.

The order of feature creation can be seen from the **Model Tree** shown in Figure 8-86.

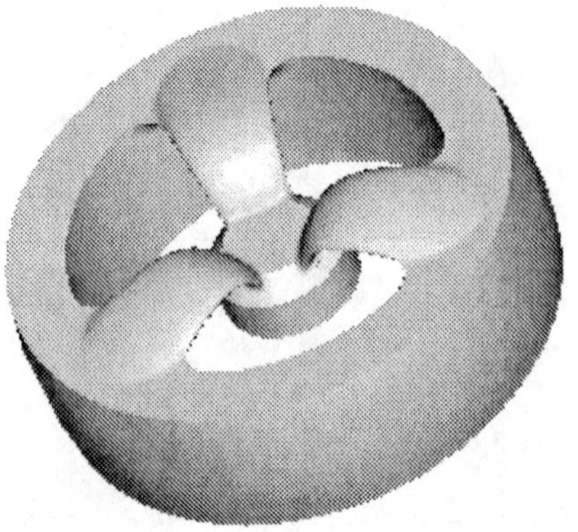

Figure 8-85 *Isometric view of the model*

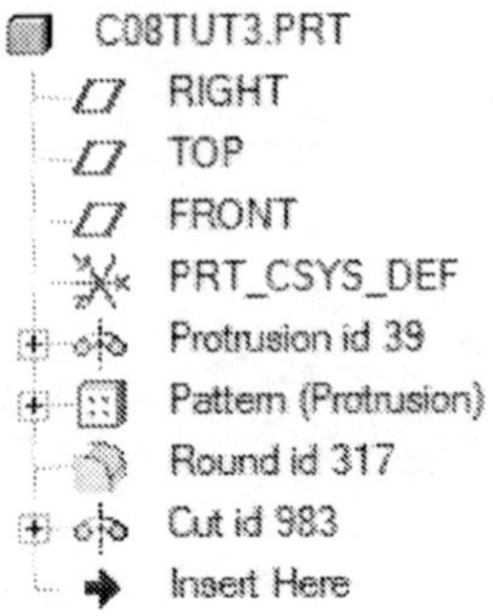

Figure 8-86 ***Model Tree*** *for Tutorial 3*

Tutorial 4

In this tutorial you will create the spring shown in Figure 8-87. Figure 8-88 shows the front view with dimensions of the spring. **(Expected time: 20 min)**

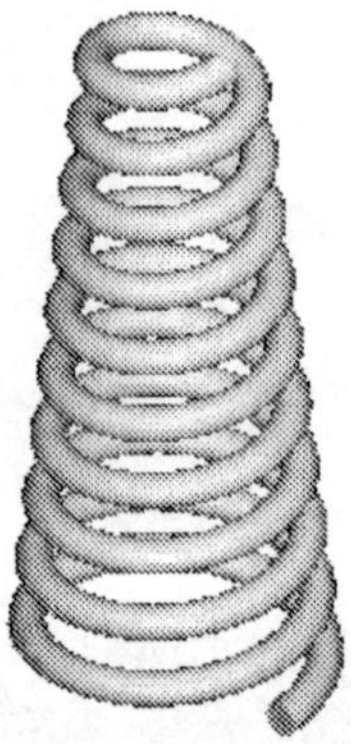

Figure 8-87 *Isometric view of the spring*

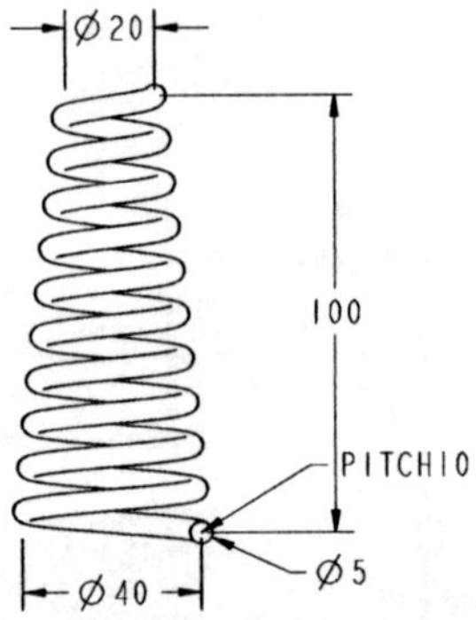

Figure 8-88 *Front view of the spring with dimensions*

The following steps outline the procedure for creating the constant pitch spring:

a. First, examine the spring and then determine the specifications of the spring. The spring is right-handed, has a constant pitch, and is created by revolving through axis.

b. Select the sketching plane, draw the trajectory using the sketcher tools, and apply dimensions, see Figure 8-89.

c. Specify the pitch of the spring.

d. Draw the section of the spring using the sketcher tools and dimension it, see Figure 8-90.

After understanding the procedure for creating the model, you are now ready to draw it. Make sure that the required working directory is selected.

Creating New Object File

1. Open a new part file and name it *c08tut4*. The three default datum planes and the **Model Tree** are displayed on the graphics screen. Close the **Model Tree**. The **Model Tree** will not appear if it has been previously closed.

Creating the Helical Sweep

The spring that you have to create is a right-handed spring of constant pitch.

1. Choose **Insert > Helical Sweep > Protrusion** from the menu bar. The **ATTRIBUTES** menu is displayed. The **Constant > Thru Axis > Right Handed** options are selected by default.

2. Choose **Done** from the **ATTRIBUTES** menu. You are prompted to select a sketching plane.

3. Select the **FRONT** datum plane from the graphics screen. The **DIRECTION** submenu is displayed. Choose **Okay**. The **SKET VIEW** submenu is displayed.

4. Select the **Top** option from the **SKET VIEW** submenu and choose the **TOP** datum plane.

5. Once you enter the sketcher environment, draw the sketch of the trajectory and dimension it as shown in Figure 8-89. As evident from Figure 8-89, the endpoint of the trajectory is aligned to the **TOP** datum plane.

   You need to draw a center line in the sketch about which the spring will be rotated. This is the axis of the spring and is the first trajectory that is drawn in the sketch.

6. After you complete the sketch of the trajectory, choose the **Continue with the current section** button from the **Right Toolchest**. The **Message Input Window** is displayed and you are prompted to specify the pitch of the spring.

7. Enter **10** in the **Message Input Window** and press ENTER.

   Now, you enter the sketcher environment to draw the section of the spring. You will notice that a yellow cross of infinite length appears on the screen. This cross consists of two perpendicular lines of infinite length. The intersection point of these lines is the start point of the trajectory.

8. Draw the section of the spring such that the center of the circle coincides with the intersection of the two perpendicular lines and dimension it as shown in Figure 8-89.

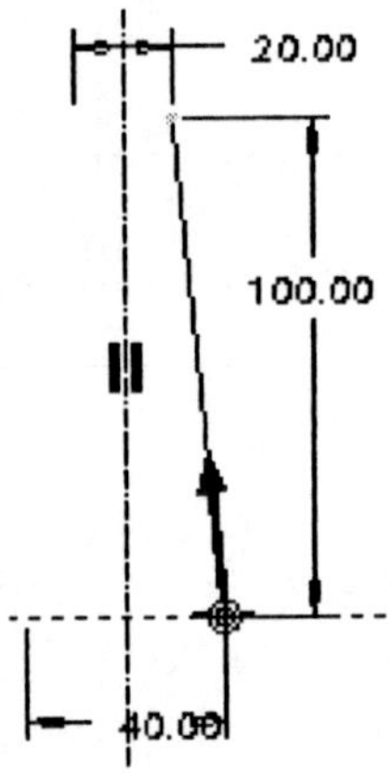

Figure 8-89 Sketch of the trajectory with dimensions

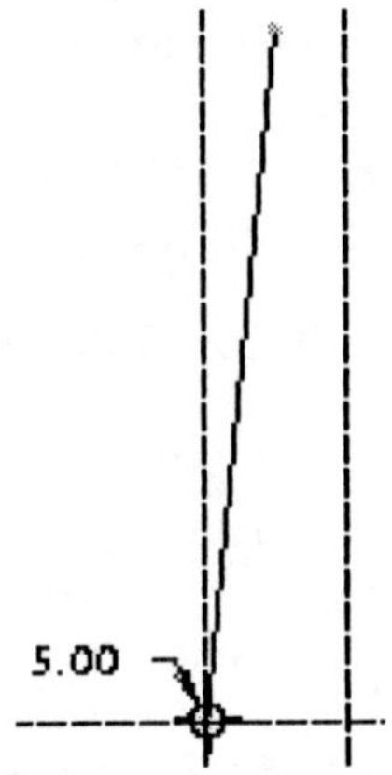

Figure 8-90 Sketch of the section with dimensions

9. After completing the sketch, choose the **Continue with the current section** button to exit the sketcher environment.

10. Choose the **OK** button from the **PROTRUSION** dialog box.

    The spring is created and is shown in Figure 8-91.

Saving the Model

1. Choose the **Save the active object** button from the **File** toolbar and save the model.

   The order of feature creation can be seen from the **Model Tree** shown in Figure 8-92.

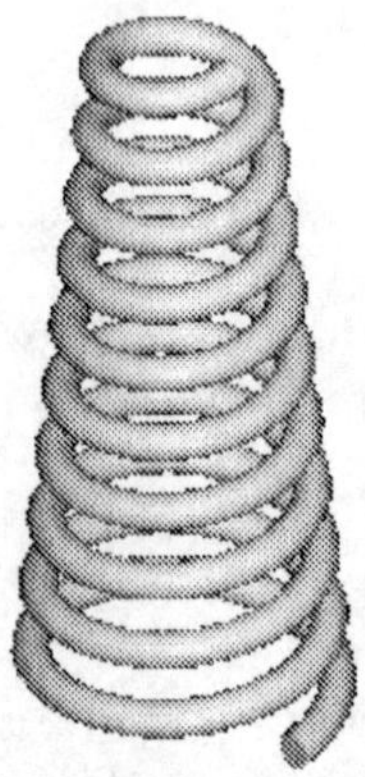

Figure 8-91 *Isometric view of the spring*

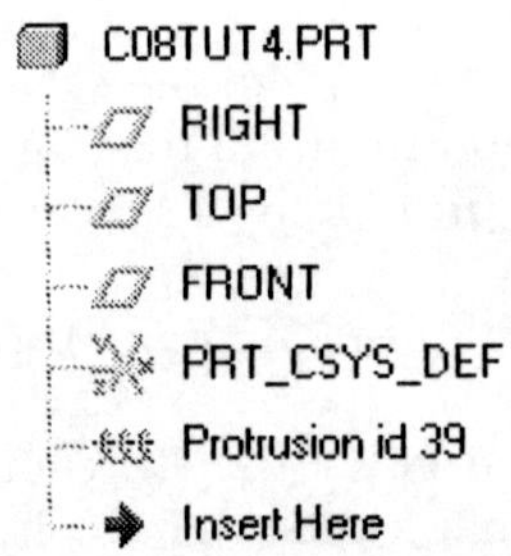

8-92 Model Tree *for Tutorial 4*

Self Evaluation Test

Answer the following questions and then compare your answers with the answers given at the end of this chapter.

1. You can create Constant pitch as well as Variable pitch helical sweep features. (T/F)
2. The **Helical Sweep** option is used to create cut as well as protrusion features. (T/F)
3. You can create a helical feature having 40 units pitch and 50 units as the diameter of the circular cross-section. (T/F)
4. You can add as well as delete points while specifying the pitch values during the creation of variable pitch helical sweep. (T/F)
5. The **Var Sec Sweep** option is available for both protrusion and cut options. (T/F)
6. While using the **Var Sec Sweep** option, the contour of the sweep feature depends on the shape of the __________.
7. While using the **Var Sec Sweep** option, when you draw the section for the sweep feature, the section should be __________ to the endpoints of the trajectories.
8. When you choose the **Done** option from the **BLEND OPTS** menu, the __________ menu is displayed.
9. When you use the **Normal to trajectory** option to create the variable sweep, the section is __________ to the origin trajectory and is swept along the X trajectory.
10. The **Swept Blend** option is used to create a model that is a combination of __________ and __________.

Review Questions

Answer the following questions:

1. Which of the following menus is displayed when you choose the **Helical Sweep** option from the **Insert** menu in the menu bar?

   (a) **ADV FEAT OPT** (b) **ATTRIBUTES**
   (c) **SWEEP OPTS** (d) None

2. Which of the following options is used to create a helical feature with constant pitch throughout the sweep?

   (a) **Right Handed** (b) **Constant**
   (c) **Thru Axis** (d) None

3. Which of the following options in the **BLEND OPTS** is used to create a sweep in which the swept section is normal to the trajectory defined as the origin trajectory?

(a) **Pivot Dir** (b) **NrmToOriginTraj**
(c) **Norm To Traj** (d) None

4. Which of the following options in the **Relations** dialog box is used to add dimensions that appear on the screen?

(a) **Insert > From Screen** (b) **Insert > Function**
(c) **Insert > From List** (d) None

5. Which of the following options is used to create a swept blend in which you need to select a plane, edge, curve, or axis to which the section to be swept is perpendicular?

(a) **Pivot Dir** (b) **Nrm To Traj**
(c) **NrmToOriginTraj** (d) None

6. The **Var Sec Swp** option can be used to create a sweep in which the section of the sweep is constant throughout the sweep. (T/F)

7. The trajectory called X trajectory is used to guide the section of the sweep feature along it. (T/F)

8. When you choose the **Edit Rel** option, the relations are not checked for errors. (T/F)

9. The **Switch Dim** option of the **RELATIONS** submenu is used to toggle the dimension value mode and the symbols. (T/F)

10. The **Helical Sweep** option is also used to create a cut. (T/F)

Exercises

Exercise 1

Create the model shown in Figure 8-93. The front view is shown in Figure 8-94.

(Expected time: 20 min)

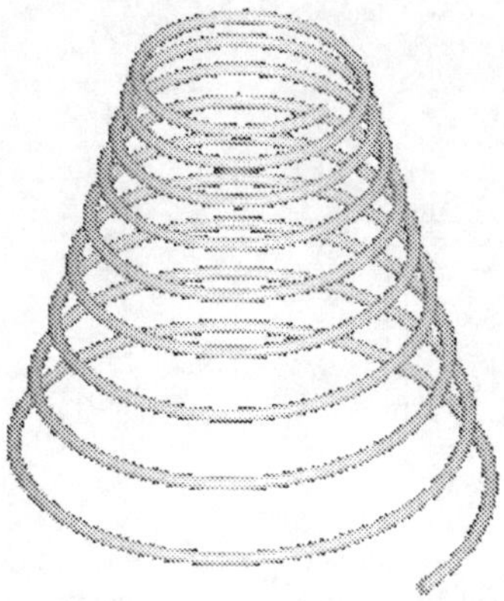

Figure 8-93 Isometric view of the spring

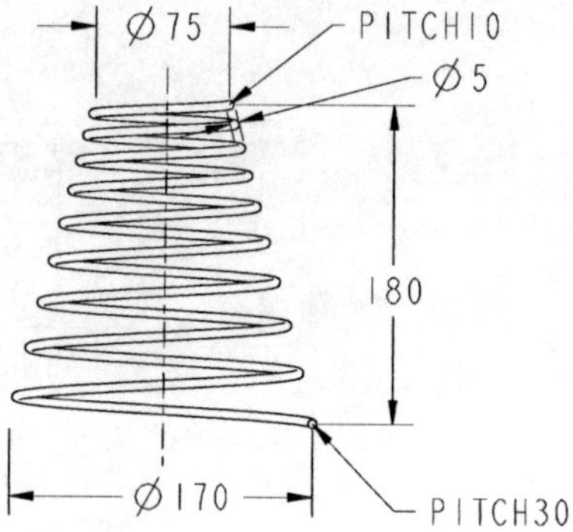

Figure 8-94 Front view of the spring

Exercise 2

Create the model shown in Figure 8-95. The front view and the top view of the model is shown in Figure 8-96. **(Expected time: 20 min)**

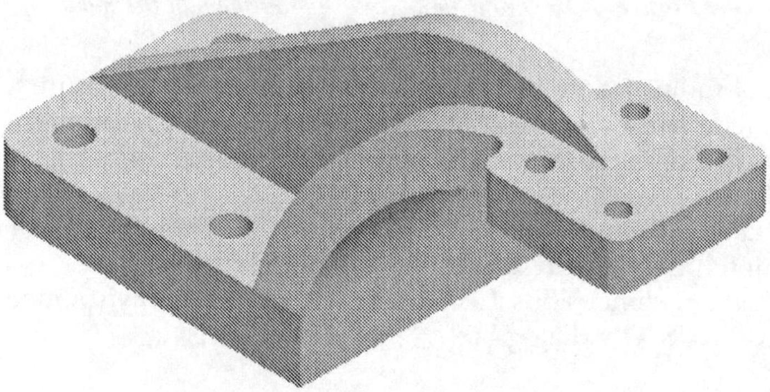

Figure 8-95 Solid model

Hints to create the swept blend feature:

1. Choose **Insert > Swept Blend > Protrusion** from the menu bar. The **BLEND OPTS** menu is displayed.
2. Accept the options that are selected by default and choose **Done**. The **SWEEP TRAJ** menu is displayed.
3. Choose the **Sketch Traj** option. You are prompted to select a sketch plane.
4. Select the **FRONT** datum plane as the sketch plane and select the top face of the base feature to be at the top while sketching.

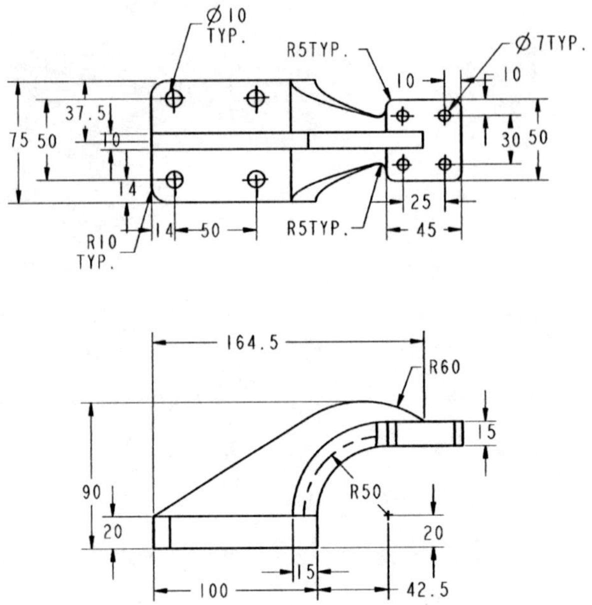

Figure 8-96 *Front view and the top view of the model*

5. Draw the arc of radius 50 and locate its center from the base feature.
6. Exit the sketcher environment by choosing the **Continue with the current section** button. You are prompted to enter the z-axis rotation angle for section 1.
7. Press ENTER.
8. Draw the rectangle of dimensions 15x50. Exit the sketcher environment.
9. You are prompted to enter the z-axis rotation angle for section 2.
10. Draw a rectangle of dimensions 15x75. Exit the sketcher environment and choose **OK** from the **PROTRUSION** dialog box.

Answers to the Self-Evaluation Test

1 - T, **2** - T, **3** - F, **4** - T, **5** - T, **6** - X trajectory, **7**- aligned, **8** - **SWEEP TRAJ** **9** - normal, **10** - sweep, blend.

Chapter 9

Assembly Modeling

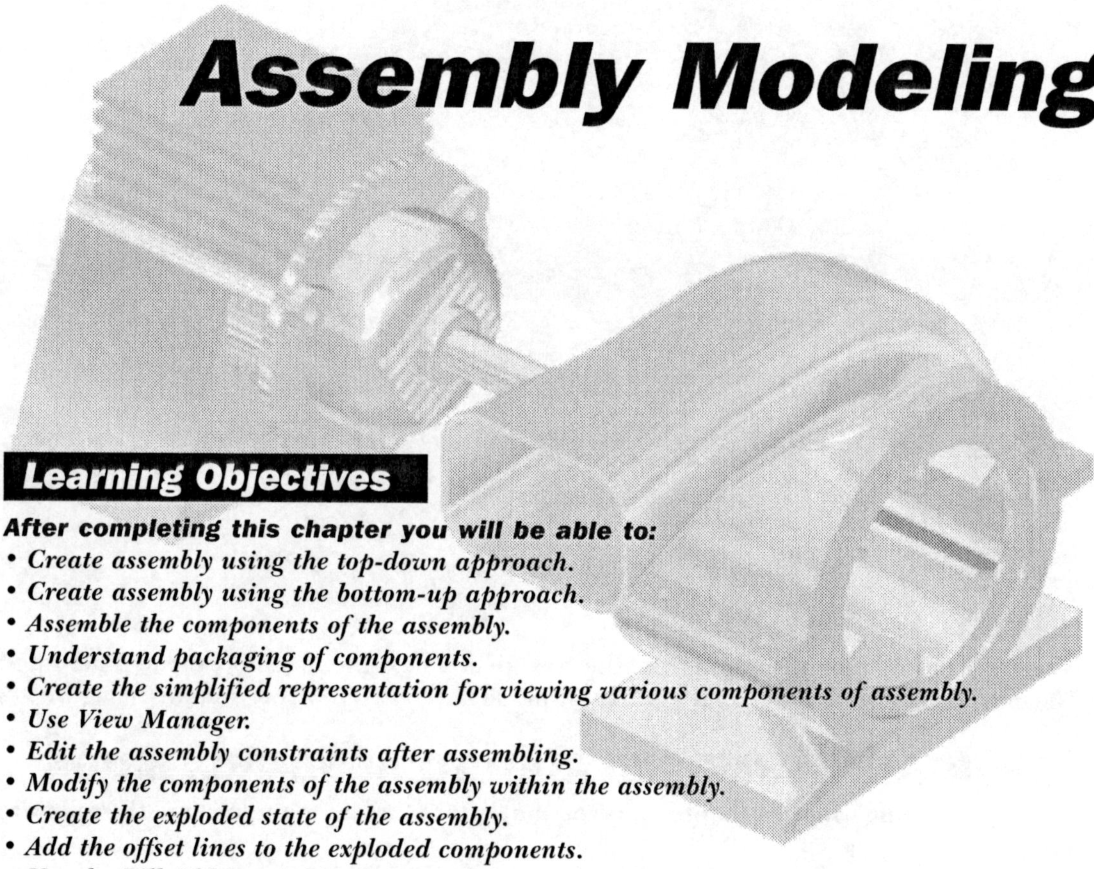

Learning Objectives

After completing this chapter you will be able to:

- ***Create assembly using the top-down approach.***
- ***Create assembly using the bottom-up approach.***
- ***Assemble the components of the assembly.***
- ***Understand packaging of components.***
- ***Create the simplified representation for viewing various components of assembly.***
- ***Use View Manager.***
- ***Edit the assembly constraints after assembling.***
- ***Modify the components of the assembly within the assembly.***
- ***Create the exploded state of the assembly.***
- ***Add the offset lines to the exploded components.***
- ***Use the Bill Of Material in the assemblies.***

ASSEMBLY MODELING

An assembly is defined as a design consisting of more than one component bonded together at their respective working positions. These assembly designs are created in the Assembly mode of Pro/ENGINEER. To proceed to the Assembly mode, choose the **Create a new object** button from the **Top Toolchest**. The **New** dialog box is displayed; select the **Assembly** radio button from the **Type** area and then select the **Design** radio button from the **Sub-type** area as shown in Figure 9-1. Specify the name of the assembly in the **Name** edit box and choose **OK**.

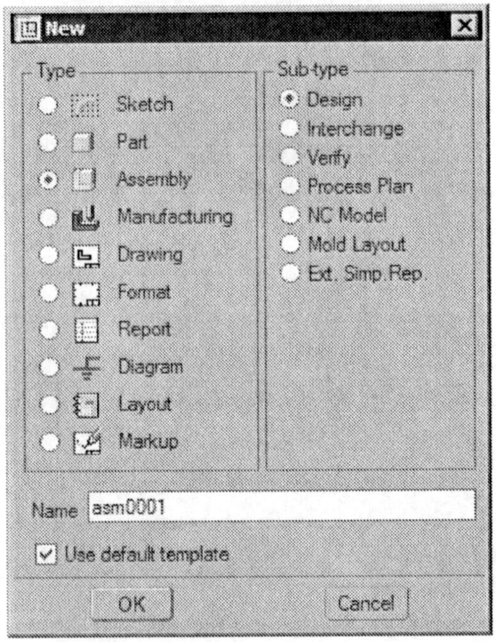

***Figure 9-1** Selecting the Assembly mode from the **New** dialog box*

When you choose the **OK** button from the **New** dialog box, the Assembly mode is activated. The initial screen appearance of the Assembly mode in Pro/ENGINEER is shown in Figure 9-2.

IMPORTANT TERMS RELATED TO ASSEMBLY MODE

Before proceeding with this chapter, it is very important for you to understand the following terms:

Top-Down Approach

This is the method of assembling the components in which the components of the assembly are created in the same assembly file. In this type of approach of assembly, the components are created in the assembly file and then assembled using the assembly constraints. The parts you create in the Assembly mode are saved as separate *.prt* files.

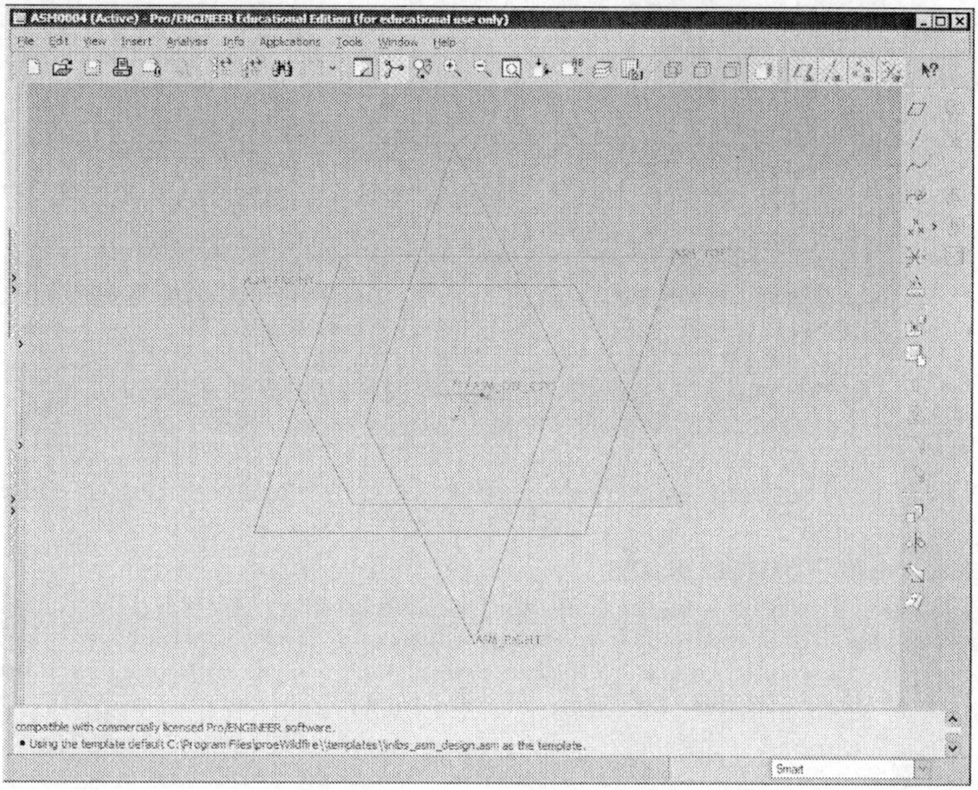

Figure 9-2 *Initial screen appearance of the Assembly mode*

Bottom-Up Approach

This is the method of assembling the components that are created as separate parts in the Part mode and are saved as *.prt* files. Once all the parts of an assembly have been created, you will create a new assembly file (*.asm*) and then assemble the parts using the assembly constraints available in the Assembly mode to create the assembly. Since the assembly file has information related only to the assembling of components, this file is not very heavy and therefore requires less hard disk space.

However, if any of the assembly component is moved from its original location on the hard disk, the assembly will not open. This implies that Pro/ENGINEER searches for the assembly component in the folder in which it was originally created. If it does not find the name of the component in that folder, it gives an error in opening the assembly file and the **RESOLVE FEAT** menu is displayed as shown in Figure 9-3. Choose the **Quick Fix** option from this menu; the **QUICK FIX** submenu is displayed as shown in Figure 9-4. You can use the options in this submenu to locate the part file.

Also, there should be no files of the same name in the current session of Pro/ENGINEER. For example, the name nut is common and is used in most of the assemblies.

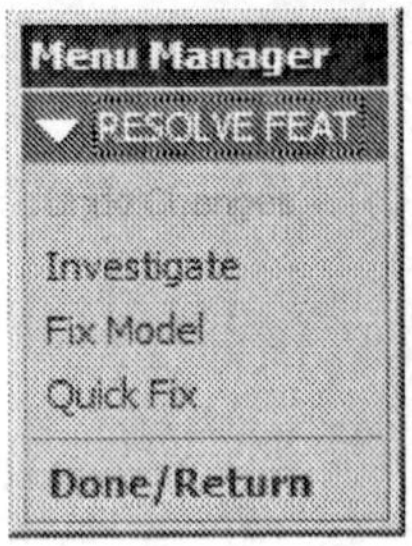

*Figure 9-3 **RESOLVE FEAT** menu*

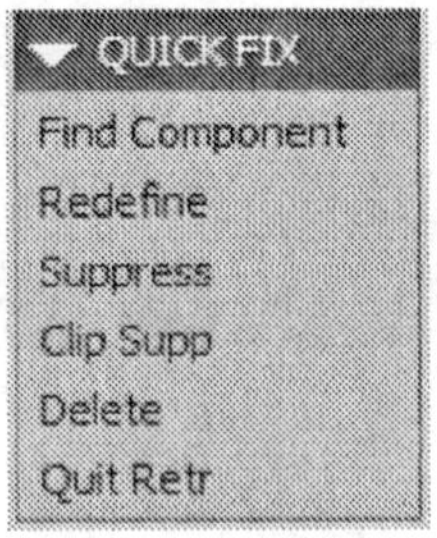

*Figure 9-4 **QUICK FIX** menu*

Tip: *It is advised that you create separate folders for all the assemblies that you create in Pro/ENGINEER.*

Placement Constraints

The placement constraints are the constraints that are used to rigidly bind the components of the assembly to their respective positions in the assembly. These constraints are also called the assembly constraints. Generally, these constraints are used in combinations in order to constrain all the six degrees of freedom of a component.

Package

It is the state in which the component that is being assembled is not fully constrained and thus is not rigidly placed at its actual location.

CREATING TOP-DOWN ASSEMBLIES

As mentioned earlier, top-down assemblies are assemblies in which all the components are created in the Assembly mode and in the same assembly file. You know that to create components you need the sketcher environment and to invoke the sketcher environment you need the Part mode. You can invoke the Part mode from within the Assembly mode and create components using the tools available in the Part mode. To create the component from within the Assembly mode, you need to choose the **Create a component in assembly mode** button. The use of this tool is discussed next.

Creating Components in the Assembly Mode

In Pro/ENGINEER, you can create components in the Assembly mode also. The components that you create in the Assembly mode are saved as separate *.prt* files when you choose to save the assembly file in which they are created.

To create a component in the Assembly mode, choose the **Create a component in assembly mode** button. The **Component Create** dialog box is displayed as shown in Figure 9-5. You can select the **Part** radio button from the **Type** area and the **Solid** radio button from the **Sub-type** area in the dialog box to create a *.prt* file. Enter the name of the component in the

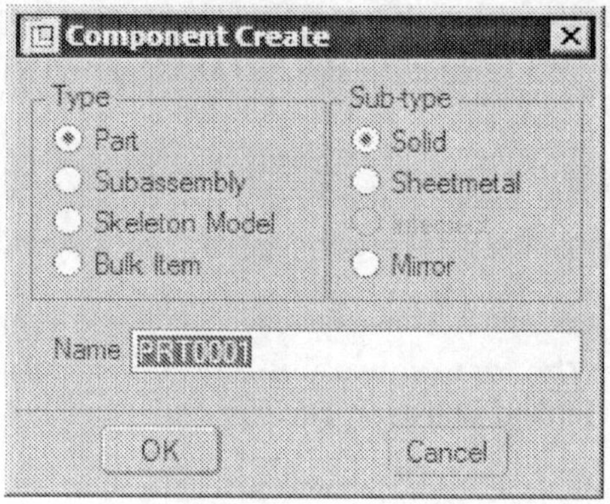

Figure 9-5 ***Component Create*** *dialog box*

Name edit box and then choose the **OK** button. The **Creation Options** dialog box is displayed. Use this dialog box to select the method of creation of the component.

CREATING BOTTOM-UP ASSEMBLIES

As discussed earlier, the bottom-up assemblies are created by inserting the various part files one at a time and then assembling them to create an assembly in the Assembly mode. The components are assembled using the assembly constraints. When the first component is inserted in the Assembly mode, its three default datum planes are placed in the same orientation as that of the default datum planes of the Assembly mode. To insert the component, you need to choose the **Add component to the assembly** button. This tool button is discussed next.

Adding Components in Assembly

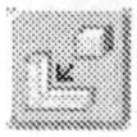

When you choose the **Add component to the assembly** button, the **Open** dialog box is displayed. Select the part that you need to add to the assembly. This dialog box is the same as the dialog box that is used to open an existing file. You can also preview the component before adding it to the assembly. After the component is displayed on the graphics window in the Assembly mode, you need to specify constraints in order to assemble them. Even if you are placing the first component, you need to constraint it using the assembly constrains. Remember that no component is placed automatically and you need to specify manually the position of the component.

PLACEMENT CONSTRAINTS

The assembly constraints are also called placement constraints and are available in the **Component Placement** dialog box which is discussed later in this chapter. The placement constraints that are available in Pro/ENGINEER are discussed next.

Automatic

When you choose this constraint, Pro/ENGINEER assumes the constraint and applies it according to the type of entity selected. For example, if you select axes of two components to assemble then Pro/ENGINEER will understand that you want to apply the **Align** constraint and the **Align** constraint will be applied to the two components.

Mate

The **Mate** placement constraint allows you to make two selected planes, datum planes, faces, or a combination of a datum plane and a face coplanar with respect to each other. The faces or datums selected may or may not be in contact with each other.

There are three options that the **Mate** constraint is used in combination with. It is used with **Offset**, **Coincident**, and **Oriented** constraints.

Mate Offset

If you need to keep the two faces or planes at some distance apart then you can use the **Mate** constraint with some offset value. The offset distance between the two coplanar faces can have a positive or a negative value. Figure 9-6 shows the faces that are selected to apply the **Mate Offset** constraint. Figure 9-7 shows the component assembled at an offset distance from the selected face.

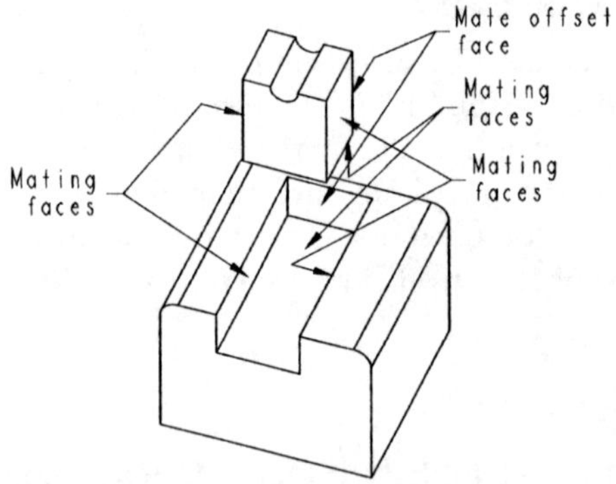

Figure 9-6 *Components before assembling*

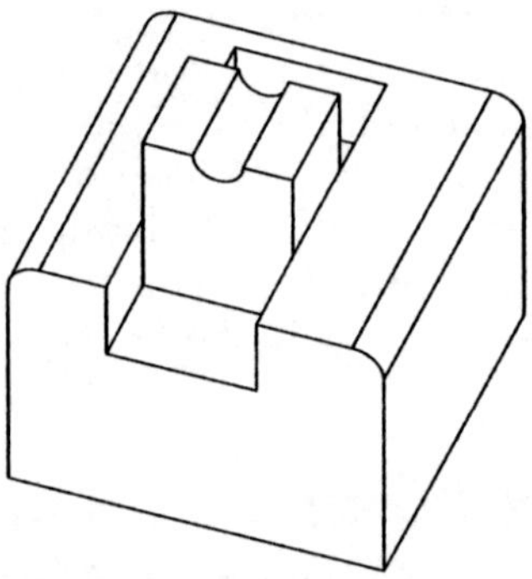

Figure 9-7 *Components after assembling*

Mate Coincident

The **Mate Coincident** combination of constraints allows you to make two selected planes or faces coplanar to each other. Figure 9-8 shows the faces that are selected to mate. Figure 9-9 shows the components after assembling.

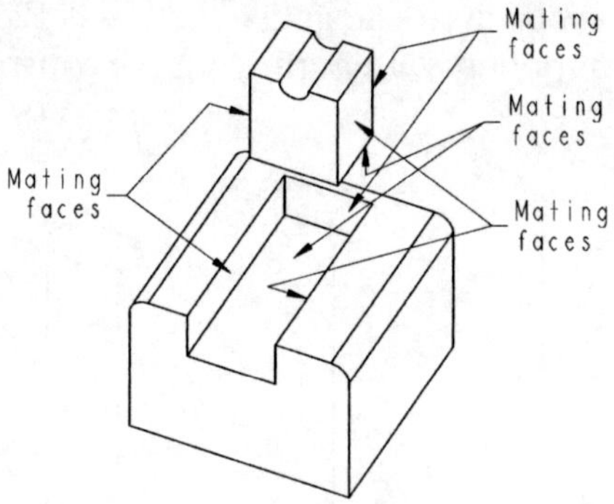

Figure 9-8 Components before assembling

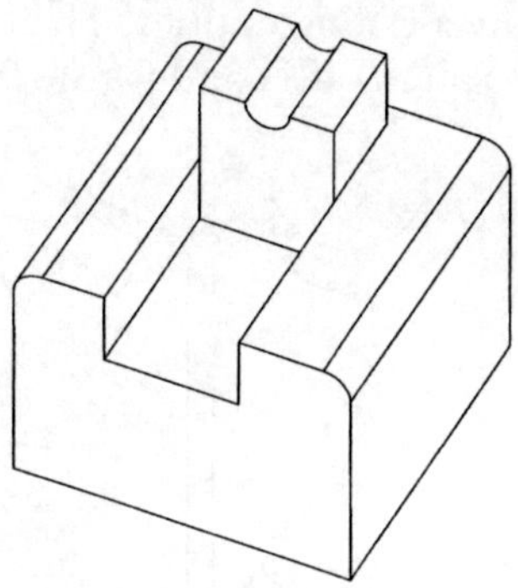

Figure 9-9 Components after assembling

Mate Oriented

The **Mate Oriented** combination of constraints is used when you want to mate two faces or planes or a combination of two and at the same time you want to orient some other faces or planes of the two components to face in the same direction. Figure 9-10 shows the faces that are selected to mate and Figure 9-11 shows the components after assembling.

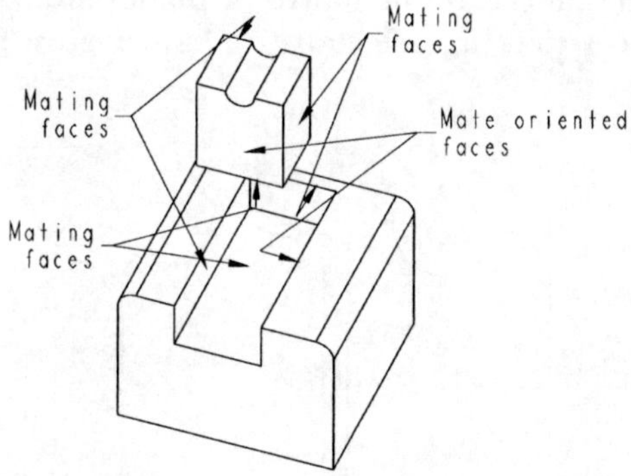

Figure 9-10 Components before assembling

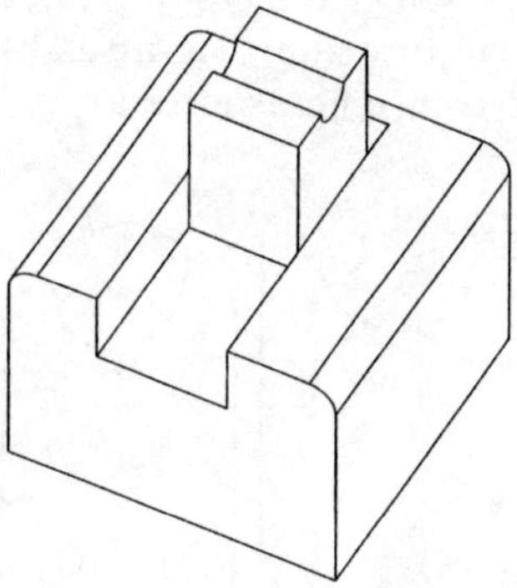

Figure 9-11 Components after assembling

Align

The **Align** constraint is used to assemble two components by making the selected faces or planes coplanar such that the aligned faces or planes are facing in the same direction. Apart from planes and faces, you can also select datum axes, datum points, edges, or vertices for applying the **Align** constraint.

There are three options that the **Align** constraint is used in combination with. It is used with **Offset**, **Coincident**, and **Oriented** constraints.

Align Offset

The **Align Offset** combination of constraints is similar to the **Align Coincident** combination of constraints, with the only difference being that you can also specify the offset distance between the aligning entities. The offset value can be positive or negative. Figure 9-12 shows the faces to align offset and Figure 9-13 shows the components after applying the constraints.

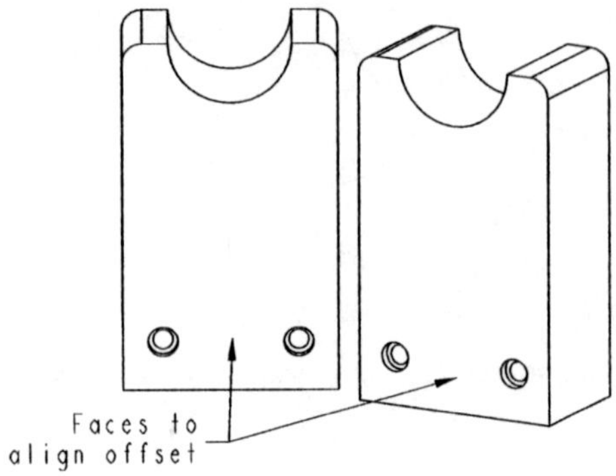

Figure 9-12 *Components before applying constraints*

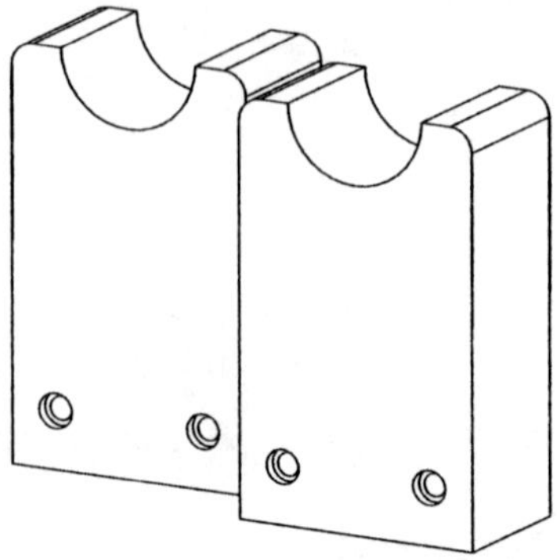

Figure 9-13 *Components after applying constraints*

Align Coincident

The **Align Coincident** combination of constraints is used to assemble two components by making the selected faces or planes coplanar such that the aligned faces or planes are facing in the same direction. Figure 9-14 shows the faces that are selected to align. Figure 9-15 shows the components after assembling.

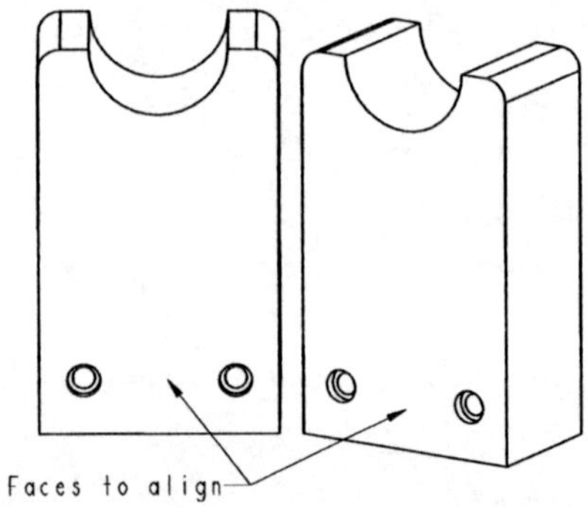

Figure 9-14 *Components before applying constraints*

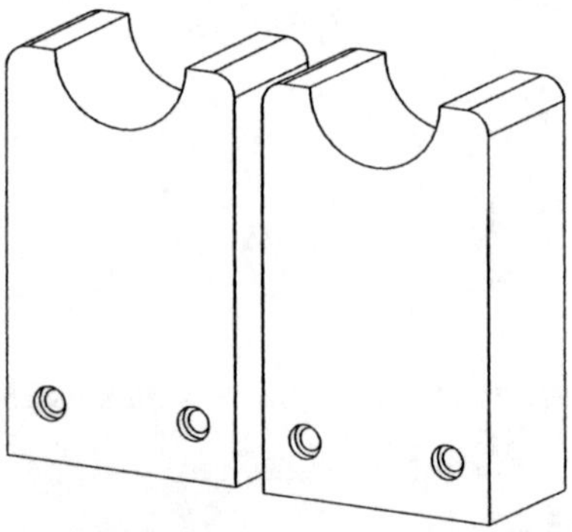

Figure 9-15 *Components after applying constraints*

Align Oriented

The **Align Oriented** combination of constraints makes the two selected faces or planes or the combination of both to face in the same direction. Figure 9-16 shows the constraints applied on the two components and Figure 9-17 shows the two components after applying the constraints.

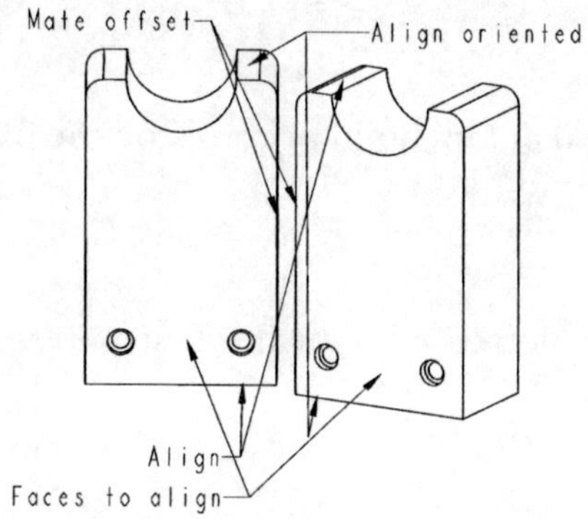

Figure 9-16 *Components before applying constraints*

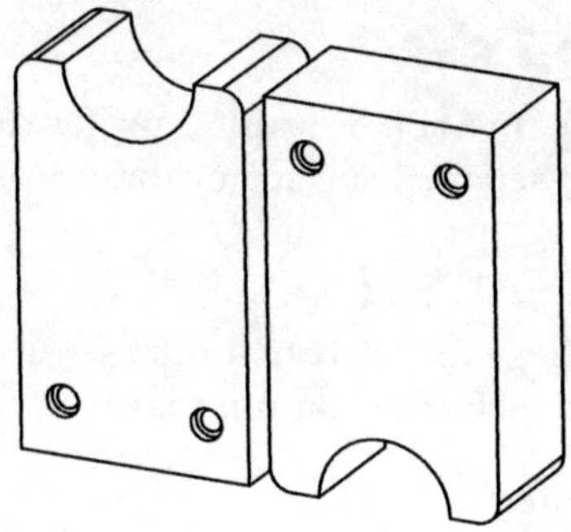

Figure 9-17 *Components after applying constraints*

Insert

The **Insert** constraint is used to assemble the revolved components. Applying this constraint allows the revolved component, holes, or a combination of both to share the same orientation of the central axis. Figure 9-18 shows the faces to insert and Figure 9-19 shows the parts after assembling.

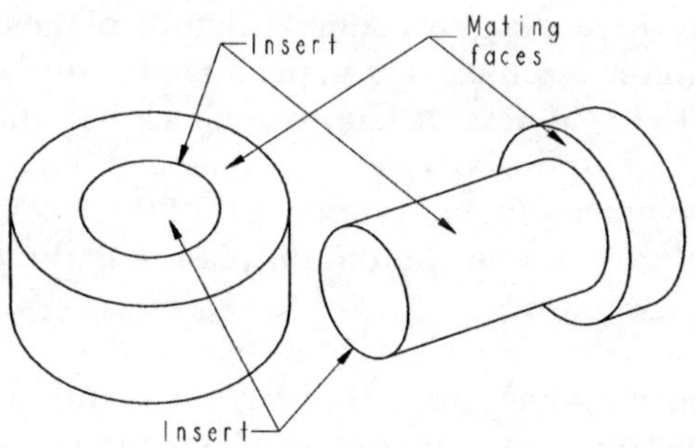

Figure 9-18 *Parts before assembling*

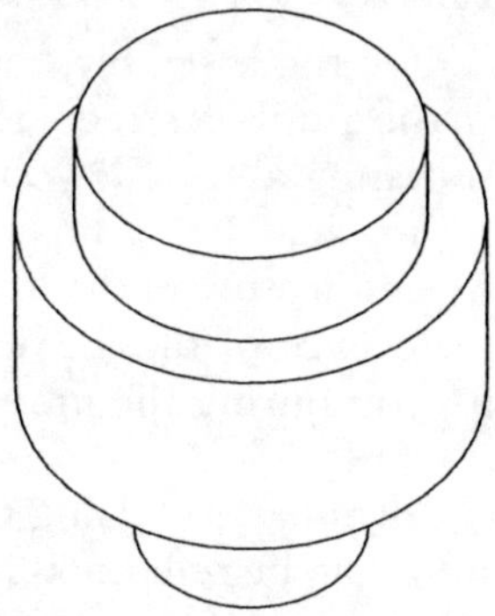

Figure 9-19 *Parts after assembling*

Coord Sys

The **Coord Sys** constraint aligns the coordinate system of the first component with the coordinate system of the second component.

Tangent

The **Tangent** constraint is used to make the selected circular face tangent to the other selected face or plane and at the same time make them coplanar.

Pnt On Line

The **Pnt On Line** constraint is used to align the selected datum point or the vertex on the first part with the selected edge, datum axis, or datum curve on the second part.

Pnt On Srf

The **Pnt On Srf** constraint is used to align the selected datum point or vertex on the first part with the selected surface or datum plane on the second part.

Edge On Srf

The **Edge On Srf** constraint is used to align the selected edge of the first part with the selected surface or datum plane on the second part.

Default

The **Default** constraint is used to assemble the component in the assembly by aligning the default coordinate system of the component with the default coordinate system of the assembly. If the assembly has the assembly datum planes then the part datum planes are aligned with the assembly datum planes.

Fix

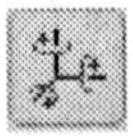

The **Fix component to current position** button packs the two components and displays that the two components are fully constrained.

ASSEMBLY DATUM PLANES

When you enter the assembly environment, the three default assembly datum planes are displayed on the graphics screen. If you do not want the assembly datum planes to be displayed then while opening a new file you need to clear the **Use default template** check box that is available in the **New** dialog box. Generally, it is advised that you use the assembly datum planes as the first feature of the assembly and then assemble the components of the assembly taking the reference of these assembly datum planes. Using the datum planes as the first feature helps you during the modification of the assembly.

Some of the advantages of using the assembly datum planes are as follows: the components of the assembly can be redefined in such a way that the component placed later can be made the first component, the first component can be replaced by some other component, and the placement constraint of the first component can be modified.

ASSEMBLING THE COMPONENTS

The components in the assembly can be placed parametrically or nonparametrically. If the components are placed using the placement constraints, it is called parametric assembly and if the components are packaged, it is called nonparametric assembly.

To assemble the components parametrically, choose the **Add component to the assembly** button. The **Open** dialog box is displayed. Once you select the part to assemble from the **Open** dialog box, the **Component Placement** dialog box is displayed as shown in Figure 9-20.

The **Component Placement** dialog box has multiple functions. This dialog box is used for

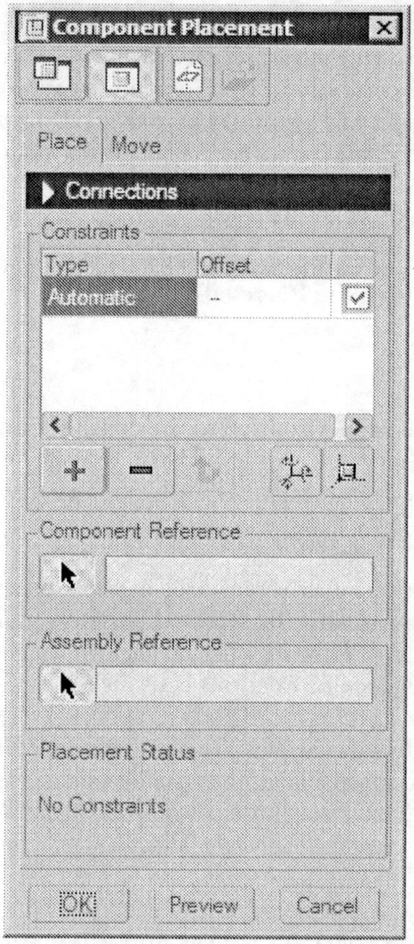

***Figure 9-20 Component Placement** dialog box*

both constrained and packaged placement of components. There are two tabs in this dialog box, the **Place** tab and the **Move** tab. The options under these tabs are discussed next.

Displaying the Components in a Separate Window

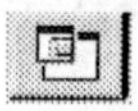

The **Show component in a separate window while specifying constraints** button is used to display the component to be assembled in a separate window. When you choose this button, a new window will open displaying the component to be assembled. It is easy to view the component to be assembled and the assembly in separate windows. This prevents cluttering of the components in a single window.

However, there is a limitation when the assembly and the component to be assembled in separate windows. If the components are in the same window they change their orientation as you apply the constraints, thus giving you an idea of the next constraint to be applied. In other words, you can easily find out which degree of freedom of the component is constrained and which needs to be constrained.

Displaying the Components in the Same Window

The **Show component in the assembly window while specifying constraints** button is used to display the component to be assembled in the assembly window itself. This button is selected by default when you invoke the **Component Placement** dialog box. As you apply constraint, the component in the assembly changes its position according to the constraints.

Place Tab

The options that are available in the **Place** tab of the **Component Placement** dialog box are discussed next.

Constraints Area

This area displays all the constraints applied to the component, along with the offset value (if any), in the order in which they were applied. In the **Component Placement** dialog box, this area is of great importance. It is recommended that you check this area after applying a constraint. Check that the constraint you intended to apply is listed in this area.

You will notice a check box on the extreme right in the **Constraints** area. After you apply a constraint, the check box is automatically selected and now you cannot move the component in the degree of freedom that is locked by the constraint applied. However, if the check box is cleared, the constrained degree of freedom is unlocked.

Moving the component after applying constraints and before fully constraining the components using the **Component Placement** dialog box as well as using the combination of mouse and keyboard is discussed later in the chapter. The options in this area are discussed next.

Type drop-down list

This drop-down list displays all the placement constraints that are available to assemble the component. You can select the constraints from this drop-down list to assemble components in an assembly. The different constraints that are available in this drop-down list were explained earlier.

Offset drop-down list

This drop-down list is used to specify the offset value (if any) for the placement constraint. This drop-down list has two more options, **Coincident** and **Orient**. The **Coincident** and **Orient** options are used in combination with **Mate** and **Align** options.

Specify a new constraint button

This button is chosen to add a constraint to assemble the components. When you choose this button, the **SELECT** dialog box is displayed. To exit this dialog box you need to choose the **Cancel** button.

Remove the selected constraint button

This button is chosen to delete the constraint that has already been applied. The constraint that is selected in the **Constraints** area is deleted when you choose this button.

Change orientation of constraint

This constraint button is used to reverse the orientation of the constraint applied. This button is used only after a constraint has been applied to the components. This button is not available initially when the **Component Placement** dialog box is invoked.

Component Reference Area

The button under this area is chosen to select the reference for specifying the placement constraint on the component to be assembled. When you choose this button, you can select the reference from the assembly or from the component.

Assembly Reference Area

The button under this area is chosen to select the reference for specifying the placement constraint on the assembly. This button is automatically chosen when you have finished specifying the reference on the component to be placed.

Placement Status Area

This area displays the placement status of the component in the assembly. The status displayed in this area is usually partially constrained or fully constrained. If you choose **OK** from the **Component Placement** dialog box when the status displayed is **Partially Placed**, then in the **Model Tree** the component assembled is displayed with a mark on it. This mark indicates that the component is packaged.

Move Tab

The options under the **Move** tab can be used only if you have selected the **Show component in the assembly window while specifying constraints** button. This tab is used to move a component along a degree of freedom that is not restricted. The **Move** tab is shown in Figure 9-21 and the options under this tab are discussed next.

Motion Type Area

The options under this area are discussed next.

Translate

This radio button is selected to move the component from its current location in the assembly. However, remember that the component can be moved only along the degrees of freedom that are not restricted.

Rotate

This radio button is selected to rotate the component in the assembly along its available degrees of freedom.

Adjust

This radio button is selected to pack a component with reference to the assembly. The message "**Select surface on packaged component to adjust to ref plane.**" is displayed in the **Message Area** when you select this radio button.

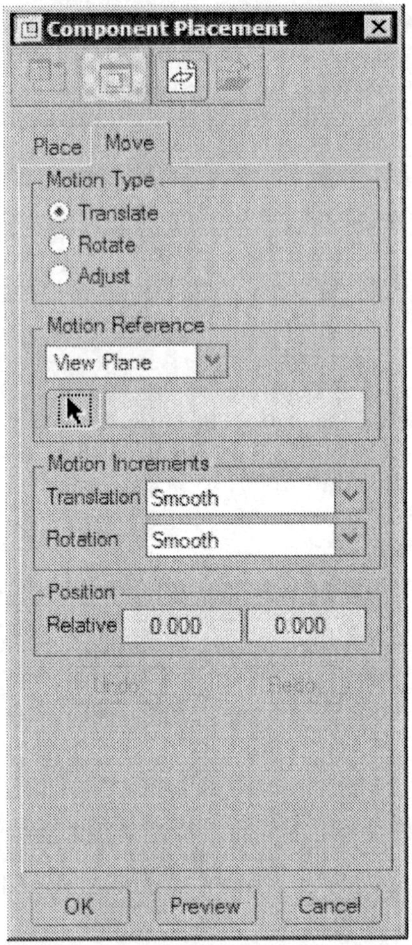

Figure 9-21 ***Move*** *tab of the* ***Component Placement*** *dialog box*

Tip: *It is recommended that you use CTRL+ALT+left mouse button to translate a component and CTRL+ALT+middle mouse button to rotate a component. It is always easy to use the above combinations to move the component rather than switching to a separate tab in the* ***Component Placement*** *dialog box and using its options to move a component. Direct manipulation of the component can speed up the process of locating the component correctly in the assembly and establishing constraints.*

Motion Reference Area

The drop-down list under this area displays the references that can be used for the motion of the component. The options provided under this drop-down list are discussed next:

View Plane

This option uses the view plane as the reference for the motion of the component.

Sel Plane
This option allows you to select a planar surface or a datum plane that you want to use as reference for the motion of the component.

Entity/Edge
This option allows you to select a straight edge, an axis, or a datum curve as the reference for the motion of the component.

Plane Normal
This option allows you to select a plane normal to which will be the allowed motion of the component.

2 Points
This option is used to select two datum points that will specify the direction for the movement of the component.

Csys
This option is used to select the coordinate system to specify the motion of the component. When you select this option, three buttons appear besides this drop-down list. These buttons are **X**, **Y**, and **Z**. These buttons are used to specify the axis of the selected coordinate system for specifying the direction of motion.

Motion Increments Area
Translation
This drop-down list is used to select the predefined increment values by which the component will be moved along the direction specified using the options in the **Motion Reference** drop-down list.

Rotation
This drop-down list is used to select the predefined increment angles by which the component will be rotated along the direction specified using the options in the **Motion Reference** drop-down list.

Position Area
Relative
These boxes display the coordinates of the current position of the component.

Undo
This button is chosen to undo the changes made using the options in the **Move** tab.

Redo
This button is chosen to reinstate the changes that were removed using the **Undo** button.

Preview
This button is used to preview the component in the assembly before it is actually assembled.

PACKAGING THE COMPONENTS

Sometimes, you do not know the exact location of a component in the assembly or you want to specify the exact location later. In such cases, you can nonparametrically place the component in the assembly. This method of nonparametrically placing the components in the assembly is called Packaging. There are two methods to package a component in an assembly. The first method is to use the **Insert** menu in the menu bar. The second method is to assemble the component using the **Component Placement** dialog box and then exit the dialog box before fully constraining it.

To package the component, choose **Insert** > **Component** > **Package** from the menu bar. When you choose this option, the **PACKAGE** menu will be displayed as shown in Figure 9-22. This submenu provides the following options:

*Figure 9-22 **PACKAGE** menu*

Add

The **Add** option is used to add the component to be packaged. You can open a new component using the **Open** dialog box, select it from the assembly, or select the last component from the assembly to assemble again.

When you select the component from the **Open** dialog box, the **Move** dialog box is displayed as shown in Figure 9-23. After you have specified the location of the component in the assembly file, choose **OK** to place the component and close the **Move** dialog box. The options under this dialog box are the same as those under the **Move** tab of the **Component Placement** dialog box. However, remember that you can move only the packaged component using this option.

Move

The **Move** option is used to move the packaged component to a new location. When you choose this option, the **Move** dialog box will be displayed as shown in Figure 9-23.

Fix Location

The **Fix Location** option is used to fix the location of an existing packaged components. When you choose this option, you are prompted to select the component to fix.

Finalize

As mentioned earlier, the packaged components are not assembled parametrically. Hence, packaging does not relate the component with its neighboring components in the assembly. Once the location of the component is decided, then it can be finalized. The **Finalize** option is used to finalize the location of the packaged component. When you choose this option you will be prompted to select the packaged component. The **Component Placement** dialog box will be displayed when you select the packaged component. You can parametrically assemble the component by adding the placement constraints to the component by using this dialog box.

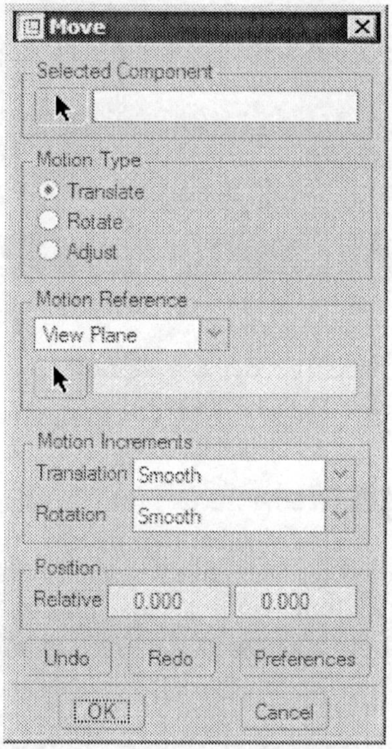

Figure 9-23 ***Move*** *dialog box*

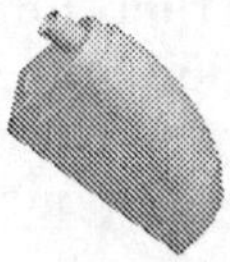

Tip: *The components that are packaged in an assembly are displayed in the* ***Model Tree*** *with a mark on them.*

CREATING SIMPLIFIED REPRESENTATIONS

As you know, the assembly designs consist of not only one part or component but a number of parts and subassemblies. The number of components in the assembly is directly proportional to the complicacy of the assembly. More the number of components, the more complicated will be the assembly. Therefore, huge assemblies like the inside of a CPU tend to become very complicated when all the chips, cables, and other parts are assembled. Also, if the outer body of the CPU is assembled then the inside components will not be displayed. So there have to be some means to temporarily remove certain components that are not desired at some point of time from the current display or force them to be displayed in the wireframe so that you can see through them. This process of removing some components or changing their type of display from the current display is called **Simplified Representation**.

Simplified representations can be created using the cascaded menu or by using the **View Manager**. The only difference is that using the **View Manager** you can give a name for the representation. To create a simplified representation using the cascaded menu, select a component from the **Model Tree** and then choose **View > Representation** from the menu bar. The cascaded menu is displayed as shown in Figure 9-24.

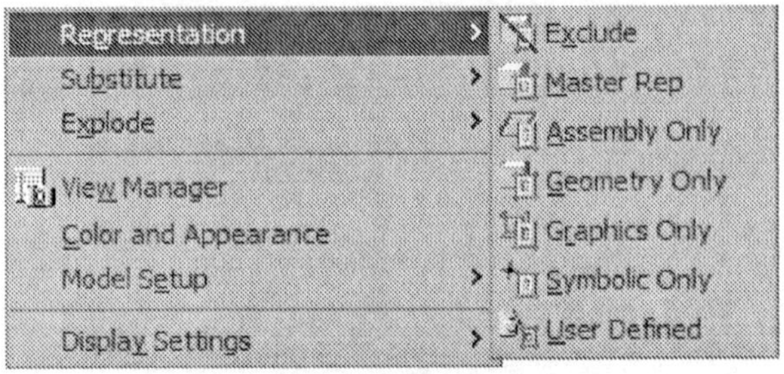

Figure 9-24 *Cascaded menu*

Using the Cascaded Menu

The options available in the cascaded menu are used to create a simplified representation without saving it. The simplified representation created reduces the regeneration time of the assembly, thus reducing the system requirements. The options in this menu are discussed next.

Exclude

When you choose this option, the selected component is removed from the display. You can even select more than one component from the **Model Tree** or from the assembly to remove them from the display. However, the component remains in the memory of Pro/ENGINEER and is even displayed in the **Model Tree**. This component can be, at any point of time, forced to be displayed again. To redisplay the components, choose the **Master Rep** option from the cascaded menu. Figure 9-25 shows a butterfly valve assembly in which the Body is removed from the current display. This kind of representation helps in assembling the components inside the body, like the plate and the screw in this case.

Master Rep

This option is used to redisplay the component in the assembly that was removed from the display using the **Exclude** option. When you choose the **Master Rep** option from the cascaded menu, the selected component is redisplayed as shown in Figure 9-26.

Assembly Only

The **Assembly Only** option is used to remove all the components from the assembly leaving the subassembly.

Tip: *When you create a simplified representation, some information regarding the type of representation is displayed on the graphics window. To remove this information, choose **View > Shade** from the menu bar.*

Geometry Only

In this type of simplified representation the complete geometry of the components is displayed

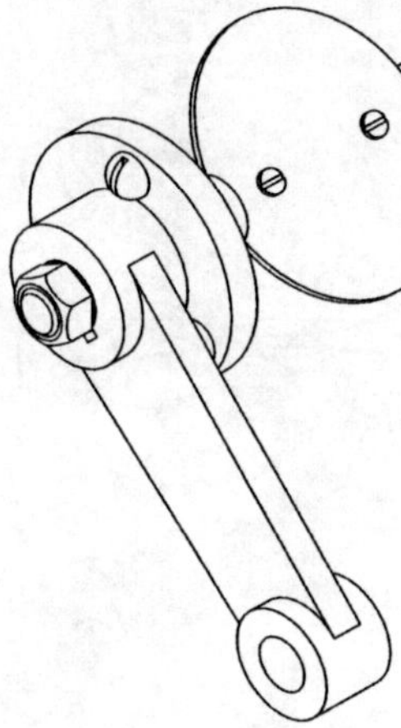

Figure 9-25 *Components excluded from the assembly*

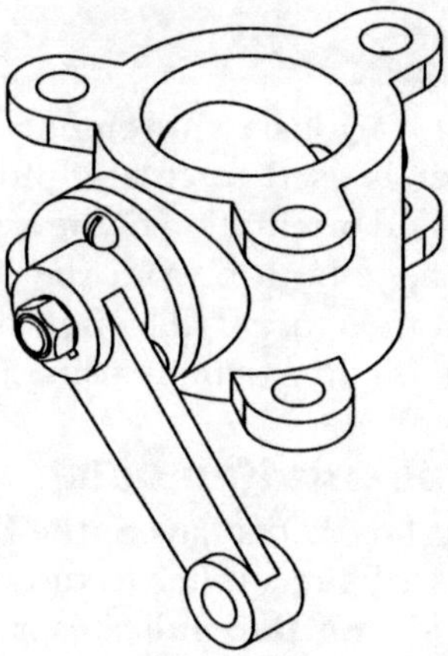

Figure 9-26 *Master representation of the assembly*

in the current model display style. This type of simplified representation takes a long time for regeneration. The components that are not selected in this type of representation cannot be modified. However, these components can be redefined.

Graphics Only

When you choose the **Graphics Only** option from the cascaded menu, all the selected components will be displayed in wireframe as shown in Figure 9-27. This type of representation is generally used for assembling the components inside some other component. This type of simplified representation also reduces the regeneration time in case of large assemblies. However, remember that the components that are displayed as the wireframe model in the graphic representation cannot be modified, redefined, or referenced.

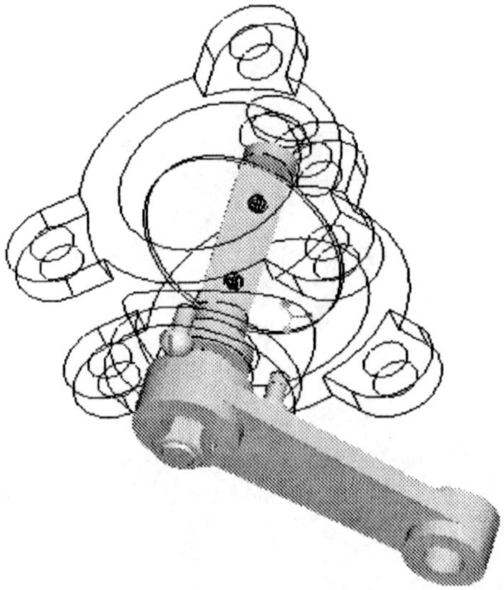

Figure 9-27 The Graphics Representation of an assembly

Using the View Manager

The **View Manager** is used for various requirements of display of an assembly and its components. The **View Manager** is used to create, modify, and switch between simplified representations, exploded views, and orientation of the assembly. To invoke the **View Manager**, choose the **Start the view manager** button from the **Top Toolchest**. The **View Manager** dialog box is displayed as shown in Figure 9-28. The various uses of this dialog box and the operations that you can perform on an assembly using this dialog box are discussed next.

Creating simplified representation using the View Manager

When the **View Manager** dialog box is displayed, the **Simp Rep** tab is selected by default. The options available under this tab are similar to those that were available in the cascaded menu. The red arrow in front of **Master Rep** indicates that this representation is set currently. To set any of the listed representation types, select the representation and right-click to invoke the shortcut menu. Choose the **Set** option from the shortcut menu to set the display of the assembly to that type of representation.

To create a user-defined representation, right-click in the dialog box and choose the **New** option. You can specify a name to the representation.

To redefine a representation type, choose the representation from the list and right-click to invoke the shortcut menu. Choose the **Redefine** option from the shortcut menu. The **EDIT** dialog box is displayed as shown in Figure 9-29. Using this dialog box you can select the components that you want to include, exclude, and substitute from the display.

Creating display style using the View Manager

When the **View Manager** dialog box is displayed, choose the **Style** tab. Two styles are available in the **Names** list of the dialog box. You can create a new style or redefine an existing style.

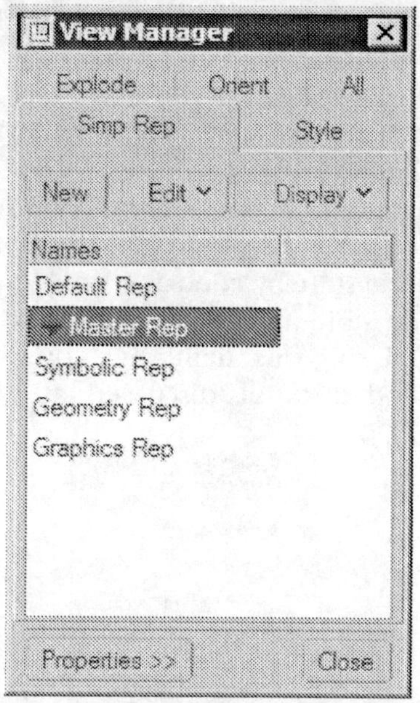

***Figure 9-28 View Manager** dialog box*

***Figure 9-29 EDIT** dialog box*

The options to do these operations are available in the shortcut menu that is displayed when you right-click in this dialog box.

Creating explode states using the View Manager

The exploded view of the assembly created using the **View Manager** can be saved and then later retrieved in the Drawing mode. Choose the **Explode** tab to view the list of explode states available. When you choose the **Explode** tab, the **Default** state is listed in the **Names** list. You can create a new explode state by choosing the **New** option from the shortcut menu. Enter the name of the explode state and then redefine it to display the **MOD EXPLODE** menu as shown in Figure 9-30. Using this dialog box you can create an exploded state of the assembly. The creation of exploded state is discussed later in the chapter.

*Figure 9-30 **MOD EXPLODE** menu*

Setting the orientation of the assembly using the View Manager

The various orientations that are given in the **Saved view** drop-down list (from **Top Toolchest**) are used to display the assembly in various views. Using the **View Manager** you can invoke the **Orientation** dialog box to set a user-defined orientation. To invoke the **Orientation** dialog box, choose the **Orient** tab and create a new orientation by selecting the **New** option from the shortcut menu. Name the view and then redefine it to display the **Orientation** dialog box. Use this dialog box to set the orientation of the assembly. You can save the orientation by giving it a name. This name is listed in the **Saved view list** drop-down list.

REDEFINING THE COMPONENTS OF THE ASSEMBLY

Sometimes, after you have assembled the components of the assembly, you may need to modify or redefine the location or the placement constraints of the components. This is done by selecting the component from the **Model Tree** and right-clicking to invoke the shortcut menu. Choose the **Edit Definition** option from the shortcut menu. When you select the component to be redefined, all the components assembled after the selected component will become invisible and the **Component Placement** dialog box will be displayed. You can modify the existing placement constraints or add new constraints using this dialog box.

You can also right-click on the component to be redefined on the graphics window, and choose the **Edit Definition** option from the shortcut menu that is displayed.

REORDERING THE COMPONENTS

The order of assembling the components can be changed by choosing **Edit > Component Operations** from the menu bar. The **COMPONENT** menu is displayed as shown in

Figure 9-31. Choose the **Reorder** option from this menu. When you choose this option, you will be prompted to select the component to be reordered. When you select a component to reorder either form the **Model Tree** or from the graphics window and choose **Done**, a message will be displayed in the **Message Area** that specifies the components before which you can insert the selected component.

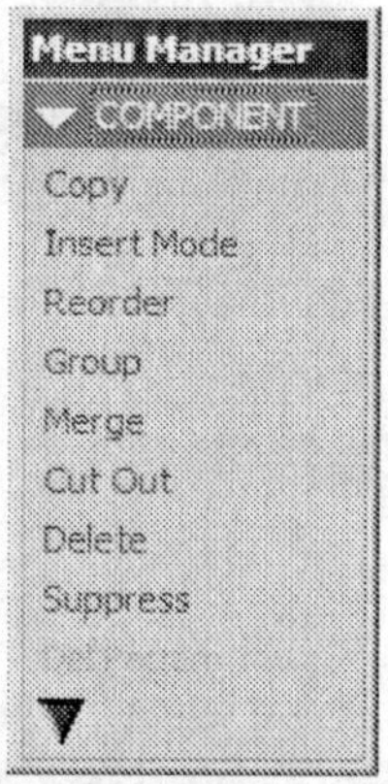

Figure 9-31 ***COMPONENT*** *menu*

Tip: *You can also reorder the components in an assembly by using the **Model Tree**. This can be done by selecting the component in the **Model Tree** and then dragging it to the position where it can be placed. It should be noted that this method of reordering does not let you know where the component can be placed and where it cannot be placed. However, the parent component of the selected child component gets highlighted when you start dragging the child component. This means that if a component has a parent then the parent is also reordered with the child. It should be noted that in a **Model Tree**, the parent component will always occur before its child component.*

SUPPRESSING/RESUMING THE COMPONENTS

If you do not want certain components of the assembly to appear in the current display as well as in the drawing views then you can suppress them. To suppress the components choose **Edit > Component Operations** from the menu bar. The **COMPONENT** menu is displayed. Choose the **Suppress** option from this menu. Similarly, the suppressed components can be resumed by choosing the **Resume** option from the same menu.

The components can also be suppressed/resumed using the shortcut menu that is displayed when you right-click on the component in the **Model Tree** or in the graphics window. In the shortcut menu, choose **Suppress** to suppress and **Resume** to resume the components.

REPLACING COMPONENTS

The existing components of the assembly that are assembled using the assembly constraints can be replaced with some other component if necessary. To replace the component, choose **Edit > Replace** from the menu bar. The **Replace** dialog box is displayed as shown in

Figure 9-32 and you will be prompted to select the model for the replacement of the component. The new component can be selected using the **Browse** button of the **Replace** dialog box. The new component selected will be displayed in the **Browse** edit box. The new component can be assembled using a family table, an interchange assembly, by layout, or manually. After you have selected one of these options, choose **OK** from the **Replace** dialog box to display the **Component Placement** dialog box for adding the placement constraints.

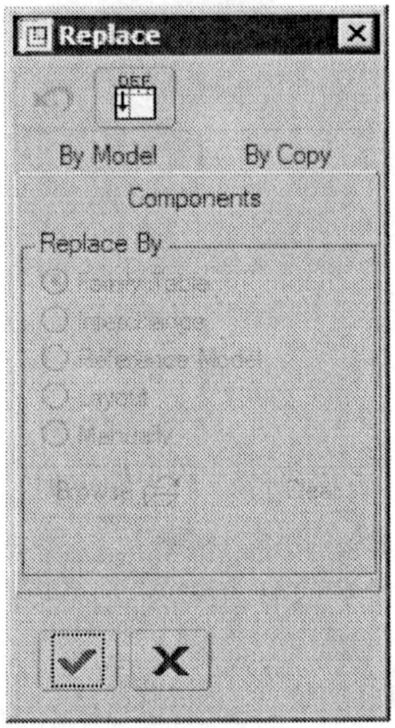

***Figure 9-32 Replace** dialog box*

ASSEMBLING REPEATED COPIES OF A COMPONENT

Sometimes, the assembly design demands the assembling of a particular component more than once. One option is that you should assemble the component every time and add the placement constraints all over again. However, this is a very tedious and time-consuming process, especially in the assemblies that have a large number of similar components. Therefore, to solve this problem, Pro/ENGINEER provides you with an option of assembling multiple copies of the components.

To repeat a component in an assembly, select the component to repeat from the **Model Tree** or from the graphics window and then choose **Edit > Repeat** from the menu bar. When you choose this option, the **Repeat Component** dialog box will be displayed as shown in Figure 9-33. The options in the **Repeat Component** dialog box are discussed next.

Component Area

The button in this area is chosen to select the component whose multiple copies are to be assembled. The name of the selected component is displayed in the edit box provided in this area.

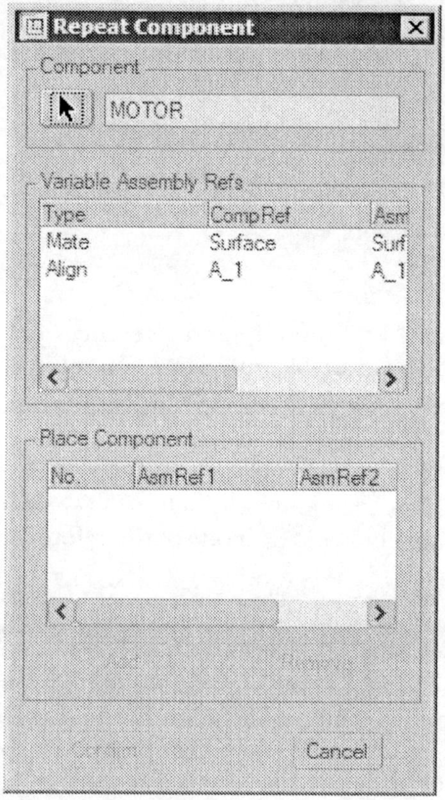

Figure 9-33 ***Repeat Component*** *dialog box*

Variable Assembly Refs Area

The **Variable Assembly Refs** area displays the placement constraints that are applied on the selected component. You can keep the placement constraints that should be present for the new component and modify the remaining constraints by selecting them from this area. The field in this area consists of the following columns.

Type

The **Type** column displays the type of placement constraints that are applied to the selected component.

CompRef

The **CompRef** column displays the references that were used to assemble the component. In case you want to change this for the new component, select this constraint and then choose the **Add** button. Now, select a reference on the assembly. A copy of the selected component will be added to the assembly at a new location. Similarly, to add another copy, repeat this procedure.

AsmRef

The **AsmRef** column displays the reference on the assembly that was used to assemble the selected component.

Place Component Area

The **Placement Component** area displays the new references added for the newly placed copies of the selected component.

Add

The **Add** button is chosen to add the required references for the new copy of the selected component. This button will be available only when you select one of the constraints from the **Variable Assembly Refs** area.

Remove

The **Remove** button is chosen to remove the selected reference from the new component. This button is available only when you select one of the references from the **Place Component** area.

MODIFYING THE COMPONENTS OF THE ASSEMBLY

Sometimes, during or after the assembly of a components, you may need to modify the dimensions of the components. The modification in the dimensions of the components can be directly made in the Assembly mode and you do not need to open the part file to make the modifications. You can even redefine the selected feature in the Assembly mode itself. The methods used to modify and redefine the features are discussed below.

Modifying the Dimensions of a Feature of a Component

The dimensions of a selected feature of the component can be modified by selecting the component from the **Model Tree** or from the graphics window. When the feature turns red or orange in color, right-click to invoke the shortcut menu. Choose the **Activate** option. The name of the component appears on the graphics window. This indicates that only the component whose name appears on the graphics window can be selected. Now, when you move the cursor on the assembly only the activated component is highlighted in cyan. Select the feature of the component to highlight it red in color and then hold down the right mouse button to invoke the shortcut menu. Choose the **Edit** option from the shortcut menu. The dimensions of the selected feature of the component are displayed on the graphics window. After modifying the dimensions, remember to regenerate the component by choosing the **Regenerate** option available in the shortcut menu. After regenerating the component you can view the effect of the modification.

Redefining a Feature of a Component

The option to redefine the features in the Assembly mode is available. This saves time in opening the part in the Part mode and then redefining the features. To redefine the feature of a component, you need to select the component from the **Model Tree** and make it active. After the component is activated, select the feature that you need to redefine from the graphics window. Hold down the right mouse button to invoke the shortcut menu. Choose the **Edit Definition** option. After choosing the option, the feature creation tool is activated and now you can redefine the feature.

CREATING THE EXPLODED STATE

Generally, the assembly designs consist of a large number of components. Some of these components are visible after they are assembled and some of them are not visible. This is because the components that are assembled outside are visible but the components that are assembled inside some other components are not visible. This is a very misleading state and may confuse the viewer as he cannot view all the components. However, to eliminate this confusing state, generally an exploded view is provided along with the assembled view. The exploded view is a state in which all the components are moved from their actual location so that they are visible as shown in Figure 9-34.

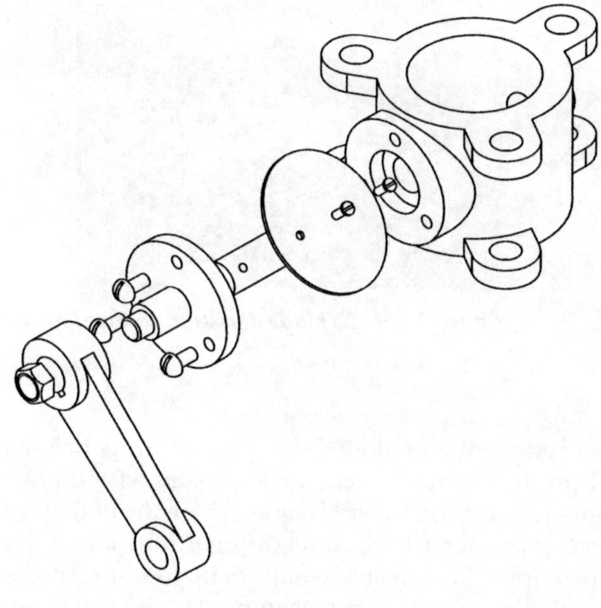

Figure 9-34 *The exploded state of an assembly*

The exploded state can be created by choosing **View > Explode > Edit Position** from the menu bar. The **Explode Position** dialog box is displayed as shown in Figure 9-35. This dialog box provides you the following options to move the components from their actual location.

Motion Type Area

Translate

This radio button when selected allows you to move a component on the graphics window from its position in the assembly.

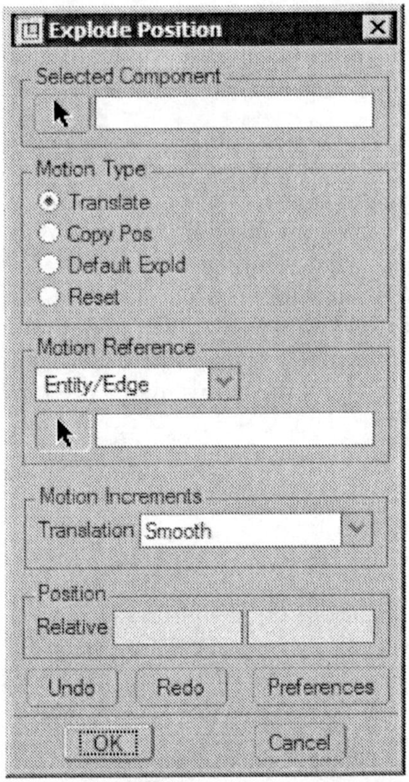

***Figure 9-34 Explode Position** dialog box*

Copy Pos

This option is used to restore the exploded component to its original location. This radio button when selected prompts you to select a component whose position you want to copy. For example, a component is exploded and is moved away from the assembly on the graphics screen and you want to assemble it back to its original position. Select the **Copy Pos** radio button; you will be prompted to select a component to copy the position from. Select a component from the assembly. Now, you are prompted to select a component to move. Select the exploded component that is moved away from the assembly. The selected component will be assembled at its original position in the assembly.

Default Expld

This radio button when selected prompts you to select a component. When you select a component from the assembly, the component is exploded from the assembly to a default position on the graphics screen.

Reset

The **Reset** radio button is used to set the selected component to its original location in the assembly.

Motion Reference Area

View Plane

The **View Plane** option is used to specify the current view plane (that is, the screen of the monitor) as the reference for exploding the selected component.

Sel Plane

The **Sel Plane** option is used to select a datum plane or a planar face for specifying the reference for the motion of the component. When you choose this option, you will be prompted to select the datum plane or a planar face and then you will be prompted to select the component to move.

Entity/Edge

The **Entity/Edge** option is selected by default and is used to select a linear edge or an axis for specifying the reference direction for exploding the component.

Plane Normal

The **Plane Normal** option is used to select a planar face or a datum plane, normal to which will be the reference for exploding the components.

2 Points

The **2 Points** option is used to select two datum points or two vertices to specify the reference direction for exploding the components.

C-sys

The **C-sys** option is used to select one of the axes of the selected coordinate system for specifying the reference direction for exploding the components.

Note

If you have not created an exploded state of an assembly, a default explode state is available when you choose ***View*** *>* ***Explode*** *>* ***Explode View*** *from the menu bar. To redisplay the unexploded state, choose* ***View*** *>* ***Explode*** *>* ***Unexplode*** *from the* ***View*** *menu in the menu bar.*

OFFSET LINES

The offset lines are the lines that are used to display the actual path and direction of the mating components. These lines are generally used for easy visualization of the exploded components of the assemblies. The exploded state of an assembly using the offset lines is shown in Figure 9-36.

Creating the Offset Lines

To create offset lines, an exploded state should exist in which the offset lines will be created. To create the offset lines, choose **View** > **Explode** > **Offset Lines** > **Create** from the menu bar. The **Entity Select** menu is displayed as shown in Figure 9-37. This menu provides you with the following options for selecting the objects for creating the offset lines.

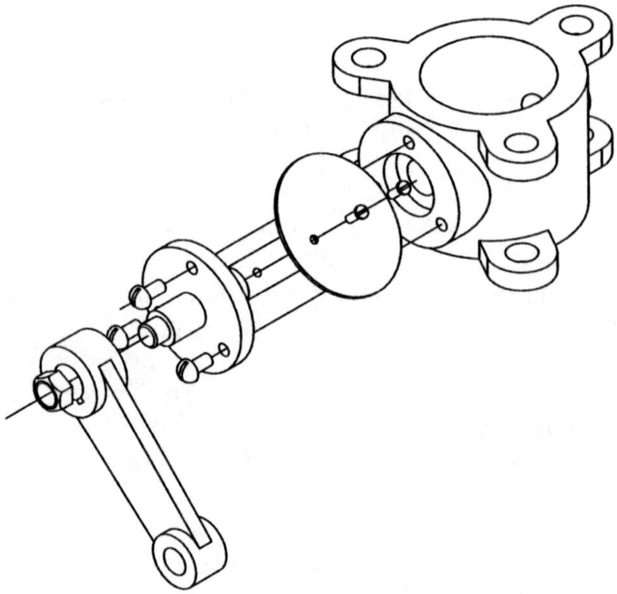

Figure 9-36 The exploded state of an assembly with offset lines

Axis

The **Axis** option is used to select the axis of the components for creating the offset lines.

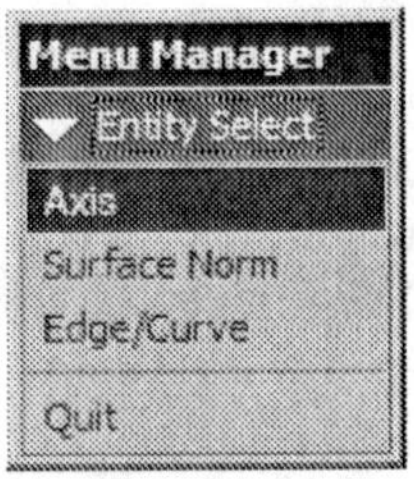

***Figure 9-37 Entity Select** menu*

Surface Norm

The **Surface Norm** option is used to create the offset lines normal to the selected surface at the specified point.

Edge/Curve

The **Edge/Curve** option is used to select an edge or a curve for creating the offset lines.

Note

The offset lines are automatically made invisible once you display the unexploded state of the assembly.

Modifying the Offset Lines

To modify the offset lines that are already created, choose **View > Explode > Offset Lines > Modify** from the menu bar. The **EXPL LINES MODIFY** menu is displayed as shown in Figure 9-38. The options in this menu are discussed next.

Move

This option is used to move the offset line that has already been created. When you choose this option, you are prompted to select the offset line segment to modify. Select the end of the offset line and move the cursor to extend or shorten the line. When you move the cursor the offset line that is being extended is not visible on the graphics window. To pause the extension of line press the right mouse button. To end the modification of the line press the left mouse button. To abort the modification process, press the middle mouse button.

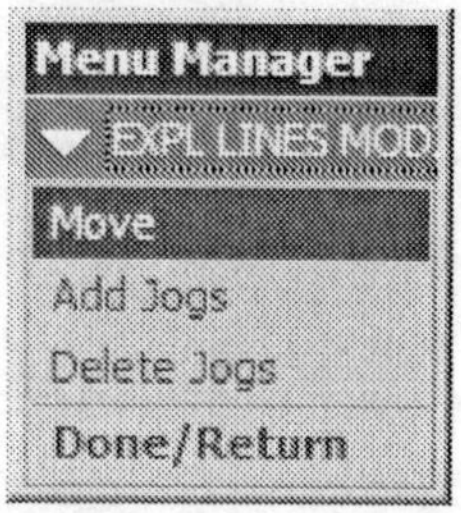

***Figure 9-38 EXPL LINES MODIFY** menu*

Add Jogs
This option is used to add a jog to an existing offset line segment. A jog is a break in the continuation of a line segment.

Delete Jogs
As the name of the option suggests, this option is used to delete a jog, if any, from a line segment.

Note

*To delete an offset line, choose **View > Explode > Offset Lines > Delete** from the menu bar. You can also set the line style for the offset lines you create in an exploded view. To set a line style for offset lines, choose **View > Explode > Offset Lines > Mod Line Style** from the menu bar.*

THE BILL OF MATERIAL
The Bill Of Material or the BOM is the tabular representation of all the components of the assembly along with the information associated with them. The information can be the material of the components, the additional note with the components, and so on. Pro/ENGINEER provides you with a readymade BOM that can be directly utilized. This BOM is automatically updated during the assembly of the components. The BOM can be viewed using the **BOM** dialog box, shown in Figure 9-39, which is displayed by choosing **Bill of Materials** from the **Info** menu in the menu bar. The options in this dialog box are discussed next.

Select Model Area
The options provided under the **Select Model** area are used to select the type of assembly whose BOM has to be displayed. These options are discussed next.

Top Level
The **Top Level** radio button is used to select the top level assembly for displaying the BOM.

Subassembly
The **Subassembly** radio button is used to select the subassembly for displaying the BOM. When you select this option, the button available under this area will be available and you will be prompted to select the subassembly.

Include Area
Skeletons
The **Skeletons** check box is selected so that the skeletons, if any, in the assembly should also be included in the BOM.

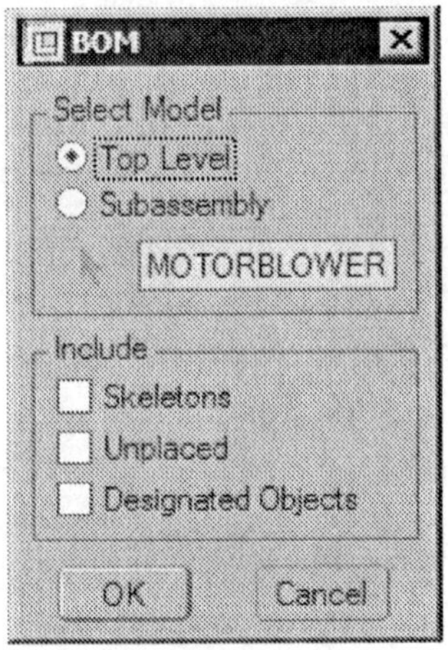

Figure 9-39 ***BOM*** *dialog box*

Unplaced

The **Unplaced** check box is selected to include the unplaced components in the BOM.

Designated Objects

The **Designated Objects** check box is selected to include the bulk items, if any, that are used in the assembly. Bulk items are items whose solid models are not created but they are used in the assembly. In Pro/ENGINEER, you can include the weight and other properties of these bulk items to calculate certain parameters related to the assembly like total weight of assembly. Examples of bulk items are paint, glue, cotton lace, and so on.

After you have selected the options in the BOM dialog box, choose **OK** from the **BOM** dialog box to open the browser that displays the BOM for the selected assembly. The file of the bill of material is automatically saved with the name of the assembly. The extension of this file is *.bom*. You can later retrieve it in the Drawing mode to display the BOM in the drawing views of the assembly.

Note

When you work on Pro/ENGINEER there are two memories you are working on. The first is the memory of Pro/ENGINEER and the second is of the computer system, which is called the hard disk. Pro/ENGINEER memory is called ***In Session*** *and can be erased as discussed in the* ***Introduction*** *chapter. The* ***In Session*** *memory is emptied once you exit Pro/ENGINEER. When you choose the* ***Save the active object*** *button on the* ***File*** *toolbar to save, the object on the graphic screen is permanently saved in the working directory and can be removed only by deleting.*

Associative BOM can be created in the Report mode of Pro/ENGINEER. You will learn to create associative BOM in Chapter 12.

TUTORIALS

Tutorial 1

In this tutorial you will create all the components of the Shock assembly and then assemble them as shown in Figure 9-40. Also create an exploded state of the assembly displaying the offset lines as shown in Figure 9-41. The BOM is shown in Figure 9-42. The dimensions of the components are shown in Figures 9-43 through 9-50. **(Estimated time: 2 hrs)**

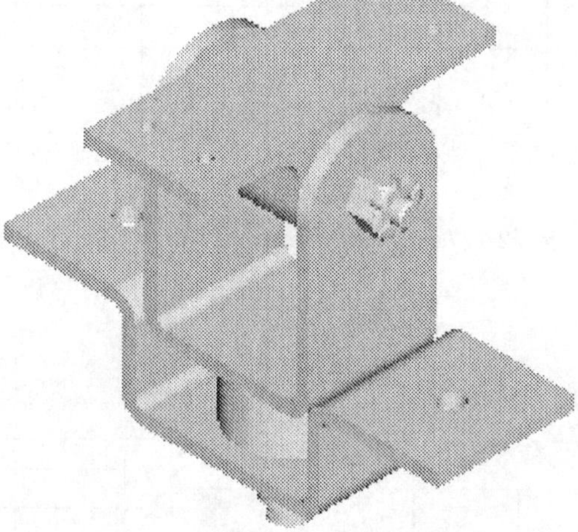

Figure 9-40 *The Shock assembly*

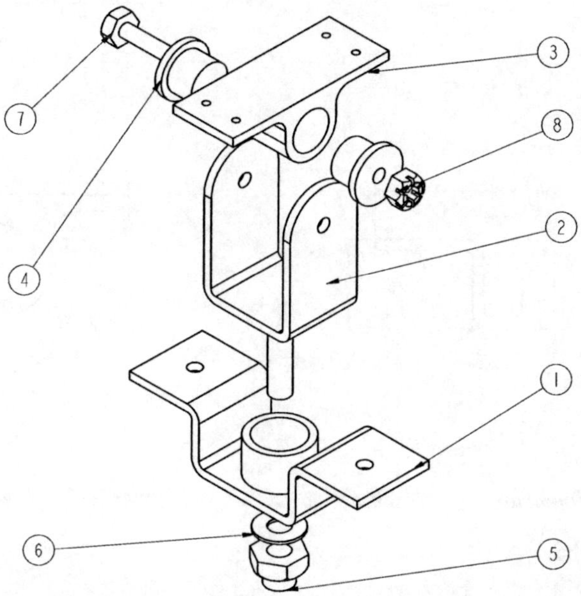

Figure 9-41 *The exploded state of the Shock assembly*

ITEM	QTY.	NAME	MATERIAL
1	1	BRACKET	STEEL
2	1	U-SUPPORT	STEEL
3	1	PIVOT	STEEL
4	2	BUSHING	BRONZE
5	1	SELF LOCKING NUT	0.625-11UNC
6	1	WASHER	USER DEFINED
7	1	HEXAGONAL BOLT	STEEL
8	1	CASTLE NUT	STEEL

Figure 9-42 *The Bill Of Material for the Shock assembly*

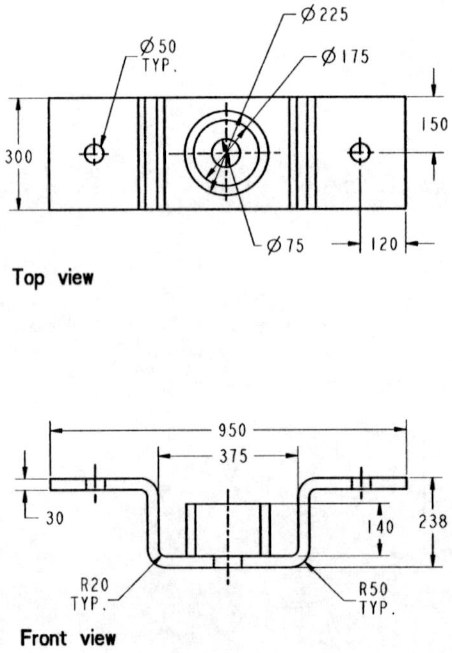

Figure 9-43 *Dimensions for Bracket*

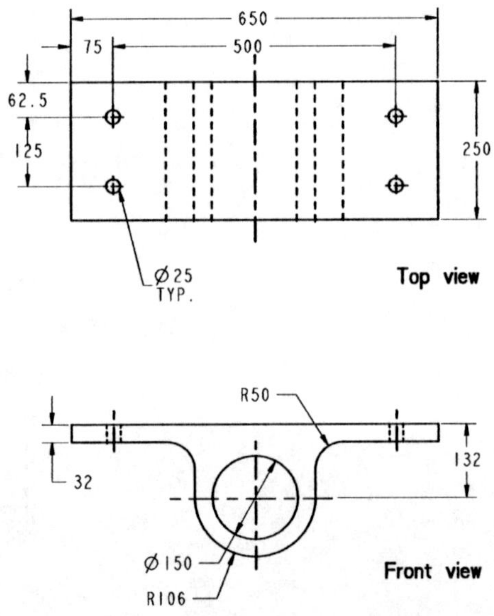

Figure 9-44 *Dimensions for Pivot*

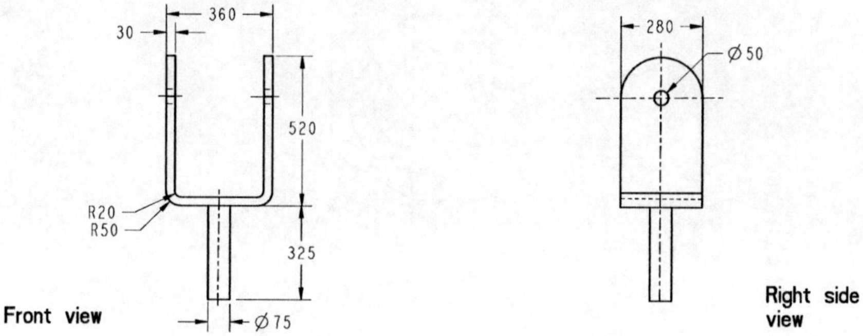

Figure 9-45 *Dimensions for U-Support*

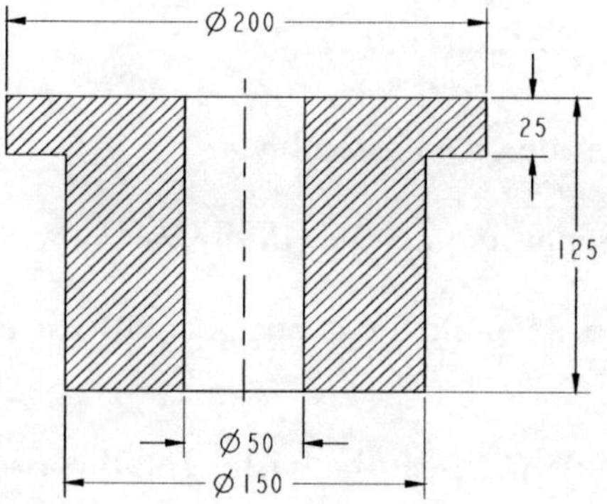

Figure 9-46 *Dimensions for Bushing*

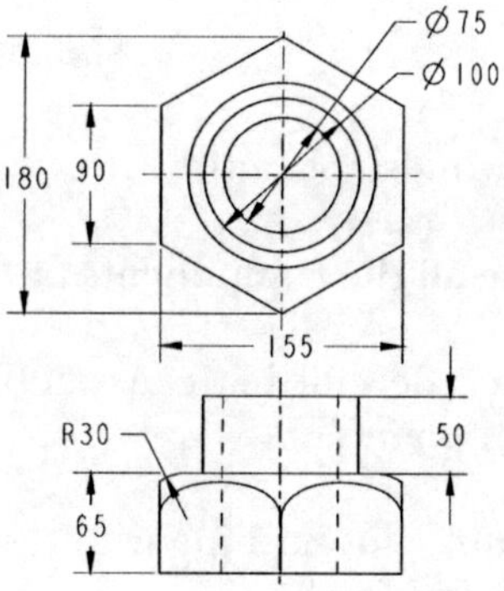

Figure 9-47 *Dimensions for Self-locking nut*

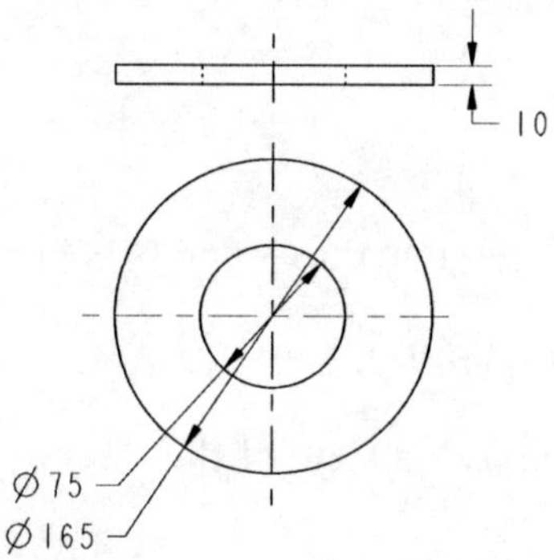

Figure 9-48 *Dimensions for Washer*

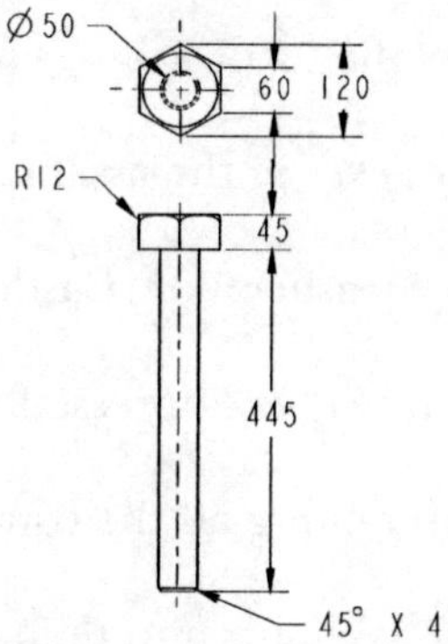

Figure 9-49 *Dimensions for Hexagonal bolt*

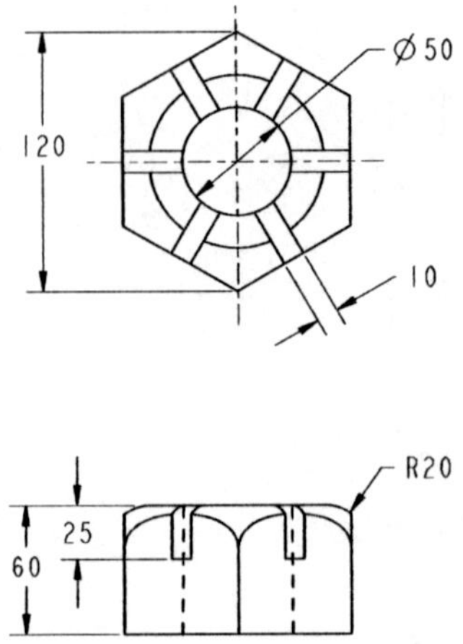

Figure 9-50 *Dimensions for Castle nut*

The following steps outline the procedure to create the given assembly:

a. Create all the components of the assembly as separate part files in the Part mode.

b. Create a new file in the Assembly mode and then assemble the U-Support with the default datum planes.

c. Assemble the Bushing with the U-Support and then repeat the bushing in the assembly, see Figure 9-53.

d. Assemble the Pivot with the Bushing, see Figure 9-54.

e. Suppress the two Bushings and the Pivot.

f. Insert Bracket in the assembly, see Figure 9-57.

g. Insert the Washer, Bolt, Castlenut, and the Self-locking nut in the assembly, see Figure 9-60.

h. Unsuppress the suppressed components, see Figure 9-60.

i. After assembling all the components, create the exploded state of the assembly.

j. Save the assembly and then exit the Assembly mode.

Before you start creating the components, set the working directory to *C:\ProE-WF\c09\Shockassembly*. Also, make sure that the c09 directory exists inside the *ProE-WF* directory. If these directories do not exist, create them and set the working directory.

Creating the Components for Assembly

The given assembly is created by using the bottom-up approach. As mentioned earlier, in bottom-up approach assemblies, the components are created as separate files and are placed in the assembly file. Therefore, you first need to create the components of the assembly.

1. Create all the components of the assembly as separate part files and save them in the current working directory.

2. Close the part files if opened.

Creating New Assembly File

All the components that you have created are assembled in an assembly file having an extension *.asm*. You need to open a new assembly file to assemble the components.

1. Choose the **Create a new object** button from the **File** toolbar to display the **New** dialog box.

2. Select the **Assembly** radio button in the **Type** area of the **New** dialog box. In the **Sub-type** area of the **New** dialog box, the **Design** radio button is selected by default. Enter the name of the assembly in the **Name** edit box as **SHOCKASSEMBLY**.

Note

*Make sure that in the **New** dialog box, the **Use default template** check box is selected. If this check box is not selected, you will not be able to use the template provided by Pro/ENGINEER. Hence, in that case you will need to create the default datum planes by choosing the **Insert a datum plane** button from the **Right Toolchest**.*

3. Choose **OK** from the **New** dialog box to proceed to the assembly environment.

Assembling the U-Support with the Default Datum Planes

When you choose **OK** from the **New** dialog box, the three default assembly datum planes are displayed on the graphics screen and the **Model Tree** is displayed on the left of the graphics window. You can close the **Model Tree** by clicking on the sash available on the right edge of the **Model Tree**. Now, you can start assembling the components.

1. Choose the **Add component to the assembly** button. The **Open** dialog box is displayed.

2. Select **u-support** from the **Open** dialog box and choose the **Open** button. The **Component Placement** dialog box is displayed with the **Show component in the assembly window while specifying constraints** button selected by default.

Since the **Show component in the assembly window while specifying constraints** button is selected by default, the U-Support is placed with the default datum planes by assuming some assembly constraints.

As discussed earlier, when you open a component to assemble, there are two display modes in which the component can be displayed. These two display modes are controlled by two buttons at the top in the **Component Placement** dialog box. You will use the display mode in which the component is displayed in a separate window called the **Component Window**.

Tip: *Viewing the components in separate windows helps the user to locate the references for the assembly constraints easily and also the user has independent viewing control over the two windows. You can spin, pan, and zoom the components in their respective windows.*

3. Choose the **Show component in a separate window while specifying constraints** button.

The subwindow is displayed on the left of the graphics screen with the U-Support. You will notice that the component is still displayed on the graphics window. This is because the **Show component in the assembly window while specifying constraints** button is still selected in the **Component Placement** dialog box.

4. Choose the **Show component in the assembly window while specifying constraints** button again from the **Component Placement** dialog box to clear it. Now, the component is not displayed on the graphics window.

You can drag the window to a desired position so that you can view the prompt messages of both the windows. You can also use the keyboard and mouse button combinations to zoom in and zoom out the component in the two windows and to spin the models individually.

5. Choose **Assemble component at a default location** button from the **Component Placement** dialog box.

As discussed earlier, the **Assemble component at a default location** button aligns the datum planes of the component with the assembly datum planes. The advantages of assembling the first component of the assembly with the default datum planes were discussed earlier.

6. The **Placement Status** area shows **Fully Constrained**. Choose **OK**. The **Component Placement** dialog box is closed and the U-Support is assembled with the default assembly datum planes.

Assembling the Bushing with the U-Support

The next component you need to assemble is the Bushing.

1. Choose **Add component to the assembly** button to display the **Open** dialog box.

2. Select **bushing** and choose **Open** to display the **Component Placement** dialog box. The component is displayed in the **Component Window**. You may need to drag the **Component Window** so that you can view the prompts in the **Message Area**.

3. Select the **Mate** option from the **Type** drop-down list in the **Constraints** area of the **Component Placement** dialog box to add the mate constraint. You are prompted to select a mating surface.

4. Select the planar surface of the Bushing shown in Figure 9-51.

5. Select the left inner face of the U-Support, shown in Figure 9-51, as the other face to add the mate constraint.

6. Select the **Align** option from the **Type** drop-down list in the second row to add the align constraint.

7. Select the axis of the Bushing, shown in Figure 9-51 to add the **Align** constraint.

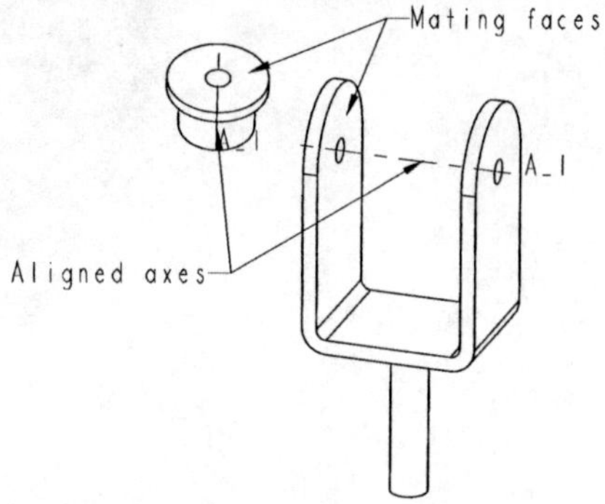

Figure 9-51 *Constraints and the location to apply them*

8. Select the axis of the hole on the vertical face of the U-Support shown in Figure 9-50. The **Placement Status** area shows **Fully Constrained**.

   Note that the **Allow Assumptions** check box is selected. This means that Pro/ENGINEER has assumed one or more constraints to fully define the assembly. If you clear the **Allow Assumptions** check box, the **Partially Constrained** message is displayed in the **Placement Status** area. Therefore, let Pro/ENGINEER assume the constraints and let the **Allow Assumptions** check box be selected.

Tip: *The arrow in the **Component Placement** dialog box under the **Component Reference** area is used to make selections. If you have made a wrong selection then you can choose this arrow to display the **SELECT** dialog box and make the selection again.*

9. You can preview the assembly of the U-Support and the Bushing by choosing the **Preview** button from the **Component Placement** dialog box. The bushing is displayed as a wireframe model in the assembly. Choose **OK**.

   The two components are assembled as shown in Figure 9-52.

10. The next instance of the Bushing can be placed directly using the **Repeat** option. To repeat Bushing in the assembly, select it from the **Model Tree** or from the graphics window and then choose **Edit > Repeat** from the menu bar. When you choose this option, the **Repeat Component** dialog box is displayed.

    In the dialog box, the **Mate** and **Align** constraints are displayed in the **Variable Assembly Refs** area.

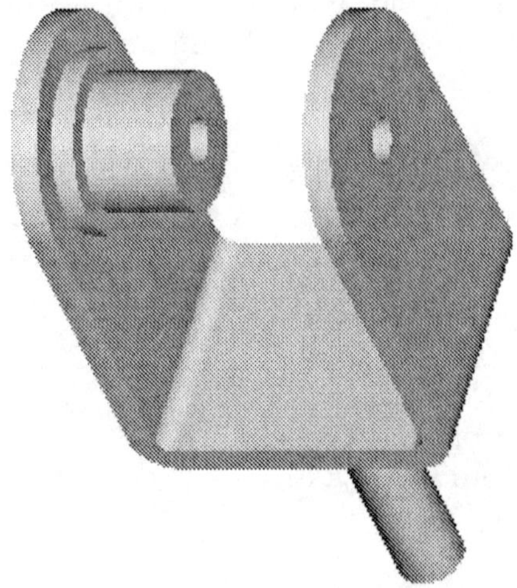

Figure 9-52 The U-Support assembled with the Bushing

11. Select the **Mate** constraint from this area and then choose the **Add** button.

12. Select the inner right face of the U-Support as the mating face. The copy of the Bushing will be assembled as shown in Figure 9-53. Choose **Confirm** from the **Repeat Component** dialog box.

Figure 9-53 Assembled Bushings and the U-Support

Assembling the Pivot with the Bushing

The next component that you need to assemble is the Pivot.

1. Choose the **Add component to the assembly** button to display the **Open** dialog box.

2. Select **pivot** and choose **Open** to display the **Component Placement** dialog box.

3. Select the **Mate** option from the **Type** drop-down list to add the mate constraint. You are prompted to select a mating surface.

4. Select the face of the Bushing, shown in Figure 9-54, to add the mate constraint.

5. Select the front face of the Pivot, shown in Figure 9-54 to add the mate constraint.

6. Select the **Align** option from the **Type** drop-down list in the second row to add the align constraint.

Tip: *If the display of datum axis is turned on, the datum axis will not be displayed in the subwindow until you choose **View > Repaint** from menu bar.*

7. Select the axis of the Bushing shown in Figure 9-54.

8. Select the central axis of the Pivot shown in Figure 9-54. The **Fully Constrained** message is displayed in the **Placement Status** area of the **Component Placement** dialog box. But note that the **Allow Assumptions** check box is selected. This suggests that Pro/ENGINEER has assumed some constraints to constrain the Pivot with the Bushing.

9. Choose the **Preview** button from the **Component Placement** dialog box. The orientation of the Pivot with the Bushing is not what is required. Therefore, you need to clear the **Allow Assumptions** check box and define one more constraint.

10. Clear the **Allow Assumptions** check box in the **Component Placement** dialog box. Now, the **Partially Constrained** message is displayed in the **Placement Status** area.

11. To add a constraint, choose the **Specify a new constraint** button from the **Constraints** area in the **Component Placement** dialog box. The **SELECT** dialog box is displayed.

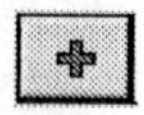

12. Select the field in the **Offset** column. The field will be changed into a drop-down list. Select the **Oriented** option from the drop-down list and select the top face of the Pivot shown in Figure 9-54.

13. Now, select the upper horizontal face of the U-Support shown in Figure 9-54. The message displayed in the **Placement Status** area is **Fully Constrained**. Choose the **Preview** button from the **Component Placement** dialog box. Choose **OK**.

Note

*If the surface is oriented toward the bottom side, choose the **Change orientation of the constraint** button from the **Component Placement** dialog box.*

The assembly after placing the Pivot should look similar to the one shown in Figure 9-55.

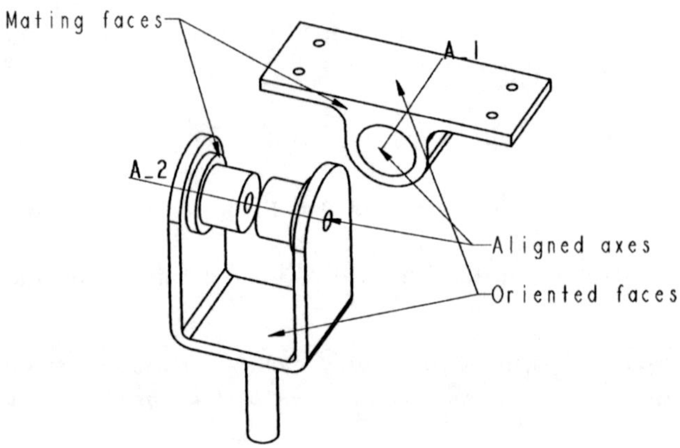

Figure 9-54 Constraints and the location to apply them

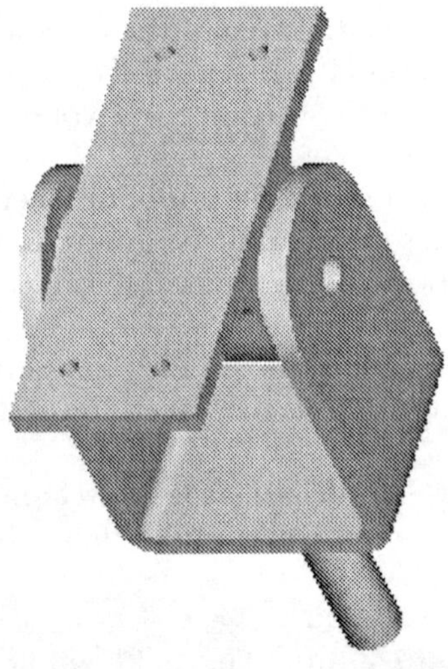

Figure 9-55 Assembled view of the Pivot, U-Support, and the Bushing

Suppressing Features

Next, you need to suppress both the Bushings and the Pivot. This is because as the components in the assembly increase, the clarity on the graphics screen reduces. Also, more the number of components, more are the datum planes. Thus, it becomes difficult to make selections on the components. However, when you suppress a component, its datum planes are also suppressed and the complications on the graphics screen are reduced.

1. Spin the model and select the first bushing from the graphics window or from the **Model Tree** and right-click to display the shortcut menu. Choose **Suppress** from the shortcut menu to suppress the component.

The **Suppress** window is displayed. Choose **OK**. The Pivot will also get suppressed because it was dependent on the first bushing.

2. Similarly, suppress the other Bushing to make the assembly of the other components easy.

Note

Very often you will need to use the datum planes and the datum axes to assemble the components. Therefore, you will need to turn their display on or off from the ***Datum Display*** *toolbar whenever required.*

Assembling the Bracket with the Assembly

The next component you need to assemble is the Bracket.

1. Choose the **Add component to the assembly** button to display the **Open** dialog box.

2. Open the **bracket** to display the **Component Placement** dialog box.

3. Select the **Mate** option from the **Type** drop-down list to add the mate constraint.

4. Select the upper face of the cylindrical feature as the first face to add the mate constraint and then select the bottom central horizontal face of the U-Support as the other mating face as shown in Figure 9-56.

5. Select the **Align** option from the **Type** drop-down list in the second row to add the align constraint.

6. Select the axis of the Bracket to add the align constraint and then select the axis of the U-Support as shown in Figure 9-56.

Note

If the ***SELECT*** *dialog box is not displayed, choose the* ***Specify a new constraint*** *button from the* ***Component Placement*** *dialog box to display the* ***SELECT*** *dialog box and to add a new constraint.*

7. Select the field in the **Offset** column in the third row. The field will be changed into a drop-down list. Select the **Oriented** option from the drop-down list and select the faces as shown in Figure 9-56.

8. Choose **OK** from the **Component Placement** dialog box. The assembled Bracket is shown in Figure 9-57.

Assembling the Washer with the Assembly

The next component that you will assemble is the Washer.

1. Choose the **Add component to the assembly** button to display the **Open** dialog box.

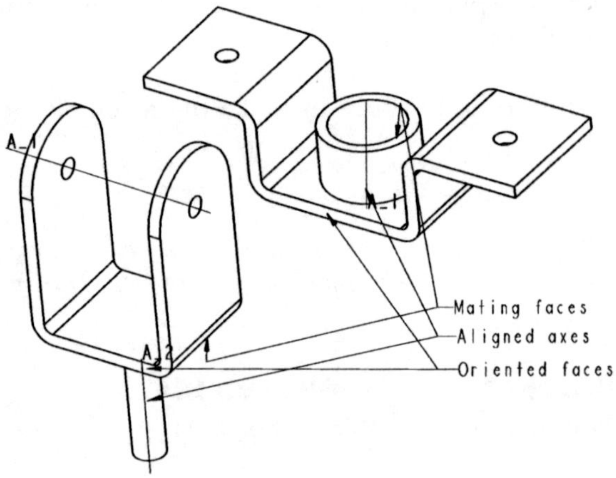

Figure 9-56 Constraints and the location to apply them

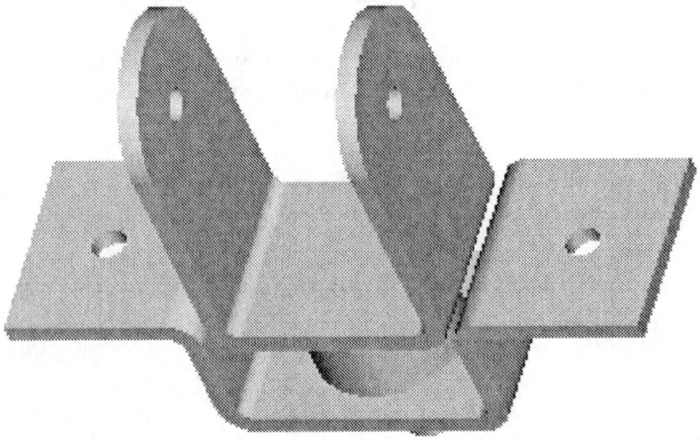

Figure 9-57 Assembled view of the Bracket with the U-Support

2. Open the **washer** to display the **Component Placement** dialog box.

3. Select the **Mate** option from the **Type** drop-down list to add the mate constraint. The **SEL/MAKE DAT** menu is displayed.

4. Select the top face of the Washer as the first face to add the mate constraint and then select the bottom face of the Bracket as the other mating face.

Note
Sometimes when you select a reference from a component to apply a constraint, the ***Message Input Window*** *is displayed and you are prompted to specify the offset value for the constraint applied. Enter a value of 0 in the* ***Message Input Window*** *if the offset distance is not required. Now, if the constraint changes from* ***Mate*** *to* ***Align****, change it back to* ***Mate*** *by selecting it from the* ***Type*** *drop-down list.*

5. Select the **Align** option from the **Type** drop-down list in the second row.

6. Select the axis of the Washer and then select the axis of the cylindrical feature on the Bracket.

7. Choose the **Cancel** button from the **SELECT** dialog box.

8. Select the **Allow Assumptions** check box. The message displayed in the **Placement Status** area is **Fully Placed**.

   You have defined two constraints for the Washer and the Bracket, but you need to define one more constraint that will constrain the rotatory degree of freedom of the Washer. That is why you selected the **Allow Assumptions** check box. You can also restrict the rotatory movement of the Washer by selecting a datum plane that is perpendicular to the direction of rotation of the Washer and orient it with a plane or a face parallel to the selected datum plane.

   The washer is assembled to the Bracket as shown in Figure 9-58.

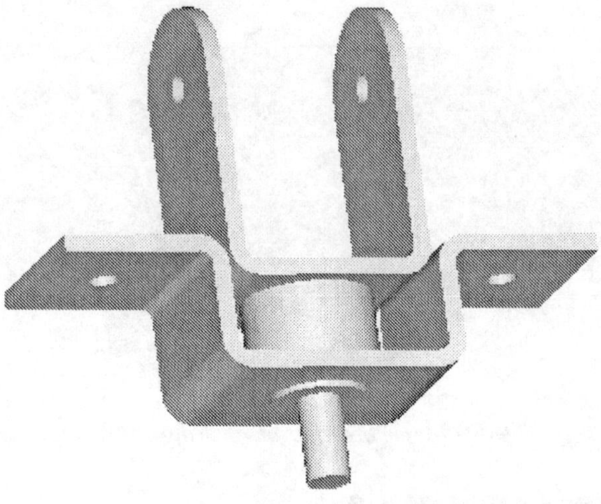

Figure 9-58 *Assembled view of the Washer with the Bracket*

Inserting the Hexagonal Bolt in the Assembly

The next component that you will assemble is the Hexagonal Bolt.

1. Choose the **Add component to the assembly** button to display the **Open** dialog box.
2. Open the **hexagonalbolt** to display the **Component Placement** dialog box.
3. Select the **Mate** option from the **Type** drop-down list to add the mate constraint. The **SEL/MAKE DAT** menu is displayed.
4. Select the inner face of the head of the Hexagonal bolt and then select the outer left face of the U-Support as the other mating face.
5. Select the **Align** option from the **Type** drop-down list in the second row.
6. Select the axis of the Hexagonal bolt and then select the axis of the hole feature on the U-Support.
7. Clear the **Allow Assumptions** check box and then choose the **Specify a new constraint** button. Select the field in the **Offset** column. The field will be changed into a drop-down list. Select the **Oriented** option from the drop-down list.
8. Out of six faces of the head of the Hexagonal bolt, select any one face and then select the top face of the Bracket. The **Placement Status** area shows **Fully Constrained**. Choose **OK**. The Hexagonal bolt is inserted in the assembly as shown in Figure 9-59.

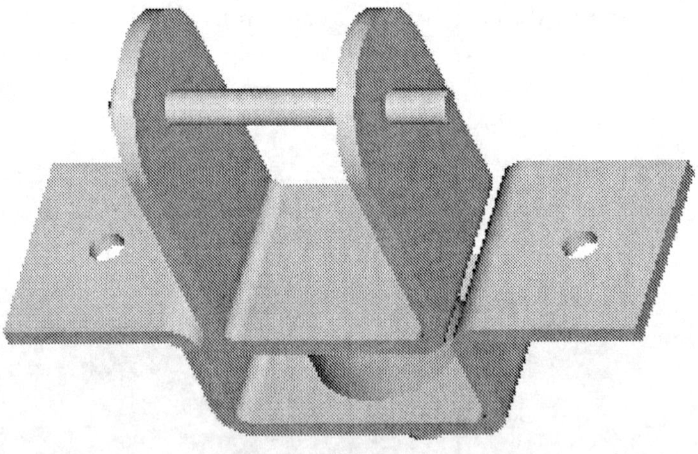

***Figure 9-59** Assembly after assembling the hexagonal bolt*

Assembling Other Components

Similarly, assemble the Castlenut and the Self-locking nut.

Unsuppressing the Components

All the components are assembled to their required positions. Now, you need to unsuppress the components that were suppressed.

1. Choose **Edit** > **Resume** > **Resume All**. The suppressed components appear in the assembly where they were constrained. The completed assembly is shown in Figure 9-60.

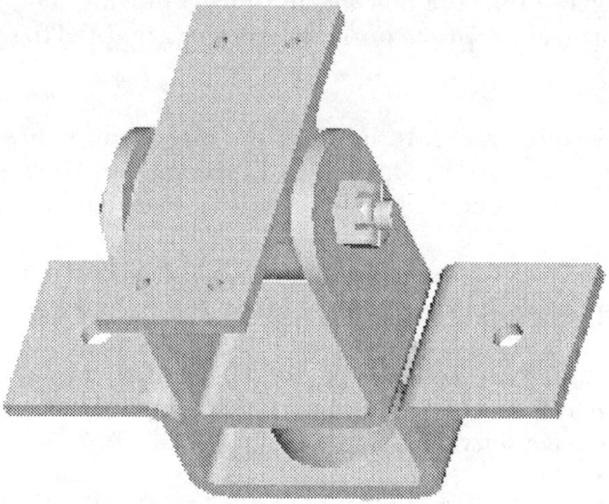

Figure 9-60 *Completed assembly with all the components*

Creating the Exploded State of the Assembly

To clearly view all the components in an assembly you need to create an exploded state of the assembly.

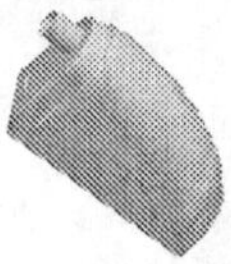

Tip: *You can also view the default exploded state of an assembly by choosing **View** > **Explode** > **Explode View** from the **View** menu in the menu bar.*

1. Choose the **Start the view manager** button from the **Top Toolchest**. The **View Manager** dialog box is displayed. Choose the **Explode** tab and right-click in the dialog box to invoke the shortcut menu. Choose the **New** option, name it **EXP1**, and press the middle mouse button.

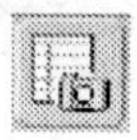

2. Right-click on the **EXP1** in the dialog box and choose the **Redefine** option. The **MOD EXPLODE** menu is displayed. Choose the **Position** option from the menu; the **Explode Position** dialog box is displayed.

   The **Translate** radio button is selected by default in the **Motion Type** area of the **Explode Position** dialog box. The **Entity/Edge** option is selected by default in the **Motion Reference** drop-down list. The **Smooth** option is selected by default in the **Translation** drop-down list. The **SELECT** dialog box is displayed and you are prompted to select an axis or a straight edge as the motion reference.

3. In the **Filter** drop-down list available in the status bar that is below the **Message Area**,

select the **Axis** filter and then select the axis of the cylindrical feature on the Bracket to specify the direction for the motion of all the components.

4. After selecting the axis, select the Self-locking nut and move the cursor down to move the component. Note that the Self-locking nut is moving along its axis only. Place the Self locking nut at an appropriate position. You are prompted to select another component to move.

5. Select the Washer and Bracket to move in the direction as shown in Figure 9-61. Next, you need to move the Castle nut, the two Bushings, the Hexagonal bolt, and the Pivot together as one component along the vertical axis selected earlier.

6. Choose the **Preferences** button from the **Explode Position** dialog box. The **Preferences** dialog box is displayed. Select the **Move Many** radio button and choose **Close**.

7. Select the Castle nut, the two Bushings, the Hexagonal bolt, and the Pivot. The selected components turn red in color. You may need to spin the assembly using the middle mouse button to make selections.

8. After selecting the above-mentioned components, choose the **Default** option from the **Saved view list** button. The assembly orients to its default orientation.

9. Press the middle mouse button and select the Pivot to place all the selected components as shown in Figure 9-61.

Figure 9-61 *Creating the exploded state*

10. After moving all the components to the desired positions, choose **OK** from the **Explode Position** dialog box. The **MOD EXPLODE** menu is displayed.

11. Choose the **Position** option from the submenu. The **Explode Position** dialog box is

displayed. Select the axis of the Hexagonal bolt and move Bushings, Hexagonal bolt, and Castle nut as shown in Figure 9-62.

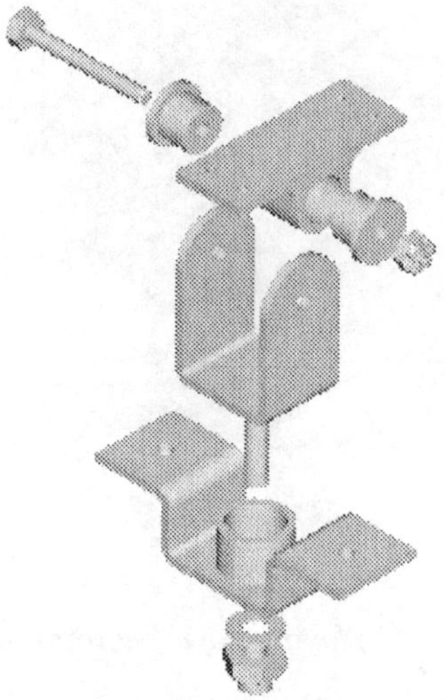

Figure 9-62 *Exploded state of the assembly*

12. After moving the components, choose **OK** from the **Explode Position** dialog box. The **MOD EXPLODE** submenu is displayed.

13. Choose the **Offset Lines** option from the **MOD EXPLODE** menu. The **OFFSET LINES** submenu is displayed. Choose the **Create** option from the **OFFSET LINES** submenu. The **Entity Select** submenu is displayed and the **Axis** option is selected by default.

14. Select the axis of the Self-locking nut and then select the axis of the shaft feature on the U-support. The green-colored offset line is created.

15. Similarly, select the axes of the other components to create the offset lines displaying the path and direction of the mating components as shown in Figure 9-63.

16. After creating all the offset lines, choose **Done/Return** from the **OFFSET LINES** submenu and then choose **Done/Return** from the **MOD EXPLODE** menu. Choose the **Close** button to exit the **View Manager** dialog box.

> **Tip**: *To unexplode the assembly, choose the* ***View > Explode > Unexplode View*** *from the* ***View*** *menu in the menu bar.*

Saving the Assembly

You need to save the assembly because you may need it later.

1. Choose the **Save the active object** button from the **File** toolbar and save the model.

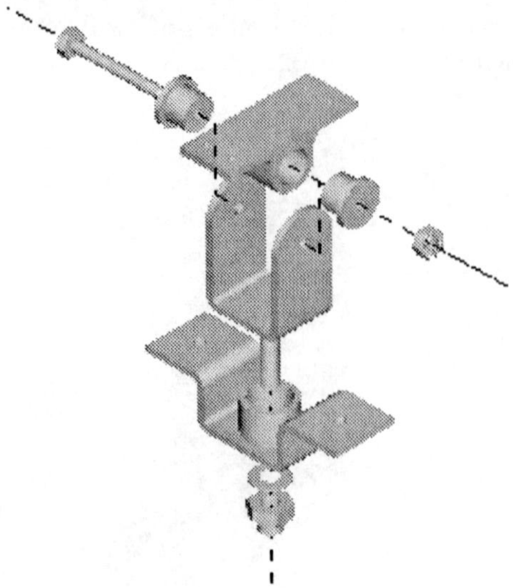

Figure 9-63 Displaying the offset lines in the exploded state

Closing the Window

Now, you have saved the assembly and the window can be closed.

1. Choose the **Close** option from the **Window** menu in the menu bar.

Tutorial 2

In this tutorial you will create all the components of the Pedestal Bearing assembly and then assemble them as shown in Figure 9-64. The BOM is shown in Figure 9-65. You will also create the exploded state, shown in Figure 9-66, displaying the offset lines. The dimensions of the components are shown in Figures 9-67 through 9-70. **(Estimated time: 2 hrs)**

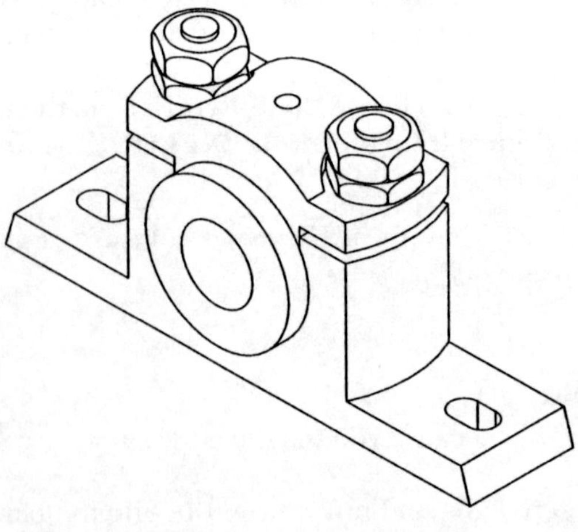

Figure 9-64 Assembly of the Pedestal Bearing

S. NO.	NO. OFF	PART'S NAME	MATERIAL
1	1	BODY	CAST IRON
2	1	CAP	CAST IRON
3	1	BRASSES	GUN METAL
4	2	SQUARE HEADED BOLT	MILD STEEL
5	2	NUT	MILD STEEL
6	2	LOCK NUT	MILD STEEL

Figure 9-65 *The Bill Of Material for Pedestal Bearing*

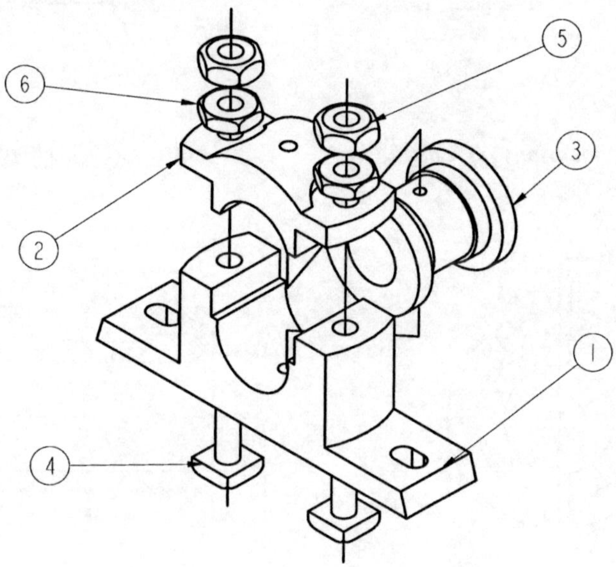

Figure 9-66 *The exploded state of the Pedestal Bearing*

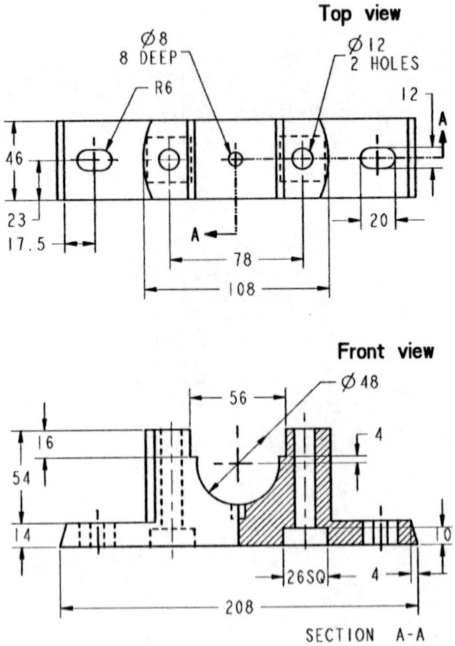

Figure 9-67 Dimensions for Casting

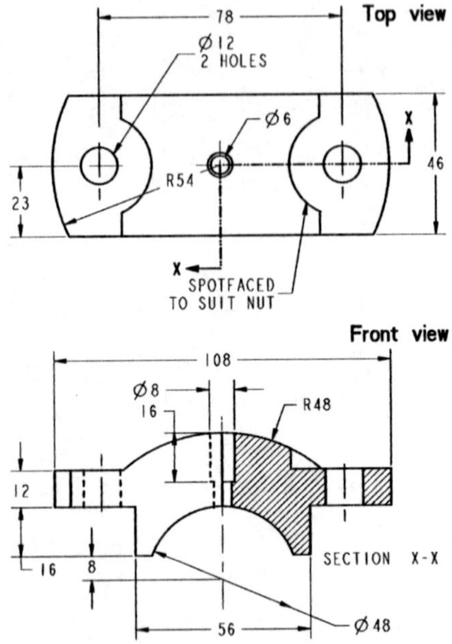

Figure 9-68 Dimensions for Cap

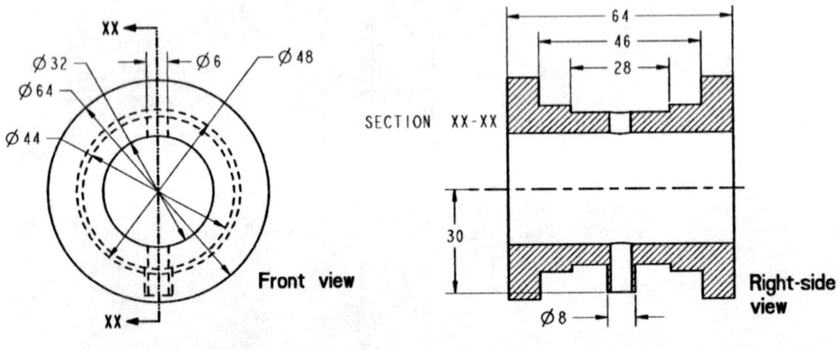

Figure 9-69 Dimensions for the Brasses

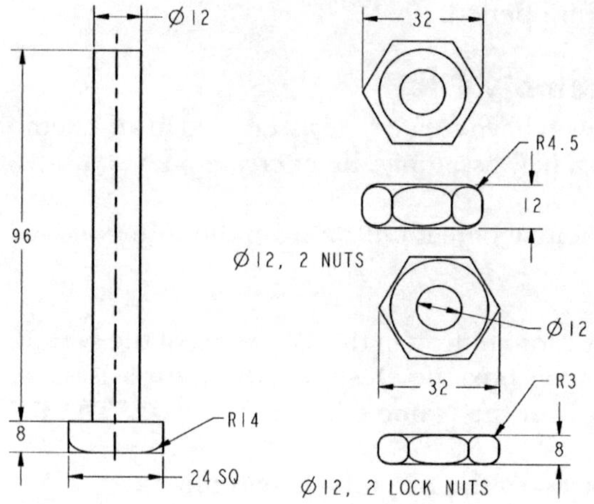

Figure 9-70 Dimensions for the components of the assembly

The following steps outline the procedure to create the given assembly:

a. Create all the components of the assembly as separate part files in the Part mode.

b. Create a new file in the Assembly mode and then assemble the Casting with the default assembly datum planes.

c. Assemble the Cap with the Casting, see Figure 9-72.

d. Suppress the Cap from the assembly.

e. Assemble the Brasses with the Casting, see Figure 9-74.

f. Insert Square headed bolt, nut, and lock nut in the assembly, see Figure 9-75.

g. Resume the suppressed components.

h. After assembling all the components, create the exploded state of the assembly (Figure 9-76).

i. Save the assembly and then exit the Assembly mode.

Set the working directory to *C:\ProE-WF\c09\Pedestalbearing*.

Creating the Components for Assembly

To create the assembly all the components must be created first and then they will be assembled at their respective positions. All the components will be created in the Part mode. You will use the bottom-up approach of assembly creation.

1. Create all the components of the assembly as separate files and save them in the current working directory.

2. Close the part files, if opened.

Creating New Assembly File

All the components that you have created are assembled in an assembly file having an extension *.asm*. You need to open a new assembly file to create an assembly of all the components.

1. Choose the **Create a new object** button from the **File** toolbar to display the **New** dialog box.

2. Select the **Assembly** radio button in the **Type** area of the **New** dialog box. In the **Sub-type** area of the **New** dialog box the **Design** radio button is selected by default. Enter the name of the assembly in the **Name** edit box as **PEDESTALBEARING**.

3. Choose **OK** to proceed to the Assembly environment.

Assembling the Casting with the Default Datum Planes

When you choose **OK** from the **New** dialog box, the three default assembly datum planes are displayed on the graphics window and the **Model Tree** is displayed on the left of the graphics window. If the display of the **Model Tree** was turned off in the previous tutorial, it will not appear. Now, you can start assembling the components.

1. Choose the **Add component to the assembly** button to display the **Open** dialog box.

2. Select **casting** from the **Open** dialog box and choose the **Open** button. The **Component Placement** dialog box is displayed with the **Show component in a separate window while specifying constraints** button selected by default.

   Since, the **Show component in a separate window while specifying constraints** button was selected in the previous tutorial, therefore, the Casting is placed in the **Component Window**.

   In this tutorial, you will use the display mode in which the component is displayed with the assembly in the same window.

Note

*If the datum planes are displayed on the graphics screen, you need to turn their display off by choosing the **Datum planes on/off** button from the **Datum Display** toolbar. This is because the datum planes clutter the graphics window and then the selection of the references for applying assembly constraints becomes difficult. You can turn their display on when the datum planes are required.*

3. Choose the **Show component in the assembly window while specifying constraints** button. Now, choose the **Show component in a separate window while specifying constraints** button to clear it.

4. Select the **Assemble component at a default location** button from the **Component Placement** dialog box. 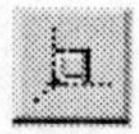

5. The **Placement Status** area shows **Fully Constrained**. Choose **OK**. The **Component Placement** dialog box is closed and the Casting is assembled with the default datum planes.

Assembling the Cap with the Casting

First the Cap will be assembled with the Casting.

1. Choose the **Add component to the assembly** button to display the **Open** dialog box.

2. Select the **cap** and choose **Open** to display the **Component Placement** dialog box.

3. Select the **Mate** option from the **Type** drop-down list. You are prompted to select a mating surface.

4. Select the top face of the Casting as the first mating face. Now, use the CTRL+ALT+middle mouse button to spin the Cap and then select the lower flat face of the Cap as the other mating face as shown in Figure 9-71.

   The CTRL+ALT+middle mouse button is used to spin the component to be assembled with the assembly and the CTRL+ALT+left mouse button is used to move the component. However, to spin the assembly you can use the middle mouse button.

5. Enter a value of **4** in the **Message Input Window**. If this window is not displayed then select the **Offset** drop-down list and select **0.0**. Now, enter offset distance of **4** and press ENTER.

6. Choose the **Specify a new constraint** button and then select the **Align** option from the **Type** drop-down list.

7. Select the axis of the left hole on the Casting as the first aligning axis and then select the axis of the left hole on the Cap as the second aligning axis as shown in Figure 9-71.

   Note that in the **Component Placement** dialog box, the **Allow Assumptions** check box is selected and the Cap is assembled with the Casting. The status displayed in the **Placement Status** area is **Fully Constrained**. This means that Pro/ENGINEER assumed some constraints and has assembled the two components. Although, you can now confirm the assembly by exiting the **Component Placement** dialog box, you will not accept the default constraints. You will define one more constraint to assemble the two components. However, while assembling the previous component if you cleared the **Allow Assumptions** check box then it will not be selected and the status will be **Partially Constrained**.

8. Clear the **Allow Assumptions** check box by selecting it. The status displayed in the **Placement Status** area is **Partially Constrained**.

The Cap is still assembled with the Casting. To know which degree of freedom of the Cap with the Casting is not constrained, perform the next step.

9. Use CTRL+ALT+middle mouse button to move the Cap. You will notice that the second hole of both the Cap and the Casting needs to be aligned.

10. If the **Automatic** option is not displayed in the **Type** drop-down list, choose **Specify a new constraint** button from the **Constraints** area in the **Component Placement** dialog box. The **SELECT** dialog box is displayed. Select the **Align** option from the **Type** drop-down list.

11. Align the axis of the right hole on the Casting with the axis of the right hole on the Cap by selecting them as shown in Figure 9-71. The **Placement Status** area shows **Fully Constrained**. Choose **OK** to exit the **Component Placement** dialog box.

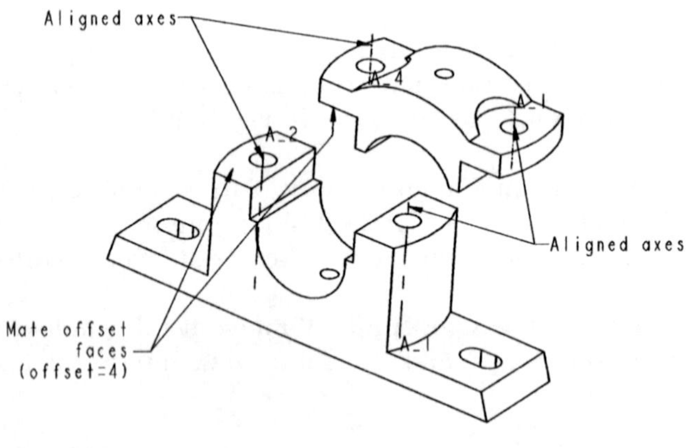

Figure 9-71 *Constraints used for assembling the components*

12. After assembling the Cap, the assembly should look similar to the one shown in Figure 9-72.

Suppressing the Cap from the Assembly

You need to suppress the Cap because the next component has to be assembled with the Casting. When you suppress the Cap from the assembly, it becomes easier to assemble the new component in the assembly. The Cap will be unsuppressed later.

1. Select the Cap from the assembly on the graphics window and right-click to display the shortcut menu. Choose the **Suppress** option. You are prompted to confirm the suppression. Choose **OK**.

Assembling the Brasses with the Casting

Next, you need to assemble the Brasses with the Casting.

1. Choose the **Add component to the assembly** button to display the **Open** dialog box.

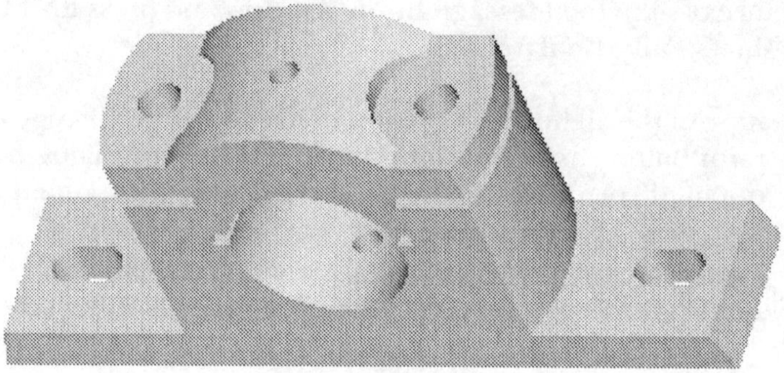

Figure 9-72 *After assembling the Cap with the Casting*

2. Select **brasses** and choose **Open** to display the **Component Placement** dialog box.

3. Select the **Mate** option from the **Type** drop-down list and mate the face of the Brasses with that of the Casting as shown in Figure 9-73.

4. Select the **Insert** option from the **Type** drop-down list and insert the oil hole of the Brasses into that of the Casting as shown in Figure 9-73.

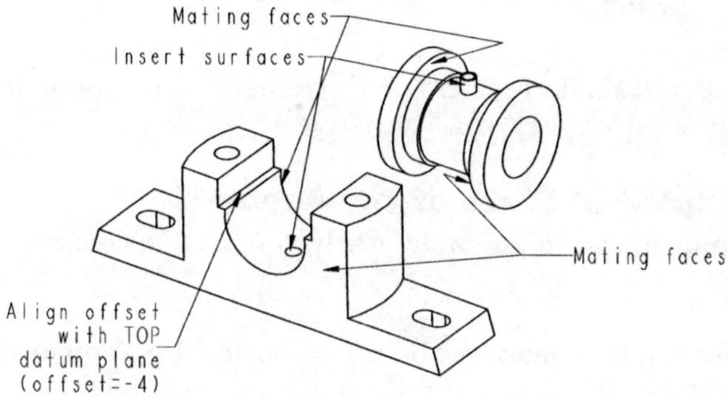

Figure 9-73 *Constraints used for assembling the components*

5. Use CTRL+ALT+left mouse button to move the Brasses and then select the **Align** option from the **Type** drop-down list.

6. Select the **TOP** datum plane from the Brasses and then select the face of the Casting shown in Figure 9-73. The red arrow is displayed and you are prompted to enter the

offset distance in the direction of the arrow.

7. Enter a distance of **-4** in the **Message Input Window** and press ENTER. The **Placement Status** area shows **Fully Constrained.**

8. If the join feature of the oil hole is at the top then choose the **Change orientation of the constraint** button from the **Component Placement** dialog box. The oil hole is now oriented correctly. Choose **OK** to exit the **Component Placement** dialog box.

After assembling the Brasses, the assembly should look similar to the one shown in Figure 9-74.

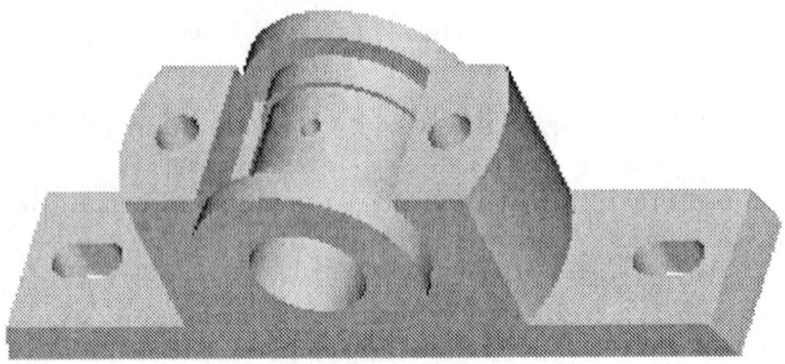

Figure 9-74 *Inserting the Brasses in the assembly*

9. Resume the Cap and similarly assemble the remaining components. The final assembly is shown in Figure 9-75.

Creating the Exploded State of the Assembly

To view all the components in an assembly clearly, you need to create an exploded state of the assembly.

1. Choose the **Start the view manager** button from the **Top Toolchest**. The **View Manager** dialog box is displayed. Choose the **Explode** tab and right-click in the dialog box to invoke the shortcut menu. Choose the **New** option, name it **EXP1**, and press the middle mouse button.

2. Right-click on the **EXP1** in the dialog box and choose the **Redefine** option. The **MOD EXPLODE** menu is displayed. Choose the **Position** option from the menu; the **Explode Position** dialog box is displayed.

   The **Translate** radio button is selected by default in the **Motion Type** area of the **Explode Position** dialog box. The **Entity/Edge** option is selected by default in the **Motion Reference**

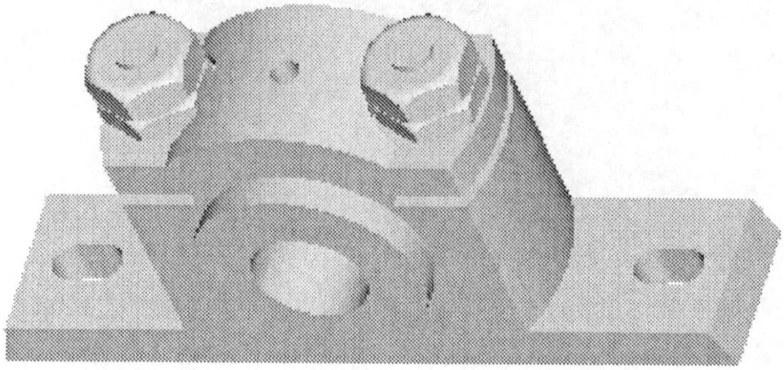

Figure 9-75 The final Pedestal Bearing assembly

drop-down list. The **Smooth** option is selected by default in the **Translation** drop-down list. The **SELECT** dialog box is displayed and you are prompted to select an axis or edge as the motion reference.

3. In the **Filter** drop-down list available in the status bar that is below the **Message Area**, select the **Axis** filter and then select the central axis of the Square headed bolt to specify the direction of motion of the components and then move the Square headed bolts, Nuts, Cap, Lock Nuts, and Casting as shown in Figure 9-76.

4. After moving the above components along the axis of the Square headed bolt, choose the arrow button in the **Motion Reference** area.

5. Select the axis of the Brasses and select it to move as shown in Figure 9-76.

6. After moving the components, choose **OK** from the **Explode Position** dialog box. The **MOD EXPLODE** menu is displayed.

7. Choose the **Offset Lines** option from the **MOD EXPLODE** menu. The **OFFSET LINES** submenu is displayed.

8. Choose the **Create** option from the **OFFSET LINES** submenu. The **Entity Select** submenu is displayed and the **Axis** option is selected by default.

9. Select the axes of the two Nuts and the Square headed bolts to create the offset lines as shown in Figure 9-77.

10. Now, spin the model and select the axis of the oil hole on the Cap from the lower curved surface and then select the axis of the upper oil hole on the Brasses to create the offset lines between them. Similarly, create the offset lines between the lower oil hole on the Brasses and that on the Casting as shown in Figure 9-77.

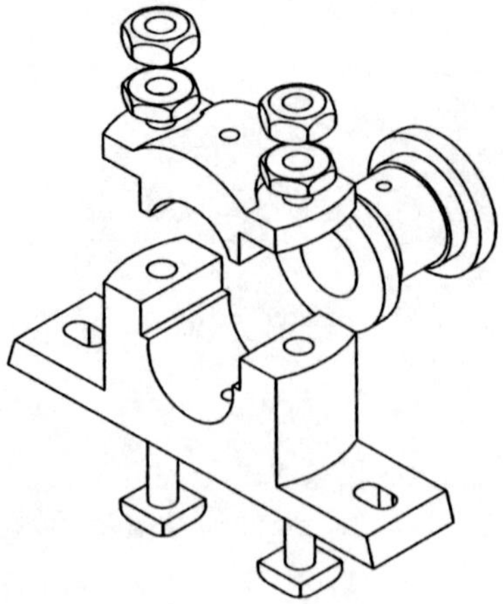

Figure 9-76 *The Exploded Pedestal Bearing assembly*

Note

You can delete or modify an offset line by using the ***Delete*** *and* ***Modify*** *options respectively from the* ***OFFSET LINES*** *submenu.*

11. The final assembly after creating the offset lines in the exploded state should look similar to the one shown in Figure 9-77.

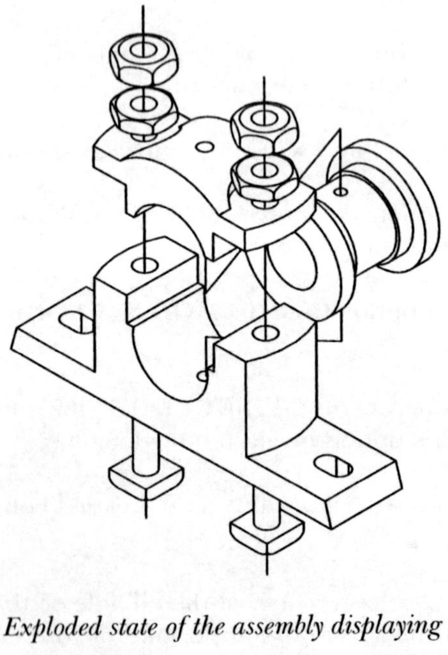

Figure 9-77 *Exploded state of the assembly displaying the offset lines*

12. After creating all the offset lines, choose **Done/Return** from the **OFFSET LINES** submenu and then choose **Done/Return** from the **MOD EXPLODE** menu. Choose the **Close** button to exit the **View Manager** dialog box.

Saving the Assembly

You need to save the assembly because you may need it later.

1. Choose the **Save the active object** button from the **File** toolbar and save the assembly.

Closing the Window

Now, you have saved the assembly and the window can be closed.

1. Choose the **Close** option from the **Window** menu in the menu bar.

Self-Evaluation Test

Answer the following questions and then compare your answers to the answers given at the end of this chapter.

1. When you select a datum plane for applying the mate constraint then you will have to specify whether you want to select the cream side or the brown side of the datum plane to be coplanar with the other face or datum plane. (T/F)

2. The exploded view is a state in which all the components are moved from their actual location so that they are visible. (T/F)

3. The dimensions of a feature cannot be modified in the Assembly mode. (T/F)

4. You can display the unexploded state of the assembly by choosing **Unexplode** from the **View** menu. (T/F)

5. The offset lines are automatically made invisible once you display the unexploded state of the assembly. (T/F)

6. In the ____________ type of simplified representation the selected component will be displayed as shaded.

7. The _____________ type of simplified representation reduces the regeneration time in case of large assemblies.

8. The ___________ constraint is similar to the **Mate** constraint, with the only difference being that this constraint allows you to specify some offset distance between the two coplanar faces.

9. The _________ constraint is used to assemble the circular components.

10. The __________ is the tabular representation of all the components of the assembly along with the information associated with them.

Review Questions

Answer the following questions:

1. Which constraint is used to align the selected datum point or vertex on the first part with the selected surface or datum plane on the second part?

   (a) **Orient** (b) **Pnt on Srf**
   (c) **Edge on Srf** (d) **Tangent**

2. When the assembly is constrained fully the message displayed in the **Placement Status** area of the **Component Placement** dialog box is

   (a) **Partially constrained** (b) **Fully constrained**
   (c) **Completely constrained** (d) None of the above

3. In which of the following types of simplified representation the complete geometry of the components is displayed in the current model display style?

   (a) Geometry representation (b) Graphics representation
   (c) Master representation (d) None of the above

4. Which constraint is used to assemble two components by making the selected faces or planes coplanar such that the mating faces or planes are facing in the same direction?

   (a) **Insert** (b) **Orient**
   (c) **Tangent** (d) **Align**

5. The unexploded state of an assembly can be displayed by choosing **Unexplode** from which of the following menu in the menu bar?

   (a) **View** (b) **File**
   (c) **Info** (d) **Window**

6. A component of an assembly that is assembled using the assembly constraints cannot be replaced with some other component. (T/F)

7. If you want some component of an assembly to disappear from the display it can be achieved. (T/F)

8. Offset lines can only be created in the exploded state of an assembly. (T/F)

9. The order of assembling the components in an assembly can be changed. (T/F)

10. Pro/ENGINEER provides you with a readymade BOM that can be directly used. (T/F)

Exercise

Exercise 1

In this exercise you will create all the components of the Crosshead assembly and then assemble them as shown in Figure 9-78. Also you will create an exploded state, shown in Figure 9-79, displaying the offset lines. The dimensions of the components are shown in Figures 9-80 through 9-84. **(Estimated time: 2 hrs)**

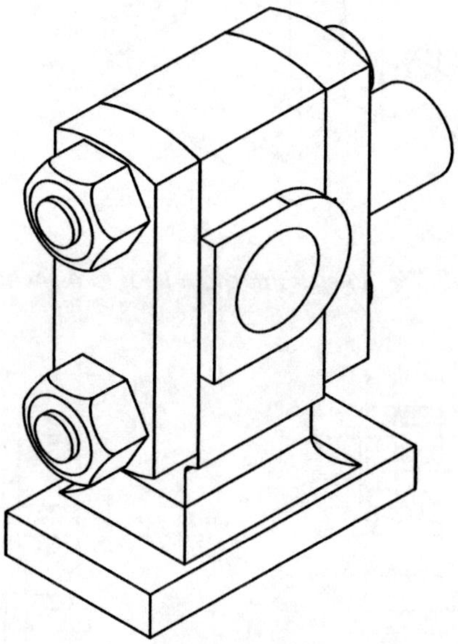

Figure 9-78 *The Crosshead assembly*

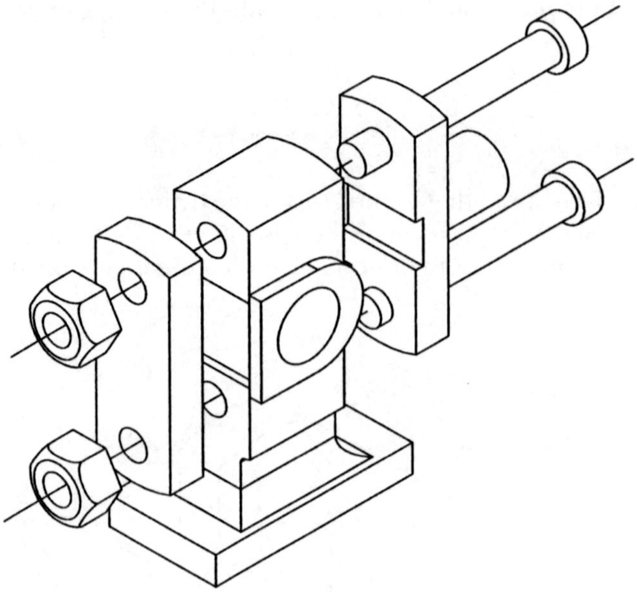

Figure 9-79 The exploded state of Crosshead assembly

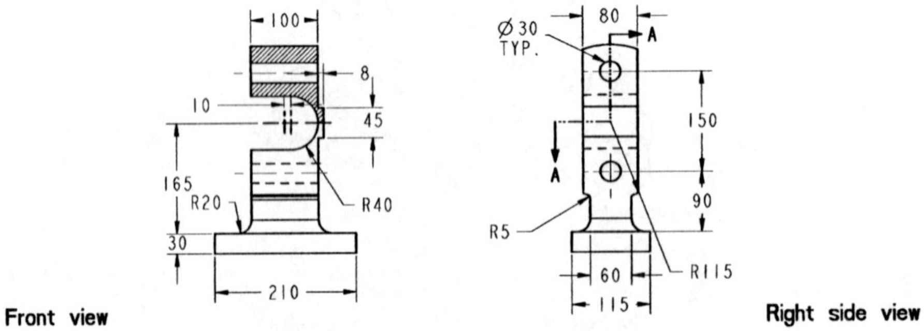

Figure 9-80 Front view and the right-side view of the Body

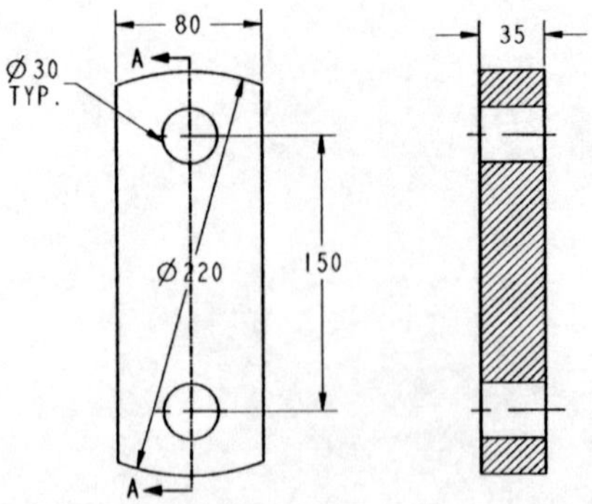

Figure 9-81 Dimensions of Keep Plate

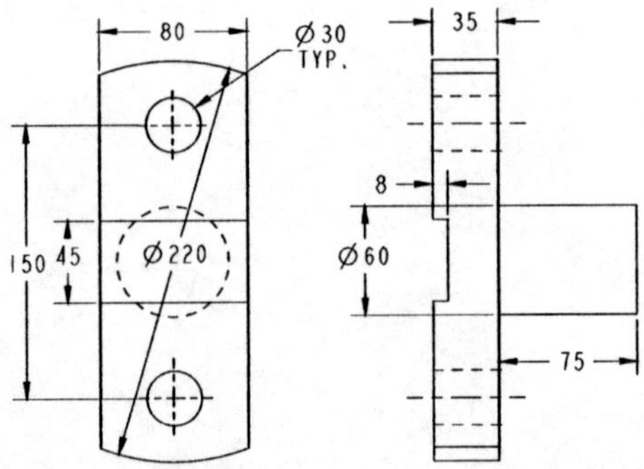

Figure 9-82 Dimensions of Piston Rod

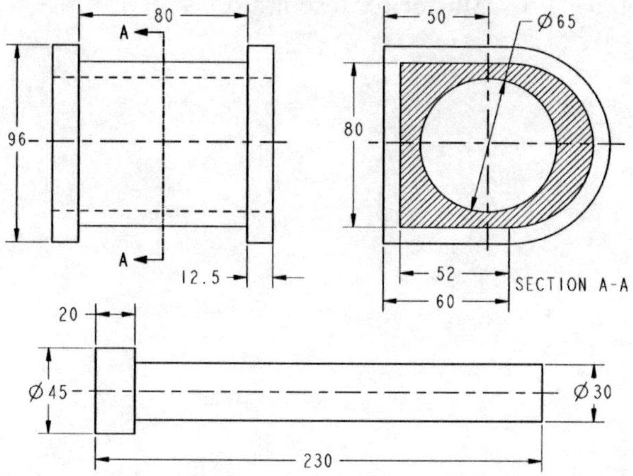

Figure 9-83 *Dimensions of the Brasses and Bolt*

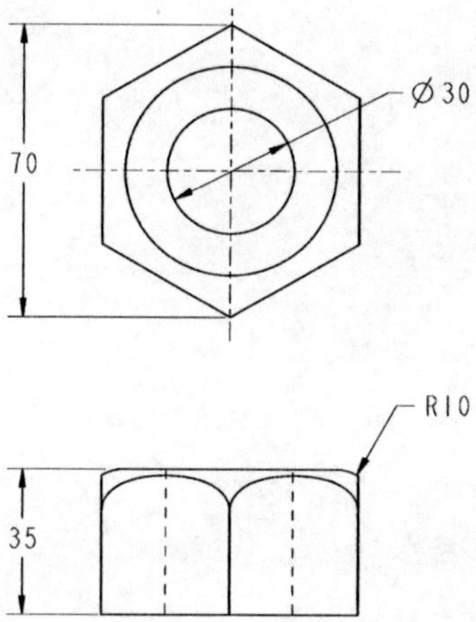

Figure 9-84 *Dimensions of Nut*

Answers to the Self-Evaluation Test

1 - F, **2** - T, **3** - F, **4** - T, **5** - T, **6** - Master representation, **7** - Graphics representation, **8** - Mate offset, **9** - Insert, **10** - BOM.

Chapter 10

Generating, Editing, and Modifying the Drawing Views

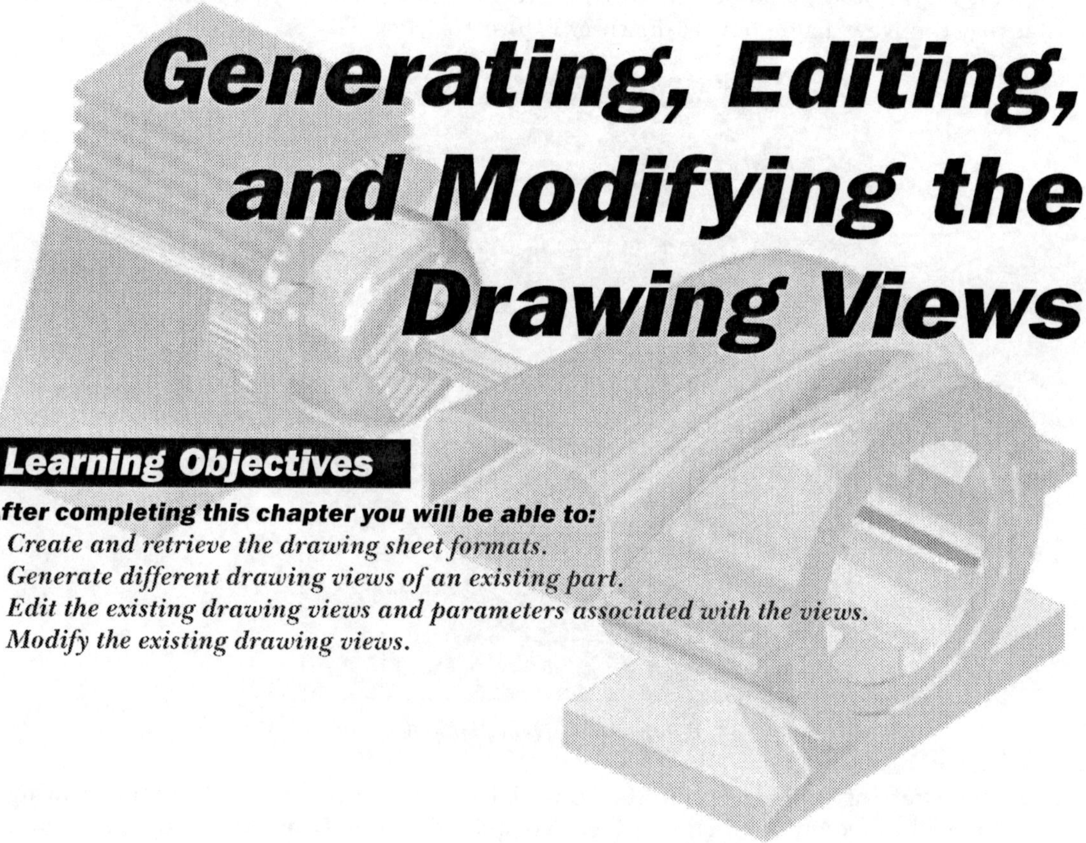

Learning Objectives

After completing this chapter you will be able to:

- *Create and retrieve the drawing sheet formats.*
- *Generate different drawing views of an existing part.*
- *Edit the existing drawing views and parameters associated with the views.*
- *Modify the existing drawing views.*

THE DRAWING MODE

In the previous chapters you have learned about creating the parts in the Part mode and assembling different parts in the Assembly mode. One of the major advantages of working with this software package is its bidirectional associative nature. This is the property that ensures that if any modifications are made in the model in the Part mode, its drawing views are updated automatically and if any modifications are made in the dimensions of the drawing in the drawing views, the model gets updated automatically. In Pro/ENGINEER there are two types of drafting methods: Interactive drafting and Generative drafting. In this chapter you will learn about generating the drawing views of the parts or the assemblies that you have created. In Pro/ENGINEER, the drawing views can be generated from the solid model or assembly in the Drawing mode. Choose the **Create a new object** button from the **File** toolbar to display the **New** dialog box as shown in Figure 10-1.

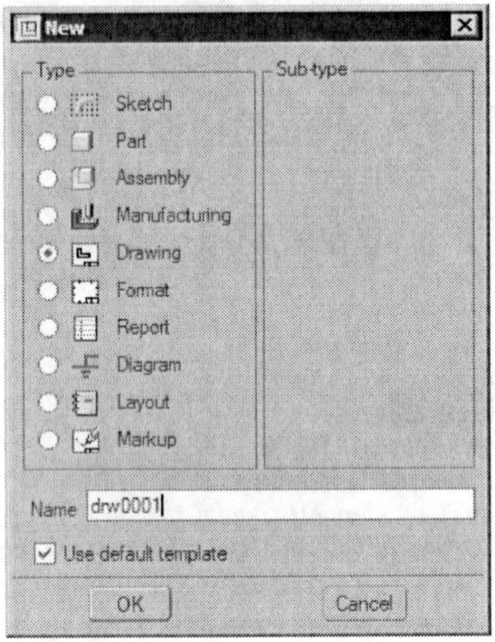

***Figure 10-1** New dialog box*

Select the **Drawing** radio button in the **New** dialog box. Specify the name of the drawing in the **Name** edit box and then choose **OK** to display the **New Drawing** dialog box shown in Figure 10-2.

New Drawing Dialog Box Options

Default Model Area

The **Default Model** area is used to specify the name of the model whose drawing views you want to generate. You can specify the name of the model in the **Name** edit box or select the model using the **Open** dialog box that is displayed when you choose the **Browse** button. If a model exists in another window then the name of the model is displayed by default in the **Name** edit box of the **Default Model** area.

Specify Template Area

The **Specify Template** area is used to specify whether you want to use the default templates available in Pro/ENGINEER, use predefined formats, or use an empty sheet. There are three radio buttons in this area. The **Use template** radio button is selected by default. When you select the **Empty with format** radio button, the dialog box changes as shown in Figure 10-3.

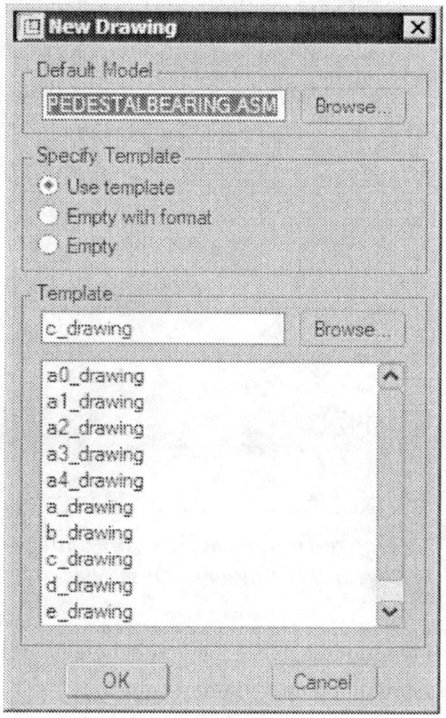

***Figure 10-2** **New Drawing** dialog box with the **Use template** radio button selected*

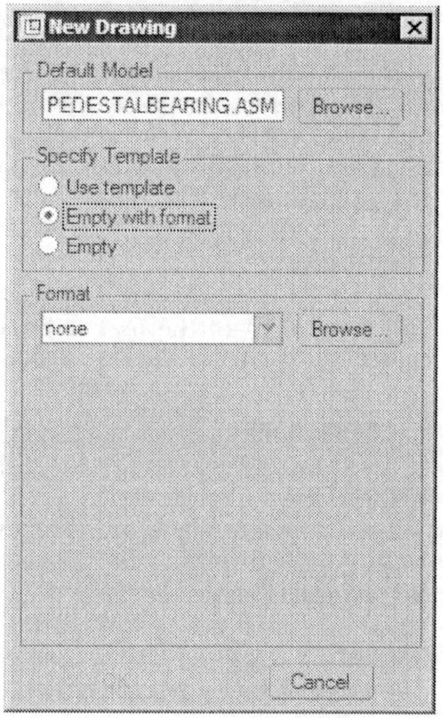

***Figure 10-3** **New Drawing** dialog box with the **Empty with format** radio button selected*

When you select the **Empty** radio button, the **New Drawing** dialog box is modified as shown in Figure 10-4.

Orientation Area

The **Orientation** area is available only when you select the **Empty** radio button from the **Specify Template** area. The buttons provided in this area are used to specify the orientation of the sheet. You can select a standard size sheet with a portrait or a landscape orientation using the **Portrait** or the **Landscape** button. You can also specify a sheet with the user-defined size by choosing the **Variable** button. The size of the sheet and its units can be set using the **Size** area.

Size Area

The **Size** area is available only when you select the **Empty** radio button from the **Specify Template** area. The drop-down list in this area is used to specify the sheet size.

Standard Size
This drop-down list is used to select a drawing sheet of standard size. This drop-down list is available only when you select the **Portrait** or **Landscape** button from the **Orientation** area.

Inches/Millimeters
These radio buttons are selected to set the standards for the user-defined sheets. You can set the size of the sheet in inches or in millimeters. These buttons are available only when you select the **Variable** button from the **Orientation** area.

Width/Height
These edit boxes are used to specify the width and the height of the user-defined drawing sheets. These edit boxes are available only when you select the **Variable** button from the **Orientation** area.

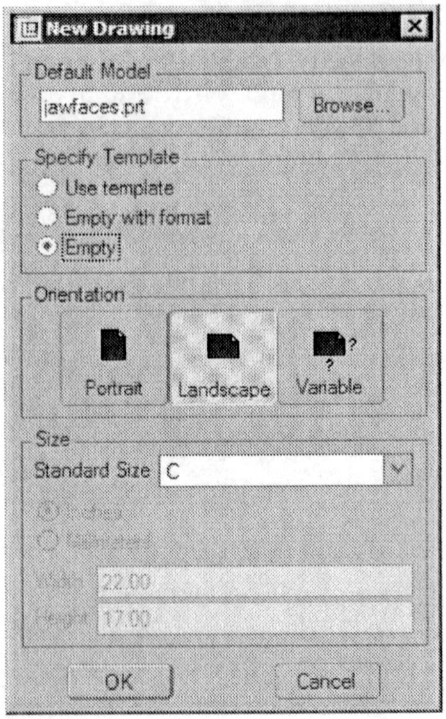

***Figure 10-4 New Drawing** dialog box with the **Empty** radio button selected*

Format Area

The **Format** area is available only when you select the **Empty with format** radio button from the **Specify Template** area.

Format
The **Format** drop-down list is used to select the available formats.

Browse
The **Browse** button is chosen to display the **Open** dialog box for retrieving the drawing formats. By default, there are only five standard system formats that can be retrieved. However, you can create your own user-defined formats that can be retrieved later.

Choose **OK** from the **New Drawing** dialog box to proceed to the Drawing mode. A drawing sheet of the specified size and orientation will be placed on which you can now generate the drawing views.

Tip: *Suppose the model in the **Default Model** area of the **New Drawing** dialog box is not selected and you choose **OK** to enter the Drawing mode. When you choose the **Insert a drawing view of the active model** button from the **Top Toolchest**, the **Open** dialog box is displayed and you can select the model from the **Open** dialog box.*

*Pro/ENGINEER automatically selects the model in the **Default Area** of the **New Drawing** dialog box. This model is selected from the memory of Pro/ENGINEER. If the model that is selected automatically is not the desired model then change the model by selecting the **Browse** button in the **New Drawing** dialog box.*

GENERATING THE DRAWING VIEWS

To generate drawing views in Pro/ENGINEER, choose the **Insert a drawing view of the active model** button from the **Top Toolchest**. The **VIEW TYPE** menu is displayed as shown in Figure 10-5. All the views that can be generated by using the options from the **VIEW TYPE** menu are discussed next.

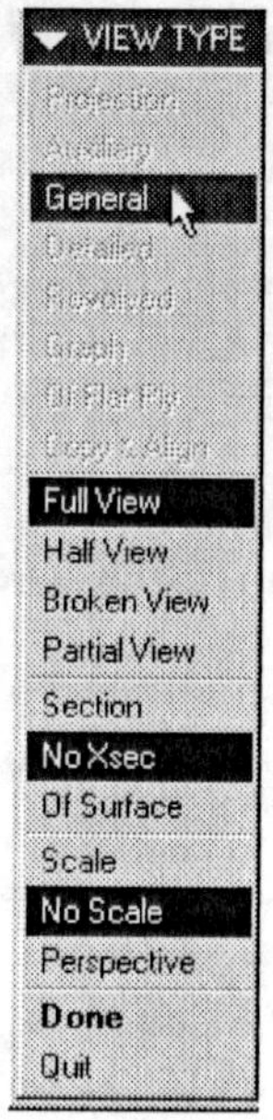

***Figure 10-5** Selecting the **General** option from the **VIEW TYPE** menu*

General View

The **General** view is generated with the help of two references and is similar to orienting the model in the Part mode. This view is also used as the parent view for generating the other views. You can specify any desired scale factor for these views. The **General** option in the **VIEW TYPE** menu and some other options are selected by default as shown in Figure 10-5. The other options are discussed later in the chapter. Accept the default options and choose **Done**. You are prompted to specify the center point for the drawing view. After specifying the center point of the view on the graphics screen, the view is first placed with the default orientation and then the **Orientation** dialog box is displayed as shown in Figure 10-6. You can orient the view using datum planes or the planar faces of the part or use a view saved in the Part mode.

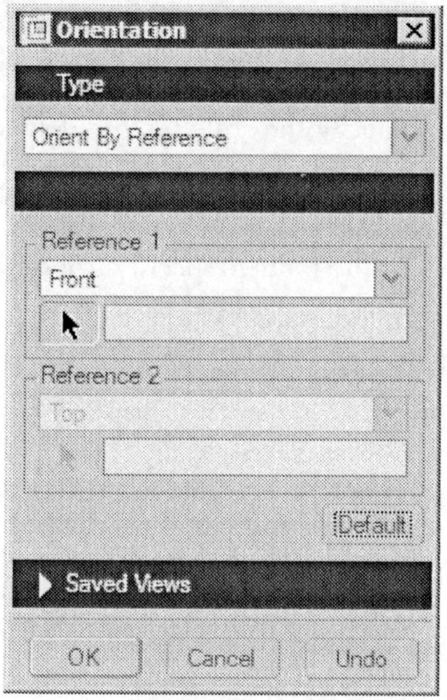

***Figure 10-6** **Orientation** dialog box to orient the **General** view*

Projection View

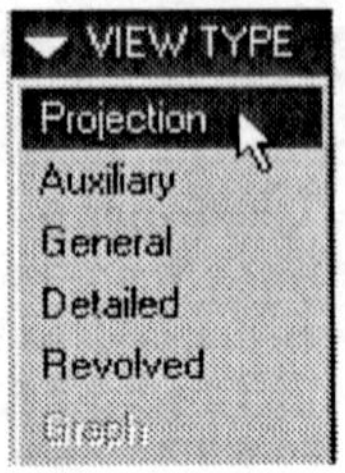

***Figure 10-7** Selecting the **Projection** option from the **VIEW TYPE** menu*

The projection views are the orthographic views generated by projecting the normal lines from an existing general view. The **Projection** option shown in Figure 10-7 is available only after you have generated a general view. The scale factor of these views will be the same as that of the parent view from which they are generated. Depending upon the center point of the projection view with respect to the parent view, the resultant view will be the top, front, left, or right view. If there exists more than one view that can be the parent view of the projection view, then you will be prompted to specify the parent view for the new view. These views move along with the parent view if their parent view is moved from its original location.

Auxiliary View

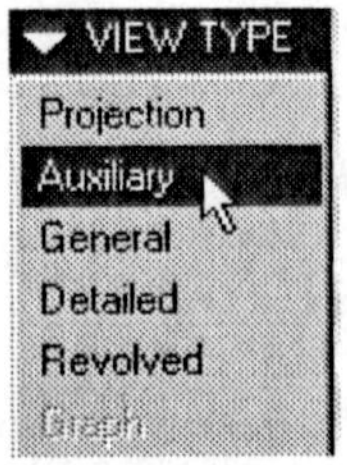

***Figure 10-8** Selecting the **Auxiliary** option from the **VIEW TYPE** menu*

Auxiliary views are mainly used where you want to show the true length of an inclined surface. The auxiliary views are created by projecting normal lines from a specified edge, an axis, or a datum plane in an existing view. The view scale will be the same as that of the parent view. The **Auxiliary** view option is available in the **VIEW TYPE** menu as shown in Figure 10-8.

Detailed View

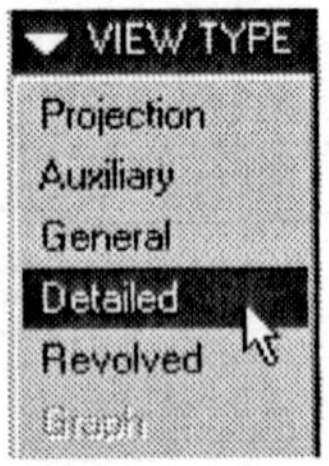

***Figure 10-9** Selecting the **Detailed** option from the **VIEW TYPE** menu*

Detailed views are used to provide the enlarged view of a particular portion of an existing view. When you choose the **Detailed** option from the **VIEW TYPE** menu as shown in Figure 10-9, you will be prompted to specify the scale for the detailed view. You will also be prompted to specify the center point and sketch a spline in the parent view to define the outline of the detailed view. You are also prompted to specify the name of the detailed view, the boundary type, and a point to specify the location for the placement of note.

Figure 10-10 shows the different drawing views.

Revolved View

Figure 10-11 shows the **Revolved** option available in the **VIEW TYPE** menu. These are the section views that are generated from an existing view by revolving the section through an angle of 90 degree about the cutting plane and then projecting it along the length. Remember that the cutting plane of the revolved section views is normal to the viewing plane.

Graph View

Figure 10-12 shows the **Graph** option that is available only when you have created a datum graph in the Part mode of the current drawing. This view is used to display the datum graphs in the drawing views.

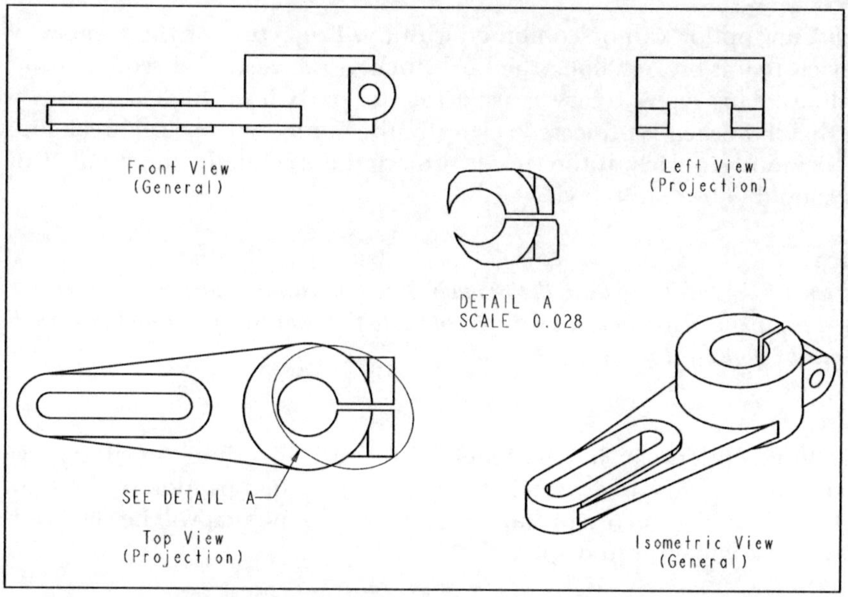

***Figure 10-10** The different drawing views*

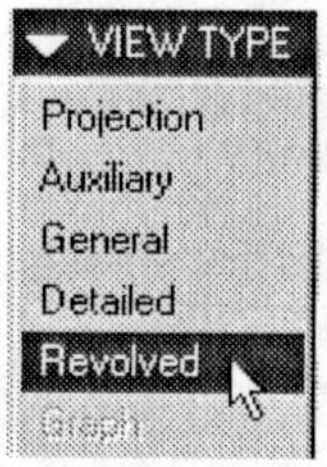

***Figure 10-11** Selecting the **Revolved** option from the **VIEW TYPE** menu*

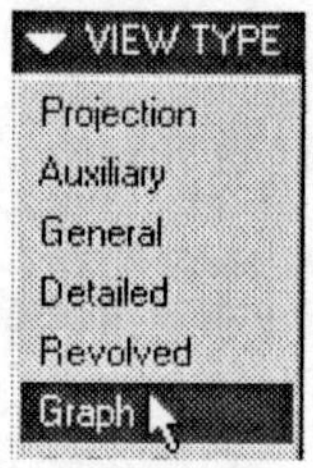

***Figure 10-12** Selecting the **Graph** option from the **VIEW TYPE** menu*

You can also combine these standard view types with some more options to generate some other views. You can use the following options and combine them with the standard view types to control their visibility. The options are discussed next.

Full View

The **Full View** option can be combined with any of the view type to generate a drawing view displaying the complete part.

Half View

The **Half View** option can be combined with the **Projection**, **Auxiliary,** or the **General** view type to create a drawing view that displays only half the part. When you place this view, you are prompted to select a reference plane that will be used to remove half of the drawing view. The reference plane can be a datum plane or a planar surface, and must be perpendicular to the screen in the new view. This option is available only for **Projection**, **Auxiliary**, and **General** views. This type of view is generated for symmetric parts.

Broken View

The **Broken View** option can be combined with the **Projection** or the **General** view type to generate a view that is broken along the horizontal or the vertical direction using horizontal or vertical lines. This type of view is used for the parts have high length to width ratio. Remember that if the view is projected vertically, the number of vertical break lines should be zero in the view and similarly, if the view is projected horizontally, the number of horizontal break lines should be zero in the view.

Note

*Remember that while creating the **Broken** view, you cannot zoom or pan even if the size of the object requires it. Therefore, it is recommended to fit the drawing size on the graphics window as required before creating the **Broken** view.*

Partial View

The **Partial View** option can be combined with the **Projection**, **Auxiliary**, **Revolved**, or **General** view type to generate a view that displays a specified portion of the view. When you choose to place this view, you are prompted to draw a spline that will be the boundary of the portion of the view you want to display.

Figure 10-13 shows the different views.

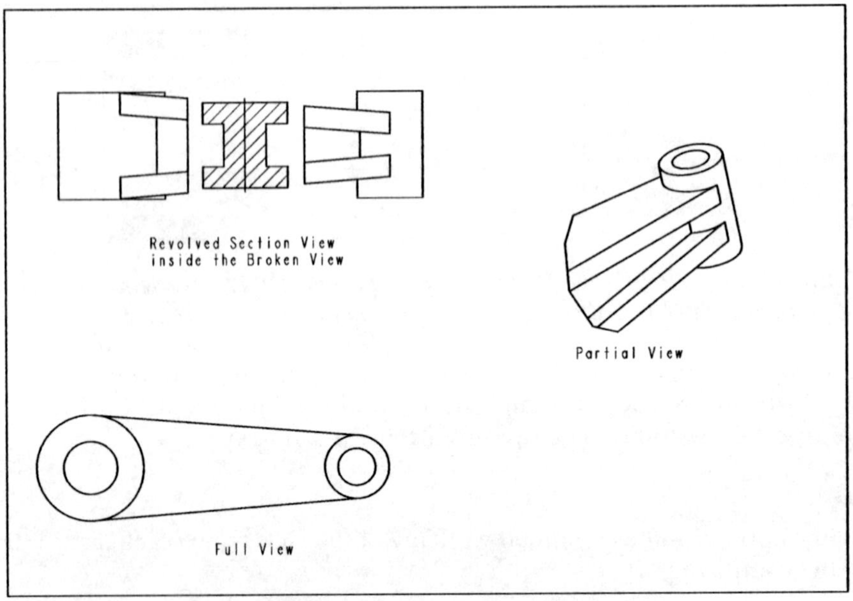

***Figure 10-13** The different drawing views*

SECTIONAL VIEWS

The sectional views are generally used for the views that are complicated from inside. As it is not possible to display the inside of the part using the conventional views, therefore, these views are cut (sectioned) using a datum plane or a planar surface and the resultant section view is displayed. Pro/ENGINEER allows you to generate the following types of section views by choosing them from the **XSEC TYPE** menu shown in Figure 10-14. The **XSEC TYPE** menu is displayed when you choose the **Section** option from the **VIEW TYPE** menu and choose **Done**.

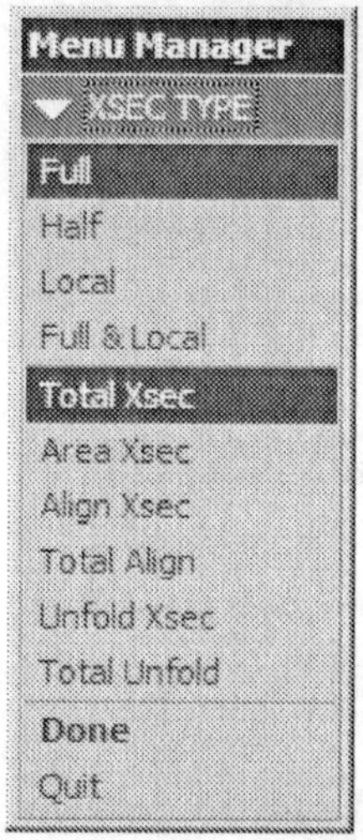

Figure 10-14 XSEC TYPE menu

Full Section View

Consider a part that is cut throughout its length, width, or height and the front cut portion (front half) is removed from the display. The remaining half portion when projected is called a full section view. When you choose to place a full section view, you are prompted to create a new cross section or retrieve an existing cross section. This is the most widely used type of section view. Figure 10-15 shows the full section view.

Half Section View

Consider a part that is cut half way through the length, width, or height and the front cut portion (front quarter) is removed from the display. This part when projected is called the half section view. In this projected view, only half of the part is displayed sectioned and the other half of the part is displayed as it is. You need to specify the half side of the view that has to be sectioned. Remember that this option is available only when you select **Full View** from the **VIEW TYPE** menu. Figure 10-16 shows the half section view.

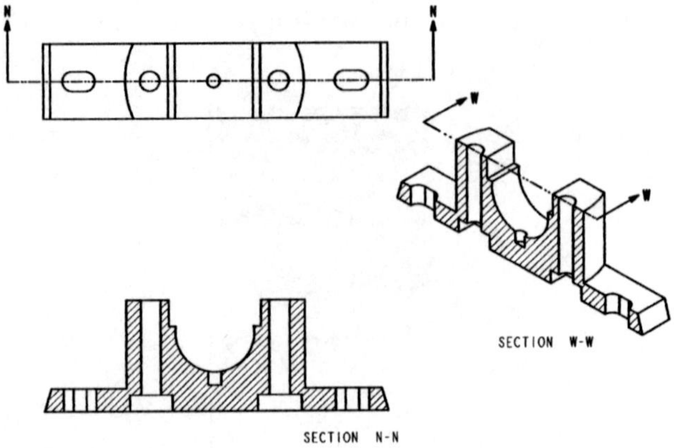

Figure 10-15 *Top view, Full section front view, and Full section isometric view*

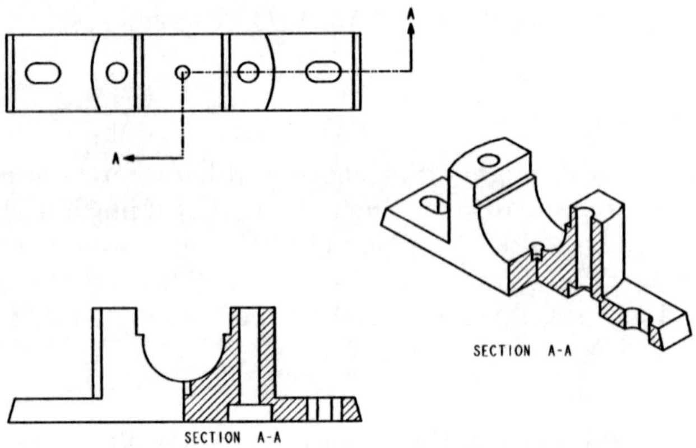

Figure 10-16 *Top view, Half section front view, and Half section isometric view*

Local Section View

The **Local Section** view is used when you want to show a particular portion of the view in section and at the same time not section the remaining view. The local section area is specified by drawing a spline around it. Figure 10-17 shows the local section view.

Note

As evident from Figures 10-15 and 10-17, to create a sectioned isometric view you have to define a section plane other than that used to generate the front section view.

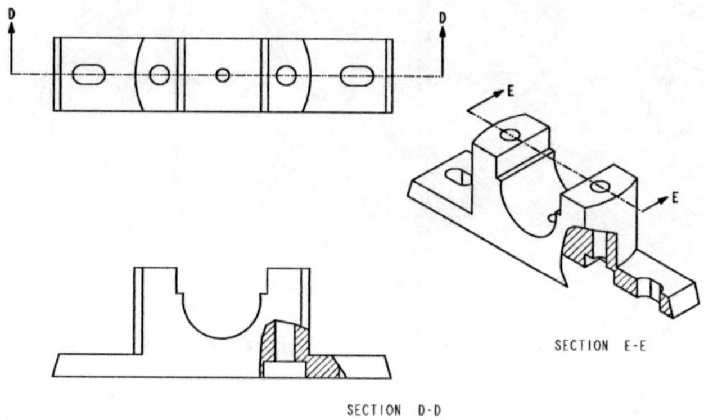

Figure 10-17 Top view, Local section front view, and Local section isometric view

Full & Local

The **Full & Local** view is the combination of the full and local section views. This option allows you to generate a full and a local section view in the same view.

Advanced Sectioning Options

You can combine the above-mentioned section types with other advanced sectioning options to generate the section views. The advanced sectioning options that can be used are discussed next.

Total Xsec

The **Total Xsec** option when selected creates a section view that displays all the visible edges of the section view in addition to the section area. This option can be combined with the full, half, local, or the full and local Xsec types. Figure 10-18 shows the front and total sectioned left view.

Area Xsec

This option when selected displays only that area of the section view that is sectioned. No other edges of the view are displayed in the area cross section view shown in Figure 10-19. This option can be combined with only the full Xsec type.

Align Xsec

These views are used to section those features that are created at a certain angle to the main section planes. Align sections straighten these features by revolving them about an axis that is

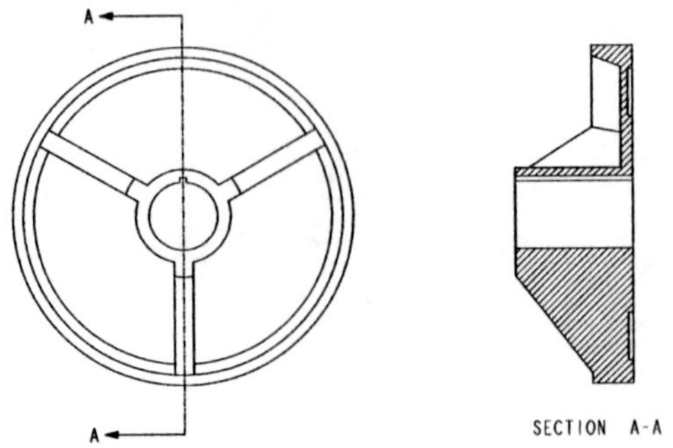

Figure 10-18 *The front view and the Total cross section view*

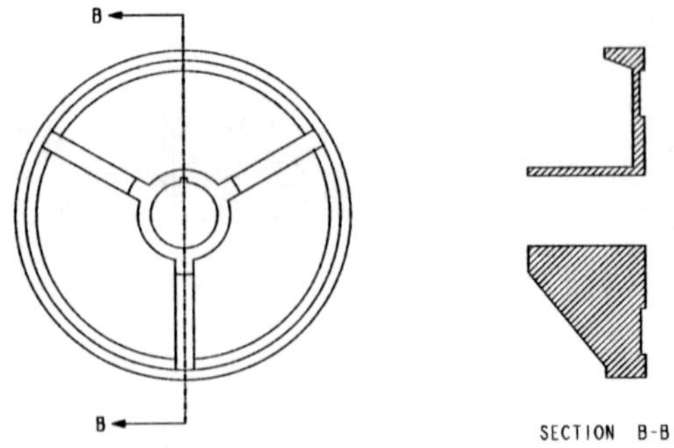

Figure 10-19 *The front view and the Area cross section view*

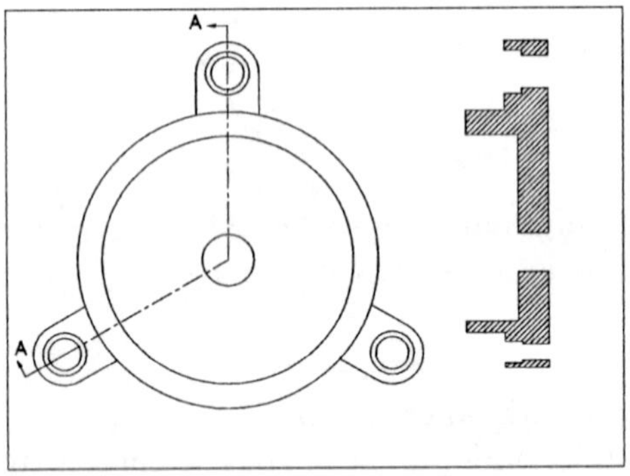

Figure 10-20 *The Area cross section view with normal lines of projection*

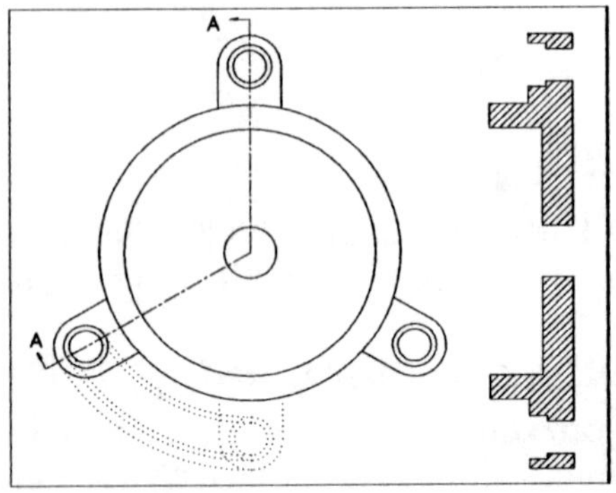

Figure 10-21 *The Align cross section view with aligned lines of projection*

normal to the parent view. Remember that the axis about which the feature is straightened should lie on all the cutting planes. This means that the lines that are used to define the section plane should pass through the axis about which the feature will be revolved. These views are similar to the area section views in which only the portion that is sectioned is displayed. Figures 10-20 and 10-21 show these views.

Total Align

These views are similar to the align section views, the only difference being that these views display all the visible edges of the section view in addition to the section portion as shown in Figures 10-22 and 10-23.

Note

The ***Align Xsec*** *and the* ***Total Align*** *options are available only when you choose the* ***Full*** *option from the* ***XSEC TYPE*** *menu.*

Unfold Xsec

The **Unfold Xsec** option is also used to generate the align section view but only for the **General** view type. Once you have placed the view and specified the cutting plane, you will

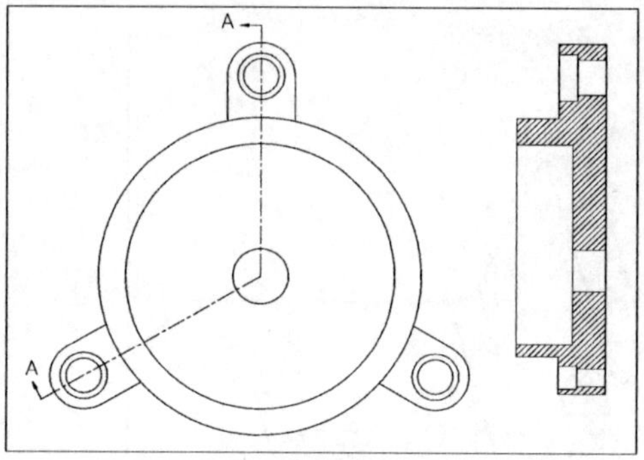

Figure 10-22 *The Total cross section view with normal lines of projection*

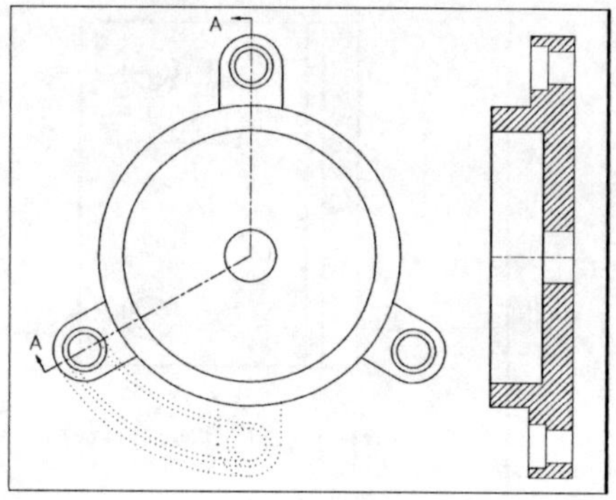

Figure 10-23 *The Total align cross section view with aligned lines of projection*

be prompted to specify the orientation of the section view.

Tip: *You can also create the cross sections in the Part mode and later retrieve those sections in the Drawing mode and use them for generating the section views.*

Total Unfold

The **Total Unfold** option is also used to generate the total align section views but only for the **General** view type.

CREATING THE SECTION PLANES

Once you have selected all the options for generating the section views and specified the location of the section view, you need to specify the cutting plane for the section view. You can either retrieve the cross section created in the Part mode or create the cross section in the Drawing mode by specifying the cutting plane. Pro/ENGINEER allows you to create two types of section planes. These are discussed next.

Planar

The **Planar** option is used to select an existing planar face or the datum plane as the cutting plane. You can also make a datum plane that can be used as the cutting plane. However, remember that you cannot draw a cutting plane using this option.

Offset

The **Offset** option is used to sketch an offset cutting plane using the sketcher options. When you invoke this option, you are prompted to specify the name of the cross section. Once you have specified the name of the cross section, another window appears displaying the part. You can then follow the Part mode options to draw a line that will act as the cutting plane for generating the section views. This option is generally used for creating the aligned section views. Figure 10-24 shows the top view, offset sectioned front view, and the sectioned general view of a part.

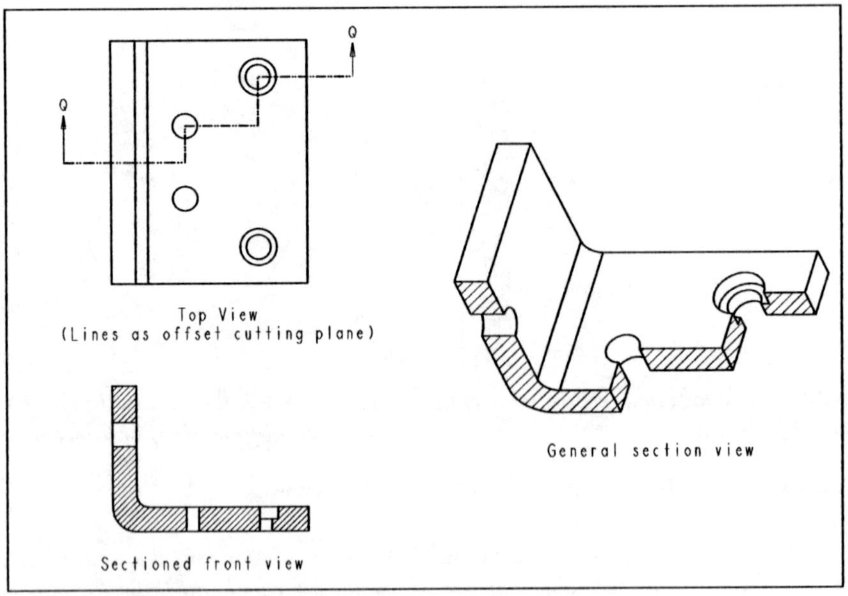

***Figure 10-24** The section views created using the offset section plane*

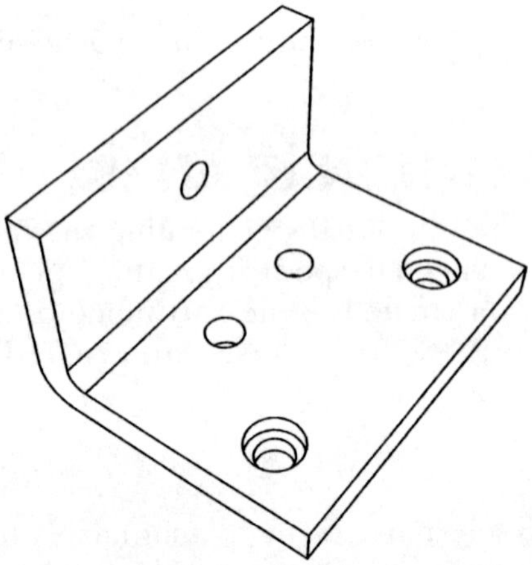

***Figure 10-25** A general perspective view*

PERSPECTIVE VIEWS

Pro/ENGINEER provides you with an option of creating the perspective drawing views of the parts or assemblies. The perspective views are generated using two parameters. These are the eye distance for the view in the model space and the diameter of the view in the paper space. The eye distance in the model space here refers to the distance of the viewer from the actual model in the Part mode and the diameter of the view in paper space refers to the size of the view in the Drawing mode. Figure 10-25 displays a general perspective view. The following points should be remembered while creating the perspective views:

1. You cannot use perspective views to generate projection or detailed views.

2. Perspective views do not have dimensions, datum planes, and axes.

3. Perspective views cannot have cross sections.

4. Only general views can be displayed in perspective.

EDITING THE DRAWING VIEWS

With this release of Pro/ENGINEER it has become very easy to edit a drawing view and the items in the drawing views. All the options of modifying the drawing views or items related to it can be chosen from the shortcut menu that is displayed when you right-click after selecting the view or the item. However, these options can also be invoked from the menus available in the menu bar. Pro/ENGINEER allows you to perform the following types of editing operations on the drawing views.

Note

By default, the drawing views generated are locked and cannot be moved by selecting them. To unlock them so that the views can be moved, choose ***Disallow the movement of drawing views with the mouse*** *button from the* ***Top Toolchest****.*

Moving the Drawing View

When a view is generated its movement is locked by default. To unlock the drawing view, choose the **Disallow the movement of drawing views with the mouse** button from the **Top Toolchest**. When you select the drawing view, it is displayed inside a boundary. To move the drawing view, select it and then drag it to the required location on the drawing sheet. Remember that if you select the view that has some child views then the child views will also move along with the parent view in order to maintain their alignment with the parent view. Also, the projected views can be moved only in the direction of projection.

Note

The ***General*** *and* ***Detailed*** *views can be moved to any new location because they are not the projections of any view.*

Erasing the Drawing View

The **Erase View** option in the **VIEWS** menu is used to temporarily remove the selected drawing view from the sheet. The **VIEWS** menu can be invoked by choosing **View > Drawing Display > Drawing View Visibility** from the menu bar. However, the view still remains in the memory of the drawing and can be resumed at any point of time. As the view is not completely removed from the memory, therefore, you can also erase a view that has some child views associated to it and it will not affect the child view. Once a view is erased, a box is displayed in place of the view displaying the name of the view. To resume the view, choose the **Resume View** option from the **VIEWS** menu.

Note

When you erase a view, the leaders and dimension values that were attached with the view are also erased. When you select the erased view to resume, the leaders and dimension values are redisplayed.

Deleting the Drawing Views

To delete a drawing view from the drawing sheet, select the view and hold down the right mouse button to invoke a shortcut menu. Choose the **Delete** option from the shortcut menu.

Deleting the Drawing Views

The **Delete selected items** button available in the **Top Toolchest** after you select the drawing view to delete is used to permanently remove a selected drawing view from the sheet. Once the view is deleted, no information related to the deleted view remains in the memory of the drawing. Remember that the view that has some child views associated with it cannot be deleted unless all the child views are deleted first. When you select the drawing view to be deleted, you are prompted to confirm the deletion process.

Adding New Parts or Assemblies to the Current Drawing

You can also add more parts or assemblies in addition to the default part or assembly for generating the drawing views. This is done by choosing **File > Properties** from the menu bar. The **FILE PROPERTIES** menu is displayed as shown in Figure 10-26. Choose the **Drawing Models** option, the **DWG MODELS** menu is displayed as shown in Figure 10-27.

Figure 10-26 FILE PROPERTIES menu

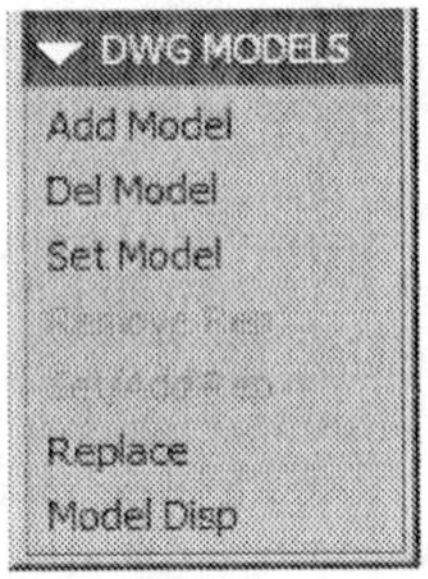

Figure 10-27 DWG MODELS submenu

Choose the **Add Model** option to add a model to the current drawing. When you choose this option, the **Open** dialog box is displayed. You can select the new model to be added using this dialog box. Remember that the latest model added will be the current model and the drawing views that will be generated now will be of this model. You can change the current model by choosing the **Set Model** option from the **DWG MODELS** menu. Similarly, you can also delete a model from the current drawing. However, only the model that does not have any view generated from it can be deleted.

MODIFYING THE DRAWING VIEWS

You can also make the following modifications in the existing drawing views.

Changing the View Type

Select the drawing view that needs to be modified and hold down the right mouse button to invoke the shortcut menu. Choose **Properties** from the shortcut menu. The **VIEW MODIFY** menu is displayed as shown in Figure 10-28. Choose the **View Type** option, the **VIEW TYPE** menu is displayed. Select the type of view from this submenu. It should be noted that only the general, projection, and auxiliary view types can be modified using this option.

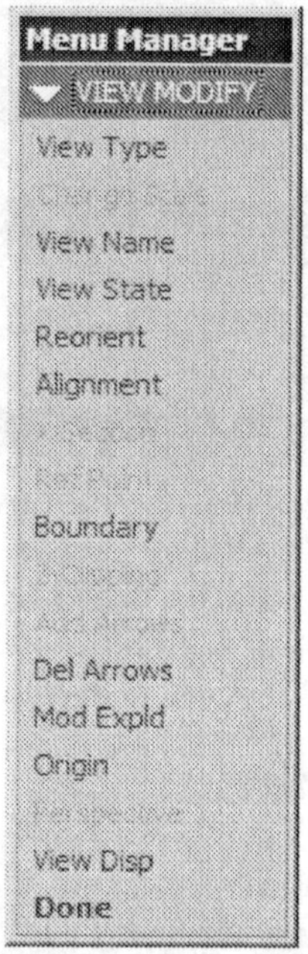

*Figure 10-28 **VIEW MODIFY** menu*

Changing the View Scale

The **Change Scale** option in the **VIEW MODIFY** menu is used to change the scale factor of the selected view. You can modify the scale factor of only those views in which the scale factor has been assigned during generation.

Note

*The scale factor represents the drawing to model scaling. For example, if the scale set is **0.25**, Pro/ENGINEER scales the drawing views to one-quarter (1/4) of the actual size of the model.*

Changing the View Name

The **View Name** option in the **VIEW MODIFY** menu is used to change the name of the selected view. View name appears inside a box when the view is erased using the **Erase** option. You can change this name by using this option. When you invoke this option, you will be prompted to select the view whose name has to be changed.

Reorienting the Views

The **Reorient** option in the **VIEW MODIFY** menu is used to reorient the general view. However, if some child view is associated with the general view then the child view will also be reoriented accordingly. The **Orientation** dialog box will be displayed when you select the general view to reorient. You can use the options from this dialog box to change the orientation of the selected view.

Modifying the Cross Sections

You can flip the side of sectioning and replace, delete, or rename the X sections in the views using the **X-Section** option in the **VIEW MODIFY** menu.

Modifying the Boundaries of Views

The **Boundary** option in the **VIEW MODIFY** menu is used to modify or resketch the boundary of the broken, partial, or the detailed views.

Adding or Removing the Cross Section Arrows

The **Add Arrows/Del Arrows** options in the **VIEW MODIFY** menu are used to add or delete the cross section arrows from the views. In case the arrows are not displayed, you will be prompted to select the section views whose arrows are to be displayed and then the view where the arrows should be displayed. In case the arrows are to be removed, you will be prompted to select the arrows to be deleted.

Modifying the Perspective Views

The **Perspective** option in the **VIEW MODIFY** menu is used to modify the perspective views. You can modify the eyepoint distance or the view diameter using this option.

MODIFYING OTHER PARAMETERS

Apart from modifying the drawing views, you can also modify other parameters related to the drawing views. For example, you can modify the size and the style of the text, scale factor of all the drawing views, cross section hatching, and so on. All this is done by selecting the item. When the item is highlighted in red color, hold down the right mouse button to invoke the shortcut menu. You can select the options from the shortcut menu to modify that item.

You can modify any parameter associated with the drawing views. Also, the selected drawing view can be moved by selecting the view and then moving the mouse. Depending upon the item selected to modify, the options related to it vary. The options in the shortcut menu that are displayed vary from item to item. For example, if you select any text item associated with the drawing views then you can move it from its original location, edit the text, edit text style, and so on. If you select a drawing view and right-click then the modifications that are possible are; changing the location of the view, changing the view scale, renaming the view, and reversing the direction of the sectioning or replacing the cross sectioning (in case of section views).

Modifying the Values

The **Value** option is used to modify any numeric value associated with the drawing views. The numeric value can be the view scale in case it is provided or changing the dimension value. When you change the dimension value, it turns white in color indicating that the dimension is changed. You can also change the scale of all the views in case it is not available by modifying the scale value provided near the lower left corner of the screen. Choose the **Value** option from the **Edit** menu in the menu bar to change the scale factor for all the views in the current sheet.

Editing the Cross Section Hatching

Select the hatching from a drawing view. When the hatching turns red in color, hold down the right mouse button to invoke the shortcut menu. Choose the **Properties** option from the shortcut menu. The **MOD XHATCH** menu is displayed as shown in Figure 10-29.

The parameters related to cross section hatching that can be modified are the spacing of the hatching, angle of the hatching lines, offset value, and the line style of the hatching lines. There are also some standard hatch patterns that are available in Pro/ENGINEER. You can retrieve these standard patterns by using the **Open** dialog box displayed upon choosing **Retrieve** from the **MOD XHATCH** menu.

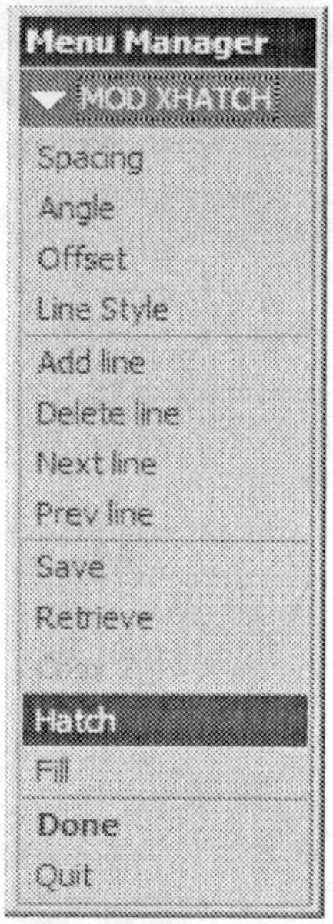

***Figure 10-29** **MOD XHATCH** menu*

TUTORIALS

Tutorial 1

In this tutorial you will generate the drawing views of the model created in Exercise 4 of Chapter 6 shown in Figure 10-30. Select the A4 size drawing sheet and generate all the drawing views shown in Figure 10-31. **(Estimated time: 40 min)**

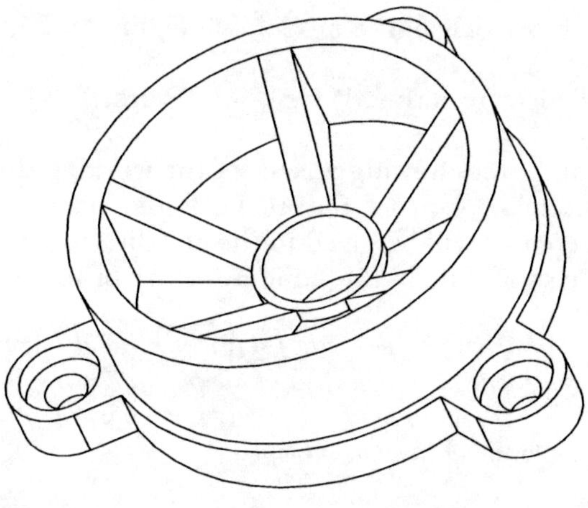

***Figure 10-30** Part for generating the drawing views*

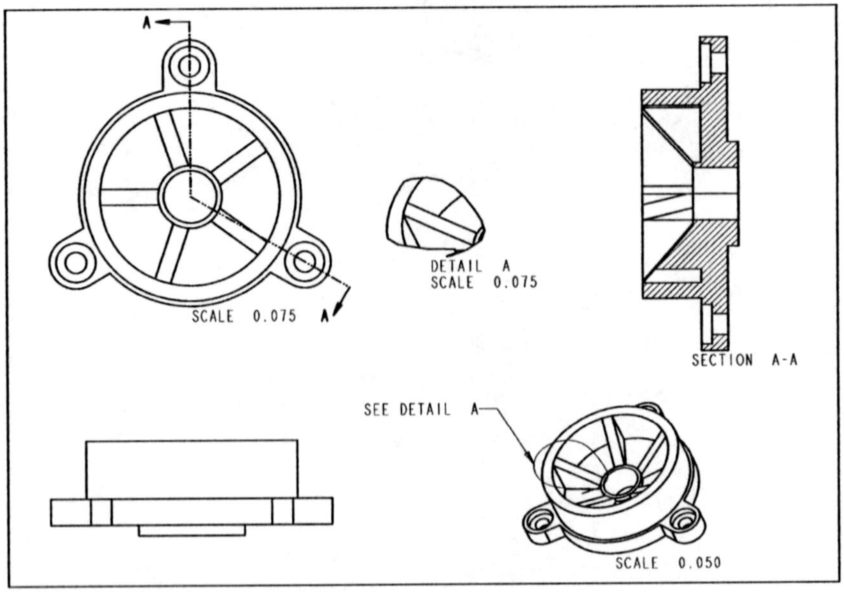

Figure 10-31 *The drawing views to be generated*

The following steps outline the procedure to generate the drawing views of the model:

a. Create a new drawing file and select the required drawing sheet.

b. Generate the top view.

c. Generate the front view taking the top view as the parent view, see Figure 10-34.

d. Generate the sectioned right view of the top view by defining a plane in the model.

e. Generate the default 3D view of the model, see Figure 10-35.

f. Generate the detail view from the 3D view, see Figure 10-36.

Before you start generating the drawing views, set the working directory to *C:\ProE-WF\c10*. Copy and paste the model *c06exr4.prt* file in the folder named *c10* from *c06* folder. The drawing that you will generate will be saved in the *c10* directory with an extension *.drw*. The *.prt* file and the *.drw* file should lie in the same directory or folder.

Tip: *If you delete the part file from which the drawing views are generated, the drawing file will not open. Also, if the part file is removed from the folder where its drawing file is placed or if you rename the part file after generating the drawing views, the drawing file will not open.*

Creating New Drawing File

1. Choose the **Create a new object** button from the **File** toolbar to display the **New** dialog box.

2. Select the **Drawing** radio button and then enter the name of the file as *c10tut1*, see Figure 10-32.

3. Choose **OK** from the **New** dialog box to display the **New Drawing** dialog box.

4. Choose the **Browse** button to select *c06exr4.prt* for generating the drawing views.

5. Select the **Empty** radio button from the **Specify Template** area.

6. Choose the **Landscape** button from the **Orientation** area if it is not chosen by default as shown in Figure 10-33.

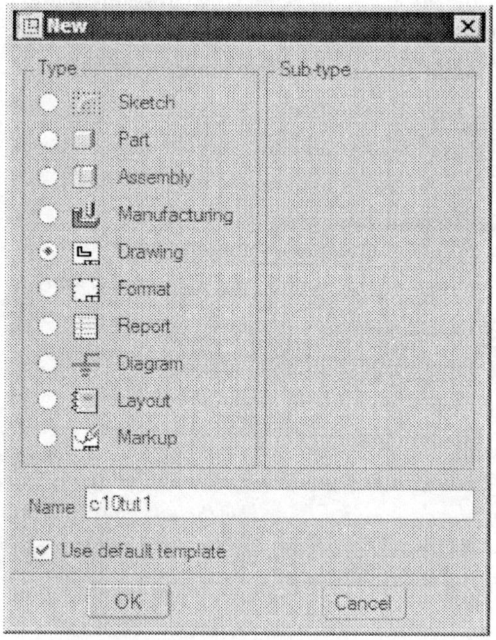

***Figure 10-32 New** dialog box*

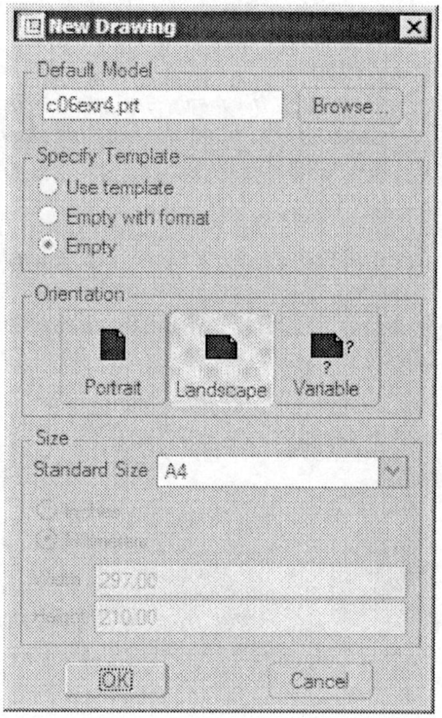

***Figure 10-33 New Drawing** dialog box with the **Empty** radio button selected*

7. Select **A4** from the **Standard Size** drop-down list. Choose **OK** to proceed to the Drawing mode.

   An A4 size sheet with landscape orientation is displayed on the graphics screen.

Generating the Top View

Generally, you can generate any view as the first view on the drawing sheet. However, in this tutorial you will generate the top view as the front view.

1. Choose the **Insert a drawing view of the active model** button from the **Top Toolchest** to display the **VIEW TYPE** menu.

2. Choose **General** > **Full View** > **No Xsec** > **Scale** > **Done**. You will be prompted to specify the center point for the drawing view.

3. Specify the placement point for the first view close to the top left corner of the sheet as shown in Figure 10-34. As soon as you specify the point, the **Message Input Window** is displayed. Enter the scale factor for the view as **0.075** and press ENTER. The default view is placed on the selected point

   The **Orientation** dialog box is displayed.

Note
Although the views once placed can be moved on the sheet, yet while specifying the placement point, try to place the drawing view inside the sheet and also leave space for the other views to be created.

4. From the **Saved Views** drop-down list, select the **TOP** option and choose the **Set** button. The default view is oriented as top view.

5. Choose the **OK** button from the **Orientation** dialog box.

6. Turn off the display of datum planes, axes, points, and coordinate system by choosing their respective buttons from the **Datum Display** toolbar. Also, change the model display to no hidden by choosing the **No Hidden** button from the **Model Display** toolbar.

Tip: *If the view that you have placed on the drawing sheet is not at the proper location on the drawing sheet then you need to move the drawing view. To unlock the movement, choose the* ***Disallow the movement of drawing views with the mouse*** *button from the* ***Top Toolchest****. Select the drawing view using the left mouse button. The selected drawing view is enclosed in an red box. Now, press the left mouse button inside the box and drag it to place it at the desired location.*

Generating the Front View

The front view of the model is generated from the top view that is already placed on the drawing sheet.

1. Choose the **Insert a drawing view of the active model** button from the **Top Toolchest** to display the **VIEW TYPE** menu.

2. **Projection** > **Full View** > **No Xsec** > **No Scale** is selected by default. Choose **Done.**

3. Specify the placement point for the front view below the top view as shown in Figure 10-34. If the **Shading** display is current, the hidden lines in the model are also displayed. Change the display type to **No Hidden** and repaint the screen.

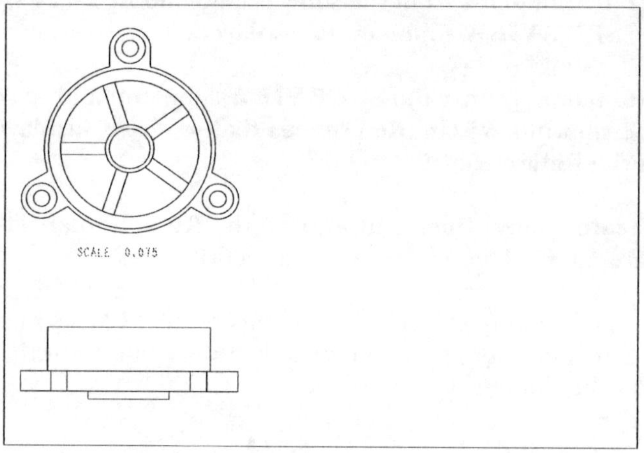

Figure 10-34 *The top and front view*

Generating the Section View

To create section view, a section must be defined on the model. You will sketch this section to generate the section view for this tutorial.

1. Choose the **Insert a drawing view of the active model** button from the **Top Toolchest** to display the **VIEW TYPE** menu.

2. Choose **Projection** > **Full View** > **Section** > **No Scale** > **Done** to display the **XSEC TYPE** menu.

3. Choose **Full** > **Total Align** > **Done** from the **XSEC TYPE** menu.

4. Specify the placement point on the right side of the top view.

5. Choose the **Create** option from the **XSEC ENTER** menu to display the **XSEC CREATE** submenu.

6. Choose **Offset** > **Both Sides** > **Single** > **Done**.

7. Enter the name of the cross section as **X** in the **Message Input Window** and press ENTER. Once you have specified the name of the cross section, a separate window will appear displaying the part. You are prompted to select a sketching plane.

Tip: *It is recommended that you use the* ***Redraw the current view*** *button available in the* ***View*** *toolbar to remove any temporary information on the graphics screen and to refresh the screen.*

8. Choose the **Datum planes on/off** button from the **Datum Display** toolbar in the original window to display the datum planes. You need to repaint the screen in the subwindow by choosing **View > Repaint** from the menu bar to view the datum planes.

9. Select the **TOP** datum plane from the subwindow. Choose **Okay** from the **DIRECTION** submenu. The **SKET VIEW** submenu is displayed.

10. Select the **Bottom** option from the **SKET VIEW** submenu and select the **FRONT** datum plane from the subwindow. The **References** dialog box is displayed. Choose the **Close** button to exit the **References** dialog box.

11. Choose the **Create 2 point lines** button from the **Right Toolchest** and draw the lines as shown in Figure 10-35. These lines create a section plane.

12. Align the lines and modify the angular dimension to **120** as shown in Figure 10-35. To modify the dimension value, press the middle mouse button in the subwindow and then double-click on the dimension value.

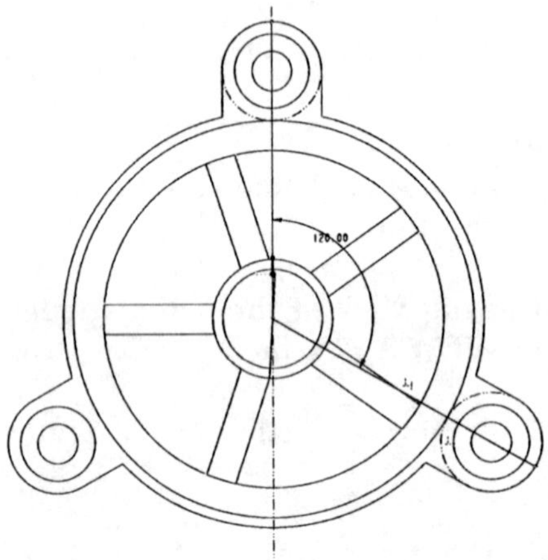

Figure 10-35 *Sketch for the Total Align section*

13. Choose **Sketch > Done** from the menu bar of the subwindow. You are prompted to select the central axis on the new view about which the section will be aligned. You can turn off the display of datum planes.

14. Select the central axis on the new view that you have created. Now, you are prompted to select a view where the section arrows will be placed.

15. Select the top view to display the arrows. You will be prompted to choose a viewing direction. Choose **Okay**. The section view is created.

    You will notice that the spacing between the hatch lines is large. Therefore, you need to modify the scale of hatching.

Tip: *If you do not want to display the section arrows in any view, you can use the middle mouse button (in case of three button mouse) to abort the creation of arrows.*

Modifying the Hatching

You need to modify the spacing between the hatch lines in order to make the hatching dense. It is not easy to select the hatching from the drawing view. Therefore you will select the filter that is available to select the hatch lines. This filter is available from the **Filter** drop-down list.

1. Choose the **X-Section** filter from the **Filter** drop-down list that is present in the status bar below the **Message Area**. Select the hatching from the drawing sheet. The hatching lines turn red in color. Press and hold the right mouse button to invoke the shortcut menu. Choose the **Properties** option from the shortcut menu.

   The **MOD XHATCH** menu is displayed.

2. From the **MOD XHATCH** menu, choose the **Spacing** option to display the **MODIFY MODE** submenu.

3. The spacing between the hatching lines has to be reduced. Choose the **Half** option twice and then choose **Done** in the **MOD XHATCH** menu. Now, the hatching appears to be more dense as shown in Figure 10-36.

Generating the General View

The **General** view is created to show a 3D view of the model, that is, the trimetric view.

1. Choose the **Insert a drawing view of the active model** button from the **Top Toolchest** to display the **VIEW TYPE** menu. 

2. Choose **General** > **Full View** > **No Xsec** > **Scale** > **Done** from the **VIEW TYPE** submenu.

3. Specify the placement point for the general view below the section view as shown in Figure 10-36.

4. Enter the scale for the new view in the **Message Input Window** as **0.05** and press ENTER.

5. Choose **OK** from the **Orientation** dialog box.

Generating the Detail View

As mentioned earlier, the detail view is required to provide details of a particular portion of the drawing view. In this tutorial you need to give the details of one of the ribs of the model.

1. Choose the **Insert a drawing view of the active model** button from the **Top Toolchest** to display the **VIEW TYPE** menu. 

2. Choose **Detailed** > **Full View** > **No Xsec** > **Scale** > **Done**.

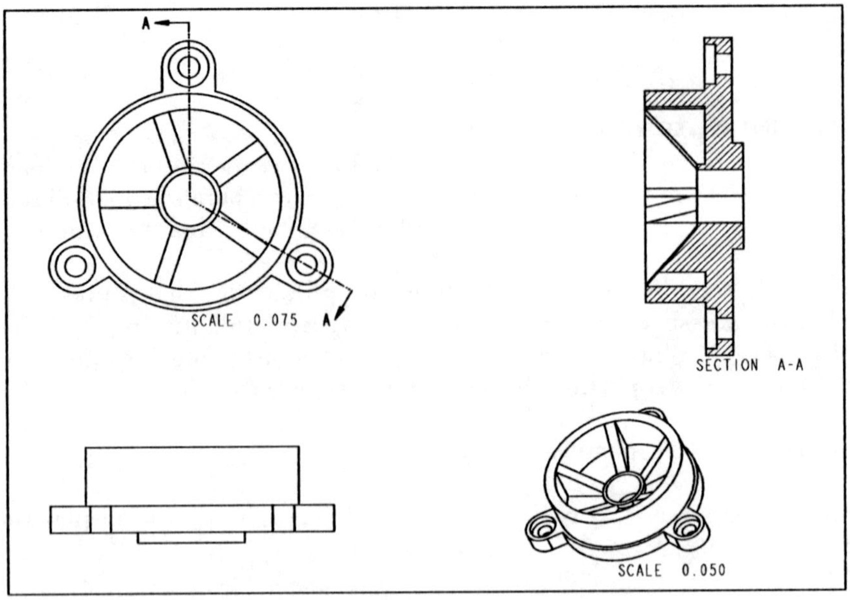

Figure 10-36 *Different drawing views*

3. You will be prompted to specify the placement point for the new view. Select a point between the top view and the sectioned side view using the left mouse button.

4. Enter the value of the view scale as **0.075** in the **Message Input Window** and press ENTER.

   Now, you are prompted to select a center point for detail on an existing view. This center point will be the center of the detailed view. Around this center point you will specify the boundary of the detail view.

5. Select the center point for the detail on one of the ribs in the trimetric view, refer Figure 10-37.

6. Sketch a spline about the center point that you selected. After the spline is drawn, press the middle mouse button to quit the spline creation.

7. The **Message Input Window** is displayed. Enter the name of the detail view as **A** in this window and press ENTER.

8. Choose **Spline** from the **BOUNDARY TYPE** menu that appears.

   Now, you are prompted to select an attachment point on the spline where the note will be attached and displayed.

9. Select a point on the sketched spline as the attachment point. You are prompted to select a location for note.

10. Select the location of the note somewhere near the spline. You may need to move the note by selecting it and then dynamically moving it. The final sheet after generating all the views should look similar to the one shown in Figure 10-37.

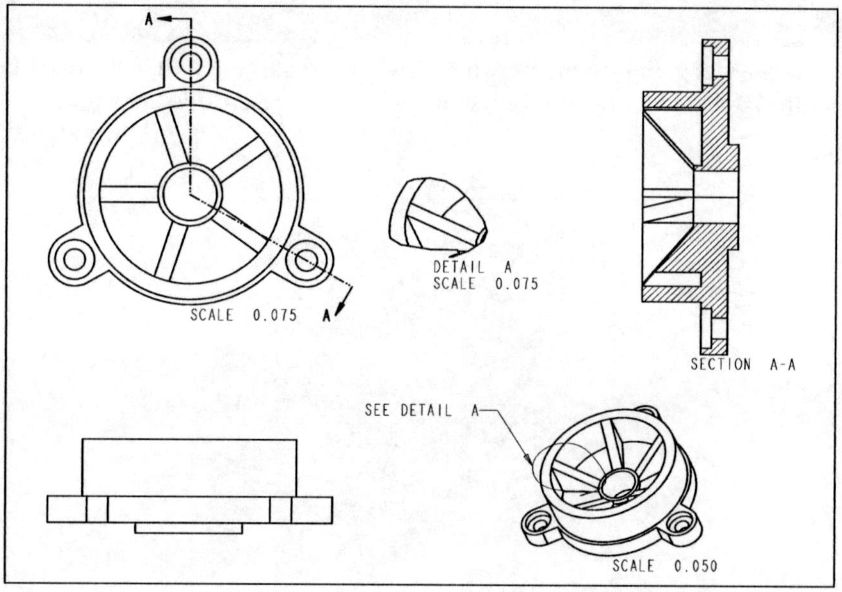

Figure 10-37 *The drawing views for Tutorial 1*

Note

To select any item or drawing view from the drawing sheet, you have to invoke the selection mode. The selection mode is invoked by choosing the ***Select items*** *button from the* ***Right Toolchest****.*

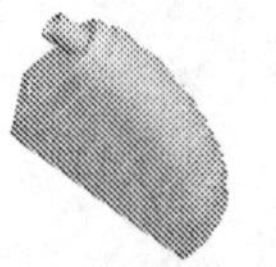

Tip: *If any of the views or the text on the drawing is overlapping or is not at the desired place on the sheet, select the view using the left mouse button and move it by dragging the mouse.*

Saving the Drawing File

You need to save the drawing file that you have created as you may need it later.

1. Choose the **Save the active object** button from the **Top Toolchest**. The **Message Input Window** is displayed with the name of the drawing file that you had entered earlier.

2. Press ENTER to confirm the saving of the file.

Closing the Drawing File

After you have saved the drawing file that you have created, you need to close the drawing file and exit the Drawing mode.

1. Choose **Window > Close** from the menu bar. The drawing window is closed and you have exited the Drawing mode.

Tutorial 2

In this tutorial you will generate the drawing views of the part created in Tutorial 1 of Chapter 7 shown in Figure 10-38. The drawing views that need to be generated are shown in Figure 10-39.

(Estimated time: 40 min)

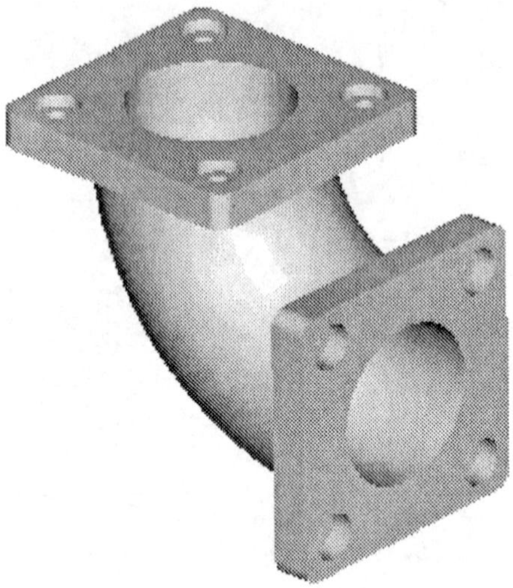

Figure 10-38 *Model for generating the drawing views*

Before you start generating the drawing views, copy the file *c07tut1.prt* from *c07* folder in the current directory.

The following steps outline the procedure to generate the drawing views of the model:

a. Create a new drawing file and select the size of the drawing sheet.

b. Generate the top view.

c. Generate the sectioned front view by defining the **FRONT** datum plane as the section plane, see Figure 10-40.

d. Generate the right-side view of the sectioned front view by using the **No Xsec** option.

e. Generate the isometric sectioned view. The section will be defined by drawing a line on the **TOP** datum plane, see Figure 10-43.

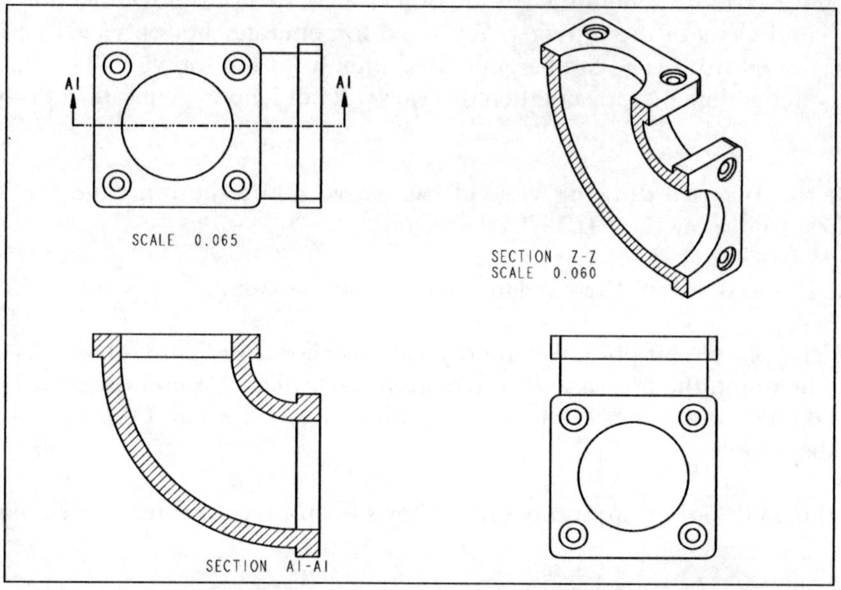

***Figure 10-39** The different drawing views to be generated*

Before starting to generate the drawing views, set the working directory to *C:\ProE-WF\c10*. The *.prt* file and the *.drw* file should lie in the same directory or folder.

Creating New Drawing File

Create a new drawing file for generating the drawing views.

1. Choose the **Create a new object** button from the **File** toolbar to display the **New** dialog box.

2. Select the **Drawing** radio button and then enter the name of the file as *c10tut2*.

3. Choose **OK** from the **New** dialog box to display the **New Drawing** dialog box.

4. Choose the **Browse** button to select *c07tut1.prt* from *c10* for generating the drawing views.

5. Select the **Empty** radio button from the **Specify Template** area.

6. Choose the **Landscape** button from the **Orientation** area.

7. Select **A4** from the **Standard Size** drop-down list. Choose **OK** from the **New Drawing** dialog box to proceed to the Drawing mode.

Generating the Top View

First the top view will be generated. All the other views, except the sectioned isometric view, will be the child views of the top view. You need to generate the top view first because the required sectioned front view can be generated only from the top view. The right-side view can be generated independently but then this view will not help to generate any other required view.

1. Choose the **Insert a drawing view of the active model** button from the **Top Toolchest** to display the **VIEW TYPE** menu.

2. Choose **General** > **Full View** > **No Xsec** > **Scale** > **Done**.

3. Specify the placement point for the top view as shown in Figure 10-40. As soon as you specify the point, the **Message Input Window** is displayed. Enter the scale factor for the view as **0.065** and press ENTER. The default view is placed and the **Orientation** dialog box is displayed.

4. Select the **TOP** option from the **Saved Views** drop-down list in the **Orientation** dialog box.

5. Choose the **Set** button. The default view is oriented as top view.

6. Choose **OK** to exit the **Orientation** dialog box.

   If necessary, move the view as shown in Figure 10-40 and choose **No Hidden** from the **Model Display** toolbar.

Generating the Sectioned Front View

The sectioned front view of the model is generated from the top view. Before proceeding further, use the **Datum planes on/off** button from the **Datum Display** toolbar to turn on the display of datum planes and repaint the screen.

1. Choose the **Insert a drawing view of the active model** button from the **Top Toolchest** to display the **VIEW TYPE** menu.

2. Choose **Projection** > **Full View** > **Section** > **No Scale** > **Done** to display the **XSEC TYPE** menu.

3. Choose **Full** > **Total Xsec** > **Done** in the **XSEC TYPE** menu.

4. Specify the placement point for the front view below the top view as shown in Figure 10-40.

5. When you place the view, the **XSEC ENTER** submenu will be displayed. Choose **Create** to display the **XSEC CREATE** submenu.

6. Choose **Planar** > **Single** > **Done**.

7. Enter the name of the cross section in the **Message Input Window** as **A1** and press ENTER to display the **SETUP PLANE** submenu. You are prompted to select a planar surface or a datum plane.

8. Select the **FRONT** datum plane (the plane that cuts the part horizontally from the center of the cylindrical feature in the top view) from the graphics screen. You are prompted to select a view where you want the arrows to be displayed.

9. Select the top view to display the arrows.

10. Turn the datum display off and repaint the screen.

Modifying the Hatching

The offset distance between the hatching lines in the front sectioned view is large. You need to reduce the distance between the hatching lines.

1. Select the **X-Section** filter from the **Filter** drop-down list present in the status bar. Select the hatching from the sectioned front view in the drawing sheet. The hatching lines turn red in color. Hold down the right mouse button to invoke the shortcut menu. Choose the **Properties** option from the shortcut menu.

   The **MOD XHATCH** menu is displayed.

2. From the **MOD XHATCH** menu, choose the **Spacing** option to display the **MODIFY MODE** submenu.

3. The spacing between the hatching lines has to be reduced. Choose the **Half** option twice and then choose **Done** in the **MOD XHATCH** menu. Now, the hatching appears to be more dense.

4. Choose **Done** from the **MODIFY XHATCH** menu and click once on the graphics window to unselect the X-hatch.

   The sheet after placing these two views should look similar to the one shown in Figure 10-40.

Generating the Right-Side View

The right-side view is the projection of the sectioned front view. However, you will use the **No Xsec** option to create a view that is not sectioned. Turn off the display of datum planes.

1. Choose the **Insert a drawing view of the active model** button from the **Top Toolchest** to display the **VIEW TYPE** menu.

2. Choose **Projection > Full View > No Xsec > No Scale > Done**. You are prompted to specify the placement point for the drawing view.

3. Specify the placement point for the view on the right side of the sectioned front view.

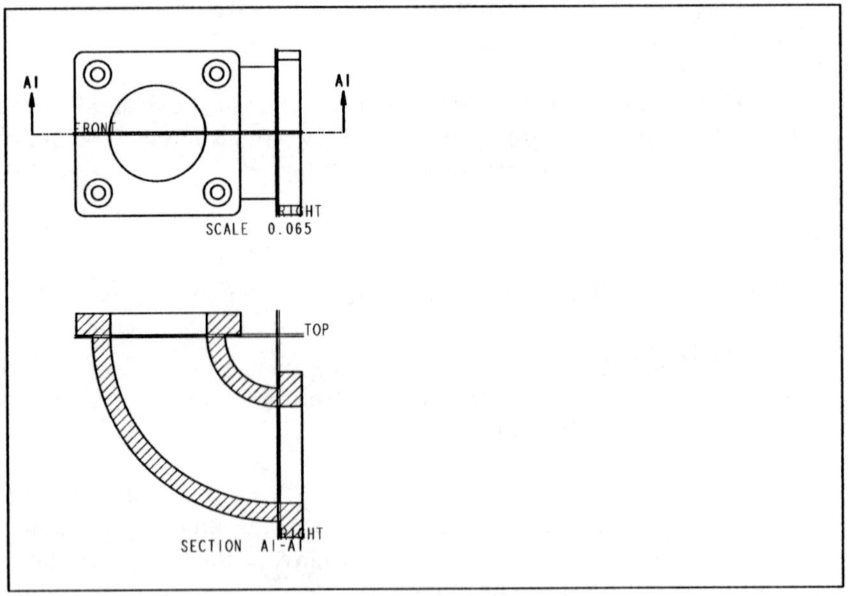

***Figure 10-40** Drawing sheet after generating the top view and the sectioned front view with the display of datum planes*

The right-side view of the model is placed as shown in Figure 10-41.

Generating the Isometric Section View

The isometric section view is an independent view and will be generated by using the **General** option.

1. Choose the **Insert a drawing view of the active model** button from the **Top Toolchest** to display the **VIEW TYPE** menu. 

2. Choose **General > Full View > Section > Scale > Done** from the **VIEW TYPE** menu. The **XSEC TYPE** menu is displayed.

3. Choose **Full > Total Xsec > Done** from the **XSEC TYPE** menu.

4. Specify a point to the right of the top view and the **Message Input Window** is displayed and you are prompted to specify the scale for the drawing. Enter **0.06** in this window and press ENTER. The default trimetric view of the model is placed and the **Orientation** dialog box is displayed.

   The default view is a trimetric view but you need the isometric view to be displayed. Therefore, you need to change the preferences of the view.

5. Select the **Preferences** option from the **Type** drop-down list.

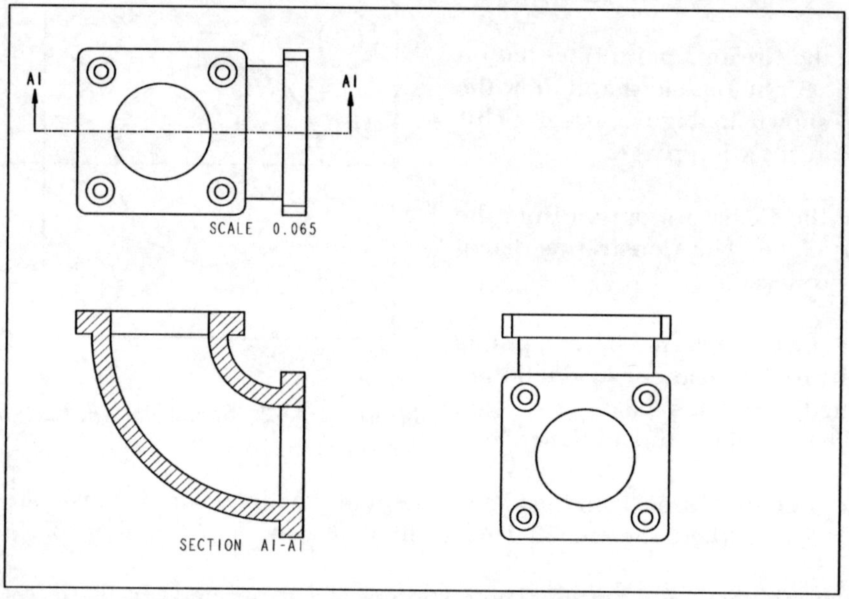

Figure 10-41 Drawing sheet after generating the top view, the sectioned front view, and the right-side view

6. Choose the **Isometric** option from the **Default Orientation** drop-down list. The isometric view of the model is displayed on the drawing sheet.

7. Choose **OK** in the **Orientation** dialog box to exit it. The **XSEC ENTER** menu is displayed.

8. Choose the **Create** option from the **XSEC ENTER** menu to display the **XSEC CREATE** submenu.

9. Choose **Offset** > **Both Sides** > **Single** > **Done** from the **XSEC CREATE** menu.

10. Enter the name of the section as **Z** in the **Message Input Window** that appears and press ENTER. Once you have specified the name of the section, a separate window will appear displaying the model.

Note

If the model in the Part mode is opened in another window, then the subwindow will not appear and you have to manually change the window. Choose the ***Window*** *menu, and select the part file that is opened. Now you can continue with the sketcher environment.*

11. Select the **TOP** datum plane from the subwindow. Choose **Okay** from the **DIRECTION** submenu. The **SKET VIEW** submenu is displayed.

12. Select the **Right** option from the **SKET VIEW** submenu and choose the **RIGHT** datum plane from the subwindow. The **References** dialog box is displayed. Choose the **Close**

button to exit the **References** dialog box.

13. Choose the **Create 2 point lines** button from the **Right Toolchest** and draw the line as shown in Figure 10-42. This line create a section plane.

14. Choose the **Constrain** option from the **Sketch** menu. The **Constraints** dialog box is displayed.

15. Choose the **Create same points, points on entity or collinear constraint** button from the **Constraints** dialog box. Align both the ends of the line to the edges.

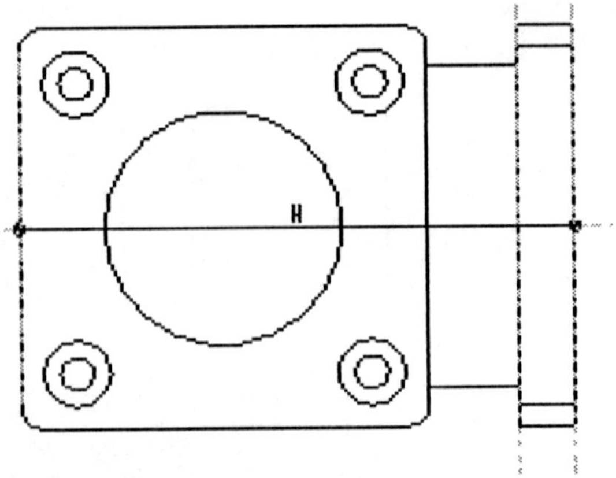

***Figure 10-42** Sketch for the section line with constraint*

16. Choose **Sketch > Done** from the **Sketch** menu in the menu bar. Now, you are prompted to select a view where the section arrows will be placed.

17. Press the middle mouse button. The section view is generated but the direction arrows are not displayed.

Note

If the drawing view is placed on the sheet such that it overlaps the boundary of the drawing sheet then you can move the drawing view. You can move the drawing view by double-clicking on it and then moving the mouse. As you move the mouse the drawing also moves.

Modifying the Hatching

The spacing between the hatching lines in the sectioned isometric view is large. Therefore, you need to reduce the distance between the hatching lines. Use the same procedure that was discussed earlier in this tutorial to modify the spacing between the hatch lines. The sheet after placing all the views should look similar to the one shown in Figure 10-43.

Saving the Drawing File

You need to save the drawing file that you have created as you may need it later.

1. Choose the **Save the active object** button from the **Top Toolchest**. The **Message Input Window** is displayed with the name of the drawing file that you had entered earlier.

2. Press ENTER to confirm the saving of the file.

Closing the Drawing File

After you have saved the drawing file that you have created, you need to close the drawing file and exit the Drawing mode.

1. Choose the **Window> Close** option from the menu bar. The drawing window is closed and you have exited the Drawing mode.

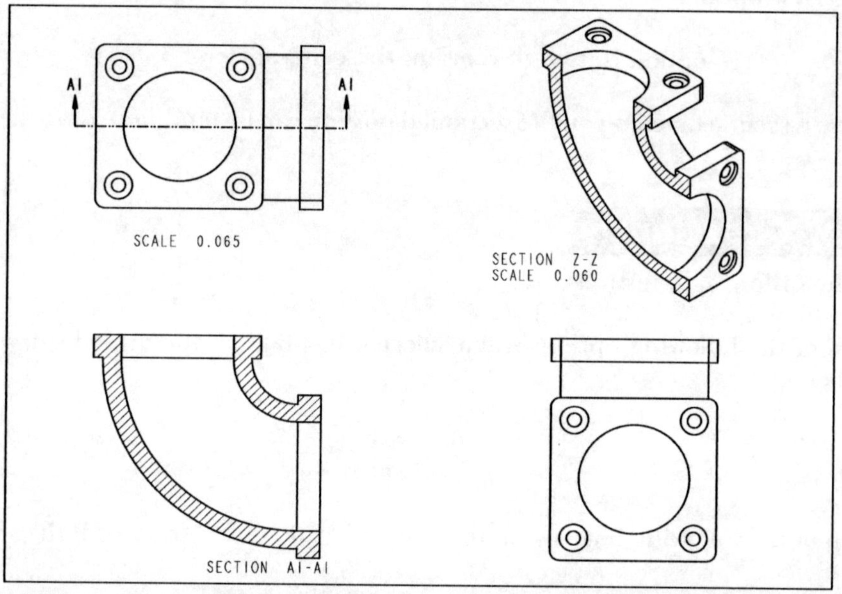

Figure 10-43 The different drawing views generated

Self-Evaluation Test

Answer the following questions and then compare your answers to the answers given at the end of this chapter.

1. The bidirectional associative nature of a software package means that when one file related to the part model is modified the corresponding modification can be seen in other related files. (T/F)

2. General view is the first view that is generated in the sheet. (T/F)

3. The Full Section view is the most widely used type of view. (T/F)

4. The sectional views are generally created for the models that have features that are not clearly visible from standard view. (T/F)

5. Broken views are used for parts that have a high length to width ratio. (T/F)

6. The __________ view is used when you want to show a particular portion of the view in section and at the same time not section the remaining view.

7. The __________ option allows you to temporarily remove the selected drawing view from the sheet.

8. The __________ option is used to redisplay the drawing views that are erased using the **Erase View** option.

9. The __________ option is used to reorient the general view.

10. The cross section hatching on the sectioned portion can be modified using the __________ menu.

Review Questions

Answer the following questions:

1. Which of the following options when selected displays all the erased drawing views on the sheet?

   (a) **Delete** (b) **Resume**
   (c) **Move** (d) **Erase**

2. Which of the following buttons on the **View** toolbar is used to refresh the screen?

   (a) **Zoom Out** (b) **Zoom In**
   (c) **Redraw the current view** (d) None

3. Which of the following options when selected displays only that area of the section view that is sectioned?

   (a) **Total Xsec** (b) **Area Xsec**
   (c) **Align Xsec** (d) **Total Align**

4. Which of the following options is used to permanently remove a drawing view from the sheet?

   (a) **Resume** (b) **Erase**
   (c) **Delete** (d) **Move**

5. Which of the following options is used to modify any numeric value associated with the drawing views?

   (a) **Xhatching** (b) **Any Item**
   (c) **Value** (d) None

6. The view type of an existing view can be changed. (T/F)

7. You can reorient only the general views. (T/F)

8. You can flip the side of the cross section views. (T/F)

9. The **Orientation** area in the **New Drawing** dialog box is available only when you select the **Empty** radio button from the **Specify Template** area. (T/F)

10. The orientation of a model saved in the Part mode can be used to orient the drawing view in the Drawing mode. (T/F)

Exercise

Exercise 1

Generate the drawing views of the model created in Tutorial 3 of Chapter 8 shown in Figure 10-44. Generate the drawing views that are shown in Figure 10-45 on an A4 size sheet.

(Estimated time: 40 min)

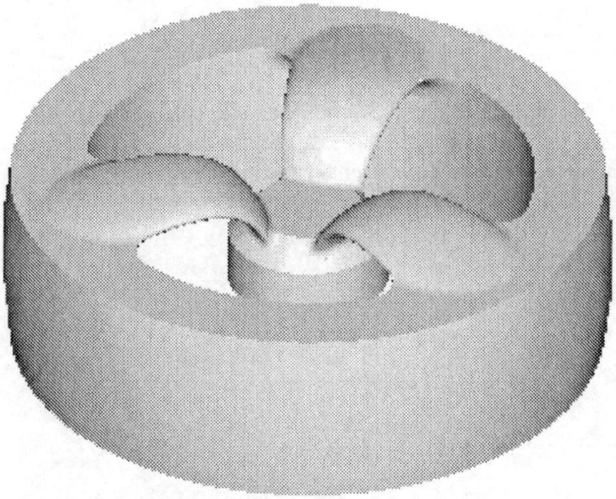

Figure 10-44 *Part for generating the drawing views*

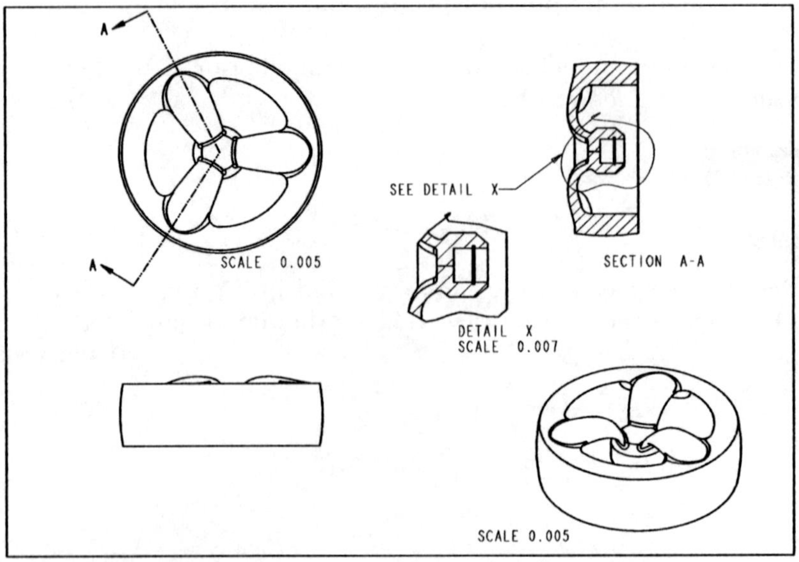

Figure 10-45 Drawing views to be generated in Exercise 1

Answers to the Self-Evaluation Test

1 - T, 2 - T, 3 - T, 4 - T, 5 - T, 6 - Local Section, 7 - Erase, 8 - Resume, 9 - Reorient, 10 - MOD XHATCH.

Chapter 11

Dimensioning the Drawing Views

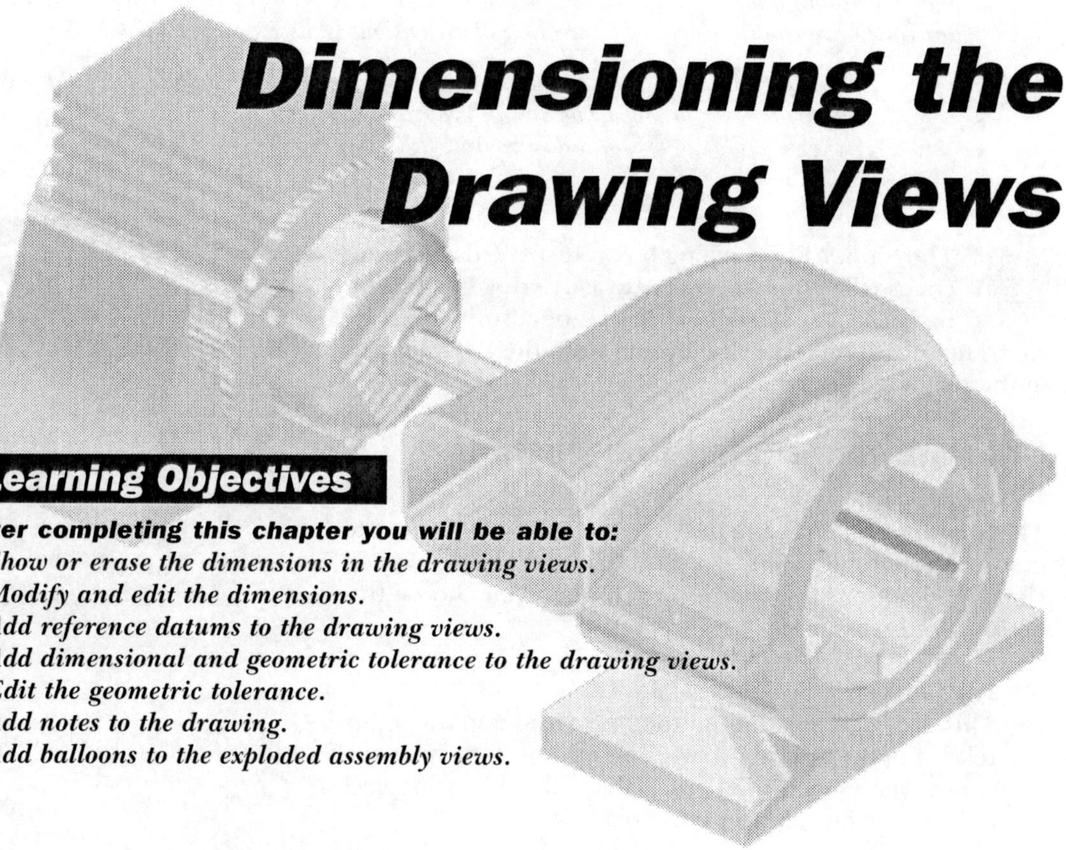

Learning Objectives

After completing this chapter you will be able to:

- *Show or erase the dimensions in the drawing views.*
- *Modify and edit the dimensions.*
- *Add reference datums to the drawing views.*
- *Add dimensional and geometric tolerance to the drawing views.*
- *Edit the geometric tolerance.*
- *Add notes to the drawing.*
- *Add balloons to the exploded assembly views.*

DIMENSIONING THE DRAWING VIEWS

Once you have generated the drawing views, you need to generate the dimensions in the drawing views, and add notes, symbols, balloons, and so on. The dimensions are the dimensions assigned to each of the entities that created the model or to the features associated with the model. These dimensions are associative in nature and hence can be used to modify or drive the dimensions of a part. Therefore, these dimensions are also called driving dimensions. Dimensions are generated using the **Show / Erase** dialog box shown in Figure 11-1.

Note

You can also create dimensions in the Drawing mode, but this type of dimension cannot drive the dimensions of the part. Therefore, in Pro/ENGINEER Wildfire the dimensions you create are called driven dimensions. The creation of dimensions is discussed in Chapter 12.

Tip: *You must have noticed that some options in the menus available in the menu bar are followed by dots. The dots indicate that after you choose any of these options, a menu, a dialog box, or a dashboard will be displayed.*

The **Show / Erase** dialog box is displayed when you choose the **Open the Show\Erase dialog box** button from the **Top Toolchest**. It can be also invoked by choosing the **Show and Erase** option from the **VIEW** menu in the menu bar.

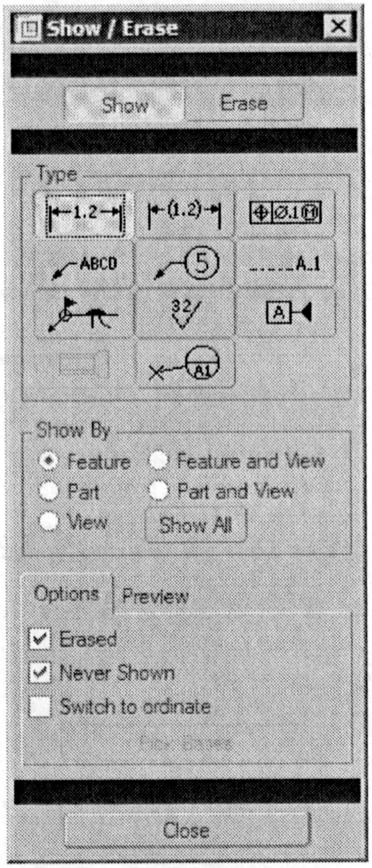

*Figure 11-1 **Show / Erase** dialog box with the **Show** button*

Show / Erase Dialog Box Options

Show

The **Show** button is chosen to select the options to generate the various parameters related to the model in the drawing views. The options that are displayed when you choose this button are discussed next.

Type Area

This area contains the buttons for various parameters related to the drawing views. The options to display the various parameters will be available only when you select at least one button from this area.

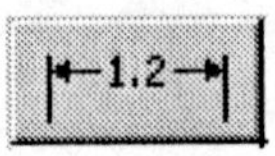

Dimension. The **Dimension** button is chosen to generate the parametric dimensions in the drawing views. The bidirectional associative nature of this software package is valid when you change the dimension value generated using this option.

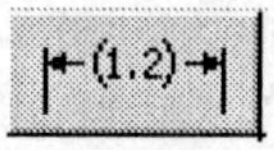

Reference Dimension. The **Reference Dimension** button is chosen to generate the reference dimensions in the drawing views.

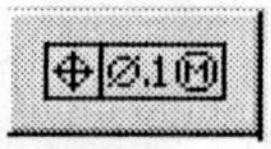

Geometric Tolerance. The **Geometric Tolerance** button is chosen to display the geometric tolerances in the dimensions.

Tip: *The dimensional tolerance will be displayed only when the* ***tol_display*** *is set to* ***Yes*** *in the configuration file. This file is displayed when you choose* ***File > Properties*** *from the menu bar. The* ***FILE PROPERTIES*** *menu is displayed. Choose the* ***Drawing Options*** *option to open the configuration file*

Note. The **Note** button is chosen to display the notes that were created in the Part mode. If any change is made to a note in the Drawing mode, the note in the Part mode is automatically modified. Similarly, if you modify this note in the Part mode, the modifications are reflected automatically in the Drawing mode.

Balloon. The **Balloon** button is chosen to display the balloons associated with an assembly in the drawing views.

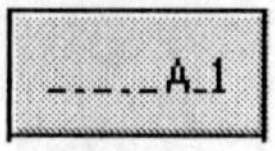

Axis. The **Axis** button is chosen to display the axis of the selected hole or cylindrical feature in the drawing views.

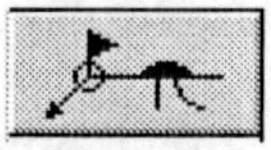

Symbol. The **Symbol** button is chosen to display any symbol associated with the model.

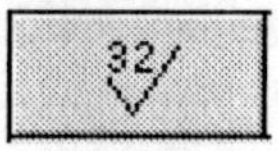

Surface Finish. The **Surface Finish** button is chosen to display the surface finish symbols associated with the part in the drawing views.

Datum Plane. The **Datum Plane** button is chosen to display the reference datum planes in the drawing views.

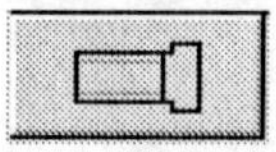

Cosmetic Feature. The **Cosmetic Feature** button is chosen to display the cosmetic features in the drawing views that are associated with the model.

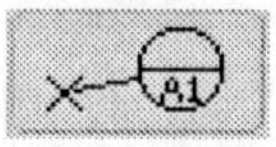

Datum Target. The **Datum Target** button is used to show the critical measurement locations on the drawing view.

Show By Area

The options in this area are discussed next.

Feature. The **Feature** radio button is selected to display the required annotation of the selected feature. If the required annotation of the selected feature cannot be displayed in the view in which the feature was selected then the required annotation will be displayed in the appropriate view.

Part. The **Part** radio button is selected to display all the required annotations of the selected part. The required annotations are displayed in all the views irrespective of the view in which the part was selected.

View. The **View** radio button is selected to display all the required annotations that can be displayed in the selected view.

Feature and View. The **Feature and View** radio button is selected to display the required annotations of a particular selected feature that can be displayed in the selected view.

Part and View. The **Part and View** radio button is selected to display all the required annotations of the selected part that can be displayed in the selected view.

Show All. The **Show All** button is selected to show all the required annotations in the drawing views. When you choose this button all the required annotations that were used in creating the model are displayed in the drawing views.

Options Tab

The options under this tab of the **Show / Erase** dialog box are discussed next.

Erased. The **Erased** check box is selected to display the annotations that were once erased. If this check box is cleared, the annotations that were once erased using the **Show / Erase** dialog box will not be displayed the next time you wish to display them.

Never Shown. The **Never Shown** check box is selected to display the annotations that are never displayed in the drawing views. At least one of the **Erased** or the **Never Shown** check box remains selected while specifying the annotation.

Switch to ordinate. The **Switch to ordinate** check box is selected to switch to the ordinate mode of dimensioning. When you select this check box, you will be prompted to select a dimension that will be the baseline dimension for the other ordinate dimensions. Ordinate dimensioning is discussed in Chapter 2.

Preview Tab

The options under this tab of the **Show / Erase** dialog box are discussed next.

With Preview. The **With Preview** check box is selected to allow you to confirm the placement of the annotation. If this check box is selected then when you show the annotations, you will be prompted to confirm the annotations that you want to keep or delete. You can select the annotations to keep, to delete, accept all, or delete all using the buttons provided under this area. The remaining options under this area will be available only after you place the annotations. If this check box is cleared then you will not be prompted to confirm the placement of the annotation. All the displayed annotations will be automatically placed in the drawing views.

Erase

The options under the **Erase** button are shown in Figure 11-2. These options are used to erase the annotations placed using the options under the **Show** button. The annotations once placed in the drawing view can also be erased by selecting it. When the selected annotation turns red in color, hold down the right mouse button and choose the **Erase** option from the shortcut menu that appears. The selected annotation gets erased from the drawing view.

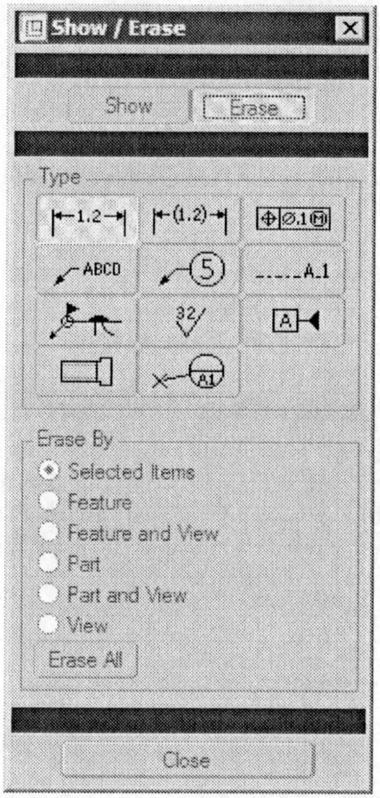

***Figure 11-2 Show / Erase** dialog box with the **Erase** tab*

Note
*The annotations erased using the **Erase** options are not deleted from the model.*

Type Area

The buttons provided under this tab are similar to those under the **Type** area of the **Show** button. The only difference is that here these buttons are used to erase the items.

Erase By Area

The options in this area are discussed next.

Selected Items. If the **Selected Items** radio button is selected then you can select the annotations to be erased using the left mouse button.

Feature. If the **Feature** radio button is selected then all the annotations of the feature that you select will be erased from all the views.

Feature and View. If the **Feature and View** radio button is selected then all the annotations related to the selected feature will be erased from the selected view.

Part. If the **Part** radio button is selected then all the annotations related to the selected part will be erased from all the drawing views.

Part and View. If the **Part and View** radio button is selected then all the annotations related to the selected part will be erased from the selected view.

View. The **View** radio button is selected to erase all the annotations displayed in the selected view.

Erase All. The **Erase All** button is chosen to erase all the annotations displayed in all the views of the current sheet. When you choose this button, you will be prompted to confirm that all the annotations can be erased.

MODIFYING AND EDITING THE DIMENSIONS

Pro/ENGINEER allows you to edit and modify the dimensions assigned to the drawing views. There are various methods of editing and modifying the driving dimensions. Figure 11-3 shows the dimensioned drawing views. But before these methods are discussed, it is important for you to learn the terminology used in these methods. Figure 11-4 shows the terminology that is used in the Drawing mode of Pro/ENGINEER.

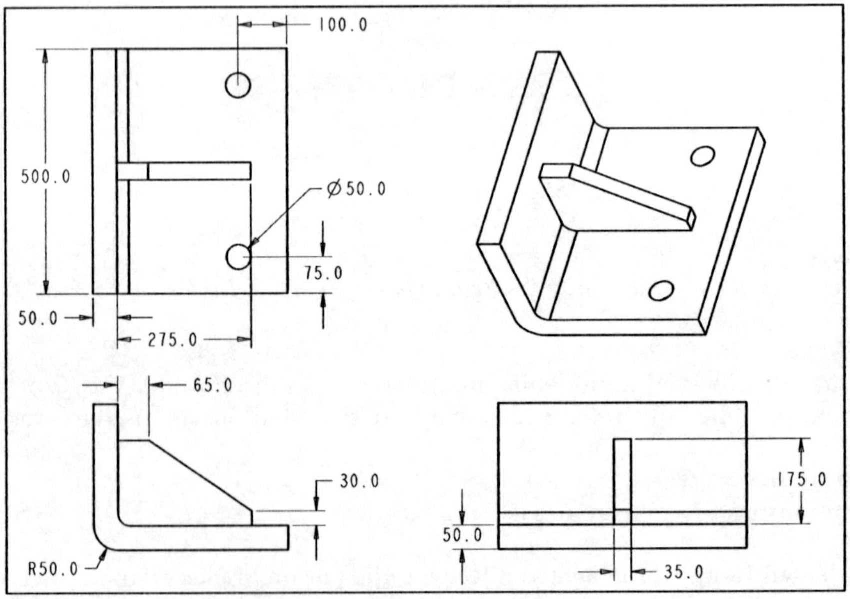

***Figure 11-3** Dimensioned drawing views*

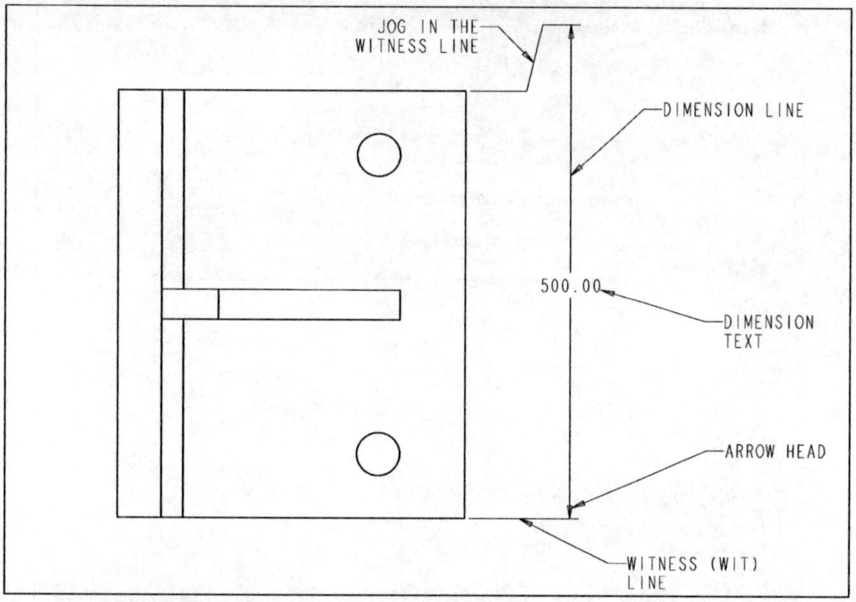

***Figure 11-4** Parameters associated with the dimensions*

Tip: *To display the dimensions in a 3D view, first you will have to set the value of **allow_3d_dimensions** to **YES** in the configuration file. If this value is set to **NO** then the dimensions will not be displayed in 3D views. The configuration file can be opened by choosing **File > Properties** from the menu bar. The **FILE PROPERTIES** menu is displayed. Choose the **Drawing Options** option to open the **Options** dialog box. This dialog box lists all the variables that can be varied for the current session.*

Modifying the Dimensions Using the Dimension Properties Dialog Box

The **Properties** option is available in the **Edit** menu only when the dimension to be modified is selected on the drawing sheet. The **Dimension Properties** dialog box is displayed when you choose the **Properties** option from the **Edit** menu in the menu bar. The second method of invoking the **Dimension Properties** dialog box is to select the dimension and hold down the right mouse button to invoke the shortcut menu. Choose the **Properties** option from the shortcut menu. The **Dimension Properties** dialog box is displayed as shown in Figure 11-5. The third method of invoking this dialog box is to double-click on the dimension value. The options in this dialog box are discussed next.

Properties Tab

Value and tolerance Area

The options provided under this area are related to the tolerance in the drawing.

Tolerance mode. This drop-down list will be available only when the value of **tol_display** is set to **YES** in the configuration file. The options under this drop-down list are used to select the type of tolerance mode. Depending upon the option selected

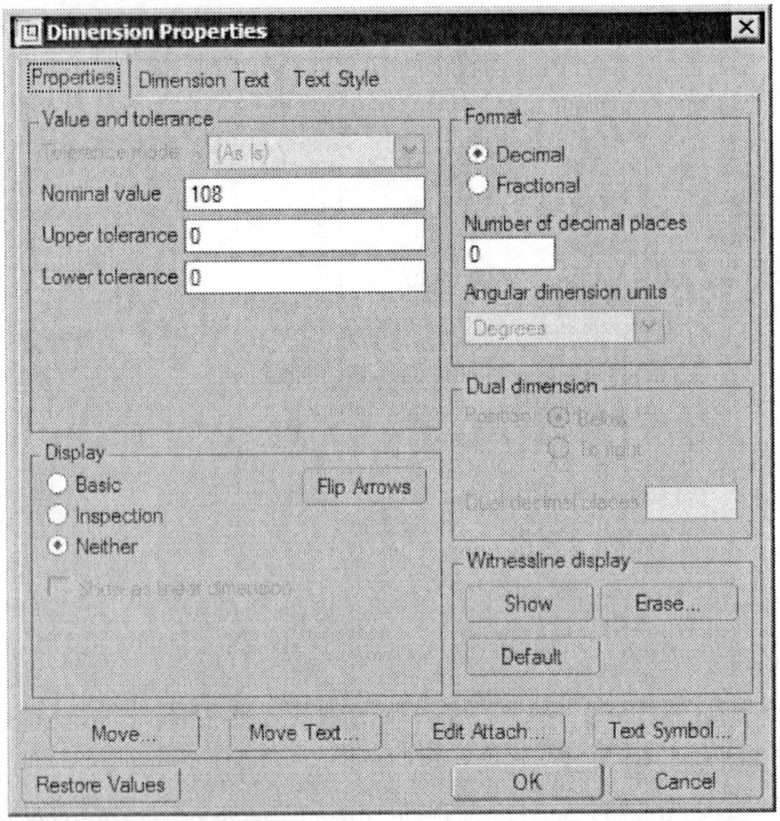

***Figure 11-5** The **Properties** tab of the **Dimension Properties** dialog box*

from this drop-down list, the other options in this area change.

Format Area

Decimal. The **Decimal** radio button is selected to display the dimensions in the decimal format. The number of decimal values can be set in the **Number of decimal places** edit box. By default, its value is 2.

Fractional. The **Fractional** radio button is selected to display the dimensions in the fractional format. The largest denominator can be set using the **Largest denominator** edit box that is displayed when you select the **Fractional** radio button.

Dual dimension Area

The options under the **Dual dimension** area will be available only when the value of **dual_dimensioning** is set to **Yes** in the configuration file. The options under this area are used to specify the placement point for the dual dimensioning and the number of decimal places.

Display Area

Basic. If the **Basic** check box is selected then the selected dimension values will be

Tip: *If you modify the nominal value of a dimension in the* ***Value and tolerance*** *area, you will have to regenerate the model using the* ***Update the display of all views in the active sheet*** *button to view the effect of modifications. The modified dimension appears blue in color.*

displayed inside a rectangular box.

Inspection. If the **Inspection** check box is selected then the selected dimension will be displayed inside an elliptical box.

Dimension Text Tab

The **Dimension Properties** dialog box with the options under the **Dimension Text** tab is shown in Figure 11-6.

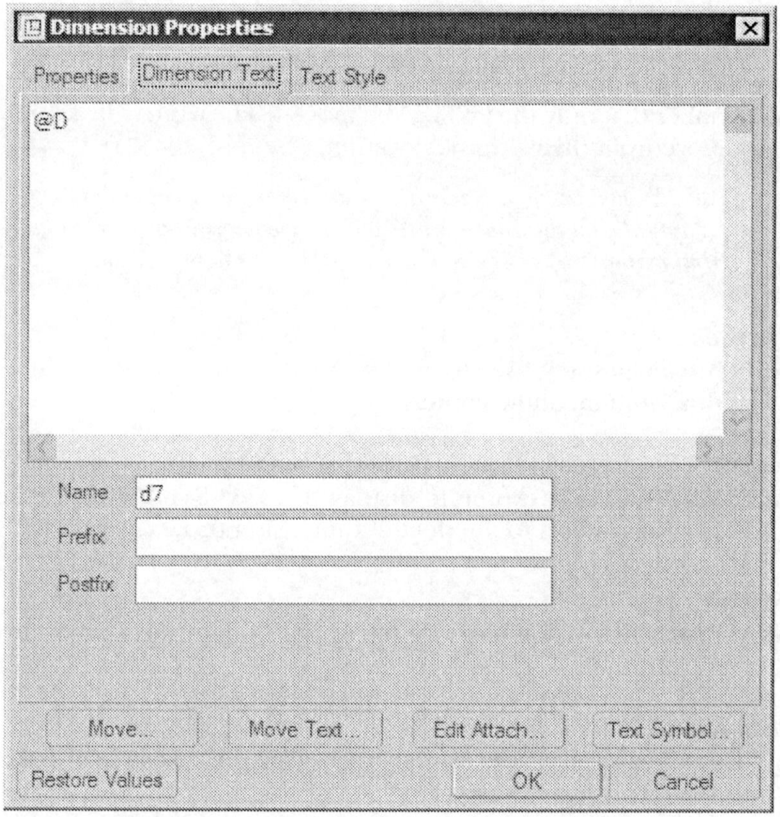

Figure 11-6 *The* ***Dimension Text*** *tab of the* ***Dimension Properties*** *dialog box*

Dimension Text Area

The **Dimension text** box displays the text of the dimension. You can edit the text using this box.

Name
The **Name** edit box display the dimension symbol. You can change the dimension symbol using this box.

Prefix
The **Prefix** edit box is used to add some additional text as prefix to the default dimension text.

Postfix
The **Postfix** edit box is used to add some additional text as suffix to the dimension text.

Move
The **Move** button is chosen to move the dimension to a new location on the current sheet. You can also move the dimension by selecting it and then dragging it by moving the mouse.

Move Text
The **Move Text** button is chosen to move the dimension text to a new location on the screen. Remember that only the text will be moved. The witness lines and the dimension lines do not move from their original location.

> **Tip**: *The modifications made using the* ***Dimension Properties*** *dialog box are applied dynamically on the dimensions. You can view this in the background by moving the* ***Dimension Properties*** *dialog box on the side of the screen.*

Modify Attach
The **Modify Attach** button is used to modify the attachment point of dimensions of some features like draft, round, and chamfer.

Text Symbol
The **Text Symbol** button is chosen to display the **Text Symbol** dialog box using which you can add special symbols to the default dimension text.

Text Style Tab
The **Dimension Properties** dialog box with the options under the **Text Style** tab is shown in Figure 11-7.

Copy from Area
The **Copy from** area displays the options to modify the text style of the dimension text. You can edit the text using this area.

Character Area
The **Character** area displays the options to modify the font style of the dimension text. Various parameters like height, font name, width of font, font style, and so on can be modified using the options available in this area.

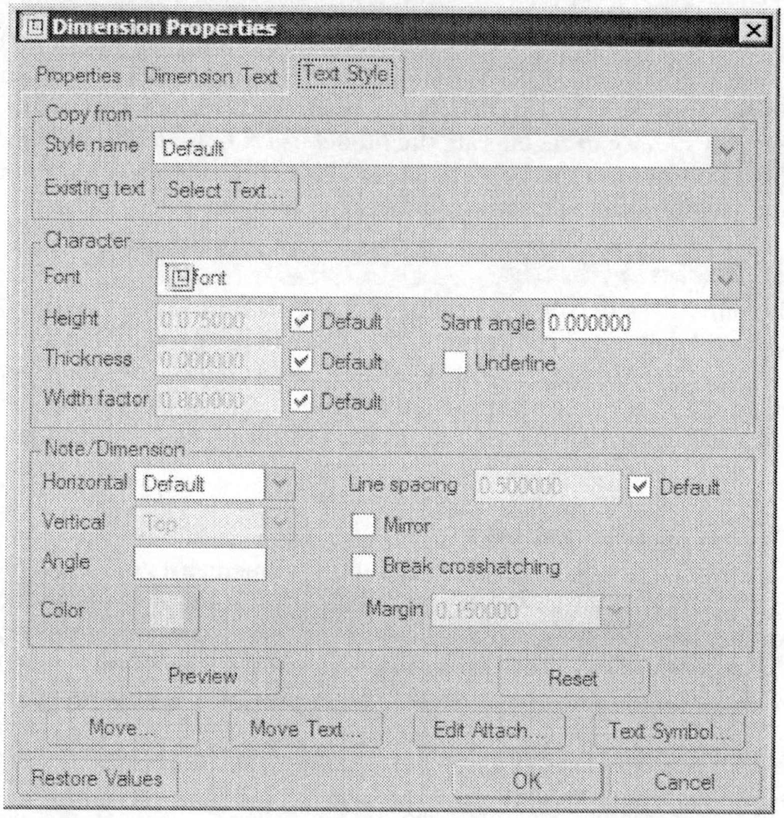

***Figure 11-7** The **Text Style** tab of the **Dimension Properties** dialog box*

Note/Dimension Area

The **Note/Dimension** area displays the options to modify the dimension or the text in the form of a note. The parameters that can be modified using this area are; text justification, color, line spacing, angle, and so on.

Apply button

The **Apply** button is used to apply the changes that are made using the **Dimension Properties** dialog box.

Modifying the Drawing Items Using the Shortcut Menu

When you select an item from a drawing view and hold down the right mouse button to invoke the shortcut menu, various options are displayed in the shortcut menu. The options in the shortcut menu depend upon the item selected to modify and vary from item to item. Some of the items related to the drawing that can be modified using the shortcut menu are discussed next.

Value

You can modify the dimension value by selecting a dimension value. When the dimension value and the arrows turn red, click on the dimension value. Notice that now only the value is highlighted in red. Hold down the right mouse button to invoke the shortcut menu. From the shortcut menu choose the **Edit Value** option. The dimension value is displayed in the edit box. Modify the value in the edit box and press ENTER. To view the effect of the modification, you will have to regenerate the model using **Update the display of all views in the active sheet** button from the **Top Toolchest**. This option is also used to modify the scale of the drawing view and the tolerance values.

Number of Decimal Places

To modify the number of decimal values for the dimension text, choose the **Decimal Places** option from the **Format** menu in the menu bar. You will be prompted to specify the number of decimal places in the **Message Input Window**. After you have specified the value, you will be prompted to select the dimensions to be modified. Select the dimension using the left mouse button and then press the middle mouse button to view the change in the decimal places.

Miscellaneous Modifications

The modifications that can be made using the shortcut menu are: move a dimension, flip the arrows of the dimensions, switch the selected dimension from one view to the other, make a jog in the witness line, modify the values or text of the dimension, modify the arrow style, break the witness line, remove the break from the witness line, erase the witness line, show the erased witness line, or erase the selected dimension from the drawing view.

Text

The parameters that can be modified are the selected text line, the entire text material, text height, text style, style library for the text, and the current style.

Cleaning Up the Dimensions

Generally, the dimensions displayed in the drawing views are scattered and improperly placed. This in turn makes the drawing views look very untidy. One method is to place each dimension manually at the appropriate location in the view. The second method allows you to place these dimensions in proper order using just one single step. The **Clean Dimensions** dialog box allows the user to set a standard for displaying the dimensions in the drawing views. When you choose **Edit > Cleanup > Dimensions** from the **Edit** menu in the menu bar, the **Clean Dimensions** dialog box is displayed as shown in Figure 11-8 and you are prompted to select the dimensions or individual views to clean. The options in this dialog box are discussed next.

Placement Tab

The options under this tab of the **Clean Dimensions** dialog box are used for placing the dimensions. The distance between the two dimensions, the distance between the view boundary and the dimension, are set under this tab.

Space Dimensions

The **Space Dimensions** option is used to maintain the distance between two dimensions.

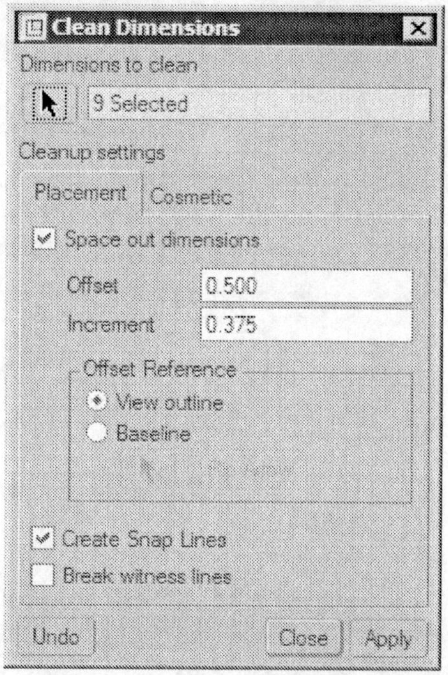

*Figure 11-8 The **Placement** tab of the **Clean Dimensions** dialog box*

Offset

The **Offset** edit box is used to specify the offset distance between the first dimension and the view boundary.

Increment

The **Increment** edit box is used to specify the offset distance between the two dimension lines.

Offset Reference Area

View Outline. The **View Outline** radio button is selected to specify the offset distances taking the outline of the view as reference.

Baseline. The **Baseline** radio button is selected to specify the offset distance taking a selected flat edge, datum plane, snap line, detail axis line, or view border as the reference. When you invoke this option, you will be prompted to select either of the above-mentioned entities as the reference.

Create Snap Lines

The **Create Snap Lines** check box is selected to create the snap lines using which the dimensions will be cleaned. The snap lines will be created at the distances specified using the **Offset** and the **Increment** edit boxes. If this check box is cleared then the snap lines will not be created.

Break witness lines

The **Break witness lines** check box is selected to break the witness lines if some dimensions are overlapping each other.

Cosmetic Tab

The options under the **Cosmetic** tab of the **Clean Dimensions** dialog box shown in Figure 11-9 are used for determining a dimension value's location with the dimension line. The arrows and the orientation of the dimension text can also be controlled using this tab.

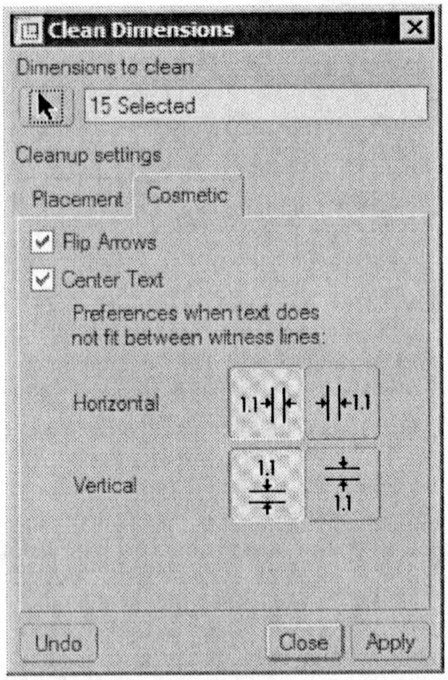

*Figure 11-9 The **Cosmetic** tab of the **Clean Dimensions** dialog box*

Flip Arrows

The **Flip Arrows** check box is selected to flip the arrows of the dimensions that are not easily adjusted inside the dimension lines.

Center Text

The **Center Text** check box is selected to place the dimensions at the center of the dimension lines.

Horizontal/Vertical

These buttons are used to specify the placement of the horizontal or vertical dimension text outside the dimension line if they do not fit.

Figures 11-10 and 11-11 explain the use of **Clean Dimensions** dialog box.

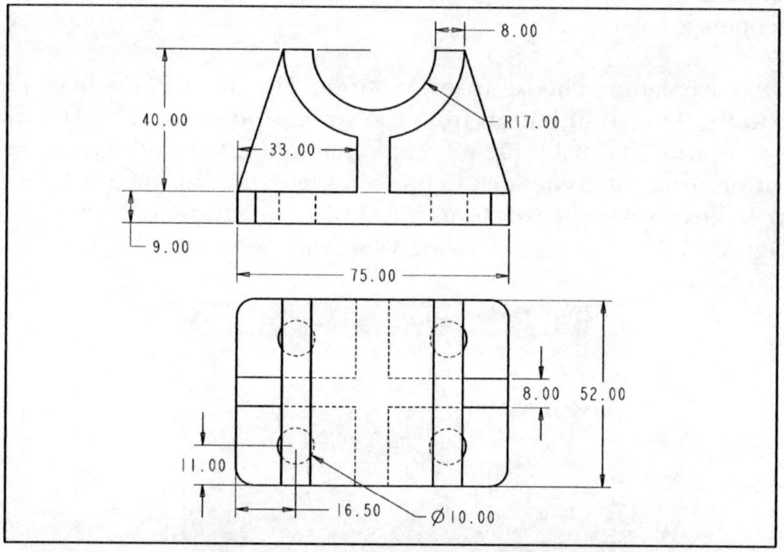

Figure 11-10 *Drawing views with dimensions before cleaning*

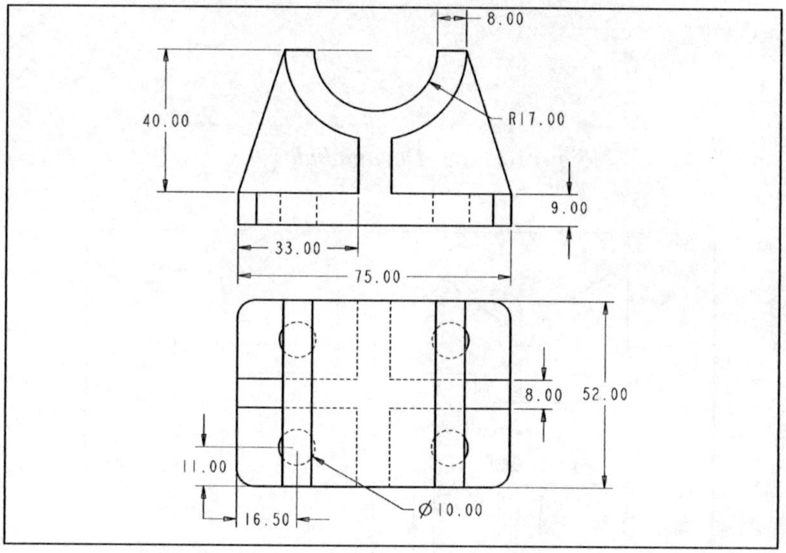

Figure 11-11 *Drawing views with dimensions after cleaning*

ADDING REFERENCE DATUMS TO THE DRAWING VIEWS

Reference datums are used as references for geometric tolerance. Datums are to be set before adding the geometric tolerance.

To create a reference datum, choose **Insert > Model Datum > Plane** from the menu bar or choose the **Datum Plane Tool** button from the **Right Toolchest**. The **Datum** dialog box is displayed, as shown in Figure 11-12. You can rename the selected datum and then choose the second button from the **Type** area to enclose the datum inside a feature control frame. You can place these reference datums freely in the drawing views or in a selected dimension. Figure 11-13 shows the drawing views with reference datums.

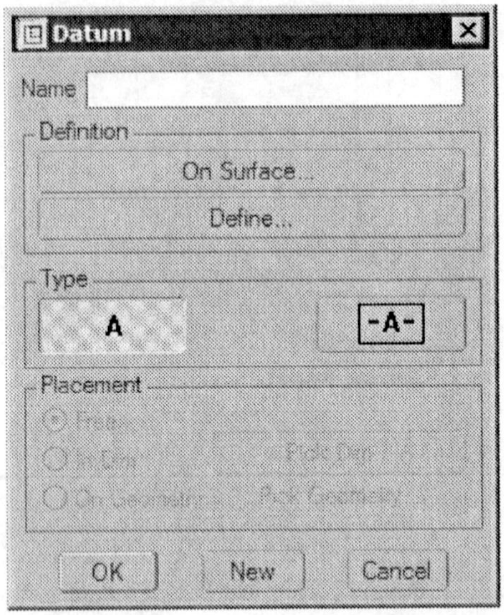

Figure 11-12 ***Datum*** *dialog box*

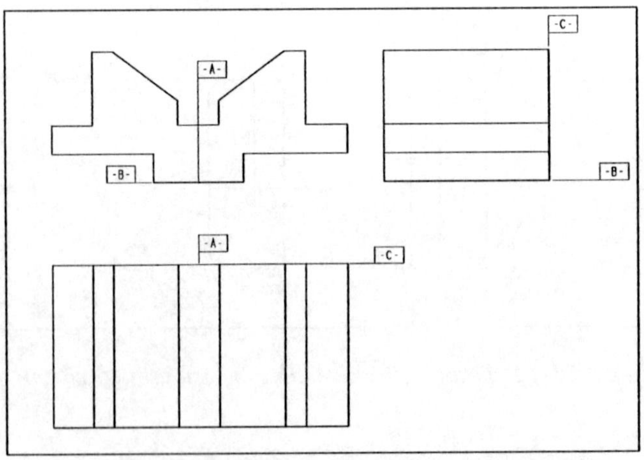

Figure 11-13 *Drawing views with reference datums*

Note

The system does not show a reference datum plane in a drawing view unless it is perpendicular to the graphics window.

ADDING TOLERANCES IN THE DRAWING VIEWS

Tolerance is defined as the difference between the maximum and minimum variations in the dimensions of the selected component. It is almost impossible to manufacture a component to the exact dimensions that have a very large precision. In such cases, the tolerance value is added to dimensions to make sure that some variation that occurs during manufacturing can be taken care of. However, when you actually send a part for manufacturing, there are other parameters along with the dimension tolerances that vary and that need some tolerances. Depending upon these factors, the tolerances are divided into two types: the dimensional tolerances and the geometric tolerances.

Dimensional Tolerances

These are the variation in the standard dimensional values. The dimensional tolerances can be easily displayed along with the dimensions in the drawing views by simply setting the value of **tol_display** to **Yes** in the configuration file. You can select and modify the type of dimensional tolerances from the **Properties** tab of the **Dimension Properties** dialog box. Figure 11-14 shows the drawing views with the tolerance added to the dimensions.

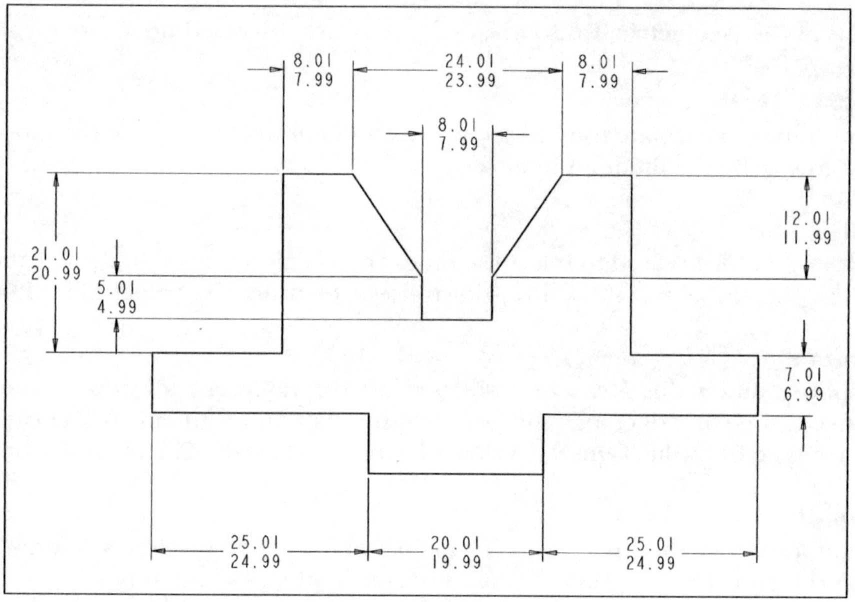

***Figure 11-14** Dimensions with tolerances*

Geometric Tolerances

When you send a component for manufacturing, you have to provide various other parameters along with the dimensions and dimensional tolerances. These parameters can be the geometric condition, the surface profile, the material condition, and so on. All these parameters are defined using the geometric tolerances. The geometric tolerances (will be called gtol henceforth) can be added to the drawing using the **Geometric Tolerance** dialog box, shown in Figure 11-15. The **Geometric Tolerance** dialog box is displayed when you choose **Insert > Geometric Tolerance** from the menu bar.

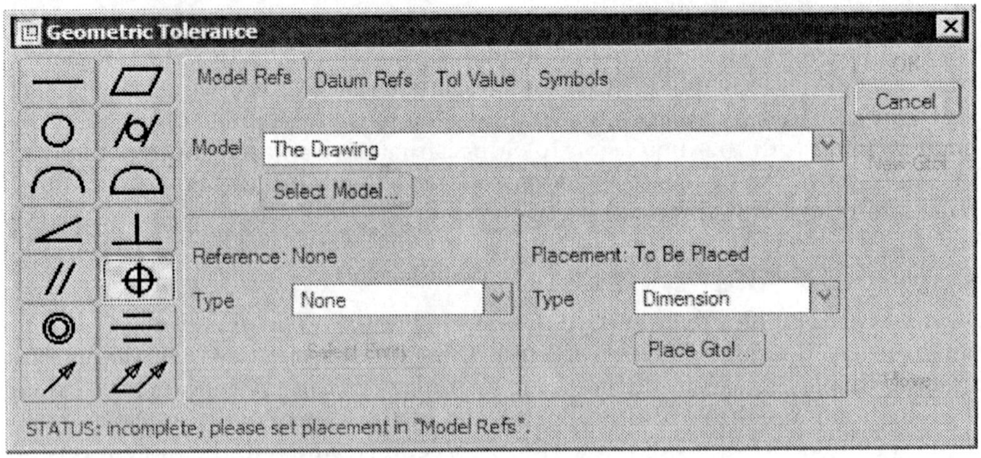

*Figure 11-15 The **Model Refs** tab of the **Geometric Tolerance** dialog box*

The options in the **Geometric Tolerance** dialog box are discussed next.

Model Refs Tab

The options under this tab are used to select the model or drawing where the gtol have to be placed and to specify the model references.

Model

This drop-down list is used to select the model on which you need to display the gtol. You can select the model from this drop-down list or by using the **Select Model** button.

Reference

The options under this area are used to select the reference for adding the gtol. The reference can be an edge, axis, surface, feature, datum, or an entity. You can select the reference type from the **Type** drop-down list or by using the **Select Entity** button.

Placement

The options under this area will be available only when you select a reference type for placing the gtol. These options are used to specify the placement type for the gtol. The placement type can be selected from the **Type** drop-down list.

Datum Refs Tab

The options under the **Datum Refs** tab, shown in Figure 11-16, are used to select the primary, secondary, or tertiary datum references for the gtol. You can also specify the composite tolerances to the primary, secondary, or tertiary datum references by using the **Composite**

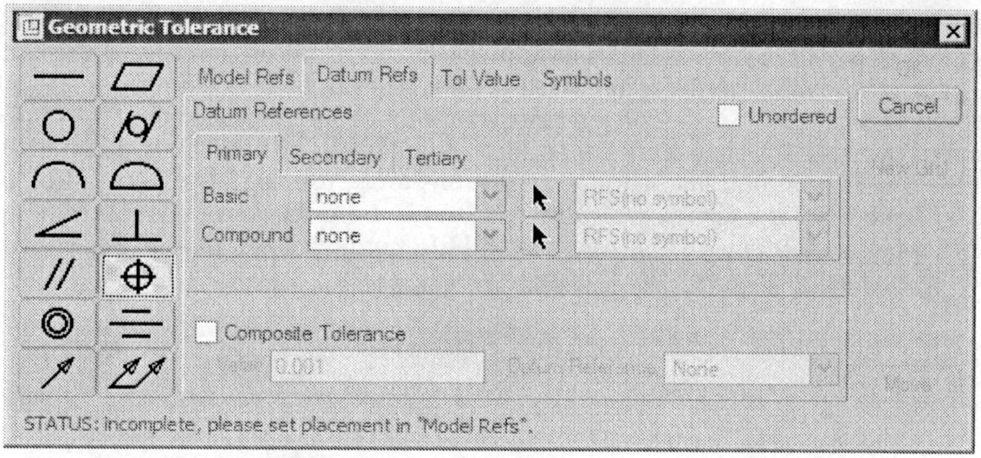

Figure 11-16 The ***Datum Refs*** *tab of the* ***Geometric Tolerance*** *dialog box*

Reference check box provided in this area. The **Composite Tolerance** check box is available only when the **Position** or the **Surface Profile** buttons from the types of gtol buttons available on the left side of the **Geometric Tolerance** dialog box are selected. The value of the composite reference can be specified in the **Value** edit box that is available only when you select the **Composite Reference** check box. The datum reference for the composite tolerance can be selected using the **Datum Reference** drop-down list that is available only when you select the **Composite Reference** check box.

Tol Value Tab

The options under the **Tol Value** tab, shown in Figure 11-17, are used to specify the tolerance value and the material condition for the gtol. You can specify the overall tolerance value or the per units tolerance value using the options under this tab. You can also specify the material condition using the **Material Condition** drop-down list.

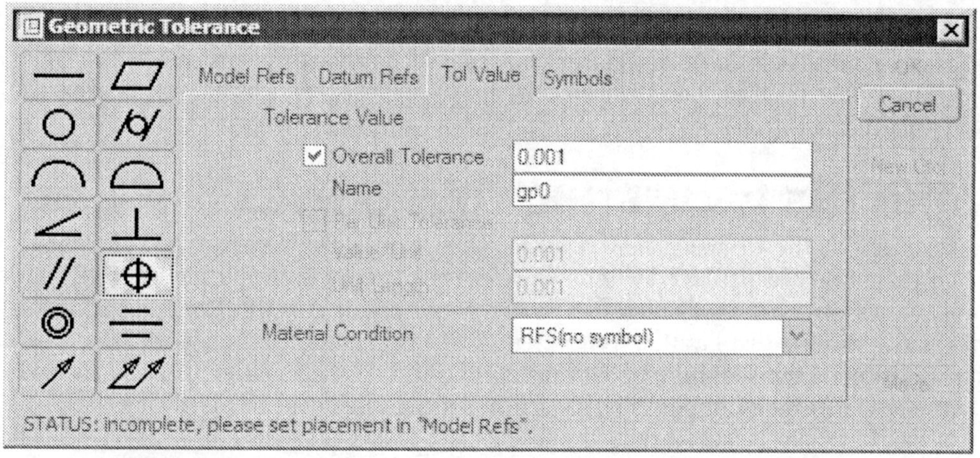

Figure 11-17 The ***Tol Value*** *tab of the* ***Geometric Tolerance*** *dialog box*

Symbols Tab

The options under the **Symbols** tab, shown in Figure 11-18, are used to specify the symbols for the gtol. The symbols and the modifiers for the gtol can be selected from the **Symbols and Modifiers** area and the projected tolerance zone symbol can be selected from the **Projected Tolerance Zone** area. The location of the projected tolerance zone symbol and its height can be specified using the options provided under this area. You can also specify the profile boundary symbol for the profile gtol from the **Profile Boundary** area that appears in the place of **Projected Tolerance Zone** area. The **Profile Boundary** area is available only if you choose the **Surface Profile or Line Profile** button.

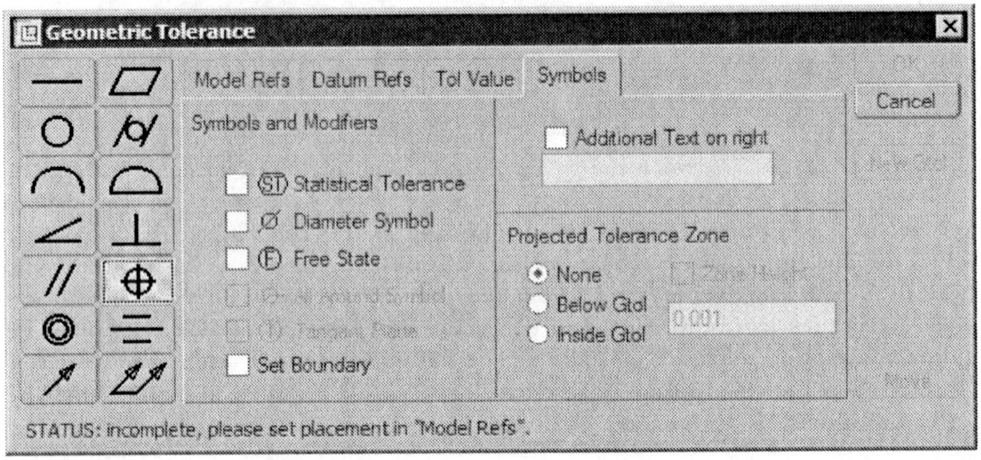

*Figure 11-18 The **Symbols** tab of the **Geometric Tolerance** dialog box*

New Gtol

The **New Gtol** button in the **Geomteric Tolerance** dialog box is chosen to accept the current gtol and place a new gtol.

Move

The **Move** button is chosen to change the location of the current gtol.

Figure 11-19 shows the drawing view with reference datums and gtol.

EDITING THE GEOMETRIC TOLERANCES

The gtol added to the drawing views can be edited using the **Geometric Tolerance** dialog box. This dialog box is displayed when you select the text of the tolerance from the drawing view and then right-click to display the shortcut menu. Choose the **Properties** option from the shortcut menu to display the **Geometric Tolerance** dialog box. All the options under the various tabs of the **Geometric Tolerance** dialog box are similar to those discussed in the **Geometric Tolerance** dialog box for adding the gtol.

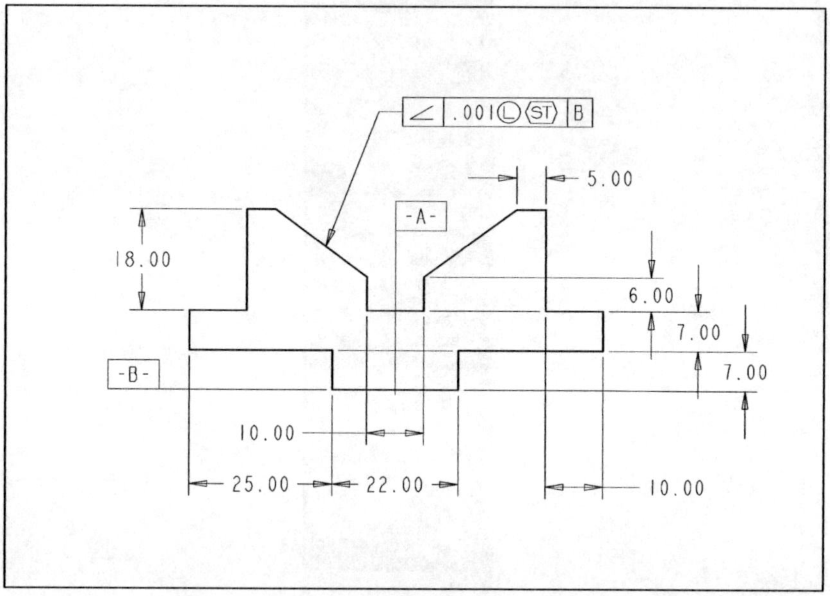

Figure 11-19 *The drawing view with reference datums and gtol*

ADDING NOTES TO THE DRAWING

The notes added to the model in the Part mode can be easily displayed in the drawing views using the **Show / Erase** dialog box. However, when you have not added notes to the model in the Part mode and still want to add them in the drawing views then you will have to create them. The notes can be created by choosing **Insert > Note** from the menu bar. When you choose this option, the **NOTE TYPES** menu is displayed, see Figure 11-20. In the **NOTE TYPES** menu, choose the type of leader to be used if required. Choose the **Make Note** option from this submenu and you are prompted to select the location for the note placement; choose a point on the screen and enter the text in the **Message Input Window** and press ENTER. The note will be placed at the selected point.

Note

You can also browse the text file saved in a .txt format using the ***NOTE TYPES*** *menu.*

ADDING BALLOONS TO THE ASSEMBLY VIEWS

The balloons can be added to the assembly drawing views by choosing **Insert > Balloon** from the menu bar. The procedure to place balloons is similar to that used while adding notes to the drawing. See Figure 11-21.

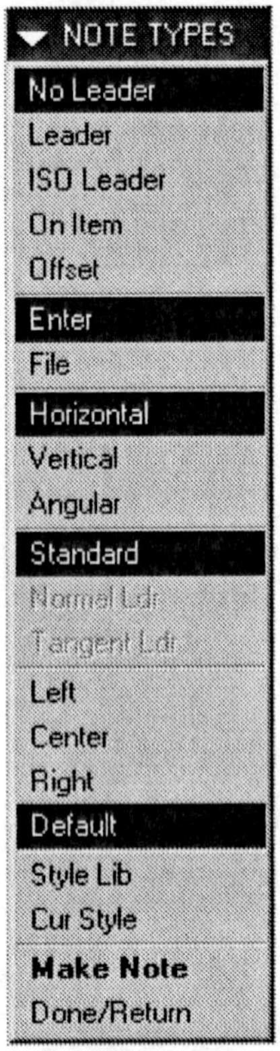

Figure 11-20 ***NOTE TYPES*** *menu*

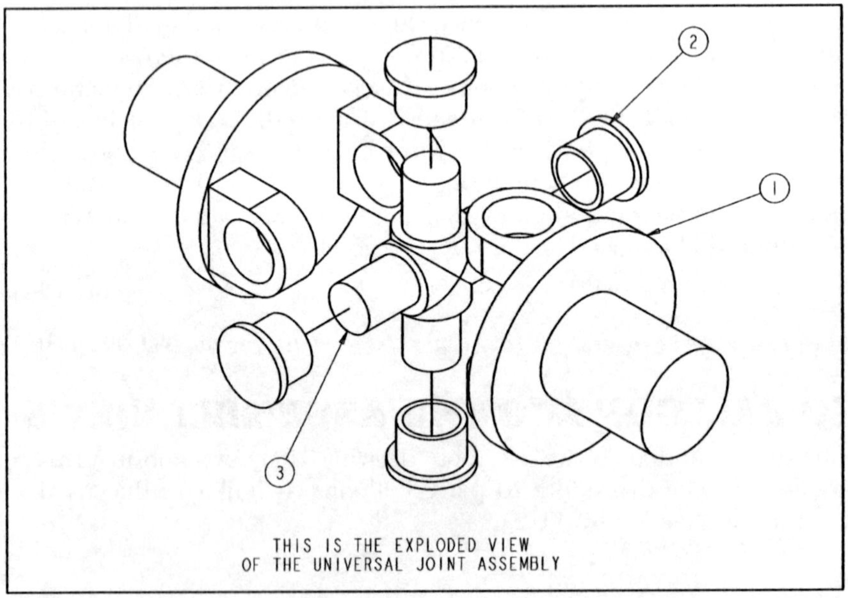

Figure 11-21 *Drawing view with balloons and note*

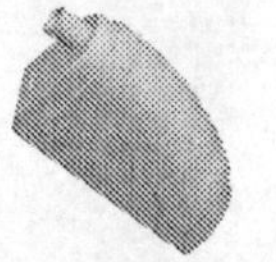

> **Tip**: *The notes or balloons can be modified and edited using the options in the shortcut menu that is displayed when you select them using the left mouse button and hold down the right mouse button.*

TUTORIALS

Tutorial 1

In this tutorial you will generate the drawing views of the part shown in Figure 11-22. Generate the dimensions in the drawing views. The views to be generated are shown in Figure 11-23.

(Estimated time: 45 min)

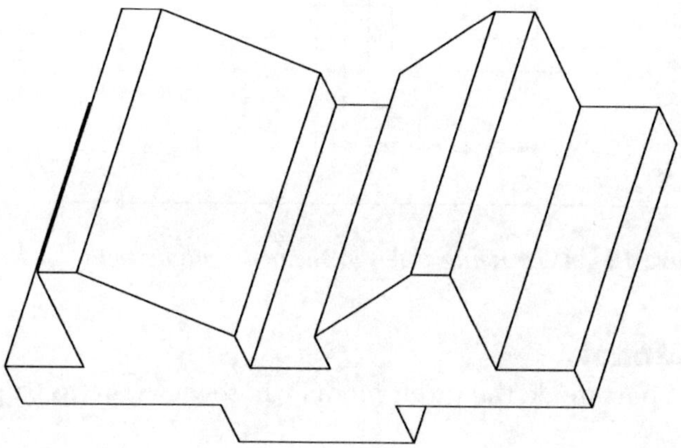

Figure 11-22 *Model for generating the drawing views*

The following steps outline the procedure to create the given drawing:

a. Create the model in the Part mode using the given dimensions and save the file.

b. Open a new drawing file in the Drawing mode and generate the top view and the front view of the model on the drawing sheet.

c. Generate the dimension to the views and clean them using the **Clean Dimensions** dialog box.

d. Add geometric tolerance to the required entities.

Before you start creating the model and the drawing views, set the working directory to *C:\ProE-WF\c11*. Make sure that the *c11* directory exists inside the *ProE-WF* directory. If these directories do not exist, create them using the **Select Working Directory** dialog box.

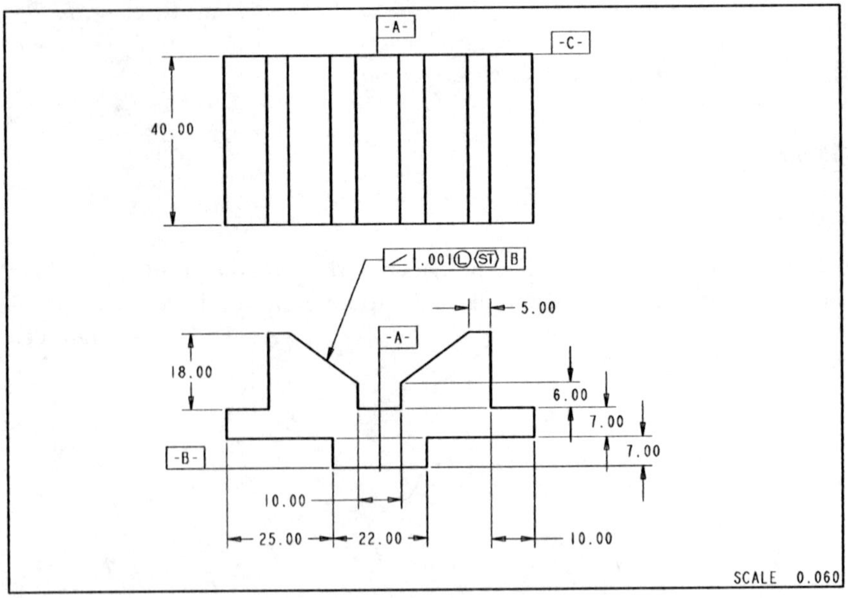

***Figure 11-23** Drawing views to be created along with the dimensions*

Creating the Model

First, create in the part mode the given model whose views are to be generated later.

1. Open a new object file in the Part mode and name it *c11tut1*.

2. Create the model in the Part mode and save this model.

Creating a New Drawing File

Create a new drawing file to generate the required views.

1. Choose the **Create a new object** button from the **File** toolbar. The **New** dialog box is displayed.

2. Select the **Drawing** radio button from the **Type** area and enter the name of the drawing as *c11tut1* in the **Name** edit box.

3. Choose **OK** from the **New** dialog box to proceed further. The **New Drawing** dialog box is displayed.

4. If the name of the model is not displayed in the **Default Model** area, choose the **Browse** button and select the model. Select the **Empty** radio button from the **Specify Template** area, **Landscape** button from the **Orientation** area, and the **A4** option from the **Standard Size** drop-down list.

5. Choose **OK** from the **New Drawing** dialog box to enter the Drawing mode.

   You enter the Drawing mode and a new A4 size drawing sheet is displayed on the screen.

Generating the Drawing Views

You need to generate the top view and the front view of the model. Any of the two views can be generated first. However, you will generate the top view first.

1. Choose the **Insert a drawing view of the active model** button from the **Top Toolchest** to display the **VIEW TYPE** menu.

2. Choose **General** > **Full View** > **No Xsec** > **Scale** > **Done**.

3. Specify the placement point for the top view as shown in Figure 11-24. As soon as you specify the point, the **Message Input Window** is displayed. Enter the scale factor for the view as **0.06** and press ENTER. The default view is placed and the **Orientation** dialog box is displayed.

4. Select the **TOP** option from the **Saved Views** drop-down list in the **Orientation** dialog box.

5. Choose the **Set** button. The default view is oriented as the top view.

6. Choose **OK** to exit the **Orientation** dialog box. Choose **No Hidden** from the **Model Display** toolbar and repaint the screen using the **Redraw the current view** button from the **View** toolbar.

   The text for scale is placed at the corner of the sheet so that it does not interfere with the other text in the drawing sheet. To move the text perform the next step.

7. Select the text; the text is highlighted in red. Select the text again and drag it to the bottom right corner of the drawing sheet using the left mouse button. After moving the text, you notice that the text is still selected. To unselect it, double-click anywhere on the sheet in the blank area.

8. Choose the **Insert a drawing view of the active model** button from the **Top Toolchest** to display the **VIEW TYPE** menu.

9. Choose **Projection** > **Full View** > **No Xsec** > **No Scale** > **Done**.

10. Specify the placement point for the front view below the top view as shown in Figure 11-24.

    The two views are shown in Figure 11-24.

Renaming the Datum Planes

The datum planes are renamed because the gtol values to the inclined surface will be applied with their reference. This is evident from Figure 11-23.

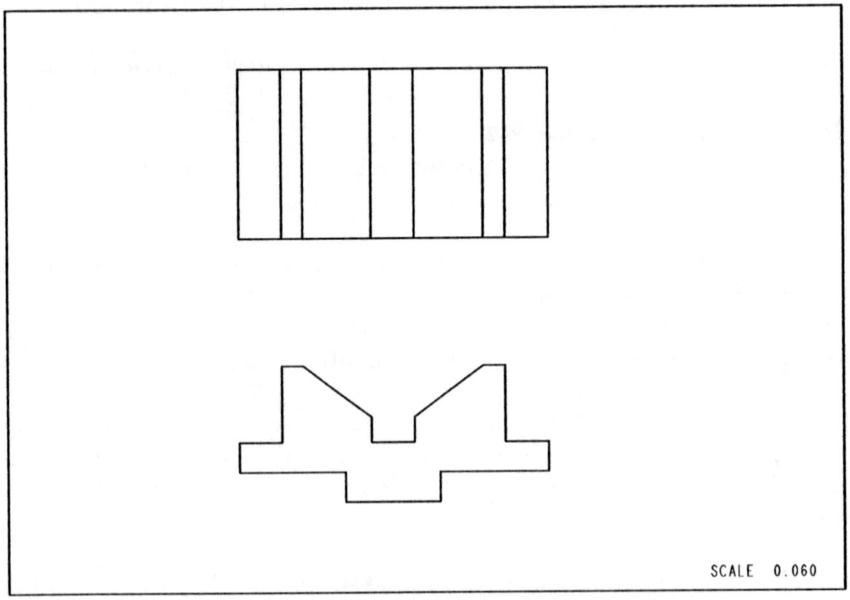

Figure 11-24 *Top view and the front view of the model*

1. Choose the **Datum plane on/off** button from the **Datum Display** toolbar to display the datum planes and then repaint the screen.

2. Select the **RIGHT** datum plane from the top view and when it turns red in color hold down the right mouse button to display the shortcut menu.

3. Choose the **Properties** option from the shortcut menu. The **Datum** dialog box is displayed.

4. Enter **A** in the **Name** edit box and then select the second button from the **Type** area to enclose the datum in the reference box. Make sure the **Free** radio button is selected in the **Placement** area.

5. Choose **OK** to exit the **Datum** dialog box.

6. Repeat steps 3, 4, and 5 to rename **TOP** and **FRONT** datum planes as **B** and **C** respectively.

Dimensioning the Drawing Views

1. Choose the **Open the Show/Erase dialog box** button from the **Top Toolchest**. The **Show / Erase** dialog box is displayed.

2. Choose the **Show** button if it is not chosen by default and then choose the **Dimension** button from the **Type** area in the **Show / Erase** dialog box.

3. Select the **View** radio button from the **Show By** area in the **Show / Erase** dialog box.

4. Select the front view (the view at the bottom) to display all the dimensions of this view.

5. Press the middle mouse button.

6. Now, select the top view and then press the middle mouse button. All the dimensions and the dimensions for this view are displayed in the selected view. Exit the dialog box.

Placing the Dimensions in Order

The dimensions displayed in the drawing views are scattered and improperly placed. This in turn makes the drawing look very untidy. You need to place the dimensions in order using the **Clean Dimensions** dialog box.

1. Choose **Edit** > **Cleanup** > **Dimensions** from the menu bar to display the **Clean Dimensions** dialog box. You are prompted to select the dimensions to clean.

2. Select all the dimensions from both the views by drawing a rectangle around them. Press the middle mouse button. You will notice that the selected dimensions are not highlighted but the display box below the **Dimensions to Clean** area displays the number of selected dimensions.

3. Enter **0.5** in the **Offset** edit box and **0.5** in the **Increment** edit box.

4. Clear the **Create Snap Lines** check box. Choose **Apply** and then choose **Close** to exit the **Clean Dimensions** dialog box.

5. After cleaning the dimensions, the drawing views should look similar to the one shown in Figure 11-25. Select and move the dimensions if necessary.

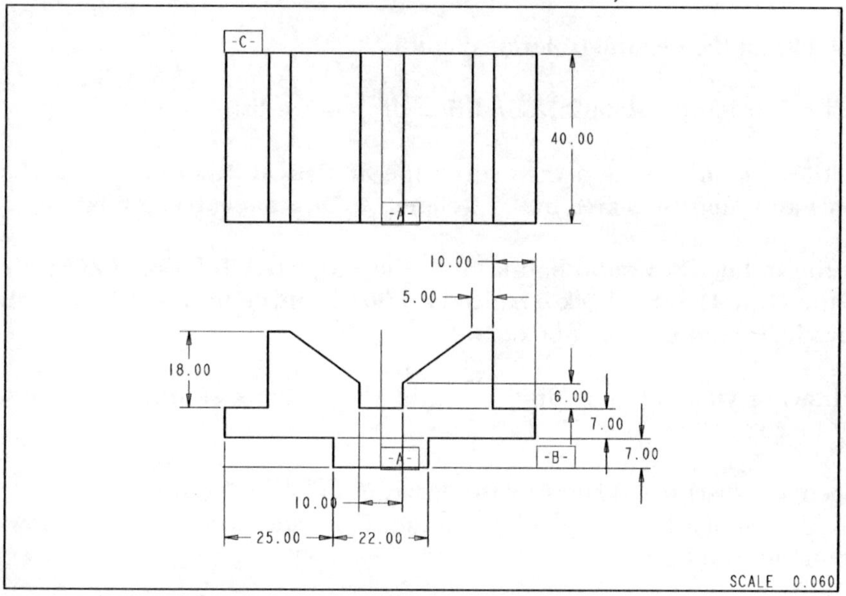

Figure 11-25 *Top and the front views of the model with the dimensions and datum planes*

Displaying the Geometric Tolerance

Actually the tolerance is created in the Part mode. However, in this chapter you will learn how the tolerances are represented in the Drawing mode.

1. Choose **Insert > Geometric Tolerance** from the menu bar to display the **Geometric Tolerance** dialog box. You will notice that the name of the model *c11tut1.prt* is displayed in the **Model** drop-down list.

2. Choose the geometric condition as angularity using the **Angularity** button from the **Geometric Tolerance** dialog box.

3. Select the **Surface** option from the **Type** drop-down list in the **Reference** area and select the left inclined edge from the front view.

4. Select **Leaders** from the **Type** drop-down list in the **Placement** area. The **ATTACH TYPE** menu is displayed and you are prompted to select an edge for attaching the leader.

5. Select the left inclined edge from the front view and then middle-click at a position on the graphics screen where you want the note to be placed. Even if the note is placed at an improper position on the graphics screen, you can move its location later.

6. Choose the **Datum Refs** tab from the **Geometric Tolerance** dialog box.

7. Select **B** from the **Basic** drop-down list under the **Primary** tab of the **Datum References** area.

8. Select the **Tol Value** tab from the **Geometric Tolerance** dialog box.

9. Enter **0.001** in the **Overall Tolerance** edit box.

10. Select **LMC** from the **Material Condition** drop-down list.

11. Select the **Symbols** tab and then select the **Statistical Tolerance** check box from the **Symbols and Modifiers** area in the **Geometric Tolerance** dialog box.

12. Select the **Below Gtol** radio button from the **Projected Tolerance Zone** area and then select the **Zone Height** check box. Enter **0.001** in this edit box. Choose **OK** to exit the **Geometric Tolerance** dialog box.

13. The drawing views after adding the gtol should look similar to the one shown in Figure 11-26.

    You need to move the note for gtol to a location shown in Figure 11-27 by selecting it and dragging it. Similarly, arrange all the dimensions and datums on the drawing sheet as shown in Figure 11-27.

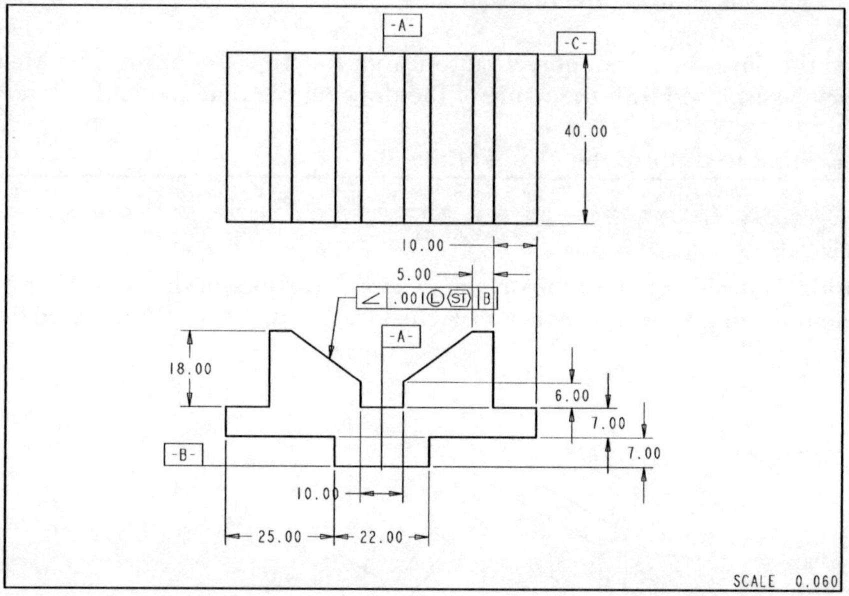

Figure 11-26 Top and the front views of the model with the dimensions, datum planes, and the geometric tolerances

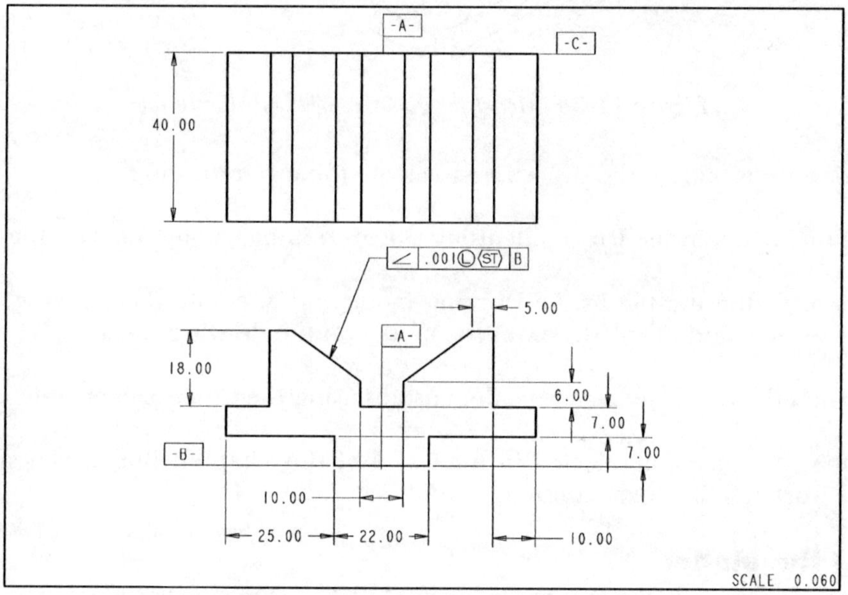

Figure 11-27 The required drawing

Saving the Drawing File

You need to save the drawing file that you have created as you might need it later.

1. Choose the **Save the active object** button from the **Top Toolchest**. The **Message Input Window** is displayed with the name of the drawing file that you had entered earlier.

2. Press ENTER to confirm the saving of the file.

Tutorial 2

In this tutorial you will generate the drawing views of the model shown in Figure 11-28. The dimensioned drawing views are shown in Figure 11-29. **(Estimated time: 40 min)**

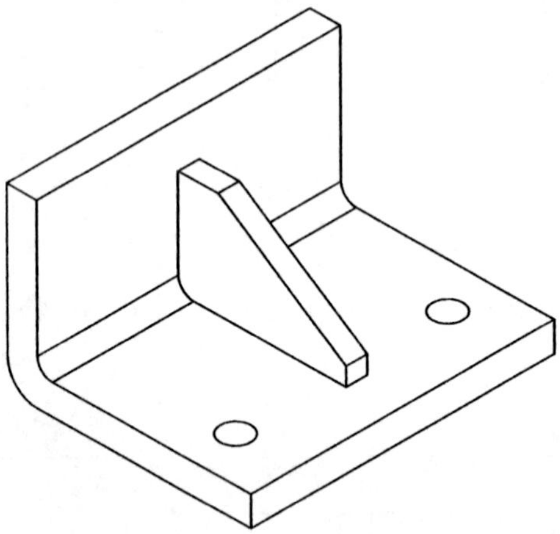

Figure 11-28 *Model for generating the drawing views*

The following steps outline the procedure to create the given drawing:

a. Create the model in the Part mode using the given dimensions and save the file.

b. Open a new drawing file in the Drawing mode and generate the top view, front view, right-side view, and detailed view of the model on the drawing sheet.

c. Dimension the views and clean the dimensions using the **Clean Dimensions** dialog box.

The working directory was selected in tutorial 1 of this chapter. But, if you still want to change the working directory set it to *C:\ProE-WF\c11*.

Creating the Model

First, you need to create in the Part mode the model whose drawing views are to be generated.

1. Create the model in the Part mode using the given dimensions.

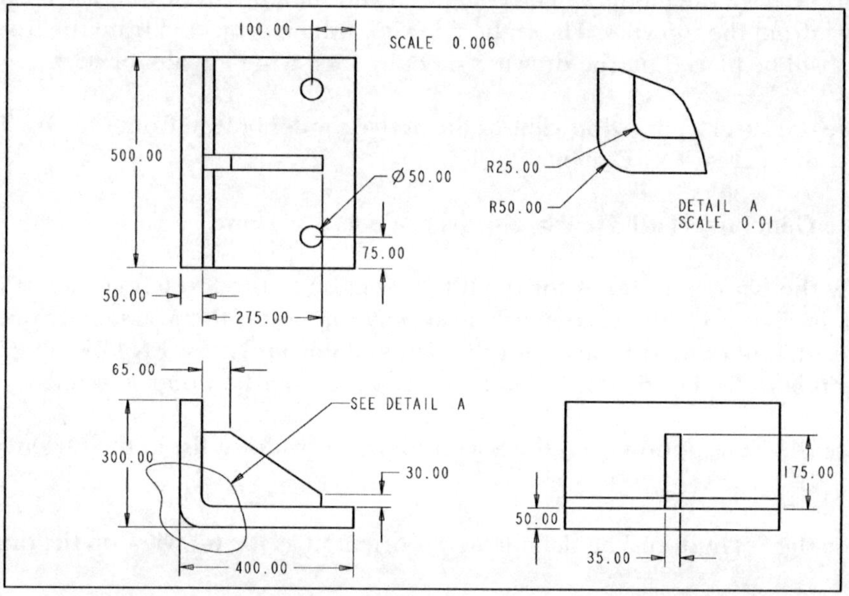

Figure 11-29 The drawing views to be generated in Tutorial 2

2. Save this model created in the Part mode as *c11tut2*.

Creating a New Drawing File

Create a new drawing file to generate the required drawing views.

1. Choose the **Create a new object** button from the **File** toolbar. The **New** dialog box is displayed.

2. Select the **Drawing** radio button from the **Type** area and enter the name of the drawing as *c11tut2* in the **Name** edit box.

3. Choose **OK** from the **New** dialog box to display the **New Drawing** dialog box. The **New Drawing** dialog box is displayed with the name of the model in the **Default Model** area.

4. If the name of the model is not displayed in the **Default Model** area, choose the **Browse** button and select the model.

5. Select the **Empty** radio button from the **Specify Template**, the **Landscape** button from the **Orientation** area, and the **A4** option from the **Standard Size** drop-down list.

6. Choose **OK** from the **New Drawing** dialog box to enter the Drawing mode. A new drawing sheet of A4 size is displayed on the screen.

Creating the Drawing Views

First, the top view of the model will be generated and then the front view will be generated by projecting it from the top view. The right-side view will be projected from the front view. The detail view will be placed on the drawing sheet by specifying a scale for it.

1. Choose the **Insert a drawing view of the active model** button from the **Top Toolchest** to display the **VIEW TYPE** menu.

2. Choose **General** > **Full View** > **No Xsec** > **Scale** > **Done**.

3. Specify the placement point for the top view close to the top left corner of the sheet as shown in Figure 11-30. As soon as you specify the point, the **Message Input Window** is displayed. Enter the scale factor for the view as **0.006** and press ENTER. The **Orientation** dialog box is displayed. The default view is placed on the point selected.

4. Choose the **TOP** option from the **Saved Views** drop-down list in the **Orientation** dialog box.

5. Choose the **Set** button. The default view is oriented as the top view on the drawing sheet.

6. Choose **OK** to exit the **Orientation** dialog box. Repaint and move the view if necessary.

   The front view of the model is generated from the top view.

7. Choose the **Insert a drawing view of the active model** button from the **Top Toolchest** to display the **VIEW TYPE** menu.

8. Choose **Projection** > **Full View** > **No Xsec** > **No Scale** > **Done**.

9. Specify the placement point for the front view below the top view as shown in Figure 11-30.

10. Similarly, place the right-side view and the detailed view on the drawing sheet as shown in Figure 11-30.

Dimensioning the Drawing Views

1. Choose the **Open the Show/Erase dialog box** button from the **Top Toolchest**. The **Show / Erase** dialog box is displayed.

2. Choose the **Dimension** button from the **Type** area and then select the **Feature** radio button from the **Show By** area. Select the feature in the detailed view. All the dimensions related to that feature (the base feature) are displayed in all the views. Press the middle mouse button.

3. Select the **View** radio button from the **Show By** area and then select all the three views one by one to display all the dimensions in the views. Press the middle mouse button and then close the **Show / Erase** dialog box by choosing the **Close** button.

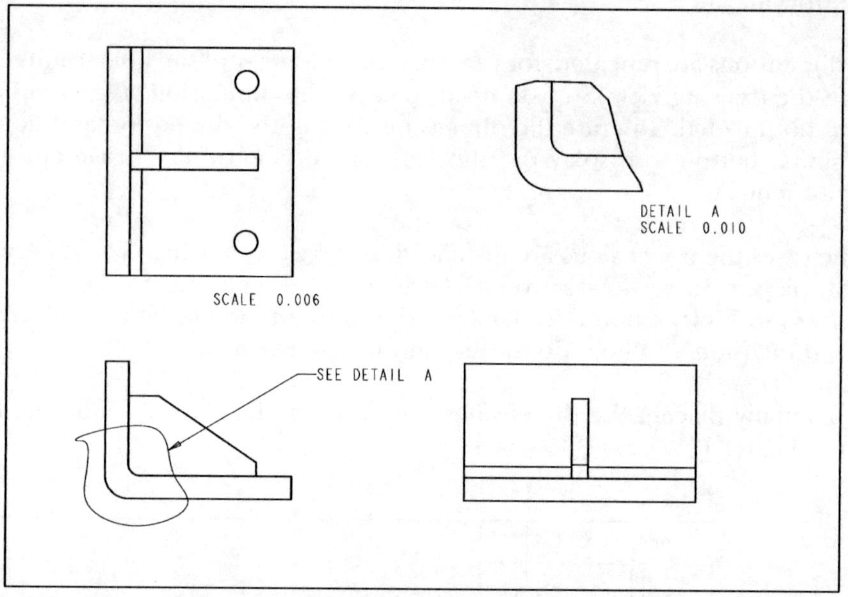

Figure 11-30 The drawing views before adding the dimensions

Placing the Dimensions in Order

The dimensions displayed in the drawing views are scattered and improperly placed. Use the **Clean Dimensions** dialog box to place the dimensions in order.

1. Choose **Edit > Cleanup > Dimensions** from the menu bar to display the **Clean Dimensions** dialog box. You are prompted to select the dimensions to clean.

2. Select all the dimensions from both the views by drawing a rectangle around them. Press the middle mouse button so that the options in the **Clean Dimensions** dialog box are available.

   When you select the dimensions, the radius dimensions in the drawing views are not selected. This means that the radius dimensions will remain unaffected by the values that you will enter in the **Clean Dimensions** dialog box.

3. Enter **0.4** in the **Offset** edit box and **0.3** in the **Increment** edit box.

4. Clear the **Create Snap Lines** check box. Choose **Apply** and then choose **Close** to exit the **Clean Dimensions** dialog box.

5. After cleaning the dimensions, you notice that the dimensions are not placed in the order that is required. Now, you need to manually place the dimensions in order.

6. To move a dimension, select the dimension and drag it to place it at the desired location

on the drawing sheet.

Some dimensions are repeated, for example, the diameter of the hole feature is displayed twice in the drawing view. So, you need to erase the dimensions that are repeated and that are not needed. To erase the dimensions, select the dimension and hold down the right mouse button to display the shortcut menu. Choose the **Erase** option from the shortcut menu.

In some cases the dimensions are displayed in the views in which you do not want them to be displayed. In such cases you can switch that dimension to the other views. The **Move Item to View** option is available in the shortcut menu that is displayed when you select a dimension and hold down the right mouse button.

After manually placing the dimensions the drawing views should look similar to views shown in Figure 11-31.

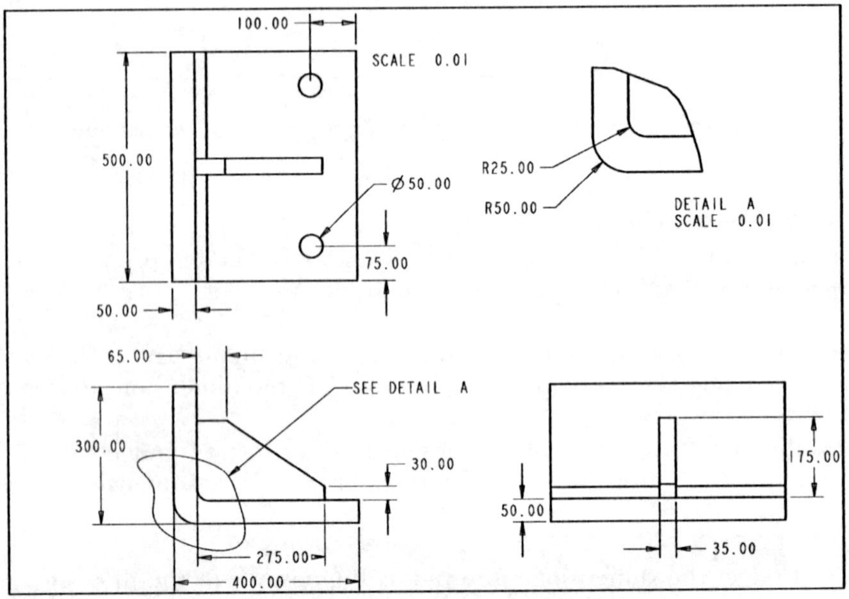

Figure 11-31 The required drawing views

Saving the Drawing File

You need to save the drawing file that you have created as you might need it later.

1. Choose the **Save the active object** button from the **Top Toolchest**. The **Message Input Window** is displayed with the name of the drawing file that you had entered earlier.

2. Press ENTER to confirm the saving of the file.

Self-Evaluation Test

Answer the following questions and then compare your answers to the answers given at the end of this chapter.

1. The reference datum planes are used to display the geometric tolerances in a drawing view. (T/F)

2. If you modify a dimension value in the Drawing mode and then regenerate the drawing view, the modification is applied to the model in the Part mode also. (T/F)

3. The notes created in the Part mode on a model can also be displayed in a drawing view in the Drawing mode. (T/F)

4. When you select a drawing view to display dimensions, all the dimensions are placed in proper order in the drawing view. (T/F)

5. Most of the editing operations in a drawing view can be done using the shortcut menu that is displayed when you hold down the right mouse button on an item to be modified. (T/F)

6. The __________ dialog box is used to display the dimensions in the drawing views.

7. The dimension values can be modified using the __________ option from the **Edit** menu in the menu bar.

8. The __________ dialog box is used to clean the dimensions in the drawing views.

9. You have to set the value of __________ to **Yes** in the configuration file to display the dimensional tolerance in the drawing views.

10. The configuration file can be opened from __________.

Review Questions

Answer the following questions:

1. Which of the following dialog boxes is used to display dimensions in a drawing view?

   (a) **Dimension Properties** (b) **Clean Dimensions**
   (c) **Show / Erase** (d) None

2. Which of the following dialog boxes is used to modify properties of text in a drawing view?

   (a) **Dimension Properties** (b) **Clean Dimensions**
   (c) **Show / Erase** (d) None

3. Which of the following dialog boxes is used to display the geometric tolerances in a drawing view?

(a) **Geometric Tolerance** (b) **Clean Dimensions**
(c) **Show / Erase** (d) None

4. Which of the following options available in the shortcut menu is used to switch a dimension from one view to another?

(a) **Erase** (b) **Flip Arrows**
(c) **Move Item to View** (d) None

5. Which of the following radio button in the **Type** area is selected in the **New** dialog box to open a new drawing file?

(a) **Assembly** (b) **Sketch**
(c) **Part** (d) None

6. The dimensions erased using the **Erase** option from the shortcut menu are deleted from the model in the Part mode also. (T/F)

7. The notes or balloons can be modified and edited using the options in the shortcut menu. (T/F)

8. The system does not show a reference datum plane in a drawing view unless it is perpendicular to the graphics screen. (T/F)

9. Generally, the dimensions displayed in the drawing views are scattered and improperly placed. (T/F)

10. You cannot move text dynamically on the drawing sheet. (T/F)

Exercise

Exercise 1

In this exercise you will generate the drawing views of the model created in Exercise 2 of Chapter 6, shown in Figure 11-32, and add dimensions to the views as shown in the figure.

(Estimated time: 40 min)

Note

In the above drawing views, the center lines are not generated but are created manually. After creating the center lines you need to define a linestyle and then apply it on the center lines.

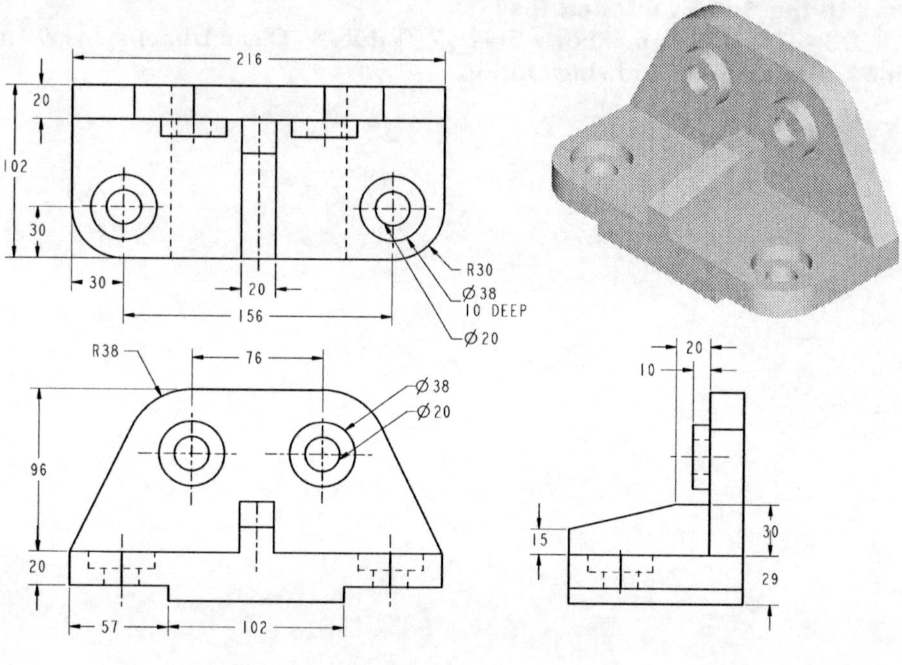

***Figure 11-32** The required drawing views with the solid model*

Answers to the Self-Evaluation Test

1 - T, 2 - T, 3 - T, 4 - F, 5 - T, 6 - Show / Erase, 7 - Value, 8 - Clean Dimensions, 9 - tol_display, 10 - File > Properties > Drawing Options.

Chapter 12

Other Drawing Options

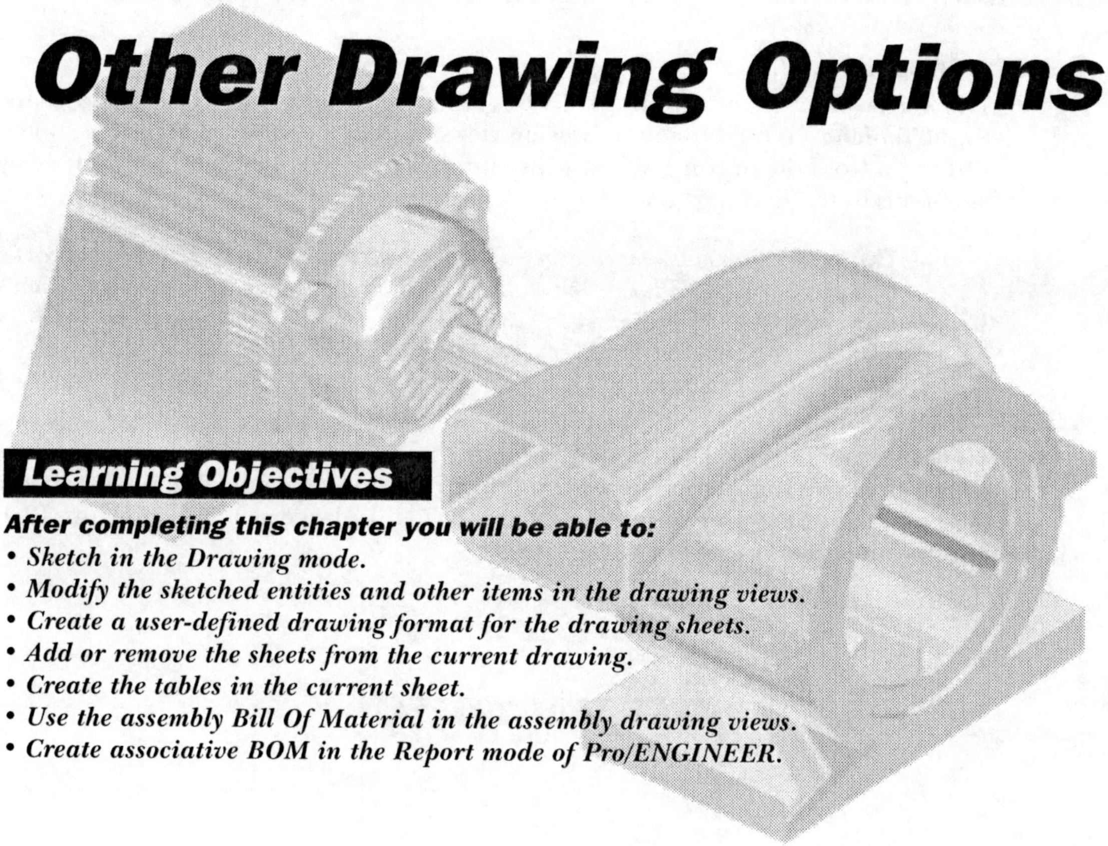

Learning Objectives

After completing this chapter you will be able to:

- *Sketch in the Drawing mode.*
- *Modify the sketched entities and other items in the drawing views.*
- *Create a user-defined drawing format for the drawing sheets.*
- *Add or remove the sheets from the current drawing.*
- *Create the tables in the current sheet.*
- *Use the assembly Bill Of Material in the assembly drawing views.*
- *Create associative BOM in the Report mode of Pro/ENGINEER.*

SKETCHING IN THE DRAWING MODE

Sketching is one of the most important tools available in the Drawing mode. Sketching in the Drawing mode is called drafting. As discussed earlier, there are two types of drafting in Pro/ENGINEER: Generative drafting and Interactive drafting. Any item on the drawing sheet that is not generated from a model is called a draft entity or a draft item. Drafting is extensively used for creating user-defined formats, drawing tables, and also for drawing the title blocks in the formats. The sketching in the Drawing mode is almost similar to the sketching in the other modes of Pro/ENGINEER. The sketching can be done by using the Sketch menu in the menu bar or the tool buttons available in the **Drawing Sketcher Tools** toolbar from the **Right Toolchest**. The tools available in the **Drawing Sketcher Tools** toolbar and the options available in the **Sketch** menu that are used to sketch in the Drawing mode are discussed next.

Select Items

The **Select one item at a time - shift to gather more than one item** button in the **Right Toolchest** is used to select drawing views, sketched entities, dimensions, notes, and so on from the drawing views. This button is used extensively to move drawing views or the items in the drawing view.

Tip: *To delete an entity or an item from a drawing view, you can select the item and then press the DELETE key. You can also select the entity and use the* **Delete** *option from the shortcut menu that is displayed when you hold down the right mouse button.*

Line

The **Line** option in the **Sketch** menu from the menu bar or the **Create 2 point lines** button from the **Right Toolchest** is used to create line segments. When you choose this option, the **Snapping References** dialog box is displayed as shown in Figure 12-1. The options in this dialog box are discussed next.

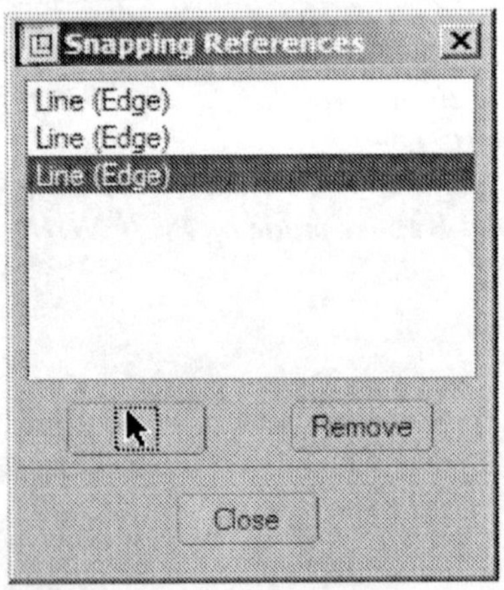

Figure 12-1 Snapping References *dialog box*

Select references button

When you choose the **Select references** button from the **Snapping References** dialog box, you are prompted to select an entity to which you want the line to be snapped.

Remove

The **Remove** button is used to remove a selected entity from the **Snapping References** dialog box.

Circle

The **Circle** option or the **Create circle** button is used to create circles. To create a circle you need to specify a center point and a point on the circumference of the circle.

Arc

The **Arc** option or the **Create an arc by 3 points or tangent to an entity at its endpoint** button is used to create arcs.

The other geometric entities that you can draw using the tools available in the **Right Toolchest** are splines, construction circles and lines, ellipses, and points. You can also create a fillet and a chamfer between two entities.

Chain

The **Enable sketching chain** button is used when the object you want to draw consists of more than one entity. For example, when you choose the **Create 2 points line** button to draw a line, you can draw a single line. But when you draw the line after choosing the **Enable sketching chain** button, you are able to draw a continuous chain of lines. The chain can be ended by pressing the middle mouse button.

Sketching Using References

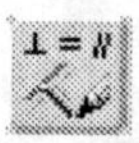

The **Remember parametric sketcher references** button is used to draw the sketches by automatically applying constraints.

MODIFYING THE SKETCHED ENTITIES

There are various options to modify the sketched entities that are termed as draft entities. The operations or the modifications that can be applied on the draft entities in Pro/ENGINEER are discussed next.

Trimming the Draft Entities

The sketched or draft entities can be trimmed by choosing **Edit > Trim** from the menu bar. When you choose this option, the cascaded menu is displayed as shown in Figure 12-2. The options in this menu are discussed next.

Divide at Intersection

The **Divide at Intersection** option is used to divide two selected entities at the point of

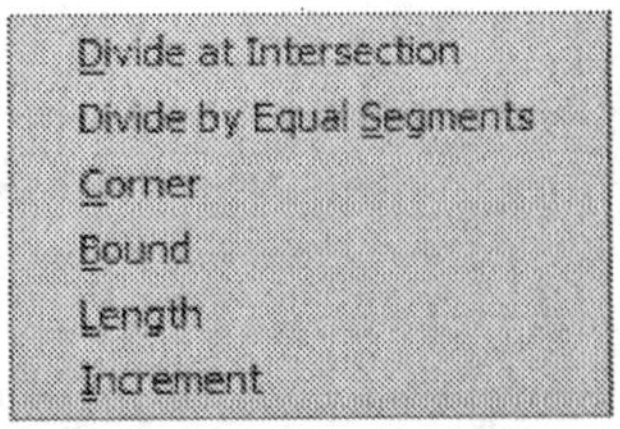

Figure 12-2 *Cascaded menu of* ***Trim*** *option in* ***Edit*** *menu*

intersection. When you choose this option you are prompted to select the two entities. Select the two entities to divide them at the intersection point.

Divide by Equal Segments

When you choose this option, you are prompted to select the entities you need to divide into equal segments. Select the entities to divide. To select more than one entity, press CTRL+left mouse button. After selecting the entity or entities, press the middle mouse button. The **Message Input Window** is displayed and you are prompted to enter the number of equal segments. Enter the number of segments into which you need to divide the entity.

Corner

This option is the same as that was discussed in the Sketch mode. This option trims the two selected entities so that a corner is formed after trimming.

Bound

This option is used for deleting or extending a portion of an entity by defining a bounding entity.

Length

This option is used to trim or extend an entity up to a specified length.

Increment

This option is used to extend or shorten an entity in specified increments. To extend the entity, enter a positive increment value and to shorten the entity, enter a negative increment value.

Transforming the Draft Entities

To translate or move the draft entities, choose **Edit > Transform**. The cascaded menu is displayed as shown in Figure 12-3. The options in this cascaded menu are discussed next.

Translate

This option is used to move a selected entity or entities on the drawing sheet. When you select the entities, press the middle mouse button to confirm the selection. After you confirm the selection, the **GET VECTOR** menu along with the **GET POINT** submenu, shown in Figure 12-4, is displayed. The options in the **GET VECTOR** menu are used to specify the direction of translation.

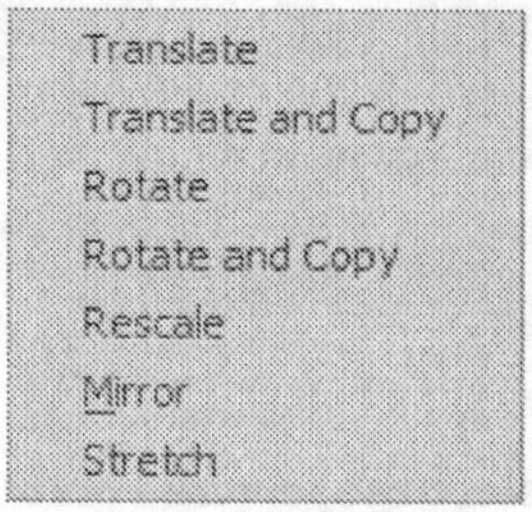

Figure 12-3 *Cascaded menu of* ***Translate*** *option in* ***Edit*** *menu*

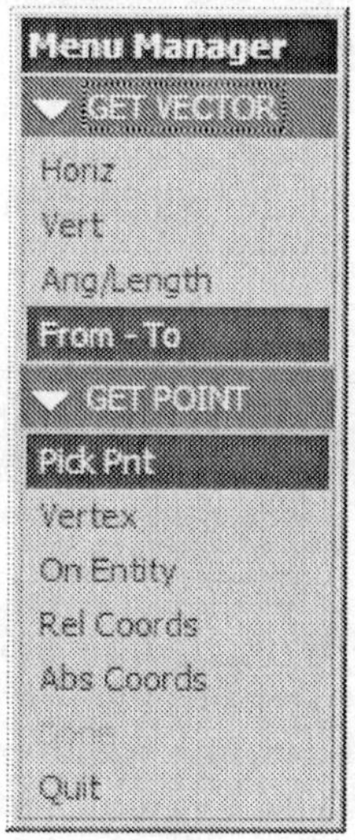

Figure 12-4 GET VECTOR *menu with* ***GET POINT*** *submenu*

Translate and Copy

The **Translate and Copy** option is used to translate the selected entity and at the same time make a copy of the entity. After selecting the entities when you press the middle mouse button, the **GET VECTOR** menu and the **GET POINT** submenu are displayed.

Rotate

The **Rotate** option is used to rotate the selected entity from its actual location. When you choose this option, you are prompted to select the entity you want to rotate. After selecting the entity or entities, press the middle mouse button. The **Message Input Window** is displayed and you are prompted to enter the rotation angle in the counterclockwise direction. The default direction of rotation will be counterclockwise. However, you can rotate the selected entity in the clockwise direction by specifying a negative rotation angle.

Rotate and Copy

The **Rotate and Copy** option is used to create a pattern of the selected draft entity. When you choose the entity after choosing this option, press the middle mouse button. You are prompted to select the center point of rotation. Select the center point of rotation. The **Message Input Window** is displayed and you are prompted to enter the rotation angle. After specifying the rotation angle, enter the number of copies in the **Message Input Window**. The rotated copies of the selected entity are placed on the drawing sheet.

Rescale

The **Rescale** option is used to rescale the selected entity. After choosing this option when you select the entity to rescale, press the middle mouse button. You are prompted to select the origin point for scaling. The new scale factor can be specified in the **Message Input Window**. A positive scale factor will increase the size of the selected entity and a negative scale factor will reduce the size of the selected entity.

Mirror

The **Mirror** option is used to mirror the selected entity using an existing sketched line.

Stretch

The **Stretch** option is used to stretch a sketched entity by selecting it using a window. The vertices of the sketched entity that are included inside the window will be moved from their original location and stretched and the remaining vertices will remain stationary.

Grouping the Entities

A draft entity can be grouped with other draft entity, with a note, a geometric tolerance (gtol), or a dimension. In Pro/ENGINEER, the note, gtol, and dimensions are called detail items. After grouping the draft entities, any operation applied on any one of them is applied to the grouped entity also. The options to group can be invoked by choosing **Edit > Group**. The cascaded menu is displayed as shown in Figure 12-5. The options in this cascaded menu are discussed next.

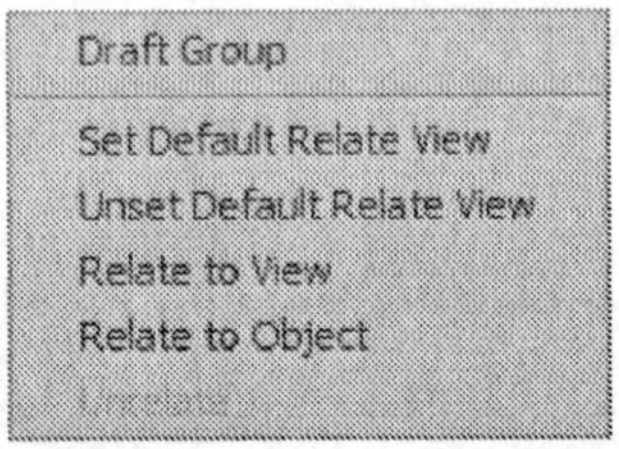

***Figure 12-5** Cascaded menu of **Group** option in **Edit** menu*

Draft Group

This option is used to form a group among draft entities like lines, arcs, circles, and so on. The operation performed on any one of the grouped entity is performed on all the entities forming that group. When you choose this option, the **DRAFT GROUP** menu is displayed as shown in Figure 12-6.

Set Default Relate View

This option is available only when the drawing view is selected. This option is used to set the

*Figure 12-6 **DRAFT GROUP** menu*

draft view as the current view. The draft entities that are drawn after setting a default draft view are grouped with the current draft view. Now, if you move the draft view then the grouped entities will also move with it.

Note

When you move a view that forms a group with draft entities, the draft entities also move. But when you move the lines, the draft view does not move.

Unset Default Relate View

This option is used to unset the current draft view. This option is available only when a current draft view exists.

Relate to View

This option is used to relate a view with a draft entity or entities. This option is available in the cascaded menu only when a draft entity is selected. When you choose this option after selecting a draft entity, you are prompted to select a view to which you want to relate the draft entity. Select the drawing view from the drawing sheet.

Relate to Object

This option is used to relate notes, dimensions, gtols, and other objects to a drawing view. This option is available in the cascaded menu only when an object is selected. When you choose this option after selecting an object, you are prompted to select a view to which you want to relate the object.

Drafting Using the Drawing Views

You can use the generated drawing views to create an entity or the whole drawing view. To use a generated drawing view, choose **Sketch > Edge**. The cascaded menu is displayed with two options; **Use** and **Offset**.

Use

The **Use** option is selected to draw a copy of the selected edge in the drawing view. This option will be available only when you have at least one drawing view in the current drawing sheet and if it is selected.

Offset

When you choose this option from the cascaded menu, the **OFFSET OPER** menu is displayed as shown in Figure 12-7. There are two options in this menu. The **Single Ent** option is used to select a single entity from the drawing view. The **Ent Chain** option is used to select a chain of entities from the drawing view. After choosing any one of the options from the **OFFSET OPER** menu, when you move the cursor over the drawing view, each edge of the model in drawing view is highlighted in cyan color.

Figure 12-7 OFFSET OPER menu

Converting a Generated View to a Draft View

To convert a generated view to a draft view, choose **Edit > Convert to Draft Entities** from the menu bar. After converting a generated view to a draft view, all the objects in the view are converted to individual entities. This option is available only when the drawing view is selected. When you choose this option, the **SNAPSHOT** menu is displayed as shown in Figure 12-8.

Figure 12-8 SNAPSHOT menu

The options in this menu is used to select either all the views or the current view.

USER-DEFINED DRAWING FORMATS

Pro/ENGINEER provides you with some standard drawing formats for generating the drawing views. These standard formats have standard sheet sizes, tables, and title blocks. However, sometimes you may need to create a user-defined drawing format that is specifically designed as per your requirements. This format can include the sheet size, tables, and title block specified as per your requirements. Choose the **Create a new object** button to display the **New** dialog box. In this dialog box, select the **Format** radio button from the **Type** area and then specify the name of the format in the **Name** edit box to proceed for creating the user-defined format. When you choose **OK**, the **New Format** dialog box will be displayed, see Figure 12-9.

You can set the size and the orientation of the format sheet using this dialog box. Choose **OK** in this dialog box after setting the parameters for the format to proceed to the **Format** mode. You can create the desired entities in this format by using the **Drawing Sketcher** toolbar. You can also add the text material to the format by using the **Note** option from the **Insert** menu in the menu bar. Figure 12-10 shows a user-defined format created using the A4 size sheet.

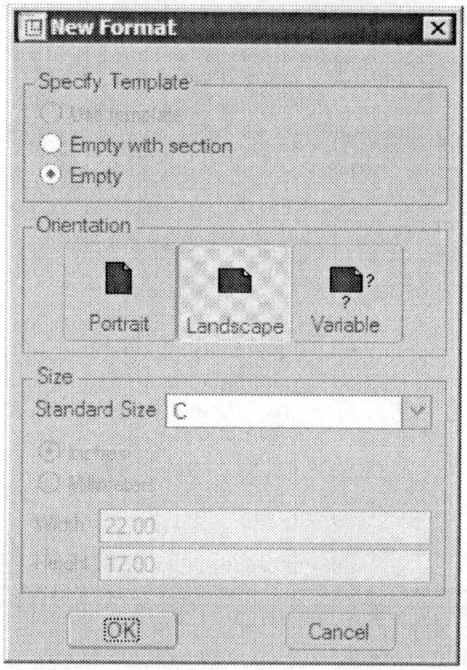

Figure 12-9 ***New Format*** *dialog box*

This format consists of the user-defined title block. Figure 12-11 shows the drawing views generated on the user-defined format.

RETRIEVING THE USER-DEFINED FORMATS IN THE DRAWINGS

For generating the drawing views you can also retrieve the user-defined format. Once you have created the user-defined formats, you can use them in the Drawing mode as sheets for generating the drawing views. To retrieve the user-defined format, select the **Empty with format** radio button and choose the **Browse** button from the **New Drawing** dialog box to invoke the **Open** dialog box. A folder named **System Formats** will be displayed in the **Open** dialog box. Some predefined formats are given in this folder. You can retrieve these predefined formats or browse the location where you have saved the user-defined format created earlier.

Tip: *The user-defined format that you create is saved in the .frm format. The location of this format will be the working directory that was selected at the time when the format was created.*

ADDING AND REMOVING SHEETS IN THE DRAWING

To add sheets in the current drawing, choose **Insert > Sheet** from the menu bar. A new sheet is displayed on the screen. At the bottom of the screen, the sheet number is displayed. You

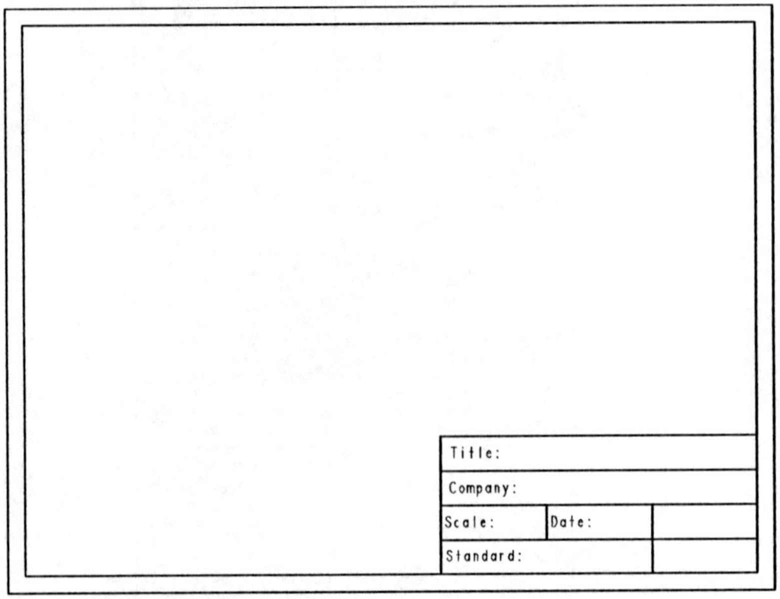

***Figure 12-10** A user-defined format*

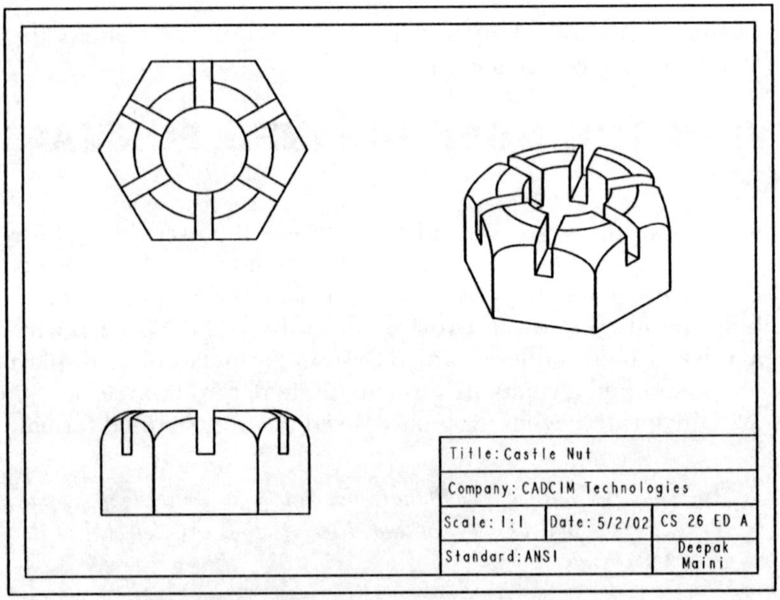

***Figure 12-11** Drawing views generated using a user-defined format*

can generate various views of a model on multiple sheets. All these drawing views on different sheets are contained in a single drawing file.

To remove sheets from a drawing, choose **Edit > Remove > Sheets** from the menu bar. The **Message Input Window** is displayed. Enter the sheet number that you want to remove.

The following points should be remembered while dealing with multisheets:

1. If you move a projected view to another sheet, the parent-child relationship of the two views is broken. That means, if you move the projected view on one sheet the source view on the other sheet does not move.

2. The scale of the drawing views on different sheets can be controlled independently.

3. If you erase a drawing view on one sheet, it can be resumed on the other sheet.

Tip: *You can add or replace the formats in drawing by using the options available in the* ***Page Setup*** *dialog box that is displayed when you hold down the right mouse button on the drawing sheet. You can add another sheet and then select the format from this dialog box.*

CREATING TABLES IN DRAWING MODE

You can easily create any kind of tabular representation in Drawing mode by using the options available in the cascaded menu that is displayed when you choose **Table > Insert** from the menu bar. The cascaded menu is displayed as shown in Figure 12-12 and the options in it are discussed next.

Figure 12-12 *Cascaded menu*

Table Option

The **Table** option is used to create a table. When you choose this option from the cascaded menu, the **TABLE CREATE** submenu is displayed as shown in Figure 12-13. The table can be created in the ascending or descending order of the rows progress and in the right or the left direction. The length and width of the cells can be specified by picking the points on the screen or by specifying the value of their length and width.

Table From File Option

The **Table From File** option is used to insert a table that exists in the *.tbl* file format. When you choose this option, the **Open** dialog box is displayed. Select the file from this dialog box to open.

Column Option

The **Column** option is used to add a column to an existing table. When you choose this

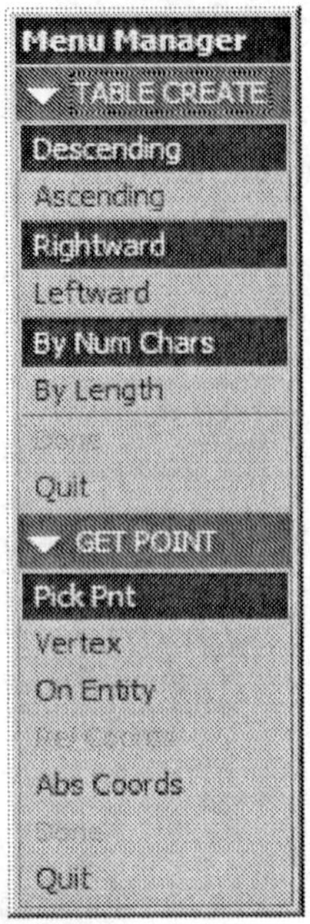

***Figure 12-13 TABLE CREATE** submenu*

option from the cascaded menu, you are prompted to select a point inside an existing table.

Tip: *To save a table with a .tbl file format, choose **Table > Save Table > As Table File** from the menu bar. This option is available only when a table exists on the drawing sheet.*

Row Option

The **Row** option is used to add a row to an existing table. When you choose this option from the cascaded menu, you are prompted to select a point inside an existing table.

Following are the editing operations that can be performed on the tables.

1. To move a table, select it and then drag it by holding it from one of the points that are highlighted on the table.

2. To enter text in any of the cells of the table you need to double-click inside the cell to display the **Enter Text** dialog box.

3. To delete a table, select the table and choose **Table > Select > Table** from the menu bar. Hold down the right mouse button and choose **Delete** from the shortcut menu.

Tip: *In Drawing mode, nearly all the modifications and editing can be done using the shortcut menu that is displayed when you select the item to modify and then hold down the right mouse button. The options in this menu vary and depend upon the item selected.*

*It is advised that you use the options under the **Table** menu in the menu bar to create the title blocks in the formats. The text can be easily added to the title block by double-clicking in the cell.*

USING THE ASSEMBLY BILL OF MATERIAL IN THE DRAWINGS

The Bill Of Material (BOM) is a representation of the components and their parameters that are used in the assembly. In Pro/ENGINEER there are two methods to create the BOM.

Creating BOM in the Drawing Mode

The BOM that is created in the Assembly mode can be directly used in the Drawing mode if it was saved in the *.bom* format. To use the assembly BOM, choose **Insert > Note > File** from the menu bar. Select the other options for placing the BOM and then enter the name of the file that was saved in the *.bom* format in the Assembly mode. You can modify the BOM and you can also sketch a table around the BOM using the sketcher tools.

The BOM created using the above method is not associative.

Creating BOM in the Report Mode

Report mode is one of the modes available in Pro/ENGINEER. All the drawing view generation options in this mode are similar to those available in the Drawing mode. To create a BOM in the Report mode, the exploded view of the assembly is created in the Report mode of Pro/ENGINEER. The table is created on the drawing sheet. Then a repeat region is defined in the table that automatically generates the BOM of the assembly. The BOM generated is associative in nature and any modification in the assembly, like addition or removal of components is automatically reflected in the BOM generated in the Report mode.

If you do not have the license for Pro/REPORT than you will not be able to create an associative BOM. In this chapter, there are two tutorials in which you will learn both the methods of creating BOM.

TUTORIALS

Tutorial 1

In this tutorial you will create a format of size A and add the title block in the format. Then you will retrieve the format in the Drawing mode and generate the exploded isometric view of the **Pedestal Bearing** assembly created in Tutorial 2 of Chapter 9. Also, add the assembly Bill of Material and the Balloons to the drawing view as shown in Figure 12-14.

(Estimated time: 45 min)

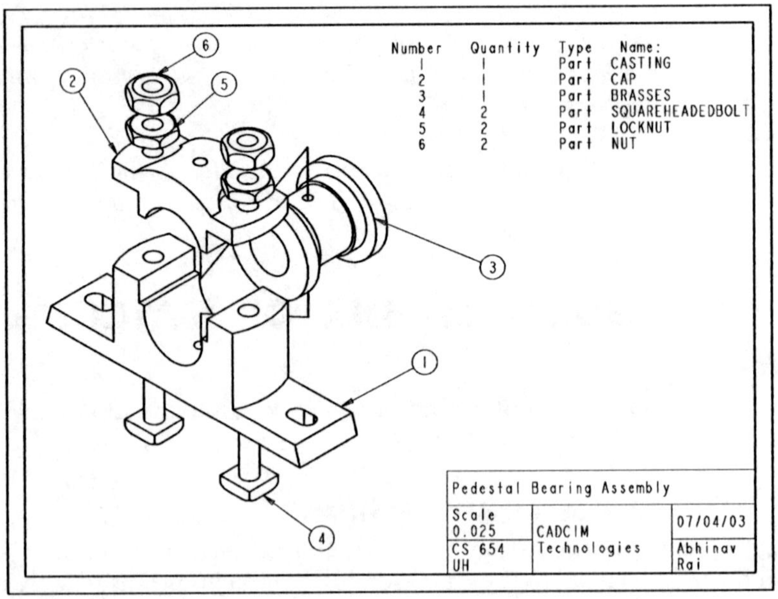

***Figure 12-14** The drawing view of the assembly showing the BOM and balloons*

The following steps outline the procedure to complete this tutorial:

a. Open a new file in the Format mode, create the format of the drawing, and add the title block in the format, see Figure 12-16.

b. Save the format file and then close it.

c. Open a new drawing file in the Drawing mode, select **Pedestal Bearing** as the model, and retrieve the format that you had created.

d. Create and save the BOM. Generate the exploded isometric view of the assembly, see Figure 12-17.

e. Add BOM and balloons to the drawing, see Figure 12-18.

Before you start creating the drawing view, you need to set the working directory. Set the working directory to *C:/ProE-WF/c12*. Since this is the first tutorial of this chapter, you need to create the *c12* folder inside the *ProE-WF* folder.

Copy all the components of the Pedestal Bearing assembly and the assembly file named *pedestalbearing.asm* from the *c09* folder in the current working directory. This is because the assembly file with all the files of the components should lie in the folder in which the drawing file will be stored.

Creating the Format File

As evident from Figure 12-14, the exploded view of the Pedestal bearing is generated on a drawing sheet with a customized format. Hence, you need to create a format before generating the drawing view. This format will then be retrieved later.

1. Choose the **Create a new object** button from the **Top Toolchest**, the **New** dialog box is displayed.

2. Select the **Format** radio button from the **Type** area in the **New** dialog box and name the file as **Format1**. Choose **OK** from the **New** dialog box.

   The **New Format** dialog box is displayed.

3. The **Empty** radio button in the **Specify Template** area and the **Landscape** button in the **Orientation** area of the **New Format** dialog box are selected by default. If they are not selected, you need to select them.

4. Select **A** from the **Standard Size** drop-down list in the **Size** area and then choose **OK** to proceed to the Format mode. Note, that a sheet of size **A** is displayed on the screen. This is evident from the text displayed below the sheet on the screen.

Creating Format

1. Choose **Sketch > Edge > Offset** from the menu bar. The **OFFSET OPER** menu is displayed. Choose the **Ent Chain** option and then select all the four border lines of the format by drawing a window around them. Press the middle mouse button after you have made the selection.

2. An arrow is displayed pointing outwards. The arrow displays the direction of offset. Since you need to offset the lines in the opposite direction, therefore, enter **-0.25** in the **Message Input Window** and press ENTER. Press the middle mouse button to exit the **OFFSET OPER** menu.

3. Choose **Table > Insert > Table** from the menu bar to display the **TABLE CREATE** menu.

4. Choose **Ascending > Leftward > By Length** from the **TABLE CREATE** menu and then choose **Vertex** from the **GET POINT** submenu.

5. Select the lower right corner of the new rectangle created by offsetting the border lines. The corner of the new rectangle is selected because the title block will be created such that it is attached to the new rectangle. Now, you need to enter the dimensions for the title block.

6. Enter the width of the first column as **1.2** in the **Message Input Window** and press ENTER. Similarly, enter the width of the second column as **2** and for the third column as **1.2** in the **Message Input Window** and press ENTER. You are prompted to enter the width for next column. Press ENTER.

Now, you are prompted to enter the height for the first row.

7. Enter the height of the first row as **0.5**, the second as **0.5,** and the third as **0.5** in the **Message Input Window**. Press ENTER twice.

   The title block created will be similar to the one shown in Figure 12-15. Note that the title block created is not as required. Therefore, you need to modify it.

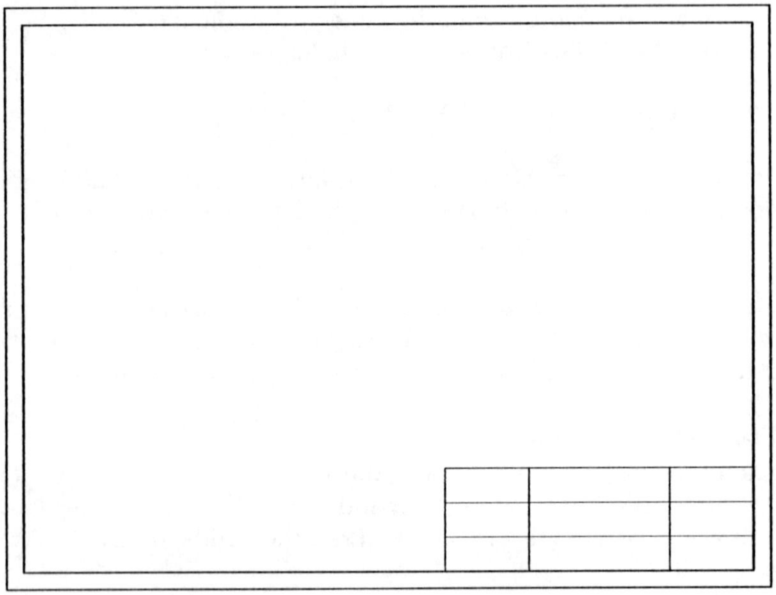

Figure 12-15 *Format with the title block*

8. Select the table by clicking on any cell. Choose **Table** > **Merge Cells** to display the **TABLE MERGE** menu. The **Rows & Cols** option is selected by default in the **TABLE MERGE** menu.

9. Select the first column and the second column of the top row to merge. Similarly, merge the third column of the top row with the new bigger column.

10. Now, merge the second column of the second row with the second column of the third row. The table after merging the rows and columns should look similar to the one shown in Figure 12-16.

11. You need to set the alignment of the text that will be entered later in the cells of the table. Double-click in a cell to invoke the **Enter Text** dialog box.

12. In the dialog box, select the **Text Style** tab.

13. In the **Note/Dimension** area of this dialog box, from the **Vertical** drop-down list, select

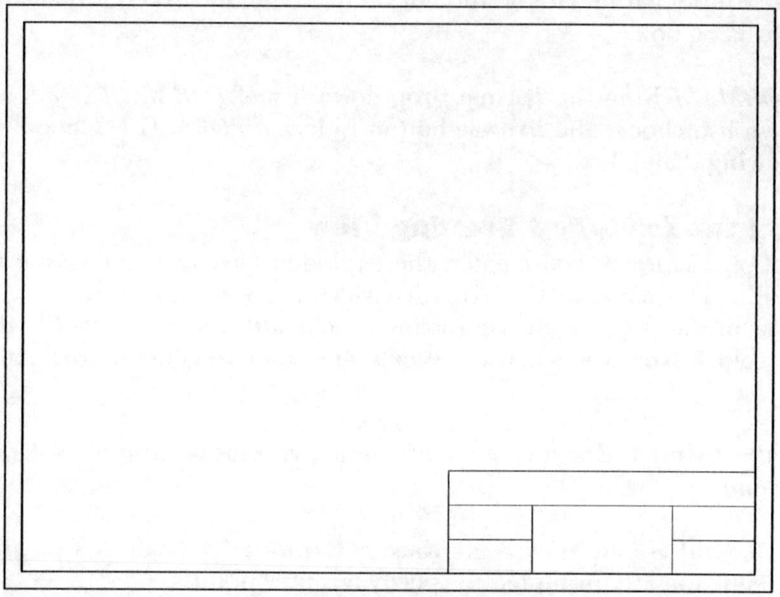

Figure 12-16 Modified title block

the **Middle** option. Exit the **Enter Text** dialog box by choosing the **OK** button.

Now, the alignment of the text is set to middle aligned vertically and left aligned horizontally. Similarly, individually set the alignment in all the cells of the table.

Saving the Format File

You need to save the format file that you have created so that you can use it as a template in the Drawing mode where you will generate the exploded drawing view of the Pedestal Bearing. The file will be stored in the *.frm* file format.

1. Choose the **Save the active object** button from the **Top Toolchest**. The **Message Input Window** is displayed with the name of the file that you had entered earlier. Press ENTER.

2. Close the current window by choosing **Window** > **Close** from the menu bar.

Creating a New Drawing File

You need to create a new drawing file to generate the exploded drawing view of the Pedestal Bearing.

1. Choose the **Create a new object** button to display the **New** dialog box. Select the **Drawing** radio button and specify the name of the drawing as *c12tut1* and then choose **OK**. The **New Drawing** dialog box is displayed.

2. Choose the **Browse** button and select the assembly file named *pedestalbearing.asm*.

3. Select the **Empty with format** radio button from the **Specify Template** area in the **New Drawing** dialog box.

4. Select *FORMAT1* from the **Format** drop-down list. If *FORMAT1* is not available in the drop-down list, choose the **Browse** button to locate *FORMAT1*. Choose **OK** to exit the **New Drawing** dialog box.

Generating the Exploded Drawing View

In Tutorial 2 of Chapter 9 you created the exploded view of the Pedestal Bearing in the Assembly mode. The name of the exploded view that you created was **EXP1**. Hence, the exploded view of the Pedestal Bearing is integrated with the assembly file that you copied from the *c09* folder. Now, you will use this exploded state to generate the exploded drawing view.

1. Choose the **Insert a drawing view of the active model** button to display the **VIEW TYPE** menu.

2. Choose **General** > **Full View** > **No Xsec** > **Exploded** > **Scale** > **Done** from the **VIEW TYPE** menu. You are prompted to specify a center point.

3. Select a point on the left of the drawing sheet. The **SEL STATE** menu is displayed. Select the **EXP1** check box if it is not selected by default and choose **Done**.

   The **Message Input Window** is displayed and you are prompted to specify the scale for the drawing view.

4. Enter a value of **0.025** in the **Message Input Window** and press ENTER. The **Orientation** dialog box is displayed with the default view of the exploded assembly. But, you need the isometric view of the exploded assembly.

5. Select the **Preferences** option from the **Type** drop-down list in the **Orientation** dialog box.

6. Select the **Isometric** option from the **Default Orientation** drop-down list. The isometric view of the exploded assembly is displayed. Choose **OK** from the **Orientation** dialog box.

7. Turn off the display of the datums from the **Datum Display** toolbar. If the model is not properly placed inside the boundaries of the drawing sheet, unlock the drawing view by choosing the **Disallow the movement of the drawing views with the mouse** button. Drag the drawing view and place it as shown in Figure 12-17.

8. Erase the text **Scale 0.025** from the drawing sheet because it is not needed. You can erase the text by selecting it and then hold down the right mouse button to display the shortcut menu. From the shortcut menu choose the **Erase** option.

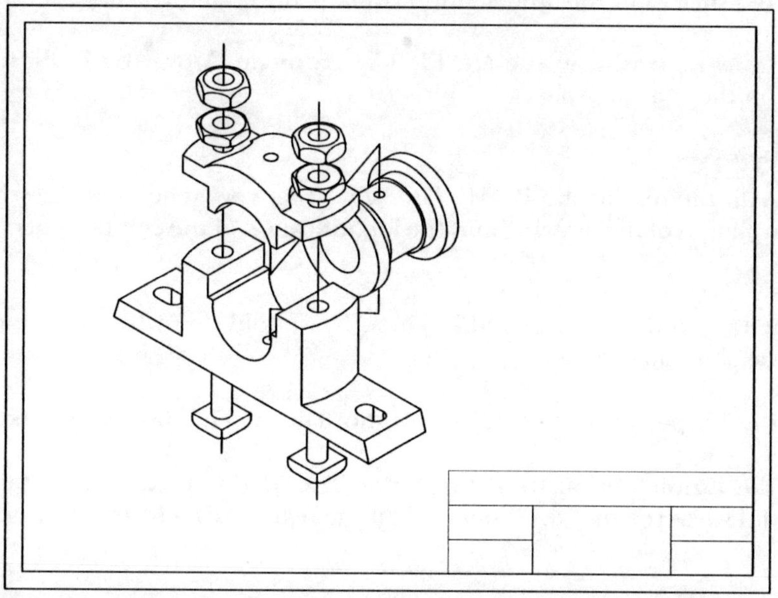

Figure 12-17 Exploded isometric view of the assembly

Creating the BOM

1. Choose **Info > Bill of Materials** from the menu bar. The **BOM** dialog box is displayed. In this dialog box, the **Top Level** radio button is selected by default. Choose **OK** to display the information of BOM in the browser.

   The message "Information written to file pedestalbearing.bom.1" is displayed in the **Message Area**.

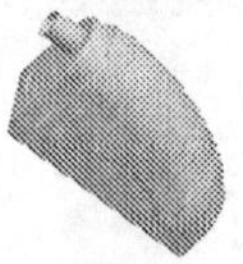

Tip: *You can also create and save the BOM in the Assembly mode and then insert it in the drawing mode.*

2. Close the browser by clicking on the sash present on its right edge.

Placing the BOM in the Drawing

1. Choose **Insert > Note** from the menu bar. The **NOTE TYPES** menu is displayed.
2. Choose **No Leader > File > Horizontal > Standard > Default > Make Note**. Specify the placement point for the BOM on the drawing sheet close to the top right corner. The **Open** dialog box is displayed.
3. Select the **All files** option from the **Type** drop-down list.
4. Select *pedestalbearing.bom* from the **Open** dialog box to open. The bom file is placed on

the drawing sheet, but this information contained in the file is not what is needed.

5. Choose **Done/Return** from the **NOTE TYPES** menu. Move the BOM to the required location on the drawing sheet.

Editing the Text

You need to edit the text in the BOM. The BOM that you generated does not convey the required information of the drawing and the identification of the components that constitutes the assembly.

1. Select the text so that it turns red in color. Now, hold down the right mouse button to display the shortcut menu.

2. Choose the **Properties** option from the shortcut menu. The **Enter Text** dialog box is displayed.
3. Choose the **Editor** button from the **Enter Text** dialog box. The **Notepad** window is displayed. Delete the first two lines and all the text starting from point number 9 in this window.

4. In line number **2** in the notepad, change **Assembly** to **Number, PEDESTALBEARING** to **Quantity**, and **contains** to **Type**. After **Type,** type and add **Name**.

5. In line number **3**, after **3:**, type **1** such that the number 1 is middle aligned with **Number**. Similarly, below **1** type **2**, below **2** type **3**, and so on.

6. Using the SPACEBAR (longest key on the keyboard) middle align the numbers under **Quantity**.

7. Using the DELETE key, middle align **Part** with **Type**.

8. Similarly, middle align the **Name** list.

9. Choose **File > Exit** to exit the **Notepad** window. You are prompted to save the changes made. Choose the **Yes** button and choose **OK** from the **Enter Text** dialog box.

   Move the BOM by dragging it if it overlaps the boundary of the drawing sheet.

Entering Text in the Title Block and Adding Balloons

The text alignment in various cells of the table was defined while creating the format. As a result, when you enter the text, it will be automatically aligned middle left.

1. Double-click the top cell in the table. The **Enter Text** dialog box is displayed.

2. Enter **Pedestal Bearing Assembly** and choose **OK**.

3. Double-click the cell that is in the first column and the second row. The **Enter Text** dialog box is displayed.

4. Enter **Scale:** and press ENTER. Now, enter **0.025** and choose **OK**. The text Scale: 0.025 is in two lines in the title block, and therefore you entered them as two lines.

   Similarly, enter the text in all the remaining cells one by one as shown in Figure 12-18.

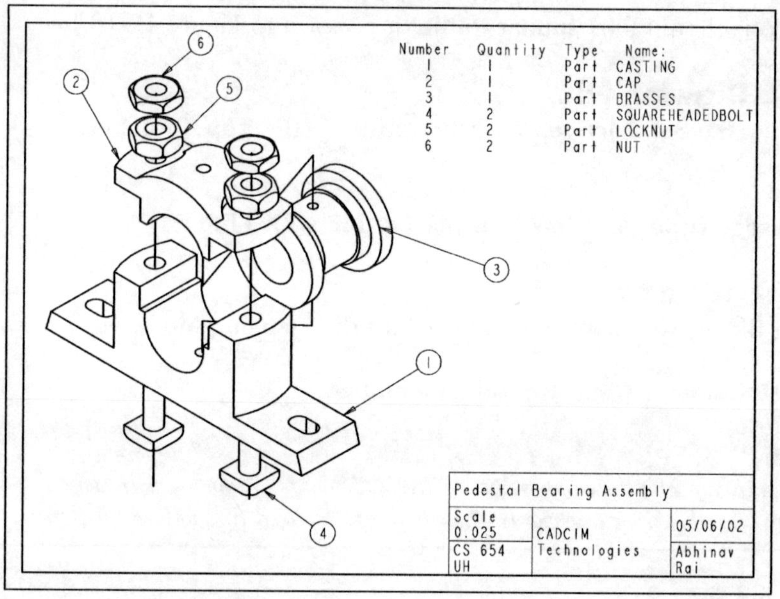

Figure 12-18 *The drawing view of the assembly showing the BOM and balloons*

5. Add the balloons to the assembly by choosing **Insert > Balloon** from the menu bar. The **NOTE TYPES** submenu is displayed.

6. Choose **Leader > Enter > Horizontal > Standard > Default > Make Note** from the **NOTE TYPES** menu in the **Menu Manager**. The **ATTACH TYPE** menu is displayed. The **On Entity > Arrow Head** option is selected by default.

7. Select an edge of the Brasses close to the back face and then press the middle mouse button on the screen where the balloon will be placed, see Figure 12-18. The **Message Input Window** is displayed. Enter **3** and press ENTER twice.

8. Choose the **Make Note** option from the **NOTE TYPES** menu. The **ATTACH TYPE** menu is displayed. The **On Entity > Arrow Head** option is selected by default. Select an edge of the Cap and press the middle mouse button on the screen where the note will be placed, see Figure 12-18. The **Message Input Window** is displayed. Enter **2** and press ENTER twice.

9. Choose the **Make Note** option from the **NOTE TYPES** menu. The **ATTACH TYPE**

menu is displayed. The **On Entity > Arrow Head** option is selected by default. Select an edge of the Casting and press the middle mouse button on the screen where the note will be placed, see Figure 12-18. The **Message Input Window** is displayed. Enter **1** and press ENTER twice.

Similarly, place the remaining balloons. The sheet after placing the BOM and adding the balloons should look similar to the one shown in Figure 12-18.

Saving the Drawing File

1. Choose the **Save the active object** button from the **Top Toolchest** to save the drawing file.

2. The **Message Input Window** is displayed. Press ENTER.

Closing the Window

The drawing file is saved and now you can exit the Drawing Mode.

1. Choose **Window > Close** from the menu bar.

Note

The occurrence of the components in BOM depends on the order in which they were assembled in the assembly. In other words, the component placed first will occur first in the BOM.

Tutorial 2

In this tutorial you will create a format of size A and add the title block in the format. You will retrieve the format in the Report mode and generate the exploded isometric view of the **Shock** assembly created in Tutorial 1 of Chapter 9. Also, add the associative Bill of Material and the Balloons to the drawing view as shown in Figure 12-19. **(Estimated time: 45 min)**

Note

This tutorial cannot be done by the users who do not have Pro/REPORT license. The users who do not have this license can complete this tutorial in the Drawing mode as was done in the previous tutorial.

The following steps outline the procedure to complete this tutorial:

a. Open a new file in the Format mode, create the format of the drawing, and add the title block in the format, see Figure 12-21.

b. Save the format file and then close it.

c. Open a new drawing file in the Report mode, select *shockassembly.asm* as the model, and retrieve the format that you had created.

d. Create the table and enter the headers.

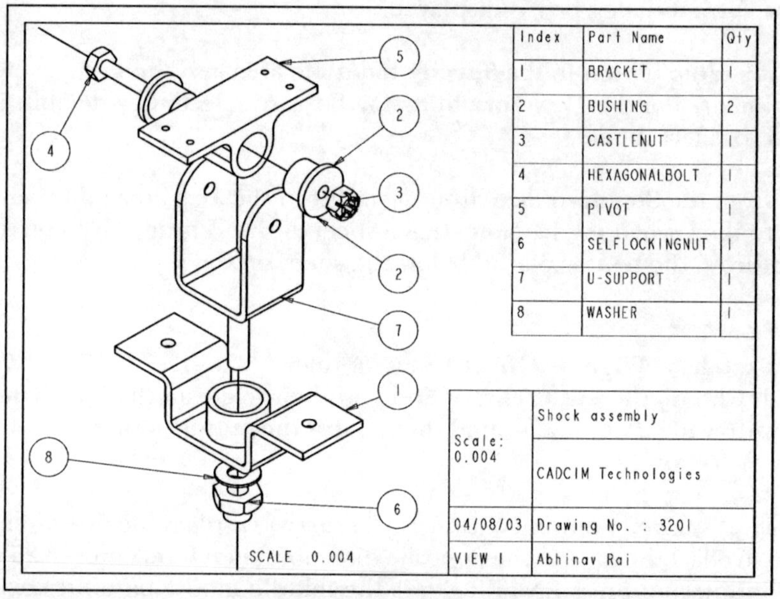

Figure 12-19 *The drawing view of the exploded assembly with BOM*

e. Define the repeat region and then enter the report symbols, see Figure 12-22.

f. Generate the exploded isometric view of the assembly and generate balloons in the drawing view, see Figure 12-25.

g. Add text in the tiltle block, see Figure 12-26.

The working directory was selected in Tutorial 1, and therefore there is no need to select the working directory again. But if you still want to select the working directory, choose the **Set Working Directory** option from the **File** menu. The **Select Working Directory** dialog box is displayed. Set the working directory to *C:/ProE-WF/c12*.

Copy all the components of the Shock assembly and the assembly file named *shockassembly.asm* from the *c09* folder in the current working directory. This is because the assembly file with all the files of the components should lie in the folder in which the drawing file will be stored.

Creating the Format File

As mentioned in the tutorial description, you need to create a format of size A that will be retrieved later to create the drawing view of the shockassembly.

1. Choose the **Create a new object** button from the **Top Toolchest**; the **New** dialog box is displayed.

2. Select the **Format** radio button from the **Type** area in the **New** dialog box and name the file as *Format2*. Choose **OK** from the **New** dialog box.

The **New Format** dialog box is displayed.

3. The **Empty** radio button in the **Specify template** area and the **Landscape** button in the **Orientation** area of the **New Format** dialog box are selected by default. If they are not, you need to select them.

4. Select **A** from the **Standard Size** drop-down list in the **Size** area and then choose **OK** to proceed to the Format mode. Note, that a sheet of size **A** is displayed on the screen. This is evident from the text displayed below the sheet on the screen.

Creating Format

1. Choose **Sketch** > **Edge** > **Offset** from the menu bar. The **OFFSET OPER** menu is displayed. Choose the **Ent Chain** option and then select all the four border lines of the format by drawing a window around them. Press the middle mouse button after you have made the selection.

2. An arrow is displayed pointing outwards. The arrow displays the direction of offset. Since you need to offset the lines in the opposite direction, therefore, enter **-0.25** in the **Message Input Window** and press ENTER. Press the middle mouse button to exit the **OFFSET OPER** menu.

3. Choose **Table** > **Insert** > **Table** from the menu bar to display the **TABLE CREATE** menu.

4. Choose **Ascending** > **Leftward** > **By Length** from the **TABLE CREATE** menu and then choose **Vertex** from the **GET POINT** submenu.

5. Select the lower right corner of the new rectangle created by offsetting the border lines. The corner of the new rectangle is selected because the title block will be created such that it is attached to the new rectangle. Now, you need to enter the dimensions for the title block.

6. Enter the width of the first column as **3.2** in the **Message Input Window** and press ENTER. Similarly, enter the width of the second column as **1.2** and press ENTER. You are prompted to enter the width for next column. Press ENTER.

   Now, you are prompted to enter the height for the first row.

7. Enter the height of the first row as **0.5**, the second as **0.5,** the third as **1**, and the fourth as **0.7** in the **Message Input Window**. Press ENTER twice.

   The title block created will be similar to the one shown in Figure 12-20. Note the difference between the title block shown in Figure 12-19 and that shown in Figure 12-20. You need to modify the title block that you have created so that it is similar to the one shown in Figure 12-19.

8. Select the table and choose **Table** > **Merge Cells** to display the **TABLE MERGE** menu.

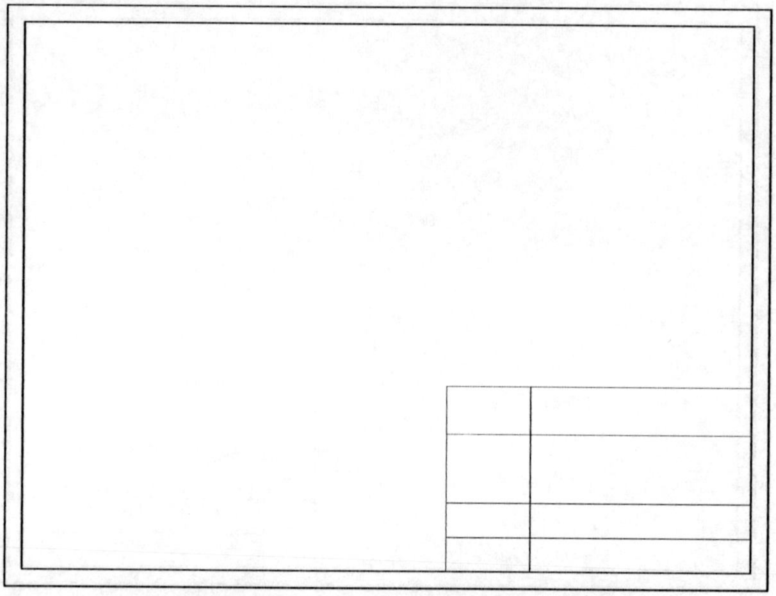

Figure 12-20 *Format with the title block*

The **Rows & Cols** option is selected by default in the **TABLE MERGE** menu.

9. Select the first row and the second row of the first column to merge.

   The table after merging the rows and columns should look similar to the one shown in Figure 12-21.

10. You need to set the alignment of the text that will be entered later in the cells of the table. Double-click in a cell to invoke the **Enter Text** dialog box.

11. In the dialog box, select the **Text Style** tab.

12. In the **Note/Dimension** area of this dialog box, from the **Vertical** drop-down list, select the **Middle** option. Exit the **Enter Text** dialog box by choosing the **OK** button.

    Now, the alignment of the text is set to middle aligned vertically and left aligned horizontally. Similarly, set the alignment in all the cells of the table individually.

Saving the Format File

You need to save the format file that you have created so that you can use it as a template in the Drawing mode where you will generate the exploded drawing view of the Shock assembly. The file will be stored in the *.frm* file format.

1. Choose the **Save the active object** button from the **Top Toolchest**. The **Message Input Window** is displayed with the name of the file that you had entered earlier. Press ENTER.

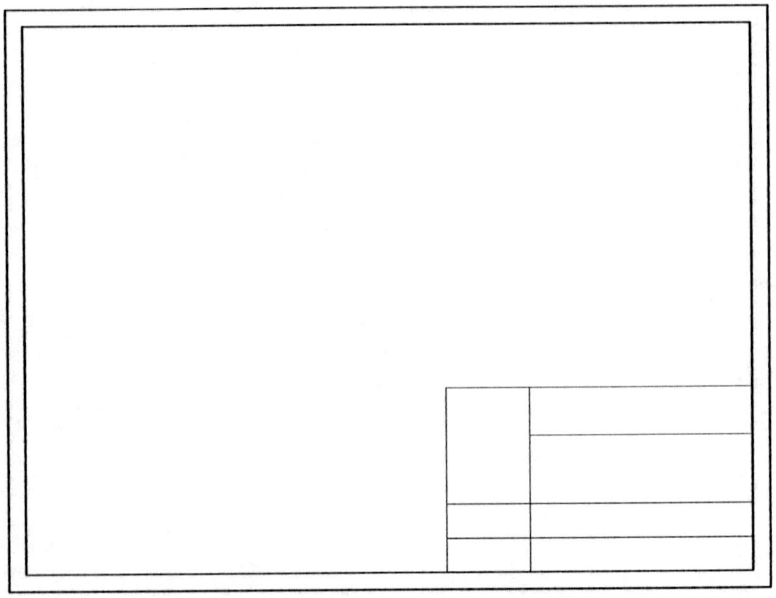

Figure 12-21 Modified title block

2. Close the current window by choosing **Window > Close** from the menu bar.

Creating a New Report File

You need to create a new report file to generate the exploded drawing view of the Shock assembly. As mentioned earlier, using the Report mode allow you to generate an associative BOM.

1. Choose the **Create a new object** button to display the **New** dialog box. Select the **Report** radio button and specify the name of the drawing as *c12tut2* and then choose **OK**. The **New Drawing** dialog box is displayed.

2. Choose the **Browse** button and select the assembly file named *shockassembly.asm*.

3. Select the **Empty with format** radio button from the **Specify Template** area in the **New Drawing** dialog box.

4. Select *FORMAT2* from the **Format** drop-down list. If *FORMAT2* is not available in the drop-down list, choose the **Browse** button to locate *FORMAT2*. Choose **OK** to exit the **New Drawing** dialog box.

Creating the Table for BOM

1. Choose **Table > Insert > Table** from the menu bar to display the **TABLE CREATE** menu. Choose **Pick** from the **GET POINT** menu.

2. Choose **Descending > Leftward > By Length** from the **TABLE CREATE** menu.

3. Select the upper right corner of the new rectangle created by offsetting the border lines. Now, you need to enter the dimensions of the BOM table.

4. Enter the width of the first column as **0.5** in the **Message Input Window** and press ENTER. Similarly, enter the width of the second column as **3** and press ENTER. You are prompted to enter the width for next column, enter **1** and Press ENTER twice.

   Now, you are prompted to enter the height for the first row.

5. Enter the height of the first row as **0.5** and of the second as **0.5** in the **Message Input Window**. Press ENTER twice.

Setting the Repeat Region

Repeat regions are the smart cells in a table that can expand depending upon the amount of data inserted in the cell. The second row will be defined as the repeat region.

1. Choose **Table > Repeat Region** from the menu bar. The **TBL REGIONS** menu is displayed.

2. Choose the **Add** option; you are prompted to locate the corners of the region.

3. Select the first cell of the second row and then select the third cell of the second row. Press the middle mouse button twice to exit the menu.

Creating the Column Headers

Next, the heading in the table will be entered.

1. Double-click in the first cell of the first row; the **Enter Text** dialog box is displayed. Enter **Index** and choose **OK** to exit the dialog box.

2. Similarly, enter **Part Name** and **Qty** in the second and the third cells.

Entering the Report Symbols in the Repeat Region

The information in the cells that are defined as repeat region contains is determined by the text that is written in the form of report symbols. These report symbols associate the information in the cells with the assembly file directly.

1. Double-click in the first cell of the second row; the **Enter Text** dialog box is displayed. Enter **&rpt.index** and choose **OK** to exit the dialog box.

2. Similarly, enter **&asm.mbr.name** and **&rpt.qty** in the second and the third cells and exit the **Enter Text** dialog box. If the text you have entered is overlapping with the other text or is extending out of the cell, ignore it. This is because the BOM will be created in the repeat region where this text exists and after the BOM is created the text of the BOM will fit inside the cells.

   The drawing sheet after adding the report symbols is shown in Figure 12-22.

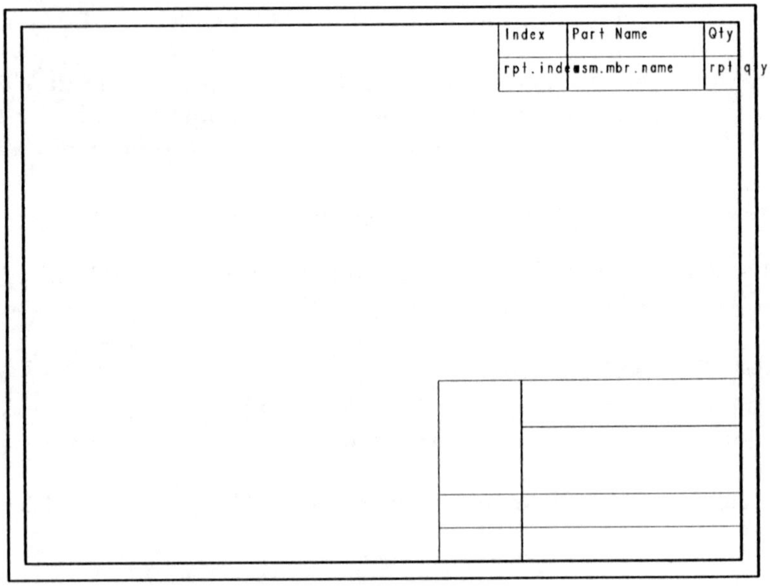

***Figure 12-22** Drawing sheet after adding the report symbols*

Generating the Drawing View

In Tutorial 1 of Chapter 9 you had created the exploded view of the Shock assembly in the Assembly mode. The name of the exploded view that you specified was **EXP1**. Hence, the exploded view of the Shock assembly is integrated with the assembly file that you copied from the *c09* folder. Now, you will use this exploded state to generate the exploded drawing view in the Report mode.

1. Choose the **Insert a drawing view of the active model** button from the **Top Toolchest** to display the **VIEW TYPE** menu.

2. Choose **General** > **Full View** > **No Xsec** > **Exploded** > **Scale** > **Done** from the **VIEW TYPE** menu. You are prompted to specify a centerpoint.

3. Select a point on the left side on the drawing sheet. The **SEL STATE** menu is displayed. Select the **EXP1** check box if it is not selected by default and choose **Done**.

   The **Message Input Window** is displayed and you are prompted to specify the scale for the drawing view.

4. Enter a value of **0.004** in the **Message Input Window** and press ENTER. The **Orientation** dialog box is displayed with the default view of the exploded assembly. In the previous tutorial the isometric view was set as the default view and therefore the view that is displayed is also isometric.

If the model is not placed properly inside the boundaries of the drawing sheet, unlock it and then drag the drawing view and place it as shown in Figure 12-23. Note that the BOM automatically appears in the table by extending the repeat region. The Qty column is empty.

Note

If the BOM is not displayed, then choose the ***Update the information displayed in tables*** *button from the* ***Top Toolchest****.*

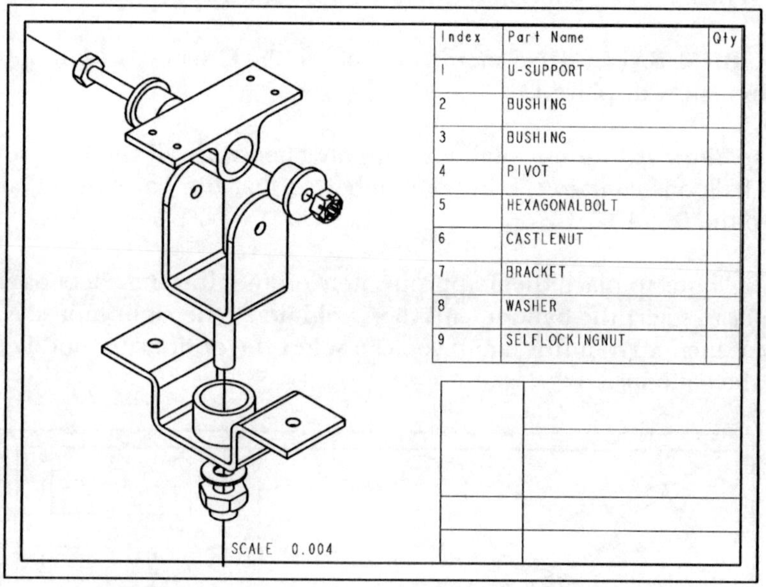

Figure 12-23 *Exploded isometric view of the assembly*

Setting the No Repeat Option

In the assembly the Bushing is repeated. Therefore, in the BOM it may get repeated. But the BOM need to be set such that no part name is repeated.

1. Choose **Table > Repeat Region**; the **TBL REGIONS** menu is displayed.

2. Choose the **Attributes** option from the menu. You are prompted to select a region.

3. Move the cursor over the table below the first row. All the rows, except the first row are highlighted in cyan. The highlighted portion is called region. Press the middle mouse button and exit the menu. The drawing sheet appears as shown in Figure 12-24.

4. Choose the **Update the information displayed in tables** button from the **Top Toolchest** if the information is not updated in the table.

Generating Balloons

The balloons are generated by using the repeat region. Before generating the balloons you need to set the repeat region for balloon generation.

1. Choose **Table > BOM Balloons**; the **BOM BALLOONS** menu is displayed.
2. Choose the **Set Region** option from the menu. The **BOM BAL TYPE** submenu is displayed with the **Simple** option selected by default and you are prompted to select a region.
3. Select the repeat region that is now extended downwards. Press the middle mouse button.
4. Invoke the **BOM BALLOONS** menu and choose the **Create Balloon** option. The **BOM VIEW** submenu is displayed.
5. Choose the **Show All** option. Balloons are attached with all the parts in the exploded view. The balloons contain the index number so that the parts in the assembly can be referred to the BOM.
6. Drag the balloons to place them appropriately on the drawing sheet as shown in Figure 12-25. You can select the balloon and then hold down the right mouse button to invoke the shortcut menu. From this menu you can select the options to modify the attachment points of the balloons.

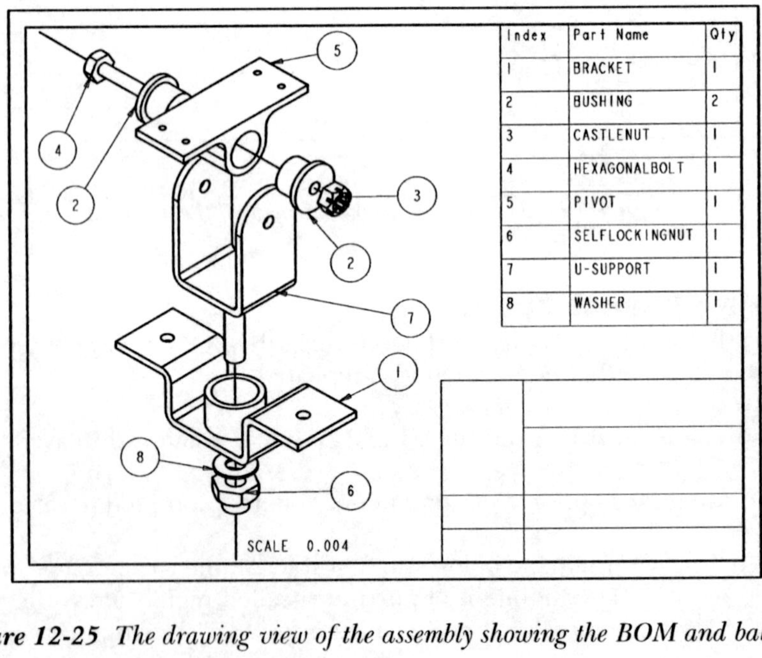

***Figure 12-25** The drawing view of the assembly showing the BOM and balloons*

Entering Text in the Title Block

1. Double-click in the cell formed by the first row and the second column. The **Enter Text** dialog box is displayed.

2. Enter **Shock assembly** and choose the **OK** button.

3. Now, select the cell from the second row and the second column. Enter **CADCIM Technologies** in the **Enter Text** dialog box and choose the **OK** button.

   Similarly, enter the text in all the remaining cells one by one as shown in Figure 12-26.

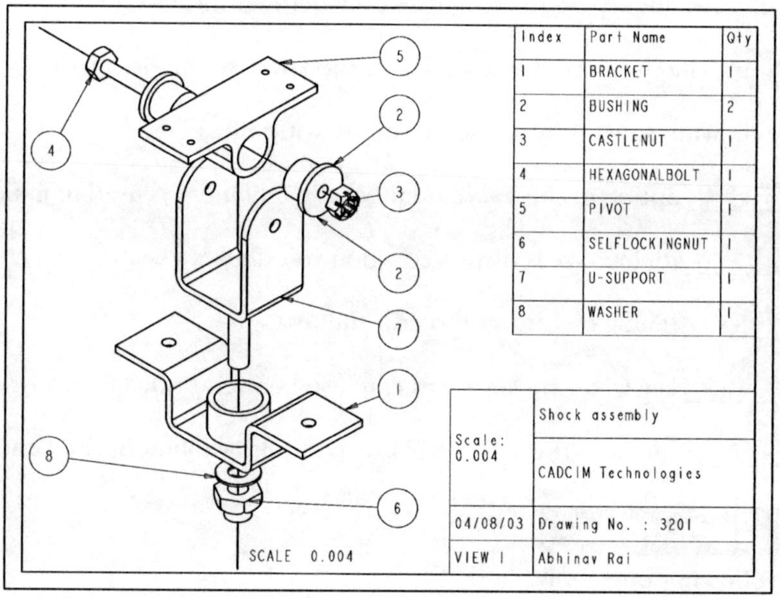

Figure 12-26 *The drawing view of the assembly showing the title block, balloons, and the BOM*

Saving the Drawing File

1. Choose the **Save the active object** button from the **Top Toolchest** to save the drawing file.

2. The **Message Input Window** is displayed. Press ENTER.

Closing the Window

The report file is saved and now you can exit the Report mode. Remember that the report file is saved with *.rep* file extension.

1. Choose **Window** > **Close** from the menu bar.

Self-Evaluation Test

Answer the following questions and then compare your answers to the answers given at the end of this chapter.

1. The **Select one item at a time - shift to gather more than one item** button in the **Right Toolchest** is used to select any item from the drawing view or the drawing view itself to modify. (T/F)

2. Pro/ENGINEER provides you with some standard drawing formats for generating the drawing views. (T/F)

3. Pro/ENGINEER allows you to create the user-defined formats. (T/F)

4. The assembly BOM can be directly used in the **Drawing** mode. (T/F)

5. Only one drawing sheet is available in one drawing file. (T/F)

6. The __________ option of the **Table** menu in the menu bar is used to generate balloons.

7. The __________ dialog box is displayed when you double-click in the cell of a table.

8. The __________ dialog box is used to edit the text.

9. The __________ key from the keyboard can also be used to delete any draft entity.

10. The __________ menu in the menu bar is used to draw tables in the Drawing mode.

Review Questions

Answer the following questions.

1. Which of the following modes in Pro/ENGINEER is used to create an associative BOM?

   (a) Drawing (b) Format
   (c) Report (d) None

2. Which of the following options in the shortcut menu is used to modify an attachment point of the leader?

   (a) **Edit Attachment** (b) **Erase**
   (c) **Move Item to View** (d) **Arrow Style**

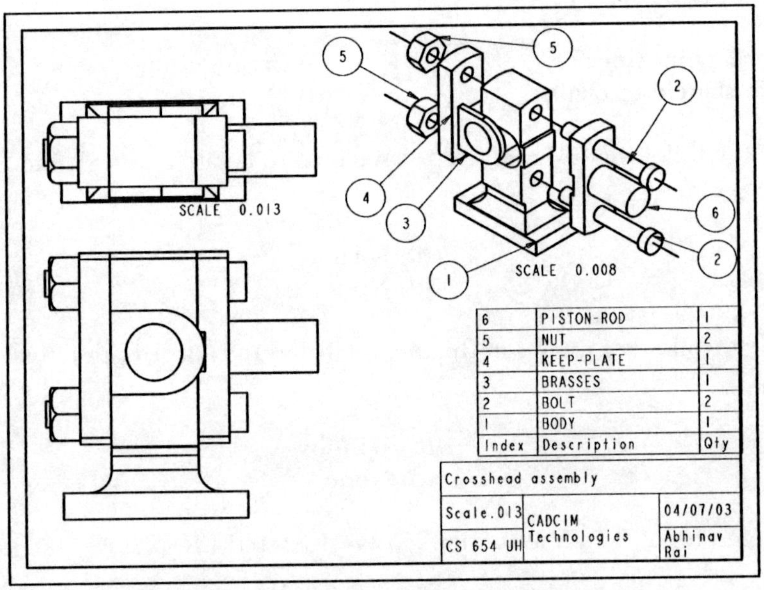

Figure 12-27 Drawing for Exercise 1

Answers to the Self-Evaluation Test

1 - T, **2** - T, **3** - T, **4** - T, **5** - F, **6** - **Bom Balloons**, **7**- **Enter Text**, **8** - **Enter Text**, **9** - DELETE, **10** - **Table**.

3. Which of the following buttons on the **Right Toolchest** enables you to draw the geometric entities continuously?

   (a) **Create 2 point lines** (b) **Create circle**
   (c) **Enable sketching chain** (d) None

4. Which of the following radio buttons is selected to create a user-defined format for the drawing sheet?

   (a) **Part** (b) **Format**
   (c) **Drawing** (d) None

5. Which of the following menus in the menu bar has the **Close** option to close the current window?

   (a) **Info** (b) **Window**
   (c) **View** (d) None

6. You can open more than one windows in Pro/ENGINEER. (T/F)

7. It is not possible to draw splines in the Drawing mode. (T/F)

8. There are two types of drafting in Pro/ENGINEER: Generative drafting and Interactive drafting. (T/F)

9. You can import a text file in the Drawing mode. (T/F)

10. The **Draft Group** option in the cascaded menu is used to add or remove notes, surface symbols, or gtols from the selected dimension. (T/F)

Exercise

Exercise 1

Create a format of size A and then retrieve it in the Report mode to place the top, front, and an exploded isometric view of the **Cross Head** assembly created in Exercise 1 of Chapter 9. Add the associative assembly BOM and the balloons to the drawing view as shown in Figure 12-27.
(Estimated time: 45 min)

Chapter 13

Projects

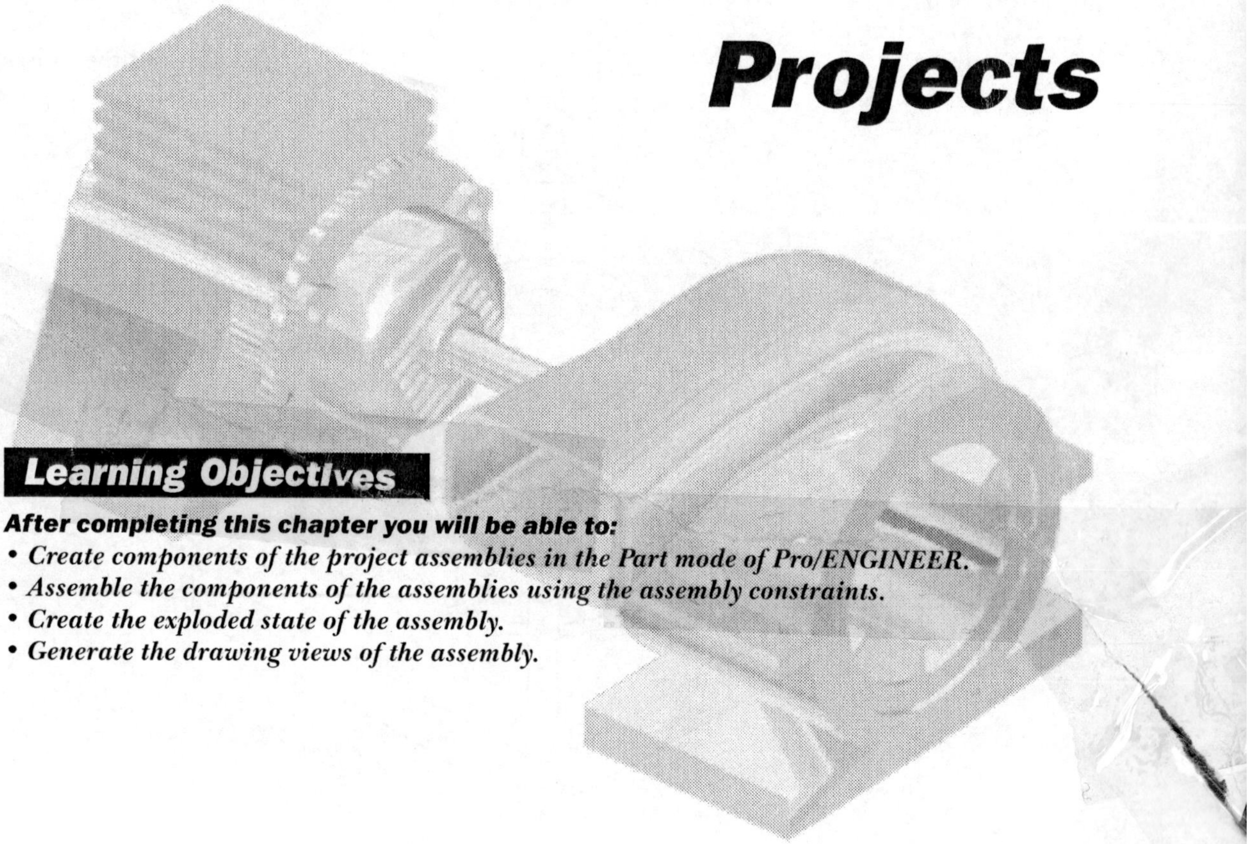

Learning Objectives

After completing this chapter you will be able to:

- *Create components of the project assemblies in the Part mode of Pro/ENGINEER.*
- *Assemble the components of the assemblies using the assembly constraints.*
- *Create the exploded state of the assembly.*
- *Generate the drawing views of the assembly.*

TUTORIAL

Tutorial 1

In this tutorial you will create the components of the Fixture assembly shown in Figure 13-1. The exploded view of the assembly is shown in Figure 13-2. The details of the components of the Fixture assembly are shown in Figures 13-3 through 13-6. After completing the assembly, create an exploded state of the assembly in the Assembly mode. Finally, generate the following drawing views of the assembly:

a. Top view
b. Front view
c. Right-side view
d. Isometric view

(Expected time: 3 hrs)

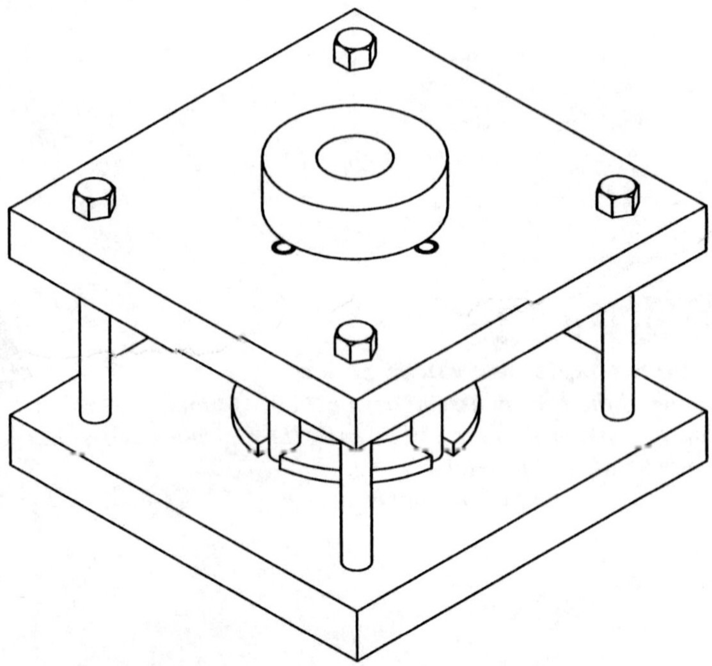

Figure 13-1 *Assembled view of the Fixture assembly*

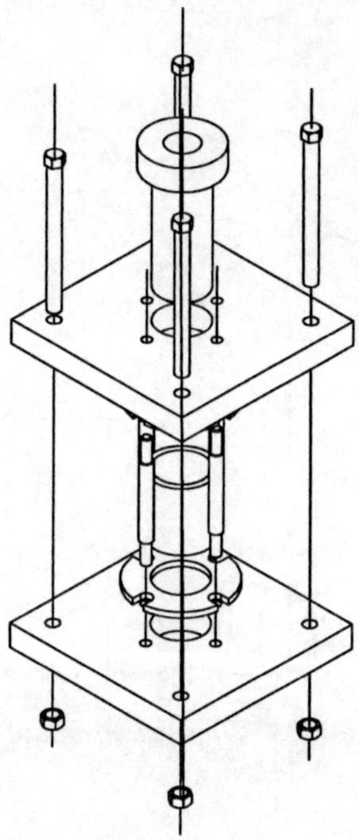

Figure 13-2 *Exploded view of the Fixture assembly displaying various components*

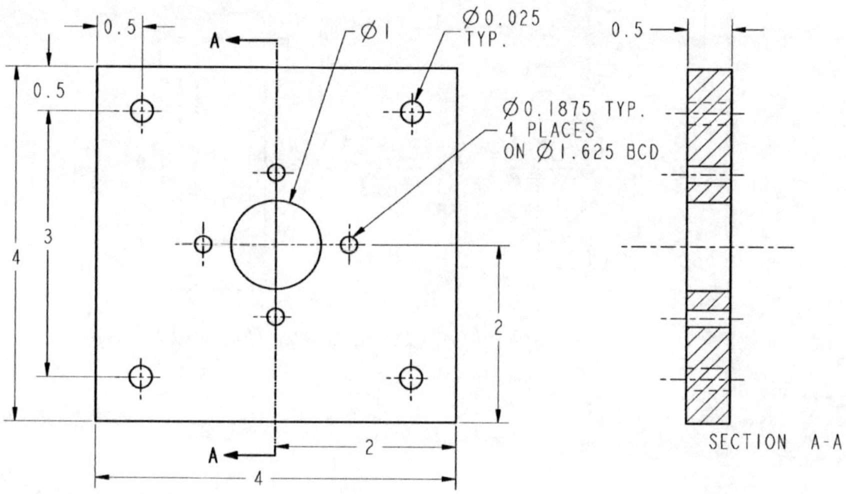

Figure 13-3 *Dimensions of the End Plate*

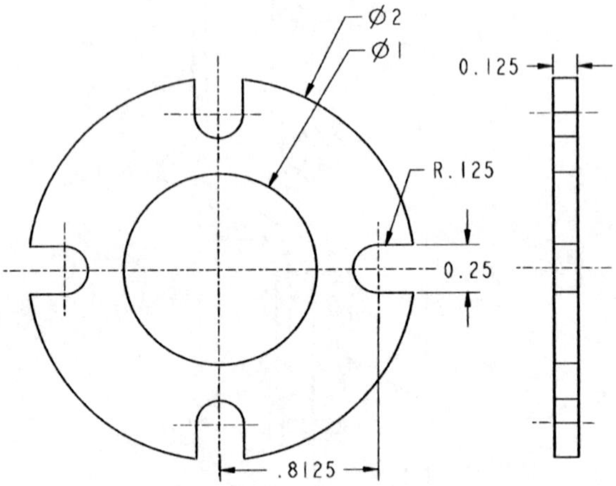

Figure 13-4 Dimensions of the Disk

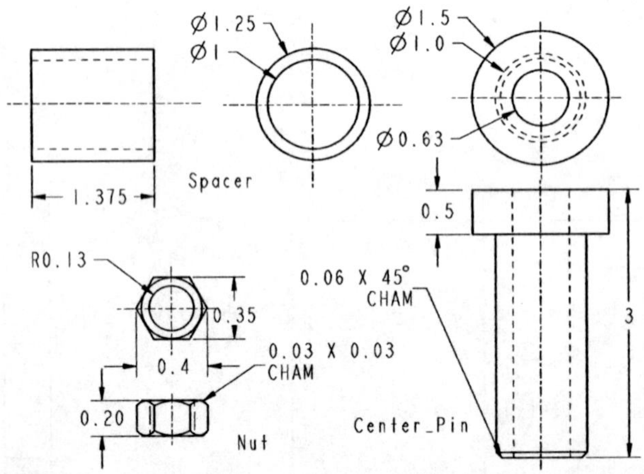

Figure 13-5 Dimensions of the Spacer, Center Pin, and the Nut

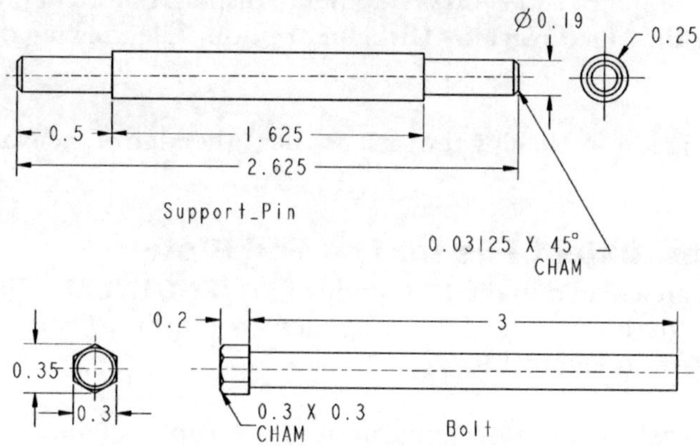

Figure 13-6 *Dimensions of the Support Pin and the Bolt*

The following steps outline the procedure for creating this assembly:

a. Set the working directory and create all the components of the Fixture assembly in the Part mode.

b. Assemble the End Plate, Disk, Spacer, Support Pin, and Center Pin in an assembly file.

c. Open a new assembly file and assemble the nut and the bolt to create a subassembly. Insert this subassembly in the main assembly and then create the reference pattern to assemble the remaining instances of the nut and the bolt.

d. Create the exploded view of the assembly in the Assembly mode.

e. Create the top, front, right-side, and isometric views of the assembly in the Drawing mode.

Setting the Working Directory

When the Pro/ENGINEER session is started, the first task is to set the working directory. A working directory is a directory on your system where you can save the work done in the current session of Pro/ENGINEER. You can set any directory existing on your system as the working directory. Before you start creating the components, you need to create a folder named *Fixture Assembly* and set it as the working directory.

1. Open the Navigator by clicking the sash on the left edge of the Pro/ENGINEER main window. The Navigator slides out.

2. Choose the **Folder Browser** button in the Navigator to view the folders.

3. Right-click on the *ProE-WF* folder to display a shortcut menu.

4. Choose the **New Folder** option. A new folder is created inside the *ProE-WF* folder. Name the folder as *Fixture Assembly*.

4. Now right-click on the *Fixture Assembly* folder to display a shortcut menu. From this shortcut menu, choose the **Make Working Directory** option. The working directory is set to *Fixture Assembly*.

5. Close the Navigator by clicking the sash on the right edge of the Navigator. The Navigator slides in.

Creating a New Object File for the End Plate

The solid models are created in the Part mode of **Pro/ENGINEER**. The extension for the file created in this mode is *.prt.* The first component is the End Plate. The dimensions of the End Plate are shown in Figure 13-3.

1. Choose the **Create a new object** button from the **Top Toolchest**.

   The **New** dialog box is displayed. The **Part** radio button is selected by default in the **Type** area and the **Solid** radio button is selected by default in the **Sub-type** area of the **New** dialog box.

2. Enter the file name as **end_plate** in the **Name** edit box and choose the **OK** button.

   The three default datum planes will be displayed on the graphics window. The **Model Tree** also appears on the left of the graphics window. Close the **Model Tree** by clicking on the sash.

Creating the Base Feature of the End Plate

The base feature will be created by extruding the sketch to a depth of 0.5. Therefore, the **Extrude Tool** will be used to create the base feature.

1. Choose **Insert > Extrude** from the menu bar or choose the **Extrude Tool** button from the **Base Features** toolbar. The Extrude dashboard is displayed below the graphics window.

   All the attributes that are needed to create the model will be defined after the sketch is drawn.

Selecting the Sketching Plane

You will select the **TOP** datum plane as the sketch plane because the direction of extrusion of this feature is normal to the **TOP** datum plane.

1. Choose the **Sketch internal section** button from the Extrude dashboard. The **Section** dialog box is displayed.

2. Select the **TOP** datum plane as the sketch plane. As you select the sketch plane, the reference plane and its orientation are set automatically. The reference plane is selected in order to orient the sketch plane.

   The yellow arrow that appears on the sketch plane indicates the direction of viewing the sketch.

In the **Section** dialog box, the **Reference** box displays **RIGHT:F1(DATUM PLANE)**. This indicates that the **RIGHT** datum plane is selected as the reference plane. In the **Orientation** drop-down list, the **Right** option is selected by default. This means that while drawing the sketch the **RIGHT** datum plane will be on the right.

3. Choose the **Sketch** button in the **Section** dialog box.

Specifying the References

After you select the planes for the orientation, the system takes you to the sketcher environment. The **References** dialog box is displayed at the top right corner of the screen and the status displayed in the **Reference status** area of the **Reference** dialog box is **Fully Placed**. This means that the references required are already defined and now you can start drawing the sketch. Exit the **References** dialog box by choosing the **Close** button from the dialog box.

Drawing the Sketch

The sketch of the base feature is a square.

1. Choose the **Create rectangle** button from the **Sketcher Tools** toolbar and draw a rectangle as shown in Figure 13-7 by defining its lower left corner and the upper right corner. The rectangle is created and weak dimensions are applied to it.

Modifying the Dimensions

The weak dimensions applied to the sketch by default are not the actual values that are required. Therefore, you need to modify the dimension values of the sketch. You can select all the dimensions at a time and then modify them by using the **Modify Dimensions** dialog box.

1. Select the sketch and all the dimensions using CTRL+ALT+A. You can also select all the entities by defining a window around them.

   You will notice that all the dimensions and entities of the sketch are selected and are highlighted red in color.

2. Choose the **Modify the value of dimensions, geometry of splines, or text entities** button from the **Sketcher Tools** toolbar. The **Modify Dimensions** dialog box is displayed.

3. All the dimensions in the sketch are displayed in this dialog box and each dimension has a separate thumbwheel and an edit box. You can use the thumbwheel or the edit box to modify the dimensions.

4. Clear the **Regenerate** check box and then modify the values of the dimensions as shown in Figure 13-8.

5. After modifying all the dimensions, choose the **Regenerate the section and close the dialog** button in the **Modify Dimensions** dialog box.

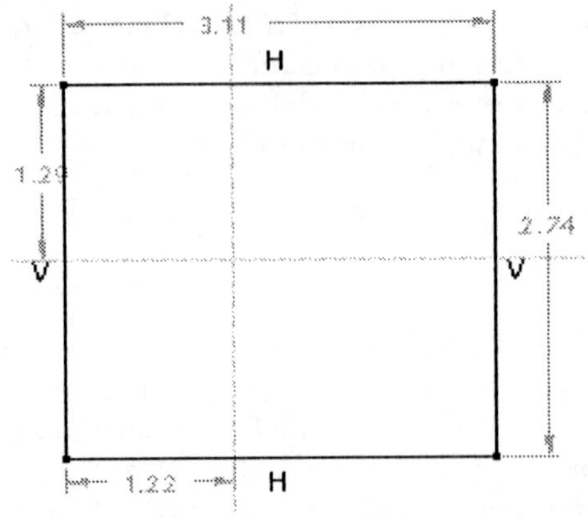

Figure 13-7 Sketch with the default dimensions

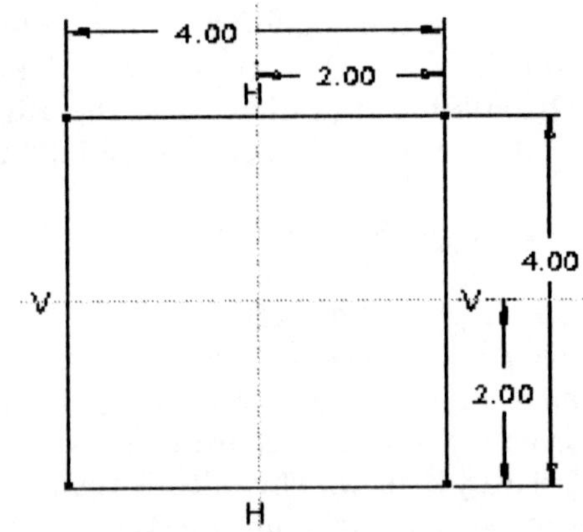

Figure 13-8 Sketch after modifying the dimensions

6. Choose the **Continue with the current section** button to exit the sketcher environment.

Specifying the Depth of Extrusion

After exiting the sketcher environment, the Extrude dashboard is redisplayed below the graphics window. Now, you need to specify the depth of extrusion for the sketch.

1. Enter a value of **0.5** in the dimension box present on the dashboard and press ENTER.

2. Choose the **Build feature** button from the Extrude dashboard to exit. Choose the **Default View** option from the **Saved view list** button in the **View** toolbar to view its trimetric view.

Creating the Hole Feature

The next feature that will be created is a hole and it will be created using the Hole dashboard. The hole is a through hole of 0.25 diameter and it will be created close to the lower left corner on the top face of the base feature.

1. Choose the **Hole Tool** button from the **Engineering Features** toolbar. The Hole dashboard is displayed.

   The **Create straight hole** button is selected by default in the Hole dashboard.

2. Select the top face of the base feature as the placement plane or primary reference.

3. Choose the **Placement** tab to invoke the slide-up panel. Click in the **Secondary References** collector.

4. Select the front face of the base feature as the first linear reference and enter a value of **-0.5** as distance in the first row of **Secondary References** collector.

5. Now, spin the model by pressing the middle mouse button and select the left face of the base feature. Enter a value of **-0.5** in the second row of the **Secondary References** collector.

6. From the depth flyout, choose the **Drill to intersect with all surfaces** button.

7. Choose the **Build feature** button. The hole is created as shown in Figure 13-9.

Creating Pattern of the Hole Feature

The hole will be patterned by specifying increment in the dimension values that are used to locate the hole feature with respect to the faces of the base feature.

1. Select the hole feature from the **Model Tree** or from the graphics window and hold down the right mouse button to invoke the shortcut menu. Choose the **Pattern** option from the shortcut menu. The Pattern dashboard is displayed.

   The dimensions of the selected hole are displayed on the graphics window and you are prompted to select the dimension to vary in the first direction.

2. Select the **Identical** option from the **Options** slide-up panel.

3. Select the dimension value **0.5** from the graphics window.

4. Enter a value of **3** in the edit box and press ENTER.

5. Hold down the right mouse button to display a shortcut menu. Select the **Direction 2 Dimensions** button from the shortcut menu.

6. Select the dimension value **0.5** that is along the longer side of the base feature. After you select the dimension in the second direction, the edit box is displayed.

   Note that in the edit boxes present on the dashboard, the number of instances **2** is specified by default.

7. Enter a value of **3** in the edit box and press ENTER.

8. Choose **Build feature** from the Pattern dashboard. The pattern of hole is created as shown in Figure 13-10.

Creating the Second Hole Feature

Next, you have to create a hole of diameter 0.1875. Before creating this hole feature, you have to create a datum axis. This axis will act as an axial reference for the hole that will be created on the Bolt Circle Diameter (BCD) of 1.625.

1. Choose the **Datum Axis Tool** button from the **Datum** Toolbar; the **DATUM AXIS** dialog box is displayed.

2. Select the **FRONT** datum plane and then select the **Through** option from the drop-down list in the first row.

3. Use CTRL+left mouse button to select the **RIGHT** datum to create the datum axis.

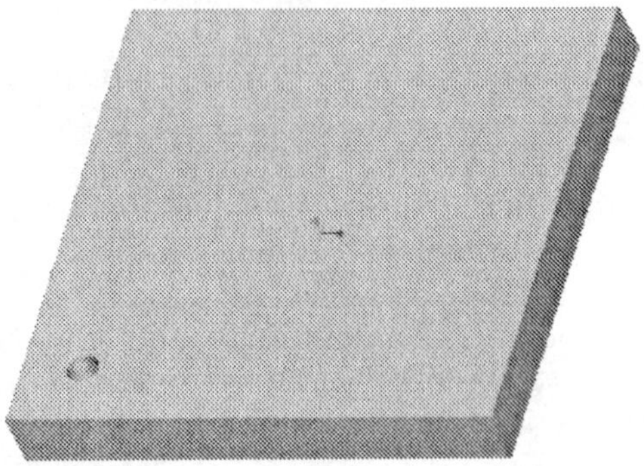

Figure 13-9 Placement of hole feature

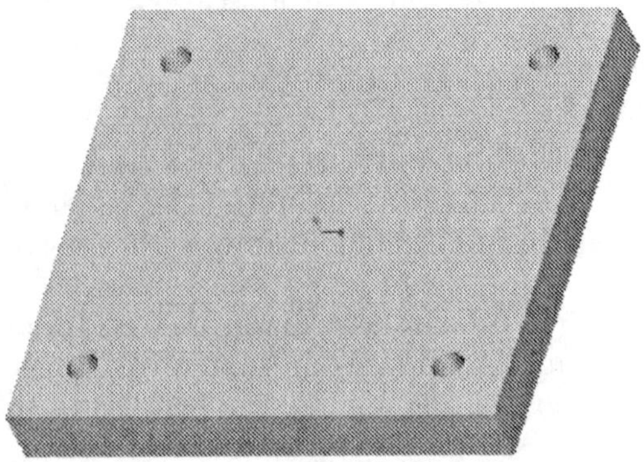

Figure 13-10 Pattern of hole feature

4. Choose the **Hole Tool** button from the **Engineering Features** toolbar.

5. Select the top face of the base feature as the placement plane.

6. Choose the **Placement** tab to invoke the slide-up panel.

7. Select the **Diameter** option from the drop-down list in the **Primary** area.

8. Click in the **Secondary References** collector and select the axis you created earlier from the graphics window. After you select the axis a default value appears in the first row of the **Secondary References** collector.

9. Click on the default value and enter **1.625**. Press ENTER. Now, you need to select a plane for the angular reference.

   You can select either the **FRONT** datum plane or the **RIGHT** datum plane as angular reference as both the datum planes intersect the placement plane and are normal to the placement plane.

10. Using CTRL+left mouse button, select the **FRONT** datum plane from the graphics window. The default angle value appears in the second row.

11. Click on the default angle value and enter a value of **90**. Press ENTER.

12. Choose the **Build feature** button from the Hole dashboard. The second hole feature is created on the base feature and is shown in Figure 13-11.

Creating the Pattern of Second Hole Feature

The second hole will be patterned by giving an increment to the angular value that was created during the placement of the second hole feature with the **Diameter** option.

1. Select the hole feature from the **Model Tree** or from the graphics window and hold down the right mouse button to invoke the shortcut menu. Choose the **Pattern** option from the shortcut menu. The Pattern dashboard is displayed.

   The dimensions of the selected hole are displayed on the graphics window and you are prompted to select the dimension to vary in the first direction.

2. Select the **Identical** option from the **Options** slide-up panel.

3. Select the angular dimension with the value of 90 from the graphics window. After you select the dimension, the edit box is displayed.

4. Enter a value of **90** in the edit box and press ENTER.

5. Enter the number of instances as 4 in the **1** edit box present on the dashboard.

6. Choose **Build feature** from the Pattern dashboard. The pattern of hole is created as shown in Figure 13-12.

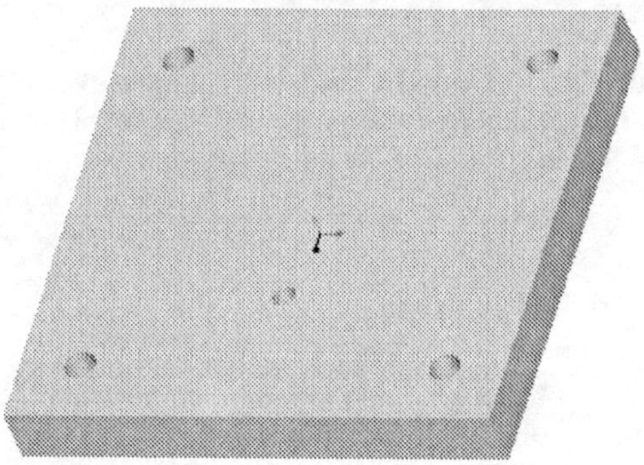

Figure 13-11 *Second hole feature*

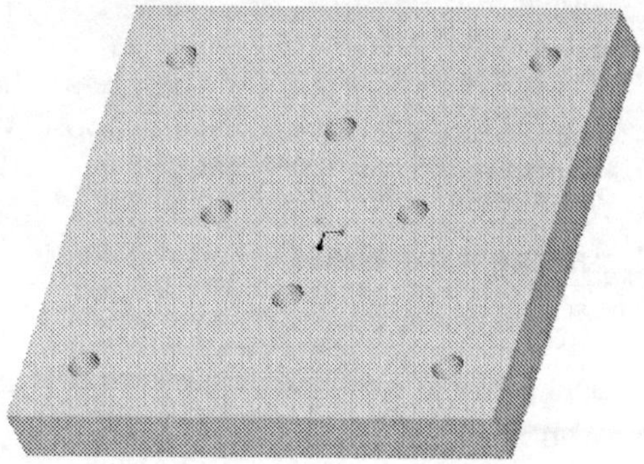

Figure 13-12 *Base feature after creation of pattern*

Creating the Third Hole Feature

The third hole feature will be created on the top face of the base feature and this hole is a coaxial hole.

1. Choose the **Hole Tool** button from the **Engineering Features** toolbar.

2. Select the top face of the base feature as the placement plane.

3. Choose the **Placement** tab to invoke the slide-up panel.

4. Select the **Coaxial** option from the drop-down list in the **Primary** area.

5. Click in the **Secondary References** collector and select the axis you created earlier from the graphics window.

6. Enter **1** in the edit box present in the dashboard and press ENTER. Select the **Drill to intersect with all surfaces** button from the depth flyout.

7. Choose the **Build feature** button from the Hole dashboard. The hole is created as shown in Figure 13-13.

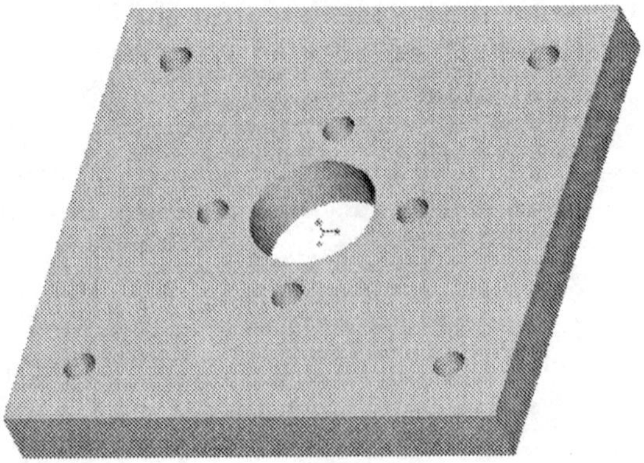

Figure 13-13 Shaded trimetric view of the End Plate

Saving the Model

You need to save this model so that you can assemble it later.

1. Choose the **Save the active object** button from the menu bar. The **Message Input Window** is displayed with the name of the file that you had specified earlier while creating a new object file.

2. Press ENTER or choose the green check mark button on the **Message Input Window**. The file is now saved. Now, choose **Window > Close** from the menu bar to close this window.

Creating the Disk

The second component is the Disk and it is an extruded feature.

Creating a New Object File for the Disk

You need to create a new file for the second component of the assembly.

1. Choose the **Create a new object** button from the **Top Toolchest**. The **New** dialog box is displayed. The **Part** radio button is selected by default in the **Type** area and the **Solid** radio button is selected by default in the **Sub-type** area of the **New** dialog box.

2. Enter the file name as **disk** in the **Name** edit box and choose the **OK** button from the **New** dialog box.

   The three default datum planes are displayed on the graphics window.

Creating the Extruded Feature

The Disk will be created by extruding the sketch to the given distance.

1. Choose the **Extrude Tool** button from the **Base Features** toolbar. The Extrude dashboard is displayed.

Selecting the Sketching Plane

To create the protrusion feature for this component you will select the **TOP** datum plane as the sketching plane. This is because the direction of extrusion for this feature is normal to the **TOP** datum plane.

1. Choose the **Sketch internal section** button from the Extrude dashboard. The **Section** dialog box is displayed.

2. Select the **TOP** datum plane as the sketching plane. As you select the sketching plane, the reference plane and its orientation is set automatically. The reference plane is selected in order to orient the sketching plane.

   The yellow arrow that appears on the sketching plane indicates the direction of viewing the sketch. The **RIGHT** datum plane is selected as the reference plane.

3. Choose the **Sketch** button in the **Section** dialog box.

Drawing the Sketch

1. Draw a circle of diameter 2 as shown in Figure 13-14.

2. Choose the **Continue with the current section** button from the **Sketcher Tools** toolbar to exit the sketcher environment. After you exit the sketcher environment, the Extrude dashboard is displayed.

3. Enter **0.125** as extrusion depth in the edit box and press ENTER.

4. Choose the **Build feature** button from the dashboard.

Creating the Cut Feature

1. Choose the **Extrude Tool** button from the **Base Features** toolbar. The Extrude dashboard is displayed.

2. Select the **Remove Material** button and then choose the **Sketch internal section** button from the dashboard. The **Section** dialog box is displayed.

3. Select the top face of the base feature as the sketch plane.

4. Choose the **Datum Plane Tool** button. Create a datum plane that passes through the axis of the base feature and is at an angle of 90-degrees from the **RIGHT** datum plane. Exit the **Datum Plane** dialog box.

5. Select the datum that you have created on-the-fly, to be oriented on the right.
   change Reference datum plane to the new

6. Choose the **Sketch** button to enter the sketcher environment.

7. Draw the sketch of the cut feature, dimension it, and modify the dimensions as shown in the Figure 13-15.

See pg 6-63 #12 #13

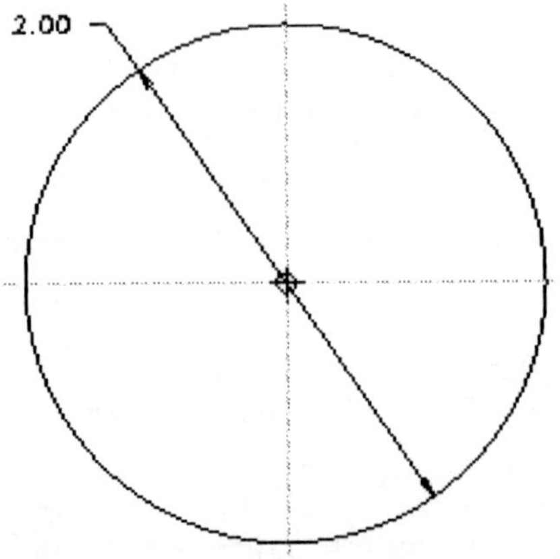

Figure 13-14 Sketch with weak dimensions

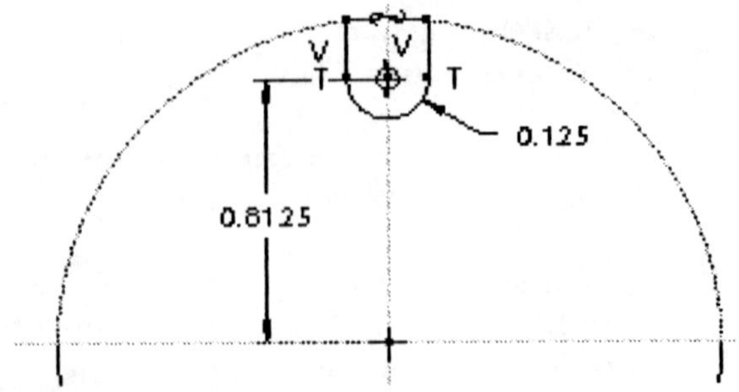

Figure 13-15 Sketch after modifying the dimensions

8. Exit the sketcher environment and specify the depth of cut from the depth flyout such that the cut is through the surface.

9. Choose the **Build feature** button to exit the Extrude dashboard. The cut feature is created as shown in Figure 13-16.

Creating the Pattern of the Cut

1. Select the cut feature from the graphics window or from the **Model Tree** and hold down the right mouse button to invoke the shortcut menu.

2. Choose the **Pattern** option. The Pattern dashboard is displayed.

3. Select the angular value of 90-degrees from the graphics window. The edit box is displayed.

4. Enter a value of **90** in the edit box. Now, you need to enter the number of instances.

5. Enter a value of **4** in the **1** edit box present in the dashboard.

6. Choose the **Build feature** to exit the dashboard. The pattern of the cut feature is created as shown in Figure 13-17.

Creating the Hole

1. Choose the **Hole Tool** button from the **Engineering Features** toolbar. The Hole dashboard is displayed.

2. Create the hole by referring to the dimensions from the drawing view of this component. The Disk after creating the hole is shown in Figure 13-18.

Saving the Part File

1. Choose the **Save the active object** button from the **Top Toolchest** to save the file.

Figure 13-16 *Disk after creating the cut*

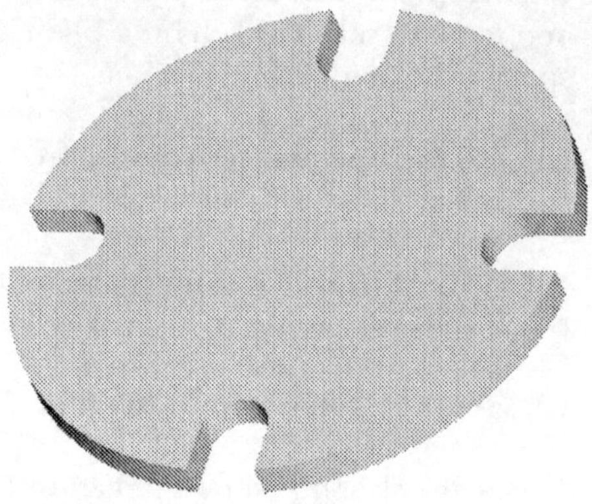

Figure 13-17 *Disk after patterning the cut feature*

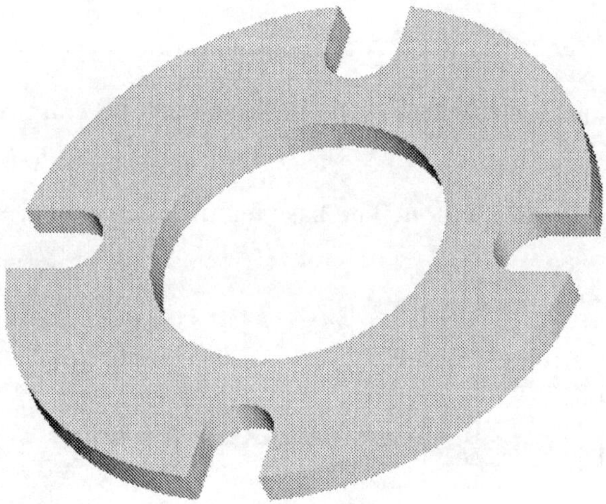

Figure 13-18 *Model of the Disk*

2. Choose **Window > Close** from the menu bar to close the file.

Creating the Center Pin

The third component is Center Pin and it is a revolved feature. The Center Pin has two features. The base feature is a revolve feature, and the second feature is a chamfer.

Creating a New Object File for the Center Pin

You need to create a new file for the third component of the assembly.

1. Create a new object file and name the file as **center_pin**.

Creating the Revolved Feature

You need to select a sketching plane to draw the sketch of the revolved feature and then specify the angle of revolution.

1. Choose the **Revolve Tool** button from the **Base Features** toolbar. The Revolve dashboard is displayed.
2. Select the **FRONT** datum plane as the sketching plane and select the **RIGHT** datum plane to be at the right.
3. Choose the **Sketch** button to enter the sketcher environment.
4. Create the sketch as shown in Figure 13-19 and then create a center line along the **RIGHT** datum plane.
5. Modify all the values of dimensions as shown in Figure 13-19.
6. Choose the **Continue with the current section** button to exit the sketcher environment.

   The Revolve dashboard is redisplayed and the angle of revolution of 360-degrees is set by default.

7. Choose the **Build feature** button. The base feature is created as shown in Figure 13-20.

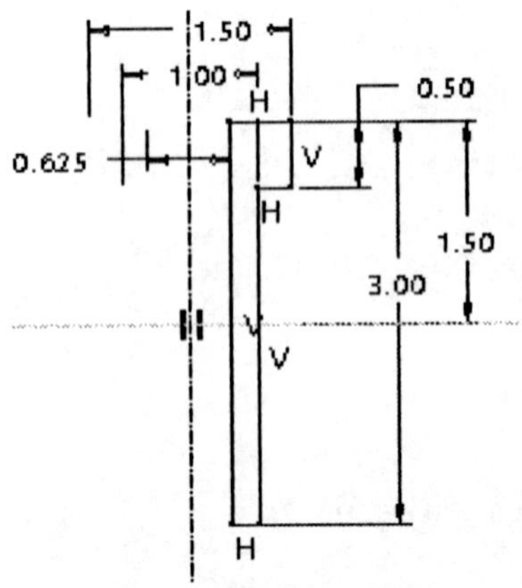

Figure 13-19 Sketch after modifying dimensions

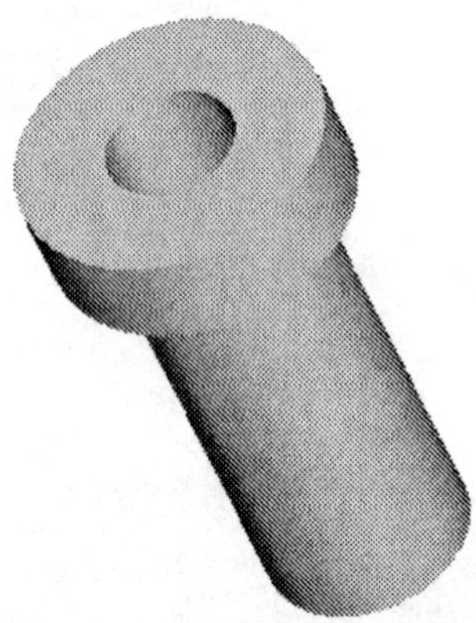

Figure 13-20 Trimetric view of the model

Creating the Chamfer

1. Choose the **Chamfer Tool** button from the **Engineering Features** toolbar. The Chamfer dashboard is displayed.
2. Choose the **45 x D** option from the drop-down list on the dashboard and then enter a value of **0.06** in the adjacent edit box.
3. Select the bottom edge of the base feature to chamfer. The preview of the chamfer is

displayed on the model.

4. Choose the **Build feature** button from the dashboard. The chamfer is created as shown in Figure 13-21.

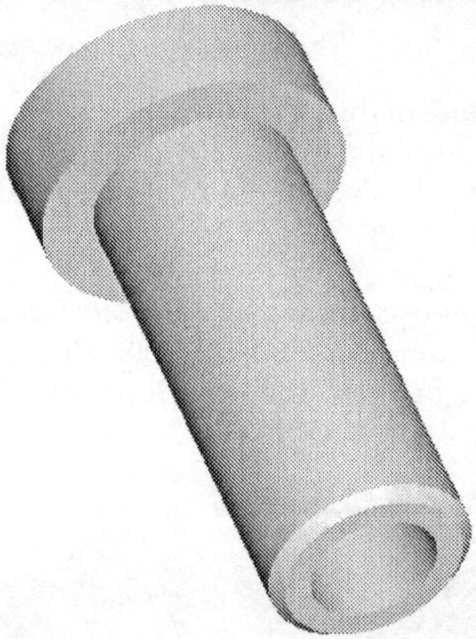

Figure 13-21 *Model after chamfering*

Creating the Remaining Components

Similarly, create the remaining components of the Fixture assembly. The components that are remaining are Spacer, Support Pin, Nut, and Bolt. Examine the given drawing views of these components and create them.

Assembling the Components

Next, you will assemble all the components of the Fixture assembly. As mentioned earlier, an assembly is defined as a design consisting of two or more than two components in which you apply some specific assembly constraints to the parts so that their **Degree of Freedom** (DOF) is controlled or restricted. Various assembly constraints, such as Mate, Align, Insert, Orient, and so on, are used to control the DOF between two components.

Creating a New Assembly File

To start with an assembly you have to create a new assembly file.

1. Choose the **Create a new object** button from the **File** toolbar. The **New** dialog box is displayed, and the **Part** radio button is selected by default in the **Type** area.

2. Choose the **Assembly** radio button from the **Type** area. The **Design** radio button is selected by default in the **Sub-type** area.

3. Enter **fixture_assembly** in the **Name** edit box and press ENTER. The three default assembly datum planes are displayed on the graphics screen.

Inserting the End Plate in the Assembly

You need to place the End Plate component first because it is the base component in the Fixture assembly.

1. Choose the **Add component to the assembly** button from the **Right Toolchest**. The **Open** dialog box is displayed.

2. Double click on the **end_plate** in the **Open** dialog box. The **Component Placement** dialog box is displayed on the right of the graphics window and you will notice that another set of datum planes, along with the default datum planes of the component, is displayed on the graphics window as shown in Figure 13-22.

3. Use CTRL+ALT+right mouse button to drag the End Plate away from the default assembly datum planes as shown in Figure 13-23. This is done in order to view the two sets of the datum planes clearly and then continue with the assembly operations.

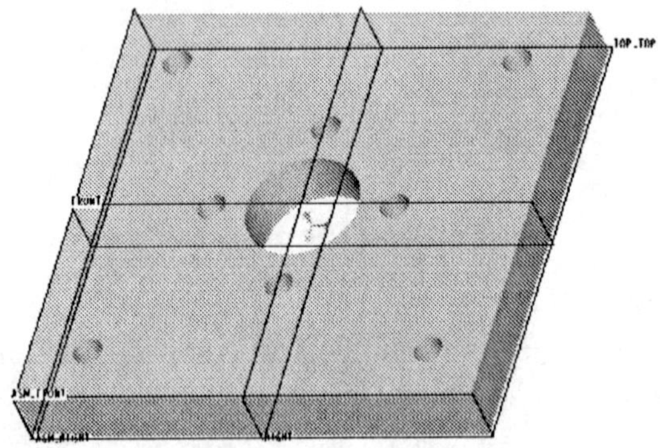

Figure 13-22 *Component placed in default position in assembly environment*

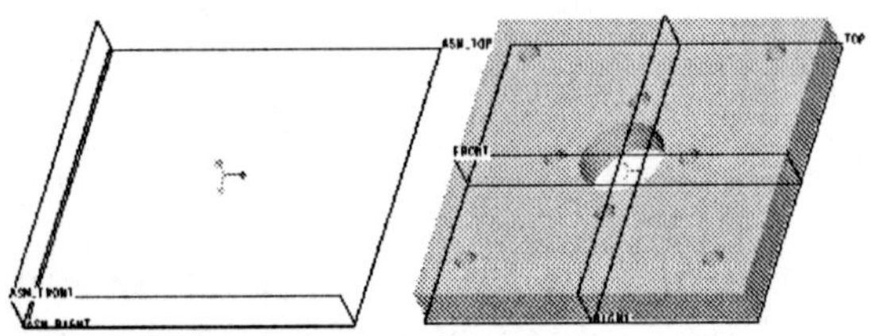

Figure 13-23 *Moving the component away from the assembly default datum planes*

4. Choose the **Align** option from the **Type** drop-down list in the **Constraints** area of the **Place** tab in the **Component Placement** dialog box. You are prompted to select a datum plane or axis.

5. Select the **TOP** datum plane from the component and then select the **ASM_TOP** datum plane from default assembly datum planes. You will notice that the **TOP** datum plane of component will align with the **ASM_TOP** datum plane. The status displayed in the **Placement Status** area is **Partially Constrained**.

6. Again, choose the **Align** option from the **Type** drop-down list.

7. Select the **FRONT** datum plane of the component and then select the **ASM_FRONT** datum plane. Enter **0** in the **Message Input Window** and press ENTER. You will see that the **FRONT** datum plane of the component and the **ASM_FRONT** datum plane will align.

8. In the same way align the **RIGHT** datum plane and the **ASM_RIGHT** datum plane with 0 offset and notice the **Placement Status** area. This area shows **Fully Constrained**. This means that the End Plate is now fully defined and all its degrees of freedom are constrained.

9. Choose **OK** from the **Component Placement** dialog box. **COMPONENT has been created successfully** will appear in the **Message Area**.

Assembling the disk

The second component that you need to assemble is the Disk.

1. Choose the **Add component to the assembly** button from the **Right Toolchest**. The **Open** dialog box is displayed. Double-click on **disk.prt** file in the **Open** dialog box.

2. The disk is placed in the assembly file but is not actually assembled at the correct position. Turn on the axis using the **Datum axes on/off** button, if needed.

3. Choose the **Mate** option from the **Type** drop-down list in the **Constraint** area of the **Place** tab. You are prompted to select the mating surface from the first part. Spin the assembly using the middle mouse button and select the bottom face of the disk as shown in Figure 13-24. Now, you are prompted to select a surface or datum plane on the other part.

4. Select the top face of the End Plate as shown in Figure 13-24. Both the faces will attain a mate constraint as shown in Figure 13-25. Next, you need to apply the align constraint to the axis of the End Plate and the disk.

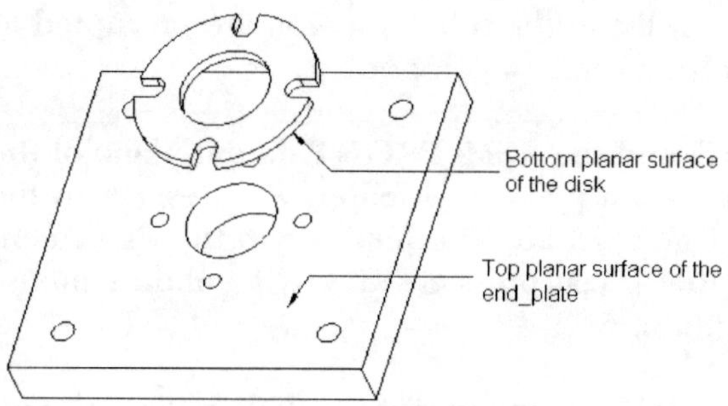

Figure 13-24 *Mating surfaces of the end_plate and the disk*

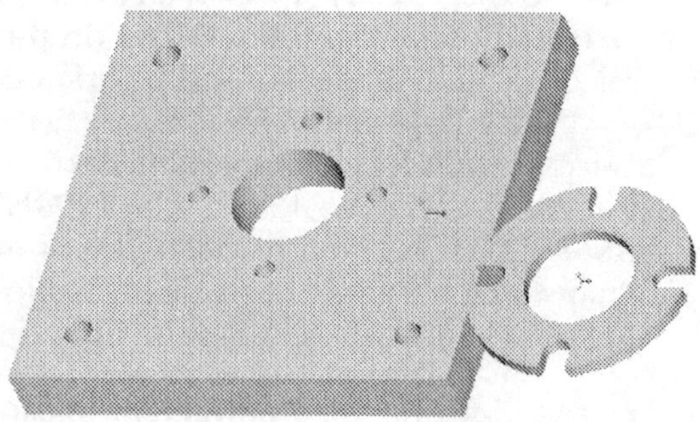

Figure 13-25 *Mate constraint applied*

5. Select the **Align** option from the **Type** drop-down list. Select the center axis of the **disk A_1** and select the center axis **A_5** of the end_plate. Both the axes of the disk and the end_plate are aligned as shown in Figure 13-26.

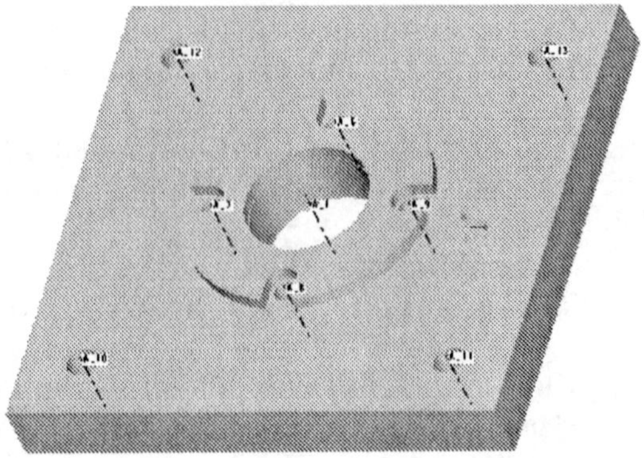

***Figure 13-26** **Align** constraint applied to the assembly*

6. In the **Placement Status** area, the message **Fully Constrained** is displayed but the **Allow Assumptions** check box is selected. This suggests that the assembly is fully constrained but you can check whether it needs some more constraints or not. If all the degrees of freedom are not specified then you can apply some more constraints. Choose CTRL+ALT+ middle mouse button and drag the disk. You will notice that the disk will revolve around the axis of the center hole. This means that the assembly is not fully constrained. So you need to apply one more constraint.

7. Turn on the datum planes using the **Datum planes on/off** button from the **Top Toolchest**, if required.

8. Clear the **Allow Assumptions** check box in the **Placement Status** area. The message in the **Placement Status** area will change from **Fully Constrained** to **Partially Constrained**.

9. Choose the **Specify a new constraint** button from the **Constraints** area. Choose the **Oriented** option from the **Offset** drop-down list in the third row. You are prompted to select a datum plane that will be oriented from the first component.

10. Select the **FRONT** datum plane of the disk and the **ASM_FRONT** datum plane of the assembly, see Figure 13-27. The **FRONT** datum plane is oriented with respect to the **ASM_FRONT** datum plane as shown in Figure 13-28. The message in the **Placement Status** area is **Fully Constrained**. Notice that if you press CTRL+ALT+middle mouse button and drag the mouse, the disk will not move.

11. Choose **OK** from the **Component Placement** dialog box to exit the dialog box. Turn off the datum planes by choosing the **Datum planes on/off** button for better visualization.

Assembling the Support Pin

The third component that you will assemble is the Support Pin.

1. Choose the **Add component to the assembly** button from the **Right Toolchest**. The **Open** dialog box is displayed. Double click on **support_pin**.prt.

2. Use CTRL+ALT+middle mouse button to orient the Support Pin to a comfortable position.

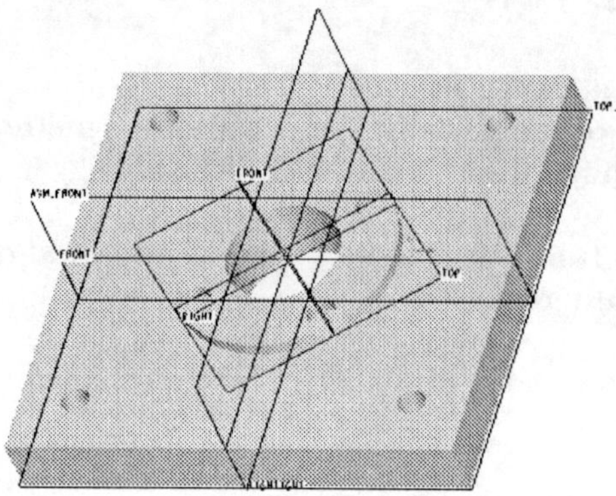

Figure 13-27 *Position of disk after dragging*

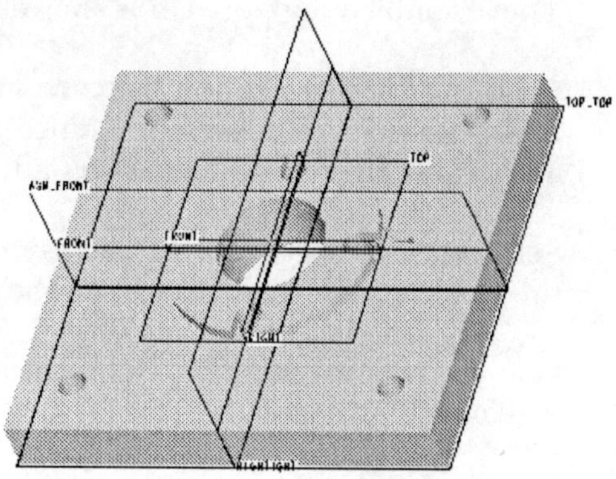

Figure 13-28 *The* ***Orient*** *constraint applied to the* ***FRONT*** *datum plane of the disk*

3. Choose the **Insert** option from the **Type** drop-down list in the **Constraints** area and select one of the small outer cylindrical surface of the Support Pin and the inner cylindrical surface of any one of the diametrically placed hole. The support_pin will be inserted inside the hole. Drag the Support Pin using CTRL+ALT+right mouse button.

4. Choose the **Mate** option from the **Type** drop-down list in the **Constraints** area and select the planar surfaces of the Support Pin and the End Plate as shown in Figures 13-29 and 13-30 respectively. Enter 0 in the **Message Input Window** and press ENTER.

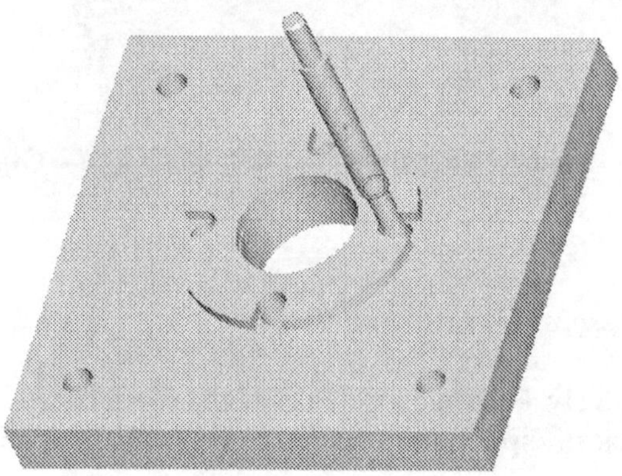

Figure 13-29 *The surface to be selected from the Support Pin*

Figure 13-30 *The surface to be selected from End Plate*

5. Clear the **Allow Assumptions** check box in the **Placement Status** area if necessary and as discussed earlier orient one datum plane of the Support Pin with one of the datum planes of the assembly.

6. Choose **OK** from the **Component Placement** dialog box and exit the assembly environment.

The assembled Support Pin is shown in Figure 13-31.

Creating the Reference Pattern of the Support Pin

Next, you need to create a reference pattern of the Support Pin because this method of assembling the components saves a lot of time of the Pro/ENGINEER users.

1. Select the Support Pin from the assembly or from the graphics window and hold down the right mouse button to invoke the shortcut menu.

2. Choose the **Pattern** option. The pattern dashboard is displayed. The **Reference** option is selected by default.

3. Choose the **Build feature** button to exit the dashboard. The reference pattern is created.

The reference pattern of the assembled Support Pin is shown in Figure 13-32.

Figure 13-31 *Assembled Support Pin*

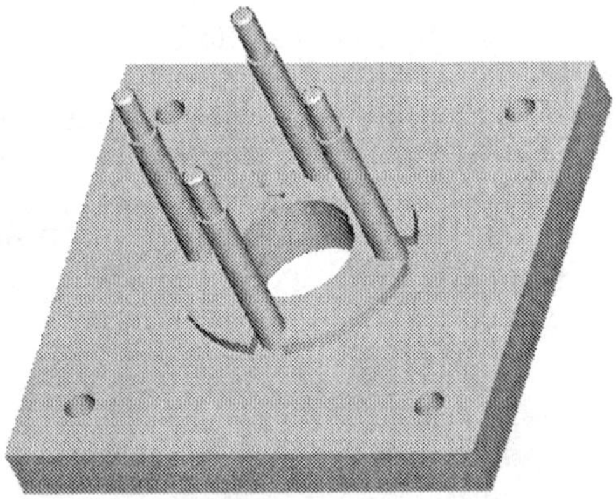

Figure 13-32 *Reference pattern of the Support Pin*

Assembling the Spacer

Next, you will assemble the spacer using assembly constraints.

1. Choose the **Add component to the assembly** button from the **Right Toolchest**. The **Open** dialog box is displayed. Double-click on **spacer.prt**. The **Component Placement** dialog box is displayed.

2. Orient the spacer using CTRL+ALT+middle mouse button to a comfortable position as shown in Figure 13-33.

3. Turn on the **Datum axes on/off** button from the **Top Toolchest** to display the datum axes.

4. Choose the **Align** option from the **Type** drop-down list in the **Constraint** area and select **A_1** axis of the spacer and **A_5** of the assembly. The **A_1** and **A_5** axes will be aligned as

shown in Figure 13-34.

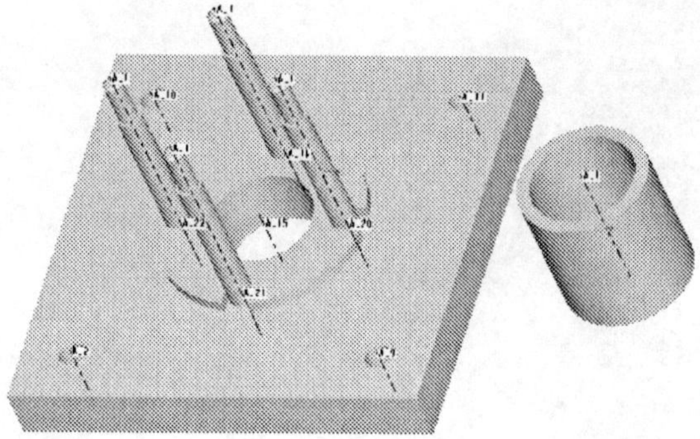

Figure 13-33 *Spacer placed in the default position*

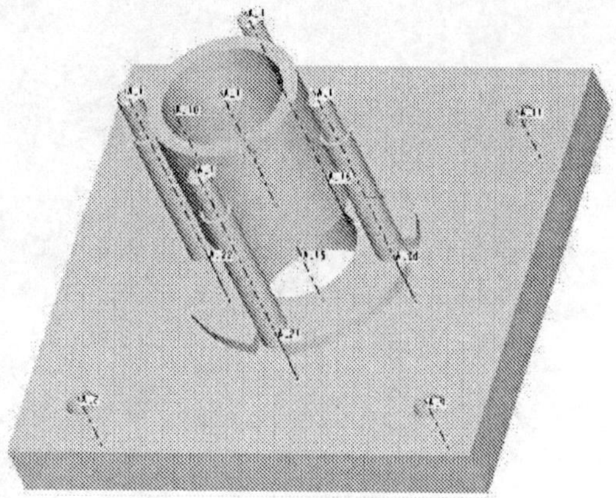

Figure 13-34 *Align constraint is applied to the spacer*

5. Use CTRL+ALT+right mouse button to drag the Spacer if necessary.
6. Now, choose the **Mate** option from the **Type** drop-down list in the **Constraints** area.
7. Select the bottom face of the Spacer and the top face of the Disk as shown in Figures 13-35 and 13-36, respectively. Enter **0** in the **Message Input Window** and press ENTER.

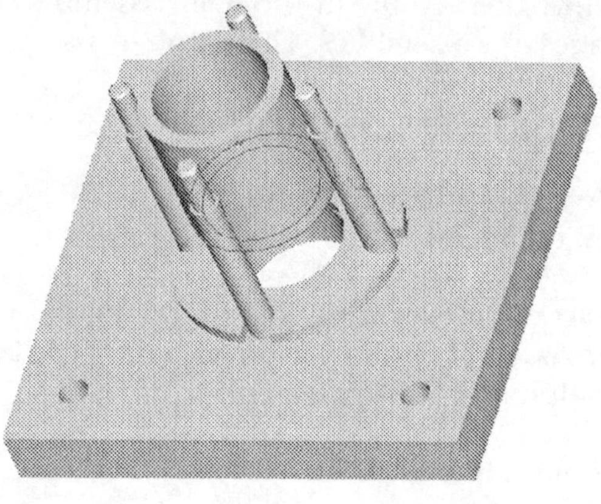

Figure 13-35 *Mating surface of the spacer*

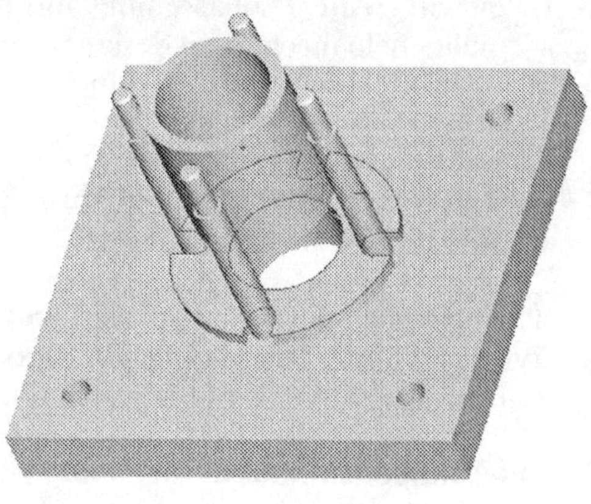

Figure 13-36 *Mating surface of the assembly*

8. Apply the **Oriented** constraint to the Spacer as discussed earlier. The **Placement Status** area will show the **Fully Constrained** message. Choose **OK** and exit the assembly environment.

   Similarly, assemble the Disk, the End Plate, and the Center Pin on the other side by using the assembly constraints as discussed earlier. Figure 13-37 shows the shaded trimetric view of the Fixture assembly. Next, you need to assemble the Nut and the Bolt.

Figure 13-37 End Plate, Spacer, Disk, and Center Pin assembled together

In this assembly you need four instances of the Nut and four of the Bolt. Assembling all of them is a very time-consuming process. Therefore, it is recommended that you create a subassembly of these components and assemble one set of the subassembly with the main assembly. The remaining 3 sets will be assembled by reference pattern.

Creating the Subassembly of Nut and Bolt

Next, you will create a subassembly and insert the subassembly in the main assembly. The subassemblies help mechanical designers to manage large assemblies. The use of subassemblies in CAD is strongly recommended in case of large assemblies.

Creating a New Subassembly File

To start with subassembly, you have to create a new subassembly file.

1. Choose the **Create a new object** button from the menu bar.

2. The **New** dialog box is displayed and the **Part** radio button is selected by default in the **Type** area of the **New** dialog box. Choose the **Assembly** radio button from the **Type** area. The **Design** radio button is selected by default in the **Sub-type** area.

3. Enter **nutboltsubasm** in the **Name** edit box and press ENTER.

   The default datum planes will appear on the graphics window.

Assembling the Components

Now, you will assemble the components to create a subassembly that will be used to complete this tutorial.

1. Choose the **Add component to the assembly** button from the **Right Toolchest**. The **Open** dialog box is displayed. Double-click on the **bolt.prt** in the **Open** dialog box.

2. You will enter the assembly environment. Assemble the datum planes of bolt with the default assembly datum planes as discussed earlier.

3. Choose the **Add component to the assembly** button from the **Right Toolchest**. The **Open** dialog box is displayed. Double-click on **nut.prt** in the **Open** dialog box. You will enter the assembly environment.

4. Choose the **Align** option from the **Type** drop-down list in the **Constraints** area and select **A_1** axis of the Nut and **A_2** axis of the Bolt from the graphics window. Both the axes are aligned as shown in Figure 13-38. Drag the Nut using CTRL+ALT+right mouse button, if needed.

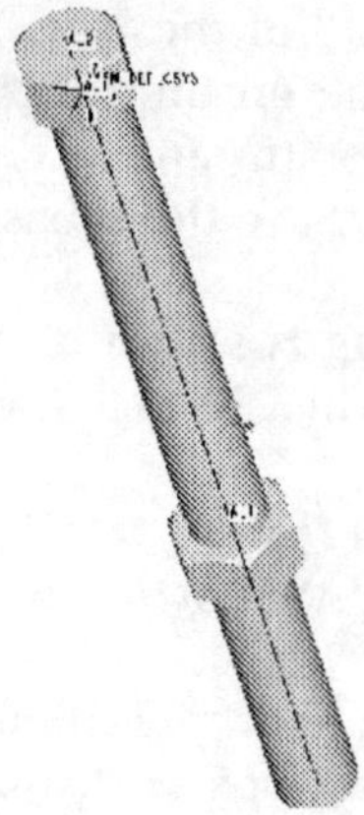

Figure 13-38 *Align constraint applied to the subassembly*

5. Choose the **Mate** option from the **Type** drop-down list. You are prompted to select a mating surface or datum plane from the first part. Select the top face of the nut as shown in Figure 13-39.

6. You are prompted to select a mating surface or datum plane of the other part. Spin the model and select the surface of the bolt as shown in Figure 13-40.

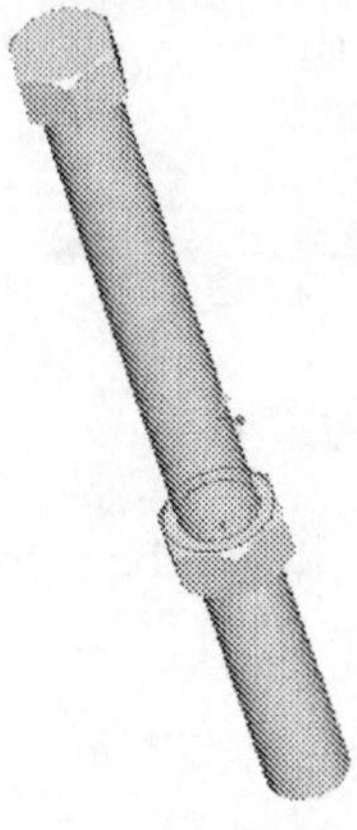

Figure 13-39 *Highlighting the mating surface*

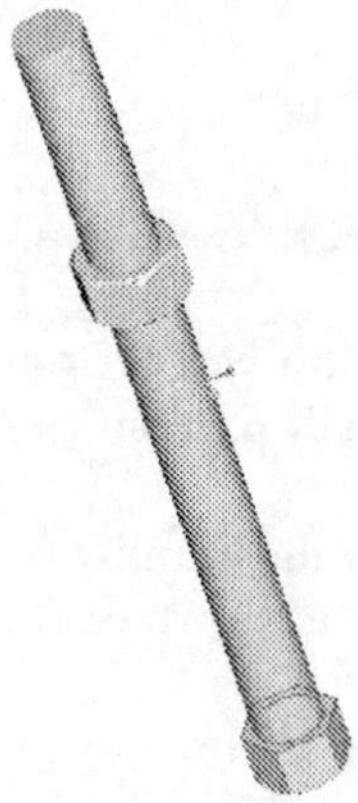

Figure 13-40 *Highlighting the mating surface*

7. Enter **2.625** in the **Message Input Window** and press ENTER.

8. Now, clear the **Allow Assumptions** check box in the **Placement Status** area and choose the **Add a constraint** button in the **Constraints** area.

9. Turn on the datum planes by **Datum plane on/off** button. Choose **Oriented** from the

Offset drop-down list in the third row. You are prompted to select the planar surface or datum plane on one part.

10. Select DTM3 of the Nut. You are prompted to select the orientation planar surface or datum plane on the other part. Select **ASM_FRONT** from the graphics window. The placement status area will show the message **Fully Constrained**. Choose **OK** to exit the dialog box. Save the subassembly.

Assembling the Subassembly with the Main Assembly

Now, you will assemble the subassembly with the main assembly.

1. Choose **Window > FIXTURE_ASSEMBLY.ASM** from the menu bar. The FIXTURE_ASSEMBLY window will be activated.

2. Choose the **Add component to the assembly** button from the **Right Toolchest**. The **Open** dialog box is displayed. Double-click on the **nutboltsubasm.asm** file from the **Open** dialog box.

3. Using CTRL+ALT+middle mouse button to orient the subassembly.

4. Turn on the axis display using the **Datum axis on/off** button from the **Top Toolchest**. Align **A_1** axis of the subassembly with **A_1** axis of the End Plate using the **Align** constraint.

5. Using the **Mate** constraint, mate the bottom planar surface of the head of the bolt with the top planar surface of the upper End Plate.

6. Orient one of the six flat faces of the head of the Bolt with the right planar surface of the End Plate.

7. Choose **OK** to exit the **Component Placement** dialog box.

Creating the Reference Pattern of Subassembly

Now, you will pattern the subassembly using the **Ref pattern** option.

1. Select the subassembly from the graphics window and hold down the right mouse button to invoke the shortcut menu.

2. Choose the **Pattern** option. The pattern dashboard is displayed. The **Reference** option is selected by default.

3. Choose the **Build feature** button to exit the dashboard. The reference pattern is created.

   The reference pattern will result in the assembling of the three remaining subassembly instances. The shaded trimetric view of the Fixture Assembly is shown in Figure 13-41.

4. Choose the **Save the active object** button from the **Top Toolchest** to save the file.

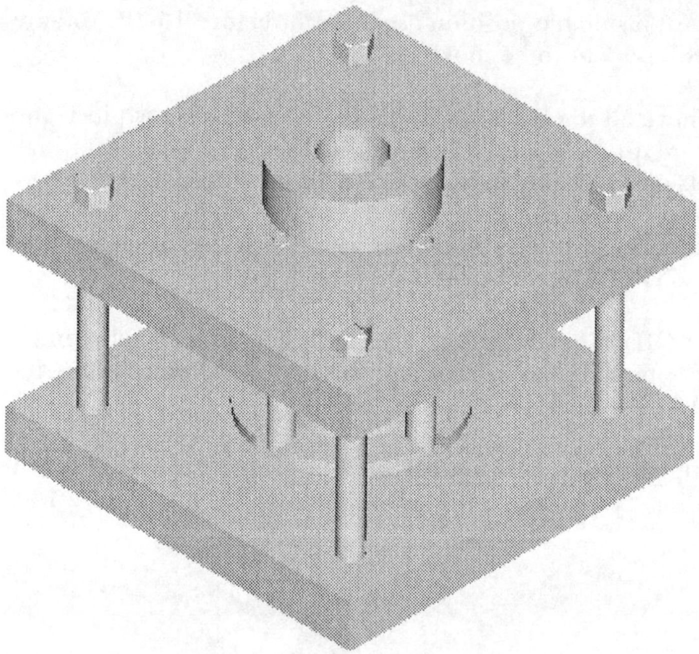

Figure 13-41 Assembled view of the Fixture assembly

Exploding the Assembly and Creating the Offset Lines

Now, you will create an exploded state of the assembly and create the offset lines.

1. Choose the **Start the view manager** button from the **Top Toolchest**. The **View Manager** dialog box is displayed. Choose the **Explode** tab and right-click in the dialog box to invoke the shortcut menu. Choose the **New** option, name it **Explode1**, and press the middle mouse button.

2. Right-click on the **EXP1** in the dialog box and choose the **Redefine** option. The **MOD EXPLODE** menu is displayed. Choose the **Position** option from the menu, the **Explode Position** dialog box is displayed.

   The **Translate** radio button is selected by default in the **Motion Type** area of the **Explode Position** dialog box. The **Entity/Edge** option is selected by default in the **Motion Reference** drop-down list. The **Smooth** option is selected by default in the **Translation** drop-down list. The **SELECT** dialog box is displayed and you are prompted to select an axis or straight edge as the motion reference.

3. In the **Filter** drop-down list available in the status bar that is below the **Message Area**, select the **Axis** filter and then select **A_1** axis at the center of the Center Pin. You are prompted to select the component to be moved.

4. Select one of the bolt from the graphics window and drag it using the left mouse button and drop it at a suitable position as shown in Figure 13-42. You are again prompted to select the component to be moved.

5. Similarly, move all the bolts to a suitable location as shown in Figure 13-42. Now, select the nuts and place them as shown in Figure 13-42. Similarly, select the other components and place them as shown in Figure 13-42.

6. Choose **OK** and exit the **Explode Position** dialog box after moving all the components. The **MOD EXPLODE** submenu is displayed.

7. Choose the **Offset Lines** option. The **OFFSET LINES** submenu is displayed. Choose the **Create** option. You are prompted to select the type of entity for first explode offset line. The **AXIS** option is selected by default.

8. Turn on the **Datum axis on/off** button from the **Top Toolchest** and select the corresponding axis to create the offset lines as shown in Figure 13-42.

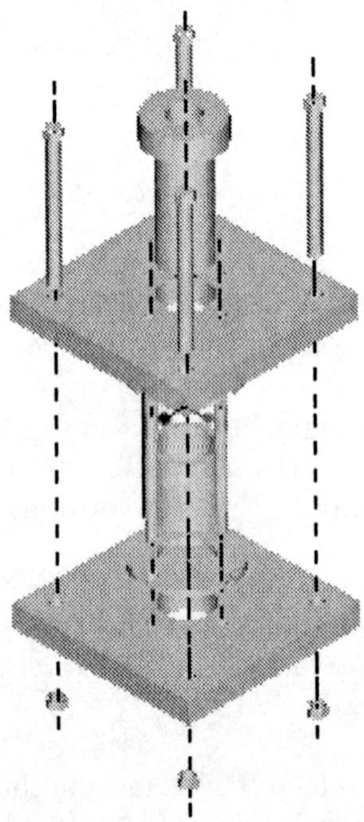

Figure 13-42 *Exploded state of the assembly*

9. After creating all the offset lines choose **Done/Return** from the **MOD EXPLODE** submenu and then choose **Done/Return** from the **EXPLD STATE** submenu.

To view the unexploded view of the Fixture assembly, choose **View > Explode > Unexplode View** from the menu bar. To view the exploded view of the assembly, choose

View > Exploded from the menu bar.

10. Choose **Window > Close** from the menu bar to close the window.

Generating the Drawing Views

Creating a New Drawing File

In Pro/ENGINEER the drawing views are generated in the Drawing mode. Therefore, you have to create a new drawing file to proceed further.

1. Choose the **Create a new object** button, the **New** dialog box is displayed. The **Part** radio button is selected by default in the **Type** area and the **Solid** radio button is selected by default in the **Sub-type** area.

2. Select the **Drawing** radio button in the **Type** area of the **New** dialog box. Enter **fixture_assembly** in the **Name** edit box and press ENTER. The **New Drawing** dialog box is displayed.

3. Choose the **Browse** button from the **Default Model area** to display the **Open** dialog box. Double-click on the **fixture_assembly** file. You will see that **fixture_assembly** will appear in the edit box that is in the **Default Model** area. The **Use Template** radio button is selected by default in the **Specify Template** area.

4. Select the **Empty** radio button. The **Orientation** and **Size** areas are displayed in the **New Drawing** dialog box.

5. The **Landscape** button is chosen by default. If not, choose the **Landscape** button. The Landscape option specifies the orientation of the drawing sheet.

6. Select **A4** from the **Standard Sheet** drop-down list in the **Size** area of the **New Drawing** dialog box.

7. Choose **OK** and exit the **New Drawing** dialog box.

   You will enter the drawing mode of **Pro/ENGINEER**. A rectangle is displayed on the graphics screen that acts as drawing sheet limits.

Generating the Drawing Views

You have to generate the drawing views from an existing solid model. This process of generating the views from an existing solid model is known as generative drafting.

1. Choose the **Insert a drawing view of the active model** button from the **Top Toolchest** to display the **VIEW TYPE** menu. In this menu, the **General** option is selected by default.

2. Choose the **Scale** option and accept all the other default options. Choose **Done**. You are prompted to select the center point for the placement of the drawing view.

3. Using the left mouse button, specify the location of the view close to the upper left corner of the drawing sheet. You are prompted to enter the scale factor value for the view. Enter **0.75** in the **Message Input Window** and press ENTER.

   The default view of the Fixture assembly will be placed on the drawing sheet. Turn off the datum display and the axis display for clear visualization. Choose the **No Hidden** button from the **Top Toolchest** so that hidden lines are not displayed. Choose **Redraw the current view** button from the **Views** toolbar in the **Top Toolchest**.

   The **Orientation** dialog box is displayed. Using the **Orientation** dialog box you can orient the drawing view placed in the default position. Therefore, you need to reorient the drawing view.

4. Choose **TOP** from the **Saved Views** drop-down list in the **Orientation** dialog box, and choose the **SET** button. The model will be oriented in the top view.

5. Choose **OK** to exit the **Orientation** dialog box.

6. Choose the **Insert a drawing view of the active model** button from the **Top Toolchest** to display the **VIEW TYPE** menu. Accept the default options and choose **Done**. You are prompted to select a center point for the placement of the drawing view.

7. Specify a point close to the lower left corner of the drawing sheet to specify the center point for the placement of the front view.

8. Choose the **Insert a drawing view of the active model** button from the **Top Toolchest** to display the **VIEW TYPE** menu. Accept the default options and choose **Done**. Specify the center point for the placement of the right-side view near the lower right corner of the drawing sheet.

   Next, you have to place the isometric view of the Fixture assembly.

9. Choose **General** > **Full View** > **No Xsec** > **Scale** > **Done**. You are prompted to specify the center point for the placement of the drawing view.

10. Specify a point close to the upper right corner of the drawing sheet as the center point for the placement of the drawing view. Enter **0.75** as the scale of the view in the **Message Input Window**. The **Orientation** dialog box is displayed.

11. Choose the **Preferences** option from the **Type** drop-down list. The **Default** radio button is selected by default in the **View Origin** area. Choose the **Isometric** option from the **Default Orientation** drop-down list and choose **OK**. The default view will be oriented to isometric view.

12. Choose the **Redraw the current view** button from the **Top Toolchest** to repaint the screen for better visualization.

Tip: *If the top view extends beyond the drawing limits, choose the* ***Disallow the movement of drawing views with the mouse*** *button from the* ***Top Toolchest*** *to unlock the drawing view. Using the left mouse button select the top view and drag and place it on the drawing sheet. Similarly, you can move any view.*

The orientation of the drawing can also be done using the following method:

In the ***Orientation*** *dialog box, the* ***Front*** *option is selected by default in the* ***Reference 1*** *drop-down list. The* ***SELECT*** *dialog box is displayed below the* ***Orientation*** *dialog box. Select the top face of the lower end plate as the first reference for drawing view orientation. Now, automatically the* ***Reference 2*** *selection button will be activated to select the second reference for the orientation of the drawing view and* ***TOP*** *will be selected by default in the* ***Reference 2*** *drop-down list. Select the right face of either of the two end plates and the drawing view that was placed in default position is oriented and now acts as the top view.*

The front view, top view, right-side view, and isometric view are generated and placed on the drawing sheet as shown in Figure 13-43.

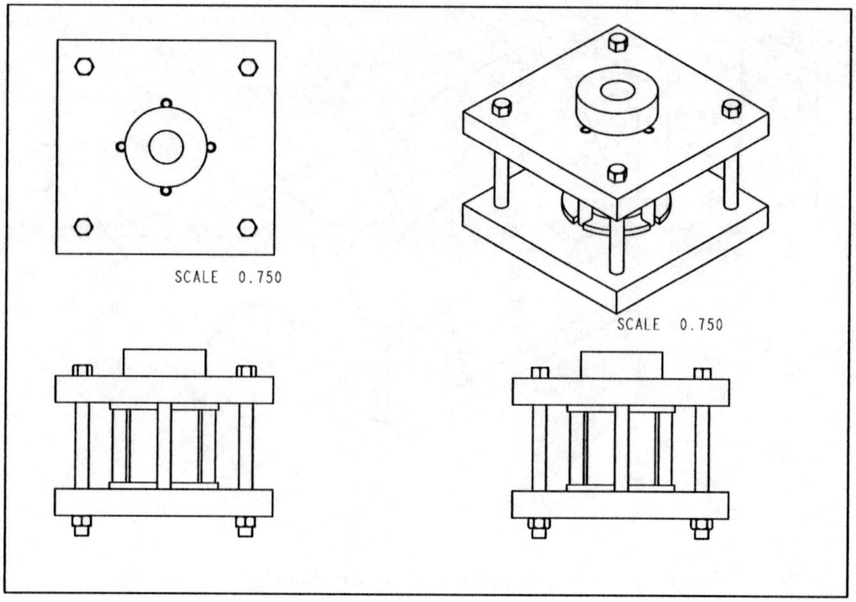

Figure 13-43 *Generated views of the Fixture assembly*

Saving the Drawing File

Now, you have to save this drawing file so that it can be referred to in the near future.

1. Choose the **Save the active object** button from the **Top Toolchest**. The **Message Input Window** displays the name of the file that you had specified earlier. Choose the green check mark on the right of the **Message Input Window** or press ENTER.

PROJECTS

Project 1

Create all the components of the Bench-Vice assembly and assemble them together as shown in Figure 13-44. The exploded view of the assembly is shown in Figure 13-45. The details of the components of the Bench-Vice assembly are shown in Figures 13-46 through 13-50. After completing the assembly create an exploded state of the assembly in the Assembly mode. Finally, generate the following drawing views of the assembly:

a. Top view
b. Front view
c. Right-side view
d. Isometric view

(Expected time: 3 hrs)

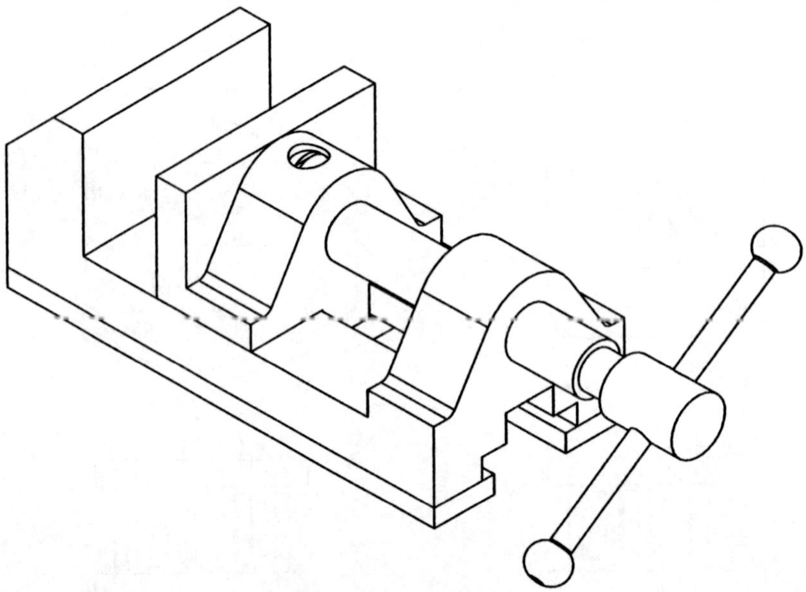

Figure 13-44 *Assembled view of the Bench-Vice assembly*

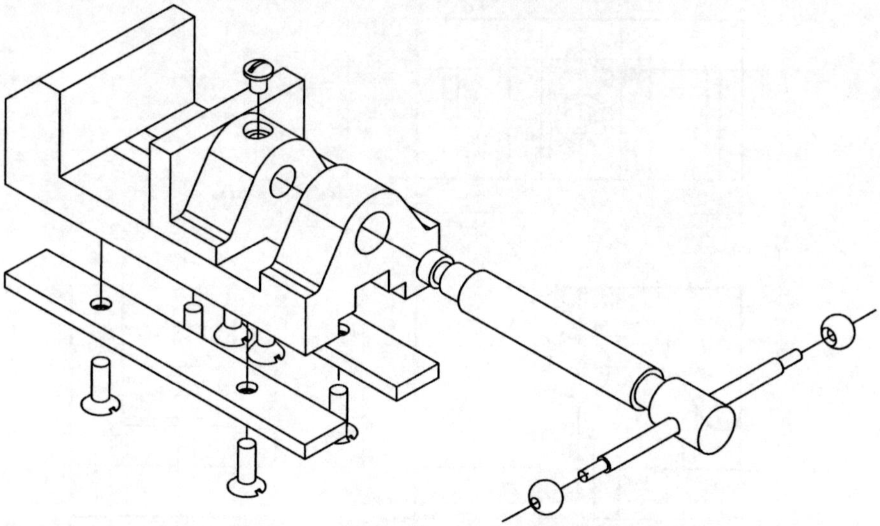

Figure 13-45 *Exploded view of the Bench-Vice assembly displaying various components*

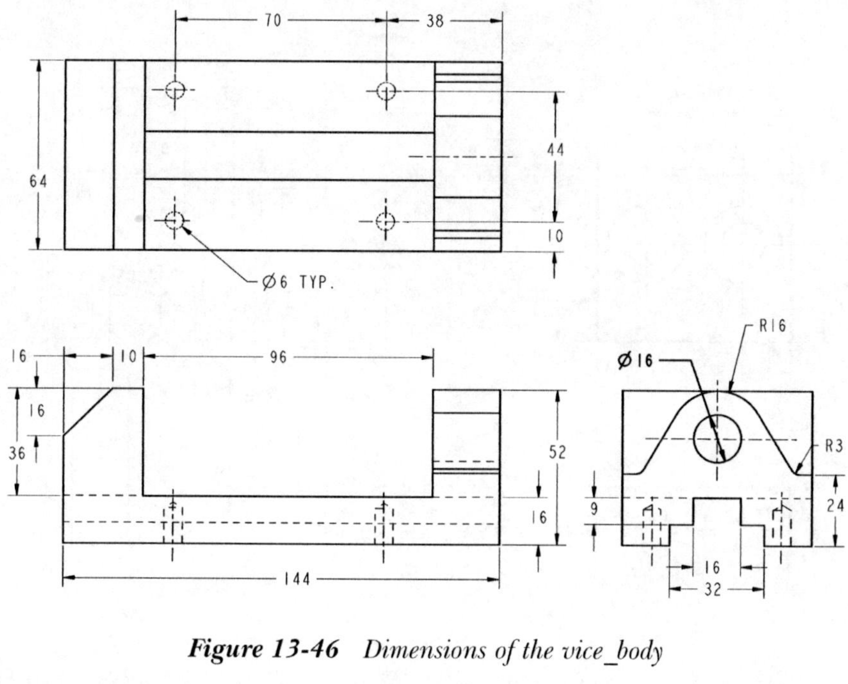

Figure 13-46 *Dimensions of the vice_body*

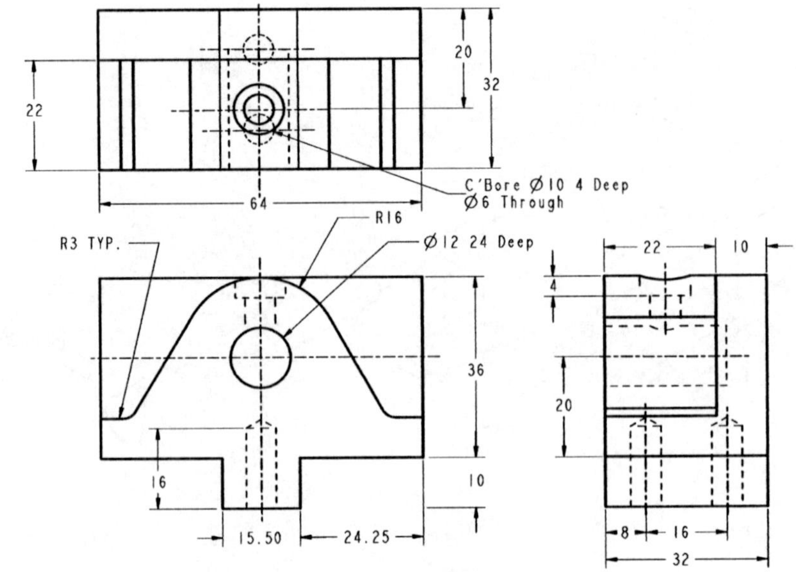

***Figure 13-47** Dimensions of the vice_jaw*

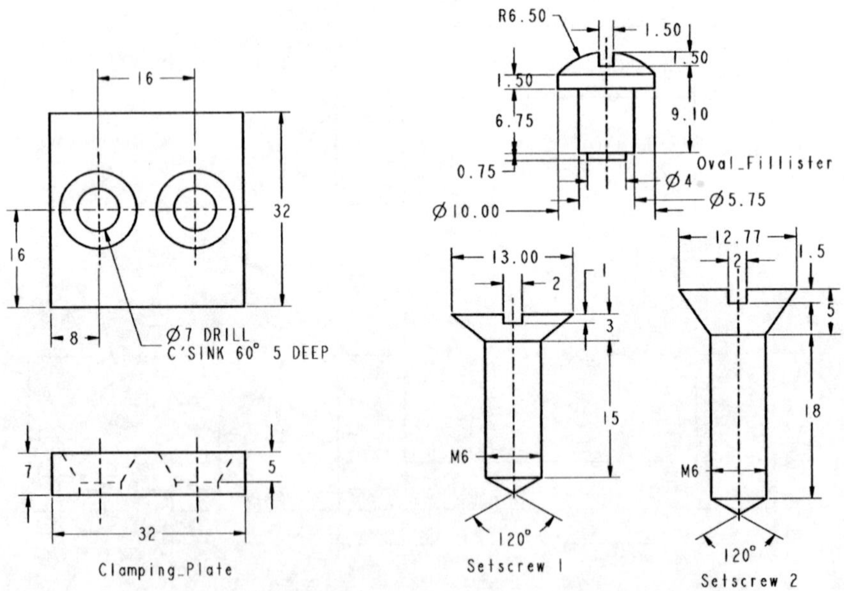

***Figure 13-48** Dimensions of the Clamping Plate, Setscrew1, Setscrew2, and the Oval Fillisters*

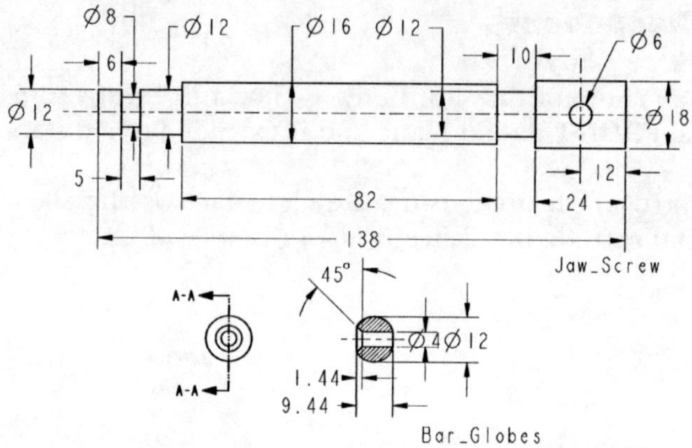

Figure 13-49 Dimensions of the jaw_screw and the bar_globes

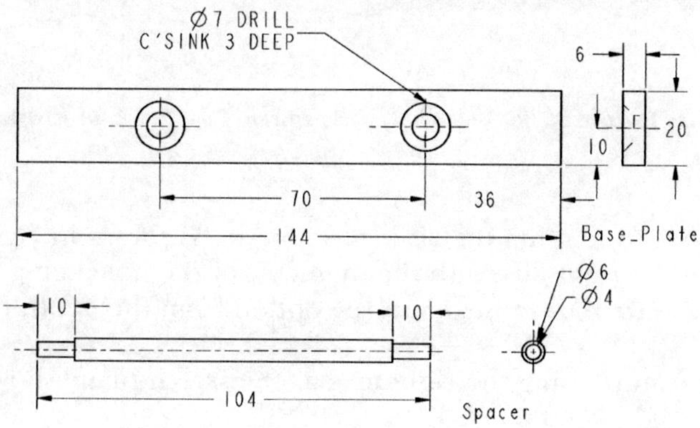

Figure 13-50 Dimensions of the base_plate and the screw_bar

The steps that will be followed to complete this project are listed below:

Creating the Components

Creating the Vice Body

1. Create the base feature of the Vice Body as shown in Figure 13-51. The base feature is sketched on the **FRONT** datum plane and is extruded to a distance of 64.

2. Create the protrusion feature using the right face as the sketch plane as shown in Figure 13-52 and extrude the sketch up to a distance of 22.

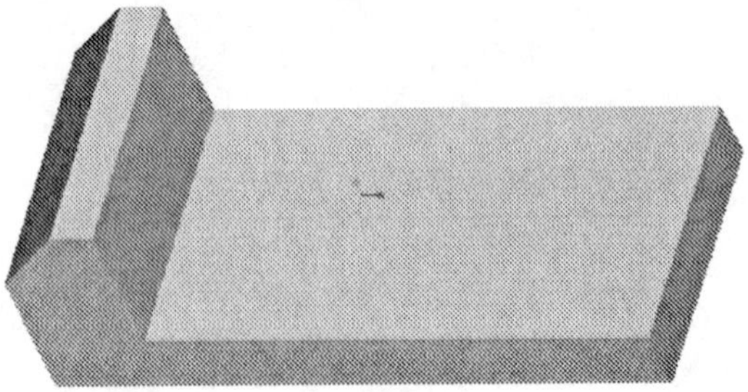

Figure 13-51 Base feature of the Vice Body

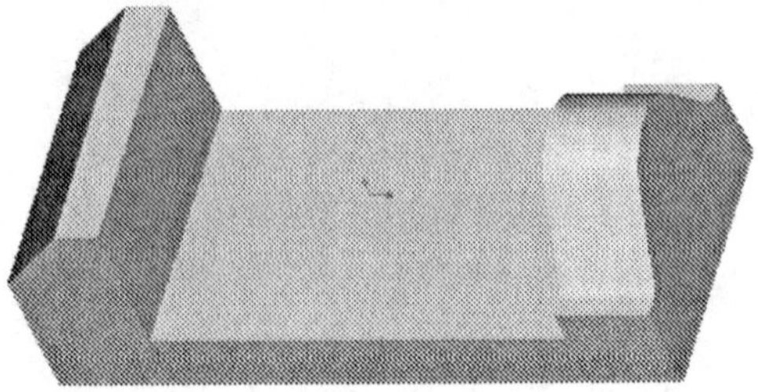

Figure 13-52 A protrusion feature added to the model

3. Create the hole feature using the Hole dashboard. As shown in Figure 13-53, the hole is to be placed on the right face with the diameter of the hole being 16 and this feature is created using the **Drill upto next surface** option from the depth flyout.

4. Create the cut feature using the right face as the sketch plane shown in Figure 13-54.

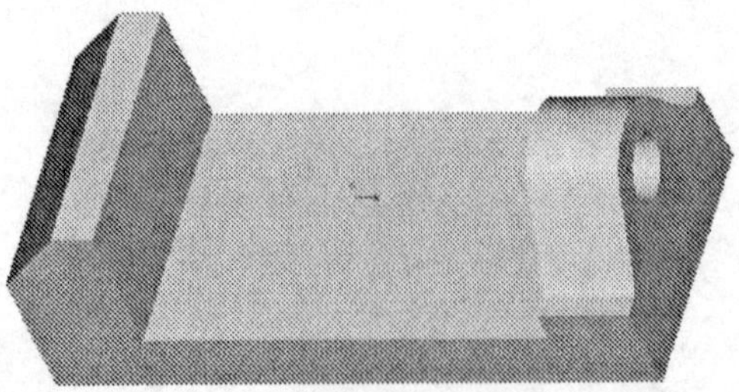

Figure 13-53 Hole created on the right planar surface of the base feature

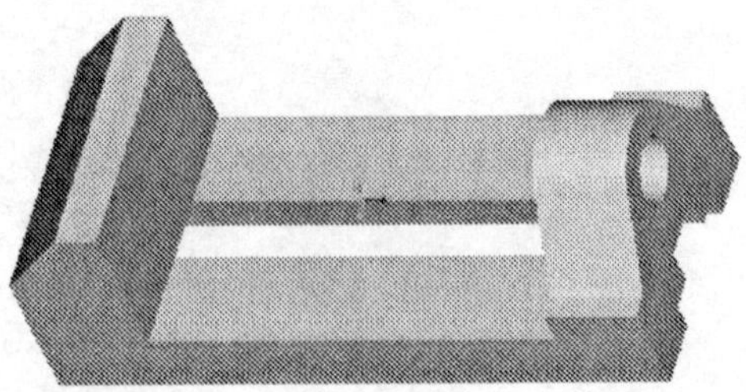

Figure 13-54 After adding the cut feature to the model

5. Create the hole feature using the Hole dashboard on the bottom face of the model and create the rectangular pattern of the same hole feature using the pattern option.

6. Now, save the model and close the file.

Creating the Vice Jaw

1. Create the base feature of the Vice Jaw as shown in Figure 13-55. For creating the base feature, the **FRONT** datum plane will be selected as the sketching plane. The sketch of the base feature is extruded to a distance of 32.

2. The second feature is a cut feature and this feature will be created using the front planar surface of the base feature as the sketching plane. The sketch of the cut feature will be extruded to a distance of 22 as shown in Figure 13-56.

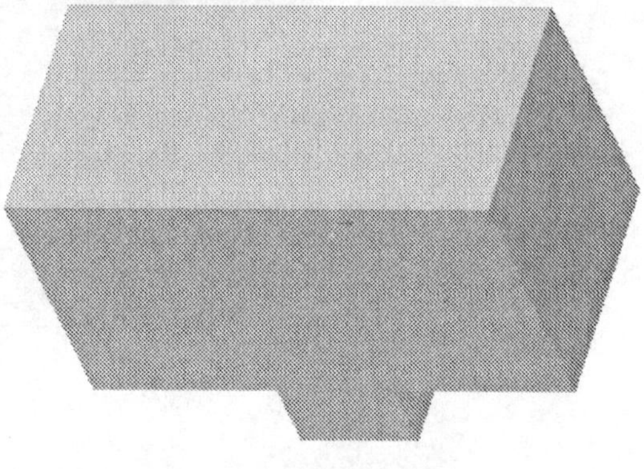

Figure 13-55 *The base feature of the Vice Jaw*

3. Create the hole feature using the Hole dashboard. The placement plane for the hole feature is the front face as shown in Figure 13-57.

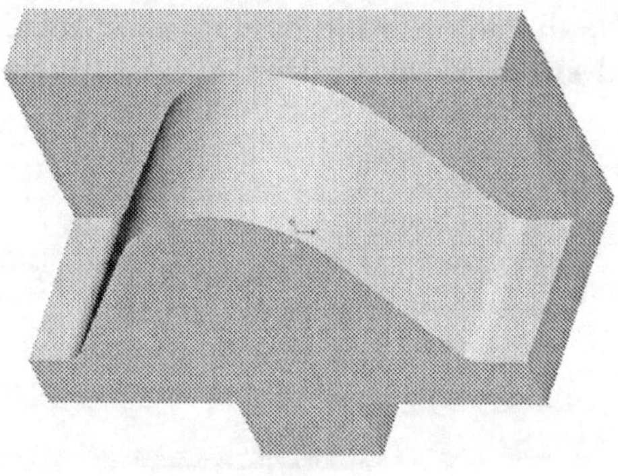

Figure 13-56 *Cut feature added to the base feature*

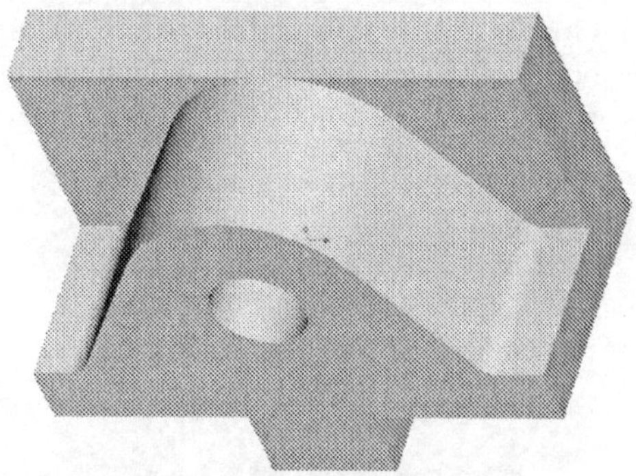

Figure 13-57 *A hole feature is placed on the front planar surface of the base feature*

4. Create a hole using the Hole dashboard on the bottom face of the base feature.

5. Create a rectangular pattern of the hole feature created previously.

6. Create a counterbore on the top face using the Hole dashboard as shown in Figure 13-58.

7. Save the model as done earlier.

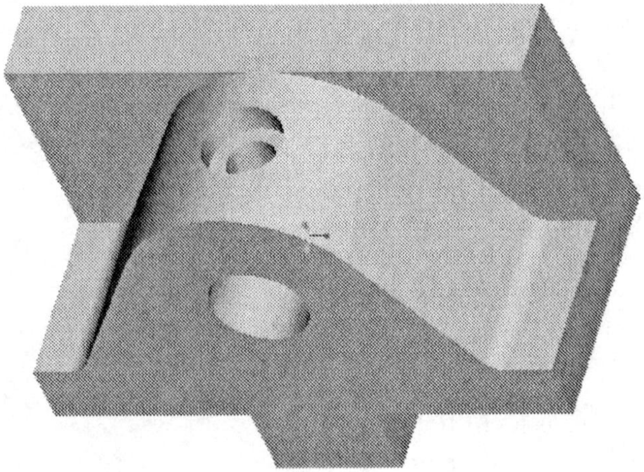

Figure 13-58 *Final model of Vice Jaw*

Create the Jaw Screw

1. Create the base feature of the Jaw Screw by sketching a section on the **FRONT** datum plane and revolving the sketch along a center line to an angle of 360 degree as shown in Figure 13-59.

2. Create a hole using the Hole dashboard. The diameter of the hole is 6 and this hole is created using the **Drill to intersect with all surfaces** button from the depth flyout as shown in Figure 13-60. Save the model.

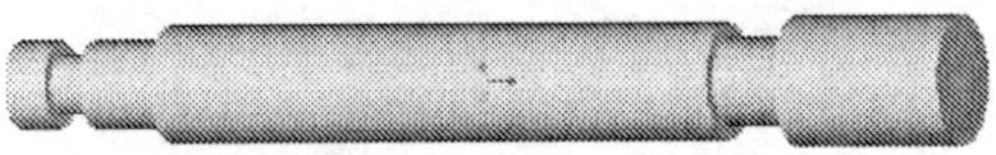

Figure 13-59 *Base feature of the Jaw Screw*

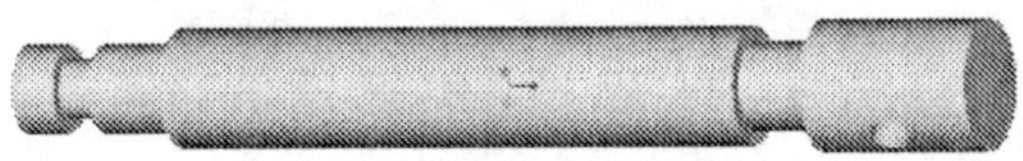

Figure 13-60 *Hole created using the Hole dashboard*

Creating the Remaining Components

Create the other components using the various tools of part modeling in Pro/ENGINEER and save them.

Assembling the Components

Assembling the Vice Jaw with the Vice Body

1. Create a new assembly file and assemble the Vice Body with the default datum planes of the assembly using the assembly constraints as shown in Figure 13-61.

2. Assemble the Vice Jaw with the Vice Body using the assembly constraints as shown in Figure 13-62.

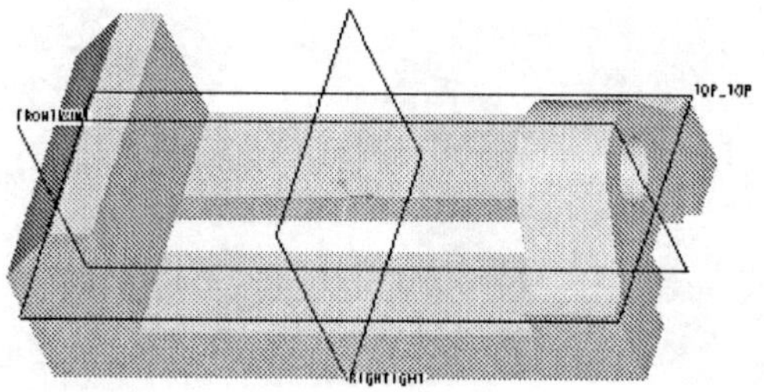

Figure 13-61 The Vice Body assembled with the default datum planes of the assembly

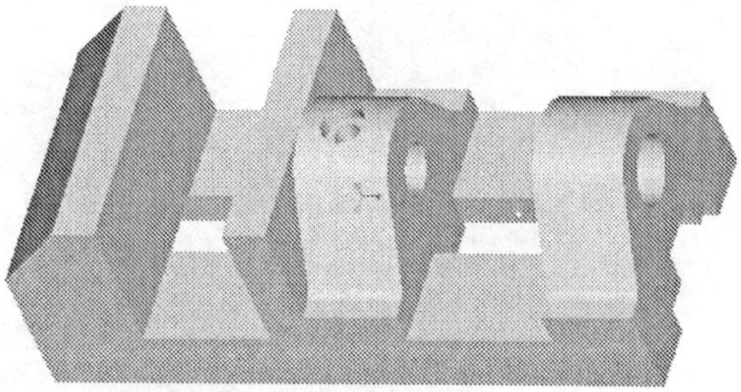

Figure 13-62 The Vice Jaw assembled with the Vice Body

3. Assemble the Jaw Screw using the assembly constraints as shown in Figure 13-63.

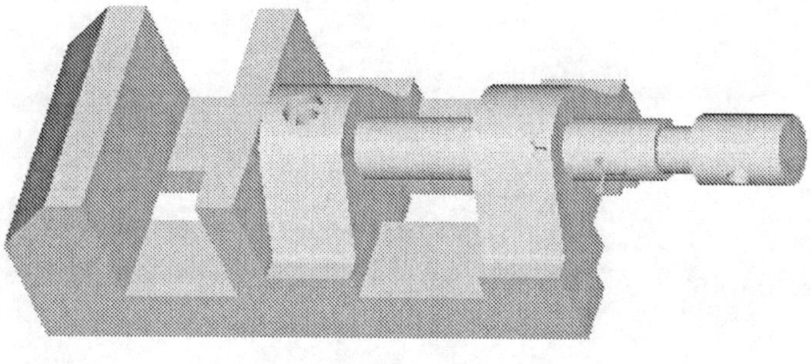

Figure 13-63 The Jaw Screw assembled to the base assembly

4. Similarly, assemble the remaining components of the Bench Vice assembly. The Bench Vice assembly after assembling all the components is shown in Figure 13-64.

Creating the Exploded State of the Assembly

1. Create the exploded view of the assembly as shown in Figure 13-65.

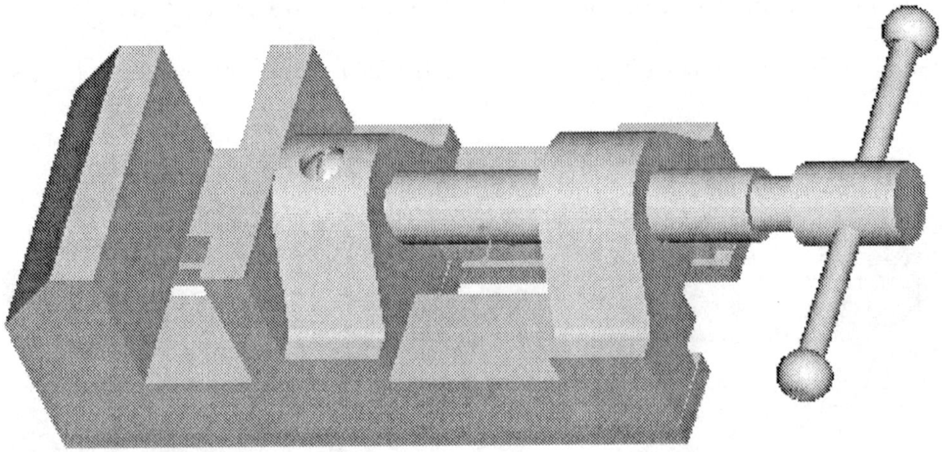

Figure 13-64 *Complete assembly of the Bench Vice assembly*

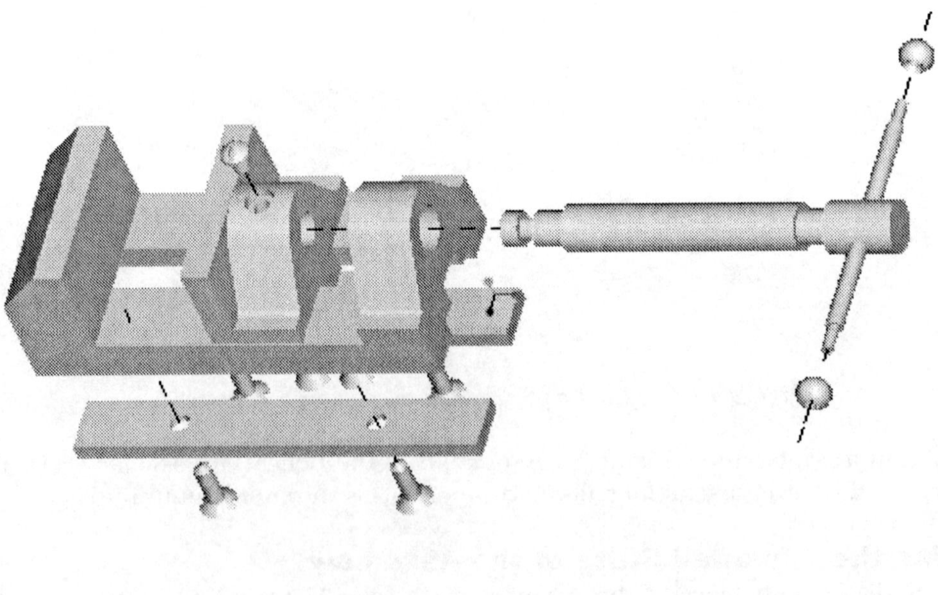

Figure 13-65 *Exploded view of the Bench Vice assembly*

Generating the Drawing view of the Assembly

1. Create a new drawing file and generate the front view, top view, right-side view, and isometric view as shown in Figure 13-66.

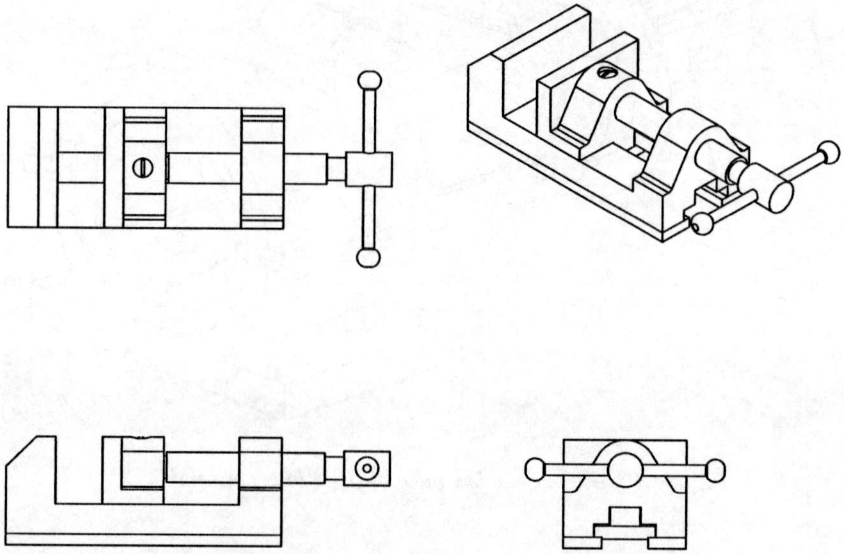

Figure 13-66 *Front view, top view, right-side view, and isometric view of the Bench Vice assembly*

Project 2

Create all the components of the Blower Assembly and assemble them together as shown in Figure 13-67. The exploded view of the assembly is shown in Figure 13-68. The details of the components of the Blower assembly are shown in Figures 13-69 through 13-74. After completing the assembly, create an exploded state of the assembly in the Assembly mode. Finally, generate the following drawing views of the assembly:

a. Top view
b. Front view
c. Right-side view
d. Isometric view

(Expected time: 4 hrs)

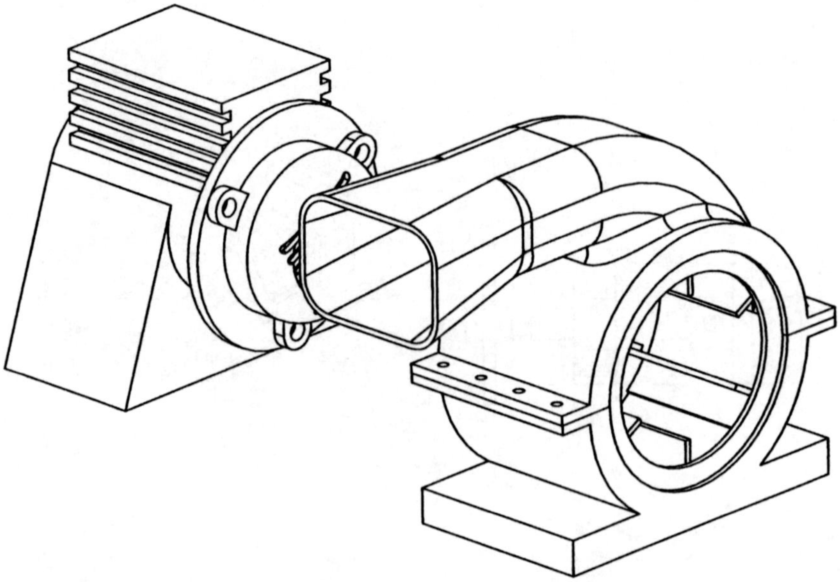

Figure 13-67 *Assembled view of the Blower assembly*

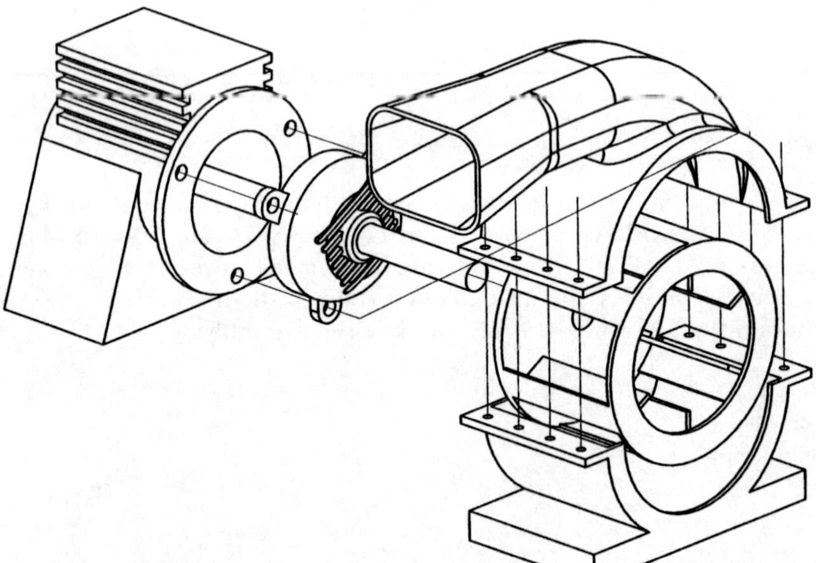

Figure 13-68 *Exploded view of the blower assembly*

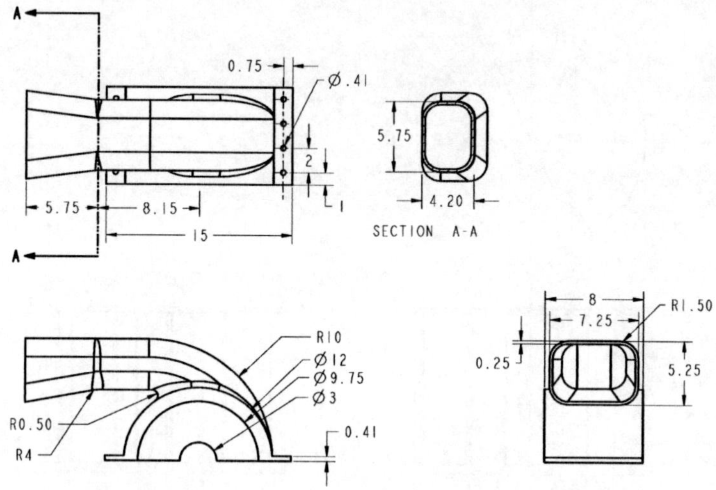

***Figure 13-69** Dimensions of the Upperhousing*

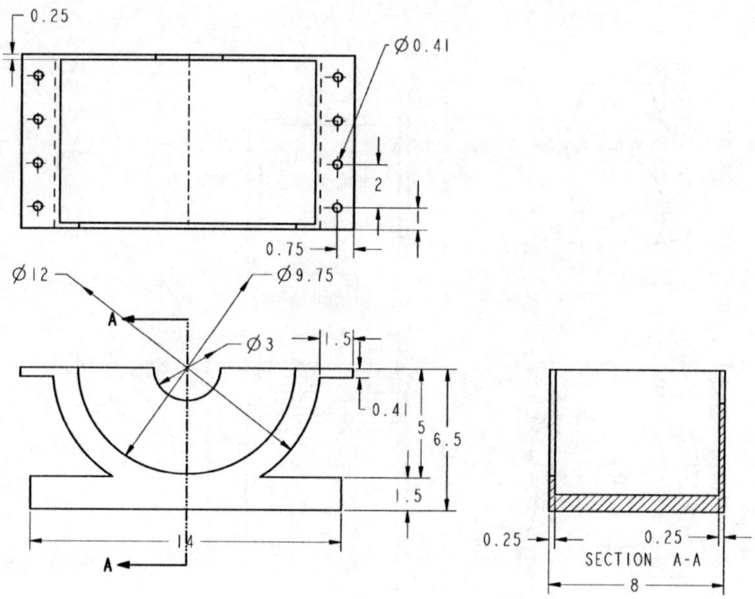

***Figure 13-70** Dimensions of the Lowerhousing*

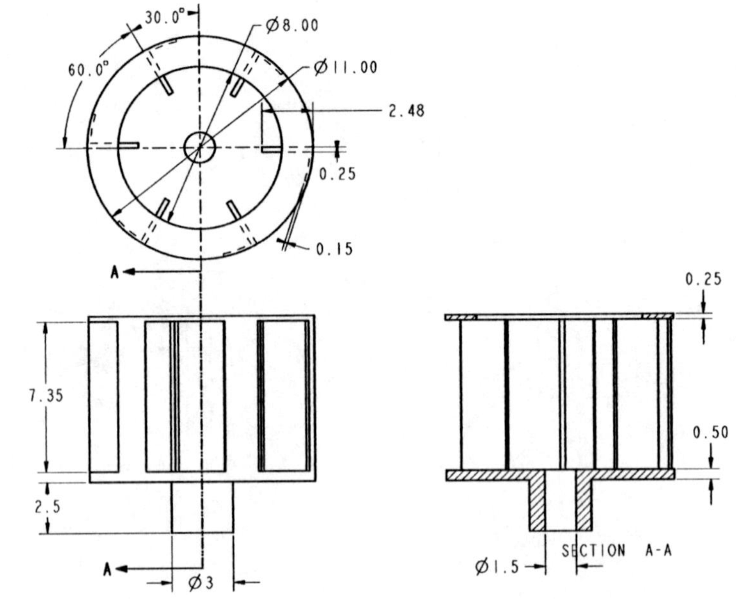

Figure 13-71 Dimensions of the Blower

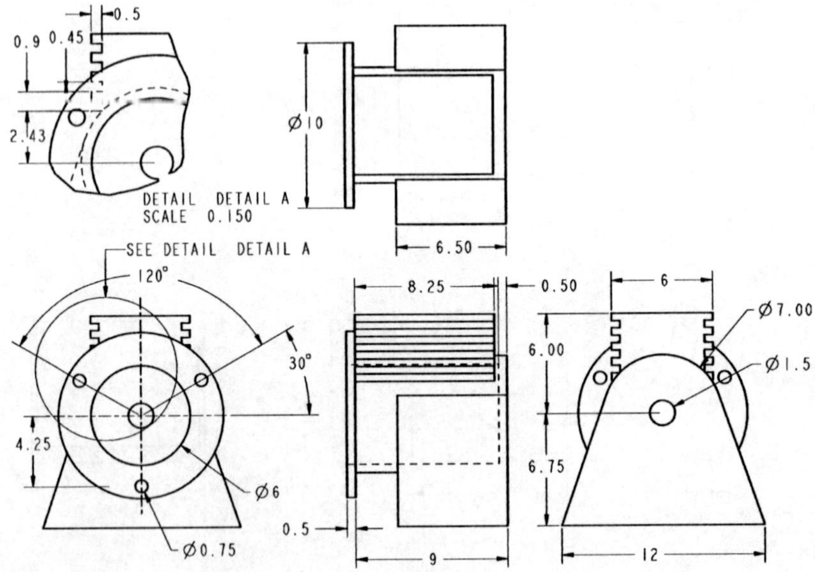

Figure 13-72 Dimensions of the Motor

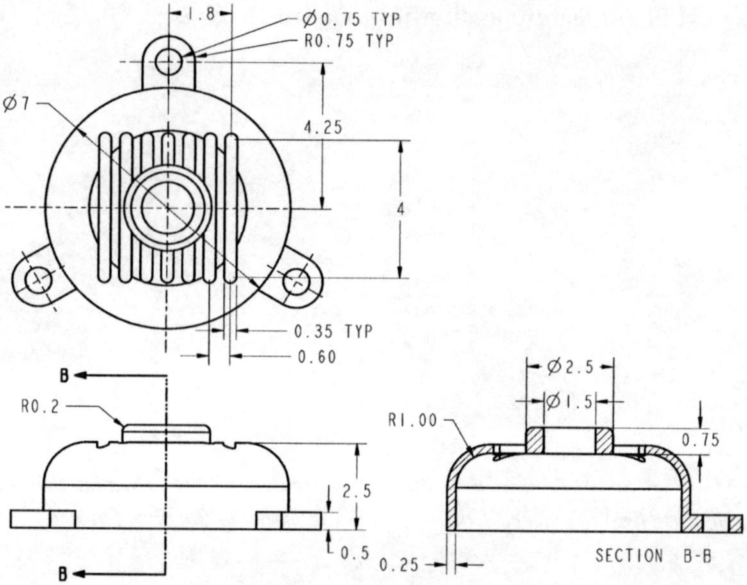

Figure 13-73 Dimensions of the Cover

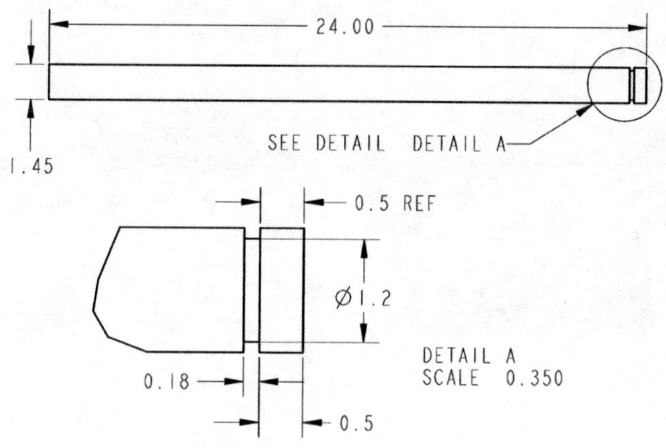

Figure 13-74 Dimensions of the Motorshaft

The steps that will be followed to complete this project are listed below:

Creating the Components

Creating the upperhousing

1. Create the base feature of the upperhousing. The base feature is a semicircular feature and is sketched on the **FRONT** datum plane. The sketch is extruded on both the sides of the sketching plane to a blind distance of 8.

2. Create the next sweep feature as shown in Figure 13-75.

3. Create the next blend feature as shown in Figure 13-76.

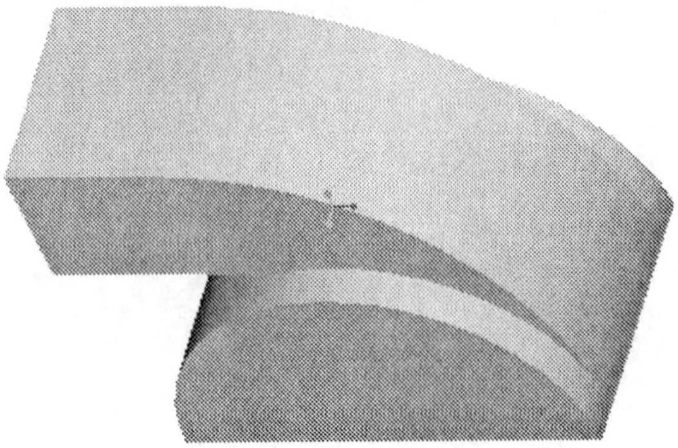

Figure 13-75 Model of the upperhousing after creating the base feature and the sweep feature

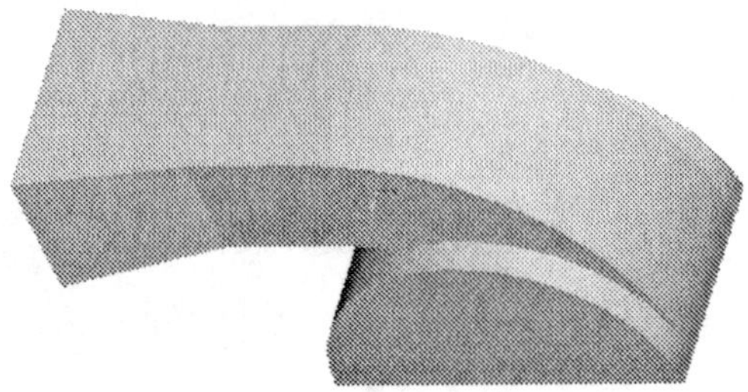

Figure 13-76 Model after adding the blend features to the base feature

4. Create the first round feature on the edges of the protrusion feature as shown in Figure 13-77.

5. Create two more round features of radius 4 and radius 0.5 as shown in Figure 13-78.

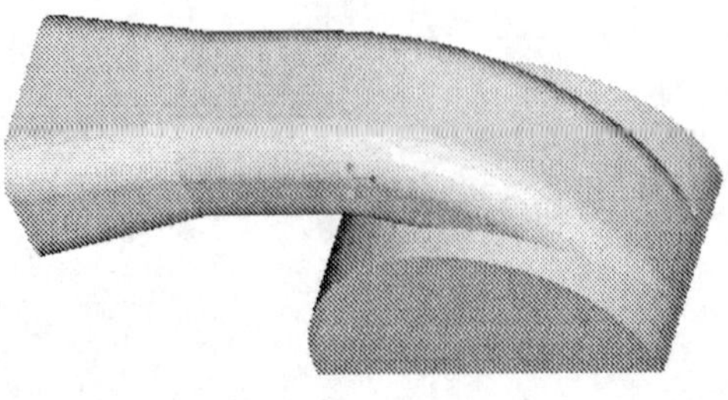

Figure 13-77 First round feature added to the model

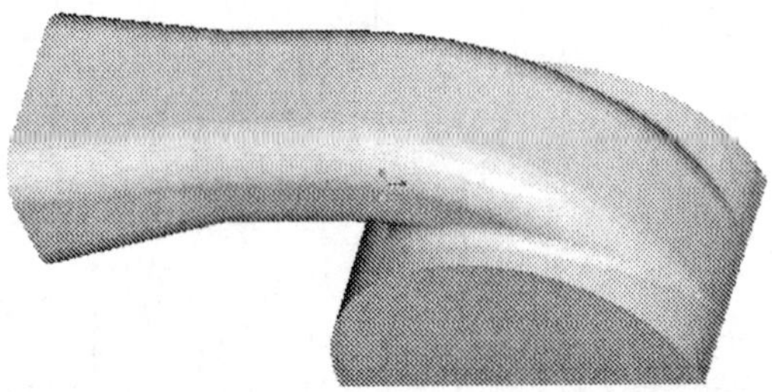

Figure 13-78 After adding the next rounds

6. Create the shell feature as shown in Figure 13-79.

7. Create the next protrusion feature. Select the front planar surface as the sketching plane and extrude the sketch to a distance of 8 as shown in Figure 13-80.

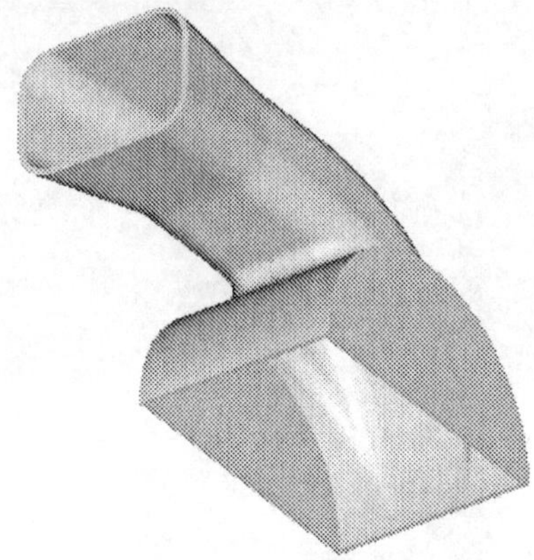

Figure 13-79 *Model after shelling*

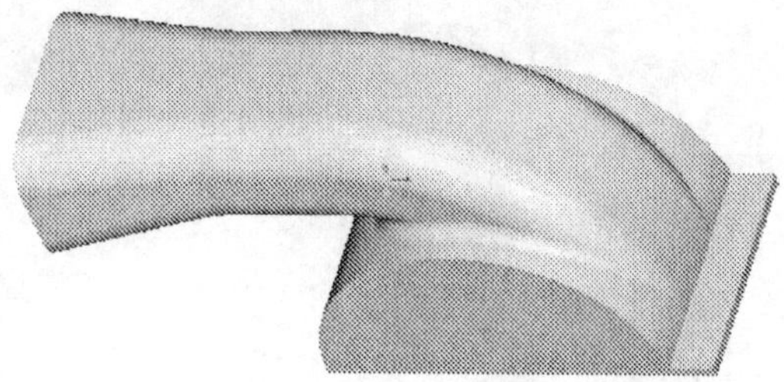

Figure 13-80 *Model after creating the next protrusion feature*

8. Create a mirror image of the protrusion feature created earlier as shown in Figure 13-81.
9. Create a cut feature on the front face of the base feature as shown in Figure 13-82.

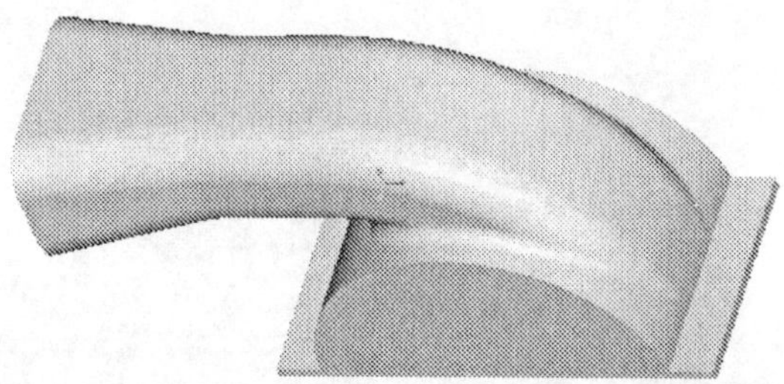

Figure 13-81 *Model after mirroring the protrusion feature*

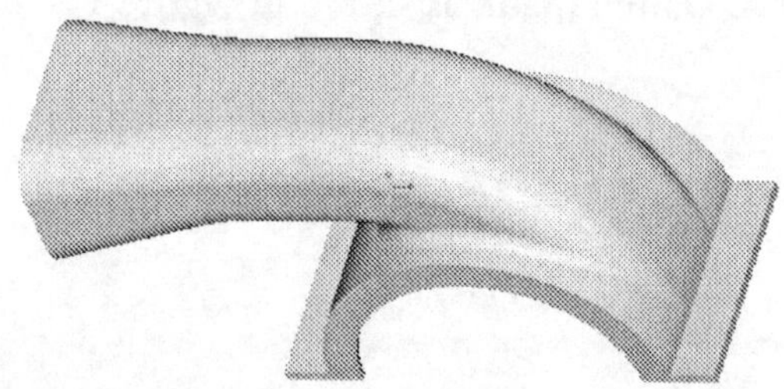

Figure 13-82 *Model after adding the cut feature on the front face of the base feature*

10. Create a cut feature on the back face of the base feature as shown in Figure 13-83.
11. Create the hole feature using the Hole dashboard and then pattern the hole as shown in Figure 13-84.
12. Save the file and close it.

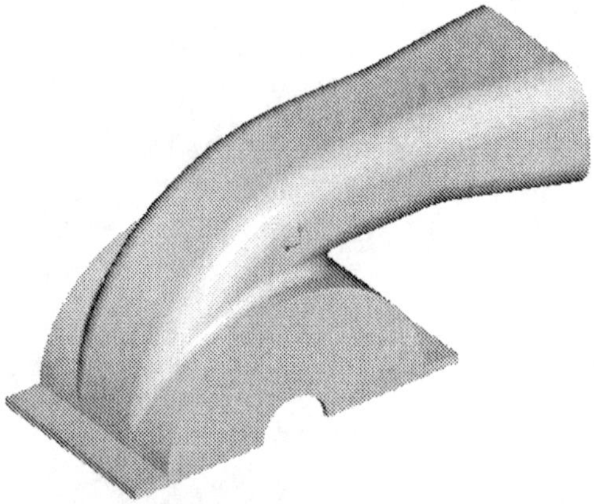

Figure 13-83 Model after adding the cut feature on the back planar surface of the base feature

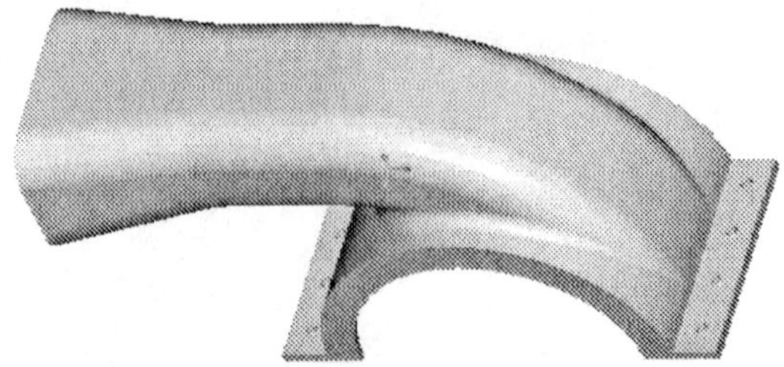

Figure 13-84 Final model of the upperhousing

Creating the lowerhousing

1. Create the base feature of the lowerhousing by extruding both sides of the sketch created on the **FRONT** datum plane as shown in Figure 13-85.

2. Create the next cut feature by extruding the sketch on both the sides of the **FRONT** datum plane as shown in Figure 13-86.

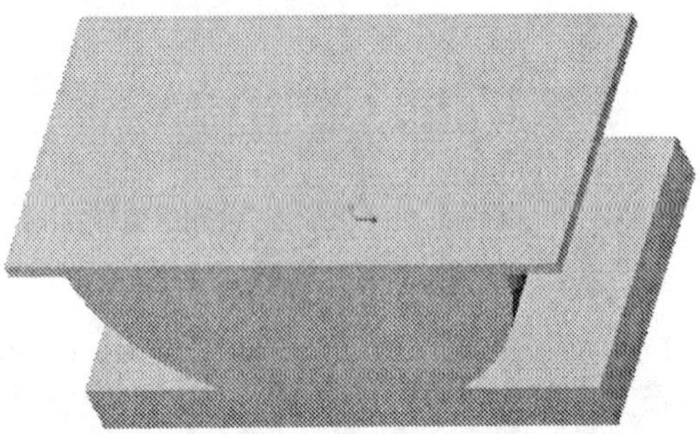

Figure 13-85 Base feature of the lowerhousing

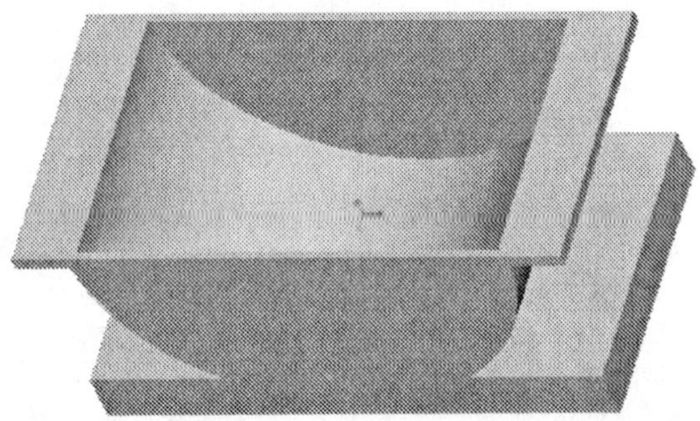

Figure 13-86 Cut feature created on the base feature

3. Create a cut feature on the front face of the base feature as shown in Figure 13-87.

4. Create a cut feature on the back face of the base feature as shown in Figure 13-88.

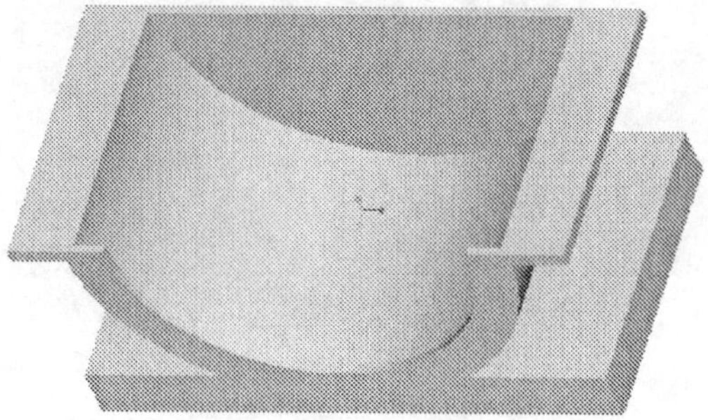

Figure 13-87 Model after creating the Cut feature on the front planar surface of the base feature

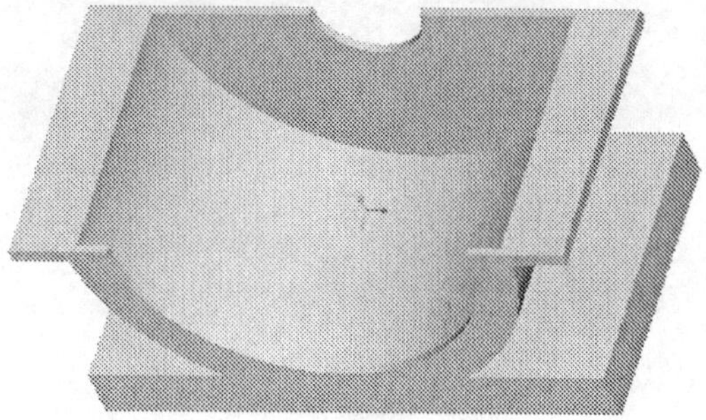

Figure 13-88 Model after creating the cut feature on the back planar surface of the base feature

5. Create a hole feature using the Hole dashboard and then pattern the hole feature as shown in Figure 13-89.

6. Mirror the holes created in the previous step as shown in Figure 13-90.

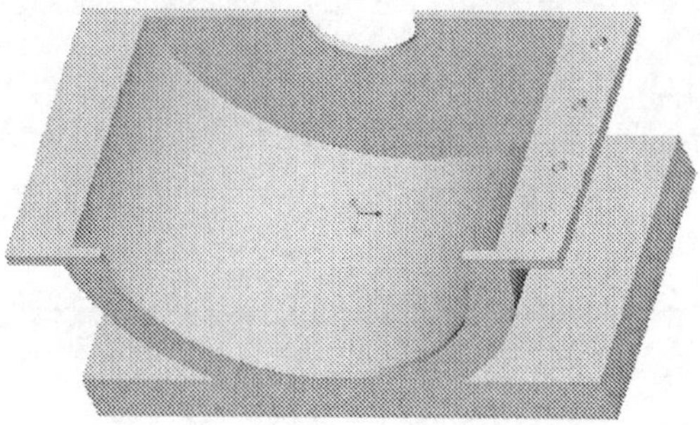

Figure 13-89 Holes created and patterned

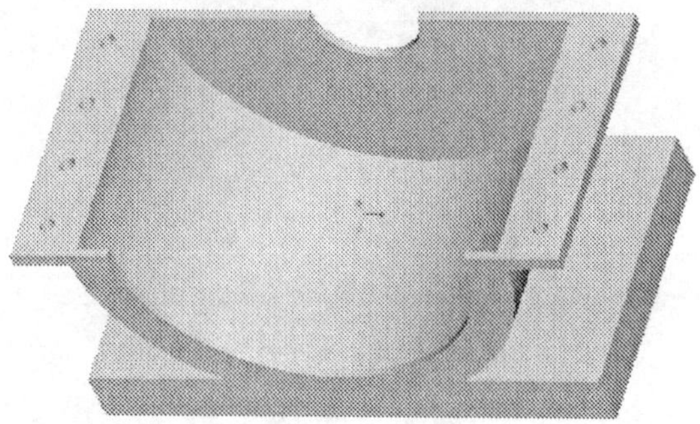

Figure 13-90 Model after mirroring the holes

7. Save the model.

Creating the Blower

1. Create the base feature on the **TOP** datum plane as shown in Figure 13-91.

2. Create the next protrusion feature on the top face of the base feature and pattern this feature as shown in Figure 13-92.

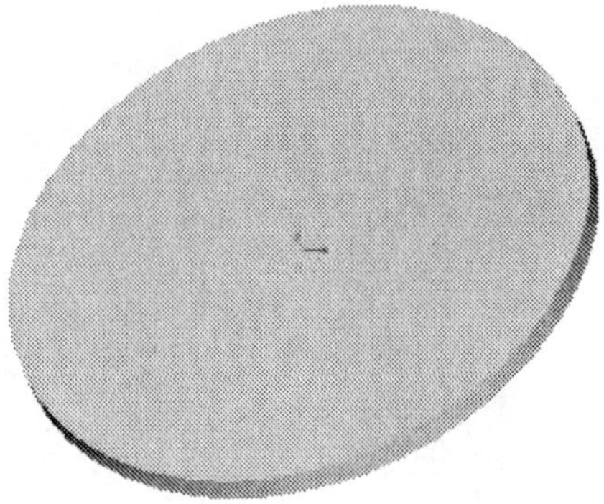

Figure 13-91 Base feature of the Blower

Figure 13-92 Patterned second protrusion feature

3. Create the next protrusion feature as shown in Figure 13-93.
4. Create the hole feature using the Hole dashboard as shown in Figure 13-94.

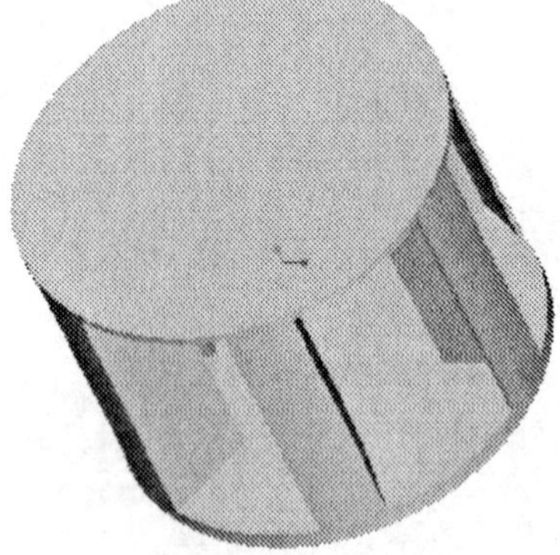

Figure 13-93 Protrusion feature added to the model

Figure 13-94 Model after creating the hole

5. Create a protrusion feature on the bottom face of the base feature as shown in Figure 13-95.
6. Create a through coaxial hole placed on the top face of the previous protrusion feature as shown in Figure 13-96.
7. Save the model.

Creating the Remaining Components

Similarly create the remaining components and save them.

Assembling the Components

1. Create a new assembly file and assemble the lowerhousing with the default datum planes

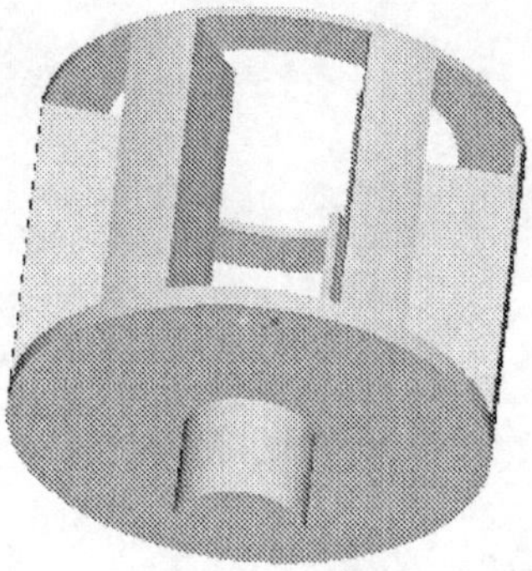

Figure 13-95 Protrusion feature created on the bottom planar surface of the base feature

Figure 13-96 A coaxial hole feature added to the model

of the assembly as shown in Figure 13-97.

2. Assemble the upperhousing with the lowerhousing using the various assembly constraints as shown in Figure 13-98.

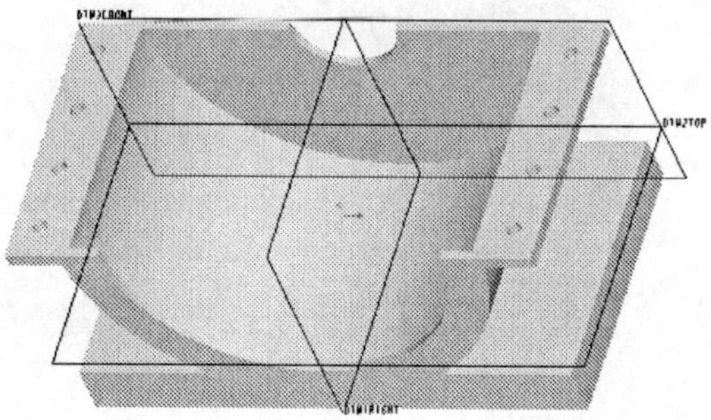

Figure 13-97 Lowerhousing assembled with the default datum planes of the assembly

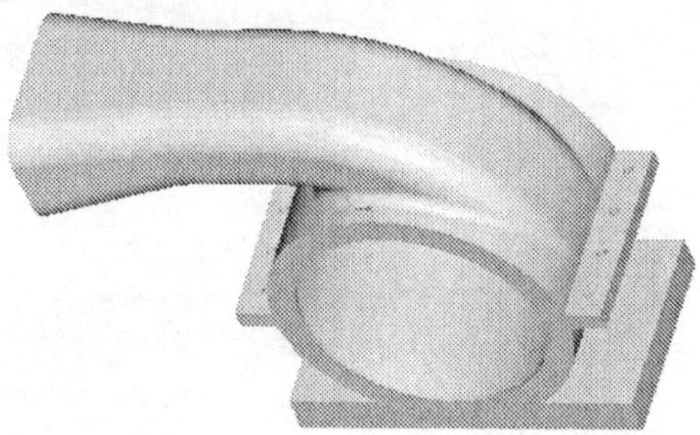

Figure 13-98 Upperhousing assembled with the Lowerhousing

3. Assemble the Blower with the base assembly as shown in Figure 13-99.

4. Similarly, assemble the other components of the Blower assembly as shown in Figure 13-100.

5. Save the assembly file.

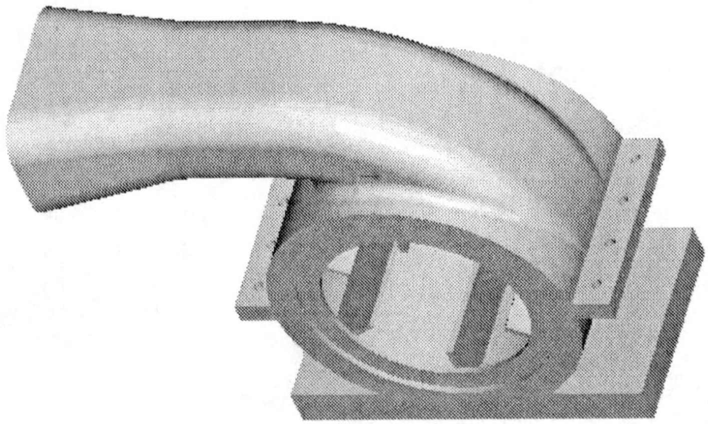

Figure 13-99 Blower assembled to the base assembly

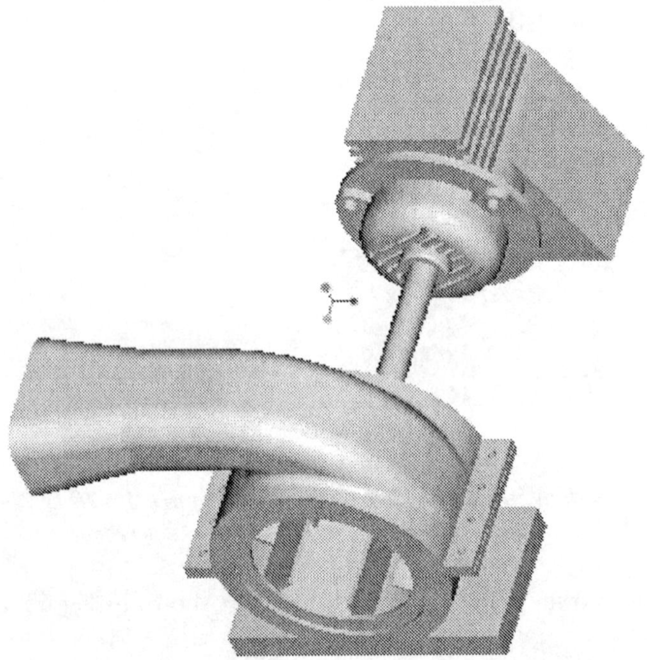

Figure 13-100 Assembled view of the blower assembly

Creating the Exploded State of the Blower Assembly

1. Create the exploded state of the assembly in the assembly mode as shown in Figure 13-101.

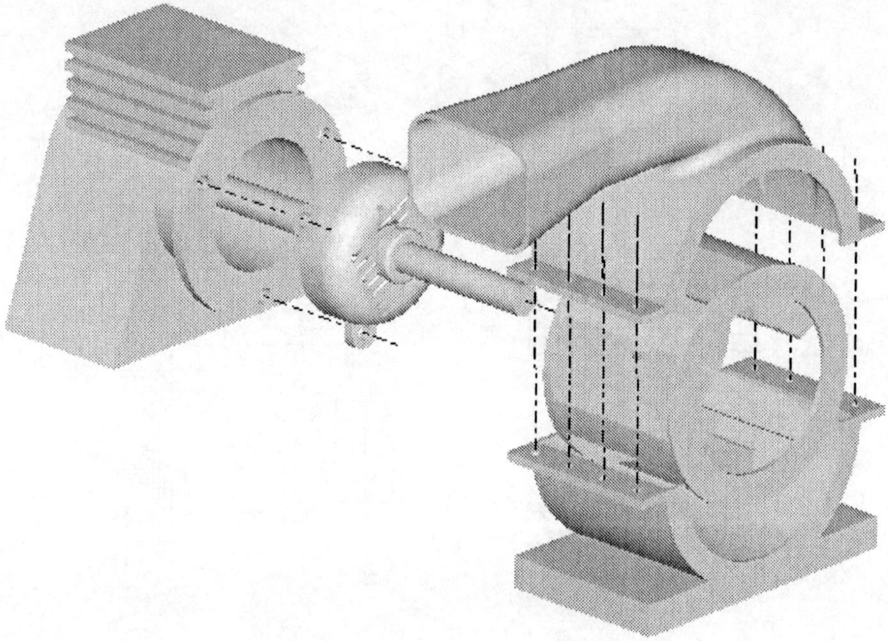

Figure 13-101 Exploded state of the Blower assembly

2. Save the assembly file after creating the exploded state.

Creating the Drawing Views of the Blower Assembly

1. Create a new drawing file and generate the drawing views as shown in Figure 13-102.

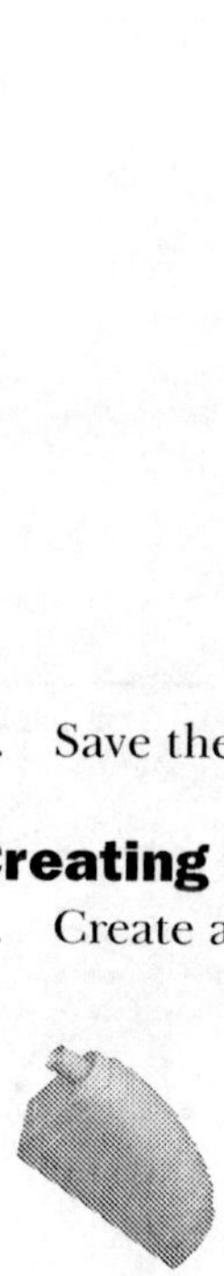

Tip: *For creating the 3D view of the Blower assembly as shown in Figure 13-102, orient the default view to isometric using the* ***Preferences*** *option in the* ***Type*** *drop-down list of the* ***Orientation*** *dialog box. Now, choose the* ***Angles*** *option in the* ***Type*** *drop-down list and select* ***Edge/Axis*** *from the* ***Reference*** *drop-down list. Select the lower vertical edge and enter a value of 90 in the* ***Angle*** *edit box and choose the apply button. Choose* ***OK*** *from the* ***Orientation*** *dialog box.*

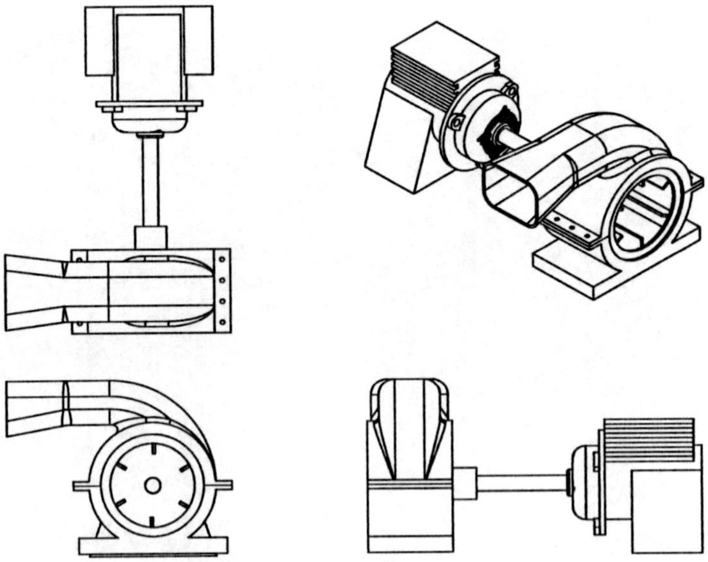

Figure 13-102 *Generated drawing views of the Blower assembly*

Appendix A

Interface

STANDARD DATA EXCHANGE

Pro/ENGINEER allows you to import or open a file created and saved in other file formats and you can also export a file created in Pro/ENGINEER using various standard file formats so that it can be opened in other design softwares. This action of importing or exporting files to or from Pro/ENGINEER is called Standard Data Exchange. The method of importing or opening some of the other format files in Pro/ENGINEER is discussed next.

Opening an AutoCAD (DWG) File in the Sketcher Environment

An AutoCAD drawing file of format *.dwg* can be imported in the sketcher environment. Note that only two-dimensional (2D) drawings can be opened in the sketcher environment. To open an AutoCAD file in the sketcher environment, choose **Sketch > Data from File** from the menu bar. The **Open** dialog box is displayed; select the *.dwg* file format from the **Type** drop-down list in the dialog box. Open an AutoCAD file.

Before placing the file in the sketcher environment, you are prompted to specify whether you want the drawing to be placed from the model space or from the paper space. You are provided with an option to change the angle of rotation and the scale of the drawing file. You can convert this sketch into a parametric sketch by adding the dimensions, relations, and constraints. After regenerating, convert the sketch into a solid model.

Note

*You can also import a 2D sketch or a drawing file saved in IGES format in the sketcher environment using the **Data from File** option. The procedure of importing the 2D IGES data is similar to the procedure of importing an AutoCAD drawing file.*

Importing Other Format Files

You can import the files stored in other formats like STEP, IGES, SAT, and so on. To import these files, choose the **Open an existing object** button from the **Top Toolchest** and select the file format from the **Type** drop-down list in the **File Open** dialog box. Figure A-1 shows the various file formats that are displayed in the **Type** drop-down list.

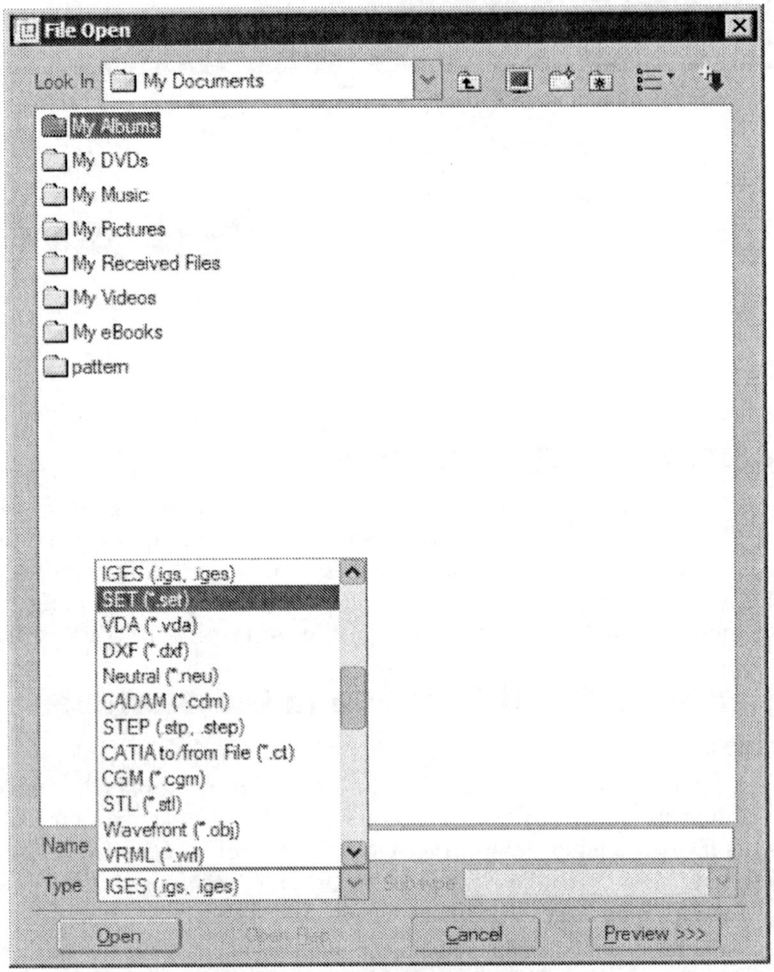

***Figure A-1 File Open** dialog box*

When you choose **OK** from the **File Open** dialog box, the **Import New Model** dialog box is displayed as shown in Figure A-2. The **Import New Model** dialog box is used to specify the mode of Pro/ENGINEER in which you want to import the file.

The individual features of the model are not imported with the solid model. This can be confirmed from the **Model Tree** after importing the model. In the **Model Tree** only one feature and a default coordinate system is displayed. This is because when a solid model is imported in Pro/ENGINEER, all the information related to its features is lost.

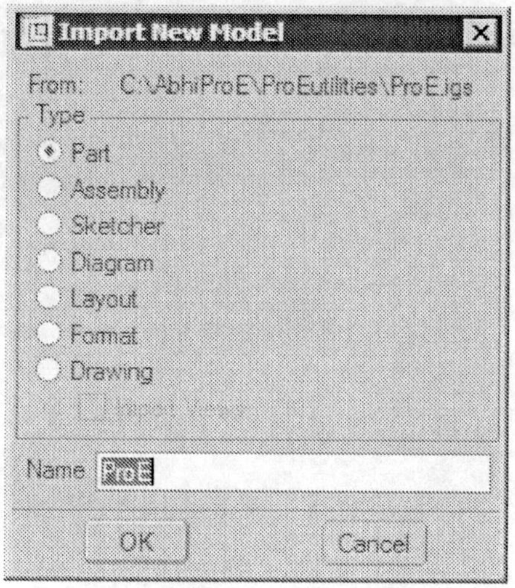

***Figure A-2 Import New Model** dialog box*

Exporting Files

Pro/ENGINEER allows you to export the Pro/ENGINEER files into various other formats such as STEP, SAT, STL, and so on. To export a file, choose **File > Save A Copy** from the menu bar. The **Save a Copy** dialog box is displayed. Select the required file format for exporting the model from the **Type** drop-down list on the graphics window. In this drop-down list you will notice a number of predefined formats in which you can export the file.

OBJECT-ACTION PARADIGM

In Pro/ENGINEER you can select a surface and then specify the action to be performed. This is called object-action paradigm in which the object is selected first and then the action is chosen. Pro/ENGINEER immediately takes you to the sketcher environment where you need to define the section. Thus, instead of using the menus you can easily select a datum curve first and then perform the operation by selecting the options. The object-action facility provided by Pro/ENGINEER reduces considerable time in going through the menus.

You can also create a protrusion feature by creating a rectangle using the datum curves and then invoking the extrude option. The feature is created without going in the sketcher environment and defining the section. When this feature is created, the yellow drag handles are displayed on the feature and the depth of extrusion can be modified dynamically. In case of creating the revolve features the rectangle created using the datum curves should contain an axis. If the axis is not present then the operation to create the revolve feature reverts to the standard interface.

MODEL PLAYER

You can replay the history of a model by using the **Model Player** shown in Figure A-3. Using the **Model Player**, you can view the features of the model one by one. You can even forward and rewind the creation of the model step by step to view its creation. The **Model Player** can be invoked by choosing **Tools > Model Player** from the menu bar.

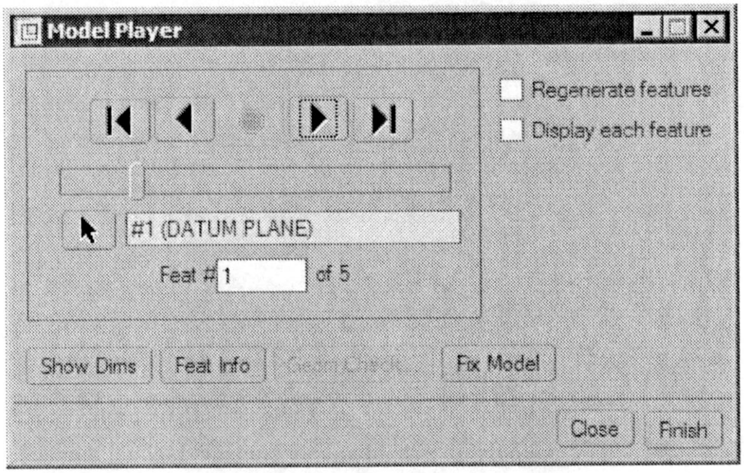

***Figure A-3 Model Player** dialog box*

APPLYING COLORS TO A SOLID MODEL

When you create a model in Pro/ENGINEER, the model is displayed in default color. However, you can apply other colors to a component using the **Appearance Editor** dialog box shown in Figure A-4. The **Appearance Editor** dialog box is displayed when you choose **View > Color and Appearance** from the menu bar.

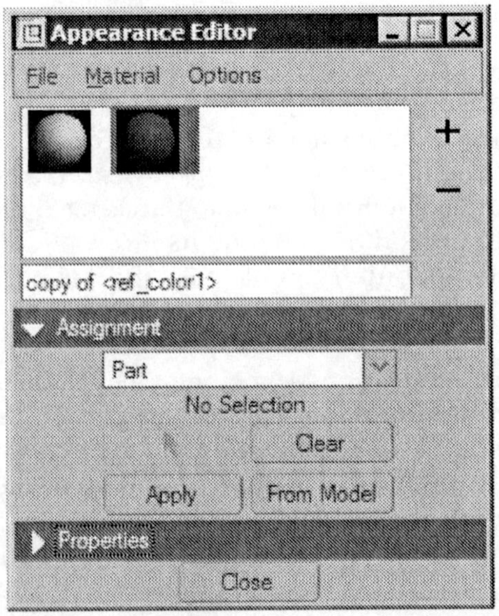

***Figure A-4 Appearance Editor** dialog box*

UDF LIBRARY

The **UDF Library** means user-defined feature library. There are certain standard features that are used in various components without any alteration in their geometry. For example, standard cuts, splines, gear teeth, and so on. Therefore, it is time-consuming if you have to create the same feature every time for different components. You can create the features and save them as a group file in a library. These stored features are known as UDF. You should be careful while referencing and dimensioning the features created as **UDF**.

USING FAMILY TABLES

The family table stores more than one instance of a model that has the same geometry but different dimensions. The family table is used extensively in industries. This is because the manufacturer can use the same model that is stored as different instances, each instance having different dimensions. Hence, considerable time is saved in creating models having different dimensions but the same geometry. An example of a component that is widely used in family tables is a bolt.

You can create a family table by choosing **Tool > Family Table** from the menu bar. The **Family Table** dialog box is displayed. Choose the **Add/delete the table columns** button. The **Family Items, Generic** dialog box is displayed as shown in Figure A-5. Select the dimensions from the model and choose **OK** to exit the **Family Items, Generic** dialog box.

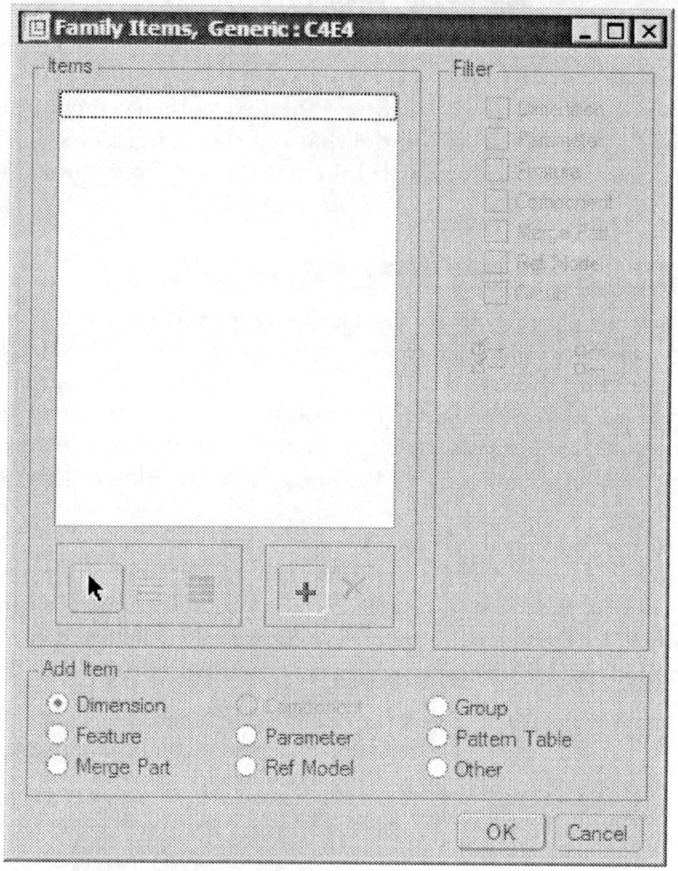

***Figure A-5 Family Items, Generic** dialog box*

When you exit this dialog box, the **Family Table** dialog box is displayed as shown in Figure A-6.

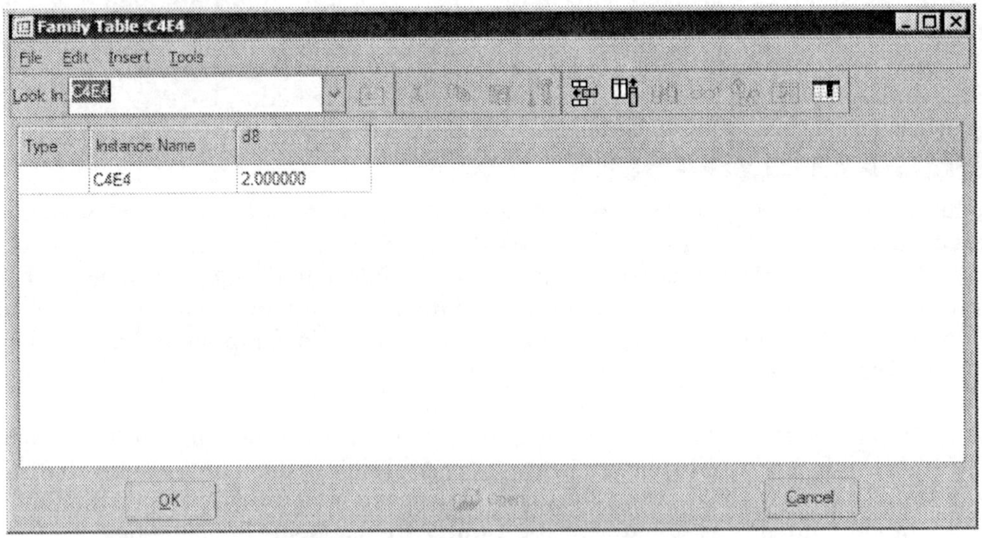

***Figure A-6 Family Table** dialog box*

Using the **Family Table** dialog box, you can create different instances of a model. You can enter different dimensions in this dialog box. When you preview any of the selected instance, a sub-window is opened in which the model that has the dimensions of the selected instance is displayed. You can even generate an excel sheet using this dialog box and edit the instances in MS Excel.

SETTING SELECTION PREFERENCES

You can set the selection preferences to select edges, surfaces, datums, and so on from a model. When you are working in Part mode, you are always in selection mode. You can pre-highlight an item and then select it. The pre-highlighted item appears in cyan color and the selected item appears in red color. When you move the mouse over the model, the items gets pre-highlighted and when you left-click on any item, it gets selected. You can also select more than one item by pressing the CTRL key and then selecting the items using the left mouse button.

The **Filter** drop-down list in the status bar is used to set the selection preferences on a model. The **Datums** option in the drop-down list is used to select the datum features like axis, planes, and points. The **Primary Items** option is used to select features, three-dimensional notes, and components. The **Geometry** option is used to select edges, vertices, and surfaces from a model.

Appendix B

Sketches

Chapter 4 > Tutorial 3

The following steps explain the procedure to create the sketch of the second feature created in Tutorial 3 of Chapter 4. The sketch in Figure 4-85 is shown in Figure B-1.

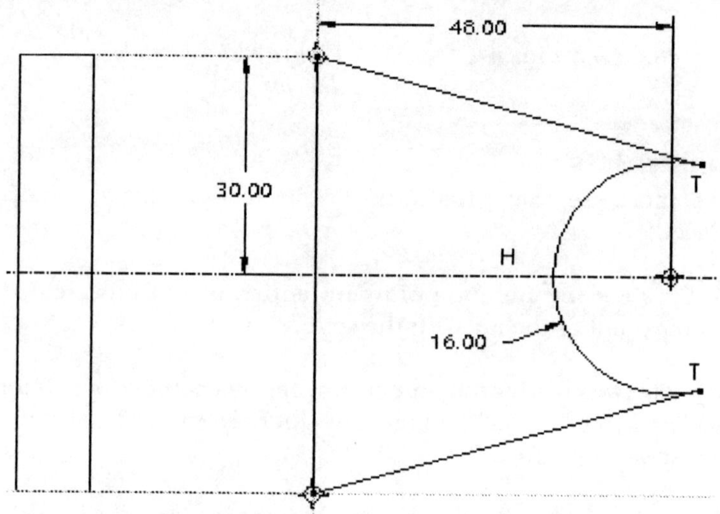

Figure B-1 *Sketch of the second feature*

Creating the Sketch

1. Choose the black arrow on the right of the **Create 2 point lines** button to display the flyout. From this flyout, choose the **Create 2 point centerlines** button. Draw a center line

such that it passes approximately from the middle of the existing feature.

2. Choose the black arrow on the right of the **Create 2 point centerlines** button to display the flyout. From the flyout, choose the **Create 2 point lines** button and draw an inclined line from the top right vertex as shown in Figure B-2. When you are prompted to align the highlighted entities, choose **Yes** from the **Confirm** dialog box. Similarly, draw the second inclined line from the bottom right vertex. Draw a line joining the top right vertex and the bottom right vertex as shown in Figure B-2.

3. Choose the **Create an arc by 3 points or tangent to an entity at its endpoint** button. Draw an arc by selecting the three points; the first point being the endpoint of the top inclined line, the second point being the endpoint of the bottom inclined line, and the third point being at the left in the sketch. The arc is created as shown in Figure B-3.

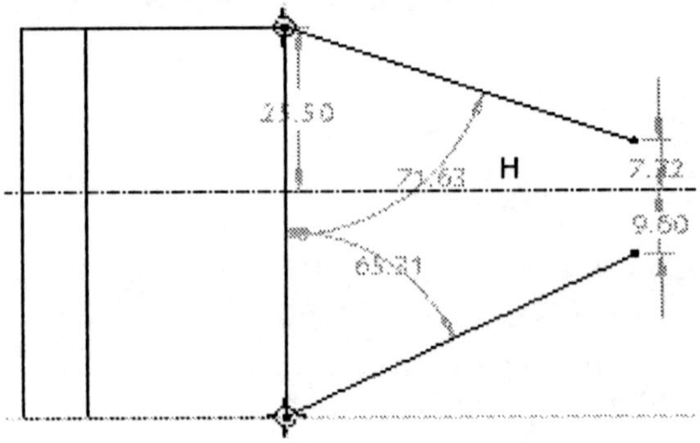

Figure B-2 Lines drawn

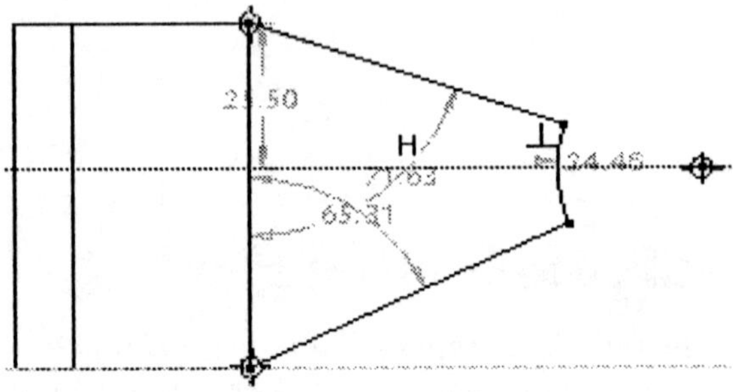

Figure B-3 Sketch before applying constraints and dimensions

Applying Constraints

1. Choose the **Impose sketcher constraints on the section** button. The **Constraints** dialog box is displayed.

2. Choose the **Create same points, points on entity, or collinear constraint** button and align the center point of the arc with the center line.

3. Choose the **Make two entities tangent** button and select the top inclined line and the arc. Similarly, make the bottom inclined line tangent to the arc. The sketch after applying the constraints is shown in Figure B-4.

   Note that the endpoints of the two inclined lines are aligned with the right edge of the base feature.

Applying Dimensions

1. Choose the **Create defining dimension** button and select the center of the arc and the right edge of the base feature and place the dimension as shown in Figure B-5.

2. Select the top edge of the base feature and the center line and place the dimension using the middle mouse button as shown in Figure B-5.

3. Select the arc and place the dimension as shown in Figure B-5.

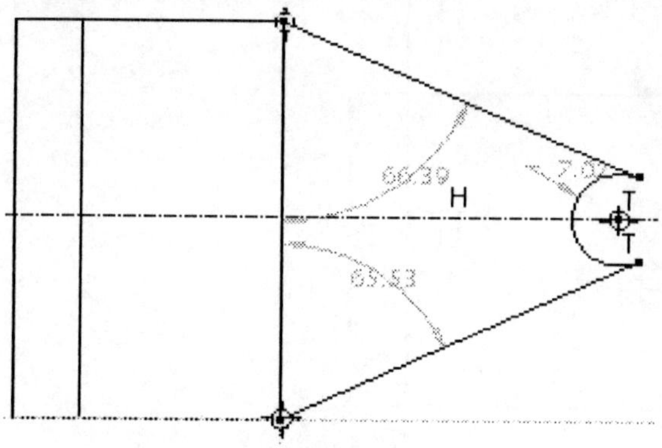

Figure B-4 *Sketch after applying constraints*

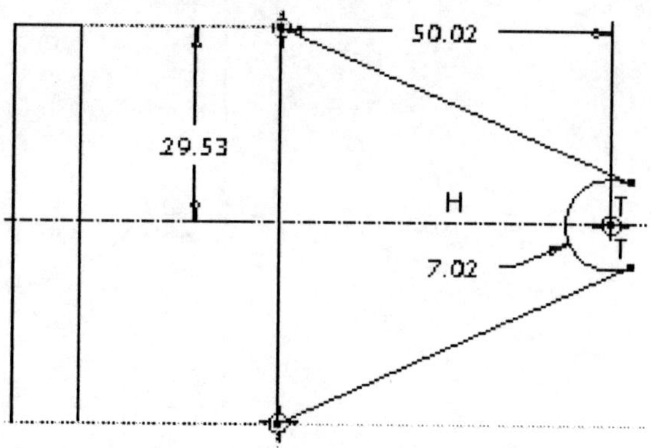

Figure B-5 *Sketch after dimensioning*

Modifying the Dimensions

1. Select all the dimensions by using CTRL+ALT+A and choose the **Modify the values of dimensions, geometry of splines, or text entities** button to display the **Modify Dimensions** dialog box. All the three dimensions are displayed in this dialog box.

2. Clear the **Regenerate** check box and then modify the values of the dimensions.

3. After modifying all the dimensions, choose the **Regenerate the section and close the dialog** button from the **Modify Dimensions** dialog box. The message **Dimension modifications successfully completed.** is displayed in the **Message Area**.

   The sketch is completed and is shown in Figure B-6.

4. Choose the **Continue with the current section** button to exit the sketcher environment.

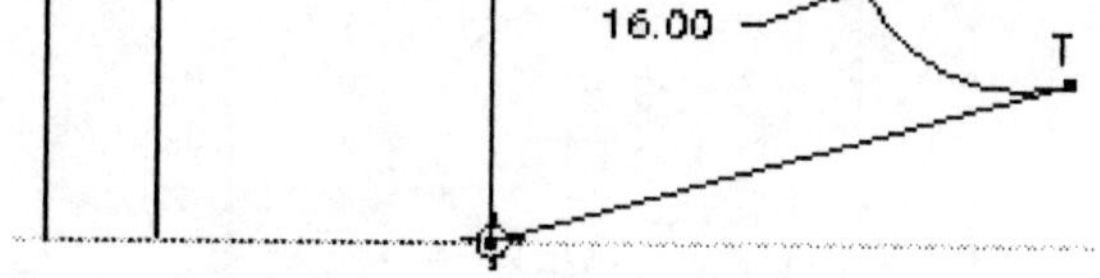

Figure B-6 *Sketch after modifying the dimensions*

Chapter 4 > Exercise 2

The following steps explain the procedure to create the sketch of the cut feature created on the right face of the model in Exercise 2 of Chapter 4. The drawing views of the model are shown in Figure 4-96 and the sketch for the cut feature is shown in Figure B-7.

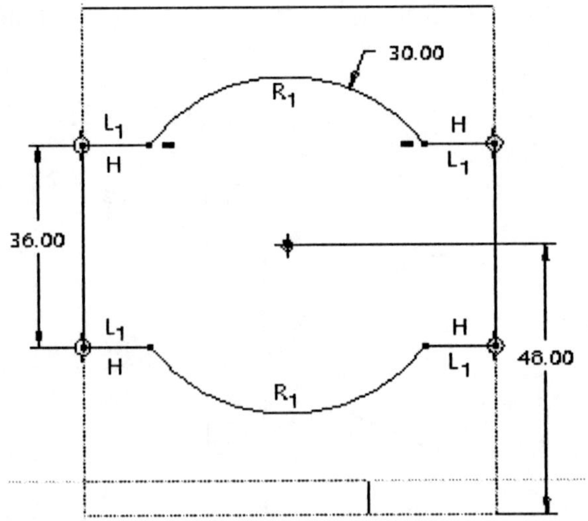

Figure B-7 Sketch for the cut feature

Creating the Sketch

1. Choose the **Create circle by picking the center and a point on the circle** button and draw a circle by selecting a center point as shown in Figure B-8.

2. Choose the **Create 2 point lines** button and draw the lines joining the circle as shown in Figure B-9. Choose **Yes** from the **Confirm** dialog box to align the endpoints of lines. As evident from Figure B-9, the lines are automatically aligned to the edge of the existing feature and to the circle. However, you will apply the constraints to the sketch manually.

3. Join the top right horizontal line with the bottom right horizontal line with a vertical line segment. Similarly, join the other two lines on the left.

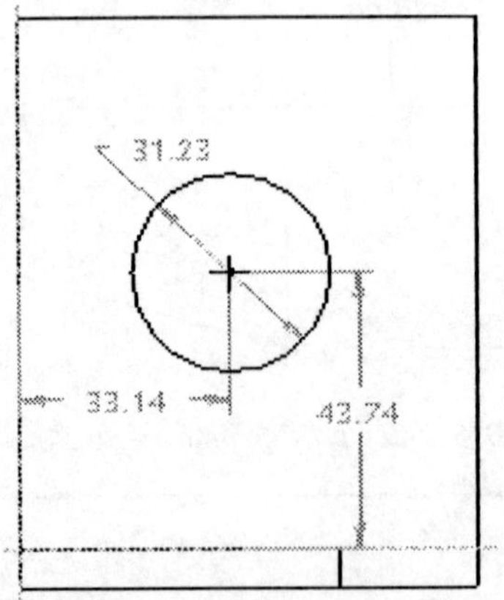

Figure B-8 Sketch of the circle with weak dimensions

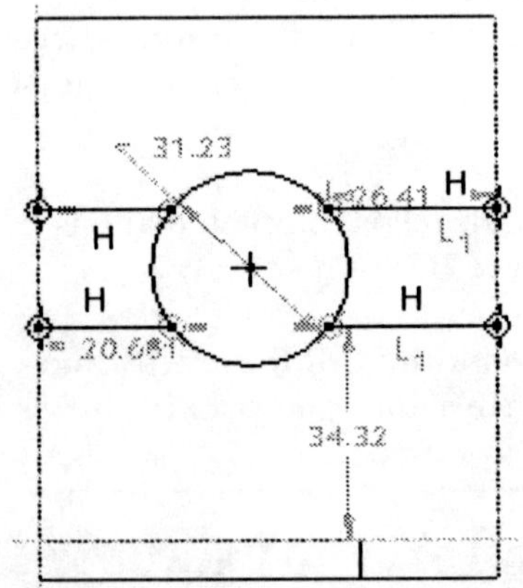

Figure B-9 Sketch of the lines joined to the circle on the circumference with weak dimensions

4. Choose the **Dynamically trim the section entities** button from the **Sketcher Tools** toolbar and bring the mouse on the right portion of the circle. You will notice that a part of the circle is highlighted in magenta color.

5. Using the left mouse button, select the part of the circle that has turned to magenta. This portion of the circle gets erased. Similarly, erase the left part of the circle that is not required. The sketch after trimming the circle is shown in Figure B-10.

Applying Constraints

1. Choose the **Impose sketcher constraints on the section** button. The **Constraints** dialog box is displayed.

2. Choose the **Create Equal Lengths, Equal Radii, or Same Curvature constraint** button and select the top right line and then the top left line to make them of equal length. Similarly, make the remaining lines of equal length.

3. Select the top arc and the bottom arc to apply the equal radii constraint.

4. Choose the **Make line or two vertices horizontal** button from the **Constraints** dialog box and select the points at the top of the sketch where the lines and the arc meet. The two points are now horizontally aligned. Similarly, horizontally align the two points at the bottom.

   The sketch after applying the constraints should be similar to the sketch shown in Figure B-11. If the **Resolve Sketch** dialog box appears, delete the constraints that are not required in the sketch.

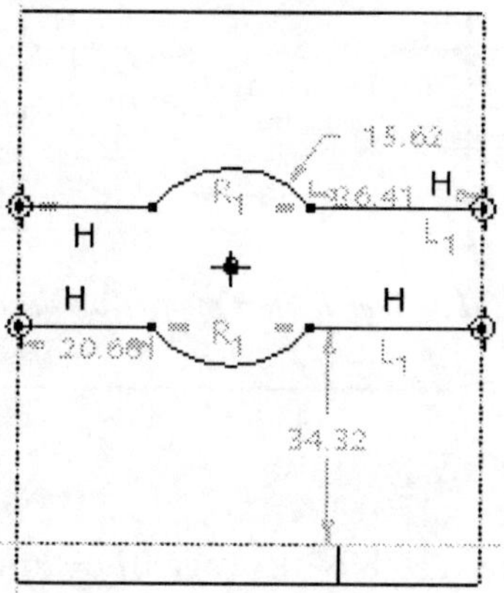

Figure B-10 *Sketch after trimming the circle*

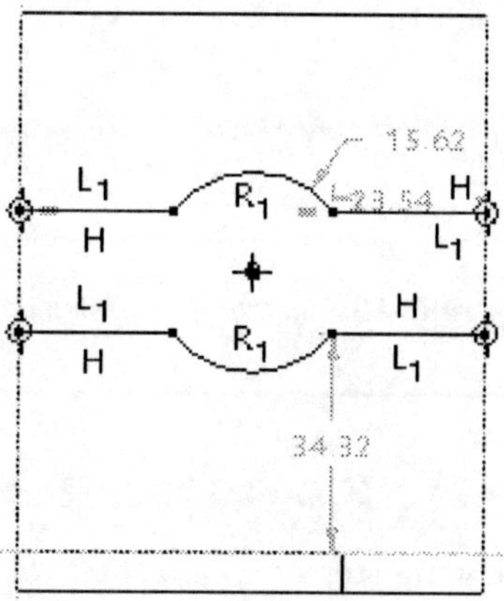

Figure B-11 *Sketch after applying constraints*

Applying Dimensions

1. Choose the **Create defining dimension** button and select the center of the circle and the bottom edge of the existing feature and place the dimension using the middle mouse button as shown in Figure B-12.

The weak dimension is already applied to the circle and therefore, you need not dimension it again.

2. Select the left vertical line in the sketch and place the dimension as shown in Figure B-12.

Modifying the Dimensions

1. Select all the dimensions by using CTRL+ALT+A and choose the **Modify the values of dimensions, geometry of splines, or text entities** button to display the **Modify Dimensions** dialog box. All the three dimensions are displayed in this dialog box.

2. Clear the **Regenerate** check box and then modify the values of the dimensions.

3. After modifying all the dimensions, choose the **Regenerate the section and close the dialog** button from the **Modify Dimensions** dialog box. The message **Dimension modifications successfully completed.** is displayed in the **Message Area**.

   The sketch is completed and is shown in Figure B-13.

4. Choose the **Continue with the current section** button to exit the sketcher environment.

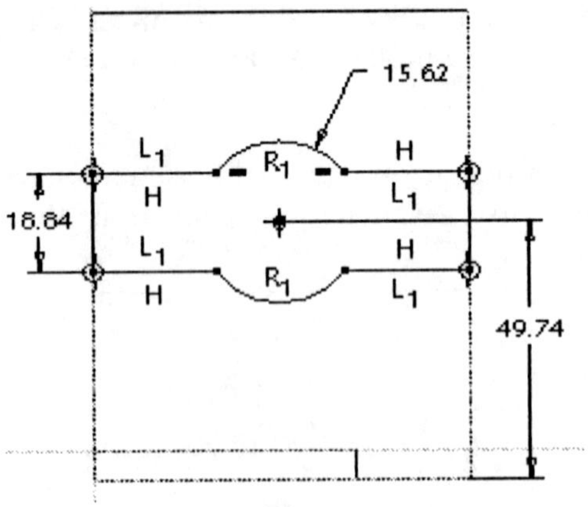

***Figure B-12** Sketch after dimensioning*

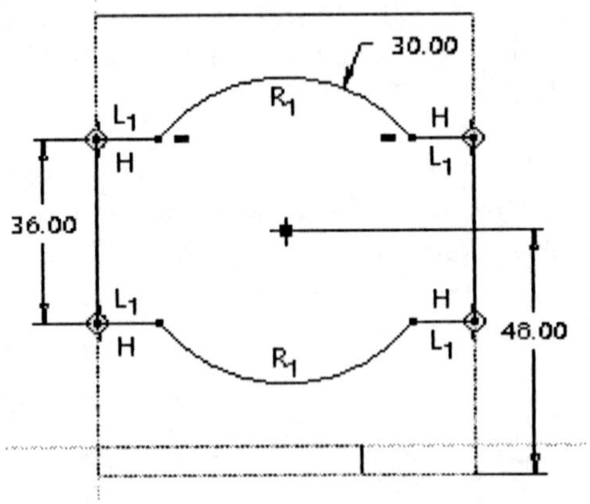

***Figure B-13** Sketch after modifying the dimensions*

Chapter 6 > Tutorial 2

The following steps explain the procedure to create the sketch of the seventh feature created on the top face of the base feature in Tutorial 2 of Chapter 6. The sketch in Figure 6-41 is shown in Figure B-14.

Creating the Sketch

1. Choose the **Create circle by picking the center and a point on the circle** button and draw a circle such that the center of the circle lies at the intersection point of the RIGHT and the FRONT datum planes.

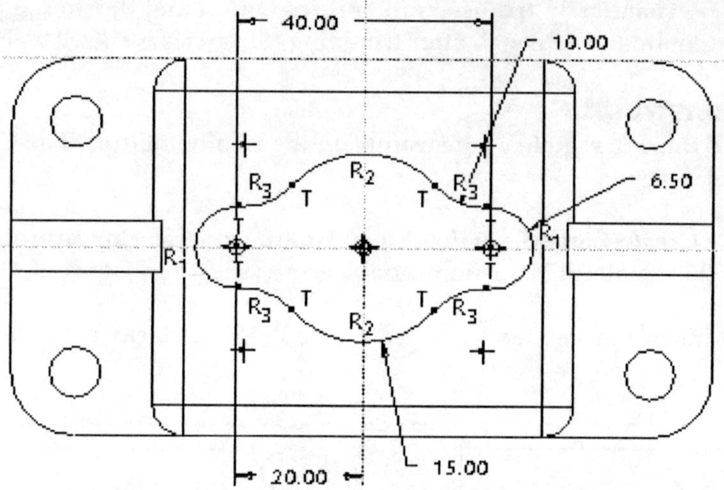

Figure B-14 *Sketch of the seventh feature*

2. Draw the second circle by specifying the center point on the **FRONT** datum plane.

3. Draw the third circle by specifying the center point on the **FRONT** datum plane and complete the circle when the equal radius constraint appears on the right and the left circles. The equal radius constraint that is applied automatically is a weak constraint and later you will need to apply the equal radii constraint manually.

   The sketch after drawing the three circles is shown in Figure B-15.

4. Choose the **Create a circular fillet between two entities** button and select a point on the top of the right circle and a point on the top of the middle circle. The fillet is created.

5. Similarly, create the remaining fillets. The sketch after creating the fillets is shown in Figure B-16.

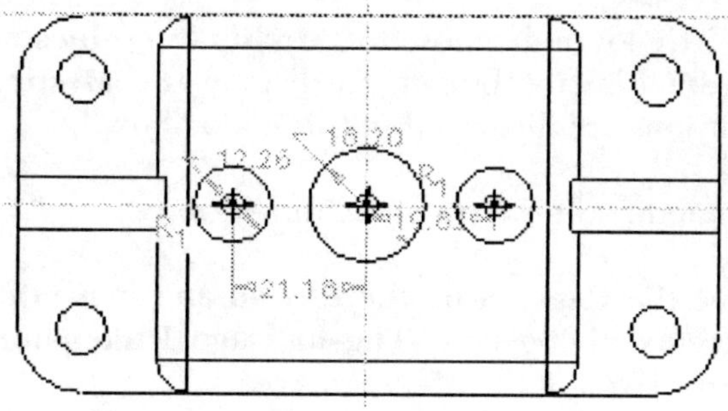

Figure B-15 *Three circles drawn*

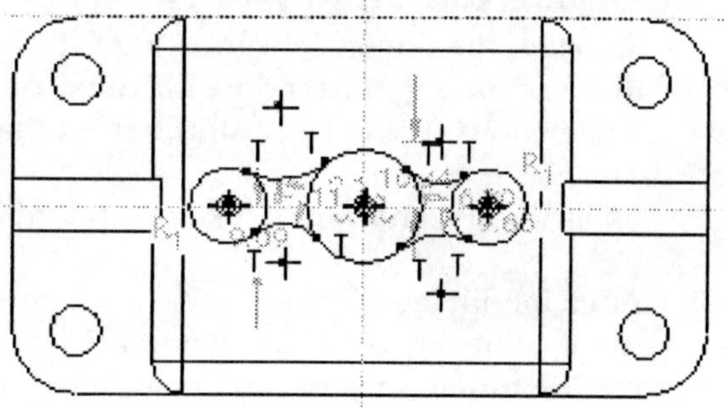

Figure B-16 *Fillets created on the circles*

6. Choose the **Dynamically trim section entities** button and delete the part of the circles that is not required. The sketch after trimming the circles is shown in Figure B-17.

Applying Constraints

1. Choose the **Impose sketcher constraints on the section** button. The **Constraints** dialog box is displayed.

2. Choose the **Create Equal Lengths, Equal Radii, or Same Curvature constraint** button and apply the equal radii constraint to the arcs as shown in Figure B-18.

   The weak dimensions are not shown in Figure B-18 for clarity of the sketch.

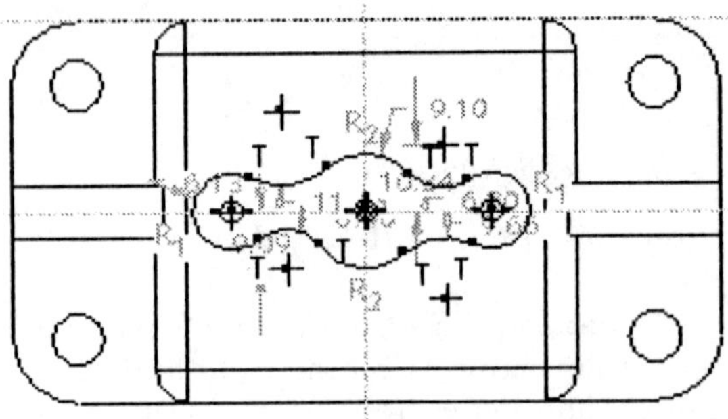

Figure B-17 Sketch after filleting

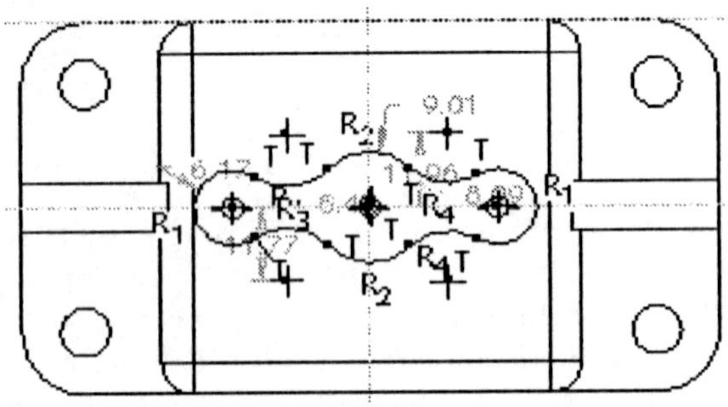

Figure B-18 Sketch after applying constraints

Dimensioning the Sketch

1. Choose the **Create defining dimension.** button and select the centers of the left circle and the right circle and place the dimension using the middle mouse button as shown in Figure B-19.

2. Dimension the remaining entities as shown in Figure B-19.

Modifying the Dimensions

1. Select all the dimensions by using CTRL+ALT+A and choose the **Modify the values of dimensions, geometry of splines, or text entities** button to display the **Modify Dimensions** dialog box. All the five dimensions are displayed in this dialog box.

2. Clear the **Regenerate** check box and then modify the values of the dimensions.

3. After modifying all the dimensions, choose the **Regenerate the section and close the dialog** button from the **Modify Dimensions** dialog box. The message **Dimension modifications successfully completed** is displayed in the **Message Area**.

   The sketch is completed and is shown in Figure B-20.

4. Choose the **Continue with the current section** button to exit the sketcher environment.

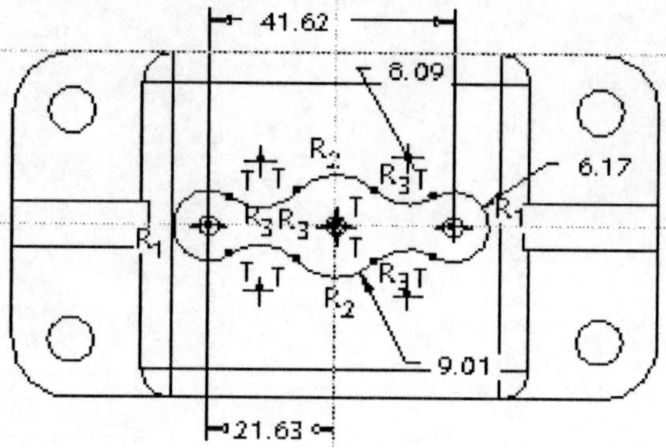

Figure B-19 *Sketch after dimensioning*

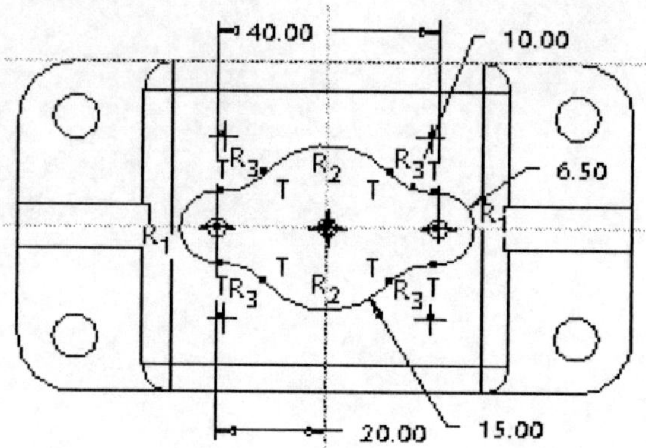

Figure B-20 *Sketch after modifying the dimensions*

Index

List of other Publications by Sham Tickoo and CADCIM Technologies

The following is the list of the other publications by Sham Tickoo and CADCIM Technologies.

SolidWorks Book

- SolidWorks for Engineers & Designers, Release 2003 (Indian Edition)
  dreamtech Press, India
- SolidWorks for Designers, Release 2003 (US Edition)
  CADCIM Technologies, USA

Autodesk Inventor Book

- Autodesk Inventor for Designers, Release 6 (US Edition)
  CADCIM Technologies, USA
- Autodesk Inventor for Engineers & Designers, Release 6 (Indian Edition)
  dreamtech Press, India
- Autodesk Inventor for Designers, Release 5 (US Edition)
  CADCIM Technologies, USA

Pro/ENGINEER Book

- Pro/ENGINEER Wildfire for Engineers & Designers, (Indian Edition)
  dreamtech Press, India
- Pro/ENGINEER for Designers, Release 2001 (US Edition)
  CADCIM Technologies, USA
- Designing With Pro/ENGINEER, Release 2001 (Indian Edition)
  dreamtech Press, India

Mechanical Desktop Book

- Mechanical Desktop Instructor, Release 5
  McGraw Hill Publishing Company

AutoCAD Books (US Edition)

- AutoCAD 2004: A Problem-Solving Approach
  Autodesk Press
- AutoCAD LT 2004: A Problem-Solving Approach
  Autodesk Press
- Customizing AutoCAD 2004
  Autodesk Press
- AutoCAD 2002: A Problem-Solving Approach
  Autodesk Press
- AutoCAD LT 2002: A Problem-Solving Approach
  Autodesk Press
- Customizing AutoCAD 2002
  Autodesk Press
- AutoCAD 2000: A Problem Solving Approach
  Autodesk Press

- Customizing AutoCAD 2000
  Autodesk Press
- AutoCAD LT 2000: A Problem-Solving Approach
  Autodesk Press
- Customizing AutoCAD R14
  Autodesk Press
- AutoCAD LT 97: Problem Solving Approach
  Autodesk Press

AutoCAD Books (Italian Edition)

- AutoCAD 2000 Fondamenti
- AutoCAD 2000 Techniche Advantaze

AutoCAD Book (Chinese Edition)

- AutoCAD 2000

AutoCAD Books (Indian Edition)

- AutoCAD 2002 with Applications
  Tata McGraw Hill Publishers
- Understanding AutoCAD 2002
  Tata McGraw Hill Publishers
- Advanced Techniques in AutoCAD 2002
  Tata McGraw Hill Publishers
- AutoCAD 2000 with Applications
  Galgotia Publishers
- Understanding AutoCAD 2000
  Galgotia Publishers
- Advanced Techniques in AutoCAD 2000
  Galgotia Publishers
- AutoCAD 14 Bible
  Galgotia Publishers
- Understanding AutoCAD 14
  Galgotia Publishers
- Advanced Techniques in AutoCAD 14
  Galgotia Publishers

3D Studio MAX and VIZ Books

- Learning 3DS Max: A Tutorial Approach, Release 4
  Goodheart-Willcox Publishers (USA)
- Learning 3D Studio VIZ: A Tutorial Approach
  Goodheart-Willcox Publishers (USA)
- Learning 3D Studio R4: A Tutorial Approach
  Goodheart-Willcox Publishers (USA)
- Learning 3D Studio MAX/VIZ 3.0: A Tutorial Approach
  BPB Publishers (India)